ALLE · ZEIT · WACH · 1842

P. Höller · V. Hauk · G. Dobmann
C. O. Ruud · R. E. Green (Eds.)

Nondestructive Characterization of Materials

Proceedings of the 3rd International Symposium
Saarbrücken, FRG October 3-6, 1988

Organized by
Deutsche Gesellschaft für zerstörungsfreie Prüfung e.V. (DGZfP), Berlin
and
Fraunhofer-Institut für zerstörungsfreie Prüfverfahren (IzfP), Saarbrücken

With 609 Figures

Springer-Verlag Berlin Heidelberg New York
London Paris Tokyo Hong Kong

Prof. Dr. Paul Höller
Fraunhofer Institut
für zerstörungsfreie Prüfverfahren
Universität, Gebäude 37
6600 Saarbrücken 11
FRG

Dr. G. Dobmann
Fraunhofer Institut
für zerstörungsfreie Prüfverfahren
Universität, Gebäude 37
6600 Saarbrücken 11
FRG

Robert E. Green, Jr.
Center for Nondestructive Evaluation
The Johns Hopkins University
Baltimore, Maryland 21218
USA

Prof. Dr. Viktor Hauk
Institut für Werkstoffkunde
Rhein. Westf. Technische Hochschule
5100 Aachen
FRG

Prof. Dr. Clayton O. Ruud
159 Materials Research Laboratory
The Pennsylvania State University
University Park, Pennsylvania 16802
USA

ISBN-13:978-3-642-84005-0 e-ISBN-13:978-3-642-84003-6
DOI: 10.1007/978-3-642-84003-6

Softcover reprint of the hardcover 1st edition 1989

2161/3020 543210

Preface

Engineering structures for reliable function and safety have to be designed such that operational mechanical loads are compensated for by stresses in the components bearable by the materials used. What is "bearable"? First of all it depends on the properties of the chosen materials as well as on several other parameters, e.g. temperature, corrosivity of the environment, elapsed or remaining serviceable life, unexpected deterioration of materials, whatever the source and nature of such deterioration may be: defects, loss of strength, embrittlement, wastage, etc. DEFECTS and PROPERTIES of materials currently determine loadability. Therefore in addition to nondestructive testing for defects there is also a need for nondestructive testing of properties.

The third type of information to be supplied by nondestructive measurement pertains to STRESS STATES under OPERATIONAL LOADS, i.e. LOAD-INDUCED plus RESIDUAL STRESSES. Residual stresses normally cannot be calculated; they have to be measured nondestructively; well-approved elastomechanical finite element codes are available and used for calculating load-induced stresses; for redundancy and reliability, engineers, however, need procedures and instrumentation for experimental checks.

Three quantitative ndt-categories should be available for the qualification and/or quality assurance of structures during fabrication and operation:

- ndt for DEFECTS (ndtd)
- ndt for PROPERTIES (ndtp)
- ndt for STRESSES (ndts)

Ndtd and ndts are much further developed and more frequently applied than ndtp. In addition, the technical communities for ndtd and ndts are much larger than that for ndtp. National and international ndt conferences deal far more with ndtd than ndts and ndtp. In November 88 the second International Conference on Residual Stresses was held in Nancy (less than

60 miles from Saarbrücken), at which 205 papers were presented and with 280 participants attending. Most contributions dealt with just one ndts-technique, X-RAY DIFFRACTION. Second place was occupied by relaxation techniques in which residual stresses are partially released by drilling holes or machining notches. Releasing strain is measured by arrays of small strain gauges (rosettes). For shallow bore holes or notches this RELAXATION TECHNIQUE is at most slightly destructive. Both techniques, X-ray diffraction and relaxation, measure strains which are directly related to the stress states to be measured by the second order elastic moduli.

Unfortunately the major mechanical properties describe the nonelastic -- i.e. nonreversible -- behaviour of materials, which physically is not correlated to elastic behaviour; however, only testing in the elastic area is nondestructive. Consequently, it is due to physical reasons that ndtp is not possible for MECHANICAL properties in a direct manner as ndtd is for defects and ndts for stresses. (Nevertheless elastic moduli have become important ndt quantities; they carry information pertaining to the stiffness and microstructure of materials. Several papers were presented on this subject during the symposium). Other physical properties such as nonelastic mechanical properties -- electric, thermal, magnetic, etc. -- can be measured using ndt techniques. They strongly correlate with the microstructure of materials.

Macroscopic nonelastic properties and behaviour of materials, especially metals, physically depend on solid solutions (alloys), microstructure (dislocations, precipitations, etc.) and stresses. Physical metallurgy deals with these dependencies and will continue research on them as long as optimization and development of materials exist.

The situation at present: we do not have nd testing methods offering direct access to macroscopic mechanical properties and describing behaviour under loads, however, we do have nd testing technology for microstructure (and microstresses). Moreover, physical metallurgists have the know-how to derive macroscopic properties from microstructural data measured nondestructively.

The scenario described above is illustrated by two diagrams. The first diagram shows microstructural parameters and defects relevant to strength and toughness of materials plotted against their linear dimensions, resulting in the resolution needed for nondestructive materials characterization. A wide band from 10^{-10} to 10^{-2} is covered; the most important area for metals and ceramics and composites as well is 10^{-9} to 10^{-5}. In the second diagram ndt techniques are plotted with the same abscissa. There are more ndt techniques for microstructures and defects above 10^{-5} than below; but those below 10^{-5} already provide access to the most important microstructures relevant for mechanical properties and early stages of deterioration.

The nondestructive characterization (ndc) of materials by electromagnetic techniques was born at least 50 years ago. A few names of pioneers, for laboratory applications as well as for industrial testing should be mentioned (in alphabetical order): W.A. Black, F. Förster, W. Gerlach, W. Jellinghaus, H. Lange. One of these pioneers, F. Förster, has accepted our invitation to present a paper on the origin of electromagnetic methods.

It is the merit of C.O. Ruud and R.E. Green that a series of symposia devoted to nondestructive materials characterization was started in 1983. This first symposium, held in Hershey Pennsylvania, was a considerable success. The same holds true for the second symposium, organized by J.F. Bussière in Montreal. During a meeting of the organizing committee and the international advisory board held during the second symposium the representatives of several countries indicated their willingness and interest to host and organize the third symposium. The decision was made in favour of Saarbrücken, FRG.

The 3rd IS-NDC, organized by the DGZfP and the FhG-IzfP, was conducted in Saarbrücken from October 3-6, 1988. 225 scientists from 14 different countries took part. 81 oral presentations were given, 15 of them were invited plenary lectures and 37 poster presentations complementing the comprehensive program. The panel discussion involving P. Adam, G. Nardoni, H. Schneider, M. G. Seitz, Ch. Thoma, F. Tonolini and R. Zeller was chaired by D. O. Thompson. R. Sharpe gave an evaluation of the conference in his concluding speech along with providing an outlook for future objectives and development trends.

We would like to especially thank the numerous ladies and gentlemen of the staffs at the DGZfP, Berlin, and at the Institut für zerstörungsfreie Prüfverfahren, Saarbrücken, for their untiring efforts in helping to organize and conduct this symposium. We would also like to extend our thanks to the members of the international advisory board and the organizing and program committees.

The present proceedings volume contains 103 articles submitted by the authors. This represents the current level of knowledge in the field of NDC of materials. The editors would like to thank the authors for adhering to the restrictions on length imposed. This means that they were forced to concentrate on essential results, dispense with details and supplementary work. We would like to thank Springer Verlag for their efforts in quickly publishing the volume.

P. Höller V. Hauk G. Dobmann

Leardimensions of microstructure and defects

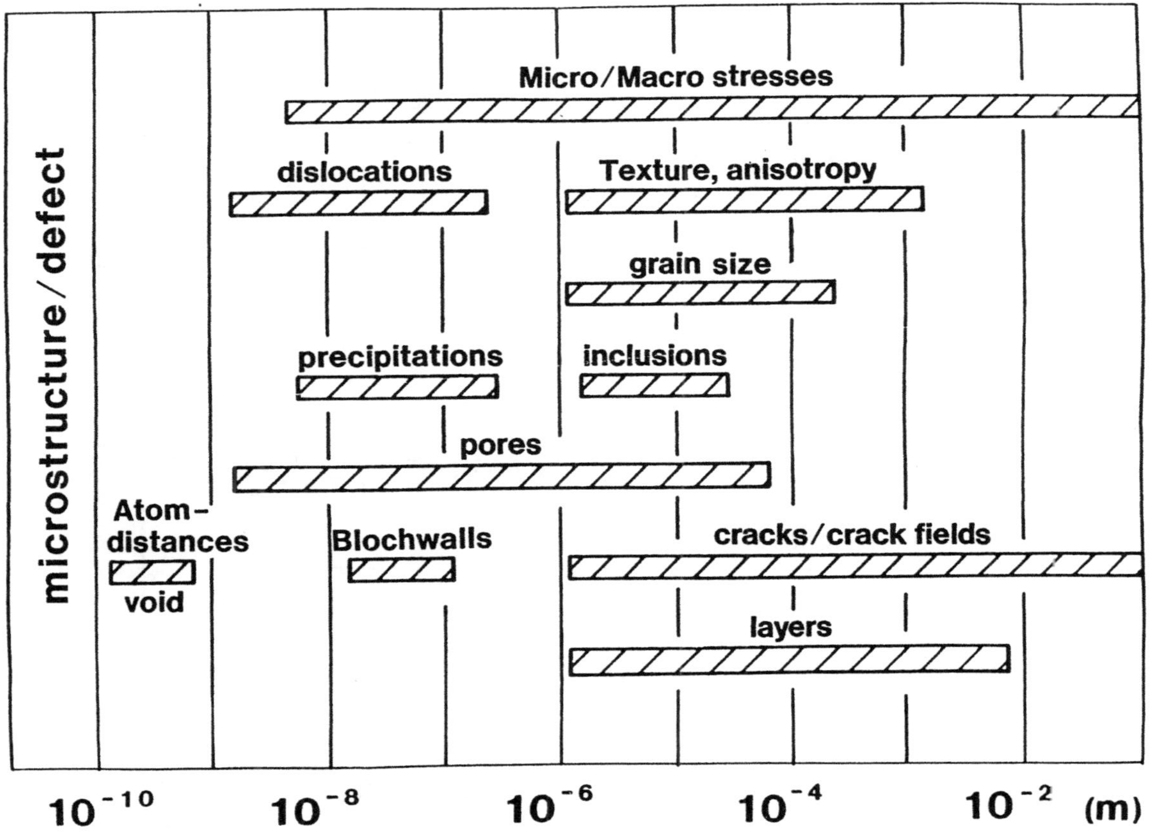

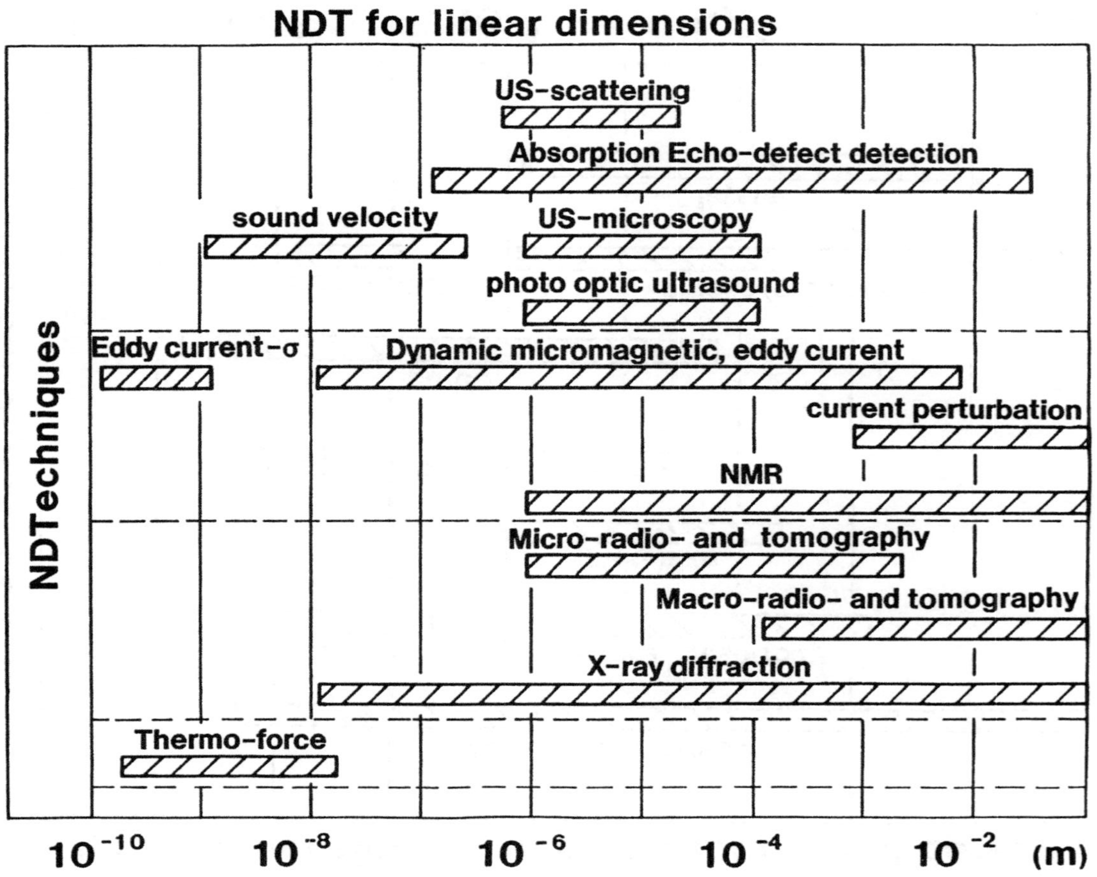
NDT for linear dimensions
NDTechniques
US-scattering
Absorption Echo-defect detection
sound velocity
US-microscopy
photo optic ultrasound
Eddy current-σ
Dynamic micromagnetic, eddy current
current perturbation
NMR
Micro-radio- and tomography
Macro-radio- and tomography
X-ray diffraction
Thermo-force
10^-10
10^-8
10^-6
10^-4
10^-2
(m)

Table of Contents

5. Texture

6. Microstructure, Stress State, Creep Damages

7. Electromagnetics

Research Programmes

Presentation of the Next BRITE-EURAM-Programme (1989-1992)

W. VAN DER EIJK - H. PERO

BRITE DIVISION
DG XII - Directorate Technological Research
Commission of the European Communities
200 rue de la Loi - 1049 BRUSSELS

Summary

The area of non-destructive testing is growing in importance due to industrial needs for better process control, assessment of materials, components and systems properties. It cannot be separated from the general trends towards the introduction of new materials and new technologies within the industry, which the next BRITE-EURAM programme will consider as a high priority.

Introduction

The Research and Technological Development programmes of the European Community aim at the scientific and technological improvement of European enterprises as regards their competitiveness.

For the first time in 1987 a Framework Programme on research has been adopted unanimously by the Council of Ministers. However, since 1985 some initial Community programmes had already been launched, like BRITE I, but a truly integrated Research and Technological Development programme was now necessary. The total Community budget for research and development has therefore been increased up to 5,4 BioECU, covering the period 1987-1991. The financial volume of this framework programme might seem very low in absolute terms (lower than 4% of total R&D budget of the member states or 2,5% of the European Community budget) but its strategic importance is fundamental for the stimulation it brings to the European enterprises.

The framework programme covers two types of actions:

- cost shared actions (including concerted actions) grouped in 8 chapters:
  - quality of life
  - towards a communication and information society
  - modernisation of industrial sectors
  - biological resources
  - energy
  - science and techniques for development
  - research and marine resources
  - intensification of European cooperation;
- direct actions carried out by the European Joint Research Centres.

BRITE-EURAM PROGRAMME (1989-1992)

BRITE-EURAM (Basic Research in Industrial Technologies in Europe)-(European Research on Advanced Materials) aims at the application of new materials and development of new industrial technologies within the European manufacturing programmes covering the research on new industrial technologies (BRITE) and on new materials (EURAM); the integration of these two programmes has revealed necessary due to their partial overlap and strong interactions.

The first BRITE programme started in 1985 consisting of a number of European collaborative projects.

Two calls for proposals have been launched, the first closed on May 1985, the second on May 1987, from which 215 projects have been selected and negotiated, corresponding to a total cost of about 360 MioECU; 50% of which are financed by the Commission of the European Communities and 50% by industry itself.

The first EURAM programme was adopted by the Council of the European Communities in June 1986; 91 projects have been selected after the 1986 call for proposals, corresponding to a total funding of 30 MioECU.

The next BRITE-EURAM (1989-1992) programme has been proposed by the Commission (COM(88) 385) to the Council of the European Communities in July 1988 and to the European Parliament; it has been defined on the basis of enquiries with the European industry, with experts, taking into account the IRDAC recommendations and following the Industrial "Strategic Planning days" on march 1988. It is essentially a "Market Pull" programme.

This programme is based on the fact that today the distinction between high-tech and traditional enterprises is disappearing. In the future, the frontier will be between those who exploit the new available technologies and those who do not.

The next BRITE-EURAM objective, as in the first programmes, is the improvement of the technological level of European manufacturing industry and through this its competitiveness in world markets. BRITE-EURAM sees it as its task to strengthen the cooperation between different sectors of industry, between enterprises, enterprises and research centres, enterprises and universities, in the areas of research towards new materials and development of new technologies.

BRITE-EURAM will cover the following areas, with, if possible, interdisciplinary and multisectorial approaches:
- technologies for advanced materials
- methodologies for design and quality assurance
- application of advanced manufacturing techniques
- technologies for manufacturing processes.

The total budget for the next 4 years will be roughly 900 MioECU, 50% of which will be financed by the Commission of the European Communities (439,5 MioECU).

90% of the budget will be dedicated to industrial applied research (financed 50% also by the industrial participants); this means actions to which 2 industrial independent enterprises from 2 different member states must participate.

Roughly 7% of the budget will be used to fund focussed fundamental research on specific themes to be carried out by universities - or associates research laboratories - from 2 different member states and supported through endorsement of their objectives by 2 independent industrial enterprises from 2 different member states. The funding on selected projects might be up to 100% of marginal costs.

Coordinated actions will also be financed in the frame of such a programme (100% of coordination costs) which offers a permanent call for proposals for such projects.

A particular attention will be given to projects involving SME's, such that their active participation might be encouraged. Feasability awards will be allocated to those SME's willing to prepare for the next call for proposals, by giving them limited funds to carry out good ideas by generating sufficient technical information and experimental evidence.

But with a limited budget this programme has to aim at the best research to assure competitiveness of the manufacturing industry. The allocated budget is not sufficient to finance all the received proposals (rate of success is 1 on 4), therefore a particular importance has to be given to the main criteria for selection which are detailed in the BRITE-EURAM information package.

Organisations from EFTA countries will be allowed to participate on collaborative research projects, under specific conditions.

It is expected that the list of the yearly calls for proposals will be published in December 1988.

Each year a new call for proposals will be launched with possible variation of priority themes. This programme would permit the funding of roughly 400 projects at the end of the reference period (1989-1992).

Non-destructive evaluation of materials

BRITE-EURAM is considering this topic as a high priority (11% of the current BRITE-EURAM projects) and technologies for non-destructive testing applied to materials characterisation and process control will be taken into account within the priority themes for the next call for proposals (1989):

2.2.1 detection and inspection of defects and inhomogeneities at the micrometer level in materials and products;

2.2.2 applications of multisensor systems for monitoring and inspection of industrial processes operating at high production rate;

2.2.3 inspection techniques, particularly when based on optical engineering methods, for use in hostile industrial environments;

2.2.5 in-process testing methods to replace destructive examination for characterisation of products;

4.1.3 new methods for the inspection and in-process quality control of surface treated components;

4.2.4 reduction in defect level through on-line process monitoring and automation of joining systems;

4.4.3 control of fine particles (1-10 micrometers) in manufacturing particle and powder processes.

These themes are not exclusive - but any non related proposal will have to be strongly argumented; priority themes will be revised every year and any new suggestion will be taken into account by the Commission.

This conference will surely define new trends for the future R&D programmes which BRITE-EURAM might consider in the next years.

Materials Research Programme of the Federal Ministry for Research and Technology (BMFT)

W. Faul

Materials and Raw Materials Research and Development Agency (PLR)
in the Kernforschungsanlage Jülich GmbH (KFA), FRG

In 1985 the Federal Ministry for Research and Technology has started the **Programme Materials Research** with a running time of ten years and a total governmental contribution of 1.100 Mio. DM.

The programme is covering five R+D-topics in the field of
Ceramics
Powder Metallurgy
Metallic High-Temperature Materials and Special Materials
New Polymers
Composites

The Materials and Raw Materials Research and Development Agency (PLR) is responsible for the programme management concerning
- advice for the preparation of R+D-proposals,
- scrutinity of submitted proposals concerning scientific-technical, financial and administrative aspects,
- preparation for the BMFT-decision relating project funding,
- supervision of supported projects concerning scientific-technical progress and transfer of funds,
- organization of seminars and workshops regarding the presentation of R+D-results and
- evaluation of new R+D-topics for the programme.

The general objectives of the programme are the support of research and development in selected key areas of material science with high innovation potential for the production of materials with considerably improved or novel mechanical and physical properties for industrial use.

To realize enhanced know-how and technology transfer, interdisciplinary joint projects between research institutes and industrial laboratories with well defined working programmes - checked one against another - in application oriented basic research, applied research and the demonstration of the principal technical feasibility are to be supported. Significant industrial contribution to the total R+D-budget (in the range of 50 %) is required.

Ceramics

In the field of high performance ceramics, including glass-ceramics and glass, structural materials with high thermal, chemical and corrosion stability, high hardness, strength and improved toughness and thermoshock resistance are to be developed for use e. g. in engines, turbines, equipment for mechanical and chemical processing

The R+D-objectives are

- development of improved or new methods for powder manufacturing: particle size, particle size distribution, crystalline and amorphous particles, new combinations and compounds, additives,
- development of new powder handling, compaction, sintering and post processing methods, e. g. machining, joining; homogeneity, cleanliness, residual stresses, shrinkage, dimensional tolerance,
- reliable and reproducible processes in order to manufacture prototypical components,
- nondestructive testing of components.

Ceramics with special physical properties, e. g. with improved or novel electrical, electronical, optical or magnetical properties are to be developed for application, e. g. in information and communication systems, microelectronic devices, energy transformation and storage systems, sensors.

The R+D-objectives are
- development of processes to manufacture bulk materials, powders or (thin) films with improved or novel physical properties, e. g. high electrical, ionic or thermal conductivity, linear and nonlinear optical properties,
- new combination of materials, new compounds, additives, doping, ion-implantation, laser/ion-beam mixing.

Powder Metallurgy

This topic is related to the powder metallurgical manufacturing of high performance components for use e. g. in engines, turbines, aircrafts, equipment for mechanical and chemical processing.

The R+D-objectives are
- the development of methods for manufacturing powders and their further processing to semifinished products and structural components; metals, mixtures of metals and hard materials,
- development of new materials for structural components having tailored, unusual properties, which cannot be achieved (at least not economically) by other methods e. g. by means of ingot metallurgy,
- ultra-rapid solidification, mechanical alloying,
- characterising and processing of fine powders,
- production of locally varied properties in structural components,
- testing methods (e. g. nondestructive).

Metallic High-Temperature Materials and Special Materials

Metallic high performance materials are needed to be used at elevated temperature e. g. in engines, power stations, turbines and aircrafts

The R+D-objectives of this topic are

- new alloys manufactured by means of ingot metallurgy with mechanical and technological properties which can be used in structural components at higher temperatures,
- improved high temperature strength,
- increased corrosion stability and resistance to wear at high temperatures,
- technical exploitation of the potential of rapid solidification processes,
- development of new manufacturing processes and (nondestructive) testing methods.

New Polymers

Advanced and new polymers as structural materials with mechanical high performance properties, e. g. high strength, stiffness, thermal stability, are to be developed for use e. g. in cars, household equipments, chemical plants, aircrafts and space

The R+D-objectives are

- synthesis of new mechanical high performance polymers and manufacturing of components,
- highly oriented molecular structure,
- liquid crystal polymers,
- polymerblends and highly ordered co-polymers,
- surface modification,
- characterization and (nondestructive) testing,

Advanced and new polymers as functional materials with high physical performance properties, e. g. electrical, electronical, optical, magnetical properties, are needed for use e. g. in electronic devices, information and communication systems, information storage, energy conversion and storage, sensors.

The R+D-objectives are
- synthesis of new monomers and polymers with physical high performance properties and processing/manufacturing of components,
- intrinsic electrical or ionic conductivity, linear and nonlinear optical properties, high transparency,
- piezo-, pyro- und ferroelectrical properties,
- photosensitivity, electron beam sensibility, electron donor and acceptor properties.

Composites

The topic covers the development of high performance materials as structural materials; reinforced polymers, metals and ceramics with high mechanical properties, e. g. high strength, stiffness, thermal and thermal-shock resistance, and low weight for use e. g. in cars, engines, turbines, aircrafts and space.

The R+D-objectives are
- development of new materials, manufacturing and (nondestructive) testing of components,
- materials for matrix reinforcement (e. g. fibres, whisker),
- development of composites with polymeric matrices (thermosets, thermoplasts), ceramic matrices, metallic matrices (light metals),
- testing methods (nondestructive),
- suitability and feasability for mass production.

Survey of joint-projects and costs
(Status Oct. 1988)

R+D-topics	number of joint-projects	number of cooperating labs in institutes and industries	total costs (Mio.DM)	BMFT-contribution (Mio.DM)
Ceramics	29	99	251	124
Powder Metallurgy	53	138	94	58
Metallic High-Temperature Materials and Special Materials	62	156	176	99
New Polymers	62	103	200	103
Composites	59	150	220	116
Total	265	646	940	501

Ceramics

Properties, Preparation and Requirements to Testing of Ceramic Materials

H. SCHMIDT

Fraunhofer-Institut für Silicatforschung
Neunerplatz 2, D-8700 Würzburg, Fed. Rep. Germany

Summary

The strength of ceramic materials is mainly defined by the defect or flaw size in the bulk as well as on the surface of the component. Defects can have various origins like raw materials, processing, moulding, sintering, and finishing. For a break through of ceramics as engineering materials the minimization of flaw size and concentration is necessary as well as the improvement of non-destructive test procedures. The test methods should be able to detect bulk and surface flaws, even in complex component shapes, to detect density fluctuations, especially in green bodies, and stresses in the final compound. They should be adaptable to industrial production processes.

Zusammenfassung

Die Festigkeit keramischer Werkstoffe hängt entscheidend von der Größe der Fehler ab. Fehler resultieren aus dem Rohstoff, dem Verarbeitungs- und Formgebungsverfahren, dem Sinterverlauf und der Endbearbeitung. Zur breiten Anwendung keramischer Werkstoffe im Maschinenbau sind Fehler an der Oberfläche und im Innern von Bauteilen sicher zu detektieren. Außerdem sollten prozeßbedingte Dichteunterschiede (besonders in Grünkörpern) sowie Eigenspannungen (auch in komplexen Bauteilen) sicher nachgewiesen werden können. Die Prüfmethoden sollten an industrielle Fertigungsprozesse anpaßbar sein.

Introduction

Ceramic materials are characterized by their brittleness, potential for high strength and high temperature stability. These basic properties have led to immense efforts in research and development to introduce ceramics into modern technologies for industrial use. But despite these efforts, the break through of structural ceramics has not yet taken place. The basic obstacles are still the same for several thousand years: The ceramic materials do not have mechanisms to decrease stresses by plastic deformation and therefore defects, at the top of which loads can create high stresses, act as fracture origins. This leads to the situation, that cracks start from defects, and under supercritical conditions catastrophic crack propagation takes place. To improve the fracture behavior (that means to improve fracture toughness, described by the K_{IC} value) various strategies have been developed (fracture toughnesses of ceramics range from about 3 up to 10; steel: 50 - 100):

- to increase fracture toughness by strenghtening concepts [1-3]: transformation toughening of ZrO_2; composites, e.g. ZrO_2/Al_2O_3

- reinforced ceramics (fibers, whiskers) [4]

- to avoid defects by improvement of processing and microstructure (e.g. better powders, agglomerate-free processing, clean room techniques).

All these efforts could not solve the problems until now and other questions like fabrication of large parts, high speed production technologies or finishing technologies are just at their beginning. Since defect-free fabrication is very difficult on a large scale production, one should, at least, be able to detect dangerous defects in ceramic components as early as possible during the production process with non-destructive techniques.

Defect origins

Defects in ceramic components can have various origins. Table 1 gives a survey over some important defect sources. The flow sheet on page 4 shows typical ceramic preparation and production procedures and the correlation to defect sources. It can be easily concluded that it is rather difficult to build up production technologies which can guarantee defect-free large scale fabrication of high performance ceramic components. Figures 1 and 2 demonstrate the formation of textures and density

Table 1. Origin of defects.

type	source
processing defects	impurities (dust, organics, grinding),
	inhomogeneities
	mixing (multicomponent systems, separation, distribution of compounds, agglomerates)
	texture (flow textures, layer formation by pressing)
	density fluctuations
	moulding (casting, pressing, injection moulding)
microstructural defects	pores, agglomerates, large crystal growth, microcracks
stresses	temperature differences during sintering, density fluctuations in the green body.

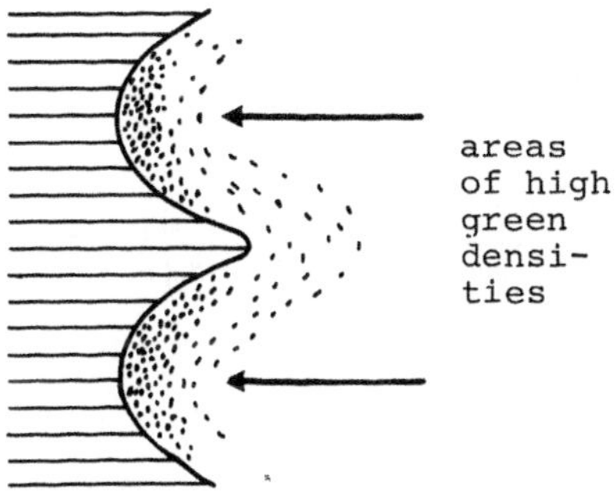

Figure 1. Density fluctuation from slip casting of complex shapes.

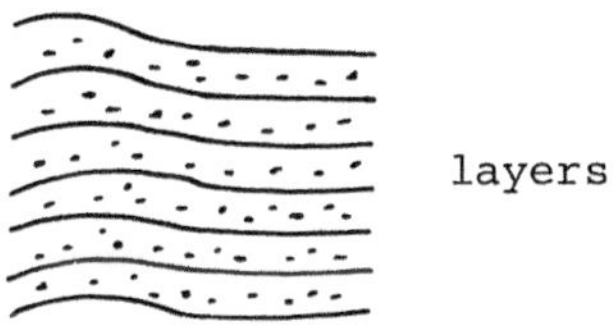

Figure 2. Layer formation resulting from dry pressing.

fluctuations of dry pressing and slip casting processes.

Ceramic composites show special features with respect to defects. In this case artificial defects are created (microcracks, "inhomogeneities" in form of whiskers, fibers or particles,

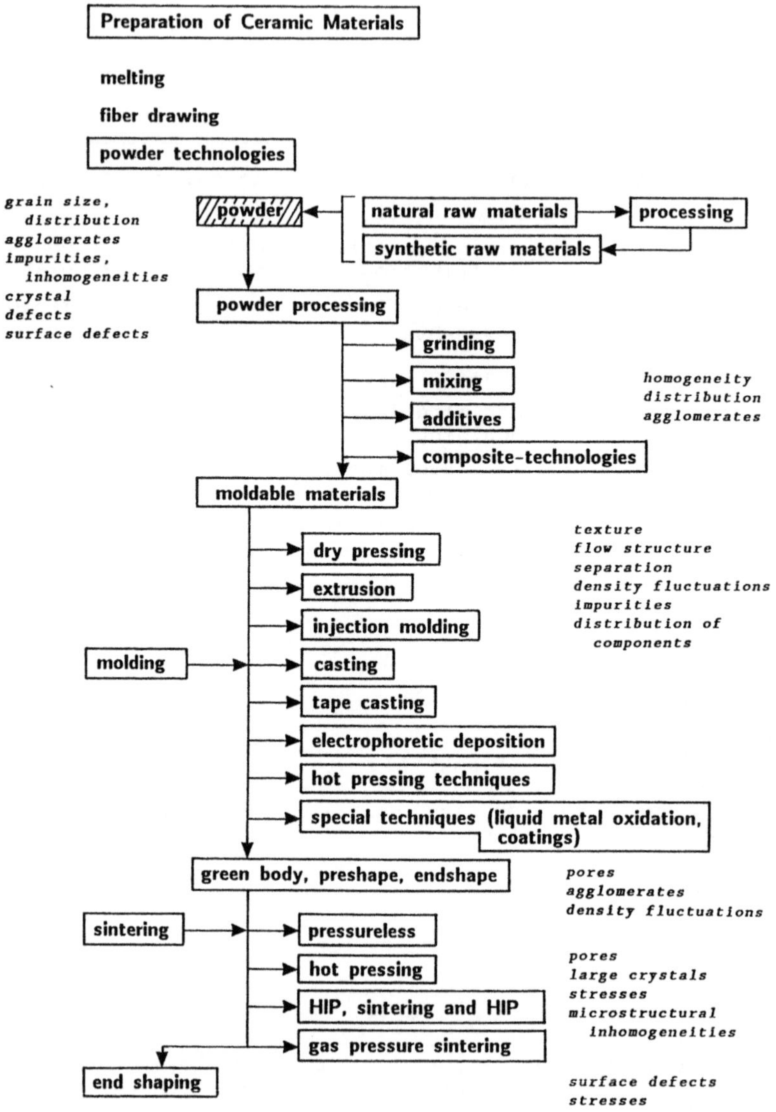

e.g. ZTA or PSZ ceramics) and are incorporated to increase fracture toughness. It is difficult to define critical defects in such systems. According to Rice [5], in BN/Al_2O_3 composites, fractography has shown, that fracture origins are only found in Al_2O_3-rich areas which are inhomogeneity type defects. Defects in fiber reinforced ceramics can be based on the pore formation between the fiber and the matrix (figure 3).

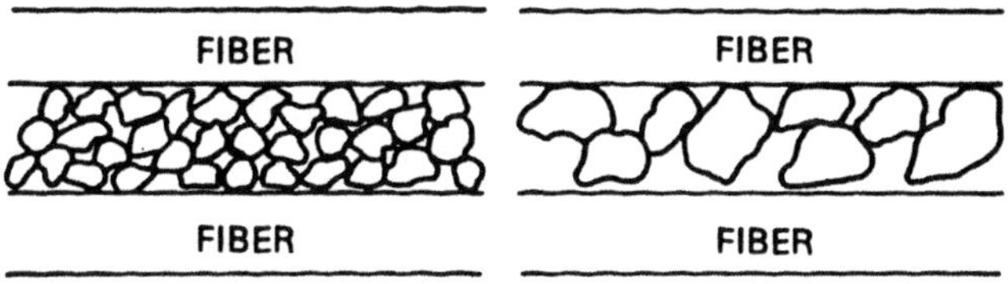

Figure 3. Defect formation during sintering in fiber reinforced ceramics.

In polycristalline ceramic materials, the mechanical strength can be correlated clearly to the defect size. Figure 4 (after Petzow) [6], shows the correlation for several types of ceramics. The effect of flaw size was determined on Al_2O_3, moulded by electrophoretic deposition (plates 8 cm in diameter, 5 mm thickness, double ring bending test, figure 5).

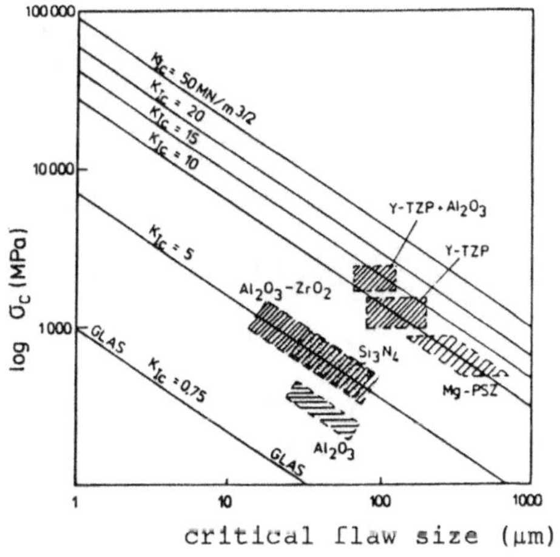

Figure 4. Dependence of strength on critical flaw size.

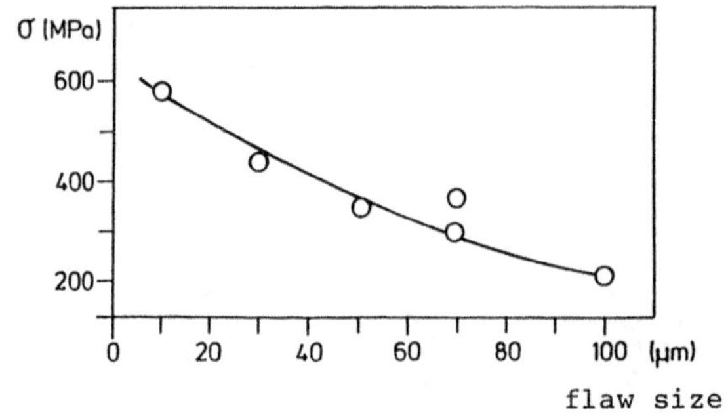

Figure 5. Strength dependence on flaw size of Al_2O_3 (electrophoretic deposition)

Figure 6 shows a large flaw, reducing strength to about 300 MPa. One can conclude, that for high performance ceramics flaw sizes should clearly be as small as 10 μm or less.

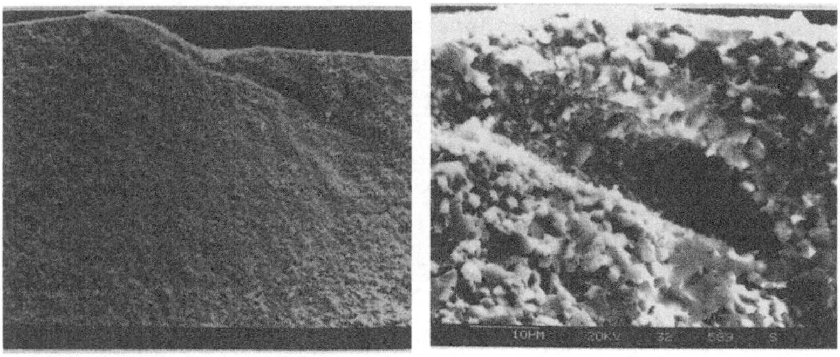

Figure 6. Microstructure of an Al_2O_3 sintered body (electrophoretic deposition).

In figure 7 [7] the fracture origin of a high strength transformation toughened ZrO_2 ceramic is shown. The defect is due to a finishing process (grinding and polishing). This demonstrates, that surface flaws have to be controlled as well as bulk flaws. Critical flaw sizes can be a result of permanent stresses as of a fatique process, too. Figure 8 shows this on

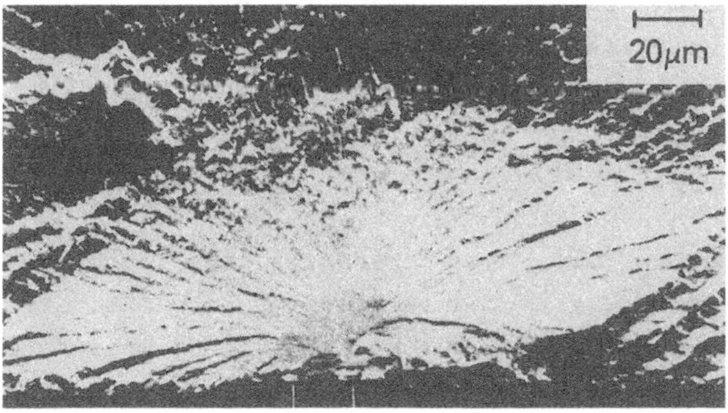

Figure 7. Surface flaw of a toughened ZrO_2 sintered body.

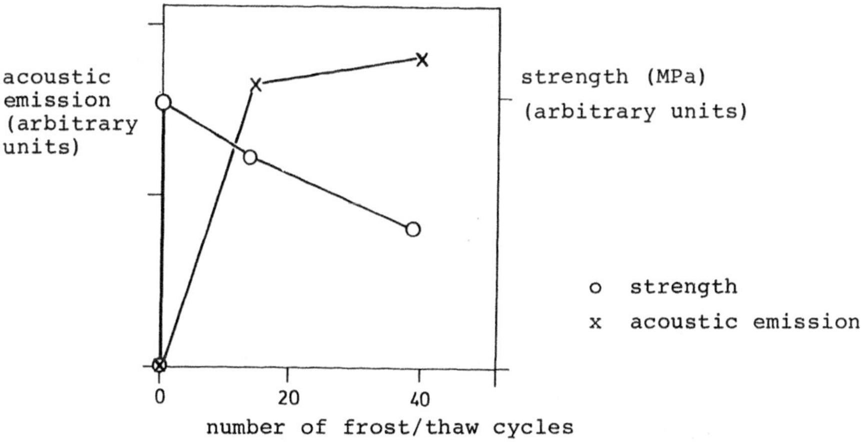

Figure 8. Fatique of a natural sandstone due to cyclic loads. Acoustic sound emission versus mechanical strength.

a "natural" ceramic, a sandstone. Frost/thaw cycles are followed by acoustic emission (summarizing the energy). Bending experiments show that crack formation reaches a critical level only after ten to twenty cycles. Therefore, it is desirable to develop test procedures which can be able to control flaw formation in ceramic components even after practical use in order to detect critical states.

Conclusions

For making better ceramics, two strategies have to be followed: Firstly, it is necessary to reduce flaw sources. One very important source seems to be processing and finishing. Therefore, the whole technology has to be improved: making better powders, improve powder processing, green body fabrication, sintering and finishing. Secondly, in order to control flaws, it is necessary to improve flaw detection procedures for bulk and surface, to detect density inhomogeneities expecially in green bodies (even with complex shapes) and to detect stresses in the endshaped components. The flaw size to be detected should be less than 10 µm. The test procedures should be able to be adapted to high speed production lines. Both improvents are an indispensable prerequirement for the break through of ceramic compounds as engineering materials.

Acknowledgement

The author wants to thank Dr. Naß and Dipl.-Phys. Storch from Fraunhofer-Institut für Silicatforschung and Dr. Riedel from Fraunhofer-Institut für Werkstoffmechanik for the experimental aid and helpful discussions, the Minister für Forschung und Technologie and several industrial companies for their financial support.

References

[1] Garvie, R.C.; Hannink, R.H.J.; Pascoe, R.T.: Nature 258 (1975) 703.

[2] Garvie, R.C.: Structural Applications of ZrO_2-Bearing Materials, Advances in Ceramics 12. Am. Ceram. Soc., Columbus, 1984.

[3] Rühle, M.; Claussen, N.; Heuer, A.H.: J. Am. Ceram. Soc. 69 (1986) 195.

[4] Li, Z.; Bradt, R.C.: Materials Science Volumes 24 - 36 (1988) 511.

[5] Rice, R.W.: Cer. Eng. and Sci. Proc. 2 (1981) 493.

[6] Petzow, G.: Keram. Z. 40 (1988) 422.

[7] Rice, R.W.: Ceramic Fracture Features, Observations, Mechanisms and Uses, Fractography of Ceramic and Metal Failures (ASTM ST 827, 1984) 5.

Material Characterization of Ceramics by Various Nondestructive Testing Methods

T. Kishi

Research Center for Advanced Science and Technology
The University of Tokyo, Japan

K. Kitadate

Japan Fine Ceramics Center, Japan

Summary

In order both to obtain the reliability of components and to understand increasing mechanism of fracture toughness, nondestructive detection, location and characterization of microcracks less than 100 μm is an essential technology. In this report experimental results, which has been carried out in Japan Fine Ceramics Center as a project of NDI, is summarized on the detectability of various detection techniques such as Ultrasonic, X-ray, Acoustic Microscope, Scanning Laser Acoustic Microscope, X-ray CT, Neutron Radiography NMR, Electron Beam, simulated Ultrasonic and Acoustic Emission. Lastly the relation between these microcracks and fracture strength is discussed [1, 2].

Materials and Defects

Standard testing materials used are sintered and hot pressed Si_3N_4, and various artificial defects are included. Shape and size of testing specimens and size, shape and position of various defects are represented in Table 1.

Testing Results

Micro Focus X-ray Nominal focus diameter of X-ray tube is 10 μm. In this equipment all the slits of which size are larger than 20 μm width and 12 μm depth are clearly detected. Fig. 1 shows the detectability of slit penetrator where specimen thickness is changed. This result shows that detectability depends on measuring conditions, and under this detective condition 0.5 ∿ 1.0% of defects size / specimen thickness are observed. In comparison with ultrasonic testing, a detectability of Microfocus X-ray for microcracks is not

Table 1. Shape, Size and Artificial Defects of Specimens

Model Specimens

	Type of Defects	Defects Size
Surface Defects (HP)	Line Defect 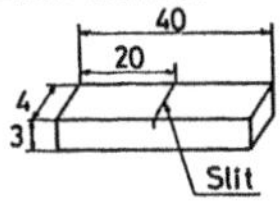 Width 20 µm	Depth 150 µm 100 70 50 30 20 10
	Indentation 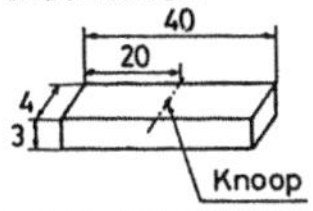	a = 100 µm 50 30 20 10
	Indentation 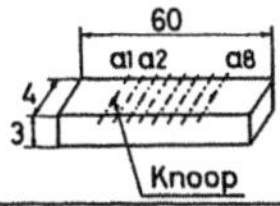	a1 = 100 µm ~ 50µm Interval a6 100µm ~ Interval a8 = 500 µm

Model Specimens

	Type of Defects	Defects Size
Internal Defects (HP)	Line Defect 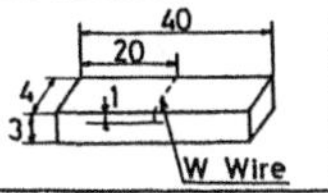	φ 150 µm 100 70 50 30
	Spherical Defect 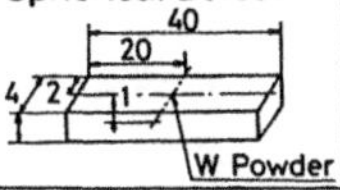	φ 150 µm 100 70 50 30
Porous Internal Defects (RS)	Line Defect 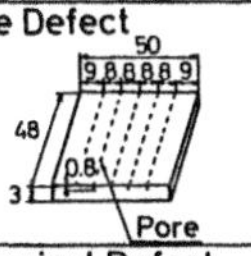	φ 75 µm 50 30 20 10
	Spherical Defect 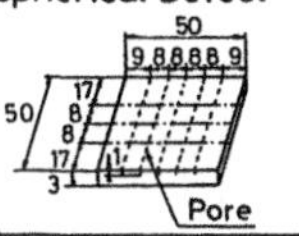	φ 190 µm 130 85 55 30

Model Specimens

	Shape	Size(mm)	Type of Defects (µm)
Green	 Sphere, Line	φ18x60 φ30x60 φ50x60 φ80x60	Nyron Wire : 74,104,235,370,570 Latex :173,273,531,920 W Wire : 50,100,200,300,500 W Sphere :100,150,200,500
Sintered	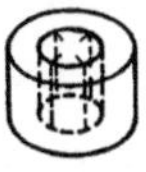	φ30x50 φ50x50 φ80x50	Slit Defect Width : 20,50,100,500 Depth :100,200,300,500,1000
	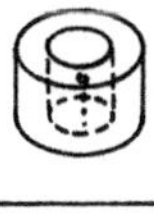	φ30x50 φ50x50 φ80x50	Drill Hole (Width = Depth) 0.3,0.5,1.0,2.0 mm
	Penetrameter 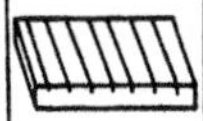	15x50x3 15x50x3 15x50x3 15x50x4	Slit Defect (Width = Depth) 50, 60, 80,100,125,160,200 100,125,160,200,250,320,400 200,250,320,400,500,640,800 400,500,640,800,1000,1280,1600

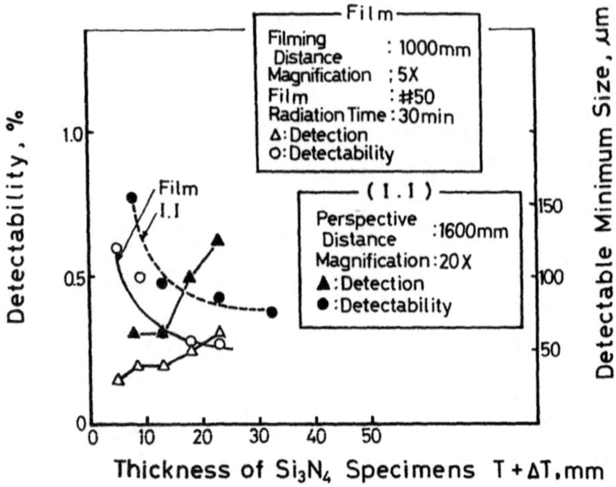

Fig. 1. Detectability of Microfocus X-ray

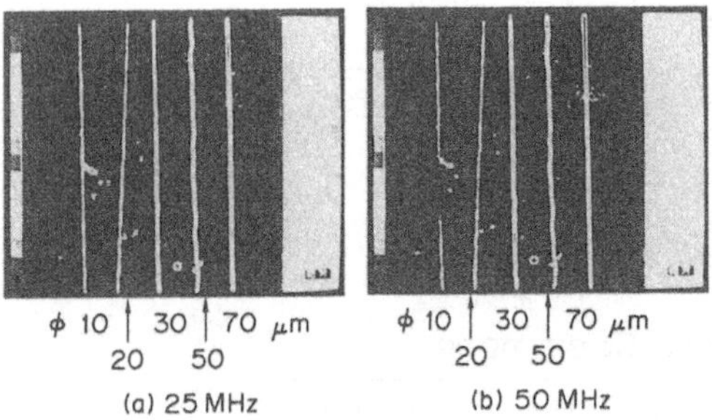

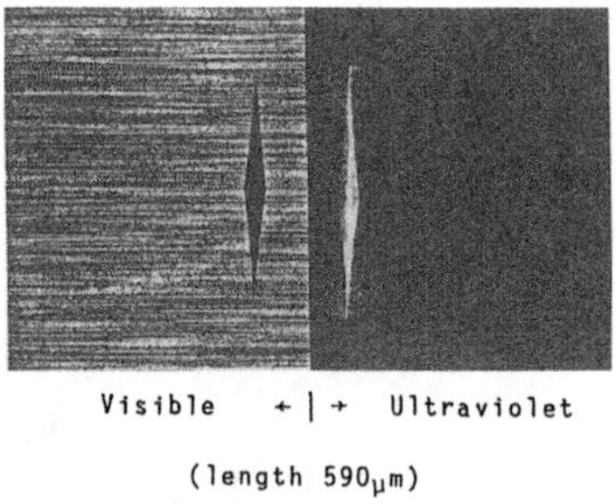

Fig. 2. Detection of Surface Slit by High Frequency UT Method

Fig. 3. Penetration Result of Knoop Indentation

dominant, but this technique does not depend on the position of defects in the specimen.

Ultrasonic Testing (UT) Fig. 2 gives the result of various surface slits detected by 50 MHz surface scattering method, and 10 μm depth minimum defect is clearly detected. Internal line defects are detected by 25 MHz angle beam technique, and minimum defect less than 10 μm is also detected. So it is concluded that UT technique is a most applicable method to detect smaller microcracks, in spite of its rather inferiority compared to acoustic microscopy. Further subjects in this technique are to evaluate shape, size and all kinds of defects.

Penetration Testing (PT) Surface line slit lager than 16 ∼ 20 μm and surface indentation larger than 260 μm (of which depth ratio is one tenth) are clearly detected as shown in Fig. 3. The detectability depends on the ratio of defects width to its depth, and relatively deeper defects can be easily observed, so Knoop indentation can not be observed so easily.

Scanning Acoustic Microscope (SAM) In this testing project, SAM showed a most excellent detectability, and could inspect all the minimum surface linear, internal linear and internal spherical defects, and 20 μm knoop indentation and 2 μm depth slip are also detected. Fig. 4 is an obtained image of 50 μm knoop indentation. Surface defects can be detected by 500 MHz ∼ 700 MHz reflection method. Due to the wave length limit of 500 MHz surface wave, the limit of resolution is 3 μm width and 0.3 μm depth. Internal defects are detected by 50MHz through transmission method, and 30 μm diameter internal pore defects are also clearly observed.

Scanning Laser Acoustic Microscope (SLAM) 100 MHz incident wave and 22 μm laser scanning are used in this test. All the surface line defects and internal pores are detected and Knoop indentations larger than 100 μm length with 10 μm depth are clearly detected. Fig. 5 shows a surface slit defect with 10 μm depth. However in the case of internal W wire, the minimum detectability is 100 μm diameter line and 50 μm diameter sphere, of which inferior results depends on the resolution principle of SLAM. So further researches concerning to the principle of scattering and attenuation of ultrasonic waves are necessary to develop the SLAM technique.

X-ray CT Third generation type equipment of 512 channel width 140 KV-300 mA pulse X-ray source is used for the test of 2 mm thickness specimens. Fig. 6 shows a X-ray CT result of 300 mm diameter cylindrical specimen, which includes slit defects, and in Table 2 the detectability of X-ray CT is summarized,

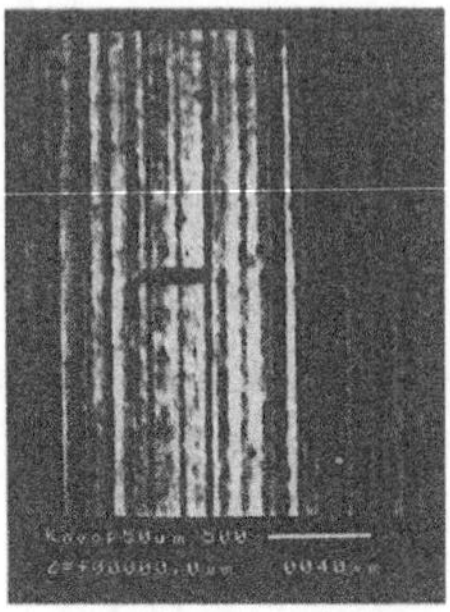

Fig. 4. Defected Photo of 50 μm Knoop Indentation by SAM Reflection Method

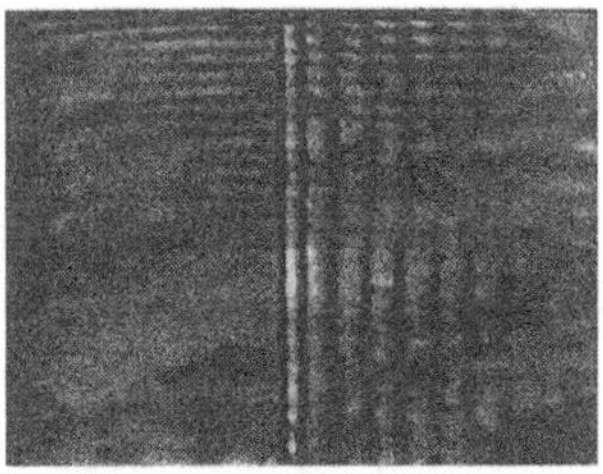

Fig. 5. SLAM Observation Result of 10 μm Depth Surface Slit

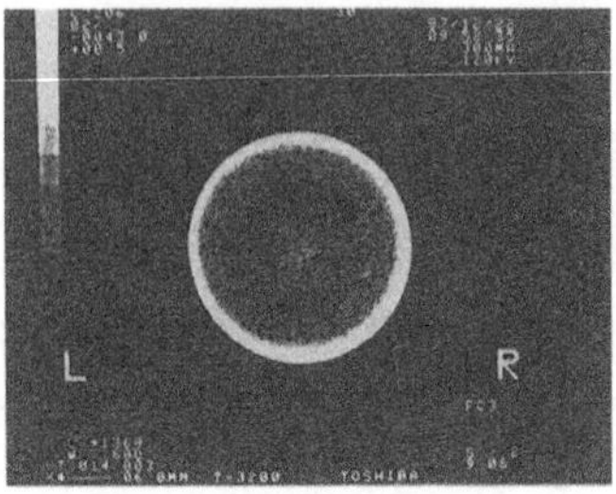

Defect depth	Defect width 20 μm	50	100	500
100 μm	×	×	×	○
200	×	×	△	◎
300	×	△	○	◎
500	×	○	○	◎
1000	△ (600)	○	○	◎

◎ : excellent ○ : good
△ : not clear × : no

Fig. 6. X-ray CT Photograph of 30 μm Diameter Cylindrical Specimen with Internal Slit, and its Detectability

which is one order larger than that expected in ceramics. So further progress such as to increase tube voltage is necessary for the application of X-ray CT to microcrack detection in ceramics, and so new microfocus X-ray CT technique has to be developed in future.

Neutron Radiography (NR) Cyclotron (CUPRIS370) and Nuclear Power Furnace are used as a source of neutron. Fig. 7 shows the detectability of NR in slit penetrometer as a function of specimen thickness. Fig. 8 represents an image photo of green body, which includes some inclusions. In this case neutron radiography is much more effective than X-ray technique and it

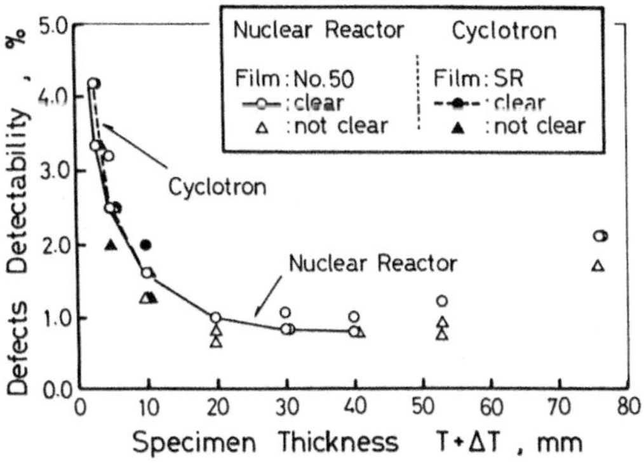

Fig. 7. Detectability of Slits in Penetrating Power Gauge by Neutron Radiography

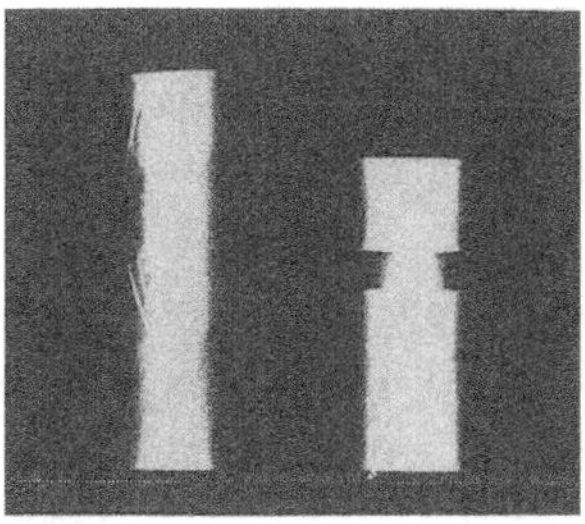

Fig. 8. Neutron Radiography of Green Specimen with Nylon and W Wires

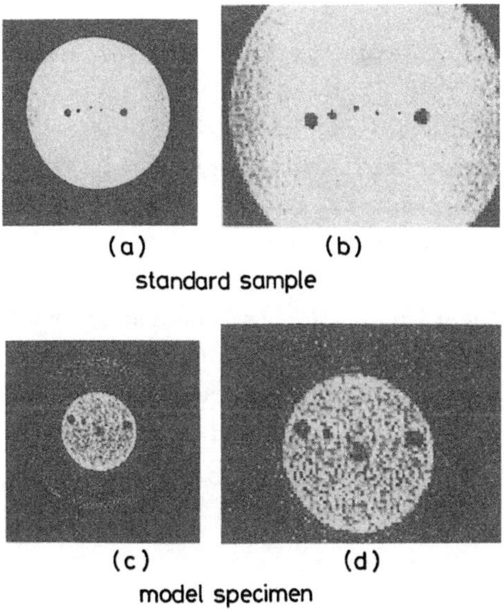

Fig. 9. NMR Projection of Line Defects in Green Body

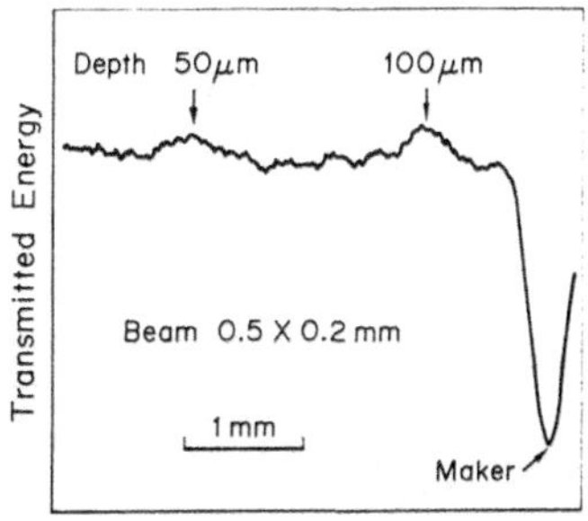

Fig. 10. Detection of Surface slit by Voltage Electron Beam

is expected that this technique will become a powerful one in the field not only of green body but also of composite materials, joining parts and thin films.

NMR-CT Fig. 9 shows a photo of green body with artificial linear defects of optional fiber of 1,000 μm, pencil lead of 50 μm and various yarn of 205 ∿ 370 μm. All the defects are clearly detected and this technique has a possibility to detect the defects of green body which includes ^{I}H proton in water.

Electron Beam Radiography 16 Mev Linac, of which electron beam energy is 10 MeV and beam size is 0.2 ∿ 0.5 mm, is applied for the test. Fig. 10 shows a result of through transmission method with the sample of 3 mm ϕ which includes both 50 μm depth × 30 μm width and 100 μm width × 100 μm depth. This indicates the possibility of EBR as a new detection method of thin film ceramics if the beam diameter become much smaller and image recording system is developed.

Hammer Drop Signal Analysis Simulated signals due to steel ball dropping becomes a source of elastic wave and the frequency analysis of this propagating wave will become a useful technique to analyze materials characteristics. Fig. 11 shows a result, and different frequency spectrum is obtained between different defects specimens.

Summary of the Advanced NDT Technique It is concluded that high frequency UT, SAM and SLAM techniques are effective under current technique level, and microfocus X-ray is a most desirable one based on future's development, and microfocus X-ray CT technique is strongly required.

Acoustic Emission

AE is a unique method to detect growing cracks, however AE signal amplitude in ceramics is relatively low because AE activity depends on the volume and rising time of each cracking. So we developed low noise measuring system. Fig. 12 show the relation between AE generating stress and bending strength σ_B. Some signals generate at the fracture stress and another signals at the much lower level. Fig. 13 shows that the specimens which include relatively smaller defects less

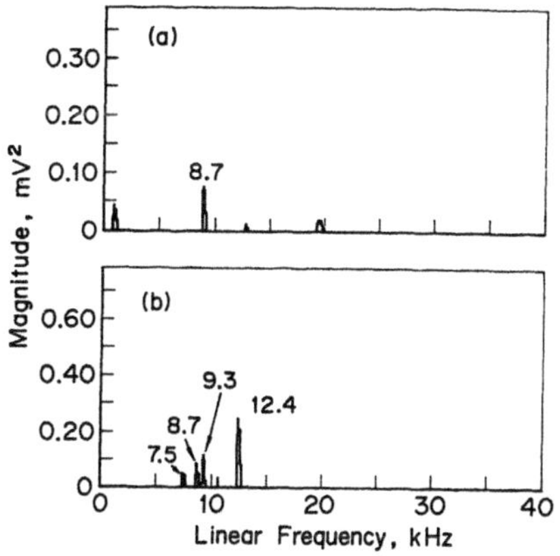

Fig. 11. Frequency Spectrum of Detected Signals due to Simulated Steel Ball

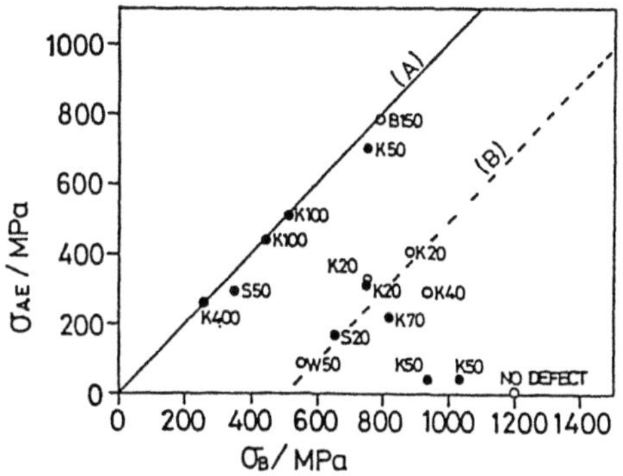

Fig. 12. Relation between AE Generating Stress σ_{AE} and Bending Strength σ_B

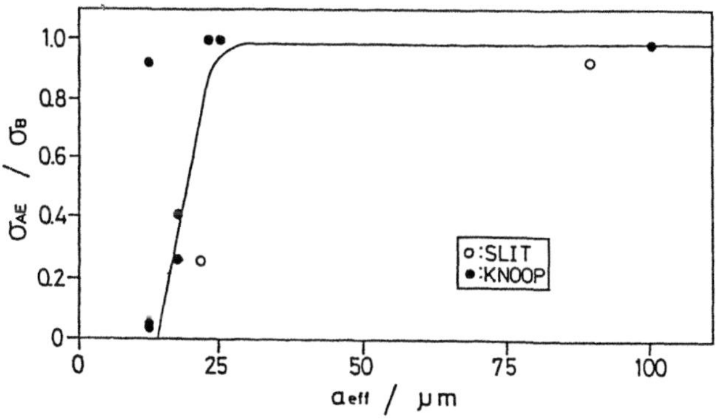

Fig. 13. Normalized AE Generating Stress as a Function of Defects Size

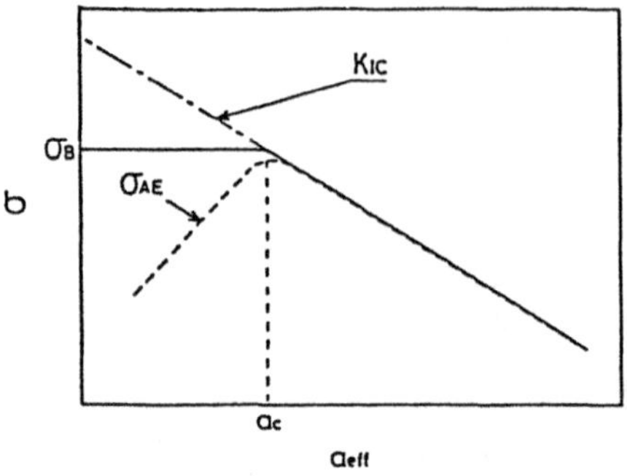

Fig. 14. Bending Strength and Microcrack Generating Stress σ_{AE} as a Function of Crack Size a_{eff}

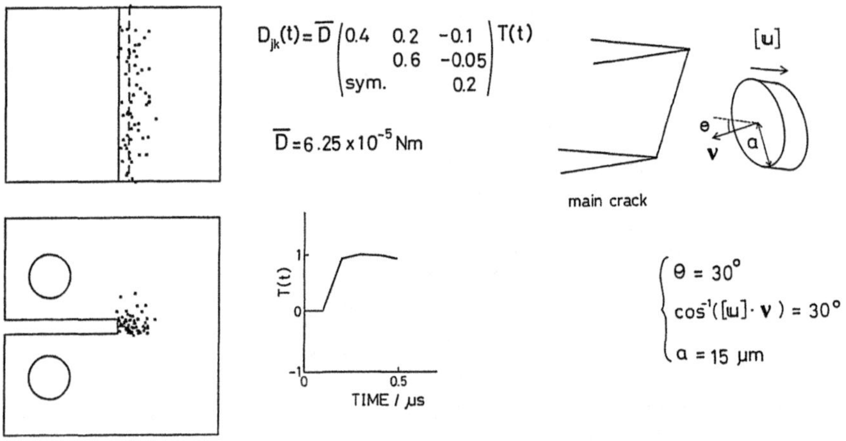

(a) (b) (c)

Fig. 15. AE Source Characterization of Si_3N_4
(a) Three Dimensional Location
(b) Moment Tensor
(c) Size, Mode and Inclination of a Crack

than 30 μm gives an lower AE generating stress. Fig. 14 shows the relation between defects size and bending strength and also shows the σ_{AE} stress. In the larger defects size than a_c, bending strength is reduced with the increase of a_{eff}, which obeys to linear fracture mechanics concept. On the contrary in the region of smaller cracks AE generate at the lower applied stress. This means that smaller defects grow from original size to critical size "a_c" during loading and fracture occurs at the same stress level of σ_B.

Lastly application of AE source characterization in Si_3N_4 materials is shown in Fig. 15 (a), (b) and (c). Fig. 15 (a) is a result of the precise three dimensional location, Fig. 15 (b) is an obtained source function tensor D_{jk} (seismic moment) and Fig. 15 (c) gives the example of cracking size, mode and inclination plane angle derived from the result of Fig. 15 (b). These D_{jk} is obtained by deconvolution integral such as

$$V(t) = S_i(x', t) * G_{ij,k}(x', X, t) * D_{jk}(x, t)$$

where V(t) is a detected signal, S_i transfer function of measuring system, D_{jk} dynamic Green's function, and * shows deconvolution integral [3, 4].

Conclusion

Detectability of various advanced NDI for the micro defects in Si_3N_4 ceramics is examined. High frequency UT, SAM, SLAM technique show an potential result and the importance of microfocus X-ray CT technique is emphasized. AE has a high detectability and also shows a close relation to the microfracture process. AE source characterization is verified to be a useful technique to evaluate a microcracking quantitatively.

Reference

1. Munz, D., Rosenfelder, O., Goebbels, K. and Reiter, H.; Fracture Mechanics of Ceramics,, Plenum Press, 7 (1986) 265-284.
2. Cannon, W. R.; Nondestructive Characterization of Materials II, Plenum Press (1987) 115-128.
3. Kishi, T. and Enoki, M.; Nondestructive Testing of High Performance Ceramics, Boston, (1987) 442-456.
4. Kishi, T. and Enoki, M.; Advanced Materials for Severe Service Application, Elsevier Applied Science (1987) 61-76.

Requirements on Nondestructive Testing Methods after Machining of Ceramics

E. Brinksmeier, H. Siemer, H.G. Wobker

Institut für Fertigungstechnik und Spanende Werkzeugmaschinen
Universität Hannover, FRG

Summary

Grinding of tool ceramics introduces microcracks and stresses into the surface layers. The intensity of the surface damage depends on the material removal mechanism in grinding. Especially very small surface damages influence the function of tool ceramics. Therefore nondestructive testing methods with high resolution and high reliability are needed.

Reasons for grinding of ceramics

High quality ceramics belong to the group of hard and brittle materials. They are expected to be used increasingly in a number of high performance applications ranging from electronical and optical devices up to heat and wear resistant parts. Especially engineering ceramics like oxides, carbides and nitrides will be increasingly used also for structural applications. Since engineering ceramics are being used in applications historically reserved for metals, close tolerances and a good surface finish are necessary. Today, there are still quite a lot of discussions whether a near net shape manufacturing of ceramics can be achieved in future without any machining operations. This could be desirable, because the high costs of ceramic parts result mainly from the very high machining costs after densification and sintering.This treatment, however, causes dimensional shrinkage of about 10 - 15 % of the ceramic parts. Therefore, up till now, a material removing operation is inevitable after molding and sintering (1).

Although, some first promising results in electro discharge machining (EDM) of ceramics have been obtained (2), the grinding with diamond wheels is usually adopted, because of the extreme hardness of the ceramics. Cutting tools with defined cutting edge cannot be used because the cutting edge deteriorates rapidly and micro cracking will be observed in the workpiece. However: grinding wheels do also consist of a numerous number of cutting edges which are represented by the active abrasive grains of the wheel. Hence, the chip removal process is achieved by the abrasion between the diamond grains and the surface of the ceramic part. Consequently, the material removal process determines the surface finish of the workpiece, and the physical state of sub-surface layers. Similar to metals, the grinding process has a considerable influence on the functional behaviour of ceramic parts (3,4). The example in fig. 1 shows how different grinding operations have its influence on the fracture probability of silicon nitride ceramics (3).

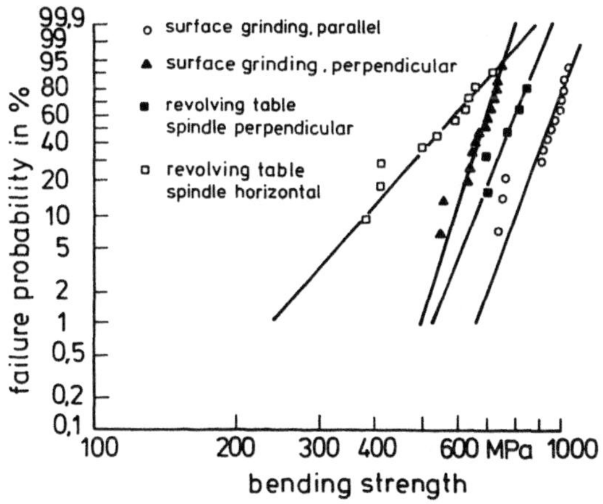

Fig. 1: Bending strength of Si_3N_4 after grinding

In general there is a great need for knowledge about the dependency of functional behaviour and machining processes. But up till now there is no widely applicable coordination of functions and processes. This is not least due to the fact that there are many functions of the different ceramic components and their surfaces, and many ways to generate these surfaces. It is therefore necessary to find geometric and physical properties of surfaces which can be correlated to the functional behaviour or the failure causes of ceramics. From this systematic approach follow the different tasks for research and investigation:

1. The material removal process of ceramic surfaces has to be investigated. This is necessary in order to estimate the order of magnitude of the local surface damage intensities.

2. Measuring methods have to be developed which are able to detect the surface and sub-surface properties of machined ceramics. It will be necessary to extend and refine those methods which meet the requirements of ceramics.

3. It has to be found out, which surfaces and surface properties are important for certain functions and how the influence of surface damages can be evaluated.

Some of these aspects are covered in this paper which concentrates on the grinding of tool ceramics, and the measurement and evaluation of the resulting surface properties.

Functional needs of tool ceramics

Ceramic cutting tools are of increasing interest due to their high temperature strength and wear resistance. The basic ceramics used for tools are oxides, carbides and nitrides (fig. 2). For practical applications especially Al_2O_3/ZrO_2, Al_2O_3/TiC and Si_3N_4 ceramics are used. They are produced by hot pressing and sintering, hot isostatic pressing and reaction bonding. Finally, all ceramic tools are surface ground in order to achieve the desired shape and surface finish.

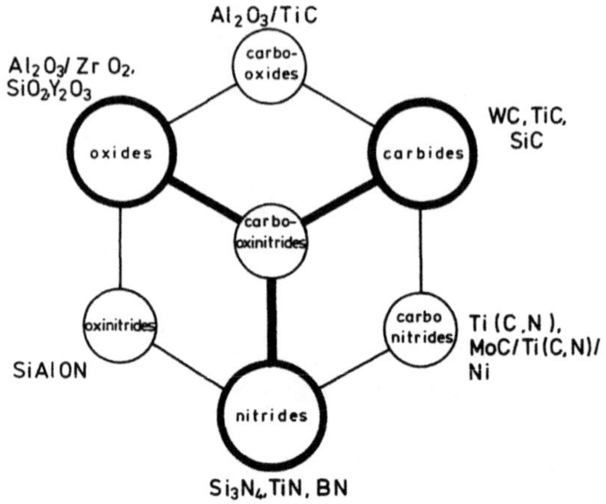

Fig. 2 Basic materials for ceramic cutting tools

Cutting tools have to resist a complex load spectrum during the cutting process (fig. 3). Cutting forces, friction induced heat and chemical interactions act upon the surface of the tool. Very often these kinds of loadings are of instationary nature, when interrupted cutting is applied.

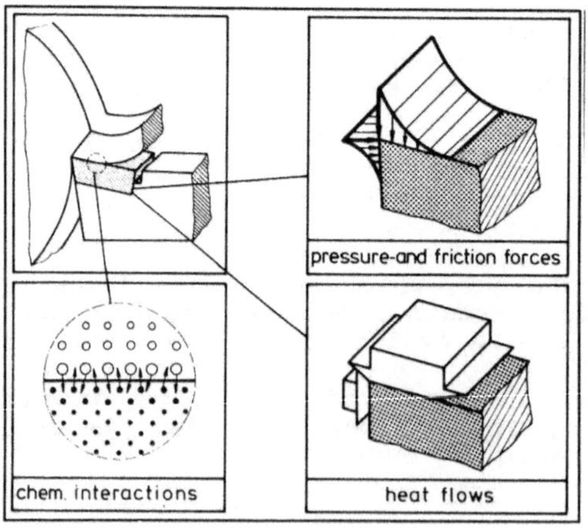

Fig. 3 Tool loading during the cutting process

It is well known that ceramics are very sensitve against tensile stresses. However, during the cutting process especially the thermal loading leads to internal tensile stresses in the tool. This has been proved by finite element calculations (5), fig. 4.

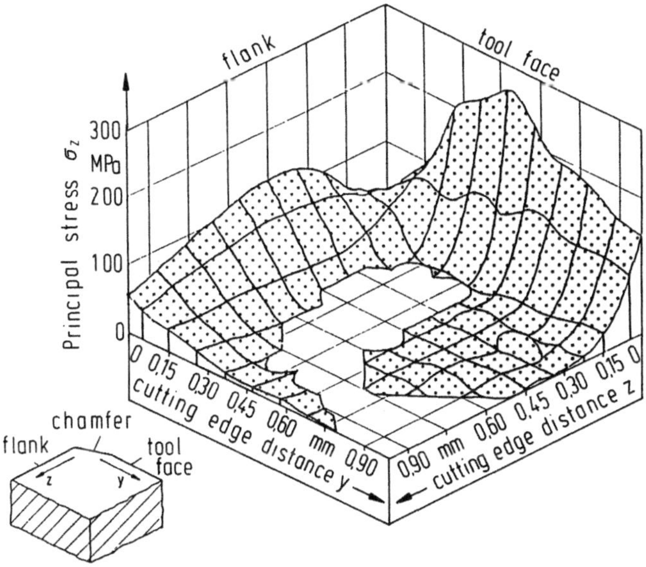

Fig. 4 Thermally induced principal tensile stresses in a ceramic insert

Considerable stresses were calculated on the surface of the tool face and the flank, which both are the critical surfaces in terms of function. As these surfaces are produced by grinding, it is of tremendous importance to know, whether resulting surface and sub-surface damages have to be expected. This is important for four reasons:

1. Shape and surface roughness of the tool determine the cutting forces and the chip flow.

2. Residual grinding stresses superimpose the thermal load stresses and may lead to early failure by cracking.

3. Material flaws and microcracks lead to local stress concentrations and thus cracking, because a stress relaxation is not possible in ceramics.

4. Surface cracks are localities for material deposits of the machined metallic workpiece. Due to the high cutting temperatures, chemical reactions take place which lead to crack growth and thus to failure (6).

A non-destructive investigation of surfaces and sub-surfaces of the tool in terms of roughness, cracks, flaws and stresses is therefore desirable.

Material removal by grinding

Up till now, only a few models are existing on how the material removal in grinding of ceramics proceeds. Marshall and Evans (7) assume brittle fracture which leads to a network of cracks. A similar concept covering plastic deformations and crack propagation is given in a paper by Inasaki (4). He assumes that material is removed in different ways depending on the size and density of defects in the material; such as flaws and cracks, and the size of stress field. When a stress field, induced by a grain cutting edge, is smaller than the defects, material will be removed mostly by plastic deformation. On the other hand, when a stress field is larger than those defects, a localized brittle microfracture must play an important part.

As the stress field depends mainly on the normal surface forces, the big importance of the normal grinding force is evident. Own investigations into the grindability of different ceramic materials showed that the ratio between normal and tangential grinding forces is much higher than it is known from metallic materials (fig. 5).

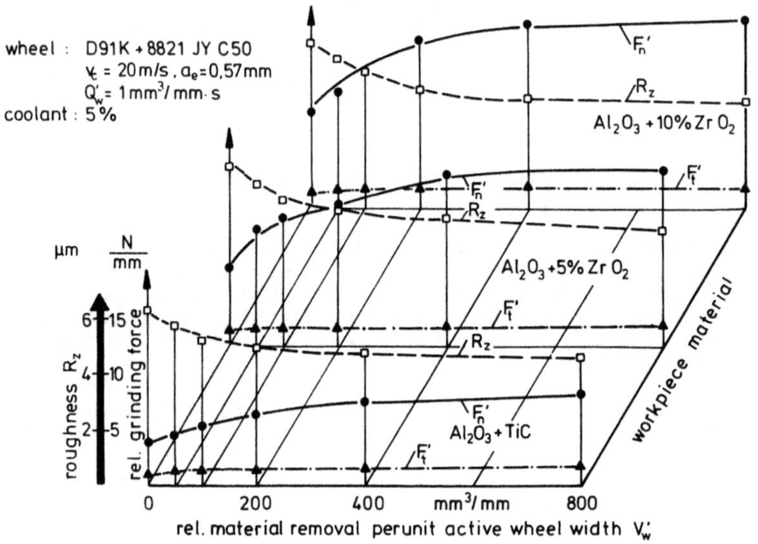

Fig. 5: Grinding forces for different ceramics

Additionally, the normal forces are considerably increasing with the material being removed. Hence, high normal forces acting upon a brittle material, lead to a material removal by micro brittle fracture and will thus induce surface cracks.

A crack-free grinding with a higher amount of plastic surface deformation can be achieved via two measures:

1. Reduction of the normal grinding forces by applying low amounts of metal removal rates.

2. Decrease of the cross sectional area of the chips by using fine grained grinding wheels, in order to keep the induced stress fields small.

In any case it is recommended to keep the cutting edges sharp and the wheel pores free from chips. Therefore Nakagawa proposes the in-process electro-discharge trueing/dressing of the wheel, if a high surface finish (mirror grinding) is desired (8). In order, to also meet the economical requirements in grinding of ceramics, it is therefore proposed to carry out a roughing with high efficiency at first. To remove the damaged surface layer, finish grinding with small cross-sectional areas, of active cutting edges is needed afterwards. The optimum combination of roughing and finishing will be a keypoint for increasing the grinding technology of ceramics in the future.

Surface damage and possibilities for detection

It has been shown that ceramic cutting tools are sensitive against cracks and residual tensile stresses. However, the inevitable grinding process produces micro cracks as far as roughing is applied. Finishing will lead to a decrease of cracks in favour of plastic surface deformations. The latter, however, will generate residual grinding stresses. It is further assumed that strain hardening or even softening within the utmost surface layers may occur due to grinding.

These are the main interactions nondestructive testing methods have to meet. The sizes of defects are generally small and so are the penetration depths of surface damages and stresses. In the following, selected examples will show the difficulties when investigating ground tool ceramics.

Some of the figures show preliminary results, taken from an international cooperative work on ceramic surfaces, which is guided by the authors (9).

The surface topography for ground tool ceramics looks different dependent on the ceramic material (fig. 6). It is important to note, that the grinding conditions were equal for both materials. Material removal by brittle fracture and little plastic deformations can be recognized on the alumina surface,whereas a considerable higher amount of plastic deformations can be noticed on the silicon nitride surface. Cracks and flaws have been found in both materials, it was difficult, however, to seperate grinding cracks from those failures resulting from manufacturing of the workpieces. Only the use of the scanning electron microscope proved that some failures in the silicon nitride are due to manufacturing. This could be seen from the growth of the Si_3N_4 sticks into pores. Dye penetrant inspection was also carried out, however with limited success. It is assumed that ultrasonic testing and the accoustic microscopy may deliver more detailed resultes (10).

Residual stress measurements have been carried out using X-ray diffraction. This method is proved on metals, it suffers, however, from low peak intensities, when used on ceramic surfaces. Unfortunately, the high penetration depth of radiation is opposed with a low depth of residual stresses. This makes it difficult to

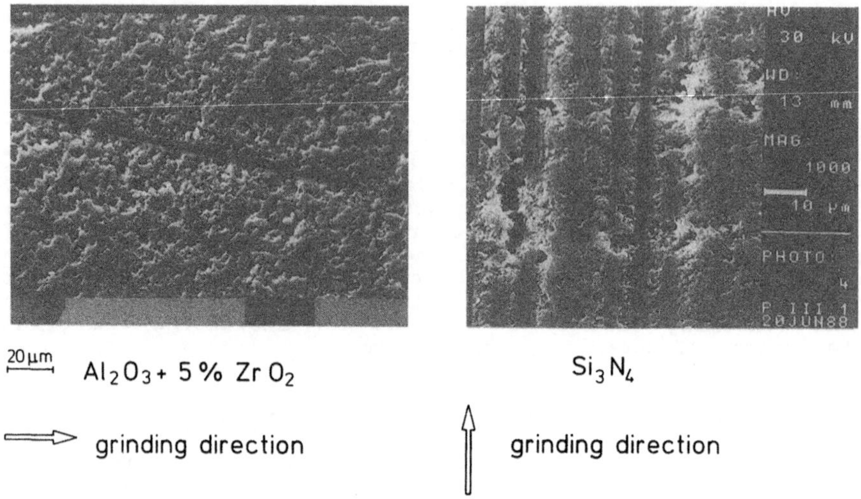

Fig. 6 Ground ceramic surfaces

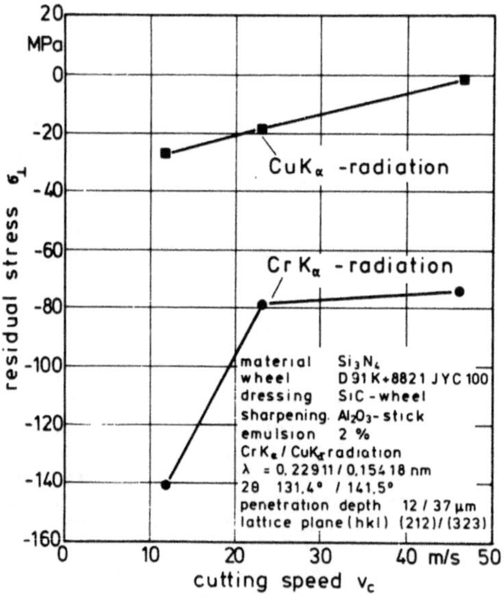

Fig. 7 X-ray radiation and residual stress

calculate the proper amount of surface stress. The low depth of stresses is proved when using different kinds of radiations (fig 7). A comparison with the penetration depth of radiation is plotted in fig. 8. A calculation of the true sub-surface stresses should be possible when using different radiations. However, a calculation of the surface stresses seems to be difficult, be-

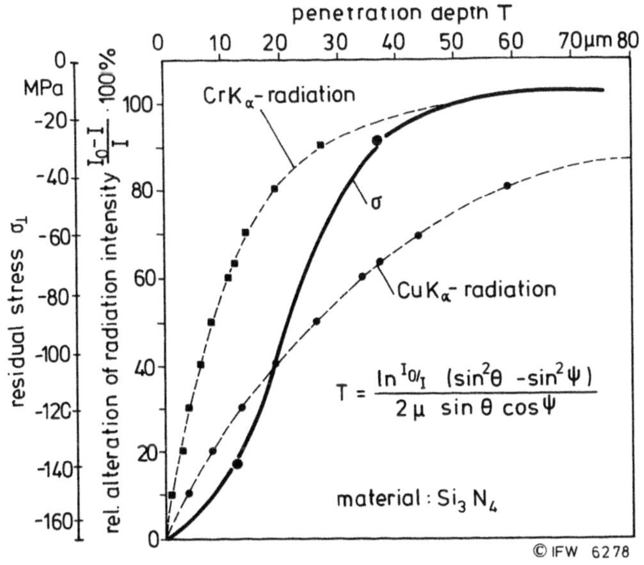

Fig. 8 Comparison of different radiations

cause steep stress gradients are evident. This is proved by the peak shift dependent on the tilt angle (fig. 9). As the penetration depth of radiation changes with the tilt angle (omega-goniometer), different amounts of the stress gradient are measured. It is assumed that the principal results, obtained on metallic surfaces (11) are also valid on ceramics.

A further problem in X-ray stress measurement is the determination of the X-ray elastic constants (REC). If they are determined in bending tests, the obtained values may differ considerably from the calculated values (12). More than this, it has to be considered that quite a lot of the ceramics are multi phase systems. Up till now, it is unknown how far this has to be considered when determining elastic constants and stresses.

When residual stresses have to be determined in sub-surfaces and in the bulk material, surface layers of ceramics have to be removed stepwise. This seems to be a big problem, as proper etching techniques are not known and a mechanical removal will induce new stresses. A new powerful tool for this task may be the use of excimer lasers. (fig. 10).

Surface layers of 1 um and less can be removed with high accuracy. No surface damage could be noticed, due to the non-thermal material processing.

Finally, this paper is concluded with a more general statement on the testing of ceramics. It is well known that the frequency of failure resp. the fracture probability of ceramics differs completely from the values for metals. Ceramic parts from the same stock may have a high stregth, but some of those may fail even under little loading. Therefore all non-destructive measuring

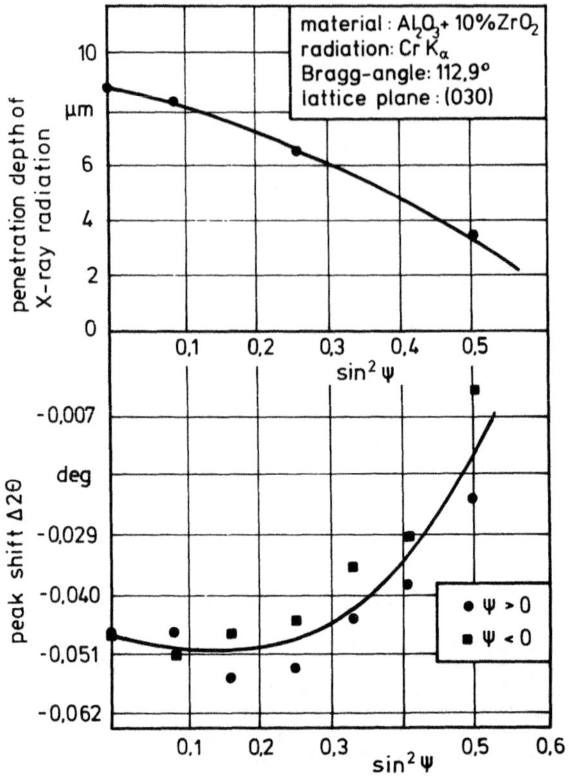

<u>Fig. 9</u> Irregularities in peak shift due to stress gradients

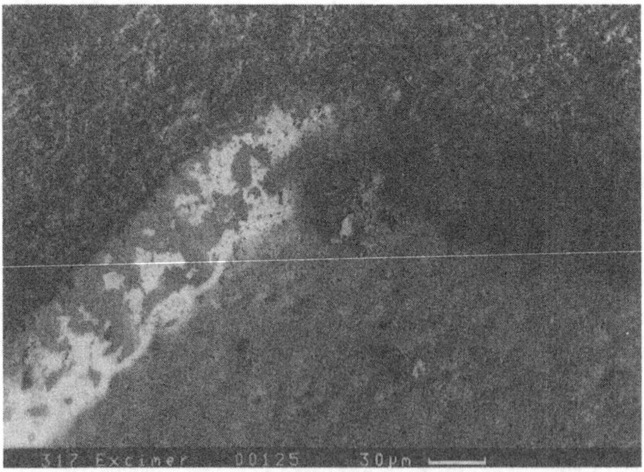

<u>Fig. 10</u> Removal of ceramic surface layers by excimer laser

methods of the future should have a high throughput in order to meet the requirements of the industry for a 100 % inspection at low testing time.

References

1. Brinksmeier, E.; Hetz, F.; Wobker, H.-G.: Oberflächen geschliffener keramischer Werkstücke. Proceedings VII Internationales Oberflächenkolloquium, TU Karl-Marx-Stadt (1987) 106-114.

2. König, W.; Daun, D.F.; Levy, G; Panten, U.: EDM-future steps towards the machining of ceramics. Annals of the CIRP (1988) Vol. 2.

3. Allor, R.L.; Baker, R.R.: Effekt of grinding variables on strength of hot pressed silicon nitride. ASME paper 83-GT-203.

4. Inasaki, I.:Grinding of hard and brittle materials. Annals of the CIRP (1987) Vol. 2, 463-471.

5. Bartsch, S.: Verschleißverhalten von Aluminium-Oxid-Schneidstoffen unter stationärer Belastung. Dr.-Ing. Diss. Univ. Hannover, 1988.

6. Tönshoff, H.K.; Brinksmeier, E.; Bartsch, S.: Notch wear and chemically induced wear in cutting with Al_2O_3-tools. Annals of the CIRP (1987) Vol. 2, 537-541.

7. Marshall, D.B. et. al.: The nature of machining damage in brittle materials. Proc. R. Soc. London A. 385 (1983) 461-475.

8. Nakagawa, T. et. al.: Developement of a new turning center for grinding ceramic materials. Annals of the CIRP (1988) Vol. 1 319-322.

9. Brinksmeier, E. et. al.: Surface layers of machined ceramics. Cooperative work between 15 research institutes. Results will be published in the Annals of the CIRP, 1989.

10. Tönshoff, H.K.; Brinksmeier, E.; Hetz, F.: Detection of microcracks. Annals of the CIRP (1987) Vol. 2, 545-552.

11. Hauk, V.; Vaessen, G.: Auswertung nichtlinearer Gitterdehnungsverteilungen. HTM Beiheft, Carl Hanser Verlag, (1982), 38-48.

12. Brinksmeier, E.; Siemer, H.: Calculation of X-ray elastic constants of aluminium oxide ceramics. Proc. of the Int. Conf. on Residual Stresses-2, Nancy, France, Nov. 23-25 (1988), to be published.

13. Dickmann, K. et. al.: Excimer Hochleistungslaser in der Materialbearbeitung. Laser Magazin 4 (1987) 34-40.

Ultrasonic NDE of Advanced Ceramics

Patrick S. Nicholson

Ceramic Engineering Research Group, McMaster University, Hamilton, Ontario, Canada

Abstract

The development of a database for the ultrasonic NDE of advanced-ceramic components is described. High frequency ($\leq$100 MHz), high power transducer development was involved and the probes are used to examine model spherical inclusions (voids ($\geq$20 μm), ZrO_2, MgO, V_2O_5 and Pt inclusions ($\geq$30 μm)) in model matrices (glass, crystallized glass and partially stabilized zirconia). Actual frequency spectra obtained from these defects as a function of defect size are compared with their calculated spectra based on scattering theory and very good agreement is obtained.

I. Introduction

The NDE of advanced ceramics is a vital step in their acceptability. It is also one of the most significant hurdles still to be cleared. Ceramic materials present unique challenges and these will be reviewed. Throughout the discussion it is assumed that NDE of advanced ceramic components means the economic evaluation of 100's to 1000's of components per day. This requirement severely limits the use of exotic techniques. Equipment should be robust and simple, cheap and reliable and above all speedy and accurate. Following metallurgical experience, this means fast ultrasonic scanning or multielement female-configuration static transducers. The required defect resolution in advanced ceramics severely limits the use of X-ray analysis. Surface flaw detection with dye penetrants may not be possible for micrometer-sized defects in ceramics. Conventional metal penetrants are recommended but unproven (1).

This paper reviews the specific problems associated with the evaluation of dense ceramic components. This approach is judged the most useful as a literature-review could leave the reader without a clear perception of the problems. The complexity of the mathematically modelling defects in real ceramics led us to develop a systematic experimental pedagogy. This involved examination of model defects (voids, spherical inclusions, indent-cracks) in model ceramic matrices (glass, crystallized glass, partially-stabilized-zirconia (PSZ)). Initial optical definition facilitated interpretation of the ultrasonic signals and a database was thus created. Regular defect shapes allowed wave physics modelling and prediction methodologies to be developed.

In parallel with the detection of smaller and smaller model defects (≤20 μm voids; ≤30 μm inclusions and cracks) the development of high frequency (≤100 MHz) high-power (≥130 dB s.n.r) transducers was pursued.

II. Unique Ultrasonic NDE Problems Posed By Ceramics

The overall requirement for the satisfactory NDE of advanced ceramics is the definition of flaws ≤10 μm, 10 mm deep in terms of their size, shape and chemistry. A 10 μm flaw in high performance Si_3N_4 can reduce the expected strength from ≃1500 MPa to ≃250 MPa. Added to these restrictions are the physical properties of covalent/ionic-bonded ceramic materials, i.e. high sound velocity and attenuation characteristics. Sound velocities are double those in metals reducing the depth of penetration of focussed ultrasound (V(longitudinal) in Al_2O_3 is 12,000 m/sec). This is equivalent to the higher refractive index effect on focussed light. To increase this depth, larger diameter transducers are needed. This requires larger electrode diameters on very thin piezoelectric diaphragms and the accompanying increase of capacitance renders the poling of these crystals difficult. The dielectric constant (d) of the piezoelectric element of the transducer must be reduced. Thickness coupling (k_t) must be maximized and radial coupling (k_p) minimized to increase element poling efficiency.

10 μm flaws require sonic wavelengths ≃60 μm i.e. a frequency of ≤100 MHz. These frequencies require very thin piezoelectric crystals (≃30 μm) (=λ/2). This thickness (t) increases as the sonic velocity in the piezoelectric crystal increases and so capacitance problems reduce. The mechanical strength requirements mean grain sizes should be ≃2 μm giving ~16 grains/section. To summarize, ceramic piezoelectrics must have low d, high k_t, low k_p, high sonic velocity, fine grain size and zero defects. The physics and methodology of the ultrasonic characterization of ceramics was recently reviewed (2).

The higher velocity in ceramics is somewhat offset by their lower density and the acoustic impedance (velocity x density) of advanced ceramics is similar to that of metals. For economic reasons, water is used as the scanning couplant in the present work. Although it will always be necessary to NDE finished components, recently the NDE of green ceramics has been investigated (3,4). Whilst making excellent economic sense, such NDE leaves some important fired defects undetected. A number of these are shown in Figure 1 (5).

Defects that develop during sintering often lead to premature failure. Agglomerates have marginally different density to their powder host so will be missed by green NDE. Their enhanced local sintering leads to porosity (Figure 1(a)), a severe flaw only detectable following sintering. Organic inclusions (such as the-fibre-that-was of Figure 1(b)) will also only evidence via ghost porosity following firing. Oxide inclusions of similar acoustic impedance to the host matrix (Al_2O_3 in PSZ, Figure 1(c)) will be undetected. In each case these were the fracture-initiating flaws. The inevitable porosity of greenware must limit the frequency of penetrating ultrasound which in turn limits its resolution. Gross defects such as density gradients and large

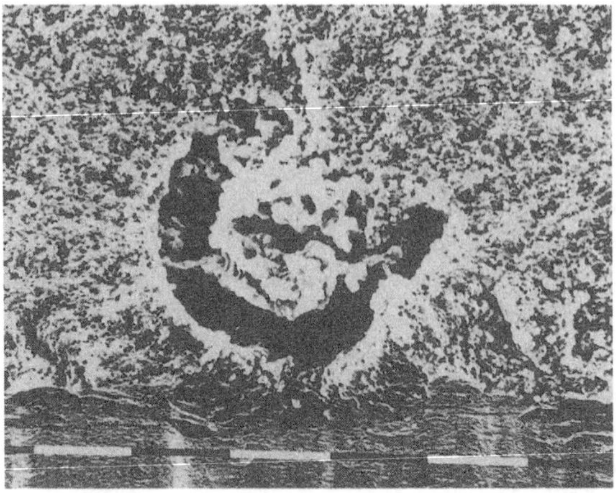

Fig. 1a

Agglomerate defect in partially-stabilized zirconia (bar = 5.6 μm).

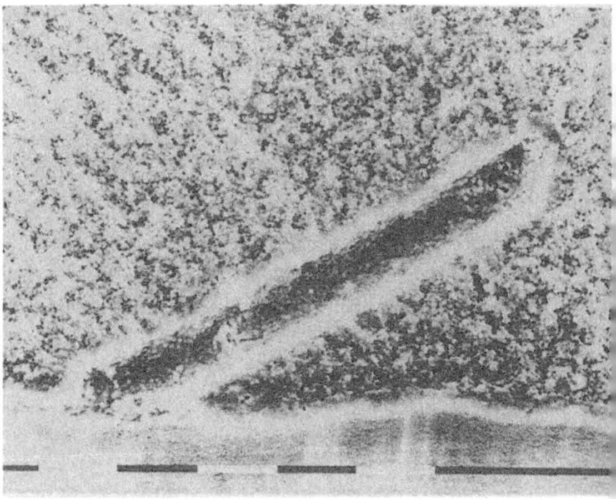

Fig. 1b

Residual pore following burn-out of an organic fibre from partially-stabilized zirconia (bar = 5.6 μm).

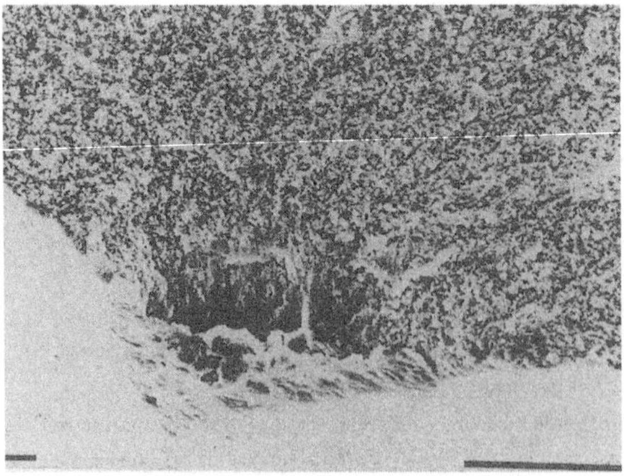

Fig. 1c

Alumina inclusion in partially-stabilized zirconia (bar = 4.4 μm).

metallic inclusions can probably be detected, making greenware NDE a limited but valuable economic tool.

Many papers report the "detection" of defects (6,7) with indications of the size to 25 µm at depths of 3 mms but the results are qualitative. Use of the reliability methodologies recently developed (8) require accurate knowledge of the flaw size and modern components require their quantitative characterization at 10 mm depth. At the end of the 1970's activity in ultrasonic NDE of ceramics reached a peak. The Stanford group reported the use of high frequency contact transducers to detect bulk defects (9). They also reported the use of wedge transducers to detect surface cracks down to 60 µm size (10). The frequencies used precluded penetration and contact transducers excluded scanning, both requirements for the economic NDE of real components.

The acoustic microscope has been used (11) to detect and characterize surface and subsurface defects in Si_3N_4 and SiC. 100 µm pores and cracks were detected. A 150 µm diameter flaw 400 µm deep in hot pressed SiC was detailed. Clearly the acoustic microscope is a useful research tool but as an on-line evaluator of up to 1000 components/day its value is limited. A recent development is scanning laser acoustic microscopy (SLAM). This technique has been used to detect 20 to 430 µm voids up to 2 mm deep. A plastic coverslip is placed on the ceramic to give a reflective coating. Voids between 50 and 400 µm diameter in Si_3N_4 were detected with 90% probability if within 1 µm and 1400 µm of the surface respectively (12). Depth, time and preparation limitations preclude this technique from on-line NDE of advanced ceramic components.

III. Wave Physics Models For Data-Base Development of Defects in Ceramics

The general failure of the "head-on" approaches to defect characterization in real advanced ceramics led to the data-base development pedagogy via model defects in model ceramic matrices described herein. The use of spherical scatterers means sensible wave-physics analysis can lead to realistic predictions.

An important piece of information is contained in the first peak of the reflected signal from a defect in the time-domain. If the reflector is a void, its acoustic impedance (density × velocity of sound) is zero so its reflective factor is negative i.e. the first peak of the time-domain reflection is phase inverted. If the reflector has an acoustic impedance (i.e. solid) the initial peak is positive and no such phase inversion occurs. Qualitative size information for an inclusion is also contained in the time signal. If the scatterer is a large inclusion, its initial reflection and that from the transmitted wave exiting later are differentiable. Their time difference can be used to estimate the defect size.

If the chemistry of the defect is known, quantitative size information is contained in the spectrum of reflected frequencies. Any time-domain signal can be Fourier transformed and a spectrum for the defect in the frequency domain thus obtained. This spectrum is characterized by its frequency at maximum amplitude (FMA) and its full-width-at-half-maximum amplitude (FWHM).

The scattering of a single frequency plane wave has been treated in detail (13). The scattering amplitudes are calculated for individual waves using the continuity of stresses and strains at the interface between a matrix and a spherical inclusion. In the reference the scattering amplitudes determine the total scattering cross-section for the wave whereas our interest lies in the scattering of the longitudinal waves which are given by the scattering amplitude A_m. These can then be used to determine the pressure at the transducer as:

$$p_r(k) = P_R\, k\, a\, (1 + i) \exp\{-ikR\} \sum_{m=0}^{\infty} (2m + 1) A_m \tag{1}$$

where the transducer to inclusion distance, R, which is also the focal length of the transducer, is much larger than the sphere radius, a. The quantity i is $\sqrt{-1}$. The pressure, p_r, is a function of the wavenumber of the longitudinal ultrasound in the matrix, k, and P_R is the pressure at the site of the sphere. The pressure amplitude at the focus of the transducer can be determined by expanding the equations of O'Neil (14)

$$p(\omega) = \partial\, \omega^2 S_f \frac{\exp[i(\omega t - k' \Delta z)]}{k'} \tag{2}$$

where

$$k' = (k/2)\{1 + [1 + (D/2A)2]1/2\} \tag{3}$$

Here D is the diameter of the transducer and ω is the angular frequency of the wave such that $\omega = C_{11}k = 2\pi f$ where f is the frequency. Δz is the distance from the center of the sphere which is at the focal point of the transducer. The quantity S_f is the amplitude of particle oscillation at the site of the sphere. This value is not easily determined due to attenuation within the couplant and reflections at the sample surface. Therefore, it is better to express the pressure p(ω), and hence $p_r(k)$, in terms of relative amplitudes.

Using these equations, it is possible to determine the relative amplitude of a wave of single frequency backscattered from a spherical inclusion at the focal point of a transducer. An incoming pulse of ultrasound consists of a distribution of frequencies with different amplitudes (15). The distribution chosen to approximate that of the pulse is Gaussian with the following form;

$$p_i(f) = p_a \exp\left\{-4\ell n(2)\left(\frac{f - f_a}{f_{FWHM}}\right)^2\right\} \tag{4}$$

where p_i is the pressure of the incoming pulse and p_a is the pressure at the maximum amplitude which will be taken as a unity. f_a is the frequency at the maximum amplitude (FMA) or center frequency of the distribution and f_{FWHM} is the frequency full width at half the maximum amplitude (FWHM) of the distribution.

The total expected pressure at the transducer for a spherical inhomogeneity in the focal zone of the transducer can be written as:

$$P_T(f) = \frac{C_{11}a}{2\pi f} \exp\left(-i\frac{C_{11}R}{2\pi f}\right) \exp\left\{-4\ln(2)\left(\frac{f - f_a}{f_{FWHM}}\right)^2\right\} \sum_{m=0}^{\infty} (2m + 1) A_m \tag{5}$$

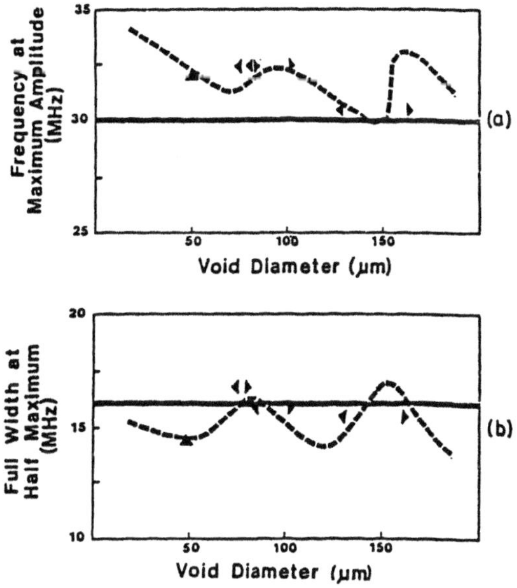

Fig. 2 Comparison of measured (triangles) and calculated (dotted lines) data for
(a) FMA vs. void diameter.
(b) FWHM vs. void diameter.

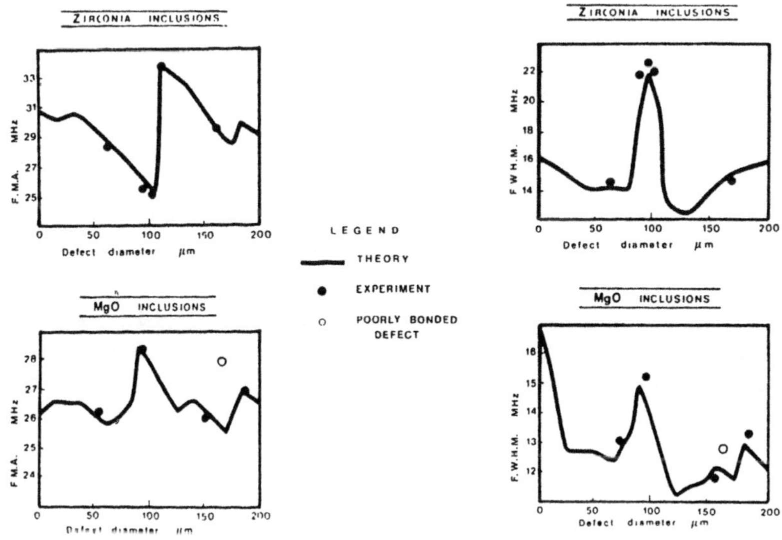

Fig. 3A FMA calculated (solid line) and experimental (points) for
(a) ZrO2 inclusions in glass
(b) MgO inclusions in glass.

Fig. 3B FWHM calculated (solid line) and experimental (points) for
(a) ZrO2 inclusions in glass
(b) MgO inclusions in glass.

where $P_T(f)$ is the relative pressure for the backscattered longitudinal wave. The summation over the scattering amplitudes will modify the Gaussian shape of the signal spectrum so the effect of density and sound velocities of both matrix and inclusion and the inclusion dimension can be measured from the change in FMA and FWHM of the magnitude spectra. The differences are attributable to resonances of the sphere with certain frequency components present in the sound pulse which physically represent creeping waves that circumnavigate the sphere (16).

The theoretical and experimental FMA and FWHM values will be compared for model defects in model ceramic matrices.

IV. Characterization of Model Defects In Model Ceramic Matrices

The following systems were examined; glass and voids (30 μm diameter minimum) (17), glass and ZrO_2, MgO and V_2O_5-inclusions (50 μm diameter minimum) (18) crystallized glass and inclusions (19) and fully dense partially-stabilized-zirconia (PSZ) containing Pt inclusions (50 μm minimum diameter) (20). Surface-breaking cracks associated with Vickers and Knoop indents in glass and Si_3N_4 were also examined. The results in comparison with the model developed in Section III will now be presented.

The calculated and measured FMA and FWHM of the frequency spectra are plotted as a function of defect size for voids in glass (Figure 2), ZrO_2 and MgO inclusions in glass (Figure 3) and ZrO_2 in crystallized glass (Figure 4). A density map for a 50 μm Pt sphere in dense PSZ with a locator loop of platinum (100 μm-diameter wire) is shown in Figure 5. A 2 minute scan detected this defect. The FMA and FWHM for Pt spheres in PSZ vs. defect size are shown in Figures 6 & 7.

The agreement between theory and experiment is good. If the type of defect is known, its size can be determined by combination of the FMA and FWHM of its frequency spectrum. The MgO point which plots off the predicted line was shown to be a decohesed defect on polishing to investigate. The depth of the 50 μm Pt inclusion (Figure 5) was 5 mm. In fact <100 μm Pt defects have been detected through 30 mms of PSZ. To use the model it is necessary to know (or estimate) the density and sound velocity of the defects. In some cases the chemistry of inclusion type defects may be unknown. Elastic modulus information is contained in the Franz surface wave that circumnavigates the defect. This wave is presently being investigated as a possible source of chemical information.

V. Summary

The development of a database for defects in advanced ceramics via model-defects in model-ceramics has been described. This work involved the development of high-frequency, high-power ultrasonic transducers and their synthesis is described starting with improved quality piezoelectric ceramics. SPN and LMN-based transducers with frequencies up to 100 MHz and power ≥ 130 dB were used to characterize voids and inclusions in glass, crystallized glass and partially-stabilized-zirconia in a scanning mode with water couplant. The characteristics of the calculated

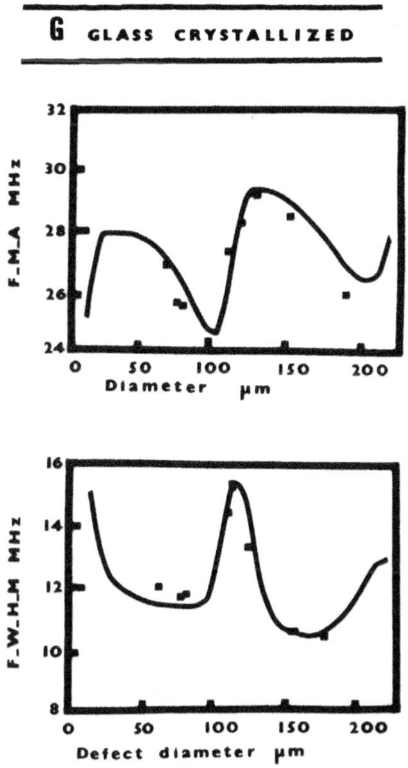

Fig. 4 FMA and FWHM for ZrO_2 inclusions in crystallized glass vs. defect size.

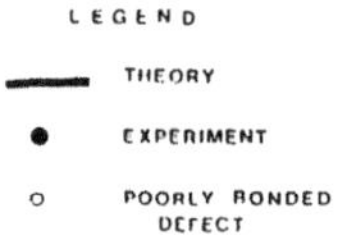

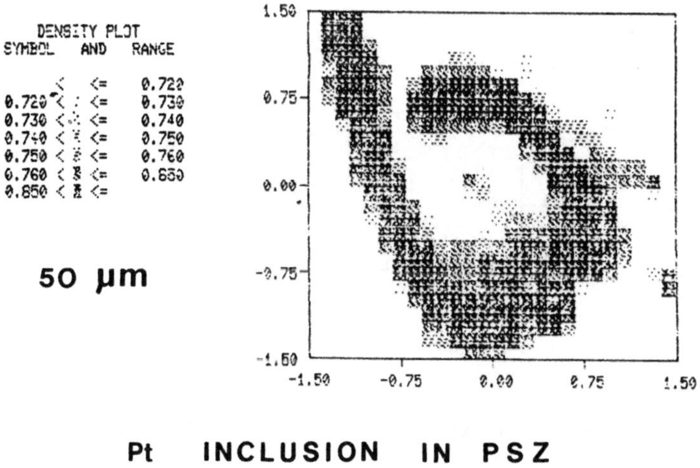

Figure 5 Density map of a 50 µm Pt inclusion and its Pt-wire locator loop, 4 mms deep in dense PSZ.

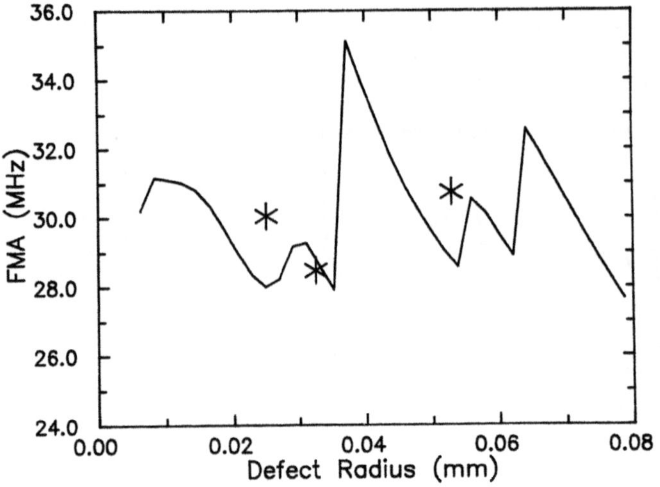

Figure 6 FMA versus defect radius for platinum particles in zirconia. Solid lines are derived from theory and asterisks are data points and error estimates.

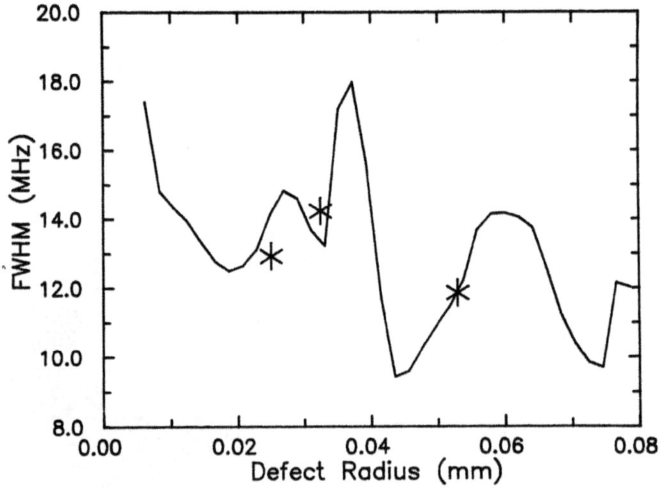

Figure 7 FWHM versus defect radius for platinum particles in zirconia. Solid lines are derived from theory and asterisks are data points and error estimates.

and experimental frequency spectra agreed closely, facilitating size prediction. The smallest defects characterized were 20 μm voids, 30 μm inclusions and 30 μm×8 μm surface-breaking cracks.

Acknowledgements

The author wishes to acknowledge the work of A. Stockman, N.D. Patel, J. Van den Andel and P. Mathieu, which is reviewed herein. The financial support of DREP-Pacific is acknowledged – W. Sturrock, Scientific Authority.

References

1. P. Emerson, Pratt & Whitney (Canada) Ltd. – private communication.
2. Bhardwai, M.C., "Principles and Methods of Ultrasonic Characterization of Materials", Adv. Ceram. Matls. 1 (4) (1986) 311-324.
3. Jones, M.P., Blessing, E.V., Robbins, C.R., "Dry-Coupled Ultrasonic Elasticity Measurements of Sintered Ceramics and Their Green States", Matls. Eval. 44 (1986) 859-862.
4. Kupperman, D.S. Karplus, H.B., "Ultrasonic Wave Propagation Characteristics of Green Ceramics", Bull. Am. Ceram. Soc., 63 (1984) 1505-1509.
5. Sung, J., Nicholson, P.S. submitted to J. Am. Ceram. Soc.
6. Schuldies, J.J., Derkacs, T., "Ultrasonic NDE of Ceramic Components", Ceramic Gas Turbine Demonstration Prog. Rev. (1978) 429-448.
7. Derkacs, T., Matay, I.M., Brentnall, W.D., "High Frequency Ultrasonic Evaluation of Ceramics for Gas Turbines", J. of Eng. for Power, 100, (1978) 549-552.
8. Marshall, D.B. and Ritter, J.E., "Reliability of Advanced Structural Ceramics and Ceramic Matrix Composites - A Review", Bull. Am. Ceram. Soc., 66 (2) (1987) 309-317.
9. Evans, A.G., Kino, G.S., Khuri-Yakub, B.T., Tittmann, B.R., "Failure Prediction in Structural Ceramics", Matls. Eval. 35 (4) (1977) 85-96.
10. Khuri-Yakub, B.T., Kino, G.S. and Evans, A.G, "Acoustic Surface Wave Measurements of Surface Cracks in Ceramics", J. Am. Ceram. Soc., 63 (1-2) (1980) 65-71.
11. Kuppermann, D.S., Pahis, L., Yuhas, D. McGraw, T.E., "Acoustic Microscopy Techniques for Structural Ceramics", Bull. Am. Ceram. Soc., 59 (8) (1980) 814-816, 839-841.
12. Roth, D.J., Baaklini, G.Y., "Reliability of Scanning Laser Acoustic Microscopy for Detecting Internal Voids in Structural Ceramics", Ad. Ceram. Matls., 1 (3) (1986) 252-258.
13. Tuell, A., Elbann, C., Chick, B.B., Ultrasonic Methods in Solid State Physics, pp. 161-179, Academic Press, N.Y. (1969).
14. O'Neil, H.T., "Theory of Focussing Radiators", J. Acoustic Soc. Am. 63 (1978) 68-74.
15. Tittmann, B.R., Cohen, E.R., Richardson, J.M., "Scattering of Longitudinal Waves Incident on a Spherical Cavity in a Solid", J. Acoust. Soc. Am., 63 (1978) 68-74.
16. Gaunard, G.C., Tanglis, E., Uberall, H., Brill, D., "Interior and Exterior Resonances in Acoustic Scattering: I-Spherical Targets", Il Nouvo Cimento 768 (1983) 153-175.
17. Stockman, A., Nicholson, P.S., "Ultrasonic Characterisation of Model Defects in Ceramics, Part I. Voids in Glass - Theory and Practice", Matls. Eval., 44 (1986) 756-761.
18. Stockman, A., Mathieu, P., Nicholson, P.S., "Ultrasonic Characterization of Model Defects in Ceramics. Part II - Spherical Oxide Inclusions in Glass - Theory and Practice", Matls. Eval. 45 (1987) 736-742.
19. Stockman, A., Mathieu, P., Nicholson, P.S., "Ultrasonic Characterization of Model Defects in Ceramics. Part III. Spherical Inclusions in Opaque Crystallized Glass - Theory and Practice", submitted to Matls. Eval. (1987).
20. Stockman, A., Nicholson, P.S., "Ultrasonic Characterization of Microspherical Inclusions in Zirconia" ASNDT/Am. Ceram. Soc. Proc. NDT Ceramics Conf., Boston, August 1987.

Tomodensitometry with X- and Gamma-Rays

J. GOEBBELS, H. HEIDT, B. ILLERHAUS, P. REIMERS

Bundesanstalt für Materialforschung und -prüfung (BAM)
Unter den Eichen 87
D-1000 Berlin 45

Summary

For materials testing computerized tomography (CT) offers the great advantage to produce nondestructively maps of the local X-ray absorption inside an object. For radiation sources in the energy range between 0.2 and 1.5 MeV the absorption is proportional to the density. Therefore with CT the density distribution can be determined. For simple cases the absolute value of the density can be measured.
All measurements to be presented are made with the two tomographs developed at BAM especially for materials testing. The density resolution is about 1 %.
Applications on concrete, powder metallurgical parts and ceramics are given. The last two groups of materials are compared for the green and sintered state.

Introduction

One of the oldest methods to characterize materials is the measurement of density. For powder metallurgical objects the knowledge of the density distribution is essential. For a determination of the local density with the Archimedes principle the objects must be destructed, a time consuming procedure.
Failure of ceramics during stressing is governed by microcracks together with residual stresses. This process is strongly affected by density fluctuations. Differences in the density of 1 or 2 % can decrease the lifetime for several orders of magnitude /1/.
This paper describes the possibility of nondestructive density measurement with CT with X- and Gamma-rays.

Experimental

CT gives a local map of tho absorption inside the object represented mostly in form of a image matrix. Each matrix element or pixel represents a volume element or voxel of the investigated slice for which the absolute value of linear attenuation coefficient can be determined averaged over this volume element.

In the energy range 0.2 up to 1.5 MeV compton scattering is predominant, e.g. the absorption is proportional to the density.
All measurements are made with the two tomographs developed at BAM especially for materials testing. The BAM-Tomograph III was installed 1984 and is designed for objects up to 1 m diameter and 1000 kg. The spatial resolution is between 0.6 and 2 mm depending on the used radiation source (X-ray, Co-60, LINAC-9 MeV) /2/.
The other tomograph is under development and is used with X-ray tubes with a maximum energy of 160 kV. For single cases a spatial resolution of about 0.05 mm was reached using a microfocus X-ray tube. The density resolution for both tomographs is about 1 %.

Absolute densitometry

Fig.1 shows the mass attenuation coefficient as function of atomic number normalized to the value of iron. As first approximation the mass attenuation coefficient is constant (+/- 2.5 %). In simple cases therefore the absolute density can be determined with CT and different materials can be classified . Fig. 2 shows a tomogram of a waste container for radioactive waste. As radiation source a Co-60 nuclid was used. Due to the density difference concrete, aluminum and steel can be distinghuised (Pseudocoloured images can show different materials much easier than grey-tone images)

Relative densitometry

Much more interesting than the absolute density is the density distribution or homogenity of a component part. The density resolution which can be reached with CT is defined as the quotient of variance of the linear attenuation coefficient to the absolute value. It is influenced by the following parameters:

- Radiation source
- Linear attenuation coefficient
- Dose rate
- Object diameter
- Scan conditions
- Filter function and reconstruction algorithm
- Artefacts

The density resolution is limited mainly by the given energy spectrum and the final specific emissivity of the used radiation sources. The scan conditions of the measurement and the algorithm of image reconstruction do not contribute in such strong manner to the density resolution.
For simple cases an analytical expression is possible and the theoretical density resolution can be calculated.
For an cylindrical aluminum part (diameter 160 mm) we have measured a density resolution of 1.0 % (Co60-radiation source) whereas the theoretical value was 0.83 %
For a precise determination of absolute density a calibration is necessary. Fig. 3 shows the correlation between linear attenuation coefficient and density for powder metallurgical components. The differences are mainly given due to inhomogenities in the PM-calibration pieces.
The tomogram of a synchronous coupling wheel is shown in Fig. 4. The density resolution is about 2 % whereas artifacts contribute. The green state shows a higher density gradient to the inner part of the wheel. Fig. 5 shows a oil-pump wheel of aluminum. This component is very homogenous over the investigated slices(1 %). The neutral zone is clearly detected.

Ceramic parts

On the contrary to the PM-components where artefacts has a non neglectible contribution to the density resolution, due to the high attenuation in these components, this effect can be neglected for the following ceramic parts. Two discs of Al2O3 should be tested on homogenity before the next process steps. The diameter is about 100 mm. Fig 6 shows the two discs in the green and in the sintered state.
The slip-casting-probe of Fig. 7 was measured with the micro-computertomograph developed at BAM, which has a spatial resolution between 0.05 and 0.3 mm.
The problems at slip-casting-ceramics are inhomogenities of the probes, due to sedimentation. Density resolution is 1 %. The density gradient (about 8 %) at right is very good visible. The same figure shows a tomogram taken with a microfocus x-ray tube. Due to the smaller focus the tomogram shows the pores sharpened in comparison to the tomogram taken with the minifocus X-ray tube.

Conclusions

The measurement of absolute density with CT with X- and Gamma-rays can be used in some cases to characterize objects, especially to determine different materials.
The great advantage is the fact that CT gives a map of the local density distribution. With carefully calibrated tomographs a density resolution of about 1 % and lower is reachable. Together with high spatial resolution CT offers the possibility to characterize advanced materials like powder metallurgical parts or structural ceramics about their density distribution even in complex geometries.

References

1. Buresch, F:E:; Meyer, W.:
   Safety Evaluation of Structural Ceramic Components
   Materialprüfung 30 (1986) 205-207.
2. Heidt, H.; Goebbels, J.; Reimers, P.; Kettschau, A.
   Development and Application of an Universal CAT-Scanner
   Proc. 11th WCNDT, pp. 664-671
   Las Vegas, Nov. 3-8, 1985

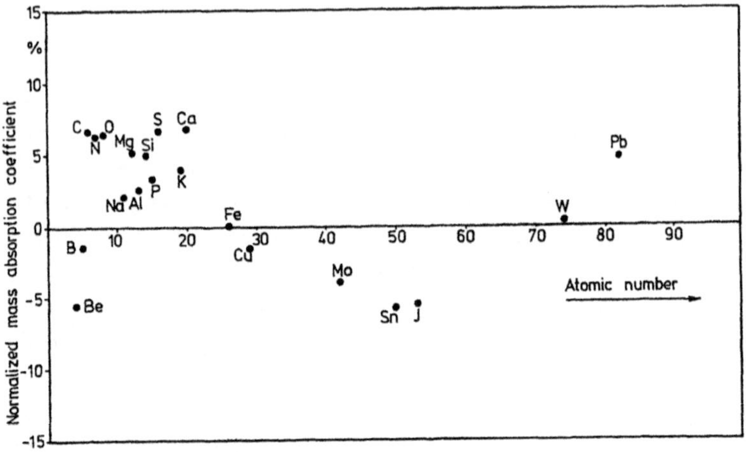

Fig. 1 Normalized mass attenuation coefficient (at 1 MeV) as function of atomic number

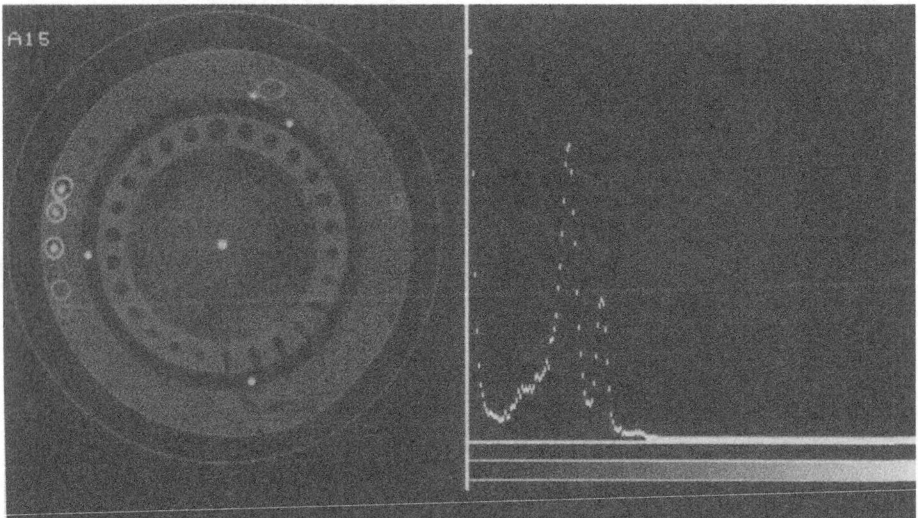

Fig. 2 Tomogram of a waste container

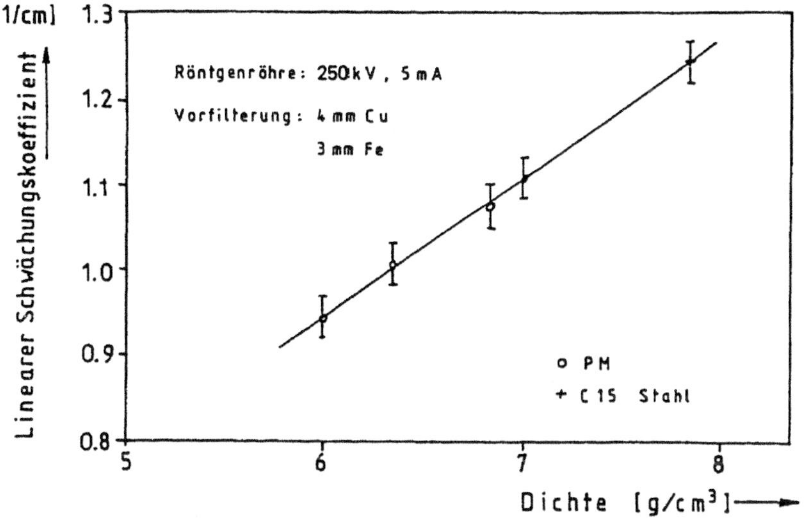

Fig. 3 Correlation between linear attenuation coefficient and density for powder metallurgical-components

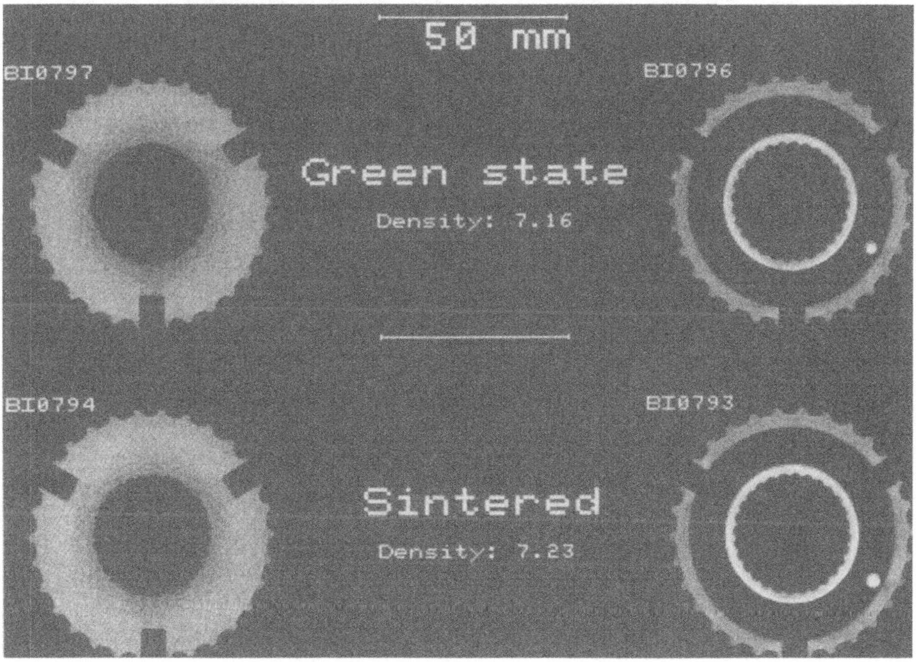

Fig. 4 Tomograms of a synchronous coupling wheel

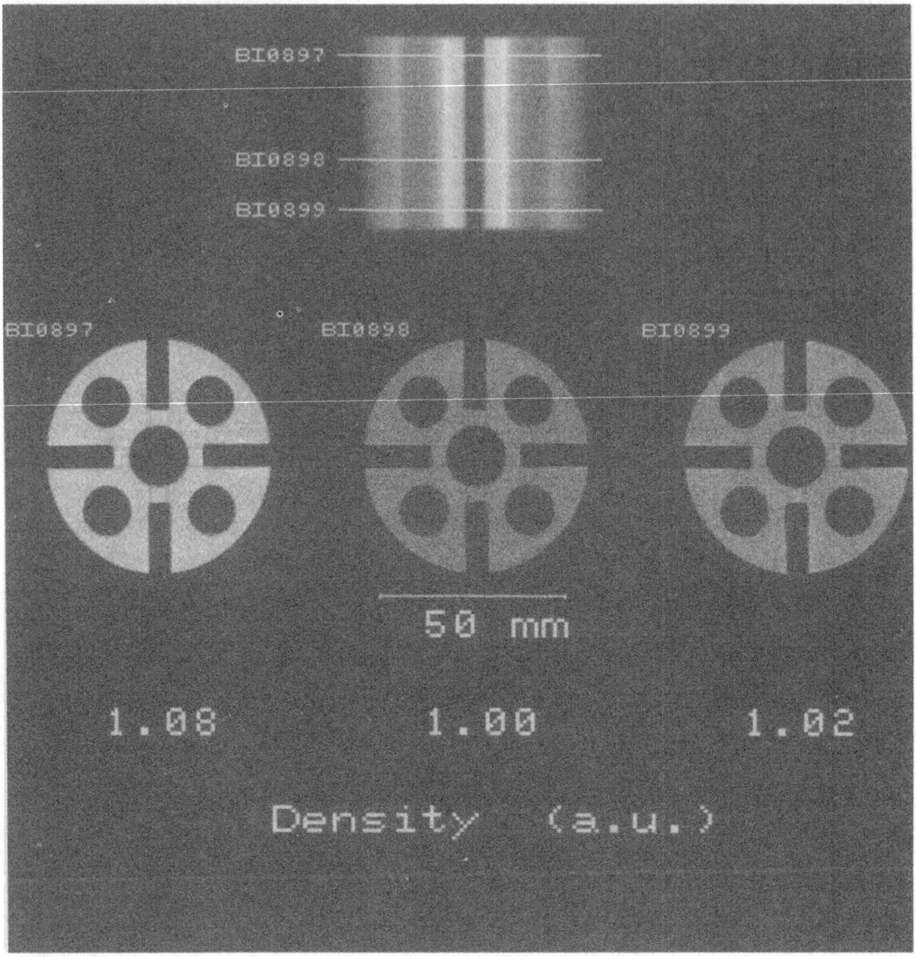

Fig. 5 Tomograms of a oil pump wheel of aluminium

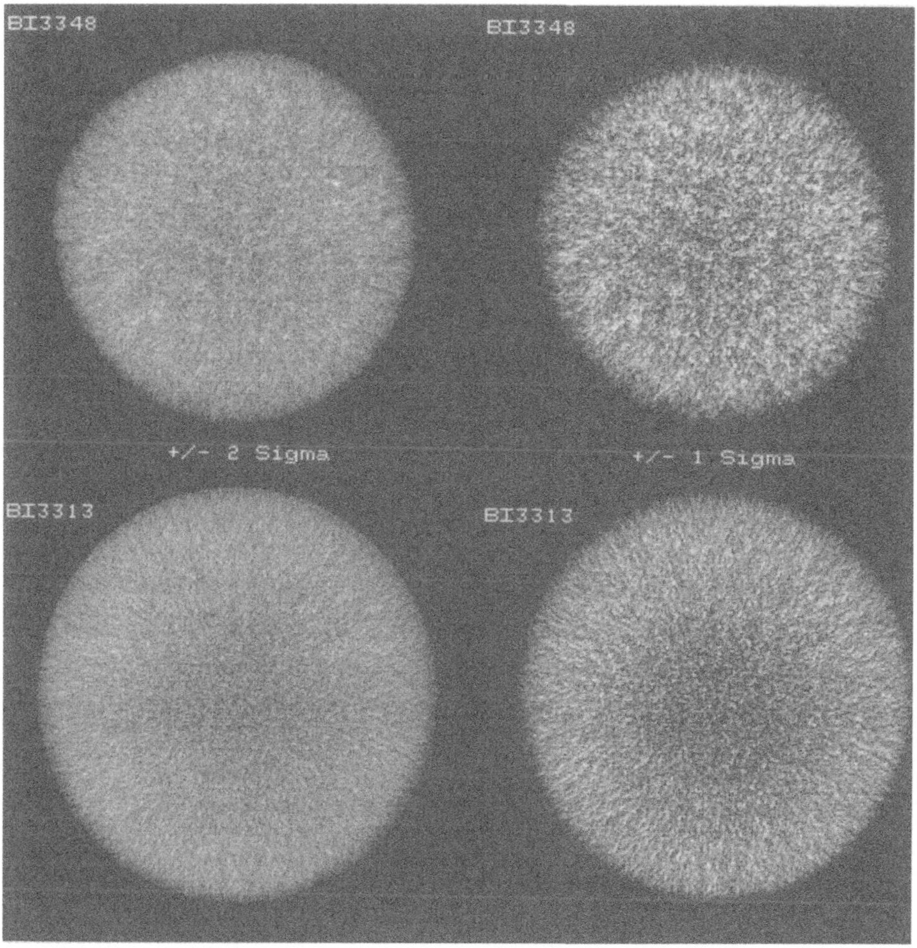

Fig. 6 Tomograms of ceramic discs for two different grey level representations (upper disc: green state, lower disc: sintered)

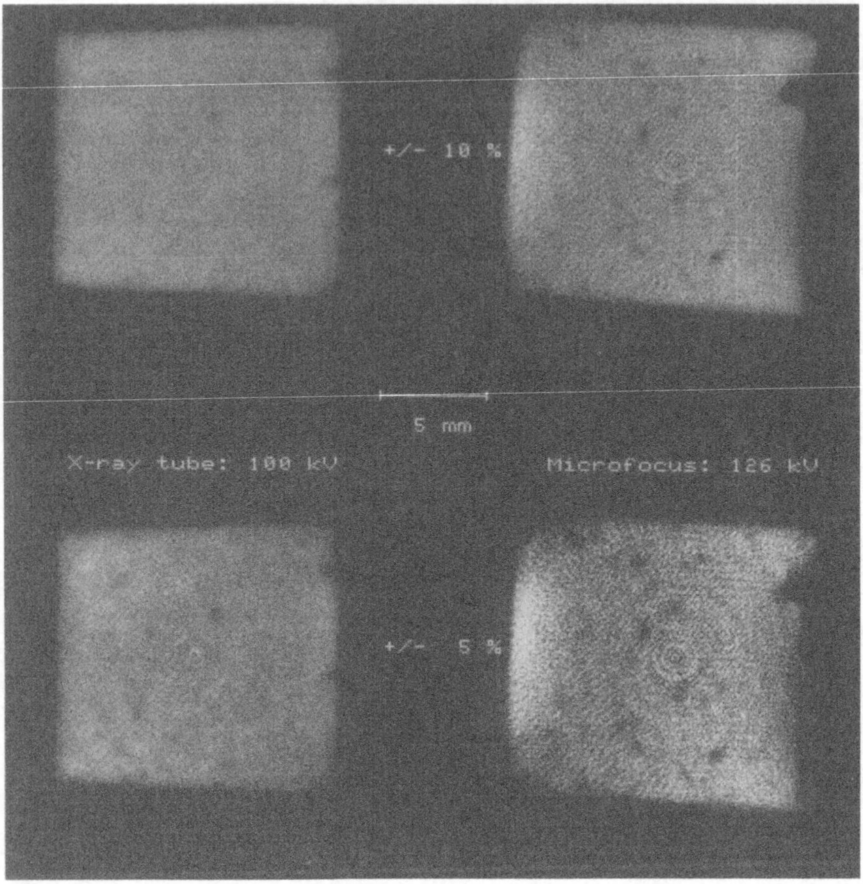

Fig. 7 Tomograms of slip-casting ceramic (left side: mini-focus x-ray tube, right side: micro-focus x-ray tube)

Nondestructive Density Measurements in Powder Metallurgy and Ceramics

G. SCHLIEPER*, V. ARNHOLD*, H. DIRKES**

*) Sintermetallwerk Krebsöge GmbH, P. O. Box 5100, D-5608 Radevormwald, West Germany

**) Dr. Heinz Dirkes, Hermann-Löns-Weg 36, D-4404 Telgte, West Germany

Summary

Absorption measurements with gamma radiation have been utilized for the determination of porosities (densities) in materials compacted or sintered from metallic or ceramic powders. The mathematical background for the assessment of this method and for evaluations of the accuracy of measurement is presented. The equipment for the practical application of density measurements in industry has been developed. Hardware and software of this computerized instrument are designed for a maximum of safety, ease of operation, reliability, flexibility, and efficiency.

Zusammenfassung

Absorptionsmessungen mit Gamm-Strahlung werden genutzt, um Porositäten (Dichten) von gepreßten oder gesinterten Werkstoffen aus metallischen und keramischen Pulvern zu bestimmen. Die mathematischen Gleichungen für die Anwendung dieses Verfahrens und zur Bestimmung der Meßgenauigkeit werden angegeben. Für den Einsatz der Dichtemessung in der industriellen Praxis wurde ein Meßgerät entwickelt. Durch einen integrierten Steuercomputer mit speziell entwickelter Software ist es sicher leicht zu bedienen, zuverlässig, vielseitig anwendbar und wirtschaftlich.

1. Introduction

Absorption measurements using gamma radiation are a well-established method to determine the thickness of sheet in rolling mills, monitor the level of goods in containers and for other purposes. Now this method was utilized in an instrument to determine the porosity of compacted or sintered parts from metallic or ceramic powders. For practical applications in quality control of a parts producing company the appropriate equipment has to fulfill various criteria, namely safety and ease of operation, reliability, flexibility, and efficiency.

The Gamma Densomat has been designed to achieve a maximum of these requirements. The characteristic features of the construction and the software concept are presented in this paper.

2. Accuracy of Measurement

The absorption of monochromatic gamma radiation in materials is described by an exponential law (Fig. 1).

$$n = n_0 \exp(-\mu \rho x) \tag{1}$$

(n = count rate of the weakened beam, n_o = count rate of the incident beam, μ = mass absorption coefficient, ρ = density, x = material thickness). If μ and x are known, ρ can be determined by measuring n_o and n.

$$\rho = \frac{1}{\mu x} \ln \frac{n_0}{n} \tag{2}$$

An important prerequisite for the application of absorption measurements is an assessment of the reliability of a density measurement. The accuracy of measurement can be influenced by four factors.

1. Errors from the measurement of absorption coefficients (calibration measurements).
2. Errors from the thickness measurement.
3. Statistical errors from the reference count rate n_o.
4. Statistical errors from the attenuated count rate n.

The total error can be written as follows.

$$\frac{\Delta \rho}{\rho} = \frac{\Delta \mu}{\mu} + \frac{\Delta x}{x} + \frac{1}{\ln \frac{n_0}{n}} \left(\frac{1}{\sqrt{n_0 t_0}} + \frac{1}{\sqrt{n t}} \right) \tag{3}$$

The statistical errors of the reference and the attenuated count rates are formally very similar. However, since the count rate and the counting time for a density measurement are always smaller than for the reference measurement, the errors from the density measurement are predominant in any case.

$$\frac{\Delta \rho}{\rho}_{stat.} = \frac{1}{\ln \frac{n_0}{n}} \frac{1}{\sqrt{n t}} = \frac{1}{\mu \rho x \sqrt{n_0 t}} \exp\left(\frac{1}{2} \mu \rho x\right) \tag{4}$$

A plot of this equation as a function of the part thickness x is shown in Fig. 2. There is an optimum part thickness x_o at which the statistical error is a minimum. This optimum thickness is

$$x_o = \frac{2}{\mu \rho} \tag{5}$$

For measurements of high accuracy the thickness of the material should be between one fifth and three times the optimum thickness. Within these limits the statistical error is less than twice the minimum value.

The error resulting from a calibration measurement can now be estimated under the assumption of a calibration sample with optimum thickness x_o, sufficiently smooth surface, and density exactly known. The counting time t_o for calibration and reference measurements is 600 sec.

$$\frac{\Delta\mu}{\mu} = \frac{1}{2}\left(\frac{1}{\sqrt{n_o t_o}} + \frac{e}{\sqrt{n_o t_o}}\right) = \frac{0.076}{\sqrt{n_o}} \tag{6}$$

The total error without consideration of the error of thickness measurement which is independent of the count rate can now be expressed as

$$\frac{\Delta\rho}{\rho} = \frac{0.076}{\sqrt{n_o}} + \frac{x_o}{2x}\left(\frac{0.041}{\sqrt{n_o}} + \frac{1}{\sqrt{n_o t}} \exp\frac{x}{x_o}\right) \tag{7}$$

In Fig. 3 the total error for a counting time of t = 240 sec has been plotted as a function of the reference count rate n_o. The lower graph is valid for the optimum thickness x_o and the upper graph for the case of $x = 0{,}2\ x_o$ and $x = 3\ x_o$.

3. Design of an Instrument for Industrial Applications

For the practical application of density measurements by absorption of gamma radiation in industry the Gamma Densomat has been developed whose schematic construction is shown in Fig. 4. As radiation sources generating monochromatic gamma radiation some radioactive elements are well suited.

The Gamma Densomat can be equipped with two different radiation sources for more flexibility with respect to the materials and material thickness to be measured. For each measurement the best suited source is automatically selected.

A collimated beam of gamma radiation emerges through a hole from the source container which is manufactured from strongly absorbing material to protect the user from the radiation. After penetrating the part to be measured the beam passes through an aperture which determines the material volume where the density measurement is taken. The intensity of the radiation is measured by a NaI detector and photomultiplier. A dial gauge is used to determine the part thickness and at the same time define the point of measurement.

An integrated computer guarantees safe and easy operation. The control program has the following duties.

- Control and supervision of all mechanical functions
- Handling of the data files
- Correction of the count rates (dead time, background)
- Storage of reference count rates
- Calculation of the density

4. Structure of the Data Files

The structure of the data files has been designed as to allow the operator to access the required material constants fast and comfortably. On the one hand the absorption coefficients for a large number of materials have to be readily available. On the other hand the amount of calibration work should be held as low as possible. Additionally the material for a measurement should be found by simply entering the commonly used designation of the part.

These requirements are met by a system of 3 data files, the part list, material list and component list, which are structured in a way that, invisible for the operator, data from the subordinate files can be referred to.

The part list allows the operator to use the commonly used designations of parts to call up the required material data. Each part code in the part list is attached to an internal material number. Different part codes can be attached to the same material number.

The material number is the internal link to the material list. This file contains the material compositions, each of which can be composed of up to 10 individual components.

The component list contains the density of each component and the absorption coefficients for the different radiation energies and apertures. A component can either be a pure element or a mixture of several elements, or a compound.

This way the calibration of the Gamma Densomat is located exclusively in the component list. Once calibrated, the components can be combined arbitrarily in the material list.

Composition changes in the material list do not require a new calibration. The calibration work is thus minimized. The calibration of just 2 components is sufficient to measure any combination of these components. For materials of more complex composition calibration measurements can either be done on the components or the materials themselves.

5. Applications

Useful applications of density measurements by gamma-ray absorption have been found in the production of powder metal and ceramic components. Fig. 5 shows a powder metal compact with different compacting levels. A uniform density distribution throughout the compact is necessary to achieve uniform material properties in the finished part. Through regular nondestructive density checks in the different sections of the parts it is possible to maintain a consistent quality without scrapping the tested parts.

Magnetic materials are also compacted from powder. Some ferrite magnet compacts are shown in Fig. 6. Densities of the small sections of these parts are very difficult to determine with accuracy using conventional methods. The gamma radiation measurement requires no cutting into sections and is more accurate than other methods.

An example for ceramic compacts is shown in Fig. 7. As the shrikage of ceramic materials during sintering depends from the density in the compact, non-uniform density distributions lead to distortions of the sintered part. To avoid scrap it is therefore important to check sectional densities regularly and adjust the press if necessary.

Further applications are expected in manufacturing cemented carbides, refractories, abrasive and friction materials etc. Substantial improvements in the control of production processes and the quality of parts are possible if the nondestructive determination of densities with gamma radiation is consequently utilized.

Literature

(1) G. Schlieper, V. Arnhold, A. Kozuch: Nondestructive Determination of Sectional Densities by the Gamma Densomat, Conference Proceedings: Nondestructive Testing of High-Performance Ceramics, Boston, 1987, The American Ceramic Society, Westerville, Ohio

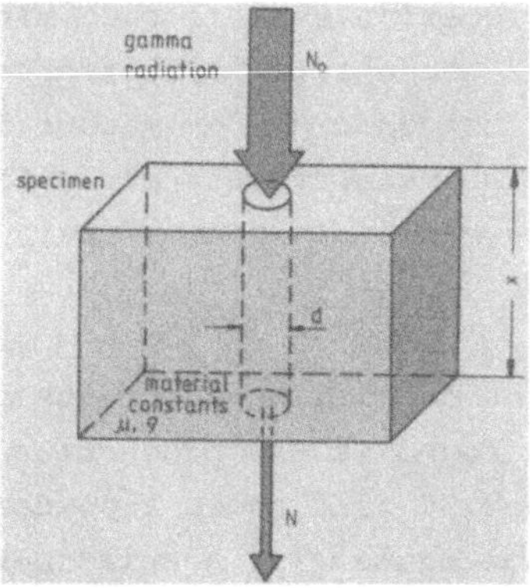

Fig. 1: Weakening of gamma radiation during penetration of material

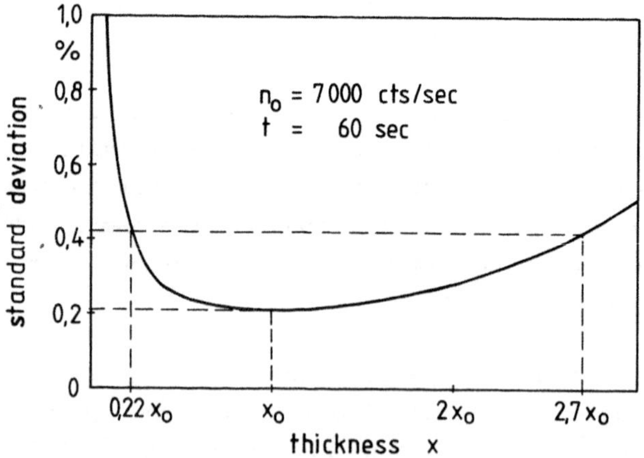

Fig. 2: Statistical error of density measurements as a function of the sample thickness

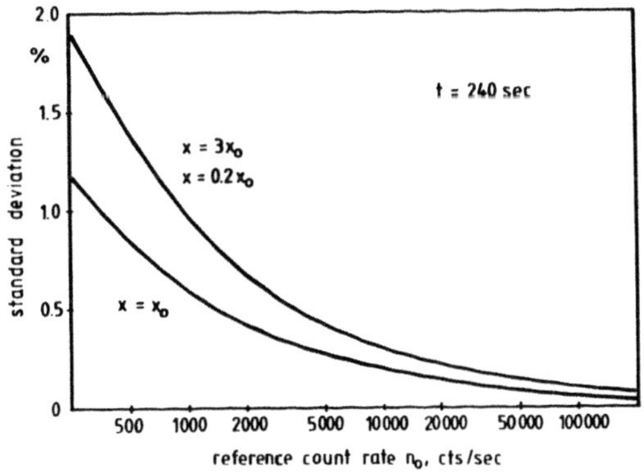

Fig. 3: Total error of density measurements as a function of the reference count rate n_o at a counting time of t = 240 sec.

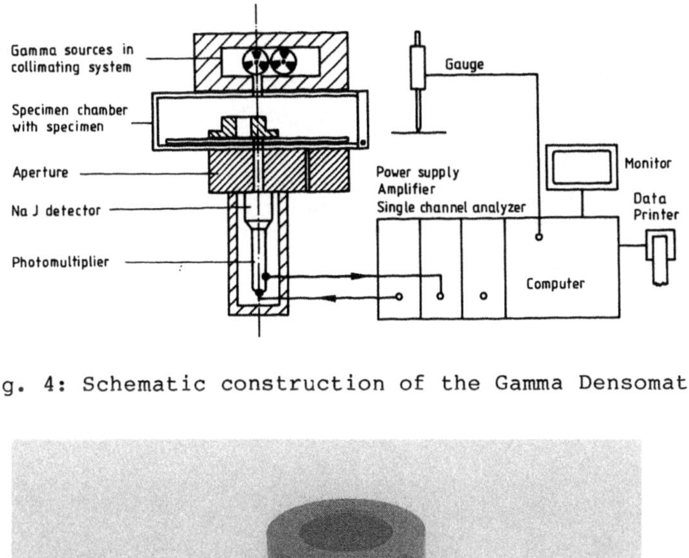

Fig. 4: Schematic construction of the Gamma Densomat

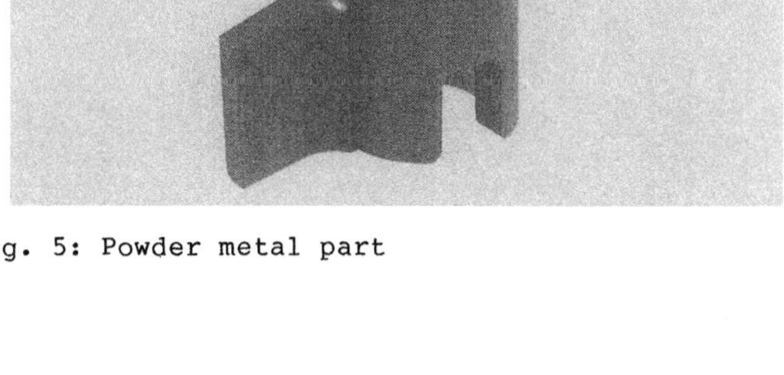

Fig. 5: Powder metal part

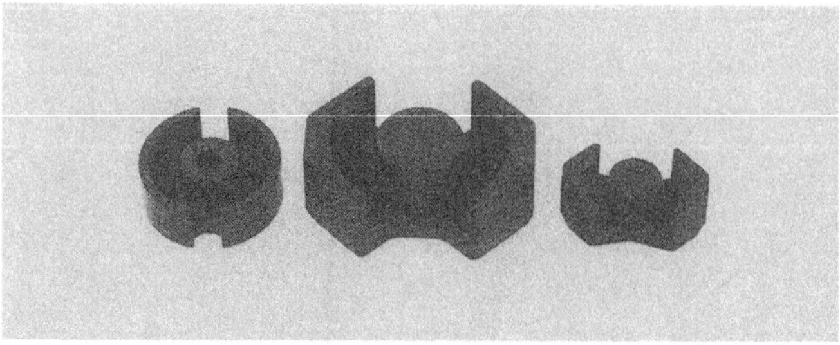

Fig. 6: Ferrite magnet compacts

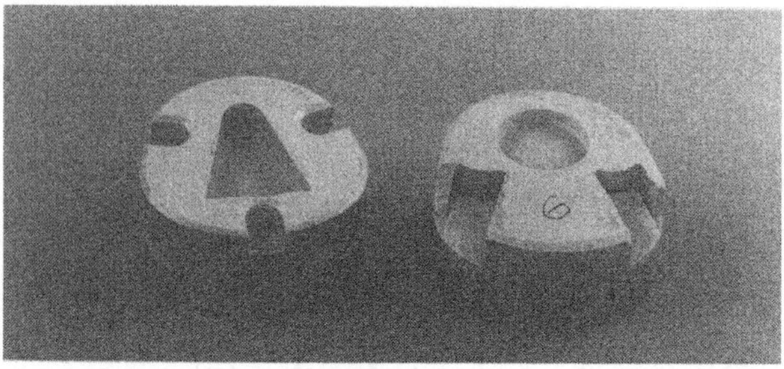

Fig. 7: Ceramic compact

New Developments for the Ultrasonic Characterization of Materials

Wolfgang Sachse and Kwang Yul Kim

Department of Theoretical and Applied Mechanics
Cornell University, Ithaca, New York - 14853 U · S · A

Summary

This paper summarizes several recent developments which are facilitating new approaches for quantitative ultrasonic measurements whose application is to the characterization of materials. These include the development of point sources and point receivers, a theory for analyzing the propagation of transient elastic waves through a bounded, dispersive and attenuative medium, and the development and implementation of appropriate signal processing algorithms. An alternative to these deterministic approaches is a processing scheme based on a simulated intelligent system which processes the signals like a *neural network*. Examples of each of these ideas are presented.

Introduction

The development of reliable non-destructive test methods for the characterization of materials, and, in particular, engineered materials such as composites, remains a critical task. Ultrasonic measurements relying on elastic waves propagating through a material often provide an ideal means for determining the global properties of the material as well as its integrity. These properties may refer to a material's mechanical properties or its composition, micro- or macro-structural features, including its microstructure, porosity and weak or failed regions or residual stresses and stress gradients. As to the detection and characterization of damage, composites differ markedly from metals. Early damage in composites is usually distributed over an extended region in a specimen and thus, for any technique to be useful, it must be capable of globally monitoring or inspecting the composite structure.

A successful ultrasonic materials characterization procedure requires the solution of two problems. One is related to the reliable determination of a particular waveform parameter such as amplitude, arrival time, spectral feature, etc. from the detected ultrasonic signals. The second is related to the evaluation of that waveform parameter to recover the material property of interest. The latter may be done by establishing a *correlation* between a particular waveform parameter with a specific property variable. More complex are the *semi-inverse methods* in which a solution to a particular forward problem of elastodynamics is studied to establish the connection between material and ultrasonic variables. Most complex is the *inverse problem* in which the ultrasonic waveform data is inverted to recover the spatially varying characteristics of the propagating

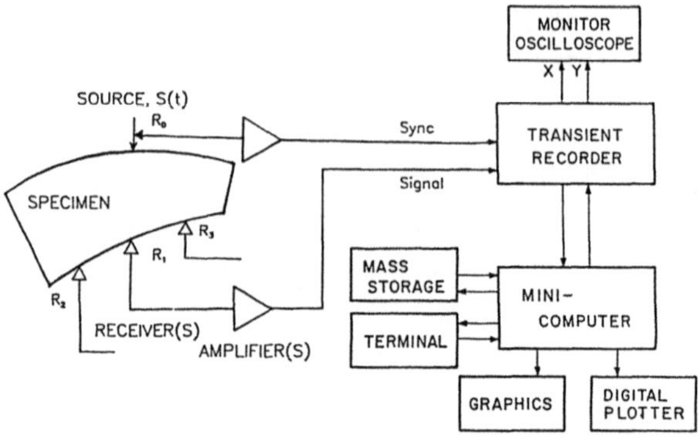

Figure 1: *PS/PR* measurement system.

medium. The focus of this paper is restricted however, to consideration of the first problem - that of reliably determining a particular wavefield parameter in a detected ultrasonic signal.

The measurement problems related to ultrasonic techniques are both material- and geometry-related. The absorption of ultrasonic waves in many materials, such as thick composites, is high and usually strongly frequency-dependent. Reflecting the effects of microstructure as well as its properties - be they elastic, viscous, geometric and/or scattering - the propagation of elastic waves through a material may be dispersive at particular frequencies. Engineered materials are also often specifically designed and fabricated to be elastically anisotropic. And they may be fabricated into large, complex-shaped parts typically possessing non-parallel and non-planar surfaces which complicate the wave propagation.

Quantitative Deterministic Ultrasonic Measurements

Such measurements utilize detailed knowledge of the *source*, *structure* and *sensor* of the measurement system combined with signal processing techniques to recover the characteristics of the unknown component - be it the source, material or sensor. The use of an ultrasonic point-source/point-receiver (*PS/PR*) measurement system which is similar to that used for quantitative acoustic emission measurements has recently been described for use as a material's characterization tool [1,2]. In the technique, which is illustrated in Fig. 1, a simulated source whose temporal and spatial characteristics are known is used as excitation and one or more well-characterized point-sensors are used to detect the signals which have propagated from the source through the medium. The essential advantage of the *PS/PR* method over conventional ultrasonic inspection methods is that absolute quantitative ultrasonic measurements are possible and the technique is capable of overcoming many of the measurement problems related to the ultrasonic characterization of materials listed above. The basis of the *PS/PR* measurement method and its application to characterize several composite materials has been given in References [1] and [2]. The requirements and operational characteristics of various sources and receivers which can be used in a *PS/PR* measurement system have also been described [3].

It was demonstrated in Ref. [1] how *PS/PR* measurements can be used to determine from just one measurement both the longitudinal and shear wavespeeds and hence the Lamé elastic constants of an elastically isotropic material. By using a non-contacting source and receiver, the measurement can be scanned, thus becoming an important materials characterization tool. By using an array of sensors to detect the signals at points equi-spaced about the source point, the orientation dependence of the wavespeeds in the material can be determined. As expected, this dependence is strongly dependent on the elastic anisotropy of the material. An algorithm was recently developed by which this information is used to recover the matrix of elastic constants for an anisotropic material [4].

The determination relies on the measurements of the P- and S-wave arrival times at each sensor comprising the array of sensors whose number is equal to or greater than the number of elastic constants to be determined. Knowing the direction cosines for each source/receiver acoustical path, the wavespeeds along these arbitrary, non-principal directions of the specimen are calculated. By constructing the characteristic equation associated with the *Green-Christoffel* tensor, a functional is found which is related to the unknown elastic constants of the material. A processing algorithm has been developed to minimize this functional using the *Newton-Raphson* method. While the procedure is general and applicable to any arbitrary symmetry, to date it has been demonstrated only with materials which are transversely isotropic.

The use of ultrasonic attenuation measurements as a basis for a materials characterization procedure has been investigated for a number of decades. The principal difficulty has been that because of a large number of parasitic effects, a procedure for the reliable determination of the absolute value of attenuation of a material is difficult to realize. However the past decade as seen the development of a theoretical basis for computing the displacement signals at a particular receiver point resulting from a given excitation in a non-attenuative material [5]. And this can be used to effect a comparison between the calculated waveforms and those measured in a real material to determine the attenuation of the real material. The frequency-dependence of the attenuation of a particular wave mode is found by windowing the corresponding wave arrival in the detected signal, transforming it into the frequency-domain, and evaluating

$$\alpha(f) = \frac{20}{d} \log_{10}\left[\frac{V(f)}{V_{theo}(f)}\right] \qquad [\text{db/length}] \tag{1}$$

Here the V's refer to the magnitude spectra of the selected signal amplitudes and d is the propagation length. The principal difficulty with this procedure is the requirement of a precise determination of the source strength and a detecting sensor whose transfer characteristics are known absolutely.

By analyzing the propagation of elastic waves in a viscoelastic material and evaluating the normal displacement signals detected at the epicentral receiver location in the frequency-domain [6], it has been shown that the Fourier phase function of particular wave arrivals can be processed to directly determine the dispersion relation for that wave in the material. To do this correctly, it is important that the effects of geometric dispersion are properly accounted for so that only the actual material-related dispersion of the material is determined. From the latter, the phase and group velocities can be directly evaluated.

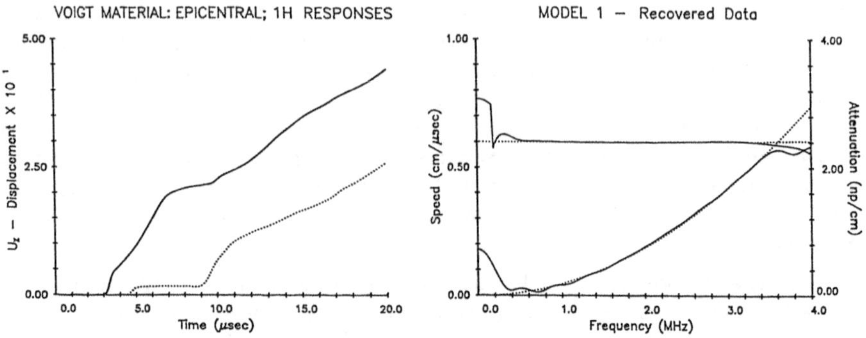

Figure 2: (a) Voigt material epicentral and $1h$ responses; (b) Input (dotted lines) and recovered (solid lines) phase velocity and attenuation data.

An extension of the above has recently been completed so that the signals at off-epicentral receiver points could also be computed for viscoelastic plates whose properties are specified in terms of arbitrary frequency-dependent wavespeeds and attenuation values [7]. A sample result is shown in Fig. 2(a), corresponding to the signals expected at epicenter and the $1h$ off-epicentral position of the plate on the side opposite from a normal force, step excitation. These responses were computed for the case of a viscoelastic Voigt-like material possessing a complex shear modulus given by $\mu = \mu_0(1 - i\omega\tau)$ where μ_0 is the static shear modulus and τ is a relaxation time, which in the example shown was $\tau = -0.017$. It is seen that the computed responses resemble those calculated for a perfectly elastic, non-attenuative material, except that in the viscoelastic case, the wave arrivals are no longer sharply delineated.

The results obtained in the forward problem have permitted the development of a more general algorithm for recovering the dispersion relation, wavespeeds and attenuation from the signals detected at two positions relative to the source location. The advantages of this algorithm include its independence of the source strength and, if similar transducers are utilized, its independence on the exact characteristics of the sensors. If the logarithm of the ratio of the Fourier-transformed epicentral and off-epicentral signals is formed and if the attenuation is not too large, the following explicit formulae are found for the phase velocity and attenuation of the longitudinal P - wave:

$$c_{P_{phase}}(\omega) \;=\; \omega\,(d-h)\left[\Im\left(\log\frac{\hat{w}_{off-epi}}{\hat{w}_{epi}}\right) - \Re\left\{\frac{c}{\omega}\left(\frac{A(0)}{h} - \frac{A(\theta)}{d}\right)\right\}\right] \tag{2}$$

$$\alpha_P(\omega) \;=\; \left[-\log\left|\frac{\hat{w}_{off-epi}}{\hat{w}_{epi}}\right| + \log\frac{h}{d} + \log\frac{G(sin\theta)\,\cos\theta}{G(0)}\right]\,[d\;-\;h]^{-1} \tag{3}$$

where d is the source/receiver path length and the A's and the G's are computed correction factors. Similar formulae have been developed for the shear and Rayleigh surface waves [7].

Numerical simulations with synthetic data shown in Fig. 2(b) have demonstrated that the above equations correctly recover the input frequency-dependent wavespeeds

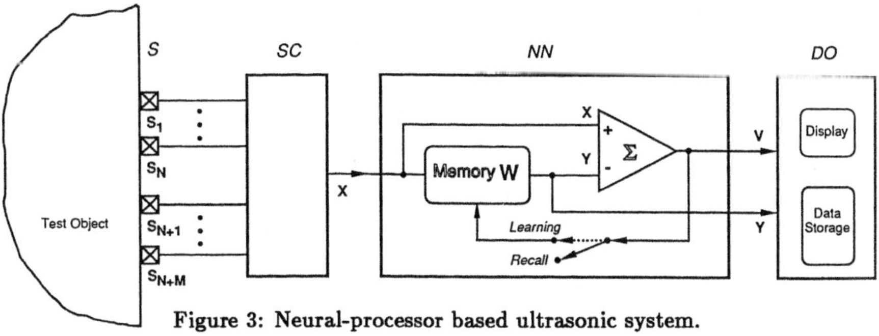

Figure 3: Neural-processor based ultrasonic system.

and attenuation values in the frequency interval between 0.5 and 4 MHz. The deviations seen to occur outside of this range are probably a consequence of the low signal amplitudes at the higher frequencies and at the lower frequencies, a breakdown of the asymptotic assumptions made in calculating the waveforms.

Neural Processor-based Ultrasonic Measurements

The basis of such processing is the teaching of a system and the development of a *memory* corresponding to known sources and wave propagation characteristics. Following this, unknown signals can then be processed using *auto-associative* algorithms to recover missing information about the source, the medium or the receiving transducer.

This approach which has been described in several publications [8,9,10], utilizes some of the fundamental principles of neural networks [11]. Our neural-like ultrasonic processing system is shown schematically in Fig. 3. It is assumed that an ultrasonic wavefield can be characterized by a finite set of data supplied from an array of sensors together with selected features of the source, structure or sensor. For one experiment in which data is collected with N sensors, a *pattern* vector can be defined as

$$X = \{v^{(1)}(t),\ v^{(2)}(t), \dots, v^{(N)}(t);\ g\} \tag{4}$$

in which the $v^{(n)}$ represents the discretized signal detected by the n-th sensor and g are M elements in which is encoded specific information about one or more of the components comprising the ultrasonic system. Such pattern vectors can be generated after signal conditioning in module SC in Fig. 3. The pattern vectors are input to the neural network processor NN.

For each fixed input vector from this ensemble, the system responds with an output vector Y which is of the same dimension as X and which can be determined from the linear matrix equation

$$Y = \mathbf{W} \cdot X \tag{5}$$

where the matrix $\mathbf{W}$ represents the response function or *memory* of the system. In order to obtain an *associative* operation of the processing system, it is assumed that the processor adapts to the input vectors such that the *discrepancy* or *novelty* between the input and output vectors, given by

$$V = X - Y = X - \mathbf{W} \cdot X \tag{6}$$

is reduced with a repetition of the inputs. This is possible with the feedback loop in the neural processor shown in Fig. 3. The adaptive law of the system which governs how the memory develops is similar to that used in other applications [8,12].

Once the *memory* of a system has been developed, then the prediction of the input by recall from memory, **W** is given by Eq. (5). The *recall*, Y, represents that part of the signal which can be specified in terms of the previously learned pattern vectors. It can be used to predict *wavefield* quantities from source data without solving the forward problem of elastodynamics. The *recall* can also be used to predict features of the *source* from wavefield data, i.e. a solution to the *inverse source problem* can be obtained. And finally, the processing system can supply missing information in a signal via an *auto-associative recall.* The output of the system is via module DO of Fig. 3.

The application of the neural-like signal processing system has been demonstrated with ultrasonic waveforms obtained in several simple source location and source characterization problems as well as ultrasonic wavefield measurement situations [8,9]. Here are only described the results of recovering wavefield data from a particular source and the detection of changes in the wavefield.

The experimental system consisted of two, miniature, broadband piezoelectric sensors mounted 20 in. apart on the surface of a 1 inch-thick square aluminum plate 0.75 m on edge. The ultrasonic signals were excited by the impact of a steel ball of diameter $\phi = 8$ mm dropped onto the plate from a height of 5 mm. The signals from the sensors were amplified by 40 db and recorded by a multi-channel waveform recording system which was triggered by one of the waveforms. The sampling rate was 5 MHz. During the learning phase, the source was activated at positions 1 inch (25 mm) apart, along the line between the two sensors. The pattern vectors consisted of 150 components of which 128 corresponded to a concatenation of the ultrasonic signals while the remaining 22 components were used to encode the source location information. The completed set of *learning vectors* is shown in Fig. 4(a).

During the iterative learning procedure, output vectors were learned. They are superimposed on the learning vectors but they are nearly indistinguishable from the original input patterns indicating a correct adaptation of the system. The memorized response matrix **W** is shown in Fig. 4(b). There are 9 characteristic regions in the matrix, corresponding to the auto- and cross-correlations between the components comprising the pattern vectors, that is, the wavefield data from each of the sensors and the encoded source information. Each portion of the memory is of importance for the associative operation of the system during the analysis of new patterns presented to the system.

After the memory has been developed, the feedback to the memory is disconnected and signals can be presented to the system for analysis via the *auto-associative recall.* If a pattern vector is presented to the system which does not include coded information about the source, the recall signals will correctly recover this information provided that it has previously been learned [9]. If only the coordinate data is presented to the system, the obtained recall, shown in Fig. 4(c) is obtained. The original signals are only partially recovered from the cross-correlation portion of the waveform signals and the source information data in the memory matrix. Since the waveform signal represents 128 out of 150 points of the total pattern length, their recall by a smaller number of coordinate points is less reliable than the recall of the source components from wavefield data. Nevertheless, the recall is able to recover the principal features of the signals.

Preliminary experiments have also been carried out with the neural-like processing approach to detect changes in the wave propagation in a specimen. In these measurements, demonstrated on an aluminum plate [10], ultrasonic signals corresponding to particular source/receiver configurations were used to develop the memory. To modify the wave propagation, a brass disk was attached at various points onto the surface of the plate. Its effect appeared as a slight variation in the measured ultrasonic wavefield relative to the original measurements. It was shown that these changes could be detected with the discrepancy vector, V. This demonstrates that an adaptive system can be used in non-destructive testing applications in which changes caused by variable wave impedances, such as the development and growth of defects in a structure are to be detected.

Conclusions

This paper has summarized several developments which are facilitating new quantitative ultrasonic measurements applicable to the characterization of materials. The use of a point source and one or more point receivers in the *PS/PR* technique provides a powerful materials testing system which, when coupled with appropriate signal processing techniques, permits the determination of the ultrasonic wavespeeds and attenuations as a function of frequency and propagation direction in a material. *PS/PR* measurements require a minimal amount of surface preparation, they can utilize non-contact sources and receivers and they can be made on specimens which are neither planar nor parallel. Measurements are also possible in ultra-attenuative ma-

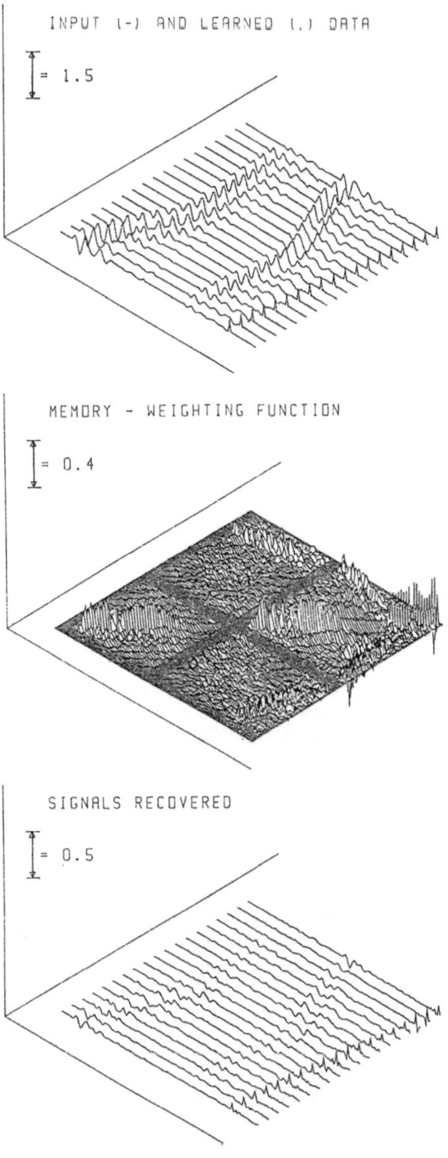

Figure 4: *PS/PR* signals: Learning, learned pattern vectors; (b) Memory matrix, **W**; (c) Recovered waveforms from source data.

terials. An important application is their application to determine the matrix of elastic constants describing the anisotropy of a material.

It was shown that an adaptive learning system comprised of an associative memory can be used to map ultrasonic source and waveform data and vice versa with the auto- and cross-correlation portions of the memory. Experiments were described utilizing such an adaptive system, to process the ultrasonic signals detected in a plate specimen to recover information about the propagating medium. The application of neural-like processing of waveforms appears to be a promising means for extracting important information from ultrasonic signals in materials and it could form the basis of a new generation of intelligent NDT/NDE signal analyzers.

Acknowledgements

It is a pleasure to acknowledge the collaboration of Profs. I. Grabec and R. L. Weaver and Dr. B. Castagnede whose results we have cited. This work has been supported by the Office of Naval Research, Solid Mechanics and Physics Programs. The use of the facilities of the Materials Science Center at Cornell University which is funded by the National Science Foundation is also acknowledged.

References

[1] W. Sachse and K. Y. Kim, "Quantitative AE and failure mechanics of composite materials', *Ultrasonics*, **25**, 195-203 (1987).

[2] W. Sachse and K. Y. Kim, "Point-source/Point-receiver materials testing", in *Ultrasonic Materials Characterization II*, J. Boussière, J. P. Monchalin, C. O. Ruud and R. E. Green, Jr. Eds., Plenum Press, New York (1987), pp. 707-715.

[3] W. Sachse, "Transducer considerations for point-source/point-receiver materials measurements", in *Ultrasonics International '87: Conference Proceedings*, Butterworth Scientific, Ltd., Guildford, Surrey, UK (1987), pp. 495-501.

[4] B. Castagnede and W. Sachse, "Optimized determination of elastic constants of anisotropic solids from wavespeed measurements", in *Review of Quantitative Nondestructive Evaluation*, **8**, D. O. Thompson and D. E. Chimenti, Eds., Plenum Press, New York (1988).

[5] A. N. Ceranoglu and Y. H. Pao, "Propagation of elastic pulses and acoustic emission in a plate: Part I. Theory; Part II. Epicentral response; Part III. General responses", *ASME J. Appl. Mech.*, **48**, 125-147 (1981).

[6] R. L. Weaver and W. Sachse, "Asymptotic viscoelastic rays in a thick plate", T&AM Preprint (June 1988). Submitted to: *Journal of Applied Mechanics*.

[7] R. L. Weaver, W. Sachse and L. Niu, "Transient ultrasonic waves in a viscoelastic plate: I. Theory; II. Applications to materials characterization", Materials Science Center Reports #6504; #6505 (July 1988). Submitted to: *J. Acoust. Soc. Am.*

[8] I. Grabec and W. Sachse, "Experimental Characterization of ultrasonic phenomena by a learning system", MSC Report #6447, Ithaca, NY (April 1988). (Submitted for publication).

[9] I. Grabec and W. Sachse, "Application of an intelligent signal processing system to AE analysis", MSC Report #6428, Ithaca, NY (April 1988). In Press: *J. Acoust. Soc. Am.*

[10] I. Grabec and W. Sachse, "Experimental characterization of ultrasonic phenomena by a neural-like learning system", in *Review of Quantitative Nondestructive Evaluation*, **8**, D. O. Thompson and D. E. Chimenti, Eds., Plenum Press, New York (1988).

[11] T. Kohonen, "State of the art in neural computing", *Proc. Intl. Conf. Neural Networks*, San Diego, CA (1987), pp. 1-77.

[12] T. Kohonen, *Neural Networks*, **1**, 3-16 (1988).

Composites

Graphite-magnesium Elastic Constants: Composite and Fiber

SUBHENDU DATTA

Department of Mechanical Engineering and CIRES
University of Colorado
Boulder, Colorado 80309-0449

HASSEL LEDBETTER

Institute for Materials Science and Engineering
National Institute of Standards and Technology
Boulder, Colorado 80303-3328

Summary

This study contains three components: measurement, modeling, and inverse modeling to get the fiber elastic constants. The studied composite consisted of 70-volume-percent continuous uniaxial graphite fibers in a magnesium matrix. By ultrasonic-velocity methods, we measured the composite's complete orthotropic-symmetry (nine-independent-component) elastic-constant tensor: C_{ij}. For a model, we used a wave-scattering method in the long-wavelength limit. The model requires two inputs: the two isotropic-matrix elastic constants and the five anisotropic-fiber elastic constants. Guessing the fiber C_{ij} gave good measurement-model agreement only for C_{11}, C_{22}, and C_{33}. Especially, the shear moduli C_{ii} (i = 4,5,6) agreed poorly (a 20-percent difference). Using inverse modeling—calculate fiber properties from measured matrix properties and measured composite properties—we estimated the anisotropic-fiber elastic constants.

Introduction

We determined the complete five-component transverse-isotropic-symmetry elastic-constant tensor for two graphite fibers: high-strength/low-modulus and low-strength/high-modulus. We did this in two steps. First, we measured ultrasonically the complete elastic constants of a metal matrix with embedded uniaxial graphite fibers. Second, we did an inverse-modeling calculation to extract the fiber's elastic constants. This calculation requires three inputs: composite elastic constants, matrix elastic constants, and fiber-matrix phase geometry, principally the fiber volume fraction. We compare the results with those expected for a random quasiisotropic graphite aggregate and for a hypothetical graphite fiber with perfectly aligned basal planes.

Graphite possesses remarkable physical properties. For example, the within-basal-plane Young modulus equals 902 GPa, four times that of iron (212 GPa), and 80% of that of diamond (1141 GPa). Also, graphite exhibits strong physical-property anisotropy: the E_1/E_3 Young-modulus ratio equals 29.8. (Here, x_3 denotes axis perpendicular to basal plane and x_1 denotes any direction in basal plane, which is isotropic.)

The fiber's elastic constants provide a valuable material characterization; they provide information on basal-plane alignment. Also, they enter many practical problems such as internal strain (residual stress).

Materials

We studied four materials produced from commercial fibers and alloys. The two fibers are categorized as high-strength/low modulus and low-strength/high modulus. The two matrices consisted of pure magnesium and 5056 aluminum alloy. For the fibers, the manufacturer reported Young moduli of 235 and 392 GPa and mass densities of 1.76 and 1.81 g/cm^3, respectively.

Composites were produced by a squeeze-casting method, where molten matrix metal infiltrates carbon-fiber bundles under high pressure. Carbon fibers were preformed with a polymer fugitive binder. Placed in a mold, the preform was heated to slightly below the matrix-metal melting temperature. The binder was burned away and the carbon fibers were preheated. Molten matrix metal was poured into the mold and pressed at 98 MPa before solidification. This pressure caused the molten metal to infiltrate the fiber bundles. After solidification, the composite was removed from the mold. Further fabrication details occur elsewhere [1].

Figure 1 shows a typical microstructure. Focusing, for the moment, on the low-modulus magnesium-matrix material, by Archimedes's method, we found a mass density of 1.771 g/cm^3. For a fiber volume fraction of 0.70, using 1.738 for magnesium, and 2.269 for graphite, we predict a mass density of 2.110 g/cm^3. Probably, the discrepancy arises from nonperfectly graphitized fibers. Indeed, the manufacturer's reported fiber density, 1.76, leads to a prediction of 1.753, within 1 percent of observed. If we assume a void-and-crack-free matrix, our results predict a fiber density of 1.79 g/cm^3, corresponding to a fiber-void fraction of 0.21.

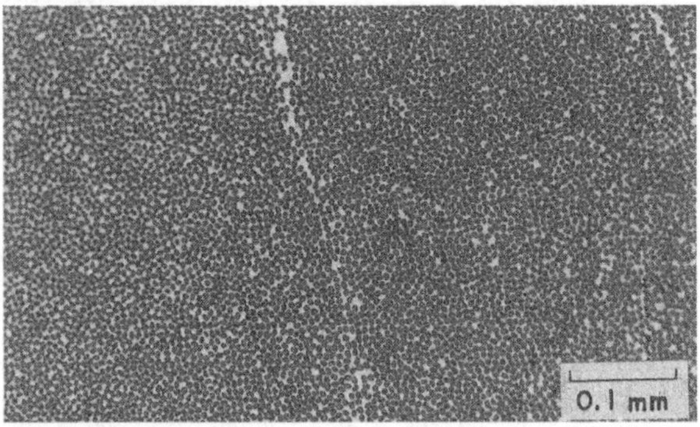

Fig. 1. Optical photomicrograph of transverse section of graphite-fiber-reinforced metal composite. The fibers, 7 μm in diameter, occupy 70 volume percent of the composite. All four studied composites show essentially the same microstructure. This case represents the low-modulus, aluminum-matrix case.

For the magnesium matrix material, a sample was prepared similar to that described above for the composite. For the matrix material, we found an Archimedes-method mass density of 1.738 g/cm^3, close to the accepted value for pure magnesium: 1.737. Similarly, for the 5056-aluminum-alloy matrix material, we found a mass density of 2.652, as expected slightly below the accepted value for pure aluminum, 2.697. Table 1 shows the measured elastic constants for these two matrix materials. The notation is C_ℓ = longitudinal modulus, G = shear modulus, B = bulk modulus, E = Young modulus, ν = Poisson ratio.

Table 1. Elastic constants of matrix materials.

	C_ℓ (GPa)	G (GPa)	B (GPa)	E (GPa)	ν
Mg	57.88	17.72	34.26	45.34	0.279
Al alloy	109.40	26.22	74.45	70.39	0.342

Measurements

We determined the nine C_{ij} by measuring eighteen sound velocities on four specimen geometries described previously [2]. For brevity, we omit further description, except for a few salient details: bond—phenyl salicylate; transducers—quartz, x-cut and ac-cut; frequencies—5 to 6 MHz; specimen size—16-mm cube, or smaller depending on specimen geometry. Previously, we reported details of the measurement method [3]. Figure 2 shows an oscilloscope display.

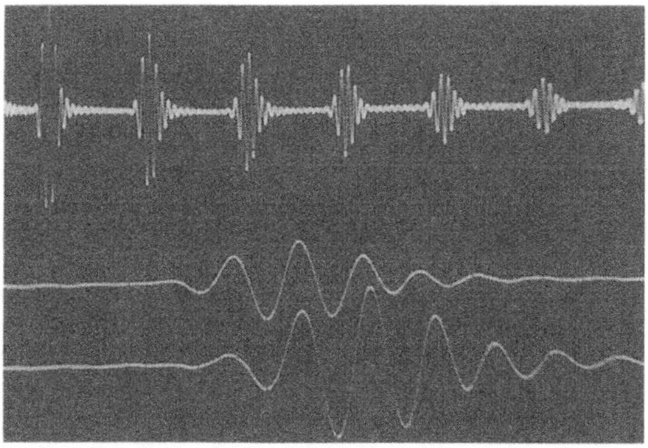

Fig. 2. Oscilloscope display of a pulse-echo pattern (top), expanded first echo (center), and expanded second echo (bottom). We measure transit time by superimposing the first nondistorted cycle of the first and second echoes. This example represents a longitudinal wave traveling parallel to the fibers.

Results

Table 2 shows principal results for one material: low-modulus fiber, magnesium matrix. Column 1 lists various elastic constants described in the previous section. Column 2 gives a set of graphite-fiber elastic constants [4]. We chose these because E_3 agrees closely with the E_3 for the present fiber. Column 3 gives elastic constants predicted by a theoretical model using the column-2 graphite-fiber elastic constants and the measured magnesium elastic constants.

Column 4 shows measured results: the nine orthotropic-symmetry C_{ij}, the principal Young moduli E_i, and the principal Poisson ratios ν_{ij}.

From the measured results and the above model, we calculated the graphite-fiber elastic constants, shown in column 5. We used the calculational sequence: C_{44}^f, C_{66}^f, $C_{11}^f - C_{66}^f$, ν_{31}^f, and E_3^f.

Table 2. Measured and calculated elastic constants for graphite-magnesium composite and calculated elastic constants for graphite fiber. Except for dimensionless ν_{ij}, units are GPa.

	Fiber[a]	Composite, Calculated	Composite, Measured	Fiber, Calculated	Composite, Recalc.	Ratio, Recalc./Meas.
C_{11}	20.02	27.28	28.19	20.99	28.19	1.00
C_{22}	20.02	27.28	27.08	20.99	28.19	1.04
C_{33}	234.77	180.63	174.68	225.17	174.30	1.00
C_{44}	24.00	21.90	17.91	17.99	17.91	1.00
C_{55}	24.00	21.90	17.70	17.99	17.91	1.00
C_{66}	5.02	7.38	8.76	6.51	8.76	1.00
C_{12}	9.98	12.52	10.66	7.98	10.67	1.00
C_{13}	6.45	9.56	12.41	9.77	12.20	0.98
C_{23}	6.45	9.56	12.41	9.77	12.20	0.98
E_1	15.00	21.38	23.65	17.79	23.81	1.01
E_2	15.00	21.38	22.70	17.79	23.81	1.05
E_3	232.00	176.04	166.64	218.58	166.64	1.00
ν_{12}	0.494	0.449	0.374	0.367	0.359	0.96
ν_{13}	0.014	0.029	0.045	0.027	0.045	1.00
ν_{23}	0.014	0.029	0.046	0.027	0.045	0.98
ν_{21}	0.494	0.449	0.358	0.367	0.359	1.00
ν_{31}	0.215	0.240	0.314	0.337	0.314	1.00
ν_{32}	0.215	0.240	0.335	0.337	0.314	0.94

[a]Ref. 3.

Column 6 shows the composite C_{ij} recalculated using the deduced C^f_{ij}. Finally, column 7 shows the ratio of column 6 to column 4, the ratio of recalculated to measured.

To calculate the predicted composite elastic constants shown in column 3, we used a model described elsewhere [5–7].

Table 3 shows the deduced graphite-fiber elastic constants for all four cases. Table 3 also includes elastic constants for two useful reference cases: an aggregate of randomly oriented graphite crystals and a fiber where all the crystallite basal planes contain the fiber axis. For this latter case, we know only C_{33} and E_3 because the appropriate averaging problem remains unsolved.

Discussion

Results in column 4 of Table 2 show that the studied composite shows orthotropic elastic symmetry, which is approximately transversely isotropic, which requires four C_{ij} interrelationships:

$$C_{11} = C_{22};\ C_{13} = C_{23};\ C_{44} = C_{55};\ C_{66} = (C_{11} - C_{12})/2. \tag{1}$$

The microstructure in Fig. 1 also suggests transverse-isotropic symmetry.

Concerning the first-guess graphite-fiber elastic-constant calculations, we see good agreement for C_{11}, C_{22}, and C_{33}; fair agreement for C_{13} and C_{23}; and poor agreement for C_{44}, C_{55}, C_{66}, and C_{12}. Thus, the criterion of choosing a graphite-elastic-constant set based on E^f_3, the axial Young modulus, succeeds partially.

One can obtain a better, complete graphite-fiber elastic-constant set by using the model inversely. If we solve the usual model equations [4–9] for the fiber elastic constants, we obtain

$$C^f_{44} = \mu^m + \frac{2\mu^m(C_{44} - \mu^m)}{2c\mu^m - (1 - c)(C_{44} - \mu^m)}\ , \tag{2}$$

$$C^f_{66} = \mu^m + \frac{2\mu^m(C_{66} - \mu^m)(k^m + \mu^m)}{2c\mu^m(k^m + \mu^m) - (1 - c)\ (C_{66} - \mu^m)\ (k^m + 2\mu^m)}\ , \tag{3}$$

$$C^f_{11} - C^f_{66} = k^m + \frac{(k^m + \mu^m)(K - k^m)}{c(K + \mu^m) - (K - k^m)}\ , \tag{4}$$

$$\nu^f_{31} = \frac{\left(\dfrac{1 - c}{K^f} + \dfrac{c}{k^m} + \dfrac{1}{\mu^m}\right)\nu_{31} - (1 - c)\left(\dfrac{1}{K^f} + \dfrac{1}{\mu^m}\right)\nu^m}{c\left(\dfrac{1}{k^m} + \dfrac{1}{\mu^m}\right)}\ , \tag{5}$$

$$E^f_3 = \frac{1}{c}\ [E_3 - (1 - c)E^m] - \frac{4(1 - c)(\nu^f_{31} - \nu^m)^2}{\dfrac{1 - c}{K^f} + \dfrac{c}{k^m} + \dfrac{1}{\mu^m}}\ . \tag{6}$$

Table 3. Deduced graphite-fiber elastic constants. Except for dimensionless ν_{ij}, units are GPa.

	Fiber 1		Fiber 2		Perfect Basal-plane Alignment	Quasiisotropic Aggregate
	Al	Mg	Al	Mg		
C_{11}	19.09	20.99	11.24	12.58	-	160
C_{22}	19.09	20.99	11.24	12.58	-	160
C_{33}	234.99	225.17	348.89	361.45	1060	160
C_{44}	19.94	17.99	14.80	14.54	-	52
C_{55}	19.94	17.99	14.80	14.54	-	52
C_{66}	5.60	6.51	2.53	3.10	-	52
C_{12}	7.89	7.98	6.19	6.39	-	57
C_{13}	10.34	9.77	6.36	11.62	-	57
C_{23}	10.34	9.77	6.36	11.62	-	57
E_1	15.66	17.79	7.81	9.25	-	130
E_2	15.66	17.79	7.81	9.25	-	130
E_3	227.07	218.58	344.25	347.22	1020	130
ν_{12}	0.399	0.367	0.546	0.493	-	0.26
ν_{13}	0.026	0.027	0.008	0.016	-	0.26
ν_{23}	0.026	0.027	0.008	0.016	-	0.26
ν_{21}	0.399	0.367	0.546	0.493	-	0.26
ν_{31}	0.383	0.337	0.365	0.613	-	0.26
ν_{32}	0.383	0.337	0.365	0.613	-	0.26

From these five equations, the graphite-fiber elastic-constant results in column 5 of Table 2 differ significantly from the first-guess values in column 2. E_3^f is 3 percent lower than the first-guess value and 5 percent less than the fiber-manufacturer's estimate (235 GPa). Among all the calculated fiber-elastic-constant values, we have most confidence in E_3^f, which is well known to follow a linear rule-of-mixture. For the fibers (columns 2 and 5 in Table 1, the notable differences occur in the shear moduli: C_{44}^f and C_{55}^f differ by 25 percent and C_{66}^f by 30 percent. For transverse-isotropic symmetry, C_{44} is the torsional modulus T_3 around the x_3 axis—for fibers, an easily measured elastic constant. Thus, measuring T_3 and E_3 for fibers should provide useful information on fiber structure. That T_3 differs while E_3 is approximately the same suggests a structural difference not related to the orientation of hexagonal graphite unit cells in the fiber. We can compute the bulk modulus:

$$B = (2C_{11} + C_{33} + 2C_{12} + 4C_{13})/9. \tag{7}$$

For the first fiber we find 35.6 GPa, for the second 33.9 GPa. Reported graphite B values range from 34 to 210 GPa; and, from monocrystal measurements, theory predicts a possible range of 37 to 163 GPa [10]. Probably, the low B values for the present fibers reflect porosity or cracks.

After matching the graphite elastic constants to the measured composite elastic constants and recalculating the composite's properties, we should observe which composite C_{ij} constants differ most from observation. From Table 2, we see three: C_{22}, C_{13}, and C_{23}. C_{13} and C_{23} hardly surprise us because these indirect, off-diagonal elastic constants almost always present problems for both the experimentalist and the theorist. The other off-diagonal elastic constant presents little problem in the transverse-isotropic-symmetry case because of the relationship $C_{66} = (C_{11} - C_{12})/2$, where one can both measure and calculate C_{11} and C_{66} directly. The C_{22} disagreement arises because we assumed that an orthotropic-symmetry material was approximately transversely isotropic.

Conclusions

From this study, there emerged seven conclusions:

(1) Using ultrasonic methods usually applied to anisotropic monocrystals, we can determine the orthotropic (nine-component) elastic-constant tensor of a uniaxially fiber-reinforced metal-matrix composite.

(2) Although orthotropic, these particular composites show approximate transverse isotropy (five independent elastic constants).

(3) For a composite containing 70-volume-percent fibers, one can use a scattered-plane-wave ensemble-average model to describe and predict the composite's elastic constants.

(4) Graphite-fiber elastic constants chosen on the basis of the axial Young modulus, E_3, the most measurable fiber elastic constant, lead to wrong composite-elastic-constant predictions, especially for the shear moduli C_{ii} $(i = 4,5,6)$.

(5) By knowing the matrix and composite elastic constants, and by using the model inversely, we can calculate the anisotropic fiber elastic constants.

(6) Graphite fibers with the same axial Young modulus, E_3, can possess different elastic constants, especially the torsional modulus $T_3 = C_{44}$.

(7) For all considered fibers, the bulk modulus computed from the C_{ij} lies near or below graphite's lower bound.

Acknowledgment

Specimens for study originated at the Toray Industries Composite Materials Laboratory in Otsu, Japan where we thank especially T. Kyono and A. Kitamura. Dr. R.D. Kriz provided the computer program. S.A. Kim made painstaking measurements. Ming Lei, guest worker from the Institute of Metal Research, Shenyang, P.R. China, provided some calculations and a critical reading. S.K. Datta received support from the Office of Naval Research, grant N00014-86-K-0280.

References

1. A. Kitamura and S. Kataoka, in Proceedings, International Carbon Conference (Bordeaux, France, July 1984), p. 216. Toray Data Sheet TY-111A (June 1983).

2. H.M. Ledbetter and D.T. Read, J. Appl. Phys. 48, 1874–1879 (1977).

3. H.M. Ledbetter, N.V. Frederick, and M.W. Austin, J. Appl. Phys. 51, 305–309 (1980).

4. R.D. Kriz and W.W. Stinchcomb, J. Exper. Mech. 19, 41–49 (1979).

5. Z. Hashin and B.W. Rosen, J. Appl. Mech. 31, 223–232 (1964).

6. S.K. Bose and A.K. Mal, J. Mech. Phys. Solids 22, 217–229 (1974).

7. R. Hill, J. Mech. Phys. Solids 12, 199–212 (1964).

8. S.K. Datta and H.M. Ledbetter, Int. J. Solids Struct. 19, 885–894 (1983).

9. S.K. Datta, H.M. Ledbetter, and R.D. Kriz, Int. J. Solids Struc. 20, 429–432 (1984).

10. H.H. Wawra, B.K.D. Gairola, and E. Kröner, Z. Metallkd. 73, 69–71. (1982).

Monitoring of Anisotropic Material Elastic Properties Using Ultrasonic Receiving Arrays

DALE W. FITTING and A. VAN CLARK

National Institute of Standards and Technology
Fracture and Deformation Division
325 Broadway
Boulder, CO 80303 USA

Introduction

A robust technique has been developed for determining the elastic constants of an anisotropic material and for on-line monitoring of changes in the the elastic properties. Use of an array of small piezoelectric elements as the receiver permits gathering information on the angle-of-arrival of the ultrasonic beam as well as the arrival time. Combining the array with a variable-angle transmitter, phase velocity measurements in an anisotropic material may be made over a range of propagation directions. The measuring and monitoring devices may be directly coupled to the material, immersion is not required.

On-line monitoring of changes in the elastic properties of a material is carried out by directing an ultrasound beam from a stationary transmitting transducer into the sample in a specified direction. Alterations in the - stiffness constants of the material (as a result of such things as moisture absorption) will cause the direction of the energy flux to deviate. These deviations are readily monitored with the array of receivers.

The arrays fabricated for monitoring elastic properties and the electronics required to process their signals are described herein. Use of the arrays for measuring the quasilongitudinal and quasitransverse wave fields in an anisotropic material (graphiteepoxy composite) is described and the importance of this measurement in correctly determining group and phase velocities is discussed.

Determining Elastic Constants of Anisotropic Materials

A technique for using ultrasonic receiving arrays to measure the elastic constants of an anisotropic material has been described elsewhere[1]. This method is based on measuring the phase velocity versus propagation direction in the anisiotropic medium. For an anisotropic plate, the measurement technique involves propagating ultrasonic waves off-axis in two orthogonal planes. Then, one solves for the moduli from a number of simultaneous equations,

----

These measurements yield 7 of the 9 elastic moduli for an orthotropic material. The remaining constants, may be measured using horizontally-polarized transverse (SH) waves.

The receiving array is used as shown in Figure 1 to gather data for calculating phase velocity. The variable-angle wedge transducer generates ultrasonic waves in the composite which may be adjusted through a range of refracted directions. The critical angle for the energy flux in the composite determines the upper limit on angle for which measurements are possible. Phase velocity measurements can be made by determining the transit time and direction of the energy flux and measuring the angle-of-arrival of the wavefront at the array.

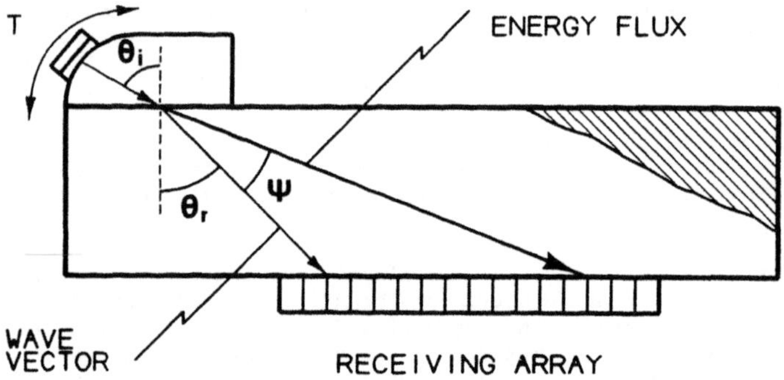

Figure 1. Use of an Array for Determination of Phase Velocity versus Angle

Array Construction

A number of linear and two-dimensional arrays have been fabricated for measuring the elastic properties of an anisotropic solid (as described above). Linear arrays are appropriate when the ultrasonic waves are constrained to the planes of symmetry in the anisotropic material. A 2-D array is useful when this condition is not met; however, the array is comprised of many more elements, with the attendent problems of additional electronics and increased processing time.

The linear arrays which were assembled contained 16 or 32 elements 0.5 mm in diameter placed on 2.5 mm centers (Figure 2). The piezoelectric material used was polyvinylidene fluoride (PVDF), which yielded transducers with a broad spectral sensitivity.

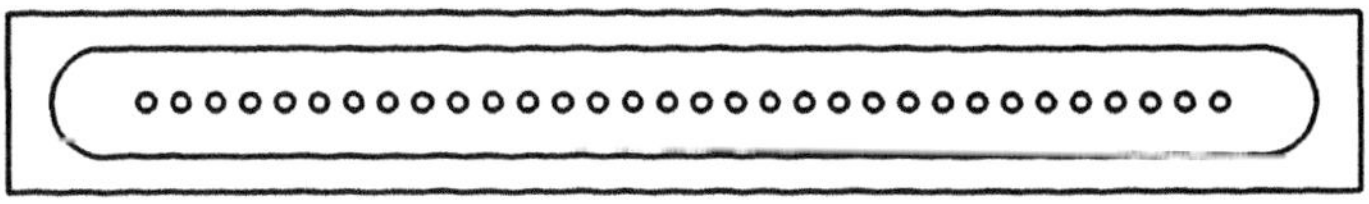

Figure 2. Linear Array Fabricated for Elastic Constant Measurement of Anisotropic Materials. Aperture: 79.2mm x 0.5mm, Element size: 0.5mm diameter, Element spacing: 2.5mm

An electronic system was designed and fabricated for preamplification and digital selection of the signal from any array element. The schematic for this system is shown in Figure 3. The device, capable of handling up to 64 array elements, consists of a two-stage multiplexer followed by an amplifier stage. A 6-bit digital word is used to select one of the 64 array elements.

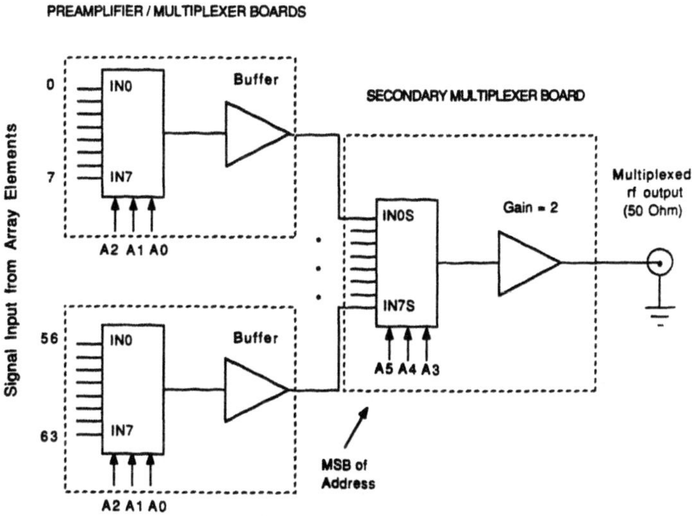

Figure 3. Electronics for Preamplification and Multiplexing of Signals from an Ultrasonic Receiving Array. The two-stage multiplexer is capable of selecting any of the 64 input channels under digital control.

Measurements were made to confirm the wide bandpass and good isolation of receiving channels in the electronics system (Table 1). Channel-to-channel crosstalk refers to isolation in the first stage of the two-stage multiplexer and board-to-board crosstalk refers to isolation in the second stage of the multiplexer.

Table 1

Specifications of the Array Electronics

Bandwidth:	200kHz to 35 MHz (-3dB)
Channel-to-Channel Crosstalk:	-58.5 to -51 dB at 5MHz (-38 dB nearest channel)
Board-to-Board Crosstalk:	-59.2 to -56.7 dB at 5MHz (-46.8 dB nearest board)

Experiments

The anisotropic material studied was a unidirectional, graphite-epoxy composite plate. The plate and specimens sectioned from this plate were used in the experiments. A wedge transducer and receiving array were coupled to the composite plate, as shown in Figure 1. The transmitting transducer, a 12.7mm diameter piezoceramic device, was shock excited producing a broadband spectrum centered about 5MHz. Waveforms at each of the elements in the receiving array were recorded (Figure 4).

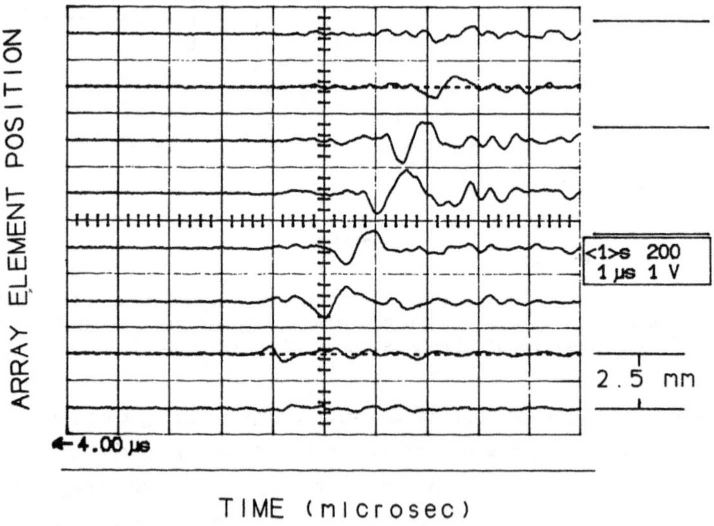

Figure 4. Waveforms at Several of the Elements in the Receiving Array Acquired During a Transmission Experiment. Experimental arrangement was as shown in Fig. 1. Horizontal Scale: 1 microsec/div, Vertical scale: 1 V/div

Although the quasilongitudinal wave was easily detected (first arrival), the ultrasonic field arriving at the linear receiving array was spatially more complex than expected and was found to vary with the angulation of the ultrasonic pulse in the anisotropic material. This complexity in the shape of the field has an effect on the accuracy of the group velocity measurement because the measured angle θ_t is affected. The angulation of the group velocity is calculated from the measured position of the ultrasound beam and the specimen thickness. It is the location of the beam 'center' which is in question. In order to further investigate these effects a set of samples was cut from the plate as shown in Figure 5 to provide a variety of fiber orientations with respect to one side of the specimen.

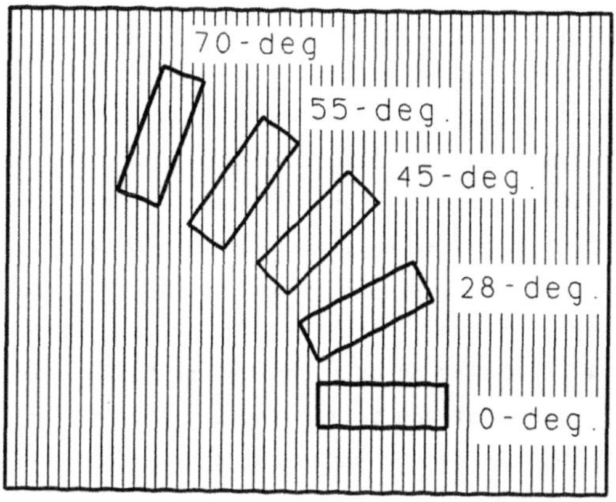

Figure 5. Specimens Were Machined from a Unidirectional Graphite-Epoxy Composite Plate to Provide Samples with a Variety of Fiber Orientations with Respect to the Side of the Sample

Ultrasonic transmission measurements were made by placing a piezoceramic transducer on one side of the sample and coupling a linear array to the opposite face (Figure 6). The wave vector for these specimens is in the direction of the transmitted beam (perpendicular to the face of the transmitting transducer). In order to obtain a spatial resolution of the ultrasonic field improved over the 2.5 mm element spacing of the array, data was recorded for each element in the array and then the array was shifted by 0.5 mm and another set of data recorded. This procedure was repeated with additional displacements until the entire data set contained information on the ultrasonic field every 0.5 mm.

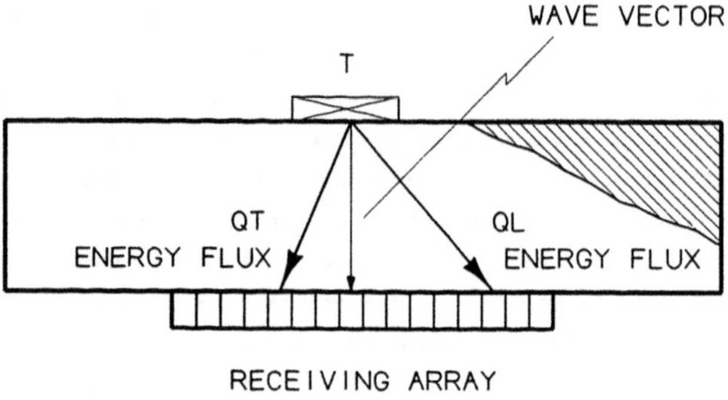

Figure 6. Measurement of Energy Flux Direction and Arrival Time in Sectioned Samples of a Unidirectional Composite

First, the ultrasonic field produced by the transmitter in an isotropic material (acrylic) was measured with the array (Figure 7). The plot is peak-to-peak amplitude of the longitudinal wave field at the array as a function of flux deviation angle (angle calculated from the position of the transmitter and each receiving element). The ultrasonic field propagating in this isotropic material is approximately Gaussian, with very little evidence of side lobes. Next, the composite specimens which had been machined from the plate (Fig. 5) were examined. Peak-to-peak avoltage at the receiving elements were recorded. The array-measured QL and QT wavefields is plotted (Figure 8) as a function of the flux deviation angle, for four of the composite samples. QT wavefields are shown as dashed lines while the QL fields are plotted as solid lines.

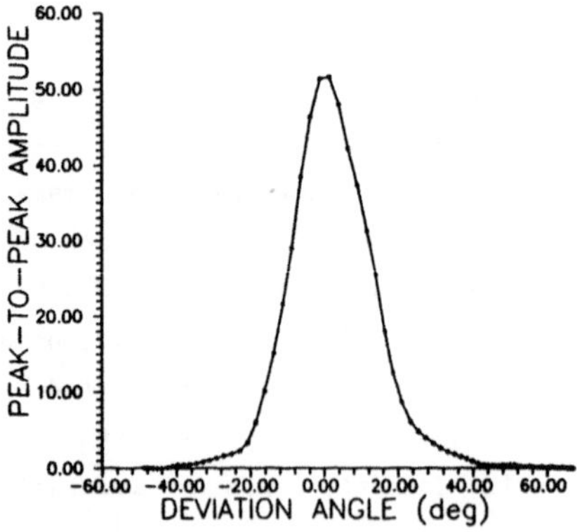

Figure 7. Ultrasonic Field Measured with a Linear Array on an Isotropic Sample (acrylic). The measurement geometry is shown in Figure 6.

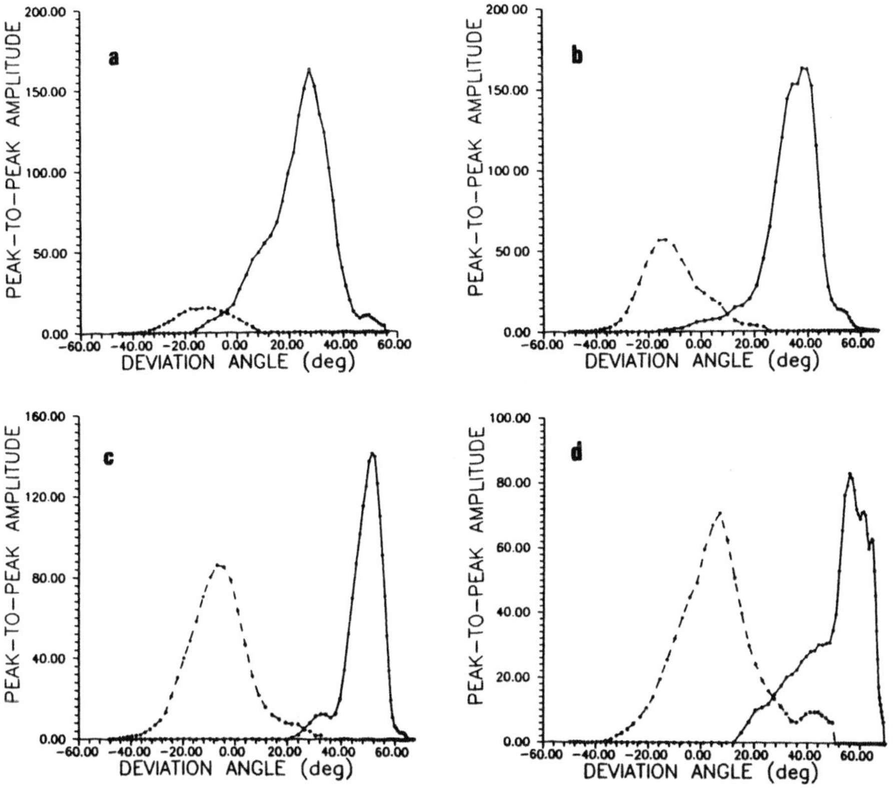

Figure 8. Ultrasonic Fields of the Quasitransverse (QT, dashed line) and the Quasilongitudinal Waves (QL, solid lines) Measured by a Linear Receiving Array. The experimental geometry was that shown in Figure 6. The samples used are those shown in Figure 5. (a) 28-degree fiber orientation, (b) 45-deg., (c) 55-deg. and (d) 70deg.

Discussion

Note that the shape of the QL and QT fields (Figure 8) are approximately Gaussian - generally the same shape as the field of the transmitter (Fig. 7). The relative amplitudes of the QL and QT fields are also seen to vary, as predicted by a bulk wave model [2]. Of interest, however, is the nonsymmetric shape of the wave fields. The question arises as to where to measure the flux deviation angle (angle at which the energy flows in the anisotropic material, θ_t). Should the angle at which the peak occurs be used, the centroid of the distribution or yet another wave field feature. The propagation of a Gaussian pulse in an anisotropic material must be examined. Norris [3] and Roberts [4] have solved aspects of this problem. Because the angular spectrum of the transmitted wave field for our experiments differed from their assumptions, a direct comparison was not easily possible.

Comparison [1] of the deviation angles calculated by the bulk wave model along with two features from the array-measured wave fields showed a slightly better correlation of the angle calculated from the measured peak in the wavefield than with the angle calculated from the spatial centroid of the field distribution. Use of either wavefield feature provides a reasonable estimate of the flux deviation angle (they vary by only a few degrees). Determination of the correct feature of the wavefield from an analysis such as presented in [4] or [5] is expected to improve the accuracy of the angle measurement and thus improve the accuracy of the velocity calculation. These analyses are now underway.

Acknowledgment

R.D. Kriz furnished the composite specimens, calculated the bulk wave flux deviation angles and provided a wealth of information on stress wave propagation in anisotropic materials. The work reported herein was partially supported by the National Research Council under their associateship program.

References

1. D.W. Fitting, R.D. Kriz and A.V. Clark, "Measuring In-Plane Elastic Moduli of Composites with Arrays of Phase-Insensitive Ultrasound Receivers," Proceedings of the Review of Quantitative Nondestructive Evaluation, 1988, Plenum Publishing Co., New York (1989).

2. R.D. Kriz and W.W. Stinchcomb, "Elastic Moduli of Transversely Isotropic Graphite Fibers and Their Composites," Exp. Mech., Vol. 19, (2), 41-49 (1979).

3. A.N. Norris, "A Theory of Pulse Propagation in Anisotropic Elastic Solids," Wave Motion, Vol. 9, 509-532 (1987).

4. R.A. Roberts, "Ultrasonic Beam Transmission at the Interface Between an Isotropic and a Transversly Isotropic Solid Half Space," Ultrasonics, Vol. 26, 139-147 (1988).

Anisotropy Measurements and Indication of Ply Orientation in Composite Materials Using Holographic Mapping of Large Amplitude Acoustic Waves

Michael J. Ehrlich
James W. Wagner

The Johns Hopkins University
Center for Nondestructive Evaluation
Maryland Hall 102
Baltimore, Maryland 21218
U.S.A.

Summary

Examination of graphite reinforced polymer matrix composite materials was performed using high speed, pulsed holographic interferometry. Holographic recordings of one face of a composite sheet were made before and immediately following mechanical excitation of the opposite side. Excitation was provided through a length of 6mm diameter aluminum rod affixed at one end to the back surface of the specimen. A pulsed laser beam was focused at the other end of the rod, launching an acoustic wave toward the specimen.

Surface displacements from the acoustic plate waves traveling along the sheet result in interference fringes superimposed on the holographic reconstruction of the specimen image. Thus variations in wavefront velocity are directly observable in the holographic fringe patterns. From the results of these experiments the effects of mechanical anisotropy of the composite material are clearly visible providing a positive indication of ply orientation.

Introduction

The testing of laminar composite materials to determine effective modulus, flexural stiffness, ply orientation, etc. is traditionally done using contact ultrasonic tests or destructive mechanical tests. However, for many composite materials, these properties are functions of both position and direction, owing to the anisotropy of the constituent layers, and inhomogeneities arising from process variation and defects.

In order to determine these properties as a function of position or direction using conventional ultrasonic testing, it is necessary to scan the testing apparatus over the surface of the material. However, using pulsed high speed holographic techniques coupled with ultrasonic excitation, full field determination of these properties is possible.

High Speed Pulsed Holography

Holography possesses the unique ability to record and subsequently reconstruct an optical wavefront. In many applications, the optical wavefront which is recorded is that from the surface of an object. When such a hologram is reilluminated, the optical wavefront of the object is reconstructed and an image of the object appears at the point in space the object originally occupied. Full field holographic methods for ultrasonic wave detection arise from this fact. In the case of high speed double exposure pulsed holography, each hologram records two wavefronts; one of the undisturbed object, the second of the same object at a later instant in time. Each exposure is of an extremely short duration, in essence a "freeze-frame" of the object at a single instant. When the hologram is reilluminated, both wavefronts are reconstructed. Any disturbance to the object surface resulting in out-of-plane displacements between the two holographic exposures results in an interference pattern superimposed on the image of the object surface. This interference pattern represents equi-displacement contours of the surface at the instant the second holographic exposure was recorded.

The displacement sensitivity for holographic interferometry depends in large part on the method used to read out the data. For example, it has been shown that displacement sensitivity for heterodyne holographic techniques approaches 1/1000 of a fringe [1]. Assuming an optical wavelength of 500nm, this sensitivity is on the order of 2.5 angstroms, which may be required in some applications. However, heterodyne techniques are inherently slow, owing to the point-by-point method of data acquisition. If, in fact, the displacements are large, adequate data analysis can be performed using simple interferometric techniques. For this reason, the disturbance of the composite plate between holographic exposures is caused by a propagating large amplitude acoustic wave. A sample interferogram is shown in Figure 4.

Experimental Procedure

Five composite sheets were manufactured from unidirectional tape prepreg consisting of Toray T-300 1, 3, 6, and 12 thousand filament graphite fibers impregnated with Fiberite 350F curing 976 epoxy resin. The specimens were manufactured with a ply orientation given by 0^{n}-90^{2n}-0^{n}, where n is an index ranging from 1 to 5, and the total superscript gives the number of plies in the 0 and 90 degree directions. One face of each specimen was painted white to improve the reflectivity of the surface.

A length of 6mm diameter aluminum rod was attached to the center of the unpainted face of each specimen, and a small amount of chemical explosive (approximately 0.2 mg silver acetelyde) was affixed to the free end of the rod. A pulsed Nd:YAG laser with output in both the visible and IR spectrum was made to double pulse with a temporal separation between pulses of 45μs and a pulse width of 9ns. The first laser pulse served to record an initial holographic exposure of the undisturbed composite surface using the visible frequency of the laser, as well as to detonate the chemical explosive using the IR beam of the laser, launching the acoustic wave down the aluminum rod. The second laser pulse recorded a second holographic exposure of the composite surface, however in the second exposure the object surface was disturbed due to the acoustic waves propagating through the composite. A schematic of this system is shown in Figure 1.

After the double exposure hologram is developed and reilluminated, the reconstructed wavefronts of the undisturbed and disturbed composite surfaces interfered and resulted in an interference fringe pattern superimposed on the holographic reconstruction of the specimen image. As stated earlier, the observed fringe pattern is essentially a topographic mapping of the out-of-plane displacements of the composite surface due to the propagating acoustic wave.

In addition, 3-point bend bar samples were cut along the two major axes of each specimen and mechanically tested (ASTM E-855) to compare with the acoustically generated holographic data.

Results and Discussion

Given a solid plate of finite thickness, there are two possible plate modes of acoustic wave propagation in directions parallel to the plate surface for acoustic wavelengths which are large compared to the plate thickness [2]. These modes are termed the symmetric and antisymmetric Lamb modes, referring to the particle displacement symmetry about the median plane of the plate, arising from the guided compressional and shear waves propagating across the plate. It has been shown for homogeneous isotropic materials that the phase and group velocities (v_p, v_g) of the lowest order antisymmetric mode are [2,3]:

$$v_p = \left\{\frac{S_F}{m}\right\}^{1/4} \omega^{1/2} \quad , \quad v_g = 2v_p \tag{1}$$

where S_F is the flexural stiffness of the plate ($= EI/(1-\nu^2)$[2]), m is the mass per unit area of the plate, and ω is the acoustic frequency.

Since $\omega = 2\pi v/\lambda$, Eq. (1) can be solved for the flexural stiffness:

$$S_F = \frac{m v_p^2 \lambda^2}{4\pi^2} \tag{2}$$

Here, the flexural stiffness is expressed as a function of mass per unit area of the plate, acoustic wavelength, and acoustic phase velocity. Both v_p and λ can be determined from the double exposure hologram for directions parallel to the plate surface. Since m is also easily obtainable, S_F can be calculated for various directions using the holographic data.

It is important to note the method with which v_p and λ are determined. Given a step input to the center of a plate, the response of the plate is initially a large non-propagating central displacement, which gives rise to a propagating wave comprised of the frequency components contained in the initial response. For a dispersive material, the velocity and attenuation of high frequency components is much greater than that of lower frequency components. As a result, the waveform which is recorded in the double exposure hologram is essentially a summation of the large amplitude, low frequency waves which were present in the initial pulse. As such, it is difficult to determine one absolute wavelength and its corresponding phase velocity. Instead, an effective wavelength is calculated for a small group of waves, and for it an effective group velocity is also determined. Since the relationship between phase and group velocity is $v_g = 2v_p$, the effective phase velocity corresponding to the effective wavelength can also be determined.

In the case of a temporal signal at a single point in space, it has been observed [4] that a technique to determine effective wavelength, frequency, and velocity which yields good correlation between theoretical and experimental results is one in which the effective frequency is calculated based on the inverse time difference between two successive maxima of the temporal signal, and associated with that frequency an effective velocity is determined using the source to detector distance and mean arrival time of the temporally separated peaks. Similarly, we propose that for a spatial signal at a single point in time, an effective wavelength may be determined using the spatial separation of two successive maxima, and that an effective velocity may be determined using the distance from the source of the signal to the spatial mean of the two successive maxima and the knowledge of the point in time at which the spatial signal is recorded and the time of the initial excitation.

Using this technique to determine an effective wavelength and an effective phase velocity, the flexural stiffness in both fiber directions were calculated for each specimen using Eq. (2). In addition, the flexural stiffness was also determined by doing 3-point bend bar tests on samples from each specimen, using [5]:

$$S_F = \frac{5PL^3}{384\delta} \tag{3}$$

where P is the applied point load, L is the distance between supporting points for the bar, and δ is the maximum deflection of the bar center. The results for both "fast" and "slow" directions of the composites are shown in table 1.

These results are also shown in Figures 2 and 3, where the best fit line is plotted through the actual data points. As can be seen, the results for the horizontal direction show a small disparity of approximately 2.6:1 between the mechanically determined flexural stiffness and that obtained acoustically, whereas in the vertical direction the results from both mechanical and holographic tests agree quite well (1.03:1).

As is evident from the actual interferograms (Figures 4,5), the difference in acoustic velocity gives a strong indication of the effective modulus in various directions, as well as an indication of ply orientation. Figure 4 shows the interferogram recorded for a composite with $n = 1$. We see that the maximum velocity occurs in the direction of outer ply alignment, with a smaller velocity in alignment with the inner plies. This serves well to indicate the relationship between the flexural stiffness in these directions. Figure 5 shows the interferogram recorded for another 4-ply composite, however the ply orientation for this sheet is 0-90-0-90. It is clear from the interferogram that the effective moduli of this composite are equal along both ply directions. In comparing the two interferograms, one can see how ply orientation and mechanical properties are coupled, as well as how full-field holographic techniques may be used to estimate these parameters.

It should be recalled that the determination of flexural stiffness based on m, v_p, and λ, was calculated using equations developed for homogeneous, isotropic media. It is quite interesting that the modulus trends for such highly anisotropic, inhomogeneous composite plates agree well with the isotropic media equations. In fact, the results are nearly identical for the case where the modulus is determined primarily by the inner ply core.

Finally, it is interesting to note (space constraints limit discussion) that the results obtained holographically for flexural stiffness (and from it effective modulus), agree with static models for laminate materials to a greater degree than do the results of mechanical tests.

References

1. Dändliker, R.: Heterodyne Holographic Interferometry, in *Progress in Optics XVII*, E. Wolf ed., North-Holland 1980
2. Meyer, E.; Neuman, E.: *Physical and Applied Acoustics*. New York, London: Academic Press 1972
3. Viktorov, I.: *Rayleigh and Lamb Waves*. New York. Plenum Press 1967
4. Dewhurst, R.J.; Edwards, C.; McKie, A.D.W.; Palmer, S.B.: Estimation of the thickness of thin metal sheet using laser generated ultrasound. Appl. Phys. Lett. **51** (14) October 1987
5. Beer, F.; Johnston, E. Jr.: *Mechanics of Materials*. New York. McGraw-Hill 1981

Acknowledgments

The authors would like to acknowledge the funding granted them by the Department of the Navy, Office of Naval Research, and the help given them by Ms. Rochelle Payne.

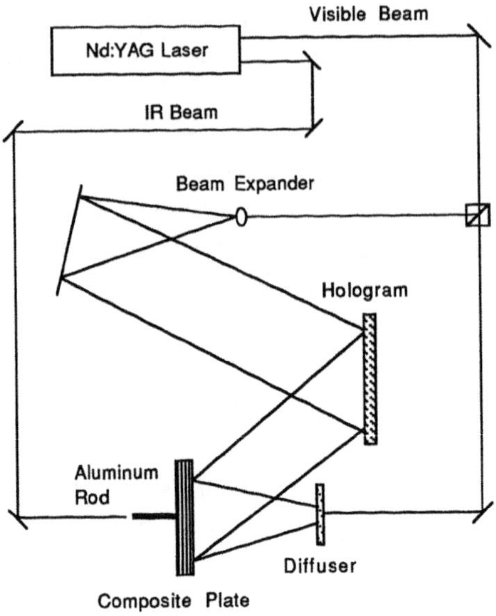

Figure 1. Schematic of Experimental Setup

	Parallel to Outer Plies		Parallel to Inner Plies	
n	S_F (acoustic)	S_F (mech.)	S_F (acoustic)	S_F (mech.)
	($Nm^2 \times 10^{-2}$)	($Nm^2 \times 10^{-2}$)	($Nm^2 \times 10^{-2}$)	($Nm^2 \times 10^{-2}$)
1	1.07	0.96	0.47	0.21
2	4.45	6.41	1.58	1.24
3	9.60	19.5	3.63	2.46
4	13.1	31.8	5.37	5.25
5	17.4	42.3	7.77	7.97

Table 1. Results for acoustic/holographic tests and mechanical tests for directions both parallel and perpendicular to outer plies.

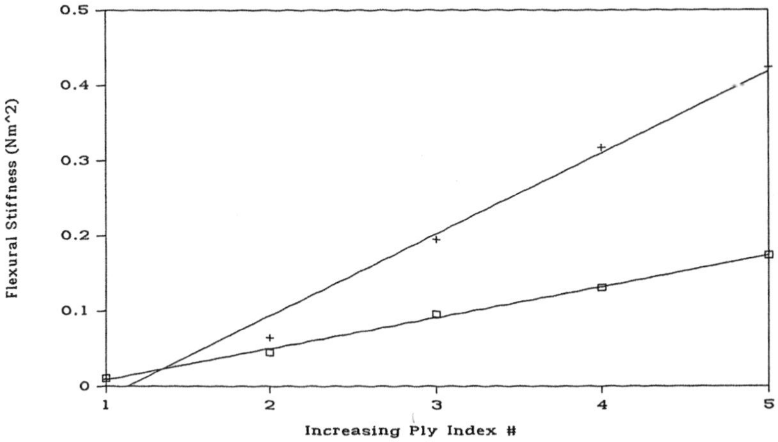

Figure 2. Flexural stiffness test results for the direction parallel to outer plies (horizontal). Mechanical test results (+) Holographic test results (□)

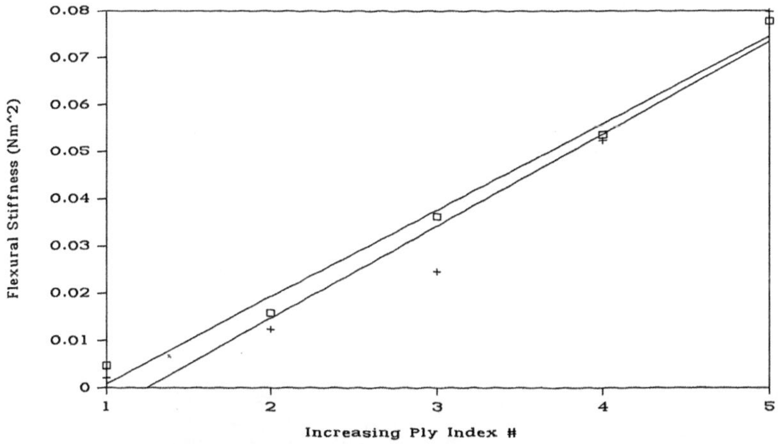

Figure 3. Flexural stiffness results for the direction parallel to inner plies (vertical). Mechanical test results (+) Holographic test results (□)

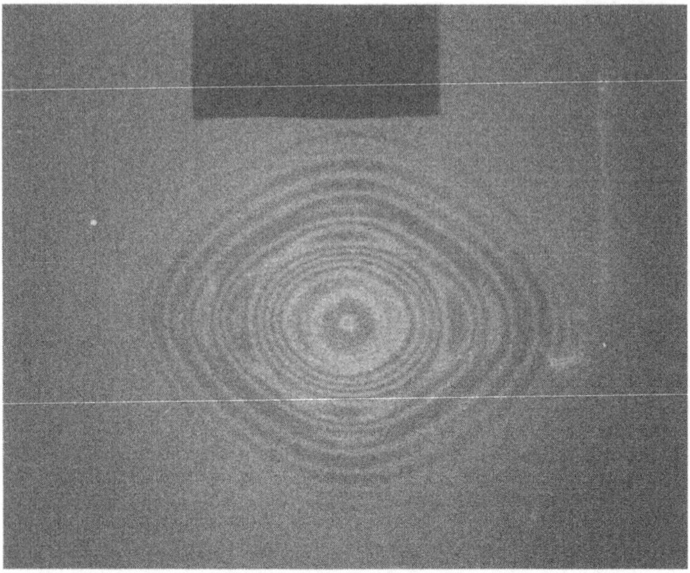

Figure 4. Holographic interferogram of 0^1-90^2-0^1 composite. (Horizontal direction parallel to outer plies)

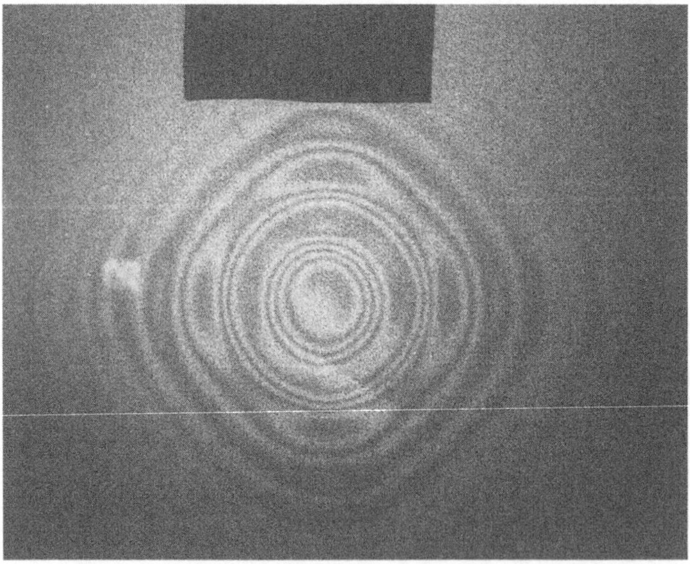

Figure 5. Holographic interferogram of 0-90-0-90 composite.

Recognition of Fracture Modes and Behaviour of Composites by Acoustic Emission

Kusuo Yamaguchi, Hirotada Oyaizu
Institute of Industrial Science, University of Tokyo
22-1, Roppongi-7, Minato-ku, Tokyo 106, JAPAN

Summary

Composite including FRP have many advantages for structural material but show more complicated fracture modes and growth behavior than homogeneous materials. Recognition of the individual mode and location by on-line measurement is very important for material evaluation and design as well as proof test. AE technology is being used for the purposes as a powerful tool. However, usual amplitude distribution analysis has a limitation because of the shortage of information. We have been developing such instrumentation that utilizes more AE waveform information including new parameters. The instrumentation systems enable more detailed recognition of fracture mechanism in composites by multi-dimensional analysis. We are applying the systems for tensile tests on GFRP specimens and obtaining results of good recognition between fracture modes.

1. Introduction

Recognition of fracture behavior in material by on-line measurement is very important for the material evaluation and design as well as structural integrity monitoring. AE technology has been being considered and used for the purposes as one of the most powerful tool. Composites including FRP have many advantages but show more complicated fracture modes and growth behavior than homogeneous materials like metal. Crack in bundle, matrix cracking, separation or delamination and fiber breakage are the typical examples of fracture modes which occur according to the growth of failure in composites depending on the material character and the stress condition. Conventional AE technique which usually used amplitude distribution analysis for such purposes has a limitation because of the shortage of information /3/4/. We have been developing advanced AE systems utilizing much more AE waveform parameters including energy moment, zero crossing count of waves and other new parameters as well as usual parameters /1/6/7/. The prototype of the systems was applied to experiments on metal components very successfully /2/6/7/. After the application to metal structures, we have been applying the system to tensile tests of GFRP specimens /5/7/ and are obtaining results of good recognition between such fracture modes as fiber breakage, matrix cracking, bundle cracking and splitting.

2. Basic concepts of AE measurement

There are variations of AE technology, however, usually AE analysis system depends on the input processing method and the input data structure. For high performance AE instrumentation, following two essential design concepts should be considered on the input processing method /6/7/.

(1) Data of all available events should be collected as far as possible to look for all possible AE sources at analysis stage.

(2) Every event data should have detailed information on the input AE as far as possible for detailed analysis on all possible sources.

However, these concepts contradict each other in cost and speed of the practical system. Therefore, compromization by good data compression as well as real-time input processing is very important.

Fig.1 shows fracture modes in FRP materials. The modes and the behaviors are more complicated than the usual case in metal. Therefore, AE technique for composites has to have more detailed analysis performances. It should utilize much more information contained in detected AE waves.

Table 1 is sorts of information which can be used in AE technology. As to AE event data, the most effective parameters from the three major categories should be selected because usually good recognition needs multi-dimensional analysis.

3. Method of multi-parameter processing

We use waveform parameters extracted from AE wave contour for good data compression. Table 2 is the current contents of the parameters called waveform microdata, and they include changeable parameters for trial and error in analysis. The system has very high speed of input processing.

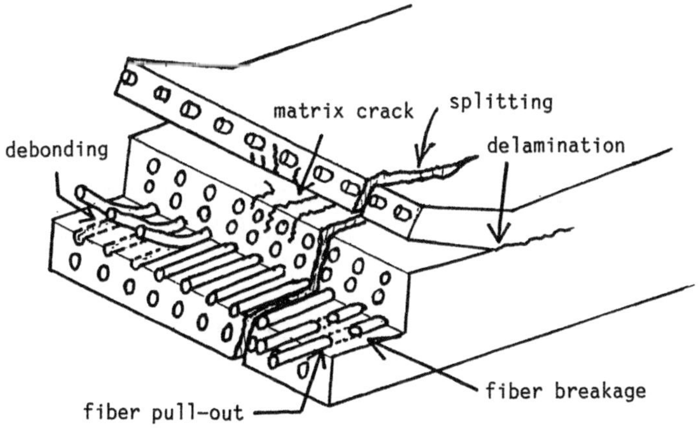

Fig.1 Illustration model of fracture modes in FRP materials.

Table 1 Sorts of information or parameter-data utilized in AE technology.

Macro data	AE count (ring-down count) stochastic distributions
Location	arrival time difference hit sequence zone
Event parameters	energy (ap, Et, etc.) time (Tem, Tr, Td, etc.) frequency (Nz, FFT, etc.)
Time trend	AE source location (intensity) event parameter values (activity)
Loading condition	load level load sequence load phase (in cyclic loading) Felicity ratio
Environmental condition	atmosphere noise process sequence plant operation

Table 2 Waveform microdata

Channel number (ch)	Wave energy (Et)
Maximum amplitude (Ap)	Zero crossing count (Nz)
Energy moment 1 (Tem1)	Wave rise time (Tr)
Energy moment 2 (Tem2)	Wave duration time (Td)
2nd order energy moment (Ed)	Arrival time (ts)

4. Example of experiments on FRP and the results

Fig.2 is the experimental setup for the tensile test of FRP material. Cyclic loading of increase-hold-decrease pattern as in Fig.3 was applied for the experiments. Specimen (A) and (B) are two examples in many tested specimens of materials and notch-shapes combination.

Fig.4 shows the distributions and the trends of energy moment (Tem) and peak amplitude (Ap) from the specimen (A).
Fig.5 and Fig.6 show the distribution-trends of energy moment (Tem) and zero crossing count (Nz) from the specimen (B).

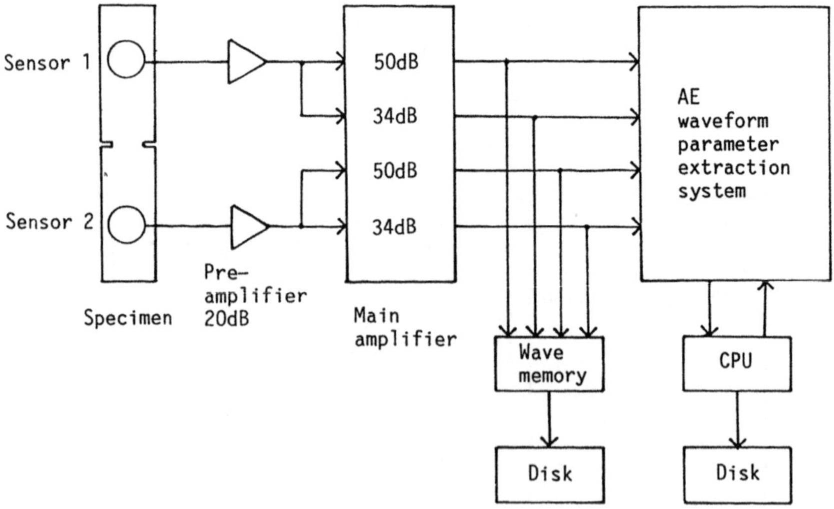

Fig.2 Experimental setup.

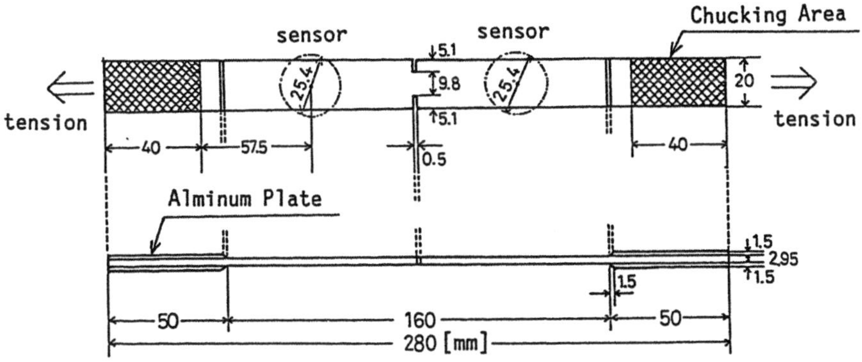

Specimen (A): Untreated woven cloth [0°/90°] / Unsaturated polyester
Specimen (B): Vinyl silane treated woven cloth [0°/90°]
/ Unsaturated polyester

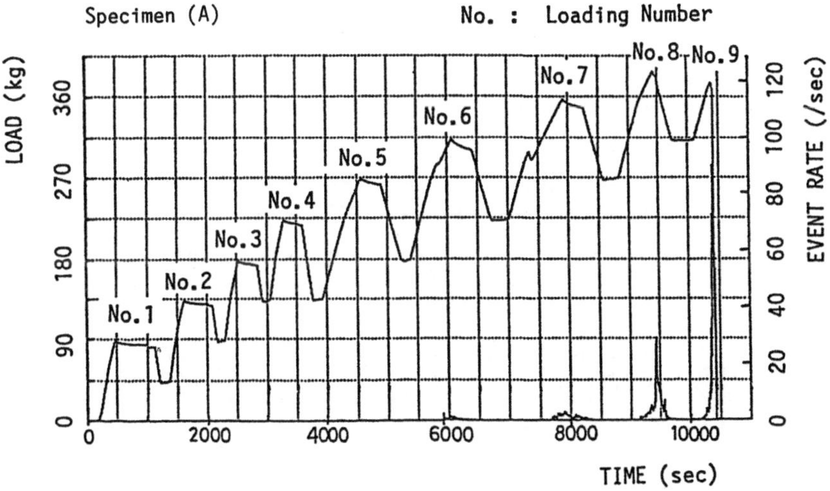

Fig.3 Example of loading pattern of the GFRP tensile test and types of the specimen.

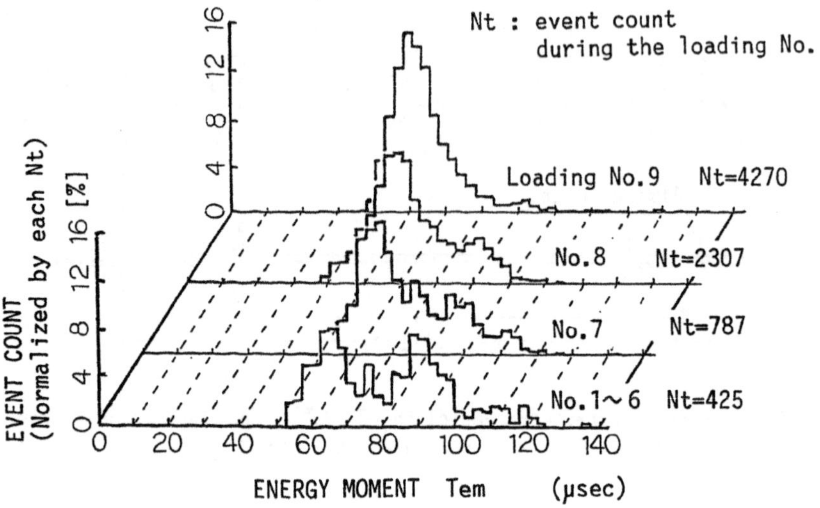

(a) energy moment Tem

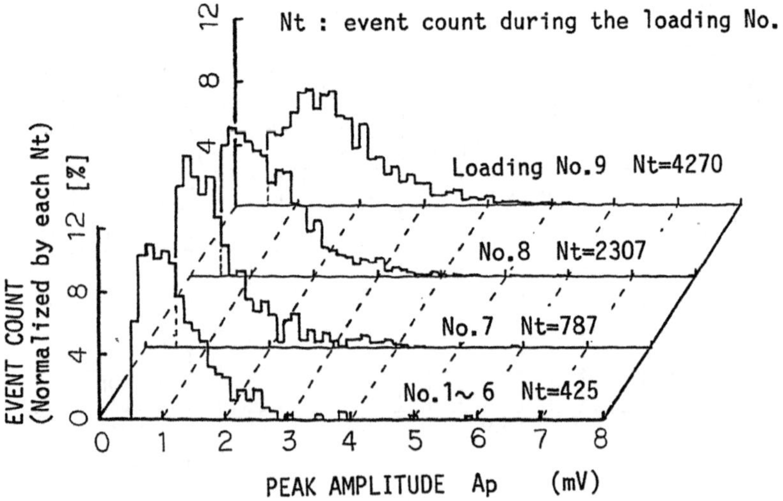

(b) peak amplitude Ap

Fig.4 Distribution of energy moment (Tem) and peak amplitude (Ap) from the specimen (A) (untreated cloth/UP) during the loading and unloading tensile test.

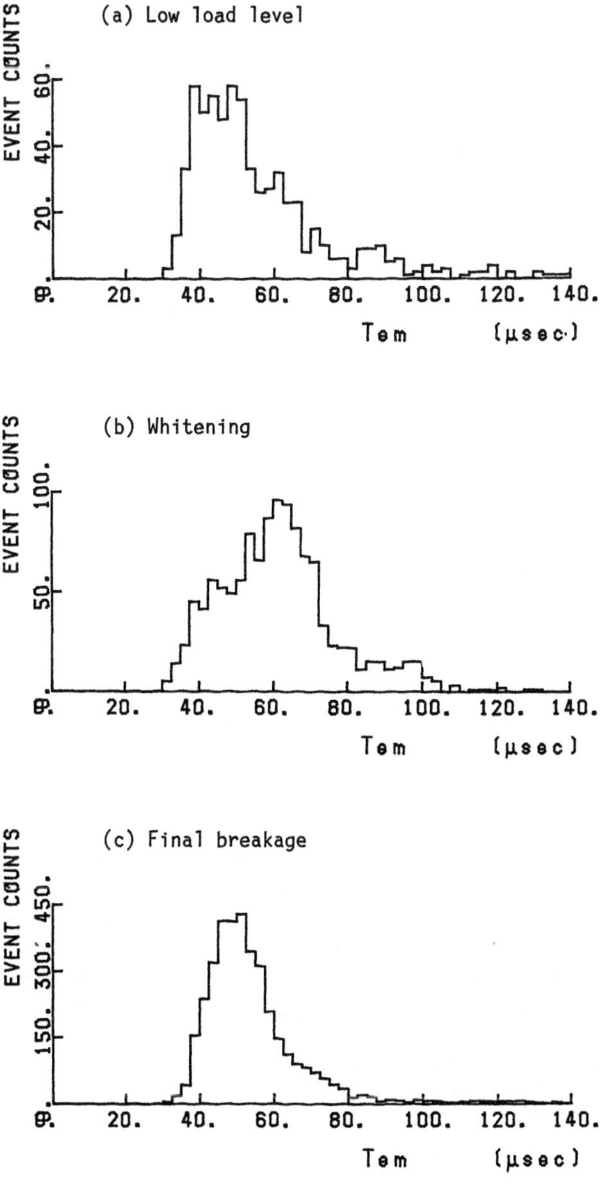

Fig.5 Distribution shift of energy moment (Tem) from the specimen (B) (silane treated cloth/UP) corresponding to loading cycle and fracture propagation.

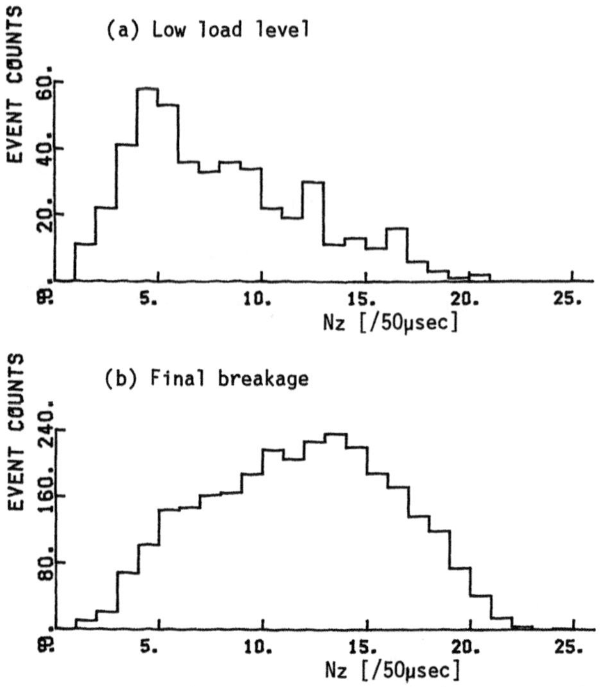

Fig.6 Distribution shift of zero crossing count (Nz) from the specimen (B) (silane treated cloth/UP). (Tem: 35 - 57.5 [μsec])

Table 3 Recognition example of GFRP fracture modes by grouping of the waveform parameters in experimental results.

Fracture mode	Tem [μsec]	Nz [/50μsec]	Location	Main phase
Debundling	40–57.5	<=9	wide	low load level
Debonding & matrix cracking	57.5–100		around notch	whitening
Fiber breakage	35–57.5	>=10	around notch	final breakage

Debundling: Cracking and dispersion in (transverse) fiber bundle

5. Discussions and conclusions

In the experiment of Fig.3 and Fig.4, as usually obtained in such tensile tests, most fiber breakages occurred at the final stages especially at No.9 loading, while matrix separation seemed to occur mainly preceding loading cycles. Amplitude distributions of AE show slight difference between the events of No.9 and other loading cycles. Higher peak amplitude events increased slightly at No.8 and a little more at No.9, however, the differences were not so clear.

On the other hand, distributions of energy moment shifted more clearly at No.8 and No.9 loading from No.1-6 and No.7 loading. In the case, about 55 - 57.5 microseconds threshold could successfully discriminate the events estimated fiber breakage from other events. The energy moment distributions from fiber breakage showed similar value at tensile tests of unidirectional FRP specimens, and the similar value of Tem threshold could apply for the discrimination/5/.

From the results of treated (silane processed) cloth specimen shown in Fig.5, Tem distribution shows very similar value at final breakage in Fig.4, but it shows two different distributions at low load level and whitening. In this treated cloth specimen, there was an AE group which shows small Tem similar to that of fiber breakage. We estimated that this AE group was generated from debundling, that is, initiation and propagation of small cracking of matrix material inside transverse fiber bundles. The reasons were as follows: wide locations of AE in the group, hair cracking by visual observation from the same period, and no detection of such group from untreated cloth specimen in which fiber bundles are not bonded by plastics. We found that zero crossing count (Nz) of this group was different from that of fiber breakage (Fig.6), and it could be used for the recognition of the modes.

Thus, we obtained a recognition table like Table 3 for fracture recognition and material characterization by using multi-parameter of AE waveform at this stage. We believe that the utilization of multi-parameter under the concepts in chapter 2 is useful and it would develop AE technology for material characterization as well as other applications. Contour shape analysis by waveform parameters with major frequency information would have advantages for such purposes.

References

1. K.Yamaguchi, H.Hamada, H.Oyaizu, H.Ichikawa, T.Kishi, H.Ishitani : "Multi-Purpose Fracture Monitoring System by Utilizing Acoustic Emission Wave Parameters", 2nd Int. Symp. on Acoustic Emission from Reinforced Composites, Montreal; SPI (1986) 213.

2. K.Yamaguchi, H.Oyaizu, Y.Matsuo, A.Yamashita, Y.Sakakibara : "Features of Acoustic Emession from Fatigue Crack in FBR Piping Component and its Generation Mechanism", 7th Int. Conf. on NDE in Nuclear Industry, Grenoble; COFREND ASM (1985) 367.

3. M.Shiwa, S.Yuyama, T.Kishi : "Acoustic Emission Signal Analysis during Fatigue Damage of GFRP", Progress in AE III (Proc. of IAES-8, Tokyo); JSNDI (1986) 554.

4. J.Awerbuch, S.Ghaffari : "Tracking Progression of Matrix Splitting during Static Loading through Acoustic Emission in Notched Uniderctional Graphite/Epoxy Composites", Progress in AE III (Proc. of IAES-8, Tokyo); JSNDI (1986) 575.

5. K.Yamaguchi, H.Oyaizu, Y.Nagata : "Acoustic Emission Waveform Characteristics from FRP during Tensile Test", Progress in AE III (Proc. of IAES-8, Tokyo); JSNDI (1986) 594.

6. K.Yamaguchi, H.Oyaizu : "Multi-Purpose Fracture Monitoring System by Flexible Processing of Acoustic Emission Micro-Data", Symposium Acoustic Emission, Bad Nauheim (West Germany); Deutsche Gesellschaft fuer Metallkunde (1987)

7. K.Yamaguchi, H.Oyaizu : "Distributed Fracture Monitoring System by High Speed processing of Acoustic Emission Micro-Data", the 16th EWGAE Conference, London (1987)

Determination of Strength Properties of Injection Moulded Parts Made from Reinforced Thermoplastics by Acoustic Emission

Bodo Brühl

Institut für Kunststoffprüfung und Kunststoffkunde
Universität Stuttgart, Pfaffenwaldring 32,
D-7000 Stuttgart 80, Germany F.R.

Summary

There is a growing scientific and economic interest in the use of non-destructive testing methods for plastics parts, mainly for expensive reinforced mouldings that are used for high-quality applications under critical load. During the production of moulded parts, the development of weld lines is possible. These lines a regions formed by the joining together of flow fronts. The properties of moulded parts with weld lines are usually examined through destructive testing methods as tensile or impact tests.
Within this framework the following acoustic emission (AE) analysis seeks correlations between acoustic activity and the properties of weld lines.

Experimental

The moulding compounds used were polyethylene terephthalate (PBTP-GF30) and polycabonate (PC-CF30) reinforced with short fibres (30% by wt.).

The specimens were loaded under a tensile force. The AE transducer was attached to the center position of the specimen as close as possible to the weld line. In all experiments, fracture occured directly in the region of the weld line.

The experiments were performed on an AE system PAC-SPARTAN (Physical Acoustics Corp.). During the experiments, the test load and increase in length of the specimen were recorded by the AE system and a separate transient recorder. The statistical evaluation of the AE experiments was performed on a microcomputer (see Figure 1).

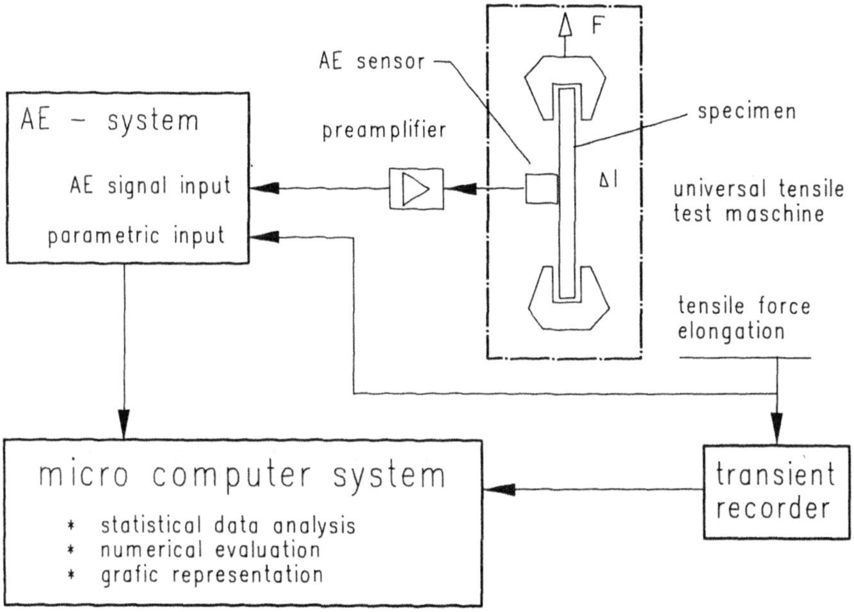

Fig. 1. Flow diagram of experimental setup

AE experimental results

The experiments provide a correlation between the strength of the weld line and the AE activity. For specimens with a weld line made from PBTP-GF30, Figure 2 shows the accumulated number of AE events at two load levels as a function of the tensile strength of the specimen. The AE activity increases with higher tensile strength.

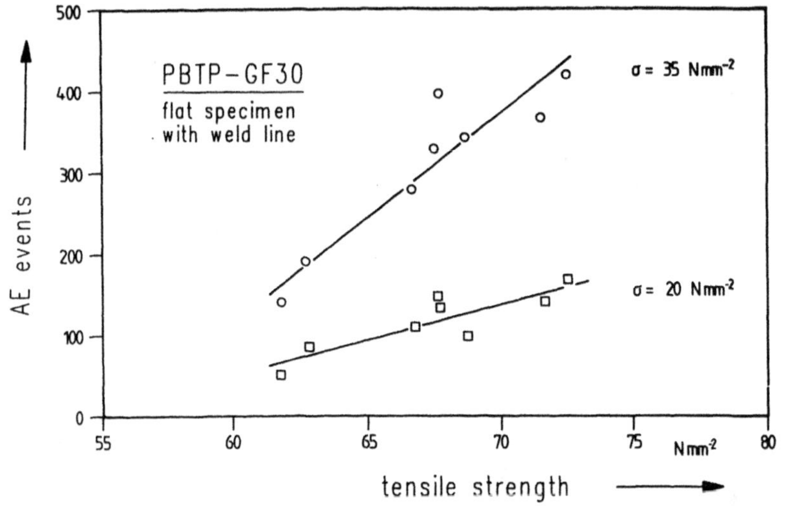

Fig. 2. Accumulated number of AE events at different load levels as a function of the tensile strength

In other experiments we investigated how the holding pressure during fabrication of specimens affects their AE behaviour. Compared to tensile testing technique the sensitivity for disclosing differences between specimens processed under various holding pressure values is higher for the AE analysis (see Figure 3). The results of impact bending tests demonstrate the influence of the holding pressure on mechanical properties.

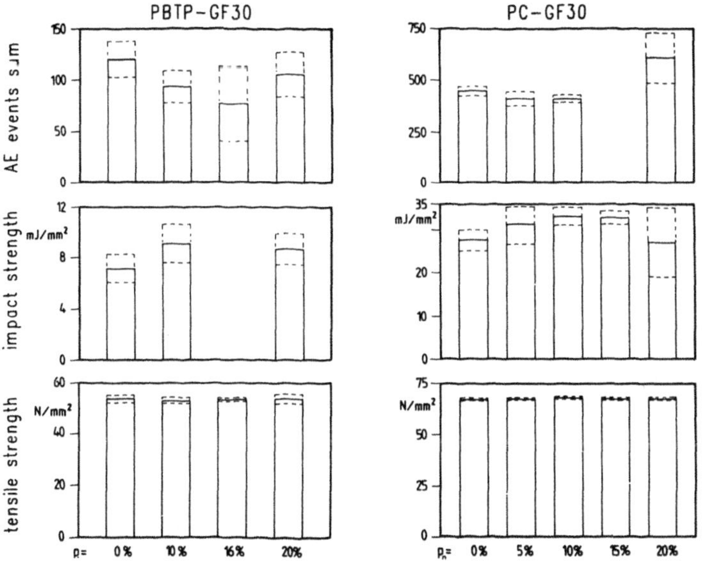

Fig. 3. Comparison of mechanical properties and acoustic activity for various holding pressure values p_n

Conclusion

Acoustic emission tests were performed on flat specimens of short fibre reinforced thermoplastics provided with weld lines. A relationship between the weld line strength and the acoustic activity is obtained for tensile forces applied below the fracture toughness value. The influence of the holding pressure on mechanical properties can be monitored by acoustic emission techniques.

Acknowledgements

The author is grateful to the Deutsche Forschungsgesellschaft (DFG) for financial support of the project, and to the companies BASF AG and BAYER AG for the injection mould and the moulding compounds.

Ultrasonic Velocity Measurements in Porous Materials

A. Jungman, L. Adler* and G. Quentin.

G.P.S. Université Paris 7

Tour 23 - 2 Place Jussieu Paris 75005 FRANCE

SUMMARY

Sintered materials with connected pores filled with fluid (air or water) has been studied using ultrasonic waves. The main applications of these materials are rocks, sediments and filters. The materials used in this study are both metallic and non-metallic structures with particle sizes ranging from 15 μ to 700 μ and porosity concentration up to 40%. Wave propagation in these two phase materials has been studied by Biot [J.A.S.A. 28, 168 (1956)] who predicted the existence of a third bulk wave - a slow compressional wave. In this work wave velocities and attenuation measurements have been carried out and related to porosity, particle size and consolidation of these composite materials. The purpose of this presentation is to report results on the ultrasonic characterization of poroelastic properties of these Biot solids.

INTRODUCTION

The purpose of this work is to study ultrasonic wave propagation in fluid-filled porous materials. These structures are two-phase materials made of a solid continuous matrix and connected pores filled with a gas or a liquid. Typical examples are given by many natural rocks or ocean sediments as well as synthetic filtration materials.

Since theories of elastic-wave propagation in fluid-saturated porous media were initiated by Biot [1,2], contributions to Biot's theory [3,6], as well as experimental confirmations [7,8], have been carried out. One fundamental aspect of these materials is their ability to support three modes of ultrasonic wave propagation : I) a fast compressional wave, II) a slow compressional wave with speed slower than that of sound in fluid, and III) a shear wave. Experiments presented in this work deal with porous materials with different grain sizes but similar porosity. The results emphasize the grain/grain contact dependence of both the fast compressional wave and the shear wave velocity.

* Permanent address: Dept. of Welding Engineering Ohio State University Colombus Ohio U.S.A.

On the other hand, attenuation of the fast compressional wave versus frequency is demonstrated to show a significant dependence with the grain size. Both results are discussed based on available theoretical analysis. In Section I, we give a description of the metallic solid porous samples, and we emphasize the relevant geometrical and physical parameters which influence the ultrasonic wave propagation.

The ultrasonic system, is briefly described in Section II. In section III, the influence of bonding between grains on velocity of fast compressional waves and shear waves is demonstrated. Section IV is an attempt to correlate the average size of the grains with attenuation measurements.

I SAMPLE DESCRIPTION

The materials used in these experiments are manufactured from metallic powders using sintering techniques. Strong intergrain bonding gives them a high stiff frame, i.e. a consolidated elastic frame, which insures the continuity of the fluid space. As filtering materials, the continuity of the fluid space is also accomplished. Typical microphotographs of bronze and stainless-steel disc samples are shown in Fig. 1

(a) (b)

Fig.1 Microphotographs of (a) bronze and (b) stainless-steel porous plates.

On the bronze sample (Fig. 1a), the pores between the spherical metallic particles are clearly visible, as well as the bonding between grains. In contrast the microphotograph of a stainless-steel sample shows irregular shaped grains. Table I gives some relevant parmeters of these samples. It is noticeable that connected porosity varies only within ± 5%, around 26% porosity for bronze and around 30% porosity for stainless-steel, while the grain size changes in a ratio 1:30. Saturation of the porous samples by water was achivèd by using either a vacuum impregnation technique or by submitting the sample to a flow of water.

SAMPLE*	DIAMETER (mm)	THICKNESS (mm)	CONNECTED POROSITY' (%)	TOTAL POROSITY (%)	AVERAGE GRAIN SIZE (mm)	DENSITY OF THE FRAME	BARRIER FILTRATION LEVEL (μm)
B3	89.18	1.98	27	32	18	6.06	2
B5	90.18	2.06	29	32	56	6.08	8
B7	90.27	2.06	31	35	84	5.75	12
B10	89.91	2.06	27	36	120	5.72	18
B15	90.52	2.15	26	38	190	5.54	25
B20	90.45	3.02	24	36	260	5.72	35
B30	89.34	3.15	24	34	370	5.91	50
B40	89.85	3.02	22	40	550	5.36	75
B60	89.60	3.25	22	41	660	5.20	100
S3	90.23	2.01	26	32		5.37	3
S5	90.37	2.01	28	34		5.17	8
S7	90.36	2.12	31	37		4.94	16
S10	90.25	2.06	32	41		4.69	20
S15	90.17	2.01	28	35		5.17	28
S20	90.72	2.11	35	41		4.67	35
S20	90.45	3.10	30	38		4.89	45
S30	90.44	3.10	33	41		4.68	75
S40	90.40	3.10	31	41		4.67	

*B = bronze, density = 8.9
S = stainless steel, density = 7.9

Table I. Geometrical and Physical Parameters of the Porous Metallic Samples

II EXPERIMENTAL SYSTEM AND MEASUREMENTS

Velocity Measurements

An ultrasonic immersion technique was used to generate bulk modes in the porous disks. Fast compressional waves are simply obtained under normal incidence by measuring the difference of time of flight between two consecutive echoes, either by using the pulse-echo method (Fig. 2A) or the through-transmission method (Fig. 2B,C).

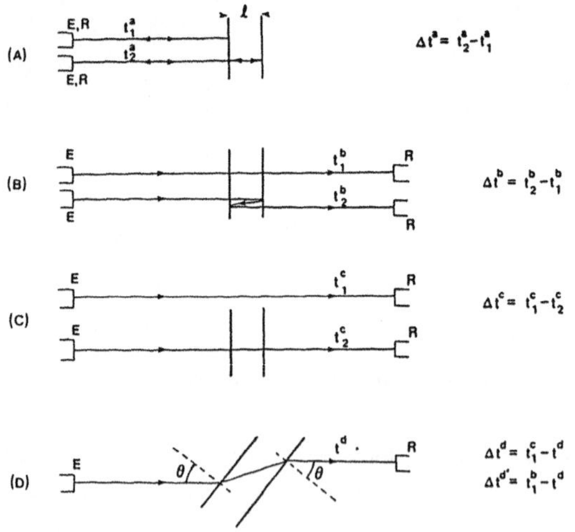

Fig. 2 Geometrical arrangement for velocity measurements.

The ultrasonic bulk speeds v for the three above situations are respectively (see Fig. 2) :

$$v = \frac{2\ell}{\Delta t^a} \quad ; \quad v = \frac{2\ell}{\Delta t^b} \quad ; \quad v = \frac{\ell . v_w}{\ell - v_w \Delta t^c} \qquad (1)$$

where v_w is the ultrasonic speed in water.
Generation of shear wave is obtained under oblique incidence by mode conversion and refraction at liquid-solid interfaces (Fig. 2D). This method, used by Plona [7] for porous materials, was first described by Hartman and Jarzynski [9]. From trigonometrical considerations, the shear velocity v_{Shear} is given by either on of the two relations.

$$v_{Shear} = v_w[(\cos\theta - \frac{v_w}{\ell} . \Delta t^d)^2 + \sin^2\theta]^{-1/2} \qquad (2)$$

$$v_{Shear} = v_w[(\cos\theta - \frac{v_w}{\ell} . \Delta t^{d''} - 1 + \frac{v_w}{v_F})^2 + \sin^2\theta]^{-1/2} \qquad (3)$$

where v_F is the fast compressional velocity,

Attenuation Measurements
The attenuation coefficient frequency dependence is deduced from the same arrangement by looking at the amplitude of the different echos. Correction for the reflection and transmission coefficients is included in the experimental results, but not the Lommel diffraction correction [10], which is neglected because of the small difference of path length between two consecutive echoes.

III VELOCITY MEASUREMENTS

Velocity and consolidation of the elastic frame
The significant character which distinguishes one sample of a given metal from the other is the grain size. Hence, in order to determine the parameter (s) of importance which influence(s) the bulk speeds, we have plotted the fast compressional and the shear speeds versus numbers proportional to the size of the grain. As shown in Fig. 3 for bronze porous samples, the frame compressional wave has velocities which do not exhibit a close dependance with the diameter of the grain. Similar observation can be made for the fast compressional and the shear velocities of stainless-steel porous samples. By considering samples B7, B10 and B15 on Fig. 3, it is seen that speeds successively decreasefrom more than 25% (samples B7 and B10) then increase 42% (samples B10 and B15), whereas grain size increases continuously (84 μm, 120 μm, 190 μm average grain diameter, respectively (see Table I)).

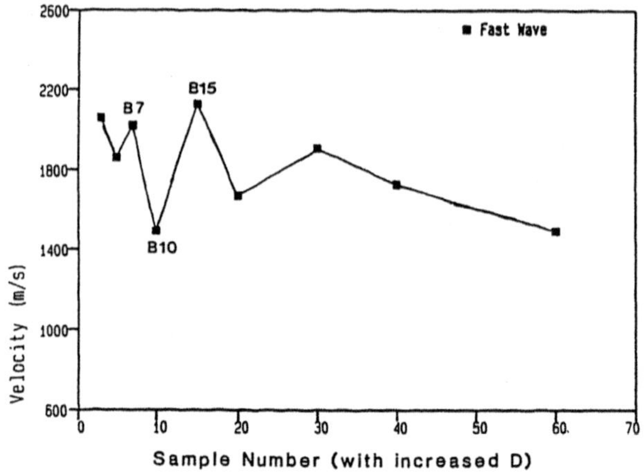

Fig. 3 Velocity of fast compressional wave in fluid-filled porous bronze.

This random behavior shows that other arguments have to be found to explain the variation of the bulk speeds in the granular porous structure. Qualitative explanations arise by examining Fig. 4, a microphotograph of the three above mentioned samples.

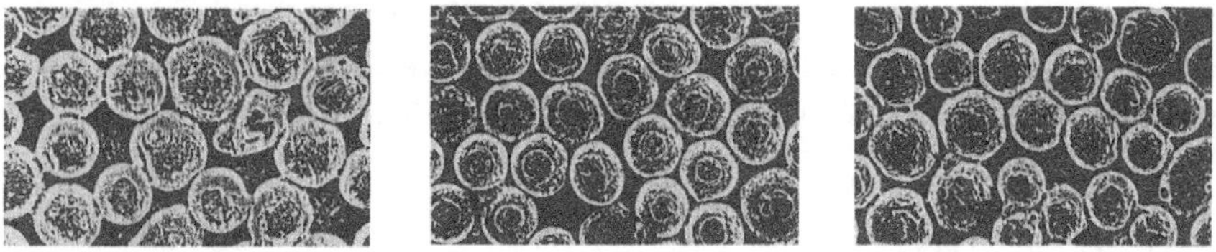

(a) (b) (c)

Fig. 4 Microphotographs of porous bronze samples (a) B7, (b) B10, (c) B15.

It appears quite evident that, to the contrary of sample B10 (Fig. 4B), with the smallest speed, samples B7 (Fig. 4A) and B15 (Fig. 4C), having much higher speed, exhibit a much stronger bonding between each sphere. As a consequence, correlation between the ultrasonic speed and the quality of grain-grain contact seems relevant, at least from a qualitative point of view. Similar observations as well as theoretical predictions have already been made by Johnson and Plona [11] and Plona and Winkler [12] on the dependance of the bulk speeds with the fusing of grains into a consolidated elastic frame. Their results show clearly how the calculated wave speeds (for the fast and slow compressional waves) change from the unconsolidated values ($K_b = N = 0$ where K_b and N are the bulk modulus and the shear modulus of the skeleton) to the consolidated stiff frame limit. These theoretical results are obtained keeping the porosity ϕ and the tortuosity α fixed, and assuming a constant proportionality between K_b and N.

Variation of fast and shear wave speed ratio of stainless-steel samples versus grain size has been carried out in Fig. 5.

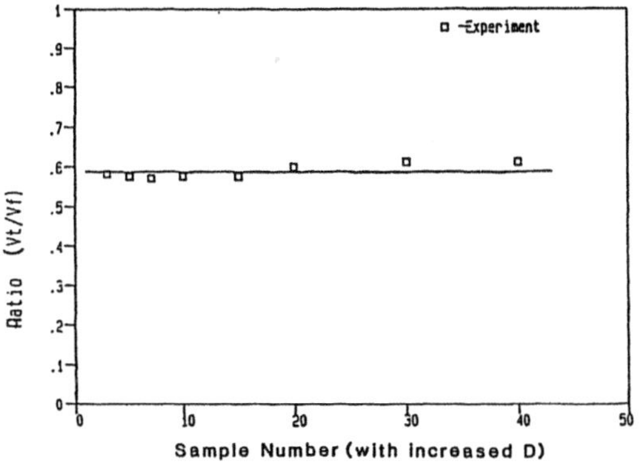

Fig.5Experimental values of the ratio of shear to fast compressional velocities.

This ratio is found constant ($\frac{v_{Shear}}{v_{Fast}} = 0.59$) for the whole range of available samples. This constant ratio can be easily obtained in terms of elastic moduli, assuming K_b, $N >> K_f$ (K_f is the bulk modulus of fluid). This asymptotic case corresponding to the stiff frame limit is valid when the pore fluid is much more compressible than the frame [13], which is relevant for water-metal systems. In this case, the fast and the shear velocities simplify greatly to the following :

$$v_{Fast} = \left[\frac{K_b + 4/3\ N}{(1-\phi)\rho_s + (1-\alpha^{-1})\phi\rho_f}\right]^{1/2} \quad ; \quad v_{Shear} = \left[\frac{N}{(1-\phi)\rho_s + (1-\alpha^{-1})\phi\rho_f}\right]^{1/2} \qquad (4)$$

where ϕ is the porosity, α is the tortuosity, ρ_s, ρ_f are the density of the solid and the fluid, respectively.

From Eqs. 4 we calculate the ratio

$$\frac{v_{Shear}}{v_{Fast}} = \left[\frac{N}{K_b + 4/3\ N}\right]^{1/2} \qquad (5)$$

This ratio is maintained constant by keeping K_b/N fixed, which is the basic theoretical assumption made by Johnson and Plona [11] in their calculations. Therefore experimental results above appear as a strong confirmation of the consolidation transition phenomenon [11].

IV ATTENUATION MEASUREMENTS

Ultrasonic estimation of pore and grain size is based on attenuation measurement. The attenuation is determined by grain scattering which disperses the energy in the travelling wave. Fig. 6 gives the attenuation curves for eight different samples of porous bronze.

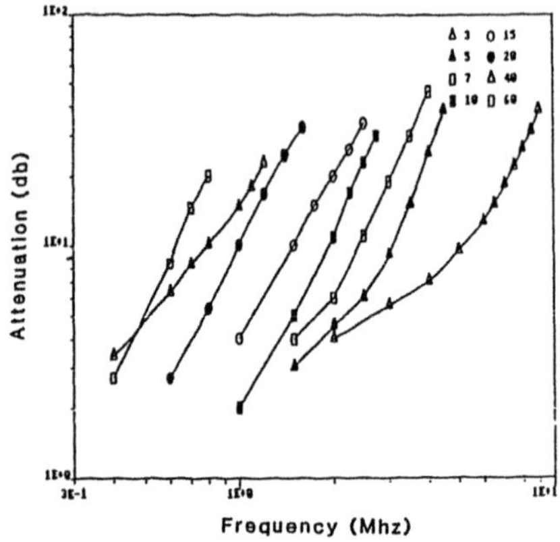

Fig. 6 Attenuation of the fast compressional wave versus grain size for fluid porous bronze (logarithmic scale).

As predicted by a simple qualitative scattering approach for a fixed frequency, attenuation increases with grain size. And, for a given sample, attenuation increases with frequency. According to the theoretical predictions, in the Rayleigh scattering domain ($kD \ll 1$, where k is the wave number and D is the diameter of the grain), the attenuation increases as the fourth power of frequency [14]:

$$\alpha = Vf^4 s \qquad (6)$$

where f is the ultrasonic frequency, V is the average grain volume, and s is a parameter of the material (depending on the density and wave velocity). The slope S of the straight lines in Fig. 6 varies from 2.7 to 3.8 which indicates, as pointed out above, that the attenuation rises as f^4, which corresponds to the upper limit of the Rayleigh scattering domain. However the above mentioned scattering theory deals with isolated spheres, which differs significantly from the present situation. The shift between the different curves gives, according to Eq. 6, indications about the average grain volume V, i.e. the size of the grains. Similar results have been obtained for stainless-steel porous specimens with same porosity but important difference in the granular structure (see Fig.1). Hence attenuation is not sensitive to the shape of the grains, but to their size.

A more surprising result is obtained by plotting the attenuation versus kD.

Fig. 7 shows that for a fixed value of 2ka (a = D/2), the attenuation increases as the grain size decreases.

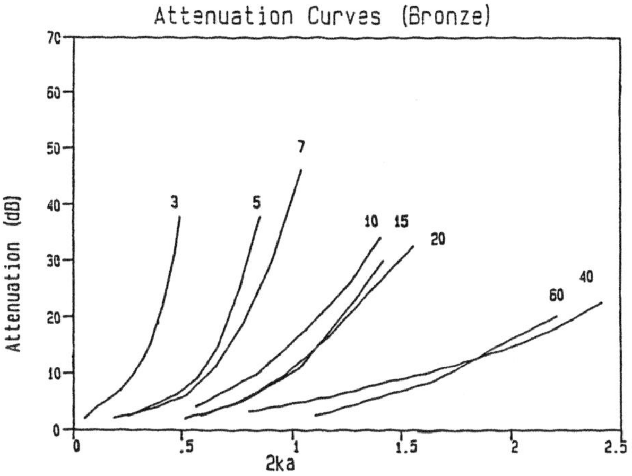

Fig. 7 Attenuation versus 2ka for the fast compressional wave in fluid - filled porous bronze.

This can be simply explained from Eq. 6 by considering spherical grain of radius a. Assuming the Rayleigh scattering domain, this equation can be rewritten

$$\alpha = [4\,\pi\, a^3/3]\; f^4 s \tag{7}$$

or

$$\alpha \simeq f^4\; a^4/a \tag{8}$$

which shows that, for fixed fa, α increases with decreasing grain size. However, the curve in Fig. 7 is plotted versus ka, which involves also the fast compressional velocity.

Hence, Eq. 8 has to be written :

$$\alpha \simeq v^4_{Fast}\; \frac{k^4\; a^4}{(2\pi)^4 a} \tag{9}$$

As mentioned above, the present situation differs significantly from the single sphere scattering problem, and the attenuation is found to vary in a different way than the fourth power of frequency. However, from a qualitative point of view it is clear that attenuation and grain size vary in an opposite way.

CONCLUSION

For a fixed prosity, we have shown that the bulk velocities vary with the consolidation of the frame, whereas attenuation is sensitive to the average grain size.

Although more quantitative measurements have to be done to be able to evaluate quantitatively parameters of interests of a porous sample such as grain and pore size, this expermental approach is a good confirmation of some theoretical predictions about ultrasonic wave propagation in porous materials.

ACKNOWLEDGMENT

This word is supported by the U.S. Department of Energy DE-FG02-- 87ER 13749, A000, by the CNRS (INT 160052) - NSF (INT 8614592) international program, and by NATO grant (0131/87)

REFERENCES

1. M. A. Biot, J. Acoust. Soc. Am. 28, 168 (1956).
2. M. A. Biot, J. Acoust. Sic. Am. 28, 179 (1956).
3. J. Geerstma and D.C. Smit, Geophys. 26, 169 (1961).
4. R. D. Stoll and G. M. bryan, J. Acoust. Soc. Am. 47, 1440 (1970).
5. R. D. Stoll, J. Acoust. Soc. Am. 66, 1152 (1979).
6. J. G. Berryman, J. Acoust. Am. 69, 416 (1981).
7. T. J. Plona, Appl. Phys. Lett. 36, 259 (1980).
8. J. G. Berryman, Appl. Phys. Lett. 37, 382 (1980).
9. B. Hartmann and J. Jarzynski, J. Acoust. Soc. Am. 56, 1469 (1974).
10. P. H. Rogers and A. L. Van Buren, J. Acoust Soc. Am. 55, 724 (1974).
11. D. L. Johnson and T. J. Plona, J. Acoust. Soc. Am. 72, 556 (1932).
12. T. J. Plona and K. W. Winkler, Conf. Proc., The Pennsylvania state University (1935).
13. D. L. Johnson, Appl. Phys. Lett. 37, 1065 (1980).
14. E. P. Papadakis, Phys. Acoust., IVB, W.P. Mason, ed. (Academic, New-York, 1968) p. 269.

Porosity Characterization in Thin Composite Plates by Ultrasonic Measurements

J.F. de BELLEVAL*, Y. BOYER*, D. LECURU**

*Division Acoustique et Vibrations Industrielles (DAVI)
Université de Technologie de Compiègne
BP. 649 - 60206 COMPIEGNE CEDEX - FRANCE

**Laboratoire Central - Aérospatiale
12, rue Pasteur
92 150 SURRESNES - FRANCE

Summary

Owing to their excellent weight/stiffness ratio, Aeronautics uses more and more composites/nomex (honeycomb) sandwich type structure. The non destructive testing of the plates used as "skin" of these structures is done by measures of attenuation of ultrasonic waves and there is no problems encountered for plates from 3 to 10mm thick. In the case of thin plates (about 1mm thick), this testing is much less precise and moreover it is better to achieve the control with the plate bonded to its nomex structure. We propose a method for the measure of ultrasonic attenuation in a plate bonded on its structure. This method is based upon a comparison of the different echoes reflected by the plate. Its uses a modelization of echoes spectra as well as ultrasonic attenuation which permits to separate two parameters, one of them characterizing the attenuation, free from frequency, due to impedance gaps at material interfaces and the other characterizing the attenuation, variable with frequency, due to porosity. We show a quite good correlation between this last parameter and the porosity measured with a destructive method, for plates bonded on nomex structure as well as for unbonded plates.

Introduction

Owing to their excellent weigth/stiffness ratio, aeronautics use more and more often sandwich type structures including one or more carbon-epoxy composite layer and nomex. The control of these pieces is done usually in two steps : first a control of the skins to determine the eventual porosity (control in a tank by double transmission) and then a control of the bonding with the nomex (control in transmission by water jets). However in some cases, the too small thickness of composite skins induces an insufficient stiffness of the sole plate which does not permit the control of porosity in a tank and this control can then be done only after the bonding.

Concerning the composites, the Laboratoire Central de l'Aérospatiale has developped a formulation permitting the correlation of the attenuation of the ultrasonic waves into the piece with the porosity ratio measured after opening and examination of these pieces. This formulation gives satisfaction for 3 to 10mm thick materials. It is much less precise for thinner or thicker plates, with or without nomex.

The control of the sandwich structures let the following problems to arise :

- the composites thickness are often small (less than 20 plies of 0.13mm each)
- the expected porosity ratio is small
- the surface quality and the size of the bond meniscus between the skin and the nomex induce amplitude signal variations which make the measures difficult to do.

The study has then for aim the research of an ultrasonic non destructive evaluation method for the measure of porosity ratio in thin carbon-epoxy composites and in sandwich type structures (thin composite/nomex/thin composite). This method consisted in a modelling of the influence of attenuation due to interfaces and due to porosity in the transfert function determined from ultrasonic echoes, and then to correlate these results with those obtained by insight examinations of samples.

The samples

The samples, as seen on Fig. 1, are composed of two composite plates (skins) bonded on honeycomb structured Nomex. The plates are unidirectional carbon-epoxy composites, whose number of plies are varying from 4 to 16. Variable porosity ratios have been obtained by a modification of polymerization cycle or by changing the number or pomping tissues.The plates have been classified into three porosity ratios : small, medium or large. The testing has been achieved first of all on unbonded plates and after on the whole samples.

Principle of the method

The method proposed is based on a comparison of two following echoes as represented in Fig. 1. The almost totality of the results will compare the echo # 1 with the plate bottom echo # 2. The method uses a modelization of the detected echoes spectra and of the attenuation in this kind of material. It permits by mean of simple analytical computations to separate two characteristic parameters of the attenuation. One of these, independant of the frequency, is due to the impedance gaps at the interfaces of

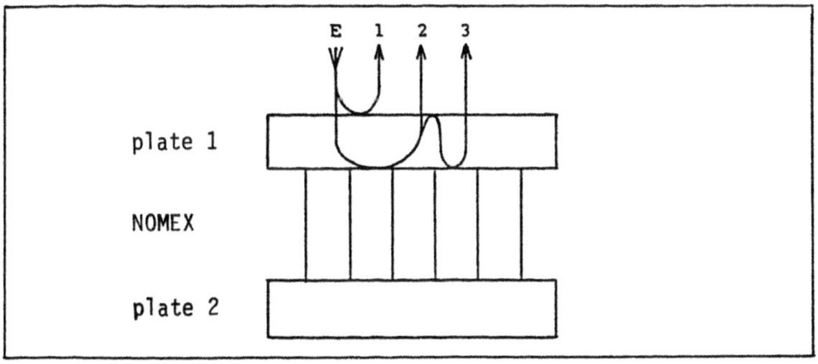

Fig. 1 : Sketch of samples

the material (water-composite, composite-nomex). The other, varying with frequency, is related to the porosity.

The#2 echoe spectrum, $S_2(\nu)$, can be written as function of the#1 echoe as following :

$$S_2(\nu) = S_1(\nu)\ f(\nu) \qquad (1)$$

where $f(\nu)$ takes into account the impedance gaps at the first interface (water-composite), at the second one (composite-water for an isolated plate, composite-nomex for a sandwich structure) as well as the attenuation in the material, including those related to porosity. The hypothesis are as following :

- the attenuation linked with reflection and transmission at interfaces is independant of frequency (perfect plane interfaces),
- the attenuation for a non porous material and those related to diffraction are neglictible (if not, a calibration with an non porous plate could take them into account),
- on the selected frequency range, the attenuation due to porosity is dependant of the square of the frequency.

The expression for $f(\nu)$ is then :

$$f(\nu) = K \exp\ (-\alpha x \nu^2) \qquad (2)$$

where : x is the distance crossed by the ultrasonic wave, i.e. two times the plate thickness in this case,

α is a coefficient depending on porosity,

K is a coefficient taking into account the attenuation independant of frequency, i.e. the interface gaps.

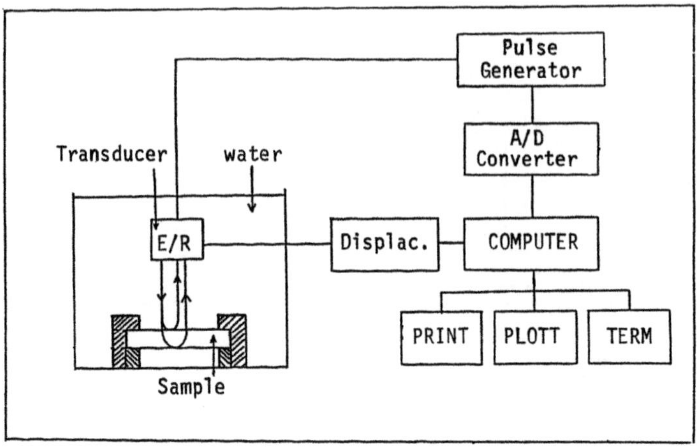

Fig. 2 : Experimental apparentus

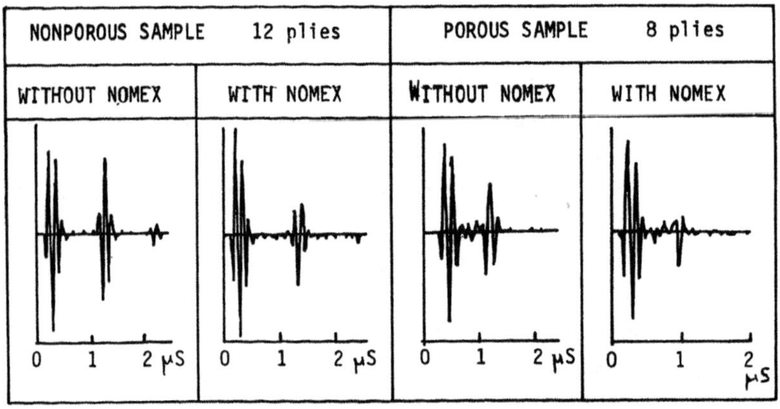

Fig. 3 : Echographic signals for two porosity ratio

$f(\nu)$ could be evaluated from formula (1) as a simple ratio between $S_1(\nu)$ and $S_2(\nu)$ but this ratio would present very strong irregularities in the frequency ranges where echoes have low energy levels and would not permit a correct evaluation of K and α. So we choosed to modelize each spectrum by an analytical function of the form (Rayleigh distribution) :

$$S(\nu) = \frac{A\nu}{\nu_m^2} \exp\left\{-\frac{1}{2}\left(\frac{\nu}{\nu_m}\right)^2\right\} \tag{3}$$

where A and ν_m characterize the considered spectrum.

If $S_1(\nu)$ can be expressed by formula (3) by mean of A_1 and ν_{m1} parameters, and $f(\nu)$ by mean of formula (2), it can be easily established that $S_2(\nu)$ can be expressed by mean of formula (3), the A_2 and ν_{m2} parameters characterizing this latter function are related to K and αx by the relations :

$$K = \frac{A_2}{A_1}\left(\frac{\nu_{m1}}{\nu_{m2}}\right)^2 \tag{4}$$

$$\alpha X = \frac{1}{2}\left(\frac{1}{\nu_{m2}^2} - \frac{1}{\nu_{m1}^2}\right) \tag{5}$$

An evaluation of A_1, ν_{m1} and A_2, ν_{m2} from $S_1(\nu)$ and $S_2(\nu)$ permits to deduce K and α X and then to separate attenuation due to interfaces from the one related to the plate porosity.

Experimental device and testing

The experimental device is sketched on Fig. 2. A transducer excited by a transient generator works in transmitter-receiver. The detected signals are sampled and processed by the computer which monitors as well the transducer moves. The transducer used, non focusing, has a 10 MHz nominal frequency. This choice is edicted by a compromise between experimental considerations (easy set up and reproducible results) and the minimal resolution required to separate echoes in the case of thin plates. Fig. 3 shows examples of signals detected for two porosity ratios and for isolated or nomex bonded plates.

First, a classical test in transmission mode has been achieved on isolated composite plates. This test revealed important local variations in porosity into a single plate. This fact has led us to evaluate precisely the porosity in the close vicinity of several measure locations by micrographic cuts. In that aim, a quantitative analysis of the image of a sample slice has been done using a CCD camera fitted microscope. With this image processing, it is possible to trigger the measures at a given grey level and to know the surface proportion overtaking the selected level by respect to the analysis window area. On the micrographies (examples Fig. 4), porosities appear in black and are quite well distinguible from the grey of resin and from the white of fibers. From this measure, we can detect the porosity ratio.

Processing and results

The computer program developped permits, for each detected signal, to plot : the experimental echoes spectra, their modelization according to formula (3), experimental (experimental spectra ratio) and modelized attenuation as well as a superposition of the second measured echoe spectrum with a reconstructed one $S_f(\nu)$ deducted from the first echoe and the attenuation model :

$$S_r(\nu) = S_1(\nu)\ K \exp\left(-\alpha X\ \nu^2\right) \qquad (6)$$

Fig. 5 depicts such results in the case of three different porosity ratios (small, medium and large). We can notice that, despite an imperfect spectra modelization, the results obtained for reconstructed spectra are quite satisfactory, which indicates a good representation of attenuation. We observe as well an increase of attenuation and more particularly of the αX coefficient versus the porosity ratio.

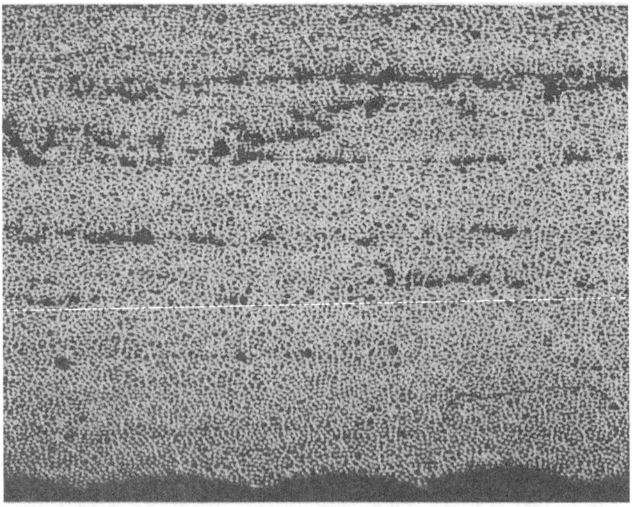

Fig. 4 : Micrography of porous sample (X 100)

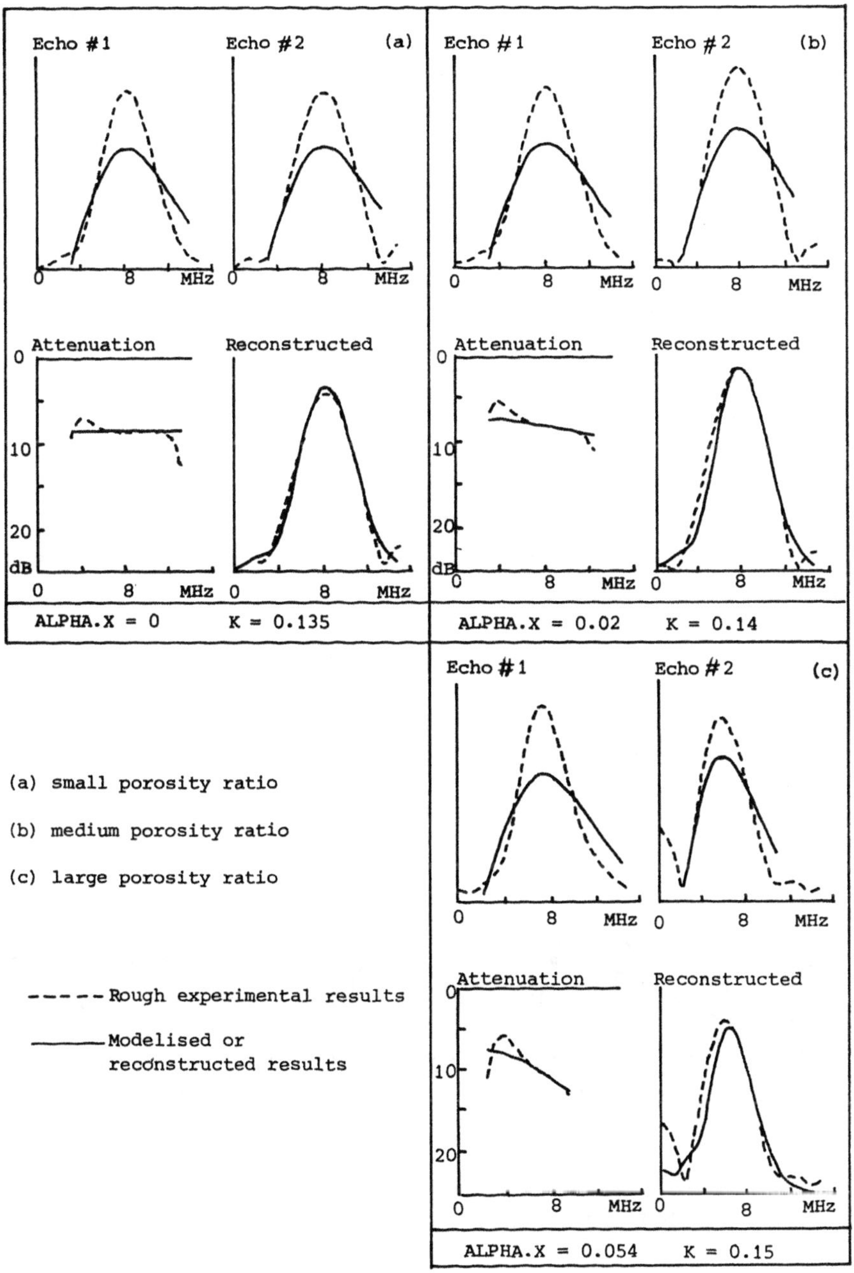

Fig. 5 : Results for plates with nomex

Fig. 6 shows the value of α X coefficient deduced from the proposed method as a function of the porosity ratio measured by micrography as well as linear regressions obtained from these points. We can notice a slight dispersion of the values around regressions curves a bit more important for the values involving nomex. However we can see that the coefficients for this latter case are relatively close for with and without nomex measures.

The results dispersion is explained by two principal problems :
- the insight investigation micrography gives very local results, at the time we constated a strong variation in porosity between close points. The more global measure using ultrasonics can then not be directly compared with micrography results. It is only by using statistics (which are done by mean of linear regressions) that the results can be compared.
- the diffusion on interface which presents visible ondulations on Fig. 5 has perturbated the results by adding a term depending of frequency which is not related to porosity. This point can be improved by taking for reference echo not the first one, the most perturbated, but a reflection echoe on a perfectly plane interface of a calibration sample.

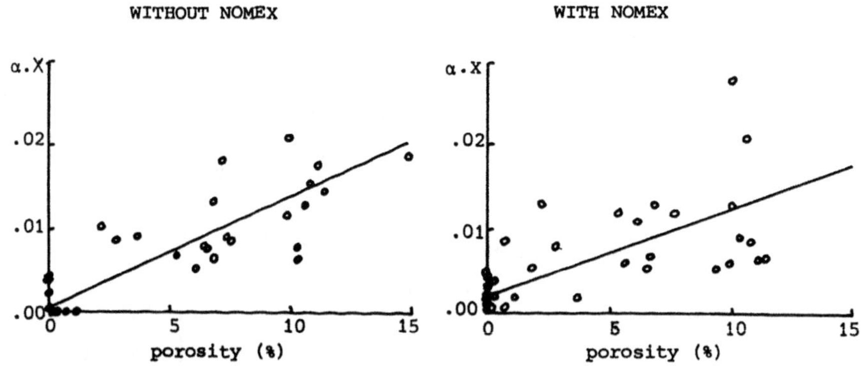

Fig. 6 : α.X coefficient versus measured porosity ratio

Conclusions

We have carried out a method permitting an evaluation of the porisity in composite plates bonded on nomex honeycomb structure. This method is based on a parametric simultaneous modelization of spectra of ultrasonic echoes reflected by the plate and of ultrasonic attenuation. The modelization permits to separate into attenuation an amplitude term independant of frequency, related to attenuation at the interfaces and a frequency dependant term related to porosity which is directly measurable by the frequency shift of the spectra of echoes after their travel into the plate.

A good correlation has been pointed out between this last term (α) and the real porosity ratio measured by micrography. This method has shown a better accuracy than the classical one (simple or double transmission without any signal processing) in the determination of porosity ratio on isolated thin plates, the classical method being not applicable to complete sandwich type structure (composite/nomex/composite).

Acknowledgement

This work was supported in part by DRET under contract N° 85-34-494

References

1. Kuc R., Schwartz M. : IEEE, trans. on Sonics and Ultra., Vol. Su 26, n° 5, sept. 79, PP. 353-362

2. Kuc R. : IEEE, Ultrason. symposium 1983, pp. 831-834

3. Dumoulin J.P., de Belleval : Ultrasonics International 85, Londres, July 85, pp. 575-580

X-Ray Diffraction Scanning Microscopy - A New Method of Nondestructive Characterization of Composites

M. P. Hentschel, A. Lange

Federal Institute for Materials Research and Testing (BAM)
Unter den Eichen 87, D-1000 Berlin 45

The nondestructive characterization of composites can be difficult if one applies traditional methods of nondestructive testing. In order to overcome this problem we have developped a new method of nondestructive characterization: "X-ray Diffraction Scanning Microscopy". As expressed in these terms it performs scanning microscopy by X-ray diffraction. The instrument is in the stage of a laboratory prototype and reveals a twodimensional picture of a section through a thick composite laminate. We explain the method and the physics behind it and demonstrate the results of two inspected composite materials. One is the carbon fibre composite with an epoxy matrix and se-

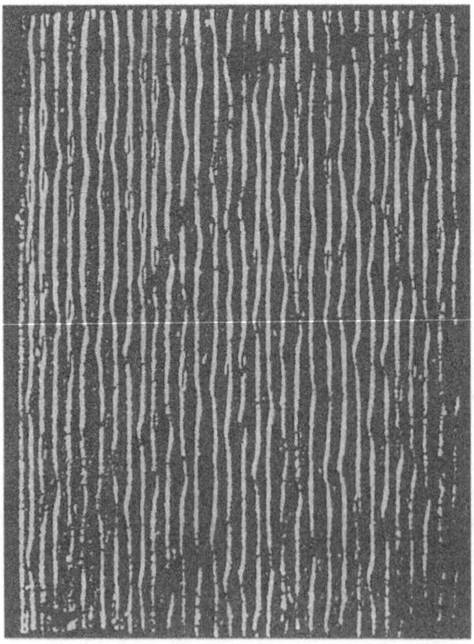

Fig. 1: Optical surface image of 25 mm carbon fibre composite bar reveals 31 vertical layer lines of in-plane fibre orientation (white lines), separated by layers of other directions (black areas)

cond is a carbon fibre ceramics. Fig. 1 shows the optical surface picture of a 25 mm thick bar of a carbon fibre composite consisting of 31 layers of vertical mean fibre orientation. These layers will be visualized inside the opaque material. The physics applied is Thomson elastic X-ray scattering as commonly used in cristallography. Fig. 2 shows the diffraction pattern of a 0.1 mm thin unidirectional carbon fibre composite. The reflection 002 is the diffracted intensity of the atomic layers of the carbon fibres. In the following only the information of this diffracted intensity is used, without interference by other substances which might be present. The selection of the diffraction angle provides a chemical contrast of the material.

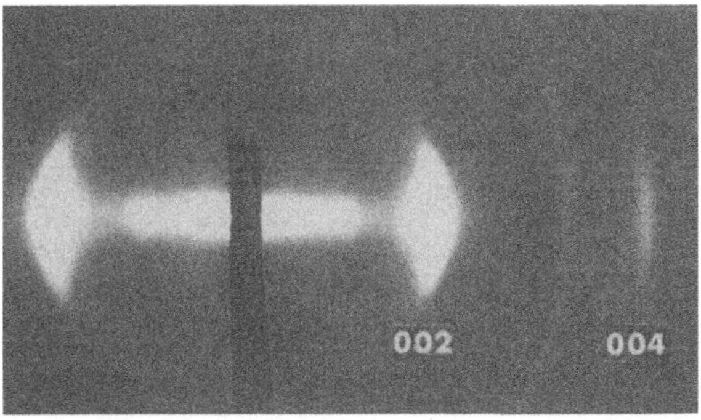

Fig. 2: X-ray diffraction pattern of thin carbon fibre composite with oriented (002)-reflection at 12° scattering angle

Fig. 3 demonstrates the change in intensity of the (002)-reflection if a thin walled composite of three layers with different fibre orientations is rotated about the axis of the incoming beam. The intensity signal of the (002)-reflection is specific to the fibre orientation. Obviously it is proportional to the density of fibre assemblies.

In the case of thick laminates all reflections of the single layers would overlapp, destroying the spatial resolution along the beam direction. This problem was overcome: Fig. 4 is a sketch of the instrument components. The necessary threedimensional resolution of X-ray diffraction from the rather amorphous

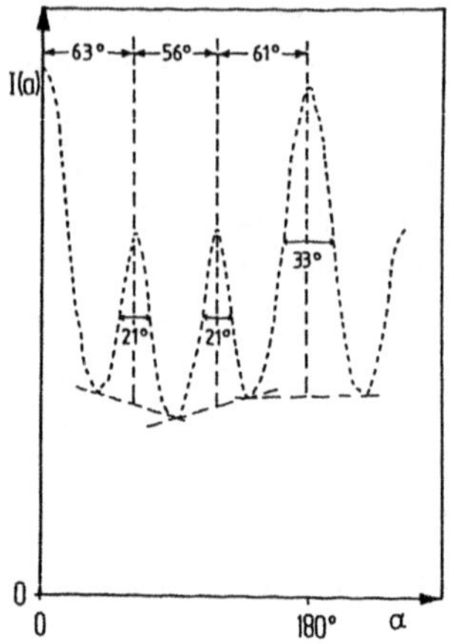

Fig. 3:
Intensity modulation of a thin 3-layer composit during rotation about the primary X-ray beam axis reveals three fibre directions

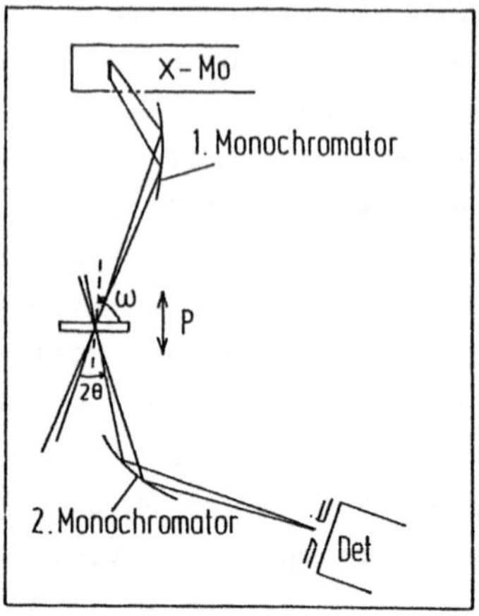

Fig. 4:
Schematic view of the X-ray focussing geometry demonstrates the principle of measurement. X-Mo: X-ray fine focus tube with Mo-target;
2 : Scattering angle; P: test sample; Det: Szintillation counter and intensity registration

materials is achieved by positioning the two focussing crystal monochromators on a common volume element where the diffraction inside the sample takes place. The upper focus is occupied by a X-ray tube target producing monochromatic Molybdenium K- radiation of 20 keV. The lower focus meets a detector slit which measures the transmitted intensity of the diffraction volume element. On moving the sample through the diffraction volume element, the changing intensities are written into a computer file which stores this information in coincidence with the position of the sample.

The geometry of the X-ray bundle passing through the composite sample gives Fig. 5 where the marked plane is the inspected area giving the twodimensional section information of the carbon fibre arrangement. P is the incoming primary beam and B the outgoing diffracted beam. The scanning direction is from front to back and than upwards following a meander pattern.

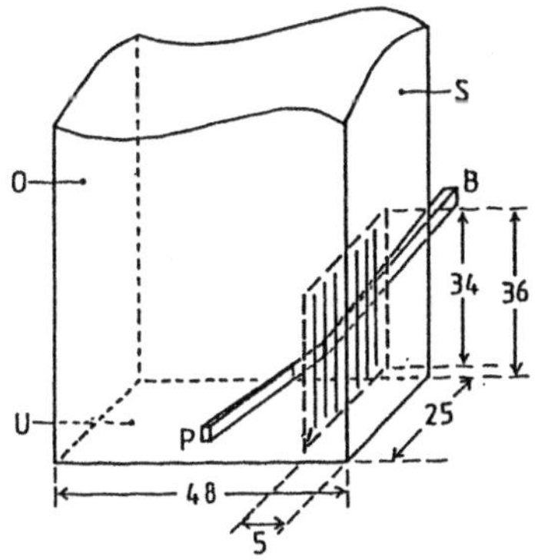

Fig. 5:
Inspected plane section through composite bar. Dimensions in mm.
P: Incoming primary beam;
B: Observed outgoing scattered beam. Carbon fibre orientation parallel to front surface O.
S: Surface of optical image (Fig. 1)

The change in intensity which occurs while stepping transversal through the sample piece is shown in Fig. 6. 31 maxima represent the 31 layers of equal carbon fibre orientation. A computer program transferres this amplitude into a grey scale. If such transversal scans are repeated and repositioned they yield a twodimensional picture as shown in Fig. 7. All the 31 layers of equal carbon fibre orientation are revealed by black and white stripes. Extraordinary contrast demonstrates regions of higher

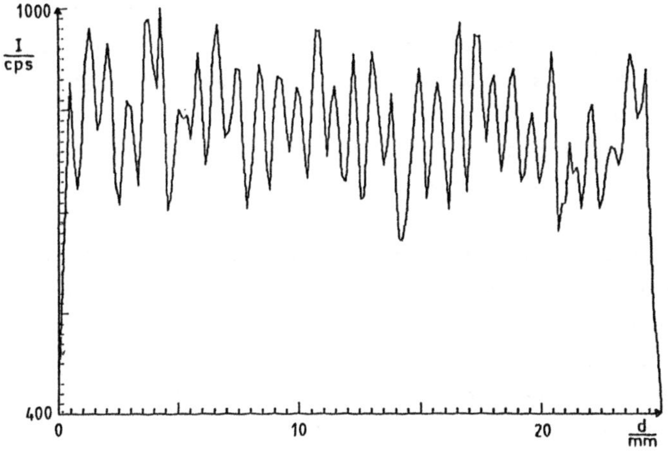

Fig. 6: Intensity modulation of transversal scan of composite bar. The 31 maxima yield the 31 carbon fibre layers of vertical orientation. Individual layers are of different shapes as determined by their internal fibre distribution

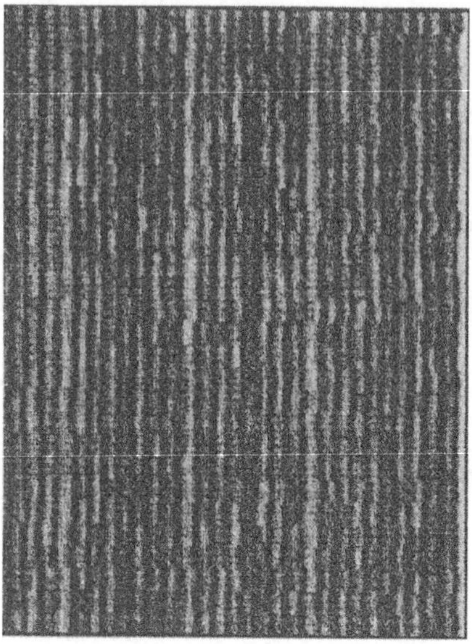

Fig. 7:
Computed section image of carbon fibre epoxy laminate by X-ray Diffraction Scanning Microscopy, 25 mm wide and 34 mm high. Note the relation to Figure 1. Dark areas are of highest fibre concentration and perfect vertical orientation. Distance variations, layer thickness, misorientation of fibres are visible and easy to analyse by the stored intensity data.

density or a misorientation inside the material of which some are systematic and known from manufacturing, namely by assembly of subunits of layer packets. The highly differentiated view through an opaque laminate consists of 5000 picture elements. The resolution is 50 µm x 200 µm x 1.5 mm. The layers of this picture are those of vertical fibre orientation. The same procedure can be repeated for any other fibre direction of the composite material. The contrast of each image reveals a fibre density distribution, but if necessary, the contrast can be optimised in order to be sensitive to fibre misorientation. In principle the resolution can be increased when higher X-ray intensities such as synchrotron radiation are applied. Few microns of resolution are feasible then. Finally we carry out the inspection of a carbon fibre ceramics plate of 4.5 mm thickness. Fig. 8 shows the optical image of the edge surface where the misalignment of the layers is visible. Fig. 9 shows a number of transversal scans revealing the changing intensities due to the 13 layers.

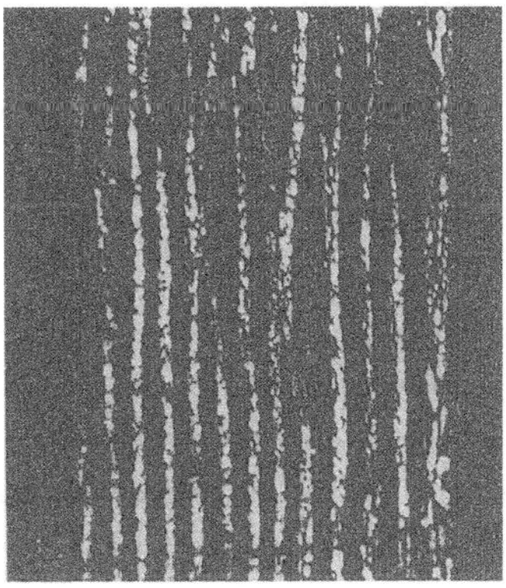

Fig. 8:

Optical surface image of tested carbon fibre ceramics panel of 4.5 mm thickness. Bright distorted layer lines are of vertical fibre orientation.

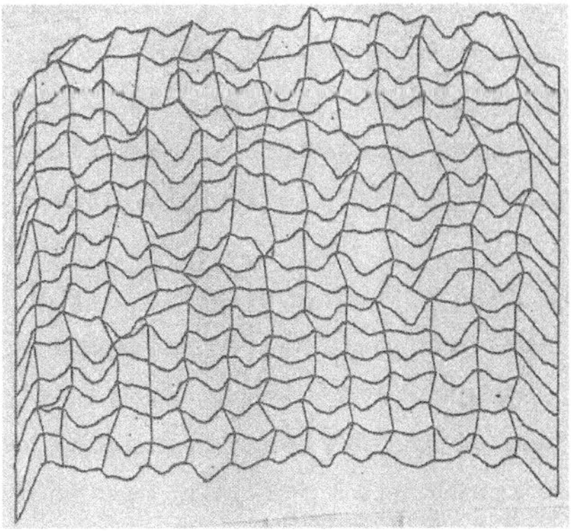

Fig. 9:

Transversal scans of carbon fibre ceramics reveal intensity amplitudes. Their maxima are marked by vertical lines according to the layer orientation.

Fig. 10: Computed section image of carbon fibre composite by X-Ray-Diffraction-Scanning-Microscopy of 4.5 mm width and 9 mm height. Layer distortions, porosity and misalignment of fibres cause the irregular pattern (compare Fig. 8)

When this information is transferred into plotter density we have the transverse section as in Fig. 10, which has similarities with the optical image (Fig. 8). The layer distances and the layer alignment as well as the fibre distribution and fibre orientation are strongly distorted.

The X-ray diffraction scanning microscopy has revealed section images of the composite samples , in which irregular and regular properties of the fibre topology become visible. It will be a matter of further research to correlate these structures to the results of destructive testing and mechanical failures.

We intend to refine the instrument, namely to improve the resolution, increase the scanning speed of now 0.5 hours per cm^2 and to introduce image refinement procedures, including an absorption correction.

X-ray diffraction scanning microscopy is generally applicable to liquids or solids of which generally any crystallographic property could be imaged. For many other "Advanced Materials", which in most cases are strongly inhomogeneous and anisotropic, the technique can be of importance.

Nondestructive Investigation of Fibre Reinforced Composites by X-Ray Computed Tomography

M. Maisl, T. Scherer, H. Reiter, and S. Hirsekorn

Fraunhofer-Institut für zerstörungsfreie Prüfverfahren,
Universität, Gebäude 37, D-6600 Saarbrücken 11

Summary

This contribution discusses the X-ray computed tomography as a tool for nondestructive testing of fibre reinforced components. A modified medical CT system is used to examine leaf springs made by glass fibre reinforced composites. The reconstructed cross sections are compared with destructively measured density distributions. Small objects were scanned by a high resolution CT system with a microfocal X-ray tube as source and selective two different detection systems. Its resolution capability is about 20 μm.

Introduction

Fibre reinforced composites are increasingly used for engineering components in automobile and aerospace industry because of their excellent mechanical properties. But because of difficulties in production process the components are not free of defects. Pores or delaminations, inhomogeneous distribution of fibres and matrix material, and dislocation of fibres reduce the strength and reliability of the components. For example, 6% porosity reduces the transversal strength about 25% [1]. Therefore, the main advantages of these materials which are light weight and high strength cannot be completely utilized. The components are overdimensioned, and the advantages in comparison to metal are reduced.

Nondestructive testing methods are required to control and thus to optimize the process of fabrication. This paper discusses the X-ray computed tomography (CT) as a tool for quality assurance and characterization of fibre reinforced components.

Medical CT System

Radiography yields a shadowgraph of the investigated object, and therefore, a superposition of the details along the penetration direction. However, the detection of plane defects like delaminations or cracks perpendicular to the penetration direction is difficult. Opposed to radiography, CT yields a shadow free image of one cross section of the object. In order to get familiar with the application of this technique for material testing, a medical CT scanner (Delta Scan 50, Ohio Nuclear) was used. Medical CT scanners are optimized to yield high contrast resolution for low absorbing (biological) materials. For the purpose of material testing the medical CT scanner was modified in order to obtain higher geometrical resolution [2]. The Delta Scan 50 is a second generation scanner (translation and rotation) with three CaF_2 detectors per slice. At maximum, the tube operates at 125 kV and 30 mA. Two cross sections are measured and reconstructed simultaneously, which takes 5 minutes at maximum. The reconstructed cross sections are displayed by 256 × 256 pixels. The reconstruction algorithm (filtered back projection) requires a complete set of data, i.e., the object diameter d must be scanned completely. Therefore, the minimum pixel width w is given by w = d/256. Some modifications improved the geometrical resolution up to 2.5 linepairs/mm (lp/mm). Depending on the diameter (at maximum 45 cm) and the absorption capability of the object the contrast resolution is about 1%.

High resolution CT system

The thickness of a cross section of an object taken by the medical system amounts to 5 mm, the smallest pixel width is 200 μm. This is not sufficient for the detection and resolution of small defects and structural details of fibre reinforced components. Therefore, we built up a high resolution CT system with a resolving capability of about 20 μm. The demanded high resolution requires special X-ray tubes and detection systems. Using the direct magnification technique in connection with microfocal X-ray sources, the resolution of the detection system can be reduced down to 1 lp/mm without reduction of the geometrical resolution of the complet CT system. As source a microfocal X-ray tube, as detection system an image intensifier are used. The object is placed on a computer controlled turntable. Its irradiation images are projected on the entrance window of the image intensifier [3]. Because of the small focus (between 10 μm and 100 μm diameter) the magnification can be varied depending on the object diameter between 3 and 30 without reduction of geometrical resolution. The cross sections are reconstructed using the filtered back projection algorithm and displayed on the monitor of the image processor.

At present, the limits of this system are given by the following points: The mechanical construction restricts the smallest pixel width to 20 µm. Depending on the pixel number of 512 x 512 points the minimum diameter for the reconstructed cross section is 10 mm. The smallest pixel size requires a power reduction of the X-ray tube (200 kV, 1 mA at maximum) down to 40 W. Otherwise, the focal spot size will increase inadmissibly. The low dynamic range of the detection system (100:1) restricts the applicability to weakly absorbing components.

The resolution of the system can be increased by the number of detector elements. In order to enable the investigation of higher absorbing samples the dynamic range of the detection system has to be increased. Figure 1 shows the set up of a high resolution CT system with a semiconductor array detector (line detector). The detection system consists of 1024 single photodiodes with a dynamic range of 1000:1. At maximum, a cross section with a diameter of 350 mm represented by 1024 x 1024 pixels can be obtained without translation. X-ray source and sample manipulation system are retained. Measurement of projections and image reconstruction are carried out simultaneously.

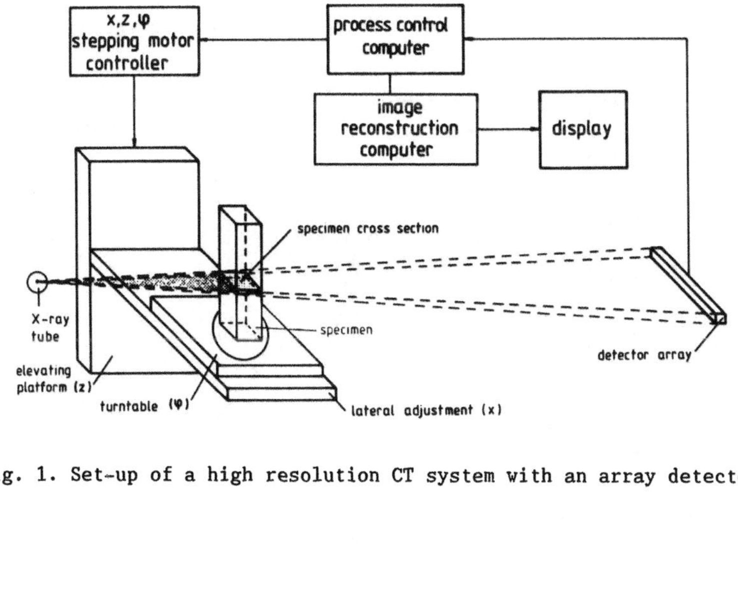

Fig. 1. Set-up of a high resolution CT system with an array detector

Experimental Results

Leaf springs for motor cars made out of fibre reinforced composites have been developed because of their excellent mechanical properties. Their quality strongly depends on the homogeneous distribution of fibres and matrix. Porosity also reduces the strength. Generally, leaf springs are produced by hot pressing of pre-pregs. During compression moulding the distribution of matrix material and fibres can become inhomogeneous because of the low viscosity of resin. On loading the inhomogeneous density distribution causes inhomogenous flow of stress and finally reduction of the life time.

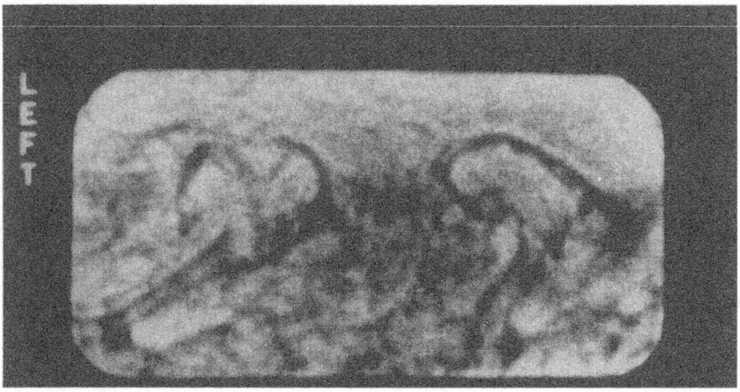

Fig. 2. Reconstructed cross section of a leaf spring

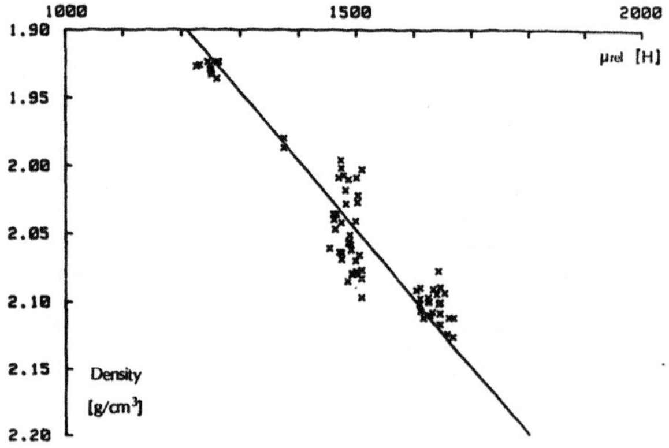

Fig. 3. Comparison between reconstructed and destructively measured density distribution

The X-ray absorption coefficient is lower for resin than for glass fibres. Therefore, X-ray CT can be used to detect the density distribution in glass fibre reinforced composites. Figure 2 shows a cross section of a leaf spring (80mm x 45mm) reconstructed by the medical CT system. In the image, the inhomogeneous distribution of fibres and resin and a resin filled meander can be seen. These meander structures are caused by the pressing process. They could be prevented by optimization of the production process. In Figure 3, the good agreement between the nondestructively measured and the destructively determined density distribution is shown [4].

In aerospace and machine building filament wound components are used. Their mechanical properties depend on the exact position and bond of rovings. Because of difficulties during fabrication these high-power components must be tested nondestructively. X-ray CT allows the detection of voids and shows up the distribution of rovings in cross section images. Moreover, size and position of delaminations can be determined. As an example, Figure 4 shows the cross section of a carbon fibre reinforced spindle of a tool machine (diameter 20cm) which is recorded by the medical CT system, too. The winding construction of the spindle, cavities (black), and layers with higher concentration of resin (grey) and fibres (white) are visible.

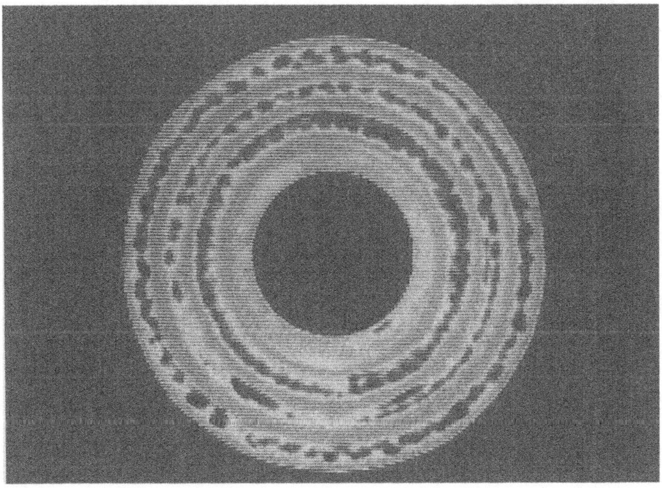

Fig. 4. Tomogramm of a tool machine spindle with voids

In order to characterize small defects and structural details such as pores and single planes of pre-pregs, respectively, the geometrical resolution of the medical CT system is not sufficient. For these investigations a high resolution CT system is necessary. Figure 5 shows a cross section reconstructed with high resolution perpendicular to the fibres of a cube cut away from a leaf spring. Because of the low dynamic range of the image intensifier the linear dimensions of the sample is limited. The edge length of the cube amounts to 10 mm. In the imaged slice, regions of different fibre and resin concentration can be distinguished. The X-ray absorption coefficient increases with fibre concentration, i. e., the lower the fibre portion the darker the image. In the middle of the imaged cross section an area of high resin concentration is located. Because of the pixel width of 44 µm single fibres (diameter about 7 µm) cannot be resolved.

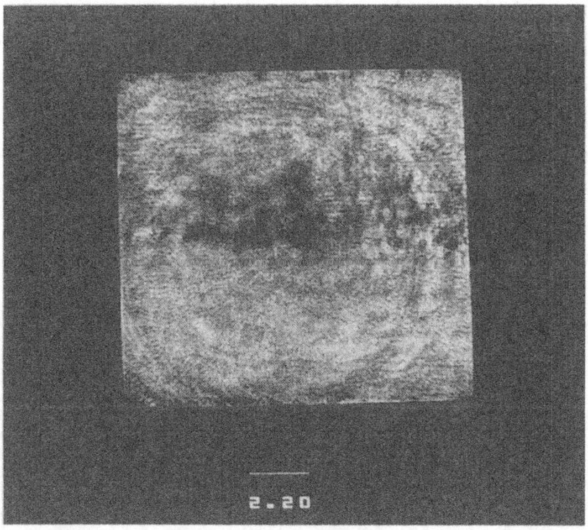

Fig. 5. Glass fibre reinforced cube, cross section perpendicular to the fibres

Complete cross sections of leaf springs can be investigated with the high resolution CT system if the array detector is used, which has a higher dynamic range than the image intensifier. Figure 6 shows a reconstructed cross section (1024 x 1024 pixels) of a slice of a leaf spring with the dimensions 80 mm x 22 mm. The image shows regions with low resin concentration at the smaller sites and a crack in the middle of the sample. Below the crack, an enrichment of resin can be seen, and the direction of pre-pregs and pores can be realized. In order to scan the whole cross section of the leaf spring the geometrical magnification was reduced so that the pixel width amounts to 90 µm. The circular artefacts result from the inhomogeneous sensitivity of the array detector. Figure 7 shows the transmitted light photo of the same slice (6 mm thickness) of the leaf spring which contains the cross section imaged in Figure 6. On this photograph, the inhomogeneous fibre distribution, the crack, and the pores are visible. But the structural details cannot be resolved as good as in the tomogramm.

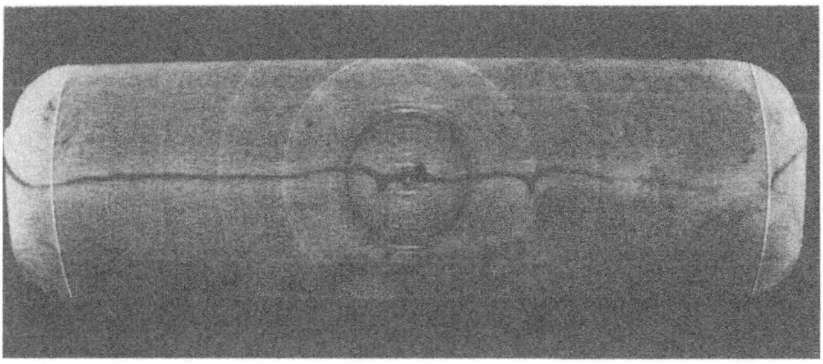

Fig. 6. High resolution CT of a leaf spring

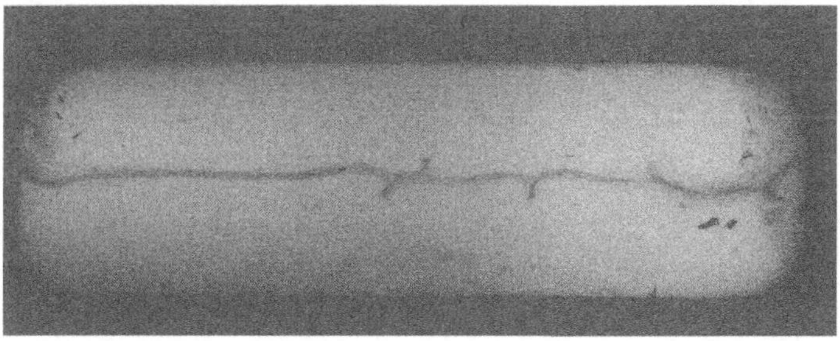

Fig. 7. Transmitted light photo of a slice of a leaf spring

Conclusion

X-ray Computed Tomography is a nondestructive imaging technique, which allows to determine the density distribution of fibre reinforced components and to detect defects such as pores or delaminations. Depending on the size of the sample to be investigated and the required resolution a modified medical or a high resolution CT system are used. The informations from a medical CT scanner are used to optimize the fabrication process of fibre reinforced leaf springs and to study the increase of delaminations during loading tests. High resolution CT can be used as a nondestructive microscopic technique to image cross sections of samples.

References

1. U. Seiler, Zur Auslegung statisch und dynamisch beanspruchter Bauteile aus Verbundwerkstoffen am Beispiel von GFK Blattfedern, Dissertation an der RWTH Aachen, 1987

2. T. Vontz, et al, NDT for small Defects by X-ray Tomography, in: Review of Progress in Quantitative Nondestructive Evaluation, Vol 7A, Plenum Publishing Corp., 1988, pp. 389

3. M. Maisl, H. Reiter, and P. Hoeller, Microradiography and Tomography for High Resolution NDT of Advanced Materials and Microstructural Components, in: New Directions in The Nondestructive Evaluation of Advanced Materials, Eds.: J.L. Rose and A.A. Tseng, ASME, New York 1988, pp. 29

4. M.Fröhlke, V.Schmitz, Nachweis von Inhomogenitäten bei der Faser-Harz-Verteilung in GFK-Federn mit Hilfe der Röntgen-Computer-Tomographie, in: Proceedings Verbundwerk 88, Demat Exposition Managing, Frankfurt 1988, 8.1

Ultrasonic Characterization of Aluminium/Epoxy Composite Materials

L.F. BRESSE, D.A. HUTCHINS* AND B. FARAHBAKHSH

Department of Physics, Queen's University, Kingston, Ontario, Canada K7L 3N6, and Alcan International R&D Centre, Kingston, Ontario, Canada, K7L 4Z4.
*Present address: Department of Engineering, University of Warwick, Coventry CV4 7AL, England.

Summary

A series of experiments have investigated the interaction of ultrasonic waves with laminated composite materials containing aluminium. It will be shown that various testing strategies are possible. In all cases, an analysis in the frequency domain allows debonding between adjacent layers to be detected.

Introduction

In recent years, there has been increased interest in the ultrasonic inspection of composite materials, to detect delaiminations at various depths within the structure. Conventional C–scans are useful in some materials, but have difficulties when thin layers are present, as reflections from each layer overlap in time. In addition, the presence of highly–attenuating layers restricts the resolution possible. Another approach is to use a separate source and receiver, as in the acousto–ultrasonic technique [1], but this often leads to only qualitative data. More recent research [2–4] has identified an ultrasonic spectroscopic approach, which determines the resonant frequencies of a particular layer within the material, and the changes in amplitude and/or frequency that occur in the presence of a delamination. To illustate this method, the cited authors presented a simplified theoretical model of a composite material, containing three layers, and showed good agreement with experimental results.

In the present work, we have extended the above by evaluating the impulse response of a composite material to a plane ultrasonic wave, where the number and properties of each layer may be defined in each case. The results have then been compared to experiment, using an analysis in the frequency domain.

Theory

The theoretical model is based upon the matrix transfer technique and leads to a solution which is easily evaluated by microcomputer. The method allows the resonant frequencies of a specific layer to be identified, and, by changing the boundary conditions within the model, the effect of a delamination may be predicted.

Consider a laminated plate, consisting of an arbitrary number (n) of elastic, isotropic layers, bounded below by a solid or liquid and above by a liquid half-space, as shown in Fig. 1. Each layer, denoted by the index (i), is characterized by its thickness $d^{(i)}$, density

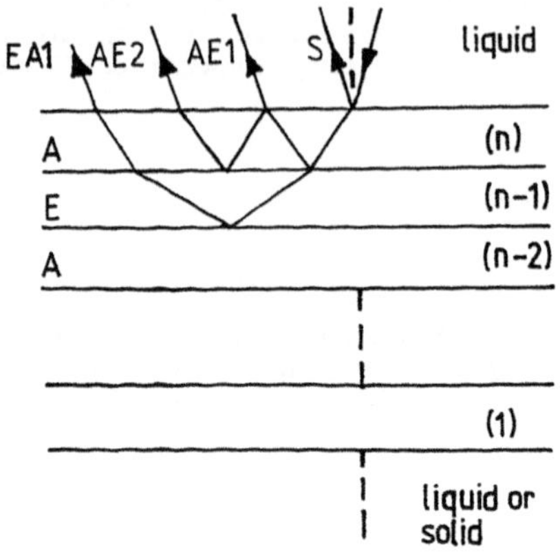

Fig. 1. Model of multilayer composite.

$\rho^{(i)}$, and elastic wave velocities $Cp^{(i)}$ and $Cs^{(i)}$. Attenuation within a given layer is treated by using a complex value for $Cp^{(i)}$ and $Cs^{(i)}$. The problem involves solving for the reflection and transmission coefficients, for a longitudinal plane wave incident at some angle θ_{ℓ}. Details of the theoretical development will be published elsewhere. Here, however, we present results of interest to the present work. As an illustration, we will examine a composite material, comprised of two 1.6 mm thick aluminium layers (Cp = 6350 ms^{-1} and Cs = 3100 ms^{-1}), separated by a 0.48 mm thick epoxy adhesive (Cp = 2580 ms^{-1} and Cs = 1100 ms^{-1}). The resulting theoretical impulse response, for reflection of a short longitudinal pulse at normal incidence, is shown in Fig. 2(a). The first echo to arrive at the transducer is from the front aluminium surface, and is marked (S). Multiple reflections within the top aluminium layer also return energy to the receiver, and the first two are shown as transients (AE1) and (AE2). Also present are reflections from the second epoxy/aluminium interface, the first of which is marked (EA1) on the figure. Although these signals are separated in time, the real waveform for a given transducer would be obtained by convolution of this impulse response with that of the transducer system; in many cases this leads to overlapping signals.

As a comparison, Fig. 2(b) shows the theoretical frequency spectrum of a reflected signal, corresponding to the impulse response of Fig. 2(a). A series of minima are present in the reflected signal. These correspond to resonant frequencies, which destructively interfere

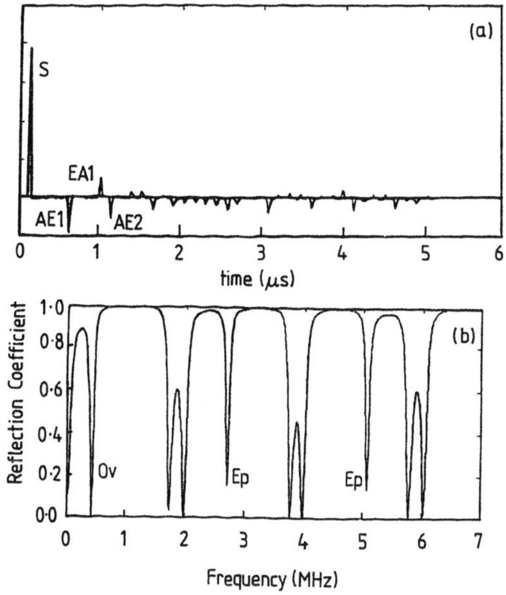

Fig. 2. Response of a three layer system to an impulsive imput in (a) time and (b)frequency domains. Layer parameters are given in the text.

with the specular echo (S) from the front aluminium surface (Note: these would be maxima in transmission, or if the echo (S) was not present). Several minima are identified. For example, frequencies Ep correspond to resonances of the epoxy layer, whereas Ov is the overall resonance of the whole structure (at the lowest frequency, as expected). Further modelling has shown that these frequencies change with layer properties (Cp, Cs, density or thickness).

It is evident from the above that specific resonances, associated with particular layers, may be identified. This may be used experimentally to detect delaminations, as will now be described.

Experimental Results

Two types of composite sample, both containing aluminium layers, have been investigated using ultrasonic techniques in the frequency domain. The first were samples similar to that described in the theory section, with aluminium plates bonded with epoxy adhesive. The second were foam panels, with 0.6 mm thick aluminium sheets attached with epoxy to an expanded closed-pore foam core of 38 mm thickness.

The instrumentation used to collect pulse-echo ultrasonic waveforms is shown in Fig. 3.

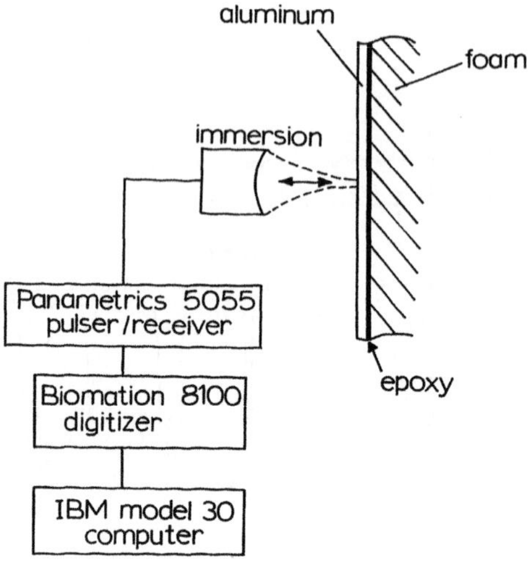

Fig. 3. Schematic diagram of apparatus

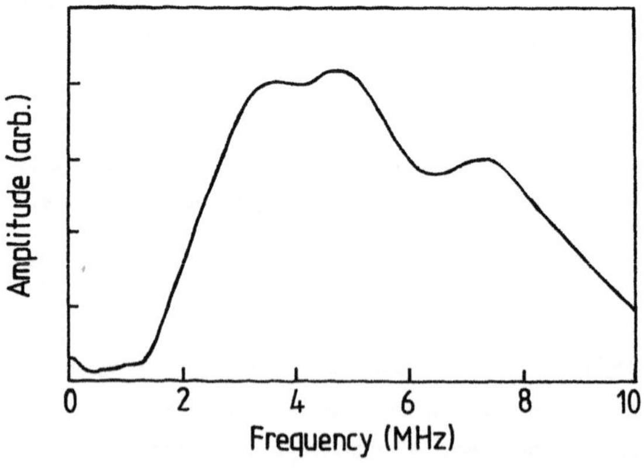

Fig. 4. Frequency response of instrumentation system

A focussed immersion transducer could be scanned over a sample, and was excited with a Panametrics 5055 pulser/receiver. The pulse–echo waveforms were recorded digitally, and processed on an IBM Model 30 microcomputer. Fig. 3 shows the foam specimen, but the same apparatus was used for all samples.

An initial experiment was conducted to determine the frequency response of the system, by recording an echo from a large flat block of aluminium. The result is shown in Fig. 4. The block of aluminium was then replaced by a composite multilayer specimen, of similar dimensions to that described in theory above. The pulse–echo spectral response is given in Fig. 5(a). Note that it contains minima, as expected, but that these are not as sharp as predicted theoretically. This is due primarily to attenuation within the epoxy layer. The computer model is, however, able to predict such spectra. An initial result for the calculated spectrum is shown in Fig. 5(b), obtained by multiplying the theoretical reflection coefficient curve with the measured spectrum of the instrumentation (Fig. 3). Attenuation may now be considered, by setting the imaginary part of the epoxy wave velocities Cp and Cs to be 4% of the real part (i.e. to use a complex velocity). The spectrum of Fig. 5(c) is then obtained, showing reasonable agreement with experiment.

Having shown that theory predicted experimental results effectively, it was now possible to identify specific minima in experimental spectra as being associated with certain layers within the sample. Any change in the structure, due to a delamination for example, would alter the amplitude and/or frequency of a particular resonance. This technique has been compared to a conventional C–scan, using a sample of two 0.9 mm thick aluminium layers, bonded by 0.6 mm of epoxy adhesive. The sample contained an artificial delamination between the epoxy and top aluminium layer. The result of the time–domain C–scan, Fig. 6(a), was in the form of a contour plot, the contours representing various levels of amplitude. A clear delamination at the centre, as well as poor bonding to the right, can be observed.

From theory and trial experiments, it was found that this sample had an epoxy resonance at 1.85 MHz. An unfocussed 2.2 MHz immersion transducer was thus selected for spectroscopic experiments, and scanned over an area of 50 mm by 90 mm, with a resolution of 5 mm. At each position of the transducer, the time waveform was transferred to the microcomputer, but only the resonant frequency and the spectrum amplitude recorded permanently. As explained in the Introduction, the resonant frequency would appear as a maximum if the specular echo (S) was removed, and this mode of detection was used in these experiments. The maximum amplitude in the 1.75 – 1.95 MHz band was plotted as a function of position, as shown in Fig. 6(b). The delamination is evident, as is the anomoly on the right side of the plate. (Note that in

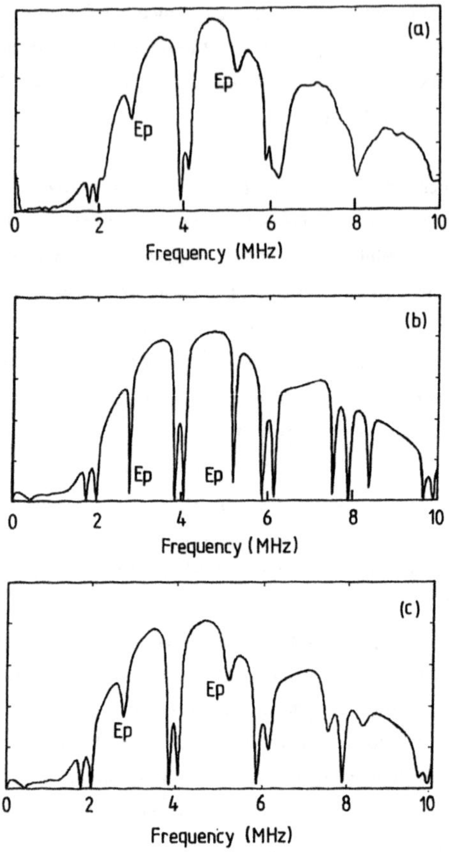

Fig. 5. Pulse-echo spectral response of an aluminium (1.6mm)/epoxy (0.48 mm) composite sample, showing (a) experimental spectrum, (b) theoretical result, with no attenuation, (c) attenuation added to theory.

Fig. 6(b), the data has been inverted, so that a lack of cohesion of the epoxy layer appears as a maximum amplitude).

Such an analysis was also undertaken on the foam composite panel, which contained a circular artificial delamination of 50 mm diameter, between the aluminium layer and the foam core. A 7 MHz immersion transducer was used, to evaluate the resonance of the upper aluminium layer. Again, the main specular echo was removed by signal processing. With the foam attached, the aluminium resonance (at ~ 5 MHz) was highly damped, as is evident from the received spectrum of Fig. 7(a). At the position of a delamination,

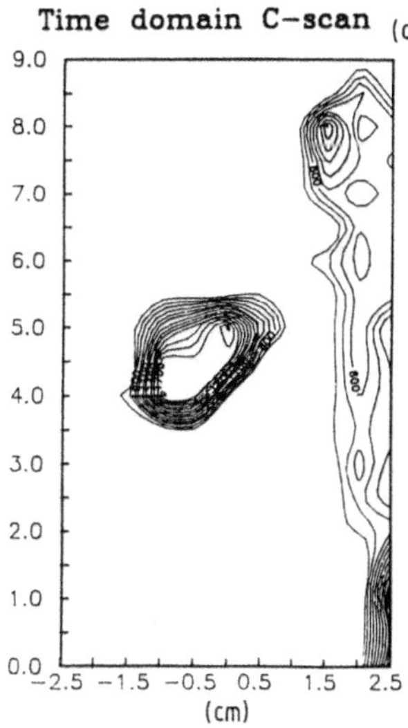

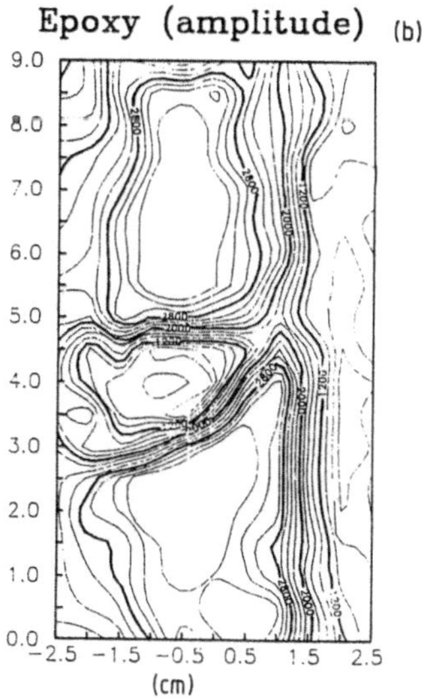

Fig. 6. (a) C-scan of an aluminium (0.9 mm)/epoxy (0.6 mm) composite sample, showing a delamination. (b) A scan of the same sample using a frequency domain technique.

however, clear resonances were observed, Fig. 7(b). A horizontal scan was again performed, the results for amplitude of resonance being presented in Fig. 8. It is evident that the circular delimination has been detected effectively.

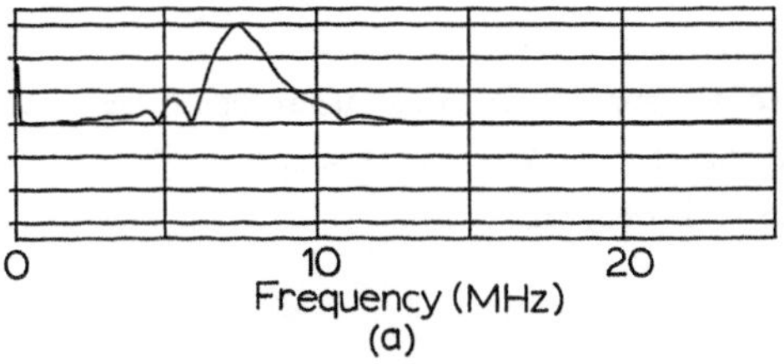

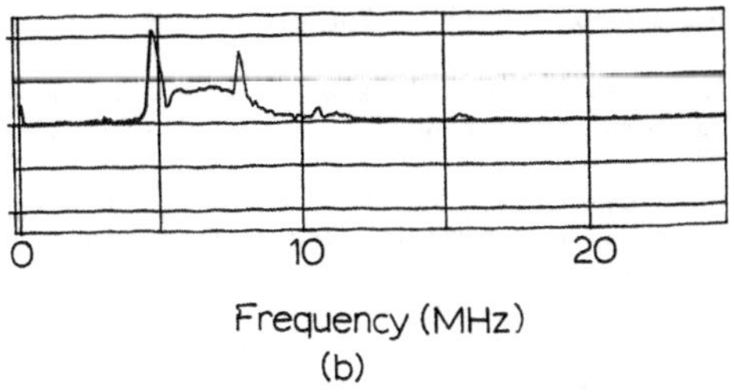

Fig. 7. Spectra for a foam composite panel, over (a) a competent bond, and (b) over a delamination.

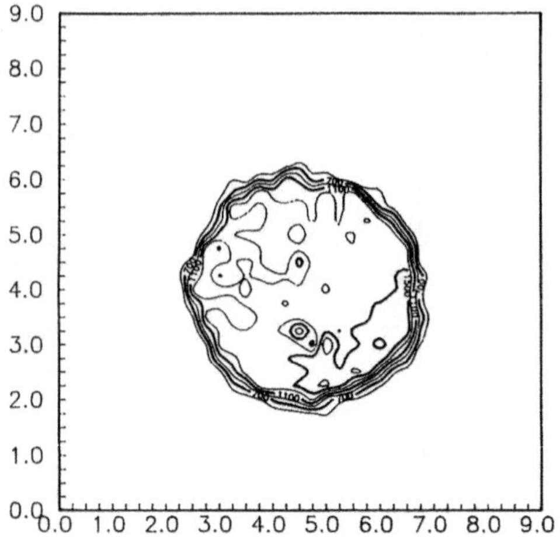

Fig. 8. Result of a scan in the frequency domain, showing an artificial circular delamination.

Conclusions

We have demonstrated that resonance techniques are able to detect delaminations in layered composite material containing aluminium. Further work will extend the analysis to shear waves, of interest in the prediction of bond strength.

Acknowledgements

This work was funded by grants from Alcan International Inc., and via a contract from the Department of Supply and Services, Government of Canada. Thanks are due to Mr. W.R. Sturrock in supporting this work.

References

1. Vary, A. and Bowles, K.J.; An ultrasonic-acoustic technique for nondestructive evaluation of fibre composite quality. Polymer Eng. Sci. 19 (1979) 373-377.

2. Gruyott, C.C.H.; Cawly, P.; Adams, R.D.: The nondestructive testing of adhesively-bonded structures: A review. J. Adhes. 20 (1986) 129-159.

3. Adams, R.D. amd Cawly, P.: Vibration techniques in non-destructive testing: In Research Techniques in Nondestructive Testing, R.S. Sharpe (ed.), Vol.8 (Academic, New York, 1985), 303-360.

4. Guyott, C.C. and Cawley, P. : The ultrasonic vibration characteristics of adhesive joints. J. Acoust. Soc. Am. 83(1988) 632-640.

Fibre and Particle Reinforced Metal Matrix Composites: Structure - Production - Properties

Gerhard Ibe, Vereinigte Aluminiumwerke AG, Bonn

1. Introduction: the principles

Anorganic solid bodies possess a theoretical lattice strength of about 10 % of the elastic modulus E, i.e. between 5 and 50 GPa. Metals however attain only about 10% of these values [1]. This is due to the defects in each solid body. These defects can be dislocations which cause plastic deformation and ductile fracture or crack nuclei which lead to premature brittle fracture in non-plastic bodies. Without such microscopic cracks brittle bodies can attain very high strengths as was shown by A.A. Griffith already in 1920 [2]. He found a strong increase of strength of thin glass fibres with decreasing thickness (Fig 1), and concluded that the number of defects per unit length decreased with decreasing diameter. This led him to the so-called fibre paradox: "the thinner the stronger". This can be demonstrated in a simple example [3]:

In a brittle cube shaped body of cage length l and of defect density the volume of one defect is in the average (Fig. 2a)

$$V_1 = 1/\rho_d \qquad (1)$$

and the average defect distance

$$l_1 = V^{1/3} = 1/(\rho_d^{1/3}) \qquad (2)$$

For a defect density of ρ_d = 100/cm^3, a very low defect free distance of

$$\underline{l_1 = 0{,}215\ cm}$$

results.

For the body in the form of a long fibre with the thickness d (Fig. 2b), the volume of one defect is

$$V_1 = (\pi/4) * d^2 * l_1 \qquad (3)$$

and with equation (1) the defect free length

$$l_1 = 4/(\pi * d^2 * \rho_d) \qquad (4)$$

For a fibre thickness of d = 5 µm and the same defect density, the defect free length is considerably higher

$$L_1 = 509{,}3\ m$$

by a factor of $2{,}5 \times 10^5$.

This simple principle is the basis of all fibre reinforced composites.

The tensile strength in length direction of a continuous fibre reinforced composite obeys a simple "rule of mixture" (ROM) [4]:

$$\sigma_{BC} = \phi_F * \sigma_{BF} + (1-\phi_F) * \sigma^*_M \qquad (5)$$

for $\phi_F >= \phi_{F,min}$

which describes a linear dependence of the composite strength from the fibre volume fraction (Fig. 3). After this a real fibre reinforcement exceeding the matrix strength occurs only for fibre fractions above the "critical" fibre fraction .

For the reinforcement with short fibres the same rule of mixtures holds, if σ_{BF} is replaced by an effective fibre strength [5]:

$$\sigma_{BF}(eff.) = \sigma_{BF} * [1-(\sigma_{BF} * d)/(2 * l * \sigma^*_M)] \qquad (6a)$$

for $l > l_c$

resp.

$$\sigma_{BF}\ (eff.) = (\sigma^*_M * l)/(2 * d) \qquad (6b)$$

for $l < l_c$

with the critical fibre length

$$l_c = (\sigma_{BF} * d)/\sigma^*_M \qquad (7)$$

Here l_c is the length below which the full fibre strength can not yet be reached by the shear stresses transferred from the matrix (Fig. 4).

For very short fibres and particles the real fibre reinforcement will be supplied and/or replaced by an induced hardening of the matrix itself. This enhancement of the flow stress is composed of several contributions [6]:

$$\Delta\sigma_{yM} = \Delta\sigma_M(\phi_F) + \Delta\sigma_M(T) + \Delta\sigma_M(O) + \Delta\sigma_M(HP) \qquad (8)$$

where the RHS terms denote the hardening due to the restriction of the macroscopic deformation to the matrix, the hardening due to deformation by quenching stresses, the hardening by particles after the Orowan mechanism, and the fine grain hardening after Hall-Petsch, resp.. This effect can be seen by comparison of the measured and the calculated strength of a short fibre reinforced material. The resulting stress enhancement is due to the induced matrix hardening.

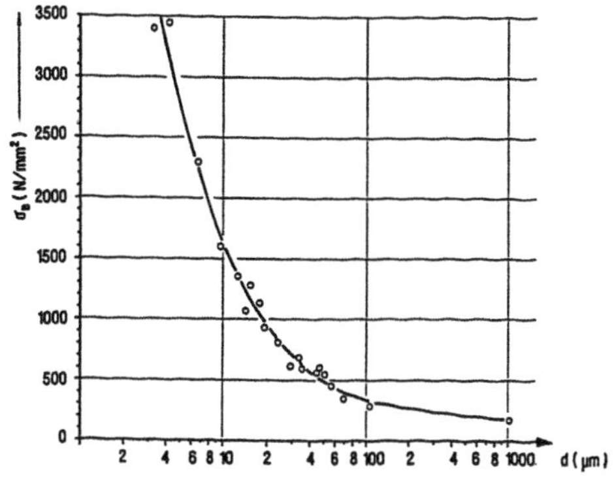

Fig. 1: Tensile strength of glass fibres as function of their thickness (after [2])

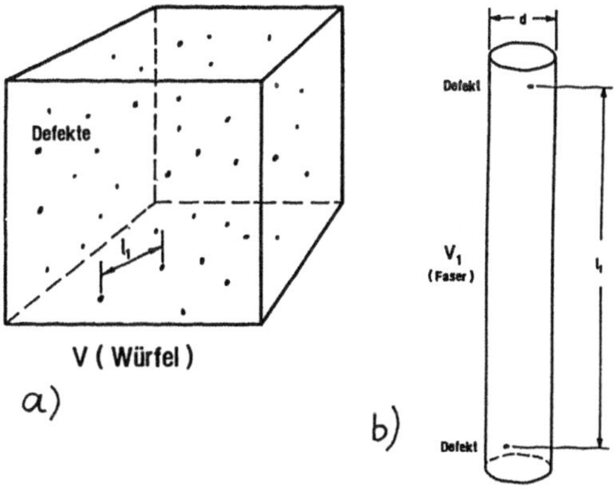

Fig. 2a: Distribution and distance of defects in compact volume (after [4])

Fig. 2b: Distance of defects in fibre volume with equal defect density (after [4])

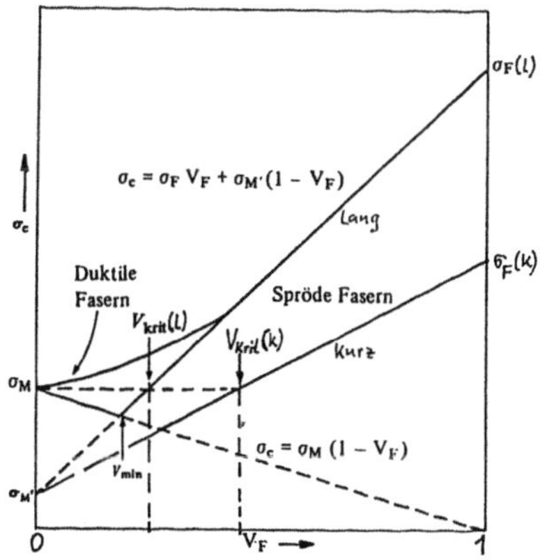

Fig. 3: Tensile strength of continuous fibre reinforced materials as function of the fibre content (rule of mixtures / ROM) (after [1])

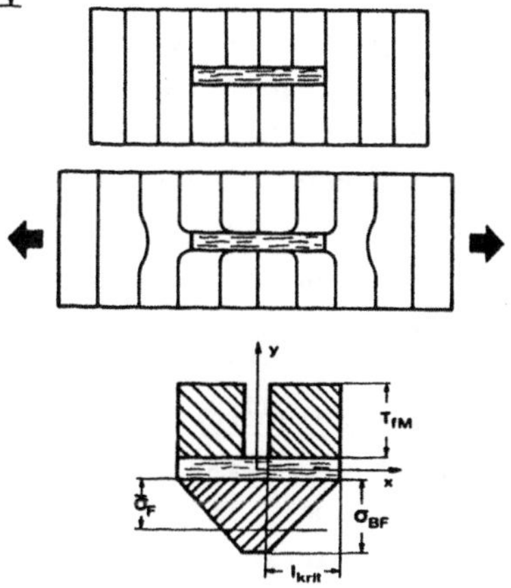

Fig. 4: Built up of tensile stress in the fibre by transfer of shear stress through the matrix-fibre interface (after [5])

After these principles many kinds of metal matrix composites have been developed, all of which can be reduced to three basic types:

-- continuous fibre reinforced sheet and tapes

-- long and short fibre reinforced cast parts

-- short fibre and particle reinforced semiproducts

Structure, production and properties of these materials will be demonstrated in the following with a few examples.

2. Fibre reinforced sheet and tapes

Sheet and tapes reinforced with continuous fibres - as a basic material for high strength and stiff spatial structures - can be produced by the process of plasma spraying a high temperature Al smelt onto a single layer winding of fibres on a thin Al supporting foil. Another method more protective for the fibres consists of hot pressing or rolling a fabric of fibres between braze plated Al sheet or tape. By hot pressing also multilayer composites in unidirectional or cross ply arrangement can be made. The roll plated braze cladding of the Al matrix sheet allows the oxide free bonding and welding of fibres and sheet (Fig. 5). This method has been converted to a continuous hot rolling technique, in which a layer of fibre fabric between braze plated Al tapes passes a preheating facility, enters the roll as a stack, and leaves it as a ready composite (Fig. 6). Because of the good directional order and good bonding of the fibres these composites reach high values of strength and elastic modules as is shown in Table 1. The strength approaches 1500 MPa, and the elastic modules 200 GPa, i.e. in the range of steel.

3. Fibre reinforced cast parts

Mechanically fixed or pre-sintered fibre preforms can be high pressure smelt infiltrated by different methods as for instance squeeze casting or high pressure die casting. In these methods the wetting behaviour is not so critical, because wetting and bonding can be achieved by a later thermal processing.

The fibre preforms for this technique were cut from sintered short fibre plates, or built up from yarn or fabric of continuous fibres, and became fixed by sintering with small additi-

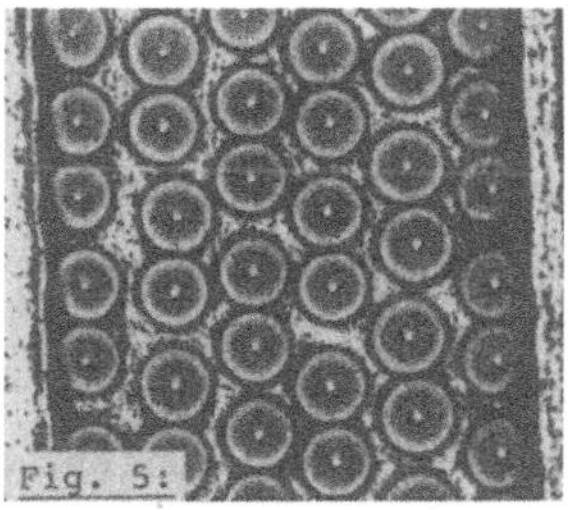

Hot pressed composite sheet with 6 layers of SiC fabric between AlMn1 sheets rollplated with AlSi10; cross section 100x (after [7])

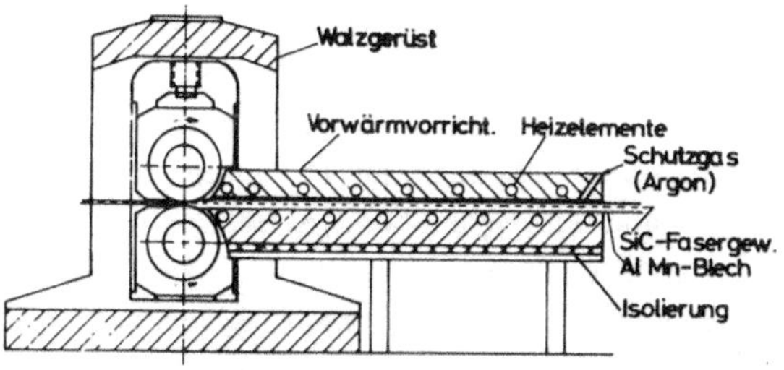

Fig. 6:
Equipment for hot rolling of fibre fabric between braze plated Al sheets, schematically (after [8])

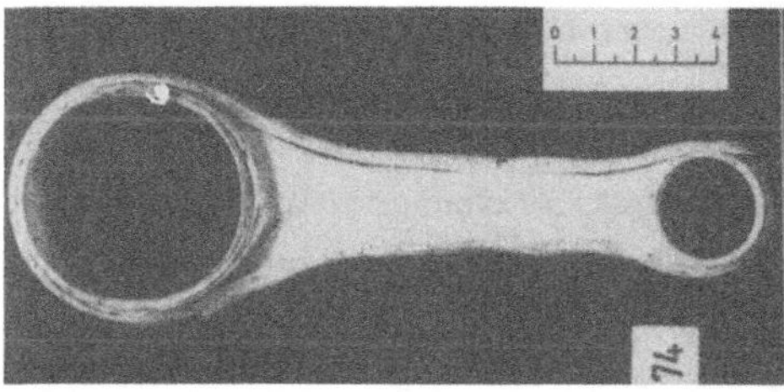

Fig. 7:
Connecting rod from AlSi casting alloy with Al_2O_3 fibre reinforcement, cast after the VACURAL PROCESS, length ca. 190 mm (after [3])

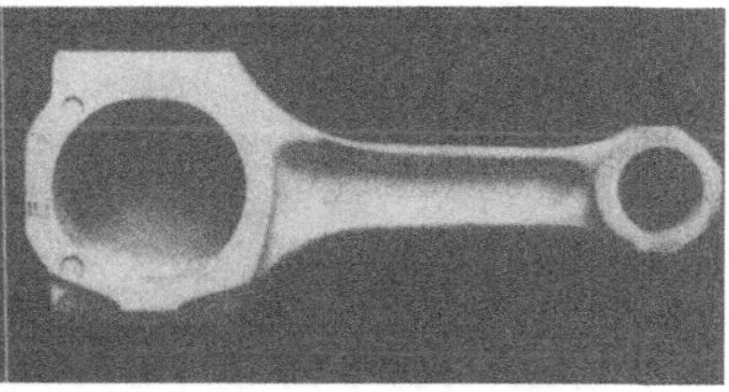

Fig. 8:
Al_2O_3 fibre preform for Al connecting rods, eyes and belt from continuous fibres, the web from short fibre preform (after [3])

Table 1

Examples of mechanical properties of Al and Mg alloys reinforced with continuous SiC,- Al_2O_3-, and C-fibres (after [12])

matrix	fibre	content (vol.%)	production	E (GPa)	R_m(l.d.) (MPa)	R_m(cr.d.) (MPa)
AlMgSiCu (6061)	SiC/C (SCS)	47	plasma spraying and hot pressing	210	1500	90
AlSi7Mg (A357)	SiC/C (SCS)	50	as cast	230	1700	105
AlMgSiCu (6061)	SiC (Nicalon)	35	"	110	900	50
Al-alloy	C	<50	"	210	700	50
AlCuMg (2024)	Al O (Safimax)	50	"	185	1000	-
AlMn1 (braze plated)	SiC (SIGMA)	50*)	hot rolled	~200	1350	ca. 120
MgZn4RE 1,2Zr (ZE 41)	SiC/C (SCS)	50	as cast	230	1350	-

*) recalculated to 50 vol.% fibers with equal "efficiency" of fibre properties

ons of anorganic binder. This technique has been applied to the production of fibre reinforced Al connecting rods (Fig. 7) by vacuum die casting in cold chamber die casting equipment. The fibre preform was built from continuous fibre belts for the shaft and the eyes, and from short fibre inserts for stiffening the shaft (Fig. 8). The bonding and the distribution of the fibres in the load bearing cross sections are good, as was proved by metallographic investigations (Fig. 9).
With this structure remarkable strength enhancements could be reached, as was proven by compression test of the connecting rods (Fig. 1o). The elastic limit was raised from 57 kN for an unreinforced part to 149 kN for a connecting rod with 22 Vol.% of Al_2O_3 fibres [3]. The fatigue strength is not yet sufficient due to the not yet finished optimization of the amount and arrangement of the fibres. The weight saving compared to steel connecting rods is about 40%.

4. Short fibre reinforced semi products

The starting material for workable semi products from fibre reinforced light metals can be produced as mentioned before, or by pressureless introduction of short fibres into the melt, for instance by stirring or injection techniques. Such material can be hot extruded to bars or simple profiles. In this process the fibres become oriented in longintudinal direction and reach a nearly unidirectional reinforcement (Fig. 11). Due to the high deformation strains and stresses in the conventional hot extrusion process the fibres exceed their critical length and pass through a multiple fracture process. The remaining fibre length is then below the critical length after equation (7), and full reinforcement is not longer possible. This can only be overcome by new forming processes in the semi liquid state with drastically reduced deformation stresses.
Despite this handicap this material shows a remarkable enhancement of the strength at higher temperatures compared with conventional Al alloys for higher temperature applications (Fig. 12). The lower reinforcement by short fibres follows directly from the rule of mixtures after equation (5 and 6b). If one replaces the fibre strength in Fig. 3 by the effective strength after equation (6b) the line of composite strength runs more

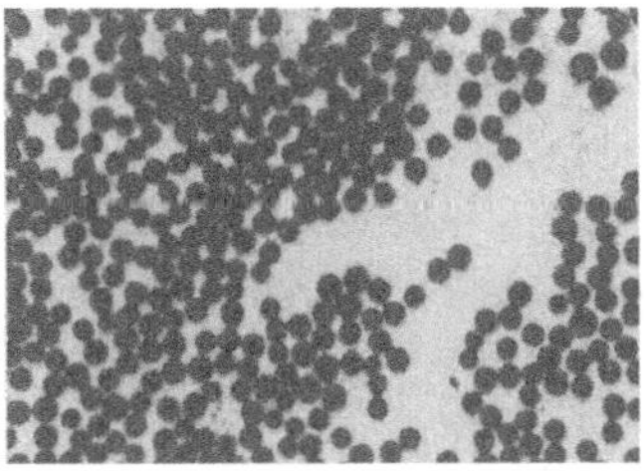

Fig. 9: Al connecting rod with Al_2O_3 continuous fibre reinforcement; cross section 100x (after [8])

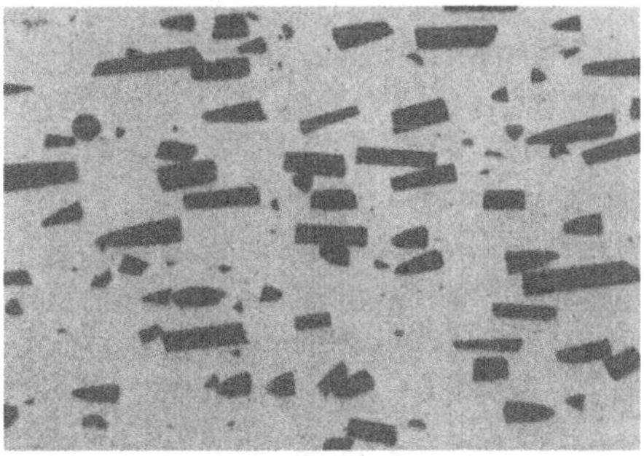

Fig. 11: Fibre distribution in an Al extrusion bar with 16 vol.% Al_2O_3 fibres (FP); length section 200x (after [8])

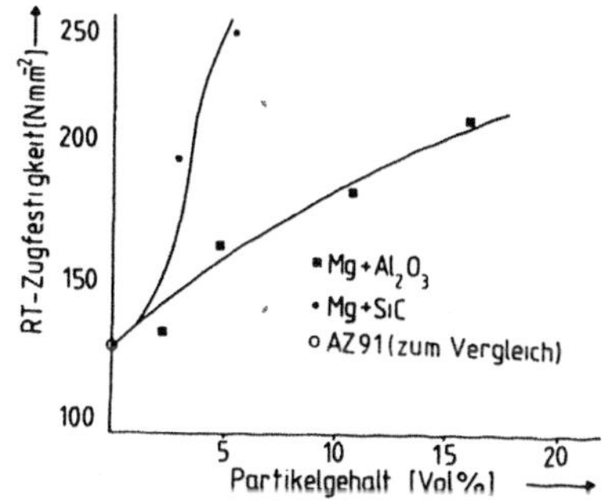

Fig. 13: RT tensile strength of particle reinforced Mg composite as function of the particle content (after [10])

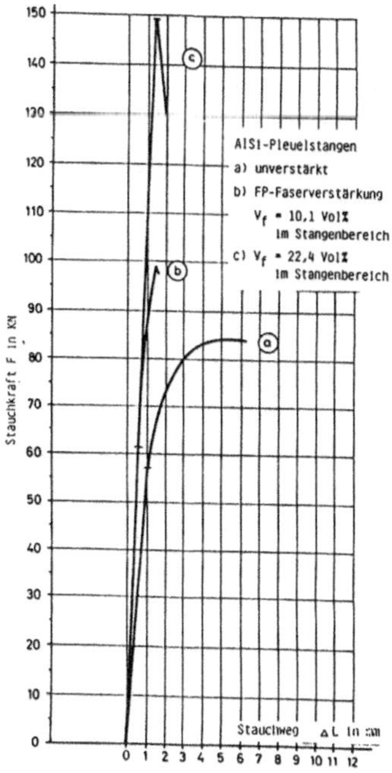

Fig. 10: Load-compression diagram of the connecting rods as in fig. 7 (after [3])

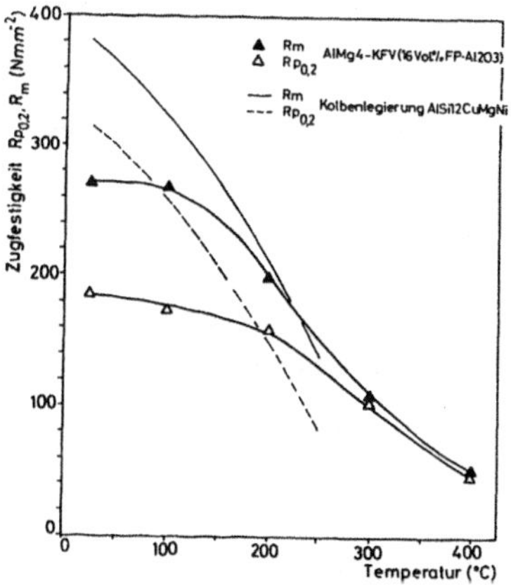

Fig. 12: Hot tensile strength of fibre reinforced Al extrusion bar after fig. 11 in comparison to a hot resistant Al piston alloy (after [7])

flatly with increasing fibre fraction. Then a much higher fibre volume fraction is necessary to reach a real reinforcement.

5. Particle reinforced semi products

Based on the experience that in conventional extrusion the short fibres undergo multiple fracture and that other strengthening mechanisms of the matrix become activated - cf. equation (8) - numerous experiments have been undertaken to reinforce Al alloys with particles of SiC, Al_2O_3 and other ceramic substances (9). The particles can be introduced into the matrix either during smelting or by powder metallurgy. Particle contents of between 10 and 30 Vol.% have been found to increase the strength up to 25%, and the elastic modules up to 40%, and to increase dramatically the wear resistance (Table 2). The particles must be extremely fine and evenly distributed and strongly bonded in the matrix. With good bonding of the particles to a matrix of Mg alloy with its higher reactivity the high temperature strength of the composite exceeds remarkably the matrix strength (Fig. 13). The elastic moduls is less sensitive to the binding and geometry of the fibre or particle reinforcement, and is in most cases more easily enhanced than the strength. An additional increase in strength is given by the induced hardening mechanisms of the matrix itself as in short fibre reinforced materials.

6. Quantitative microstructure analysis

The properties of a composite do not only depend on the material of the matrix and the kind and amount of the fibres but also on their spatial arrangement and orientation. These parameters should therefore influence the resulting signals of non destructive materials testing methods, e.g. with ultrasonics. Therefore first efforts for the microstructure evalution by the methods of quantitative metallography have been undertaken [11]. The distribution of fibres in the cross section of an unidirectional reinforced sample (cf. Fig. 9) has been compared with theoretical distributions of full ordering, of full disordering, and with extreme clustering. The result was a small cluster-parameter of 5.9%, i.e. a corresponding deviation from a statistical distribution in the direction of cluster formation.

7. Final remarks

Fibre reinforced metal matrix composites represent high performance materials with very heterogenuous microstructure. They are characterized by large difference in the thermo-elastic and transport properties of their components, e.g. for the elastic moduli, the thermal expansion coefficients, the electrical and the thermal conductivity. These differences form a good basis for the testing of the microstructure and its defects with non destructive testing methods.

8. References

[1] A.Kelly, N.H.Macmillan: "Strong Solids"; Clarendon Press, Oxford, 3. Aufl. 1986

[2] A.A.Griffith; Phil.Trans.Roy.Soc. 221A (1920) 163

[3] G.Ibe, J. Penkava; METALL 41 (1987) 590

[4] A.Kelly, W.R.Tyson; J.Mech.Phys.Solids 13 (1965) 329

[5] G.Ibe: "Grenzen der Kurzfaser- und Partikel-Verstärkung von Aluminiumwerkstoffen"; in DGM-Tagung "Verbundwerkstoffe und Stoffverbunde in Technik und Medizin", 8.-10.6.1988; will be published

[6] P.M. Kelly: "The Quantitative Relationsship between Microstructure and Properties in Two-Phase Alloys"; Int.Met. Reviews 18 (1973) 31

[7] G.Ibe, W. Gruhl, H.Gebhardt: "Untersuchungen zur Verstärkung von Aluminiumlegierungen mit keramischen Lang- und Kurzfasern"; in W.Bunk, J.Hansen (Herausg.): Faserverbundwerkstoffe (Vol.1)"; Springer-Verlag Berlin 1980, P. 27

[8] G.Ibe, J.Penkava: "Entwicklung von Verfahren zur Herstellung von faserverstärkten Aluminiumlegierungen"; in J.Hansen (Herausg.):"Faserverbundwerkstoffe, Band 2"; Springer-Verlag Berlin 1985,P.77

[9] S.V.Nair, J.K.Tien, R.C.Bates,Int.Met.Reviews 30(1985)275

[10] G.Ibe, J.Penkava, E.Tank, K.U.Kainer, F.Hage: Entwicklung faser- und partikelverstärkter Leichtmetalle"; in "Symposium Materialforschung 1988" des BMFT, Hamm 12.-14.9.1988, KFA Jülich,PLR,1988, Vol.2, P.1099

[11] M.Clark, H.Cordier, G.Ibe: "Kontrolle der Abstandvertei-teilung von Fasern in Aluminiumwerkstoffen mit automatischen Bildanalysemethoden"; in "6. Arbeitstagung Quantitative Bildanalyse" der KFA Jülich, 20.-21.6.1988, will be published

[12] Producer Information: AVCO Speciality Materials Division, Lowell, Mass., USA; Du Pont Co., Textile Fibers Dept., Wilmington, De., USA; Nippon Carbon Ltd., Tokyo, Japan; ICI, Chemicals & Polymers Group, Runcorn, Cheshire, UK

[13] Producer Information: DWA Composite Specialties Inc., Chatsworth, Ca., USA; DACC Dural Aluminium Composites Corp., San Diego, Ca., USA

Table 2

Examples of mechanical properties of Al and Mg alloys reinforced by particles (after [13])

matrix	fibre	content (vol.%)	pro-duc-tion	E (GPa)	R_m(l.d.) (MPa)	A (%)	R_m(cr.d.) (MPa)	K_{IC} ($N/mm^{3/2}$)
AlCuMg (2014-T6)	-	0	I/M	72	490	13	-	780
AlCuMg (2014-T6)	SiC-P	20	"	108	500	2,5	-	600
AlZnMgCu (7075-T6)	-	0	"	71	580	11	-	780
AlZnMgCu (7075-T6)	SiC-P	15	"	100	610	2,5	-	630
AlMgSiCu (6061-T6)	SiC-P	15	"	96	355	3,0	363	-
AlMgSiCu (6061-T6)	SiC-P	15	P/M	98	460	7,5	-	-
AlMgSiCu (6061-T6)	SiC-P	25	"	116	540	3,5	-	-
AlMgSiCu (6061-T6)	SiC-P	40	"	147	560	1,75	-	-
AlCuMg (2124)	SiC-P	25	"	116	615	4,5	-	-
AlZnMgCu (7090)	SiC-P	25	"	117	805	2,0	-	-

Ultrasonic Characterization of SiC-Reinforced Aluminium

D. F. Lee, K. Salama and E. Schneider*

Mechanical Engineering Department, University of Houston
Houston, TX 77004, USA

*Fraunhofer Institute for Nondestructive Evaluation, Saarbrücken, FRG

ABSTRACT

The nondestructive characterization of the elastic and the anelastic behavior of metal-matrix composites is essential to their manufacturing processes. The effects of volume fraction of second-phase SiC-particles on the second- and third-order elastic constants have been investigated in the aluminum alloys 8091, 7064, and 6061. The influence of second phase on the ultrasonic absorption and dispersion has also been studied. For relatively small amount of reinforcement (20 volume percent), the second-order elastic constants are found to increase linearly with SiC content. The measurements also show a high degree of anisotropy along the extrusion direction where the constants have the highest values. The third-order elastic constants, l and m are found to increase with increasing SiC content, while n decreases with SiC content. As expected, the presence of SiC increases the absorption coefficient in these composites, and indicates that the reinforcement is a dominant factor in the anelastic behavior of these materials. It is also shown that the dispersion measurements can be used to estimate the size of the second-phase scatterers which influences the physical and mechanical properties of the composites.

1. INTRODUCTION

Metal-matrix composites are emerging as high-performance materials which combine the advantages of lesser weight and improved mechanical and thermal properties. The properties of these composites are controlled by the mechanical and chemical interactions between the two phases. In addition, the interfacial bonds also influence macroscopic properties such as elastic constants, ultrasonic velocities and scattering. The properties of the composite are also controlled by the properties of the two phases as well as the percentage, size, shape, distribution and orientation of the reinforcement.

In the development and manufacturing of MMC, it is essential that elastic and mechanical properties can be determined nondestructively. Ultrasonic techniques seem to hold high promise in the characterization of two-phase materials. The measurement of ultrasonic velocities enables the evaluation of second-phase content, and ultrasonic scattering reveals informations on the size and shape of the reinforcement. Third-order elastic constants and the temperature dependence of ultrasonic velocities are under investigation in order to characterize the residual stress states. Improvements on the absorption measurement also enable the investigation of the anelastic interaction between the ultrasonic wave and the material.

The objective of this investigation is to study the influence of SiC-particles on the second- and third-order elastic constants and on the anisotropy of the composites. In addition, ultrasonic dispersion and absorption are studied in order to demonstrate their potential on the characterization of MMC.

2. EXPERIMENTAL

2.1 Samples

The experiments are performed on specimens of the Al-alloys 8091, 7064 and 6061 containing up to 20% volume fraction of SiC particles. The chemical compositions of the aluminum alloys and the volume percentage of SiC in specimens used in this study are given in Table 1. The original materials are received as extruded rods of 25 mm in diameter and extruded bars of 20 mm in height and 60 mm in width. Prismatic specimens of 15 to 18 mm wide and 30 mm long are machined such that the length direction is parallel to the extrusion direction. Opposite faces are made flat and parallel to within 0.025 mm accuracy. Figure 1 displays a sketch of the specimen. Also shown in the figure are the directions of applied stress and propagation and polarization of the ultrasonic waves used in the measurements.

	Alloying Elements									
Alloy	Si	Fe	Cu	Mg	Zr	Li	Zn	Cr	Co	Al
8091	0.02	0.01	1.90	0.80	0.11	2.70	—	—	—	rem
7064	0.05	0.10	2.00	2.30	0.20	—	7.10	0.12	0.22	rem
6061	0.71	0.29	0.21	0.86	—	—	0.10	0.06	—	rem

Specimens		
8091 + 0% SiC	8091 + 10% SiC	8091 + 15% SiC
7064 + 0% SiC	7064 + 15% SiC	7064 + 20% SiC
—	6061 + 15% SiC	6061 + 20% SiC

Table 1: Chemical compositions of aluminum alloys and volume percentage of SiC reinforcement of the specimens.

2.2 Second-Order Elastic Constants

In order to evaluate the second-order elastic constants, the density and the ultrasonic velocities are to be determined. The density is measured using the Archimedes' method, and the results are found to be reproducible to within 0.3%. Ultrasonic velocities are determined from measurements of the time-of-flight of 10 MHz longitudinal and shear waves using the pulse-echo-overlap technique. A detailed description of this method and the experimental set-up used in this investigation is given elsewhere.[1]

The second-order elastic constants are calculated using the relationship $C_{ij}=\rho V_{ij}^2$ where ρ is the density, and V_{ij} is the velocity of a wave propagating in the direction i and polarized in the direction j.

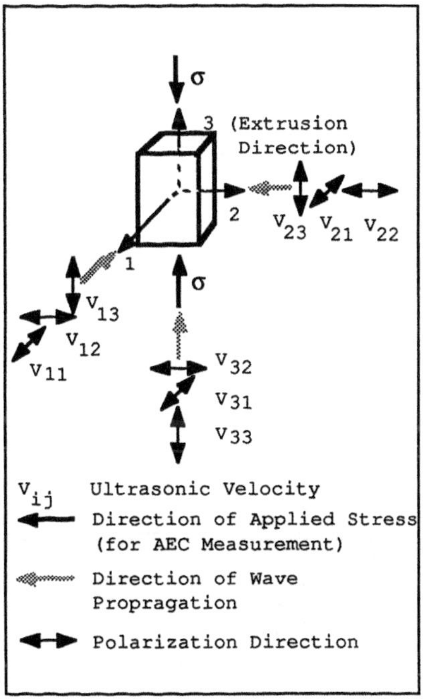

Figure 1: Velocity designations and direction of applied uniaxial compressive stress used in velocity measurements.

The bulk modulus K is calculated using the relationship $K = [C_{11} + C_{22} + C_{33} + 2(C_{12} + C_{13} + C_{23})] / 9$ where $C_{12} = C_{11} - 2C_{66}$, $C_{13} = C_{33} - 2C_{55}$, and $C_{23} = C_{22} - 2C_{44}$. Table 2 contains values of the stiffness constants C_{11}, C_{22}, C_{33}, C_{44}, C_{55}, and C_{66}. Also included in Table 2 are the values of K. Using the constants given in the table, the Young's moduli for the directions perpendicular and parallel to the extrusion direction are calculated using the relationship $E=[\mu(3\lambda+ 2\mu)]/(\lambda+\mu)$ where μ and λ are the Lamé constants given by $\mu = \rho V_s^2$, $\lambda = \rho V_l^2 - 2\mu$, and V_l and V_s are the longitudinal and shear velocities, respectively. Values of the Young's modulus for the composites are shown in Table 3, and the variation of the Young's modulus in Al-7064 specimens is shown in Figure 2 as a function of volume percentage of SiC.

2.3 Elastic Anisotropy

In order to characterize the elastic anisotropy between the direction

Stiffness (GPa)	Specimens							
	8091+ 0%SiC	8091+ 10%SiC	8091+ 15%SiC	7064+ 0%SiC	7064+ 15%SiC	7064+ 20%SiC	6061+ 15%SiC	6061+ 20%SiC
C_{11}	102	114	118	108	127	129	116	128
C_{22}	102	114	118	108	128	132	123	138
C_{33}	105	120	129	107	132	143	131	142
C_{44}	30	35	38	27	35	38	33	38
C_{55}	30	36	38	27	35	38	33	39
C_{66}	30	35	37	27	34	38	32	38
K	63	69	71	72	83	84	80	85

Table 2: Second-order elastic constants of the 8091, 7064, and 6061 Al-MMC specimens.

Modulus (GPa)	Specimens							
	8091+ 0%SiC	8091+ 10%SiC	8091+ 15%SiC	7064+ 0%SiC	7064+ 15%SiC	7064+ 20%SiC	6061+ 15%SiC	6061+ 20%SiC
$E_{\perp}$	77	90	94	71	90	98	83	98
$E_{//}$	78	92	98	72	93	100	88	104
$G_{\perp}$	30	35	37	27	34	38	32	38
$G_{//}$	30	36	38	27	35	38	33	39

Table 3: Young's and shear moduli of the 8091, 7064, and 6061 Al-MMC specimens.

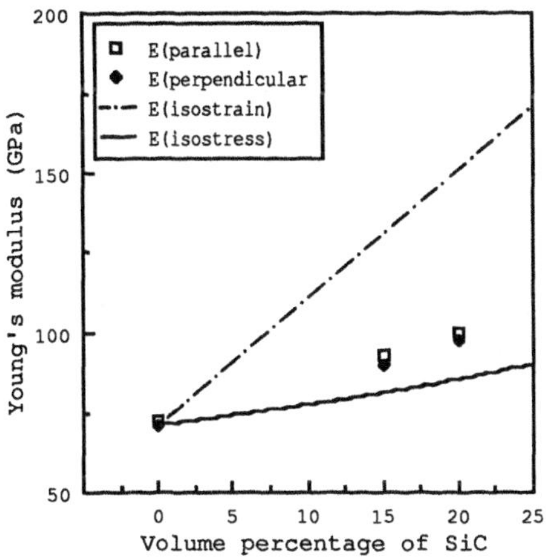

Figure 2: Young's modulus for the directions perpendicular and parallel to extrusion in the Al-7064 MMC specimens.

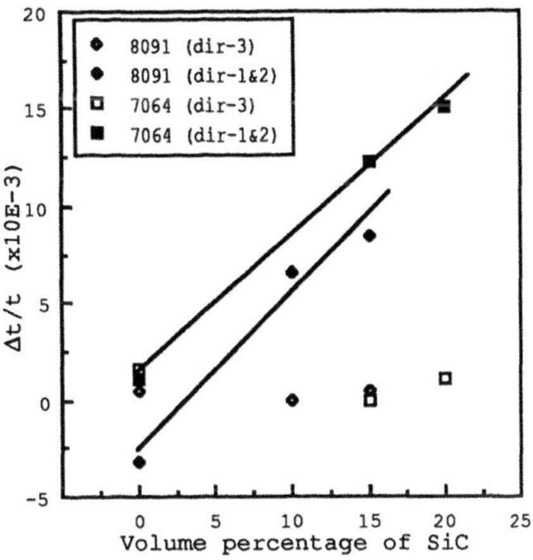

Figure 3: Birefringence effect in SiC-reinforced aluminum alloys.

of extrusion (direction 3) and those in the perpendicular plane, measurements of the time-of-flight of linear polarized 5 MHz shear waves are performed. The propagation directions are adjusted perpendicular to the extrusion direction while the polarization is kept along and perpendicular to the extrusion direction, respectively. The relative changes in the time-of-flight are measured with an accuracy better than 1 in 10^4 by rotating the transducer and thus the polarization direction by 90°. Since the two linear-polarized shear waves are propagated along the same path, there are no variations in the path length. Similar measurements are also performed with shear waves propagating along the direction of extrusion. The birefringence effect in the MMC's are displayed in Figure 3.

2.4 Third-Order Elastic Constants

The third-order elastic constants (TOEC) of the composites are determined from measurements of the acoustoelastic constants (AEC), namely, the changes of sound velocities as function of stress. Compressional stress is applied in the direction of extrusion, and ultrasonic waves are propagated in a direction perpendicular to the extrusion direction. The polarization directions are parallel and perpendicular to the extrusion direction, respectively. Assuming that the conditions for their evaluation are satisfied [2], the TOEC's l, m and n are calculated using the following relationships:

$$l = [3\ K(\lambda + 2\mu)\ /\ AEC_1] + [\lambda(m + l + 2\mu)/\mu] \quad (1)$$

$$m = [6K\ \mu\ /\ AEC_2] - [\lambda n\ /\ 4\mu] - \lambda - 2\mu \quad (2)$$

$$n = \{[24K\ \mu^2\ (AEC_2^{-1} - AEC_3^{-1})] - 12\lambda\ \mu - 8\mu^2\}\ /\ [3\lambda + 2\mu] \quad (3)$$

where AEC_1: AEC of longitudinal waves with polarization direction perpendicular to that of applied stress.

AEC_2: AEC of shear waves polarized in the direction of applied stress.

AEC_3: AEC of shear waves polarized in a direction perpendicular to that of applied stress.

The values of the TOEC's are listed in Table 4 where the subscript indicates the direction of wave propagation, and Figure 4 displays the variation of the TOEC's with SiC content in the Al-8091 MMC specimens.

2.5 Ultrasonic Absorption Coefficient

The ultrasonic absorption coefficients are determined using a new method based on ultrasonic reverberation in the specimens investigated. Ultrasonic pulses with a center frequency of 6 MHz and a length of about 15 cycles are transmitted into the specimen. After a given amount of time, amplitudes of the receiving signals are used to calculate the absorption coefficient. The accuracy in measuring the absorption coefficient is found to be about 10%. The technique is described in detail elsewhere.[3] Two A-scans showing the absorption effect in samples of 7064-Al alloy with 0%

TOEC (GPa)	Specimens							
	8091+ 0%SiC	8091+ 10%SiC	8091+ 15%SiC	7064+ 0%SiC	7064+ 15%SiC	7064+ 20%SiC	6061+ 15%SiC	6061+ 20%SiC
l_1	-54	-18	19	-169	-14	-52	-3	37
l_2	-14	86	47	103	100	99	171	200
m_1	-325	-332	-291	-404	-362	-349	-322	-325
m_2	-314	-293	-285	-313	-324	-269	-264	-292
n_1	-439	-498	-463	-557	-550	-519	-478	-547
n_2	-436	-434	-445	-472	-481	-452	-408	-467

Table 4: Third-order elastic constants of the 8091, 7064, and 6061 Al-MMC specimens.

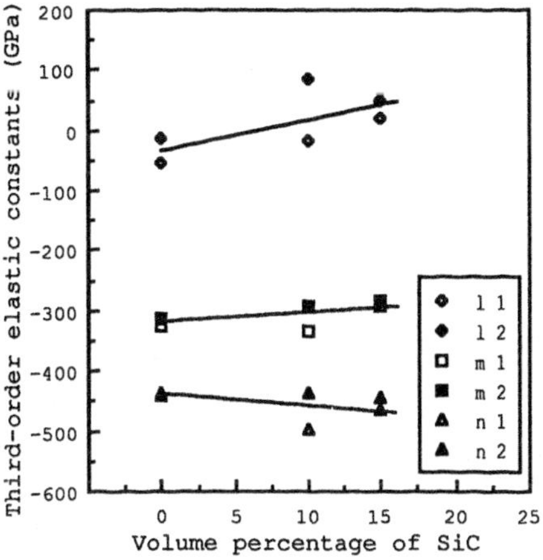

Figure 4:Variations of the TOEC's as functions of SiC content in the Al-8091 MMC specimens.

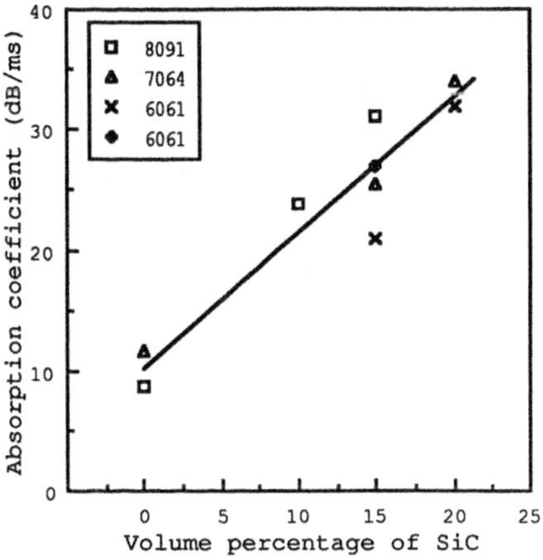

Figure 6:Absorption coefficient as a function of SiC content in Al-MMC specimens.

and 20% SiC reinforcement are shown in Figure 5. The absorption coefficients of the MMC specimens are evaluated and shown in Figure 6 as a function of SiC content.

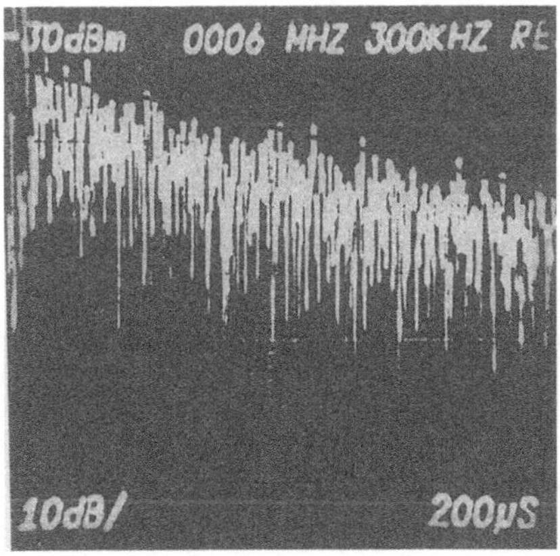

Al-7064 with 0 vol. % SiC

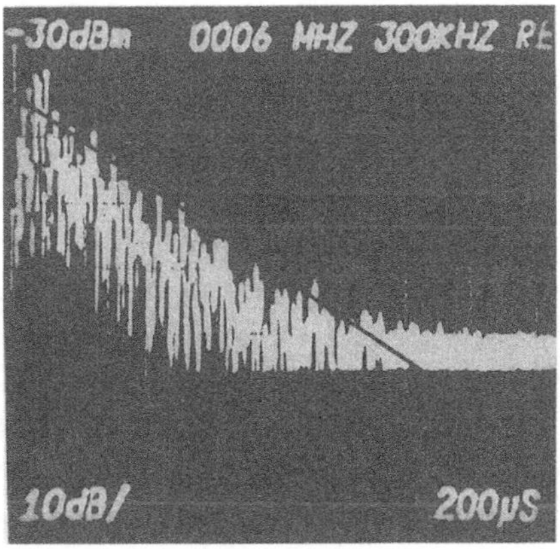

Al-7064 with 20 vol. % SiC

Figure 5:Absorption measurements in SiC-reinforced Al alloys.

2.6 Ultrasonic Dispersion

Ultrasonic scattering from the SiC-particles in the matrix results in the dependency of ultrasonic velocities on frequency. In order to estimate the scattering effect of SiC-particles in aluminum, the phase velocity of a longitudinal wave is calculated as a function of the wave number times the radius of the SiC-scatterers.[4] In this calculation, the matrix as well as the scatterers are assumed to be isotropic and homogeneous. The phase velocity, shown in Figure 7, is normalized to the longitudinal wave velocity of the matrix.

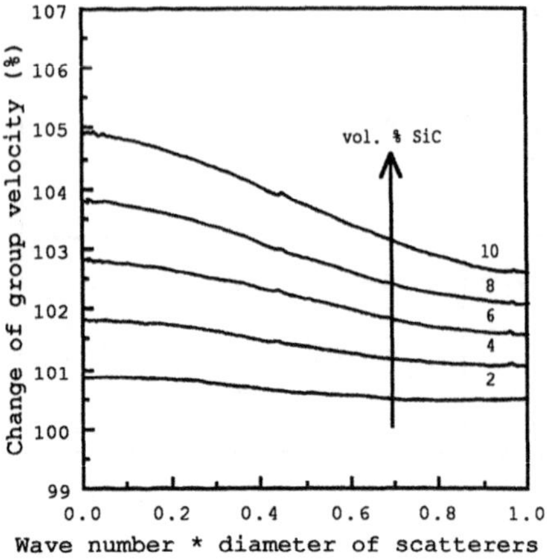

Figure 7: Ultrasonic dispersion in Al-SiC MMC specimens.

3. RESULTS AND DISCUSSION

3.1 Elastic Constants and Elastic Anisotropy

The elastic constants of a two-phase material depend on the elastic constants and the volume percentage of each phase in the composite, as well as the size, the shape, and the orientation distribution of the second phase. As seen from Table 2, the stiffness constants measured in the 1- and 2-directions are equal within the measurement's inaccuracy of 0.5%. The stiffness constants in the 3-direction (extrusion direction), however, are found to be 3 to 10% larger. This indicates that the samples are stiffer in the 3-direction than in the other two directions, and this tendency increases with the increasing amount of SiC in the composite.

From the engineering point of view, the Young's modulus and its change with volume fraction of SiC is of greater interest. From Figure 2, one can see that the Young's moduli in the alloys exhibit a linear relationship with the SiC content. The figure also shows the calculated Young's modulus considering isostrain, $E=E_{Al}F_{Al}+E_{SiC}F_{SiC}$ (upper limit) and isostress, $E=[E_{Al}E_{SiC}]/[E_{SiC}F_{Al}+E_{Al}F_{SiC}]$ (lower limit) bounds between the matrix and the reinforcing phase. The experimental results are found to be close to the lower bound condition of isostress and confirm earlier results [5] and numerical calculations.[6]

The results of the anisotropy measurements, shown in Figure 3, indicate a rotational symmetry around the extrusion axis. The birefringence effect (obtained using shear waves with polarizations parallel and perpendicular to the extrusion direction) is found to increase with the volume percentage of SiC. This indicates that the reinforcement is the dominant factor in influencing the anisotropy. This finding is in agreement with the results obtained in a recent investigation on the texture effect in SiC-reinforced Al composites.[7] In this study, it is found that the presence of SiC-particles changes the expansion coefficients of the orientation distribution function of the specimens, and the reinforcing particles influence the developement of the texture of the aluminum matrix.

3.2 Third-Order Elastic Constants

Third-order elastic constants are a measure of the nonlinear behavior of a material. The results given in Table 4 and Figure 4 shows linear changes in the TOEC's with increasing SiC content. l and m increase with an increasing amount of SiC, while n decreases with SiC content. Furthermore, l appears to be more sensitive to changes in SiC content. It is also seen from the figure that the TOEC's, calculated using ultrasonic waves propagating along the direction perpendicular to the extrusion, are equal within the measuring inaccuracy of about 10%.

3.3 Ultrasonic Absorption and Dispersion

The steady increase of ultrasonic absorption coefficient with SiC content (Fig. 6) is found in all the specimens. This is an indication that the content of reinforcement is the dominant factor in influencing the anelastic interaction between the ultrasonic waves and the SiC-particles and the metal-matrix.

Figure 7 shows the calculated dispersion curves for aluminum reinforced with 2 to 10% volume percentage of SiC-particles. It is found that the dispersion of the longitudinal phase velocity changes between 1 and 5%. This effect enables the evaluation of the SiC-particle size by measuring the phase velocity as a function of the ultrasonic frequency.

This scattering theory has been expanded sucessfully to account for the influence of SiC-ellipsoids on the dispersion of longitudinal and shear waves.[8]

4. CONCLUSION

In this investigation, the second- and the third-order elastic constants are determined in the SiC reinforced aluminum 8091, 7064, and 6061 composites. In addition, the influence of second-phase reinforcement on the ultrasonic absorption and dispersion has also been studied. For relatively low percentage of SiC reinforcement (up to 20 volume percent) in the aluminum alloys 8091, 7064 and 6061, the following conclusions can be drawn:

1. The second-order elastic constants increase linearly with the SiC content. Furthermore, the elastic moduli can be approximately described using the isostress condition.
2. Anisotropy measurements show that the MMC's are reinforced to a larger extend in the extrusion direction. Anisotropy in the MMC's increase with SiC content by the same amount in all the composites investigated. This steady increase with volume fraction of SiC indicates that the content of reinforcement is the dominant factor in influencing the anisotropy.
3. The third-order elastic constants, l, m and n, are sensitive to changes in the volume percentage of SiC. The constants l and m increase with the increase of SiC, while n decreases with the SiC content. In addition, l is more sensitive to changes in the SiC content.
4. The absorption coefficient in the composites increase in the presence of the SiC reinforcement, and the increase is of the same magnitude in all of the specimens investigated.
5. The ultrasonic dispersion changes in terms of the product of the SiC-particle size and the wave number. This enables the evaluation of the particle size by measuring the ultrasonic velocity as a function of frequency.

REFERENCES

1. K. Salama and C. K. Ling, "The Effect of Stress on the Temperature Dependence of Ultrasonic Velocity", J. Appl. Phys., 51(3), p.1505, 1980.
2. D. S. Hughes and J. L. Kelly, "Second-Order Elastic Deformation of Solids", Phys. Rev., 92, p. 1145, 1953.
3. H. Willems, "A New Method for the Measurement of Ultrasonic Absorption in Polycrystalline Materials", Proc. Review of Progress in Quantitative NDE, 6A, p.473, 1987.
4. S. Hirsekorn, (private communication).
5. D. Lee, S. Razvi, K. Salama and E. Schneider, "Nondestructive Characterization of Metal-Matrix Composites", Proc. 16th Symposium on NDE, NTIAC, San Antonio, 1986.
6. F. Corvasce, P. Lipinski and M. Bervciller, "Thermomechanical Behavior of Metal-Matrix-Composites", 3rd Int'l Symposium on Nondestructive Characterization of Materials, Saarbrücken, FRG, Oct. 1988.
7. M. Spies and K. Salama, "Ultrasonic Evaluation of Texture in Metal-Matrix Composites", Proc. Review of Progress in Quantitative NDE, La Jolla, CA, 1988.
8. S. Hirsekorn, "Ultrasonic Propagation in Metal Matrix Composites", 3rd Int'l Symposium on Nondestructive Characterization of Materials, Saarbrücken, FRG, Oct. 1988.

Ultrasonic Propagation in Metal-Matrix-Composites

Sigrun Hirsekorn

Fraunhofer-Institut für zerstörungsfreie Prüfverfahren,
Universität, Geb. 37, D-6600 Saarbrücken

Abstract

Ultrasonic scattering at phase and grain boundaries in composites causes attenuation of sound waves and dispersive sound velocities. This effect can be used for nondestructive materials characterization if it can be described quantitatively. This paper presents a theory which allows calculation of the velocities and scattering coefficients of compressional and shear ultrasonic waves in fibre reinforced composites as function of frequency. To begin with, the polycrystalline structure of matrix and fibres is neglected and an exactly parallel alignment of the fibres is assumed. The structure anisotropy causes directional dependent sound velocities and scattering coefficients. Numerical evaluation was done for SiC-fibre reinforced aluminium.

Inhaltsangabe

In Verbundwerkstoffen führt die Streuung von Ultraschallwellen an den Phasen- und Korngrenzen zur Schwächung von Schallstrahlen und zu dispersivem Verhalten der Schallgeschwindigkeiten. Dieser Effekt kann zur zerstörungsfreien Gefügeanalyse ausgenutzt werden, wenn er quantitativ beschrieben werden kann. In der vorliegenden Arbeit wird eine Theorie vorgestellt, die die Berechnung von Schallgeschwindigkeiten und Streukoeffizienten von longitudinalen und transversalen Ultraschallwellen in faserverstärkten Verbundwerkstoffen als Funktion der Frequenz erlaubt. Zunächst wurde die polykristalline Struktur von Matrix und Fasern vernachlässigt und eine exakte Parallelausrichtung der Fasern angenommen. Die Anisotropie in der Struktur bedingt richtungsabhängige Schallgeschwindigkeiten und Streukoeffizienten. Die analytischen Ergebnisse wurden für SiC-Fasern verstärktes Aluminium numerisch ausgewertet.

1. Introduction

Ultrasonic scattering at phase and grain boundaries in composites causes attenuation of sound waves and dispersive sound velocities. The scattering coefficients which describe the attenuation by scattering are frequency dependent, too. In materials with macroscopic anisotropy, the ultrasonic propagation constants also vary with direction. These effects can be used for nondestructive materials characterization if they can be described quantitatively.

This paper presents a theory which allows calculation of the velocities and scattering coeffcients of plane ultrasonic waves in fibre reinforced composites in dependence on their structure properties as function of frequency and direction. Only two-phase materials are considered. To begin with, the polycrystalline structure of matrix and fibres is neglected. This is a good approximation if the single-crystal anisotropy of both phases is small so that the scattering at grain boundaries can be omitted in comparison to the scattering at the phase boundaries. The fibres are described as prolate ellipsoids. Analytical calculations are carried out for composites with exactly parallel aligned fibres. This structure ansiotropy causes directional dependence of sound velocities and scattering coefficients. The purpose of this work is to provide with the theoretical basis needed for the development of ultrasonic techniques to determine nondestructively size, shape and distribution of the fibres in metal-matrix-composites.

2. Mathematical description of the structure in MMC

Metal-matrix-composites are two-phase materials the secondary phase of which are fibres being embedded in the primary phase (matrix). The fibres approximately can be described by prolate ellipsoids of resolution. If the single-crystal anisotropy of both phases is small, the scattering at grain boundaries within a phase can be neglected in comparison to the scattering at phase boundaries. In this case, both phases are approximately homogeneous and isotropic.

Let V_α be the volume part, ρ_α the density, and $C^{(\alpha)}_{ijkl}$ the elastic constants of phase α. From this, the mean values of density and elastic constants of a two-phase composite follow to be

$$\langle\rho\rangle = V_1\rho_1 + V_2\rho_2 \, , \tag{2.1}$$

$$\langle C_{ijkl}\rangle = V_1 \, C^{(1)}_{ijkl} + V_2 \, C^{(2)}_{ijkl} \, . \tag{2.2}$$

Because both phases are assumed to be homogeneous and isotropic, their elastic constants are related to their densities and sound velocities by

$$C^{(\alpha)}_{iiii} = \rho_\alpha \, v^2_{L\alpha} \, , \; C^{(\alpha)}_{ijij} = \rho_\alpha \, v^2_{T\alpha} \, , \tag{2.3}$$
$$C^{(\alpha)}_{iijj} = \rho_\alpha \left(v^2_{L\alpha} - 2v^2_{T\alpha}\right) \, , \; C^{(\alpha)}_{ijkl} = 0 \text{ otherwise} \, .$$

$v_{L\alpha}$ and $v_{T\alpha}$ are the compressional and shear wave velocity, respectively, within phase α.

The matrix is indicated to be phase 1, the fibres are phase 2. With the probability $W_2(\underline{r}-\underline{r}')$ of two positions $\underline{r}$ und $\underline{r}'$ to be in the same fibre if one of these positions is located in a fibre one gets an approximate expression for the correlation function of two material constants f and g as follows:

$$<f(\underline{r})\ g(\underline{r}') = V_2 W_2(\underline{r}-\underline{r}')\ (f_2-<f>)\ (g_2-<g>) + <f><g>. \qquad (2.4)$$

Because of the shape anisotropy of the fibres $W_2(\underline{r}-\underline{r}')$ not only depends on the distance $|\underline{r}-\underline{r}'|$ but also on direction. In the case of exactly parallel alignment of the fibres the directional dependence is only determined by the shape of the fibres, that is, $W_2(\underline{r}-\underline{r}')$ holds ellipsoidal symmetry. If Poisson statistic is assumed for size and shape distribuion [1] one has

$$W_2(\underline{r}) = \exp\left(-\frac{r}{l(\theta,\phi)}\right) = \exp\left(-\sqrt{\sum_{i=1}^{3}\frac{x_i^2}{l_i^2}}\right), \qquad (2.5)$$

where the aspect ratio $l_1:l_2:l_3$ has to be equal to that of the effective semiaxes a_1, a_2, and a_3 of the fibres. The effective fibre volume is given by the two different equations

$$V_{eff} = \int d^3r\ W_2(\underline{r}) = 8\pi\ l_1 l_2 l_3\ , \qquad V_{eff} = \frac{4}{3}\pi\ a_1 a_2 a_3\ , \qquad (2.6)$$

which leads to relations between the constants l_i and the effective semiaxes a_i. An ellipsoid of resolution requires two axes to be equal. To fix the coordinate system, $a_1=a_2$ is chosen to be the lateral and a_3 the longitudinal effective semiaxis of the fibres.

3. Calculation of complex propagation constants of ultrasonic waves in MMC

3.1 Description of the approach

As in the case of the description of ultrasonic propagation in single-phase polycrystals [1-6] the calculations start with the equation of motion of the displacement vector $\underline{s}$,

$$L\underline{s} = \sum_{j,k,l=1}^{3}\frac{\partial}{\partial x_j}\left(\underline{C}_{jkl}\frac{\partial s_k}{\partial x_l}\right) + \rho\ \omega^2\ \underline{s} = 0, \qquad (3.1)$$

$$\underline{C}_{jkl} = \left(C_{1jkl},\ C_{2jkl},\ C_{3jkl}\right)\ .$$

In inhomogeneous media the elastic constants C_{ijkl} and the density ρ are position dependent. In general, the elastodynamic equation is not solvable exactly, and an approximate equation of motion of the ensemble average of the displacement vector $<\underline{s}>$ is required.

The second order Keller approximation of the elastodynamic equation [1,6] is given by

$$L_o \langle \underline{s} \rangle = \left\{ \varepsilon \langle L_1 \rangle + \varepsilon^2 \left(\langle L_1 L_o^{-1} L_1 \rangle - \langle L_1 \rangle L_o^{-1} \langle L_1 \rangle \right) \right\} \langle \underline{s} \rangle , \qquad (3.2)$$

if the operator L describes the inhomogeneous material of interest (equation (3.1)), L_o leads to the equation of motion of a fictitious homogeneous reference material, and the operator

$$\varepsilon L_1 = L_o - L , \qquad (3.3)$$

which represents the difference between the actual and the fictitious reference material, can be considered as a small perturbation in comparison to L_o. Ensemble averaging of the operators L_1 and $L_1 L_o^{-1} L_1$, as required in equation (3.2), can be reduced to the calculation of average values and correlations of the material constants.

The fictitious homogeneous reference material has to meet the following conditions:

a) The Green's function of the operator L_o, and therewith the inverted operator L_o^{-1}, must be known.

b) The operator εL_1 has to present a small perturbation compared to L_o.

c) The analytical und numerical evaluation of the perturbation terms must not require additional approximations.

It is obvious that these conditions cannot be fulfilled independently of each other. It is convenient to use the elliptical coordinates defined by

$$\tilde{x}_i = \frac{x_i}{l_i} ; \quad i = 1,2,3 \qquad (3.4)$$

which are adapted to the geometric symmetry of the composite. The fictitious reference material is chosen to be homogeneous and isotropic with respect to the elliptical coordinates, that is, the operator L_o is defined by

$$L_o \underline{s}^o = \sum_{j,k,l=1}^{3} \underline{\tilde{C}}^o_{jkl} \frac{\partial^2 s^o_k}{\partial \tilde{x}_j \, \partial \tilde{x}_l} + \rho_o \omega^2 \underline{s}^o = 0 \qquad (3.5a)$$

$$\underline{\tilde{C}}^o_{jkl} = \left(\tilde{C}^o_{1jkl}, \tilde{C}^o_{2jkl}, \tilde{C}^o_{3jkl} \right),$$

$$\tilde{C}^o_{iiii} = \frac{\rho_o \omega^2}{\tilde{k}^2} , \quad \tilde{C}^o_{ijij} = \frac{\rho_o \omega^2}{\tilde{\kappa}^2} , \qquad (3.5b)$$

$$\tilde{C}^{o}_{iijj} = \rho_o \omega^2 \left(\frac{1}{\tilde{k}^2} - \frac{2}{\tilde{\kappa}^2} \right) \; ; \quad i,j = 1,2,3; \quad i \neq j;$$

$$\tilde{C}^{o}_{ijkl} = 0 \quad \text{otherwise} \; , \qquad \rho_o = \langle \rho \rangle$$

It is evident that this reference material is anisotropic when described in the original coordinates x_i. The density ρ_o of the reference material is set equal to the mean value $\langle\rho\rangle$ of the density of the composite. The Green's function of L_o which obeys the equation

$$L_o \, G_{ij}(\underline{\tilde{r}}-\underline{\tilde{r}}') = \delta_{ij} \, \delta(\underline{\tilde{r}}-\underline{\tilde{r}}') \tag{3.6}$$

can be taken from the literature [7]. If the ensemble average of the displacement vector $\langle \underline{s} \rangle$ is the plane wave

$$\langle \underline{s} \rangle = \underline{e}_P \, e^{-i\underline{\beta}\cdot\underline{r}} \; , \qquad \underline{\beta} = \beta \underline{e}_\beta , \tag{3.7}$$
$$\underline{e}_\beta = \left(n_{\beta 1}, n_{\beta 2}, n_{\beta 3} \right), \qquad \underline{e}_P = \left(n_{P1}, n_{P2}, n_{P3} \right)$$

with the complex propagation constant β, the direction of motion $\underline{e}_\beta$, and the polarization $\underline{e}_P$, the approximate elastodynamic equation (3.2) becomes

$$\sum_{q=1}^{3} \left\{ \sum_{j,l=1}^{3} \langle C_{ijql} \rangle n_{\beta j} n_{\beta l} - \frac{\rho_o \omega^2}{\beta^2} \delta_{iq} + V_2 \frac{(\rho_2 - \rho_o)\omega^4}{\beta^2} \int d^3\tilde{r} \; G_{iq}(\underline{\tilde{r}}) W_2(\underline{r}) e^{i\underline{\beta}\cdot\underline{r}} - \right.$$
$$- iV_2 \frac{(\rho_2-\rho_o)\omega^2}{\beta} \sum_{j,k,l=1}^{3} n_{\beta i} \frac{1}{l_l} \left[\left(C^{(2)}_{klqj} - \langle C_{klqj} \rangle \right) \int d^3\tilde{r} G_{ik}(\underline{\tilde{r}}) \frac{\partial}{\partial \tilde{x}_l} \left(W_2(\underline{r}) e^{-i\underline{\beta}\cdot\underline{r}} \right) + \right.$$
$$\left. + \left(C^{(2)}_{ijkl} - \langle C_{ijkl} \rangle \right) \int d^3\tilde{r} \; G_{kq}(\underline{\tilde{r}}) \frac{\partial}{\partial \tilde{x}_l} \left(W_2(\underline{r}) \; e^{-i\underline{\beta}\cdot\underline{r}} \right) \right] - \tag{3.8a}$$
$$- V_2 \sum_{j,k,l=1}^{3} \; \sum_{s,t,v=1}^{3} n_{\beta j} n_{\beta v} \frac{1}{l_l l_t} \left(C^{(2)}_{ijkl} - \langle C_{ijkl} \rangle \right) \left(C^{(2)}_{stqv} - \langle C_{stqv} \rangle \right) \cdot$$
$$\left. \cdot \int d^3\tilde{r} \; G_{ks}(\underline{\tilde{r}}) \frac{\partial^2}{\partial \tilde{x}_l \partial \tilde{x}_t} \left(W_2(\underline{r}) \; e^{-\underline{\beta}\cdot\underline{r}} \right) \right\} n_{pq} = \sum_{q=1}^{3} B_{iq} \, n_{pq} = 0 \; .$$

For arbitrarily fixed directions of motion $\underline{e}_\beta$, the condition

$$\mathrm{Det}(B_{iq}) = 0 \tag{3.8b}$$

yields the complex propagation constants of ultrasonic waves. $B = (B_{iq})$ is a three-dimensional matrix, so that one gets three solutions for each direction of motion which stand for one compressional and two shear waves. The polarization $\underline{e}_P$ of these waves can be determined by (3.8a) after fixed values of β and $\underline{e}_\beta$ are inserted.

In general, purely longitudinal and purely transverse waves do not exist in macroscopically anisotropic media but each mode has compressional and shear parts. The only exceptions are directions of motion parallel to symmetry axes of the macroscopic anisotropy, that is parallel or perpendicular to the fibres in the case of fibre reinforced composites with exactly parallel aligned fibres. The following calculations are restricted to these particular directions because a general consideration will become too extensive.

Restriction to pure compressional and shear waves propagating parallel or perpendicular to the fibres essentially simplifies equation (3.8a). The remaining integrals are to find in previous papers if the elliptic coordinates defined above are used. The results show that the wave numbers k and κ of the fictitious reference material have to obey the relations

$$\tilde{k} = l_1 k, \qquad \tilde{\kappa} = l_1 \kappa, \qquad k^2 = \frac{\rho_o \omega^2}{\langle C_{iiii} \rangle}, \qquad \kappa^2 = \frac{\rho_o \omega^2}{\langle C_{ijij} \rangle}. \tag{3.9}$$

to ensure small perturbation terms in comparison to the zeroth order approximation.

The resulting determination equations for complex propagation constants of ultrasonic waves with direction of motion parallel or perpendicular to the fibres can be solved only numerically. For this purpose, a modified Newton's method is used [8]. The real and imaginary part of the complex propagation constants are the wave numbers and scattering coefficients, respectively. The phase and group velocities follow from the wave numbers.

3.2 Validity limitations of the theory

The presented method to calculate complex propagation constants of ultrasonic waves in MMC yields reasonable results only if the used expansion with respect to the inhomogeneity is convergent and if the terms up to second order describe the ultrasonic propagation with sufficient accuracy. The formal Keller approximation of arbitrary order of the equation of motion shows that the expansion parameters are given by the relative mean quadratic deviation of the actual wave numbers k_{ci} and κ_{cij} from the corresponding Voigt approximations. Fundamental transformations lead to the following validity limitations of the theory:

$$\varepsilon^2_{\text{long } i} = \sum_{\alpha=1}^{2} V_\alpha \left(\frac{C^{(\alpha)}_{iiii} - \frac{\rho_\alpha}{\rho_o} \langle C_{iiii} \rangle}{2\langle C_{iiii} \rangle} \right)^2 \ll 1 , \tag{3.10a}$$

$$\varepsilon^2_{\text{trans } ij} = \sum_{\alpha=1}^{2} V_\alpha \left(\frac{C^{(\alpha)}_{ijij} - \frac{\rho_\alpha}{\rho_o} \langle C_{ijij} \rangle}{2\langle C_{ijij} \rangle} \right)^2 \ll 1 . \tag{3.10b}$$

4. Numerical evaluation for SiC-fibre reinforced aluminium

Numerical evaluation of the analytical results was done for SiC-fibre reinforced aluminium. The single-crystal elastic constants of both phases are to find in Table 1. Aluminium has cubic, SiC hexagonal crystal-symmetry. The anisotropy factors follow from the single-crystal constants to be

$$A_i = C_{iiii} - C_{jjkk} - 2C_{jkjk};\ i,j,k = 1,2,3;\ i \neq j \neq k;\ i \neq k . \tag{4.1}$$

Table 1: Elastic single-crystal constants of Al and SiC [9]

$$[C_{ij}] = [A_i] = 10^9 \frac{N}{m^2}$$

	Symmetry	C_{11}	C_{33}	C_{12}	C_{13}	C_{44}	$C_{66} = \frac{1}{2}(C_{11}-C_{12})$	A_1	A_2
Al	cubic	108		62.2		28.4		-11	
SiC	hexagonal	500	506	100	60	170	200	100	60

For both phases, the ratios of anisotropy factors and single-crystal constants reach values distinct below 1, that is, the single-crystal anisotropy of both phases is small.

Table 2 contains the densities and sound velocities as well as the elastic constants of macroscopically isotropic aluminium and SiC in the homogeneous isotropic approximation. The listed values show that the differences between the elastic constants of the two phases are significantly larger than those which follow from different grain orientation within each of the phases. That is, for the example considered the assumption that grain scattering can be omitted in comparison to the scattering at phase boundaries is fulfilled.

Table 2: Material constants of macroscopically isotropic Al and SiC

	$\rho\left[\frac{kg}{m^3}\right]$	$v_L\left[\frac{km}{s}\right]$	$v_T\left[\frac{km}{s}\right]$	$c_{iiii}\left[10^9\ \frac{N}{m^2}\right]$	$c_{ijij}\left[10^9\ \frac{N}{m^2}\right]$	$c_{iijj}\left[10^9\ \frac{N}{m^2}\right]$
Al	2700	6.568	3.149	116.47	26.77	62.93
SiC	3200	12.5	7.6	500.00	184.83	130.34

The material constants of Table 2 are used for the further calculations. In Table 3, the expansion parameters for volume parts of fibres from 1 to 50% are listed. It is to be seen that the convergency of the used expansion with respect to the inhomogeneity may be not good enough in the region of 10%. This may cause invalid numerical results in second order perturbation theory, and therefore, may require terms of higher order.

Table 3: Expansion parameter for SiC-fibre reinforced aluminium in dependence on volume part of fibres (V_2)

V_2	0.01	0.02	0.03	0.04	0.05	0.1	0.15	0.2	0.5
ε^2_{long}	0.022	0.0414	0.0576	0.0713	0.0830	0.1185	0.1306	0.1306	0.0722
ε^2_{trans}	0.0719	0.1273	0.1698	0.2025	0.2273	0.2805	0.2776	0.2558	0.1096

As an example, figure 1 shows the normalized phase velocities and scattering coefficients of compressional waves with direction of motion perpendicular to the fibres as function of wave number (k_1) of the incidend wave in the matrix (aluminium) times effective lateral semiaxis of the fibres (a) for volume fractions of secondary phase from 0.01 to 0.05. Normalization was done with the sound velocity v_{L1} and wave number k_1, respectively, of the matrix. The aspect ratio was chosen to be 5:1. The scattering coefficients of the three wave modes with direction of motion perpendicular to the fibres show a maximum the position and height of which is dependent on the volume fraction of fibres and which coincides with strong variations of the corresponding sound velocities as function of frequency. The velocities of compressional waves additionally run through a relative minimum at $k_1a \approx 1$. Numerical evaluation for several aspect ratios has shown that the velocities and scattering coefficients of waves propagating perpendicular to the fibres are nearly independent of fibre length.

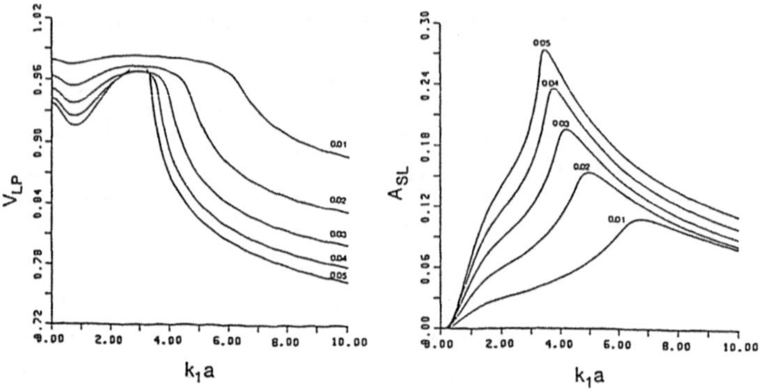

Fig.1. Normalized phase velocities and scattering coefficients of compressional waves propagating perpendicular to the fibres for different volume fractions of fibres, aspect ratio 5:1

The scattering coefficients of both wave modes with direction of motion parallel to the fibres are orders of magnitude smaller than those of waves propagating perpendicular to the fibres. The velocity dispersion decreases appropriately but still reaches orders of magnitude of per cent. The scattering coefficients as function of frequency have a maximum the position of which depends on fibre length but not on the volume part of the fibres. The height is determined by both of these parameters. The scattering coefficients decrease rapidly with increasing fibre length, e.g. about two orders of magnitude if the aspect ratio is changed from 5:1 to 10:1. Figure 2 shows the normalized phase velocities and scattering coefficients of shear waves with propagation along the fibres.

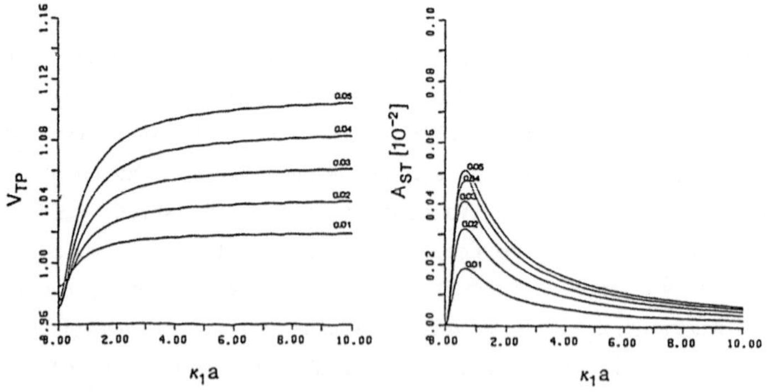

Fig.2. Normalized phase velocities and scattering coefficients of shear waves propagating parallel to the fibres for different volume fractions of fibres, aspect ratio 5:1

5. Concluding remarks

Numerical evaluation of the analytical results for SiC-fibre reinforced aluminium has shown that the velocities and scattering coefficients of ultrasonic waves as function of frequency, direction of motion, and polarization contain informations from which volume fraction, size, and shape of the fibres can be determined. It is to pay attention to the fact that the presented scattering theory for MMC supposes an exactly parallel alignment of the fibres and therefore, only can be an approximation of real composites. Additionally, grain scattering is omitted in comparison to scattering at phase boundaries which presumes small single-crystal anisotropy in both phases. The relations between ultrasonic propagation constants and structure properties are rather complicated, so that in the present form the theory cannot yet be used for quantitative structure characterization of fibre composites. Further research is necessary.

6. References

1. Stanke, F.E.; Kino, G.S.: A unified theory for elastic wave propagation in polycrystalline materials. J. Acoust. Soc. Am. 75, (1984) 665-681.

2. Hirsekorn, S.: The scattering of ultrasonic waves by polycrystals. J. Acoust. Soc. Am. 72, (1982) 1021-1031.

3. Hirsekorn, S.: The scattering of ultrasonic waves by polycrystals. II. shear waves. J. Acoust. Soc. Am. 73, (1983) 1160-1163.

4. Hirsekorn, S.: The scattering of ultrasonic waves in polycrystalline materials with texture. J. Acoust. Soc. Am. 77, (1985) 832-843.

5. Hirsekorn, S.: Directional dependence of ultrasonic propagation in textured polycrystals. J. Acoust. Soc. Am. 79, (1986) 1269-1279.

6. Hirsekorn, S.: The scattering of ultrasonic waves by multiphase polycrystals. J. Acoust. Soc. Am. 83, (1988) 1231-1242.

7. Lifshits, I.M.; Parkhomovskii, G.D.: Ultraschallabsorption in Polykristallen. Uc. Zap. Charkov Gos. Univ. Im. 27, (1948) 25-36.

8. Flügge, S.: Handbuch der Physik, Mathematische Methoden II. Springer, 1955.

9. Landolt-Börnstein: Gruppe III: Kristall- und Festkörperphysik, Band 1+2, Springer 1969.

Thermomechanical Behaviour of Metal Matrix Composites

F. CORVASCE, P. LIPINSKI, M. BERVEILLER

Laboratoire de Physique et Mécanique des Matériaux (UA CNRS)
ISGMP - ENIM
Ile du Saulcy 57045 Metz Cedex

Summary

This paper deals with the modelling of the inelastic thermomechanical behavior of Metal Matrix Composites (MMC) within the framework of the classical uncoupled thermo-elasto-plasticity without damage. At first, the local behavior of materials is presented, followed by the global or overall description. A general approach to the determination of concentration tensors from the integral equation is proposed and next specialized for the case of the self-consistent scheme. A few applications are presented concerning elasticity (evaluation of the overall elasticity tensor), thermoelasticity (local stresses, thermal expansion coefficients,...) and elastoplasticity (stress-strain curves, residual stresses...).

Introduction

The evaluation of the thermo-elastic (linear) behavior of microinhomogeneous materials such as polycrystals or composites was studied by many authors, for example Levin [1], Hill [2], Kröner [3]. The theoretical solution of this type of problems is known. The question becomes different when dealing with Metal Matrix Composites. The utilization of short fibers makes more difficult the problems of concentration and averaging of thermomechanical fields and the ductility of the matrix may lead to plastic deformations of the material.

The aim of this study is to evaluate the thermomechanical and inelastic behavior of MMC. The Medium is considered as continuum microinhomogeneous and macrohomogeneous. The uncoupled thermo-elasto-plastic behavior is supposed and damage is not taken into consideration. The mechanism of crystallographic slip is adopted in order to describe the plastic deformation of the matrix.

The general form of local relations is reviewed in first section. In this approach, the temperature acts by the intermediary of thermal expansion coefficients. Introducing the concentration tensors linking local fields with the overall ones and applying the usual averaging operations, the overall behavior of the MMC may be determined.

In the next section, an integral equation is presented allowing the determination of concentration tensors. The self-consistent approximation of these tensors is deduced from the integral equation. Finally, theoretical and numerical results concerning the behavior of MMC are presented. The overall elastic moduli and thermal expansion coefficients as well as concentration tensors are determined for Al-SiC composites. Elasto-plastic behavior and residual stresses are discussed for the same composite.

Local constitutive equations

The infinite microinhomogeneous medium of volume V undergoes a thermomechanical loading defined by the state of overall stresses Σ, the temperature θ and the corresponding rates $\dot{\Sigma}$ and $\dot{\theta}$.

At a point $\vec{r}$ of the medium, various physical mechanisms contribute to the total strain rate $\dot{\varepsilon}^T(r)$ i.e. :

* The elastic deformation defined by the local stiffness s(r) and stress rates $\dot{\sigma}(r)$

$$\dot{\varepsilon}^e_{ij}(r) = s_{ijkl}(r)\,\dot{\sigma}_{kl}(r) \tag{1}$$

* The thermal deformation described by local thermal expansion coefficients $\alpha_{ij}(r)$ and temperature rate $\dot{\theta}$ assumed to be homogeneous through the considered volume V

$$\dot{\varepsilon}^{th}_{ij}(r) = \alpha_{ij}(r)\,\dot{\theta} \tag{2}$$

* The plastic deformation modelled by the local plastic stiffness tensor P(r) submitted to the local yield conditions

$$\dot{\varepsilon}^p_{ij}(r) = P_{ijkl}(r)\,\dot{\sigma}_{kl}(r) \tag{3}$$

$$f(\sigma_{ij}(r)) = 0 \text{ and } \frac{\partial f}{\partial \sigma_{ij}}\, d\,\sigma_{ij} > 0$$

The total deformation rate $\dot{\varepsilon}^T_{ij}(r)$ is then :

$$\dot{\varepsilon}^T_{ij}(r) = \dot{\varepsilon}^e_{ij}(r) + \dot{\varepsilon}^{th}_{ij}(r) + \dot{\varepsilon}^P_{ij}(r) \tag{4}$$

A more general case of a constitutive equation including phase transformation was given by Patoor et al [4] and the form of P and f in the case of crystallographic slip can be found in [5].

Introducing (1), (2) and (3) into (4), one has :

$$\dot{\varepsilon}^T_{ij}(r) = g_{ijkl}(r)\, \dot{\sigma}_{kl}(r) + n_{ij}(r)\, \dot{\theta} \tag{5}$$

where
$$g_{ijkl} = s_{ijkl} + P_{ijkl} \qquad n_{ij} = \alpha_{ij} \tag{6}$$

Global thermomechanical constitutive equations

The global thermomechanical constitutive law takes a form analogous to (5)

$$\dot{E}^T_{ij} = G_{ijkl}\, \dot{\Sigma}_{kl} + N_{ij}\, \dot{\theta} \tag{7}$$

where the overall stress and strain rate $\dot{\Sigma}$ and $\dot{E}^T$ may be deduced from the local ones by the usual averaging operations

$$\dot{\Sigma}_{ij} = \frac{1}{V} \int_V \dot{\sigma}_{ij}(r)\, dV \equiv \overline{\dot{\sigma}_{ij}(r)} \tag{8}$$

$$\dot{E}^T_{ij} = \frac{1}{2} \int_V \dot{\varepsilon}^T_{ij}(r)\, dV \equiv \overline{\dot{\varepsilon}_{ij}(r)} \tag{9}$$

The dual form of equations (5) and (7) may be written as

$$\dot{\sigma}_{ij}(r) = l_{ijkl}(r)\, \dot{\varepsilon}^T_{kl}(r) - m_{ij}(r)\, \dot{\theta} \tag{10}$$

$$\dot{\Sigma}_{ij} = L_{ijkl}\, \dot{E}^T_{kl} - M_{ij}\, \dot{\theta} \tag{11}$$

where

$$l = g^{-1} \qquad m = l : n \tag{13}$$

$$L = G^{-1} \qquad M = L : N \tag{14}$$

In order to obtain the overall behavior of the material using the local relations and knowing the microstructure of the composite, two concentration relations are introduced

$$\dot{\sigma}_{ij}(r) = B_{ijkl}(r)\, \dot{\Sigma}_{kl} + b_{ij}(r)\, \dot{\theta} \tag{14}$$

$$\dot{\varepsilon}^T_{ij}(r) = A_{ijkl}(r)\, \dot{E}^T_{kl} + a_{ij}(r)\, \dot{\theta}$$

To obtain the overall thermomechanical behavior i.e. L and M tensors, one has

$$\dot{\sigma}_{ij}(r) = l_{ijkl}(r)[A_{klmn}(r)\dot{E}^T_{mn}+a_{kl}(r)\dot{\theta}]-m_{ij}(r)\dot{\theta} \quad (15)$$

Applying the averaging operations

$$\dot{\Sigma}_{ij} = \overline{l_{ijkl}(r)\ A_{klmn}(r)}\ \dot{E}^T_{mn} + \overline{l_{ijkl}(r)\ a_{kl}(r) - m_{ij}(r)}\dot{\theta} \quad (16)$$

it follows

$$\begin{aligned} L_{ijmn} &= \overline{l_{ijkl}(r)\ A_{klmn}(r)} \\ M_{ij} &= \overline{l_{ijkl}(r)\ a_{kl}(r) - m_{ij}(r)} \end{aligned} \quad (17)$$

Similar calculations allow to determine G and N in function of g, n, B and b.

In order to find the concentration tensors A and a, several methods may be proposed. Nevertheless, they constitute only an approximation of the integral equation proposed in the following section.

Determination of Concentration Tensors

The general approach is to consider the inhomogeneous solid as a continuum medium with a microstructure satisfying

- The equilibrium equation

$$\dot{\sigma}_{ij,j}(r) = 0 \quad (18)$$

- The constitutive equation

$$\dot{\sigma}_{ij}(r) = l_{ijkl}(r)\ \dot{\varepsilon}^T_{kl}(r) - m_{ij}(r)\dot{\theta} \quad (19)$$

- The compatibility equation

$$\dot{\varepsilon}^T_{ij}(r) = \frac{1}{2}\ (\dot{u}^T_{i,j} + \dot{u}^T_{j,i}) \quad (20)$$

Introducing a homogeneous medium (without a microstructure) and characterized by its tangent moduli tensor L° such that

$$l(r) = L^\circ + \delta l(r)$$

equations (18) (19) and (20) take the form

$$L^{\circ}_{ijkl}\dot{u}^T_{k,lj}+[\delta l_{ijkl}(r)\dot{\varepsilon}^T_{kl}(r)-m_{ij}(r)\dot{\theta}]_{,j} = 0 \quad (21)$$

Using the Green tensor technique, (21) may be transformed into an integral equation

$$\dot{\varepsilon}^T_{ij}(r) = \dot{\varepsilon}^{\circ}_{ij}+ \int_V \Gamma_{ijkl}(r-r')[\delta l_{klmn}(r')\dot{\varepsilon}^T_{mn}(r')-m_{kl}(r')\dot{\theta}]dV' \quad (22)$$

where

- $\dot{\varepsilon}^{\circ}_{ij}$ is the strain rate of the homogeneous medium submitted to the same boundary conditions as the effective medium

- $\Gamma_{mnij} = \frac{1}{2} (G_{mi,jn} + G_{ni,jm})$
- G_{mi} is the Green tensor of the infinite medium with L° moduli tensor.

Relation (22) constitutes an integral equation linking the local strain rate $\dot{\varepsilon}^T$ and the thermomechanical loading characterized by $\dot{\varepsilon}^\circ$ and $\dot{\theta}$. To exploit this complex equation, various approximations have to be applied leading to numerous models. In this work, the self-consistent approach is developed and applied. This method may be shortly characterized as follows :

* the interactions between constituents, in the case of one-site scheme, are modelled by interactions between an ellipsoïdal inclusion and the homogeneous equivalent medium L^{eff} considered as matrix
* the particular choice of $L^\circ = L^{eff}$ is adopted.

The multiple-site version of this approach was established by Berveiller et al [6] and for elastoplastic metals at large strains, the model was generalized by Lipinski and Berveiller [7].

Let us now consider a granular medium and suppose that l, $\dot{\varepsilon}^T$ and m are piece-wise constant. The following holds

$$\delta l(r) = \sum_{I=1}^{N} \delta l^I \, y^I(r)$$

$$m(r) = \sum_{I=1}^{N} m^I \, y^I(r) \tag{23}$$

$$\dot{\varepsilon}^T(r) = \sum_{I=1}^{N} \dot{\varepsilon}^{TI} \, y^I(r)$$

where $y^I(r)$ represents N characteristic functions such that

$$y^I(r) = \begin{cases} 0 \text{ if } r \notin V^I \\ 1 \text{ if } r \in V^I \end{cases} \tag{24}$$

In order to simplify the considerations, suppose for the moment that $\dot{\theta} = 0$. In this case, the integral equation is rewritten in the form

$$\dot{\varepsilon}^{TI}_{ij} = \dot{\varepsilon}^\circ_{ij} + T^{II}_{ijkl} \, \delta l^I_{klmn} \, \dot{\varepsilon}^{TI}_{mn} \tag{25}$$

where

$$T^{II}_{ijkl} = \frac{1}{V_I} \int_{V_I} \int_{V_I} \Gamma_{ijkl}(r-r') \, dV dV' \tag{26}$$

By definition, $L^{\circ} = L^{eff}$ which leads to

$$\dot{\varepsilon}^{\circ} = \dot{E}^{T} \quad (27)$$

and the concentration tensor A^{I} for the Ith inclusion is

$$A^{I}_{mnkl} = (I_{mnkl} - T^{II}_{mnij}\ \delta l^{I}_{ijkl})^{-1} \quad (28)$$

The tangent moduli tensor L^{eff} may be obtained from (17) and (28)

$$L^{eff}_{ijkl} = \sum_{I=1}^{N} F^{I}\ l^{I}_{ijmn}\ A^{I}_{mnkl} \quad (29)$$

where F^{I} denotes the volumic fraction of the considered phase. In the similar manner, one may determine a^{I} and using (17)

$$M^{eff}_{ij} = \Sigma\ F^{I}\ (l^{I}_{ijkl}\ a^{I}_{kl} - m^{I}_{ij}) \quad (30)$$

Expressions (29) and (30) are rather complex ones because of their implicit character.

The proposed theory has been applied to elasticity, thermo-elasticity and elastoplasticity of Al-SiC metal matrix composites.

Application to Al-SiC composites

The composite is considered as a two-phased material composed of a matrix and reinforcements. The first application corresponds to an experimental study by Schneider et al [8]. The properties of the matrix and reinforcements have been deduced from this study [8] and are presented in Table 1.

	Young's Modulus E	Poisson's ratio ν
Al	77.5	0.3
SiC	430	0.3

Table 1 : Elastic properties of Al-Matrix and SiC reinforcements.

The material was extruded and exhibits a small anisotropy. The reinforcements presented an ellipsoïdal aspect elongated in the extrusion direction. The numerical calculations of Young's modulus in the transversal direction using the self-consistent model are shown in Fig 1. The results agree very well with the experimental measurements. An accurate modelling of the elastic anisotropy has been stated.

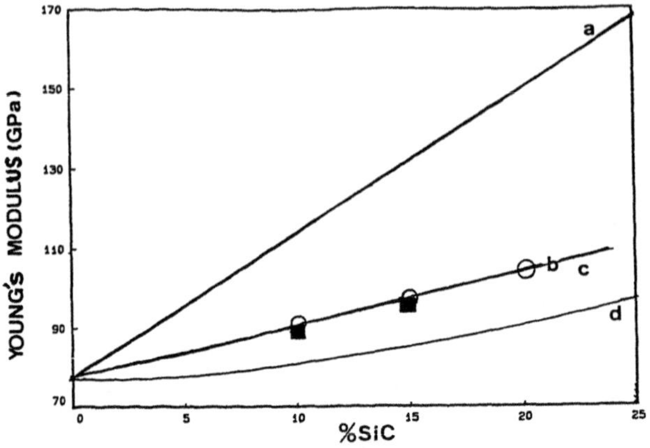

Figure 1 : Comparison of the transversal Young's modulus (O) calculated by the self-consistent model (■) experimentally obtained by Schneider et al [8] and determined using (a) Voigt and (b) Reuss models

Next, the thermoelastic behavior of Al-SiC composites the local thermal expansion tensors have been supposed isotropic both for matrix and SiC ($\alpha_{Al} = 21\ 10^{-6}\ K^{-1}$, $\alpha_{SiC} = 4\ 10^{-6}\ K^{-1}$). Fig 2 presents the evolution of α_{33}^{eff} component of the effective thermal expansion tensor for four reinforcement shapes and various F^I.

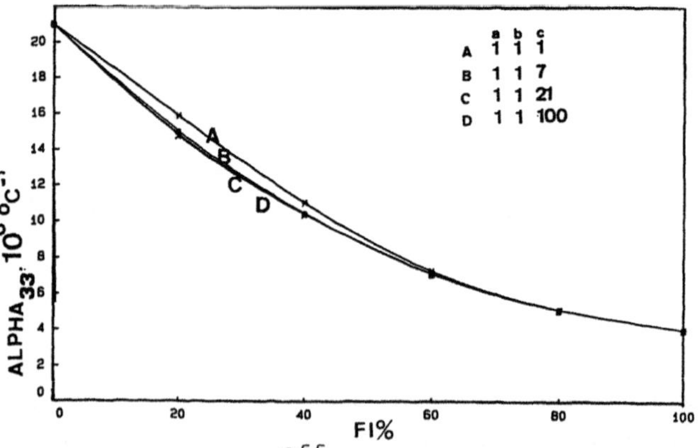

Figure 2 : Evolution of the α_{33}^{eff} component of the effective thermal expansion tensor as a function of the volumic fraction

The components of the concentration tensors B_{3333} and b_{33} are plotted in Fig 3 and 4. A very strong stress concentration is observed in both cases depending on the volume fraction and reinforcements shape.

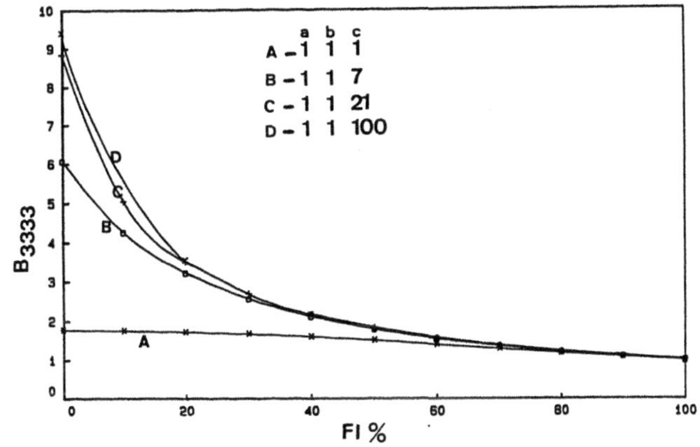

Figure 3 : Evolution of B_{3333} component of the concentration tensor as a function of the volumic fraction and various fiber aspects

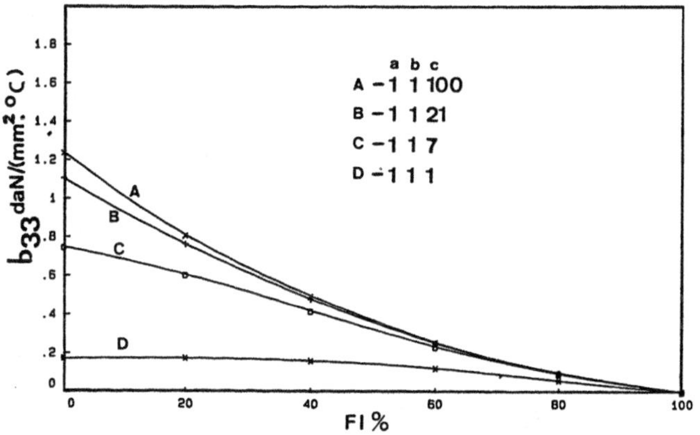

Figure 4 : Evolution of b_{33} component of the concentration tensor as a function of the volumic fraction and various fiber aspects

The last class of applications concerns the elastoplastic behavior of Al-SiC MMC. The material properties are summarized in Table II.

In this case, the composite is considered as a microinhomogeneous medium with two constituents

- the elastic and isotropic reinforcements

- the microinhomogeneous polycrystalline matrix composed of grains crystallographically mis oriented.

		SiC	Al
Elasticity	μ	$\mu = 27559\ daN/mm^2$	$2692\ daN/mm^2$
	ν	0.27	0.3
Plasticity	τ_o	∞	$14\ daN/mm^2$
	slip systems	–	$(111) < 110 > (12x)$
	Hardening Matrix	–	$H_1 = \mu/250$ $H_2 = H_1 \times 3$
Microstructure	Volumic Fraction	F_I	$(1 - F_I)$
	form and orientation	ellipsoïdal (a,b,c)	spherical
		same orientation	100 crystallographic orientation (random)

Table II : Mechanical properties and microstructure of the studied composite

The imposed loading corresponds to a tension test in the direction of the fibers axis c. Fig 5 presents four tension curves for two volume fractions and two fiber aspects. The influence of the reinforcements shape is very pronounced.

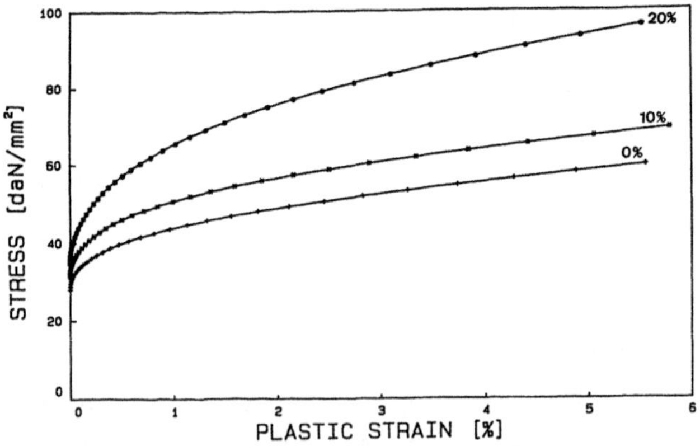

Figure 5 : Tensile curves for 0, 10 and 20% of fibers of spherical shape (a = b = c)

The evolution of residual stresses (σ_{33} component) for some particular grains of the matrix and SiC fiber as a function of the volume fraction and plastic strain is plotted in Fig 6. The intragranular inhomogeneities appear negligible with respect to the difference of the mechanical behavior between matrix and fiber.

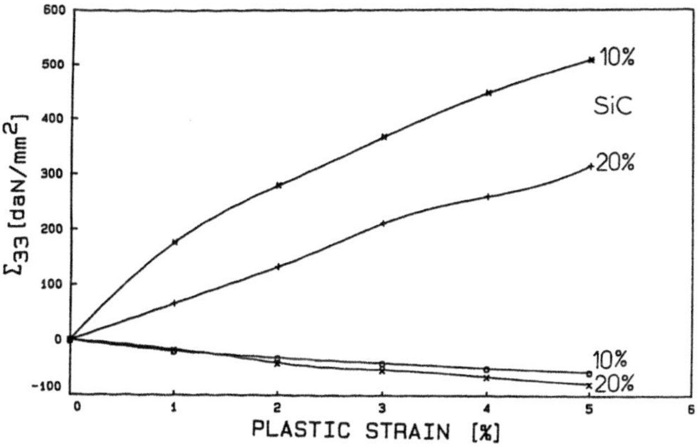

Figure 6 : Residual stress (σ_{33}) inside the reinforcements and particular grains of the matrix as a function of the overall plastic strain and volumic fraction of fibers for ellipsoïdal reinforcements ($a = b = \frac{c}{10}$)

References

1. Levin V.M. : Thermal expansion coefficients of heterogeneous materials. Mekhanika Tverd. Tela (1967), 88-94.
2. Hill R. : Elastic properties of Reinforced Solids. J. Mech. Phys. Solids (1963), 11, 357-372.
3. Kröner E. : Zur klassischen Theorie statistisch aufgebauter Festkörper. Int. J. Engng. Sci. (1973), 11, 171-191.
4. Patoor E., Eberhardt A., Berveiller M. : An integral equation for the polycrystalline thermomechanical behavior of shape memory alloys. Int. Seminar Mecamat Besançon (France), 1988, 319-330.
5. Beradai C., Lipinski P., Berveiller M. : Plasticity of metallic polycrystals under complex loading path. Int. J. Plasticity (1987), 3, 143-162.
6. Berveiller M., Hihi A., Fassi Fehri O. : Multiple site self-consistent scheme. To appear in Int. J. Engng. Sci.
7. Lipinski P., Berveiller M. : Elastoplasticity of microinhomogeneous metals at large strains. To appear in Int. J. Plasticity.
8. Schneider E., Lee D., Razvi S., Salama K. : Non destructive characterization of metal-matrix composites. Proc. 16th Symposium of Non Dest. Evaluation. San Antonio USA (1987).

Polymers

Correlation Between Manufacturing Parameters and Residual Stresses of Injection-Molded Poly-Propylene: An X-Ray Diffraction Study

V. HAUK, A. TROOST, D. LEY

Institut für Werkstoffkunde (Lehrstuhl A)
Rheinisch-Westfälische Technische Hochschule Aachen
D-5100 Aachen, FRG

Summary

The characteristics of lattice strain measurements by X-rays on amorphous and semicrystalline polymeric materials will be described. In amorphous polymers, a metallic powder serves as the crystalline phase. In the case of semicrystalline polymers, the crystalline phase is being measured.
The residual stresses (RS) of injection molded plates of polypropylene were determined using the usual reflection technique and the recently developed transmission technique. The RS state of an injection molded plate consists of compressive RS in the near-surface regions; in the middle of the cross-section, tensile RS maintain equilibrium.
The RS states of polypropylene plates, 5-mm thick injection molded, by varying the manufacturing parameters in wide limits were determined. All the varied manufacturing data and the measured RS on the surface were correlated by a multiple-linear-regression analysis. Thus, the nondestructively determined RS in injection molded polypropylene plates are characteristic material data compiled from the parameters of manufacturing.

Zusammenfassung

Die Merkmale der Gitterdehnungsmessung mit Röntgenstrahlen an amorphen und teilkristallinen Kunststoffen werden vergleichend dargelegt. Bei amorphen Polymeren dient ein Metallpulver als kristalline Phase, und bei teilkristallinen Polymeren wird der kristalline Anteil vermessen.
Die Eigenspannungen (ES) in spritzgegossenen Polypropylen(PP)-Platten wurden mit Hilfe der üblichen Reflexions- und der kürzlich entwickelten Transmissionstechnik ermittelt. Der ES-Zustand spritzgegossener Platten ist durch Druck-ES nahe der Oberfläche gekennzeichnet, die von Zug-ES in Querschnittsmitte kompensiert werden.
Die ES-Zustände der 5 mm dicken PP-Platten, gefertigt unter Variation der Herstellparameter in weiten Grenzen, wurden bestimmt. Die Fertigungsparameter und die ermittelten ES an der Oberfläche wurden mit Hilfe einer multiplen linearen Regressionsanalyse korreliert. Somit stellen die zerstörungsfrei an den Platten ermittelten ES charakteristische Materialkennwerte dar, die von den Herstellparametern herrühren.

The X-ray stress analysis on amorphous polymeric materials was introduced by Barrett and Predecki /1-6/. The crystalline phase to measure the lattice strain was realized by metallic fine-grained powder mixed with the carbon-fibre reinforced epoxy resin before curing. Hauk, Troost, Vaessen /7/ and Hauk, Troost, Ley /8/ used this method on polystyrene with Al-powder. Wörtler and Schnack /9/ made three-dimensional studies with carbon-fibre reinforced epoxy laminates. Hauk, Troost, Vaessen /7/ and Hauk, Troost, Ley /8, 10-13/ used the crystalline α-phase of polypropylene (PP) itself to measure the lattice strain caused by residual or load stresses.

Characteristic properties of both methods are listed in Fig. 1. These are the crystalline phase, the location of the interference line in back or in front reflection, the X-ray elastic constants (XEC), the calibration of the diffractometer, the accuracies of strain measurement and of stress determination, the coupling between the metallic powder and the matrix to be taken into account, and the penetration depth. A surprising aspect is the fact that the accuracy of the stress determination is of the same order of magnitude for both methods. The accuracy of the determination of the strain of the crystalline phase of polymers is poor, but the Young's moduli of the metallic powder and the crystalline phase of the polymer are very different.

property parameter		amorphous material	semi-crystalline material
crystalline phase		≤ 5 vol.% metal powder	≥ 10 vol.% phase of the material
peaks		back reflection	front reflection
XEC		known	to be determined
calibration of the diffractometer		metal powder in amorphous material	powder of the semi-crystalline material
accuracy of measurement	strain	2×10^{-5}	5×10^{-4}
	stress	2 MPa	1 MPa
coupling with matrix		factor = 2, φ = 0° factor = -0.3, φ = 90°	
penetration depth for $\sin^2\psi = 0.3$		0.2, 0.6 mm	0.1, 0.2 mm

Fig. 1: Characteristics of lattice strain measurements by X-rays on amorphous and semicrystalline polymeric materials

The coupling of the single crystals and the elastic constants of a polycrystalline material using the compliances were evaluated by Eshelby /14/, Kröner /15/, and Kneer /16/ supposing the model of an anisotropic sphere within an isotropic matrix. From these results Bollenrath, Hauk, Müller /17/ have calculated the X-ray elastic constants (XEC) and their orientation of (hkl) plane dependence; for noncubically crystallizing metals see /18,19/. The XEC of heterogeneous and composite materials have also been evaluated /20/. If the Young's modulus of the inclosed sphere is much higher than that of the isotropic matrix, as in the case of metallic powder in a polymeric material, the stress on the inclusion is two times larger than that of the PP matrix parallel to the load stress and is in transverse direction -0.3 times the load stress.

The penetration depth, though the polymer with metallic powder has a higher attenuation factor as an unfilled polymeric material, is bigger since the incident angle is smaller according to the higher Bragg's angle. The range of values of penetration depth (1/e intensity decrease for $\sin^2\psi = 0.3$) is caused by different Bragg angles using CrK_α- or Cu K_α-radiation. The angle ψ, representing the measuring angle, is the angle between the normals to the surface of the specimen and to the reflecting plane of the crystallites. The penetration depth as defined above amounts, for the (130) peak of PP, to 0.1 mm using Cr-radiation and to 0.2 mm for Cu-radiation.

The two methods were tested with plates 3-mm thick of injection molded PP filled with 10 wt.% Al-powder, Fig. 2. The basic formula of X-ray stress analysis is the following /21, 22, 23/:

$$\sigma_\varphi - \sigma_3 = \frac{1}{\frac{1}{2}s_2} \frac{1}{D_\circ} \frac{\partial D_{\varphi,\psi}}{\partial \sin^2\psi} \qquad (1)$$

To determine the load or the residual stress (RS) from the measured lattice strain distribution, slope of $D_{\varphi,\psi}$ vs. $\sin^2\psi$ dependence, the value of the XEC $1/2\ s_2$ must be evaluated since the compliances of the α-phase of PP are not determined. The XEC of Al-powder is determined /21, 22/. The meaning of the symbols is as follows: $D_{\varphi,\psi}$ the lattice parameter, φ azimuth, ψ the measuring angle as defined above, and $D_\circ$ being the lattice parameter of the stress-free state.

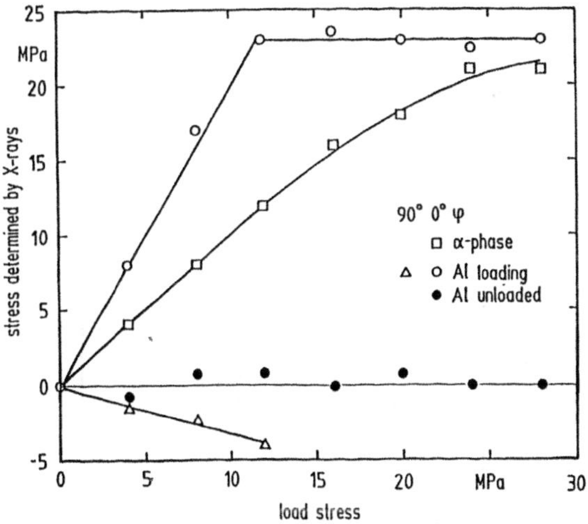

Fig. 2: Stresses evaluated by X-rays in dependence on the mechanical load stress for PP filled with 10 wt.% Al-powder. Al {511+333} peak with CuK_α-radiation, α-phase (130) peak with CrK_α-radiation.

In Fig. 2 the stress determined by X-ray measurements of the lattice strain on the Al-powder and on the α-phase crystallites of PP is plotted against the mechanical load stress. Up to a mechanical load stress of approximately 12 MPa, both dependences are linear; but there is a factor of approximately 2 according to the theory mentioned. The stresses in the transverse direction are compressive in sign with a factor 0.3 to the load stress according to the theory. Unloading the sample yields no RS as measured on the Al-powder. The results of the tests shown in Fig. 2 enable us to evaluate the load stress and the RS by both methods of using the crystalline phase of the Al-powder or the α-phase of PP. Of course there is a third phase in PP (β-phase) with a content of 10 vol.%. The influence of this relatively small amount on the result of stress values may be neglected.

In the following the lattice strain measurements were made on the α-phase with CrK_α- and CuK_α-radiations. Both techniques, the usual reflection technique and the recently developed transmission technique /8, 11/, were used, Fig. 3. The reflection technique requires a Ψ-diffractometer while the Ω-diffractometer reveals the possibility after very careful calibration and using a narrow and long vertical Soller slit to measure the lattice strain at different points of the cross-section. Typical results of both techniques are plotted in Fig. 4.

Fig. 5 represents the result of the RS state of an injection molded plate 5 mm thick on both surfaces, in the near-surface region, and in the bulk ma-

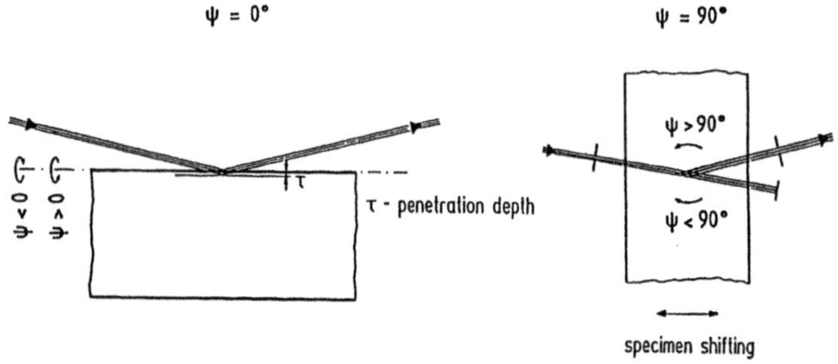

Fig. 3: The usual reflection technique (left, Ψ-diffractometer) and the newly developed transmission technique (right, Ω-diffractometer)

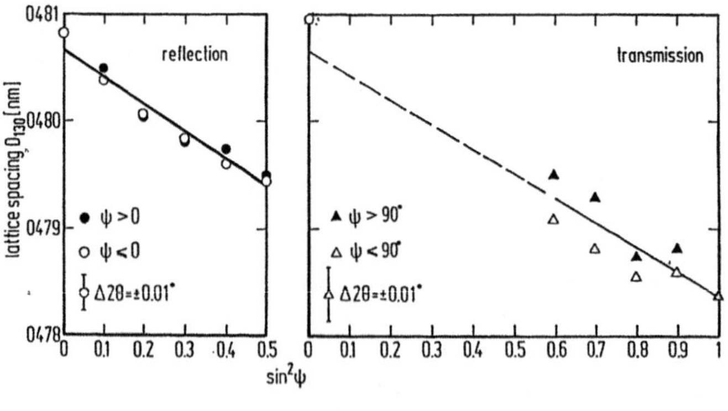

Fig. 4: Lattice spacing vs. $\sin^2\psi$ evaluated by reflection (CrK_α-radiation, Ψ-diffractometer) and transmission (CuK_α-radiation, Ω-diffractometer) techniques; outer surface of a longitudinal cut of a PP tube, 110-mm outside diameter and 10-mm wall thickness. The error bar represents the calibration accuracy.

terial. There are compressive RS in the near-surface regions; in the middle of the cross-section tensile RS hold the equilibrium. From Fig. 5 also follows that the torques are compensated. Since the RS distribution over the cross-section of injection molded plates of PP has the shape as shown in Fig. 5 the amount of compressive RS on the surface characterize the manufacturing procedure.

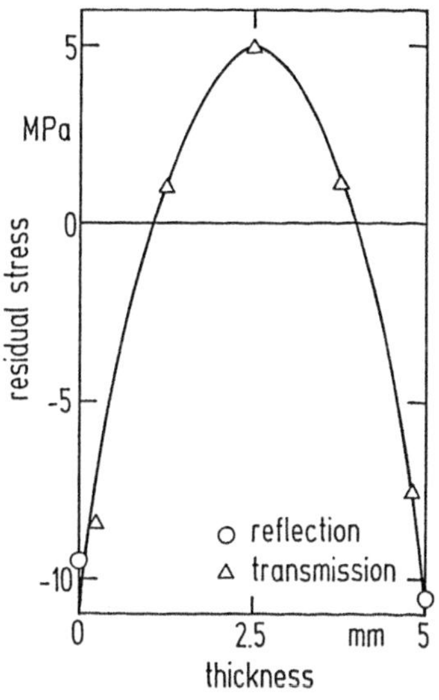

Fig. 5: The RS state of an injection molded plate of PP

Fig. 6 shows the parameters of injection molded plates and the data chosen that are of a larger range as usually used. In detail, the wall temperature was chosen between 20° and 80° C, the mass temperature between 200° and 280° C, and the volume stream between 30 and 100 cm^3/s whereas the packing pressure was taken always the same 835 bar and the thickness of the plates with 5 mm. In total 25 combinations of manufacturing parameters were used to produce plates.

All the varied manufacturing data and the measured RS on the surface using CrK_α-radiation and the Ψ-diffractometer were correlated by a multiple linear regression analysis taking into account the least square method /24/. The result of this evaluation shows the following formula that holds for this special manufacturing process:

$$RS = 0.082\,\vartheta_W + 0.005\,\vartheta_M - 0.148\,\dot{V} - 0.9 \quad \text{in MPa} \tag{2a}$$

with ϑ_W, ϑ_M in ° C, $\dot{V}$ in cm^3/s. This formula was normalized by the minimum quantities:

$$RS = 1.7\,\frac{\vartheta_W}{\vartheta_{W\,min}} + 1.0\,\frac{\vartheta_M}{\vartheta_{M\,min}} - 4.4\,\frac{\dot{V}}{\dot{V}_{min}} - 0.9 \tag{2b}$$

The compressive RS are generated by the nonuniform cooling of the plate. The volume stream has a major influence. The effect of the mass temperature is small relative to the effect of the wall temperature /25/. This formula should not be considered as a general law but as a representation of these measurement results.

Fig. 7 shows good correlation between the measured RS and the RS calculated from the manufacturing data. The confirmity of the experimental results and the calculated values is regarded as very good, recognizing that the accuracy of RS determination is 1 MPa. Thus the nondestructively determined RS in injection molded plates are characteristic material data compiled from the parameters of manufacturing.

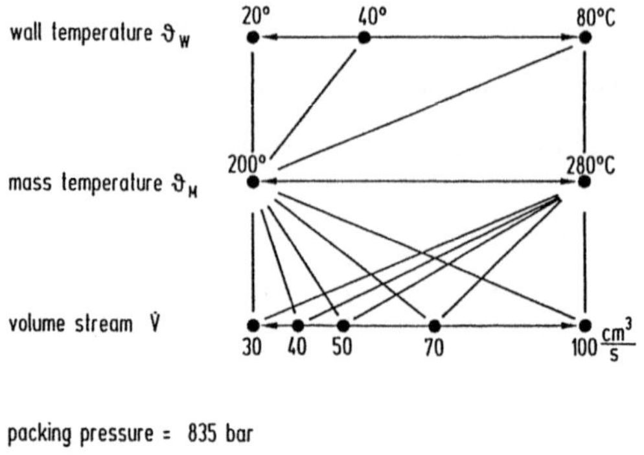

Fig. 6: 25 combinations of manufacturing parameters of injection molded PP-plates 5 mm thick

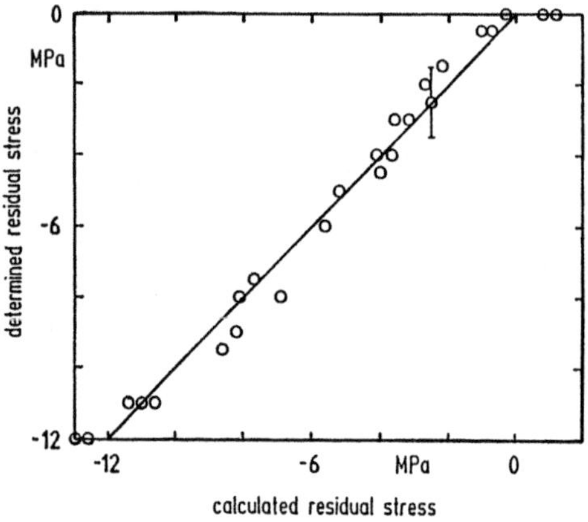

Fig. 7: Còrrelation of RS in injection molded PP-plates 5 mm thick and the manufacturing parameters compiled in formula (2)

References

1. Barrett, C.S.; Predecki, P.: Stress measurement in polymeric materials by X-ray diffraction. Polym. Eng. Sci. 16 (1976) 602-608.

2. Barrett, C.S.: Diffraction technique for stress measurement in polymeric materials. Adv. X-Ray Analys. 20 (1977) 329-336.

3. Barrett, C.S.; Predecki, P.K.: Measuring triaxial stresses in embedded particles by diffraction. Adv. X-Ray Analys. 21 (1978) 305-307.

4. Predecki, P.; Barrett, C.S.: Stress measurement in graphite/epoxy composites by X-ray diffraction from fillers. J. Comp. Mat. 13 (1979) 61-71.

5. Barrett, C.S.; Predecki, P.: Stress analysis in graphite/epoxy. Adv. X-Ray Analys. 23 (1980) 331-332.

6. Barrett, C.S.; Predecki, P.: Stress measurement in graphite/epoxy uniaxial composites by X-rays. Polymer Comp. 1 (1980) 2-6.

7. Hauk, V.; Troost, A; Vaessen, G.: Zur Ermittlung von Spannungen mit Röntgenstrahlen in Kunststoffen. Materialprüf. 24 (1982) 328-329.

8. Hauk, V.; Troost, A.; Ley, D.: Lattice strain measurements and evaluation of residual stresses on polymeric materials. In "Residual stresses in science and technology". Macherauch, E.; Hauk, V. (eds.) DGM-Informationsgesellschaft Verlag, Oberursel 1 (1987) 117-125.

9. Wörtler, M.; Schnack, E.: Röntgenographische Spannungsermittlung an Faserverbunden. VDI Berichte Nr. 631 (1987) 163-174.

10. Hauk, V.; Troost, A.; Ley, D.: Röntgenographische Ermittlung von Eigenspannungen in Polypropylen. Materialprüfung 27 (1985) 98-100.

11. Hauk, V.; Troost, A.; Ley, D.: Evaluation of (residual) stresses in semicrystalline polymers by X-rays. Adv. Polym. Techn. 7 (1987) 389-396.

12. Hauk, V.; Troost, A.; Ley, D.: Röntgenographische Dehnungsmessung und Spannungsermittlung an kohlenstoffaserverstärktem PEEK. Kunststoffe 78 (1988) 10, in the press.

13. Hauk, V.; Troost, A.; Ley, D.: Ermittlung des Eigenspannungszustands extrudierter Polypropylen-Rohre mit Röntgenstrahlen. Kunststoffe 79 (1989) 1, in the press.

14. Eshelby, J.D.: The determination of the elastic field of an ellipsoidal inclusion and related problems. Proc. Roy. Soc., Ser. A 241 (1957) 376-396.

15. Kröner, E.: Berechnung der elastischen Konstanten der Vielkristalls aus den Konstanten des Einkristalls. Z. Physik 151 (1958) 504-518.

16. Kneer, G.: Die elastischen Konstanten quasiisotroper Vielkristallaggregate. Phys. Stat. Sol. 3 (1963) K 331 - K 335.

17. Bollenrath, F.; Hauk, V.; Müller, E.H.: Zur Berechnung der vielkristallinen Elastizitätskonstanten aus den Werten der Einkristalle. Z. Metallkde. 58 (1967) 76-82.

18. Evenschor, P.D.; Hauk, V.: Berechnung der röntgenographischen Elastizitätskonstanten aus den Einkristallkoeffizienten hexagonal kristallisierender Metalle. Z. Metallkde. 63 (1976) 798-801.

19. Behnken, H.; Hauk, V.: Berechnung der röntgenographischen Elastizitätskonstanten (REK) des Vielkristalls aus den Einkristalldaten für beliebige Kristallsymmetrie. Z. Metallkde. 77 (1986) 620-626.

20. Hauk, V.; Kockelmann, H.: Berechnung der Spannungsverteilung und der REK zweiphasiger Werkstoffe. Z. Metallkde. 68 (1977) 719-724.

21. Hauk, V.; Macherauch, E.: Die zweckmäßige Durchführung röntgenographischer Spannungsermittlungen. In HTM-Beiheft "Eigenspannungen und Lastspannungen". Hauk, V.; Macherauch, E. (eds.) Carl Hanser Verlag, München, Wien (1982) 1-19.

22. Hauk, V.M.; Macherauch, E.: A useful guide for X-ray stress evaluation (XSE). Adv. X-Ray Analys. 27 (1984) 81-99.

23. Hauk, V.; Macherauch, E.: The actual state of X-ray stress analysis. In "Residual stresses in science and technology". Macherauch, E.; Hauk, V. (eds.) DGM Informationsgesellschaft Verlag, Oberursel 1 (1987) 243-255.

24. The authors would like to thank M. Schaefer (Institut für Statistik und Wirtschaftsmathematik, RWTH Aachen) for calculations.

25. Menges, G.; Thienel, P.; Targiel, G.: Verteilung der Eigenspannung und Verarbeitungsschwindung thermoplastischer Spritzgußteile. Maschinenmarkt 84 (1978) 240-242.

Acknowledgement

The financial support from the Deutsche Forschungsgemeinschaft is gratefully acknowledged.

Characterization of Varnish Layers Using Optically Generated Thermal Waves

G. BUSSE and D. VERGNE

Institut für Kunststoffprüfung und Kunststoffkunde
Universität Stuttgart, Pfaffenwaldring 32,
D-7000 Stuttgart 80, Germany F. R.

Summary

Thermal wave radiometry is applied to the inspection of paint on polymer substrates. Thickness changes and surface effects like contamination and pretreatment are monitored.
The investigation of curing, drying and temperature effects is also possible.

1 INTRODUCTION

The general idea of nondestructive evaluation (NDE) is that parts with no defects can still be used after being inspected and that their quality is not reduced.
This concept is presently not applicable to paint on polymers, if one uses the conventional techniques. However, thermal waves seem to be a very promising method for the inspection of polymer coatings, especially since there is no physical contact required.

The principle of operation is based on heat transport. If a laser beam with modulated intensity is focussed on the surface of a sample (fig.1), then periodical deposition of optical energy generates a temperature modulation which propagates from the focus as a heavily damped wave.
In a one-dimensional model its propagation behaviour $\Theta(z,t)$ with respect to the focal point is described by

$$\Theta(z,t) = \exp(-z/\mu)\ \sin(\omega t - z/\mu) \tag{1}$$

with the thermal diffusion length μ given by

$$\mu = (2\ k/(\omega\ \rho\ c))^{-1/2} \tag{2}$$

(temperature Θ, modulation frequency ω, depth z, time t, thermal conductivity k, density ρ, specific heat c).

While remote generation of thermal waves is optically achieved with a modulated laser beam, non-contacting detection of thermal waves is possible whith photothermal radiometry [2]. As the phase of the signal does not depend on the optical absorption, we use this significant information in our experiments to reveal thermal structures.

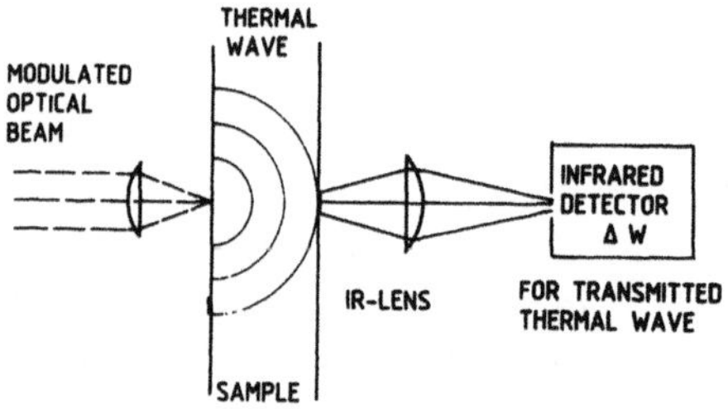

Fig.1. Transmission arrangement

2 RESULTS

2.1 Transmission arrangement

In this experiment and the following [1,3,4] the thermal wave transmission arrangement (Fig.1) was used.
We monitored remotely the drying process of paint. The phase changes as a function of time (Fig. 2). Refered to such a cali-

bration curve a single phase measurement will indicate the stage of drying during similar processes.

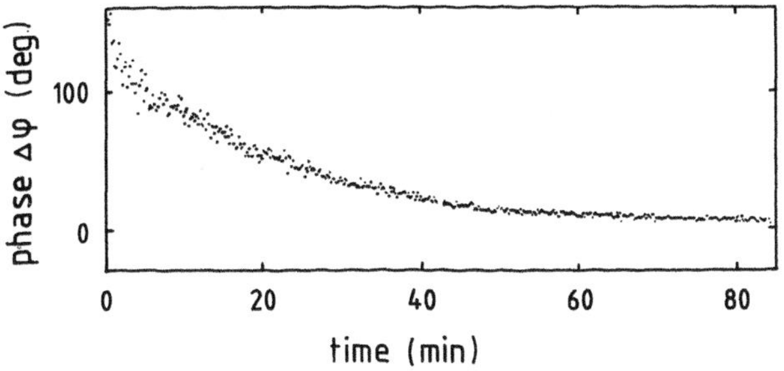

Fig.2. Drying of paint on metal

As another example for a time dependent process the curing of an epoxy adhesive between two metal plates was measured [1,5]. The rapid change of phase angle (Fig.3) shows the macroscopic solidification. After that the phase keeps changing slowly for a very long time. So monitoring the phase reveals processes going on in paint and polymers.

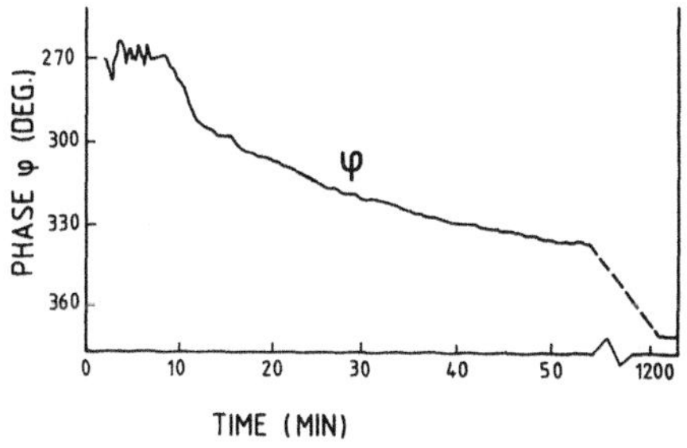

Fig.3. Effect of curing on phase lag

2.2 Front surface inspection

The transmission arrangement may be a problem for some applications. An alternative is a single-ended arrangement (Fig.4) where both generation and detection of thermal waves are performed on the same side of the sample. If there is a boundary in the sample the detector will observe the superposition of two waves: one direct wave from the laser focus to the detector point, and in addition a reflected wave from the boundary to the detector point. In the gaussian plane the superposition of these two waves is represented by the addition of the two corresponding vectors. Thus the contribution of a reflection modifies the signal vector by a certain angle.

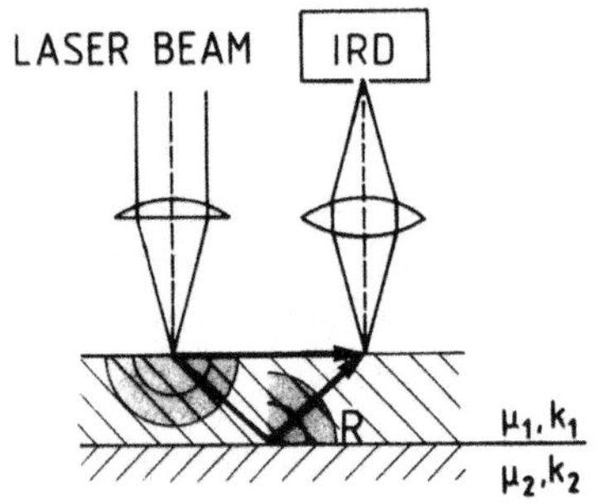

$$R = \frac{1 - \frac{\mu_1 k_2}{\mu_2 k_1}}{1 + \frac{\mu_1 k_2}{\mu_2 k_1}}$$

Fig.4. Front surface arrangement

The more the materials are similar the less reflection occurs. Therefore the boundary paint/polymer has only a rather weak reflection and thickness variations of coatings on metal provide stronger signal changes than on a polymer substrate [1,4]. In fig.5 curve "a" shows how the signal phase depends on the paint thickness for an aluminium substrate, curve "b" corres-

ponding on a polymer substrate. The reduced slope makes the calibration more difficult giving a thickness accuracy of nearly ± 2 μm at 50 μm and ± 6 μm at 75 μm [1,4].
Due to the boundary reflection the calibration curves need to be made for each polymer substrate separately.

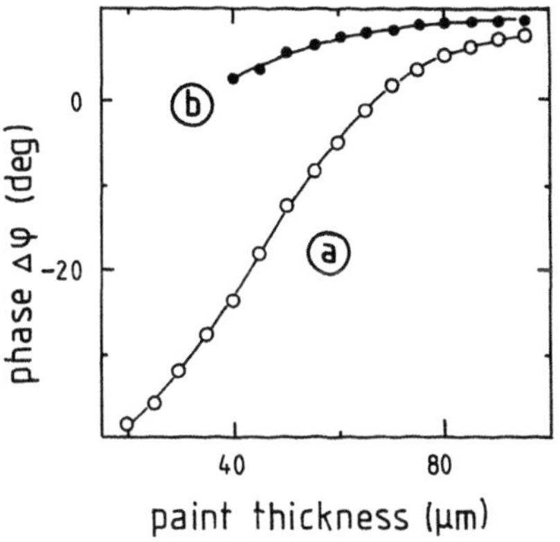

Fig.5. Thickness dependent phase for metal (a), and polymer (b) substrates

In the special case of very similar materials like polyurethane paint on polyurethane substrate the method is still suited for coating thickness measurement (Fig.6). The variation of phase angle is still correlated with a thickness change from 1 μm to 20 μm. The structure on the signal is reproducible, it can be attributed to pigment effects or even to the roughness of the substrate surface.

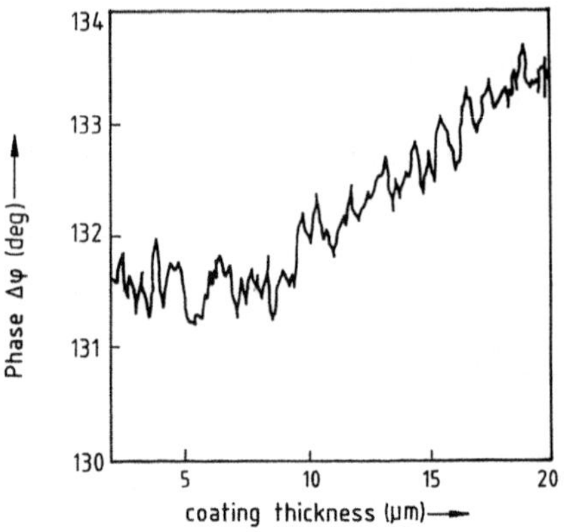

Fig.6. Calibration curve for similar substrate and paint (polyurethane) [1]

As paint undergoes certain processes with a temperature dependence, we monitored signal variations caused by temperature changes (Fig.7).

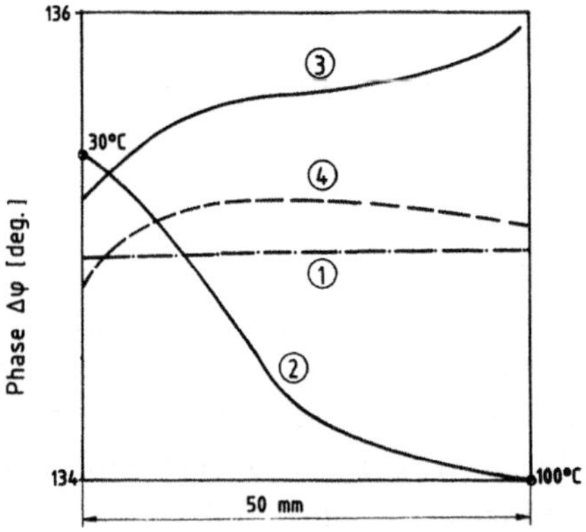

Fig.7. Temperature effects [1]: (1) untreated, (2) scan across heated region from 30 °C to 100 °C, (3) heating switched off, (4) after long cooling

A constant phase angle is observed during the scan across a polyurethane sample with a homogeneous layer of polyurethane paint (curve 1). Heating the right end of the sample to 100 °C while the left end is kept at 30 °C produces a temperature gradient which results in a signal gradient (curve 2). After this temperature gradient is switched off this curve changes to a curve with an opposite sign of slope (curve 3). After a long time curve 4 is measured which is much closer to the initial curve 1. Thus we can look at slow processes which are correlated with temperature.

The signal may be affected by surface pretreatment and contaminations too [1]. Figure 8 is the result of a raster scan across a polymer substrate partially covered with grease (line from top left to bottom right) and thereafter provided with a stripe of black paint (top right to bottom left).
These simulated defects were hidden under 40 μm thickness of the same black paint.
So this method is very sensitive to boundary situations and reveals defect regions in polymer systems.

Fig.8. Raster scan across black paint on a polymer substrate

So thermal wave radiometry is well suited for the inspection of paint on polymer substrates. Therefore we think that this method is applicable to a lot of industrial problems, e.g. in the car manufacturing process.

References

[1] Vergne D., Busse G., Infrared Physics (to be published)

[2] Nordal P.-E. and Kanstad S.O., Phys. Scr. 20, 659 (1979)

[3] Busse G., Infrared Physics 20, 419 (1980)

[4] Busse G., Vergne D., and Wetzel B., in " Photoacoustic and photothermal phenomena " (eds. P. Hess and J. Pelzl), 427, Springer, Heidelberg (1987)

[5] Busse G., Eyerer P., Appl. Phys. Lett. 43, 355 (1983)

Application of Dielectric Spectroscopy for Nondestructive Investigation of Expoxy Curing

P. ELSNER

Institut für Kunststoffprüfung und Kunststoffkunde
Universität Stuttgart, Pfaffenwaldring 32, 7000 Stuttgart 80
Germany F.R.

Summary

The monitoring of curing processes is of interest for industrial applications. However, there is presently a real need for methods that can be used in a nondestructive way.

The method suggested here is dielectric spectroscopy where the sample under investigation is exposed to a modulated electric field. One measures the admittance as a function of modulation frequency.

As only electrical quantities need to be measured and controlled, such experiments can be conveniently coupled to a computer which also handles data evaluation and presentation. This way one obtains real-time information. This method is also nondestructive, since the required arrangement of the electrodes can often easily be adapted to the sample or the component geometry.

From the frequency shift of the dielectric loss curves during the cure process we determine the the time constant of the curing process. As the values found this way agree with known data, we conclude that the method is generally applicable to real life samples.

Theory

When a voltage is applied across an insulating material a polarization occurs. The polarization produces a permittivity in where the imaginary part can be split into a dipol and a conductivity term /1/,

$$\epsilon(\omega) = \epsilon'(\omega) - i\{\epsilon''(\omega) + \sigma_0/\omega\} \qquad (1)$$

with ϵ = complex permittivity; ϵ' = dielectric constant; ϵ'' =dielectric loss modulus; σ_0 = dc conductivity; ω = circular frequency .

From an admittance measurement one can calculate the loss factor

$$\tan\delta = \epsilon''/\epsilon' \qquad (2)$$

which is described elsewhere /1/. In this case the loss factor contains the dipolar and the conductive part.

Experimental

Polymeric resins and prepregs were placed between electrodes (e.g. metal plates, metal foils) of a plate capacitor (Fig. 1) and the admittance was measured as a function of frequency with a dielectric bridge (HP 4192A) that is controlled by a computer (HP 9000). Depending on the polymeric system the samples were cured by different temperatures. The frequency range from 10 Hz to 10 MHz can be swept in less than 10 seconds with a reading rate of ten points per frequency decade. The curing time of these systems are about on hour. Therefore we can assume that the measured admittance spectrum characterizes the system at a certain process stage.

Results and Discussion

Figure 2 shows subsequent frequency spectra obtaunde for the curing process of an epoxy dicyandiamide (EP-Dicy) prepreg reinforced with glass fibers. With inreasing curing time the spectra shift to lower frequencies. Each frequency sweep contains ten measurement points. These points are connected with a full line. No mathematical fit was done.

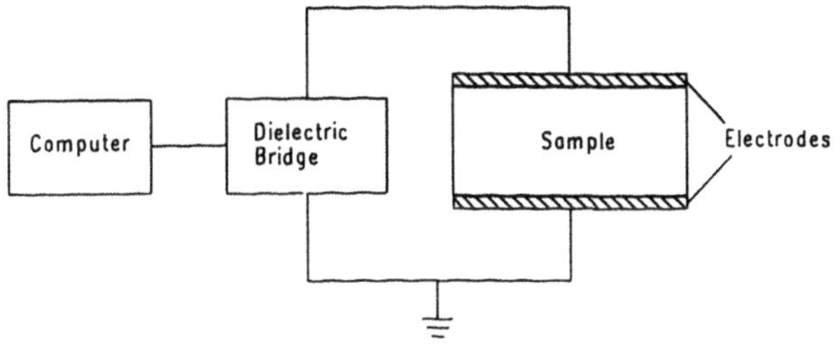

Fig.1. Experimental setup for measuring dielectric properties

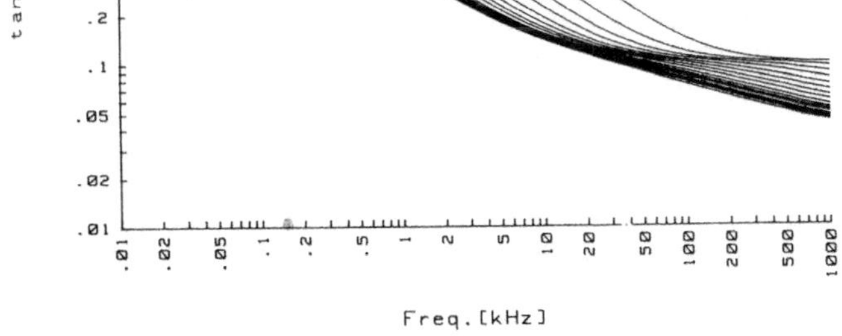

Fig.2. Dielectric tan δ during cure as a function of frequency and curing time of an EP-Dicy prepreg.
Isothermal curing temperature: 150 °C

To eliminate the contribution of the conductivity term a thin polymer foil (Mylar 4 μm) was placed between the electrodes and the prepreg. The result is seen in figure 3. By this technique the dipole peaks are well separated from the conduction part of the loss factor . Usually mathematic peak cleaning methods or physical assumptions are used: However they often fail because of oversimplyfying theories /2/. A peak reversal predicted from theory /3/ cannot be found.

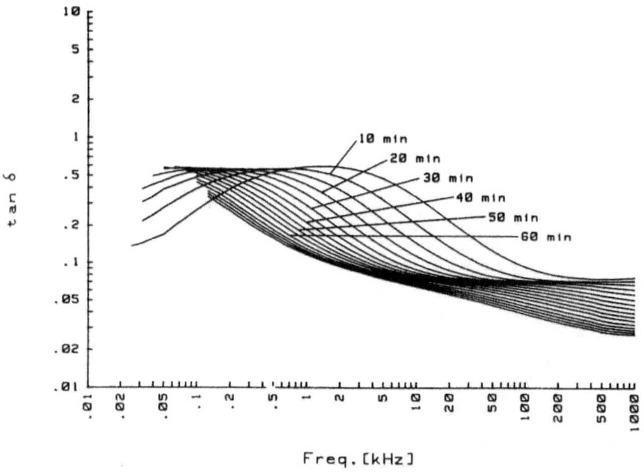

Fig.3. Dielectric tan δ during cure as a function of frequency and curing time of an EP-Dicy prepreg with a thin foil between the electrodes and prepreg. Isothermal curing temperature: 150 °C

In figure 4 the frequency shift of the curves in figure 3 is plotted against the curing time. The shift was done with regard to a constant value of 0.2 of the loss factor. The full line is an exponential fit

$$y(t)=A * \exp^{-Bt} + E. \qquad (3)$$

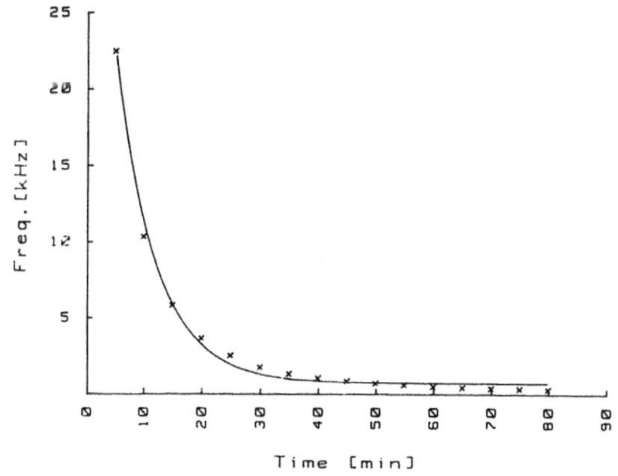

Fig.4. Frequency shift of the dielectric loss factor (tan δ=0.2) as a function of curing time. Data from fig.3.
Full line: Fit with eq. 3

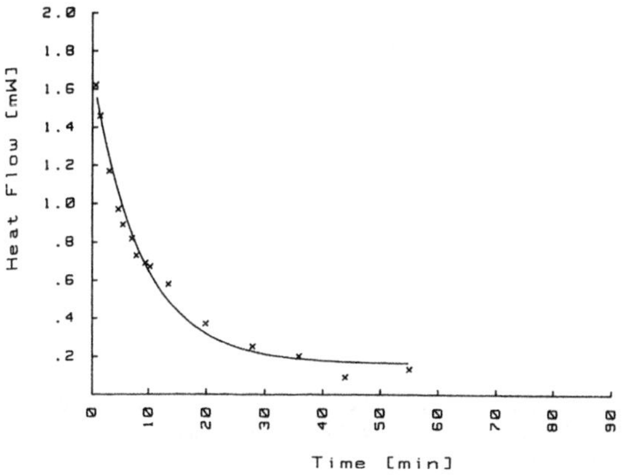

Fig.5. Heat Flow as function of curing time

In reaction kinetics a similar relation exists where the term 1/B is the time constant of the reaction /4/. The values found with dielectric measurements are equal to the values found by DSC (Dynamic Scanning Calorimetry). Figure 5 shows the curve obtained by

DSC measurements. It is very similar to the curve in figure 4 and gives the same B-value.

As a difference to the DSC measurements the values of the dielectric measurements display a nonstatistical deviation from the simple mathematical fit. Considering that EP-Dicy is a system with two distiguishable reaction steps /5,6/, a double exponential fit

$$y(t) = A * \exp^{-Bt} + C * \exp^{-Dt} + E \qquad (4)$$

performs significantly better (figure 6).

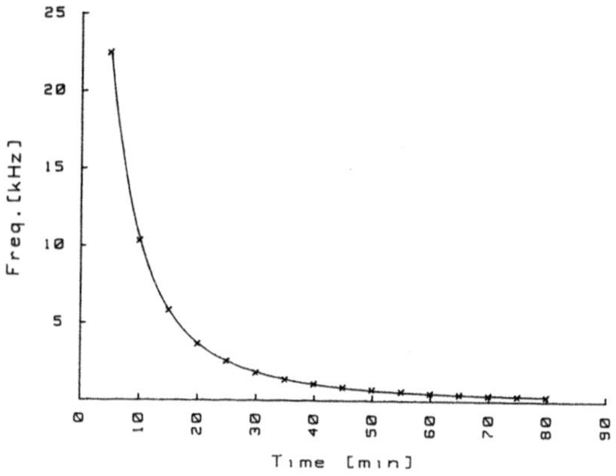

Fig.6. Frequency shift of the dielectric loss factor (tan δ=0.2) as a function of curing time. Data from fig.3.
Full line: Fit with eq. 4

Assuming that both exponential terms 1/B an 1/D are the time constants of both reaction steps (a polymerisation (chain growing) and a polyaddition (crosslinking) reaction) we perform experiments with the EP-Dicy at different curing temperatures.

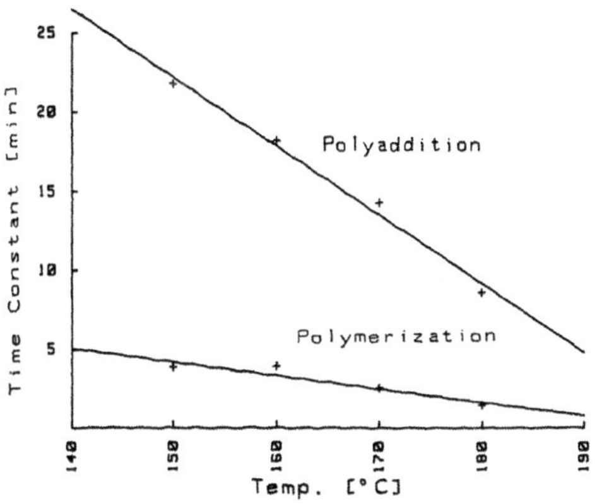

Fig.7. Time constants for the EP-Dicy System as a function of curing temperature (eq. 4).

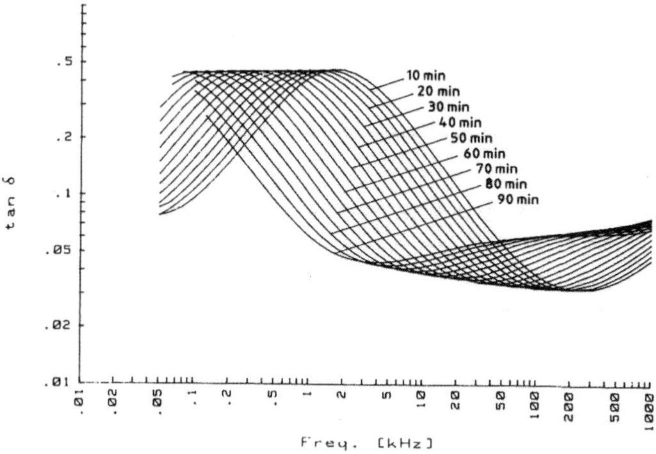

Fig.8. Dielectric tan δ during cure as a function of frequency and curing time of an EP-Diamine prepreg with a thin foil between the electrodes and prepreg. Curing temperature: 23 °C

Figure 7 shows how the calculated time constants depend on temperature . After three time constants the degree of crosslinking is about 95% . These results correlate well with practical experiences at the used system. In addition, one can use the second time constant to describe the annealing effects of these epoxy systems.

Other epoxy systems (e.g.diamine systems) display similar results. Even carbon fiber prepregs can be measured in the same way. Figure 8 shows the dielectric spectra of a epoxy diamine system cured by room temperature.

Conclusion

Reaction kinetics parameters can be derived from dielectric measurements. "Frequency - curing time" shift allows both controlling and optimizing the curing process of polymeric resins using the . Advantages of the dielectric method are

- nondestructive evaluation
- short measurement time
- no special sensors required
- adaption to nearly every geometry
- absolute values are not nessecary (no calibration !)
- also usable for carbon fiber reinforced prepregs
- automatization capability
- annealing effects can be estimated.

The disadvantages are

- empirical interpretation
- application to other polymeric resins are not yet tested.

References

/1/ Jonscher, A.K.: "Dielectric relaxation in solids". Chelsea Dielectrics Press London 1983

/2/ Senturia, S.D., Sheppard, N.: Dielectric Analysis of Thermoset Cure. In : Advances in Polymer Science 80, (1986), 3-50

/3/ Day, D.R., Lewis, T.J., Lee, H.L., Senturia, S.D.: The Role of Boundary Layer Capacitance at Blocking Electrodes in the Interpretation of Dielectric Cure Data in Adhesives. In: J. Adhesion, 18, 73 (1985)

/4/ Hemminger/Höhne : "Grundlagen der Kalorimetrie" Verlag Chemie Weinheim New York 1979

/5/ Eyerer, P. : "Eine zerstörungsfreie elektrische Prüfmethode zur Überwachung von Aushärtevorgängen an Duromeren". Dissertation, Universität Stuttgart 1972

/6/ Saunders, M.F., Levy, M.F., and Serino, J.F. In: J. Polym. Sci. Mechanism of the Teriary Amine-Catalysed Dicyandiamide Cure of Epoxy Resins. A-1, 5, (1967) 1609-1617

The Investigation of Mass Transport Through Polyethylene with Concentration Waves

F. TWARDON, O. LEITZBACH, and G. BUSSE

Institut für Kunststoffprüfung und Kunststoffkunde, Universität Stuttgart, 7000 Stuttgart 80, Germany F.R.

Summary

The investigation of mass transport is of interest since it provides information on the parameters involved in the process of permeation, diffusion and solution and so also on the structure of polymers. Applications are in the field of material characterization and nondestructive testing. Transport processes can be analysed both by the reaction to step functions (the conventional method) or by the stationary response to a periodical input. The latter is the general concept of concentration waves. In this paper we report on measurements where the concentration of the diffusing gas was modulated in a sinusoidal way. From the observed phase shift between the input (the concentration variation) and the output (the permeation rate) the diffusion coefficient can be calculated.

Zusammenfassung

Die Untersuchung des Stofftransports liefert Informationen über die Kenngrößen der Permeation, Diffusion und Sorption und damit auch über die Struktur von Polymeren. Anwendungen liegen im Bereich der Materialcharakterisierung und zerstörungsfreien Prüfung. Transportprozesse können zum einen über die Auswertung der Systemantwort auf ein stufenförmiges Eingangssignal (die herkömmliche Methode) oder der stationären Antwort auf ein periodisches Eingangssignal untersucht werden. Letzteres ist die Basis der Konzentrations-Wellen-Methode. Die vorliegende Arbeit behandelt Messungen bei denen der Partialdruck und damit die Konzentration des diffundierenden Gases sinusförmig moduliert wird. Aus dem beobachteten Phasenwinkel zwischen dem Eingangssignal (der Partialdruckvariation) und dem Ausgangssignal (dem Permeationsstrom) ist der Diffusionskoeffizient berechenbar.

1. Introduction

Permeation of polymers means that atoms or molecules of a gas have a certain probability to enter polymers at one surface, to diffuse through this solid phase, and to leave them at the opposite surface. Therefore, if the partial pressure is different on both sides, a "permeation flux" results where atoms or molecules are transported through the polymer material. The investigation of these processes is of interest for a broad spectrum of applications ranging from food or beverage packages to membranes, utilized as devices to separate liquid or gas mixtures. In these cases the permeability itself is of main interest to give informations about the barrier properties of packages or about the selectivity of membranes.

The mass transport is influenced by the structure of the solid material, the polymer. So the improvement of permeability properties reqires the investigation of mass transport parameters as a function of structure parameters. The relationship between mass transport and structure allows to control the quality of materials.

2. Measuring Principle

The Permeation through a polymer exists on several partial processes. While the permeation coefficient P determines the whole transmission of molecules through the polymer the diffusion is characterized by the diffusion coefficient D. The solubility S describes the equilibrium between fluid and solid phase. These parameters are connected by the equation

$$P = D \cdot S \quad (1)$$

D and S depend directly on the structure of the solid polymer. To investigate the relationship between structure and mass transport one determines these coefficients.

This is usually done by measuring the response to a step function input /1/ where at a time t=0 a sample is exposed to a gas or vapor phase and the permeation rate I_P is determined as a function of time (fig 1). The permeability P results from the asymptotic value I_{Pmax}. The time-lag allows to calculate the diffusion coefficient D. In case of sorption measurements the mass uptake by the material is depicted as a function of time. The solubility S results from the asymptotic value and D from the initial slope of the function.

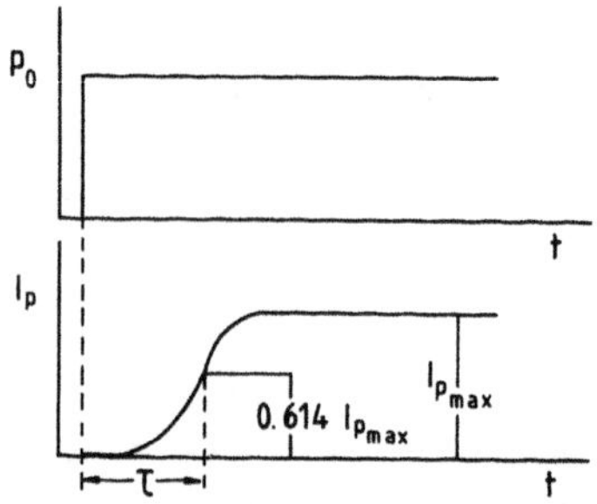

Fig 1: Input and output of the convential permeation measuring method

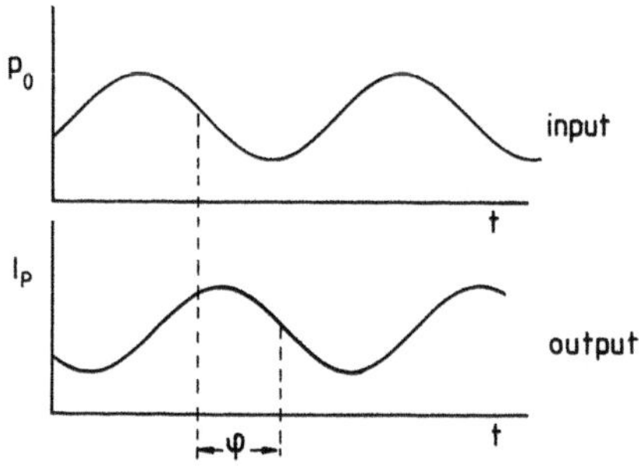

Fig 2: Input and output of concentration wave method

As the diffusion coefficient D describes the dynamic behaviour, one can determine this value only from insteady state measurements. As an alternative to the step function input one can use as well a periodical nonequilibrium situation. Therefore in our experiments the partial pressure is modulated (fig. 2). In the steady state the output of the system (permeation rate) oscillates with a constant phase shift that depends on the diffusion coefficient /2/.

The equation describing the kinetics of this process is FICK's second law

$$\frac{\partial c}{\partial t} = D\cdot\frac{\partial^2 c}{\partial x^2}. \tag{2}$$

Satisfying the boundary conditions

$$c = S\cdot(p_0+p_0^m\cdot\omega t), \quad x = 0, \quad t \geq 0 \tag{3}$$

$$c = 0, \quad x = L, \quad t \geq 0 \tag{3a}$$

for semi-infinite samples and for asymptotic behaviour (steady state) the solution of equ. (2)

$$c(x,t) = S\cdot p_0 - \frac{S\cdot p_0}{L}x + S\cdot p_0^m\cdot e^{-x\sqrt{\omega/2D}}\cdot e^{i(\omega t-x\sqrt{\omega/2D})} \tag{5}$$

determines the concentration as a function of location and time. p_0^m is the amplitude and p_0 the average of the modulated partial pressure. The dependence on x and t is like in a heavily damped wave. Therefore equ. (5) describes a concentration wave. As we measure the phase lag between partial pressure and permeation rate we have to differentiate equ. (5) and derive from Fick's first law

$$I_p = D\cdot S\cdot\frac{p_0}{L} + D\cdot S\cdot p_0^m\cdot e^{-x\sqrt{\omega/2D}}\cdot e^{i(\omega t-x\sqrt{\omega/2D}+\pi/4)} \tag{6}$$

With equation (5) and (6) we obtain

$$\varphi = \frac{L}{\sqrt{2D}}\sqrt{\omega} - \pi/4 + \varphi_{ap} \tag{7}$$

where L is the thickness of the sample and ω the angular frequency of modulation. The linear relationship between phase lag and the square root of frequency has in reality an additional intersect φ_{ap} caused by the apparatus. However, this can be eliminated by measuring the phase shift at two or more frequencies. Then D can be calculated from the slope

$$m = L/\sqrt{2D}. \tag{8}$$

By applying equ. (6) the solubility S is obtained. Also by evaluating the amplitude data one can determine the solubility.

3. Experimental Apparatus

The apparatus (fig. 3) has been described previously /2/. One

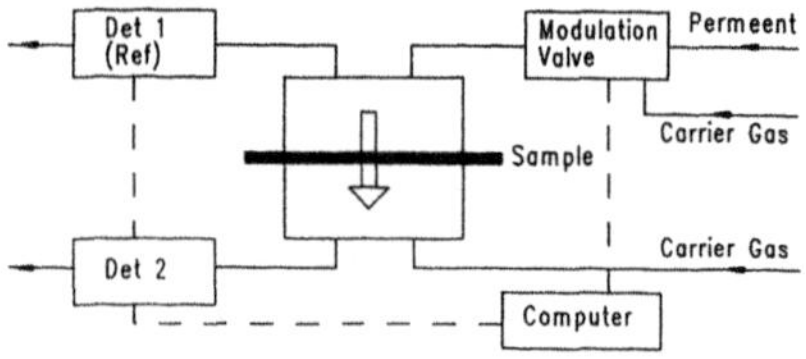

Fig. 3: Experimental arrangement for concentration wave analysis

difference as compared to other equipments is that it is a permeation and not a sorption method /3-6/. The other difference is that carrier gas is used and the permeation rate is measured. So we can use the equation (7) instead of the mathematical formalism used by Evnochides and others /7/. Another advantage is the variability of the measurement cells which are adaptable to a lot of different measuring problems.

4. Experimental Results

To verify the method we investigate the mass transport of methane and ethane through low density polyethylene (LDPE). Fig 4 shows the results for a 0.15 mm thick PE sheet. The diffusion coefficients (table 1) calculated from the slopes (see equ. 7 and 8) are very close to data published by Yasuda /8/.

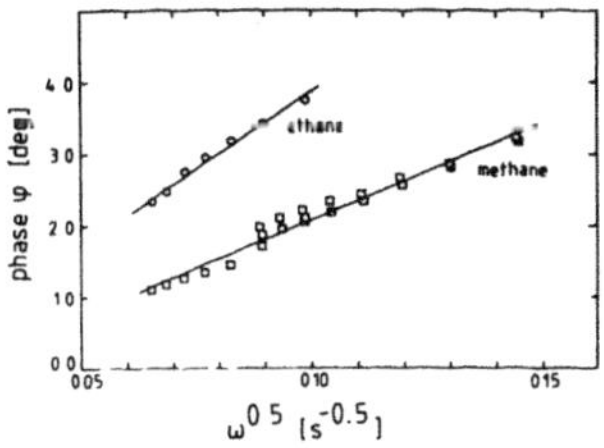

Fig. 4: Frequency dependent phase shift of methane and ethane in polyethylen (LDPE)

permeent	D in 10^{-12} m^2/s	
	literature	own results
CH_4	19	21
C_2H_6	6,8	9
temperature	25°C	20°C

Table 1: Comparison of own results and data published by Yasuda /8/

5. Conclusion

The concentration wave method has been used to investigate the mass transport through polymers and to determine the diffusion coefficient and solubility of gases and vapors. The advantages of the method are accuracy and the possibility to observe changes in the structure of the sample by continuous measurements. So the dynamic concentration wave method may be used for quality control. Another possible advantage is that by using the frequency response technique results are obtained in a shorter time.

6. References

1. H. Schuch: Kunststoffe, 33, 705 (1980)
2. F. Twardon, G. Busse, R. Müller: Photoacoustic and Photothermal Phenomena. Berlin, London: Springer 1988, page 329-332
3. D.L. Cummings, R.L. Reuben, D.A. Blackburn: Metallurgical Transactions A, 15A, 639 (1984)
4. R. Paterson, P. Doran: J. of Membr. Sci., 27, 105 (1986)
5. J.S. Vrentas, J.L. Duda, S.T. Ju, L.-W. Ni: J. of Membr. Sci., 18, 161 (1984)
6. Ju Shiaw-Tzuu: Thesis, Pennsylvania State Uni. 1981

7. S.K. Evnochides, E.J. Henley: J. of Poly. Sci., Part A-2, 8, 1987 (1970)
8. H. Yasuda, V. Stannett: Polymer Handbook. New York: Interscience Publishers, John Wiley & Sons 1975

Acknowledgement: The authors are grateful to S. Weiss for her skillful assistance.

Texture

Texture Analysis - A Method of Nondestructive Characterization of Materials I

H.J. Bunge
Department of Physical Metallurgy
Technical University of Clausthal
FRG

ABSTRACT

Anisotropic properties of polycrystalline materials of any kind may depend on the orientation distribution of the materials crystallites — the *texture*. Hence, texture is one of the basic structural parameters of such materials. Textures can be measured by optical and electron-optical methods as well as by diffraction methods e.g. x-rays, neutrons, electrons. This can be done by individual orientation determination or by pole figure measurement followed by pole figure inversion. For property control, only partial texture analysis is required which can be carried out by a fixed angle texture analyzer or by anisotropy measurements (e.g. Young's modulus, magnetic anisotropy). When mathematical models of texture formation will have been developed, an automatic feed-back property control may finally be possible.

INTRODUCTION

It is a trivial knowledge that any industrial product can only be as good as the materials it is made of. This holds for a saucepan, a motor car or the most advanced space vehicle equally well as for any other product. Hence, materials testing is one of the most important tasks in industry. By now, we know already millions of different materials and their number increases steadily and rapidly. First of all, it is mechanical strength which is required of virtually any material. Besides this, materials are bing used, however, for various other properties of theirs, e.g. specific weight, elastic stiffness, electric conductivity, magnetic permeability or coercive force, thermal expansion, corrosion resistance, colour and many many more. Hence, all these properties must be tested after, or better during production of the material, with the aim of controlling the production process appropriately.

By the term "property of a material" we understand the relationship between any external influence acting upon the material and the material's response to it:

Influence	⟵	Property	⟶	Response
Stress		Young's Modulus		Strain
Voltage		Conductivity		Current
Temperature		Expansion coefficient		Strain
⋮		⋮		⋮

The properties of a material depend on the materials "structure". This latter term is a very complex one. It is the sum of a great manifold of various structural parameters, and virtually any of them may have an influence on any of the material's properties (see e.g. /1/).

Structure		Property
Chemical composition		Specific weight
Crystal structure(s)		Young's modulus
Partical size, shape		Strength
Grain size, shape	⟶	Fracture toughness
Texture = Orientation distribution		Electric conductivity
Residual stresses		Thermal expansion
Ordering		Permeability
		Coercive force

Hence, the task of "testing materials" may be divided into:

— Testing the material's **properties**
— Testing the material's **structure**

If we are sure of *all* structural parameters influencing a particular property then the "direct" property test can be replaced by an "indirect" test i.e. a test of all relevant structural parameters. Under these premises *texture analysis* can be applied as a method of materials *property testing.*

Structure and properties of a material are ideally assumed to be homogeneous, i.e. any sample taken from any part of the material should have the same properties. This is, however, not the case. It is then convenient to split a structural parameter into a homogeneous basic structure and local inhomogeneities:

Structure	=	Overall structure	+	Local deviations
		Chemical composition		Segregation
		Specific weight		Holes, voids, cracks
		Mean grain size		Local deviations
		Global Texture		*Local Textures*
		⋮		⋮

What is a "local deviation", depends strongly on the scale at which it is being looked upon. Hence, local quantities may be defined anywhere between "macro" and "micro"

	Macro ⟵				⟶ **Micro**
Segregation	Ingot ($m - dm$)		Crystal (mm)	Grainboundary (μm)	
Stresses	1.Kind ($cm - mm$)		2.Kind ($1 - 100\mu m$)	3.Kind ($< 1\mu m$)	
Cracks Holes Voids	(cm)				(μm)
Local Texture	Band length (km)	Band width (m)	Deepdrawing- Forging- parts ($dm - cm$)	Band thickness Stampings (mm)	Shear bands (μm)

The methods of materials testing may be divided into destructive and non–destructive ones:

Destructive
— The testing procedure changes the tested quantity
— The "identical" sample can only be measured once

Tensile test
Fracture toughness
Fatigue testing
Corrosion tests
⋮

Non–destructive
— The testing procedure does not change the tested quantity
— The "identical" sample can be repeatedly measured

Young's modulus
Magnetic properties
Electric conductivity
Texture
⋮

The non–destructive methods may be further divided into methods which require samples to be cut out of the material (destructive sample taking) and methods which can be applied to the undestroyed work piece (Truely non-destructive methods). The methods of texture analysis can be grouped into both of these categories.

Destructive sample taking
(Depending on the size of the investigated material or work piece)

Texture goniometer ("large" parts)
Electron diffraction
texture measurement
Anisotropy measurements

Truely non-destructive

Texture goniometer ("small"parts)
Neutron diffraction
Fixed angle texture analyzer
Anisotropy measurements

TEXTURE ANALYSIS

The term "texture analysis" in the most comprehensive sense comprises a number of "orientational" structural parameters which may be classified as follows (see e.g. /2/):

Texture (narrower sense)	Orientation distribution of the cristallites of a single phase
global	(whole material or workpiece)
local	(various dimensions) Inhomogeneous Textures Texture topography
multiphase textures	(Texture of each phase in a multiphase material)
Generalized textural quantities	
Orientation correlation	Composites
Shape anisotropy	Eutectics
Orientation – size – relation	
Orientation – strain – relation	
Directional ordering	

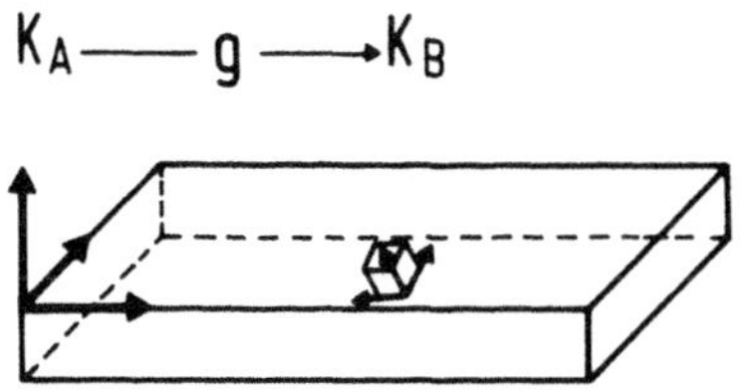

1) The orientation g is defined by the sample coordinate system K_A and the crystal coordinate system K_B

Definition of the texture

The texture of a material is defined as the orientation distribution function (ODF) of its crystallites (see e.g. /3/). Hence, it is, at first, necessary to define the orientation of an individual crystallite in a polycrystalline material. For this purpose, we specify a sample coordinate system K_A (in a rolled sheet for instance by rolling direction, RD, transverse direction, TD and normal direction, ND) and a crystal coordinate system K_B in each individual crystallite (in a cubic material for instance by the cube axes [100][010][001]). The orientation of the crystallite is then specified by the rotation g which transforms the sample coordinate system K_A into the crystal coordinate system K_B as is shown schematically in Fig.1

$$K_B = g \cdot K_A \tag{1}$$

The orientation g can be specified, in detail, by various parameters

Representation of an orientation g

		Number of parameters
metallurgical	$(uvw)[hkl]$	(6)
matrix	$[g_{ik}]$	(9)
Euler angles	$(\{\varphi_1, \Phi, \varphi_2\})$	(3)
Direction-angle	(α, β, γ)	(3)
Axis–angle	$(r_{\psi\vartheta}, \omega)$	(3)

Three of these sets of parameters are shown in Fig.2. The last column of the above table gives the number of parameters. It is seen that the minimum number needed for complete specification is three. If there are more than three parameters then they are interdependent. Hence, crystal orientation may be represented by a point in a three dimensional space (orientation space) as is shown in Fig.3a. The texture is then defined

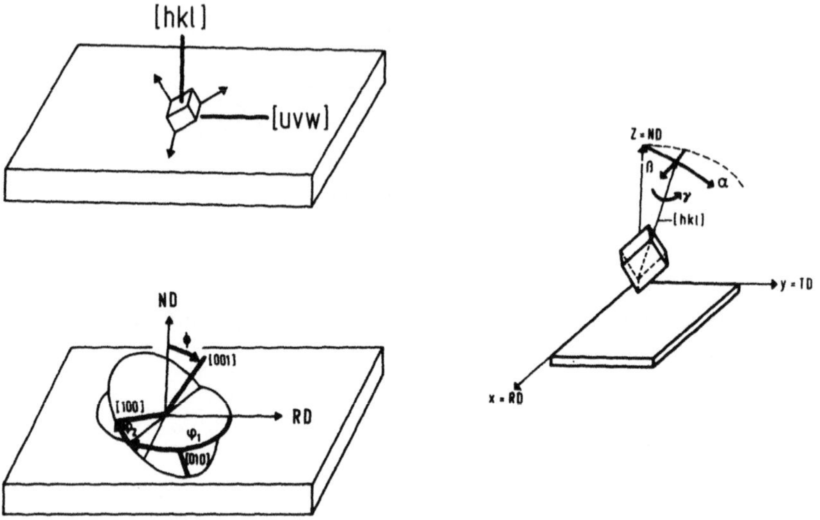

2) Three different representations of a rotation
   a) metallurgical representation
   b) Euler angles
   c) Direction–angle

by the volume fraction of the material having an orientation g within the limits dg.

$$f(g) = \frac{dV/V}{dg} \tag{2}$$

If the material consists of individual, well defined crystallites then the continuous Orientation Distribution Function (ODF) has to be constructed by smearing out the cloud of orientation points as is shown in Fig.3b,c.

The texture is a statistical quantity. It is only defined within the *statistical relevance* $\pm\Delta f$ which depends on the number of crystallites in the sample and on the *angular resolving power* Δg which has to replace dg in eq.2. Furthermore, it is important to consider crystal symmetry and sample symmetry with respect to which the texture function $f(g)$ must be invariant (see e.g. /4/,/5/,/6/,/7/).

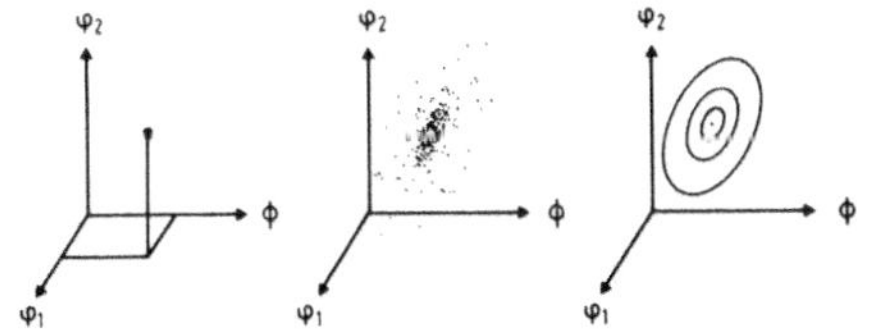

3) Construction of the orientation distribution function (ODF) in the orientation space
   a) individual crystallite
   b) all crystallites
   c) orientation density

Texture measurement

Textures can be measured by a variety of different methods (see e.g. /8/):

1. INDIVIDUAL ORIENTATION MEASUREMENTS (Resolving power < grain size)

In this case individual grains can be resolved and their orientation can be measured by:

1.a. Imaging methods

Thereby some crystallographic features (slip lines, twins, etch pits) must be visible

a) Optical microscopy

b) Electron microscopy

1.b. Diffraction methods

By a radiation with a wavelength comparable to the lattice parameters, individual diffraction diagrams can be obtained i.e. by:

a)	X-rays	Laue Patterns
b)	Electrons	Spot Patterns
		Kikuchi Patterns

Individual orientation measurements can be taken in equidistant scans or in individual crystallites as is shown in Fig.4.

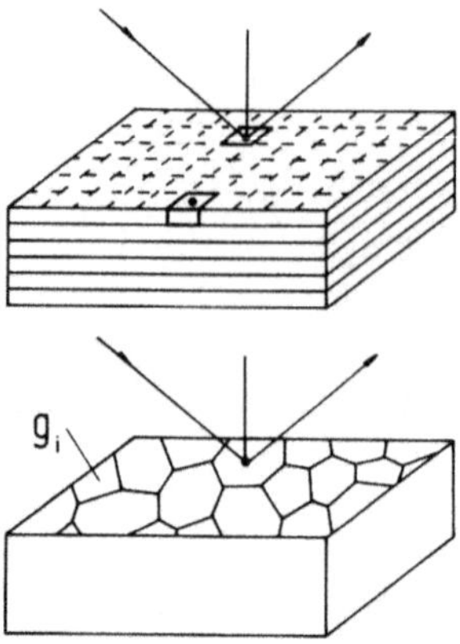

4) Individual orientation measurements
   a) in equidistant scans
   b) in individual grains

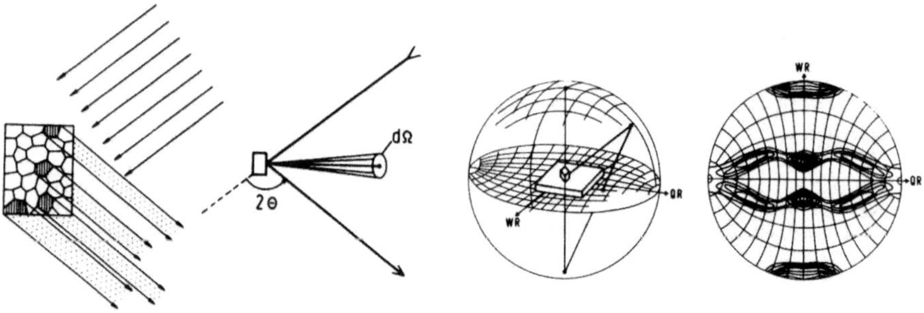

5) Pole figure measurement
   a) Diffraction in a polycristal
   b) Representation in the pole figure

2. <u>POLE FIGURE MEASUREMENTS</u> (Resolving power $\gg$ grain size)

In this case polycrystal diffraction measurements are being made in a large number of sample orientations as is shown in Fig.5. The pole figure is the volume fraction of crystallites the (fixed) crystal direction h of which is parallel to the (variable) sample direction y

$$P_h(y) = \frac{dV/V}{dy} \quad ; \quad y = \{\alpha\beta\} \tag{3}$$

The crystal direction h is selected by the Bragg-angle ϑ_h using Bragg's equation

$$n \cdot \lambda = 2 \cdot d_h \cdot \sin\vartheta_h \tag{4}$$

This requires a powder diffractometer. The sample direction y is brought into the diffraction direction s (the bisectrix between incident and reflected beam) using, for instance, a Eulerian cradle, Fig.6. A diffractometer equipped with a Eulerian cradle is called a texture goniometer

Texture goniometer = Diffractometer + Eulerian cradle

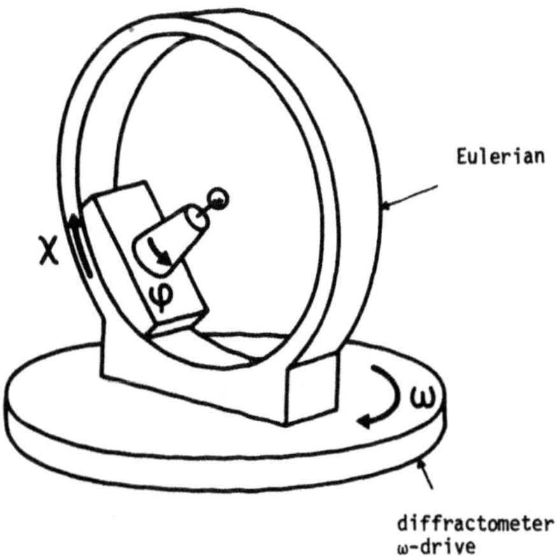

6) Eulerian cradle

Modern texture goniometers are stepping-motor operated and computer controlled as is shown schematicall in Fig.7a (see e.g. /9/). A particular texture goniometer is shown in Fig.7b.

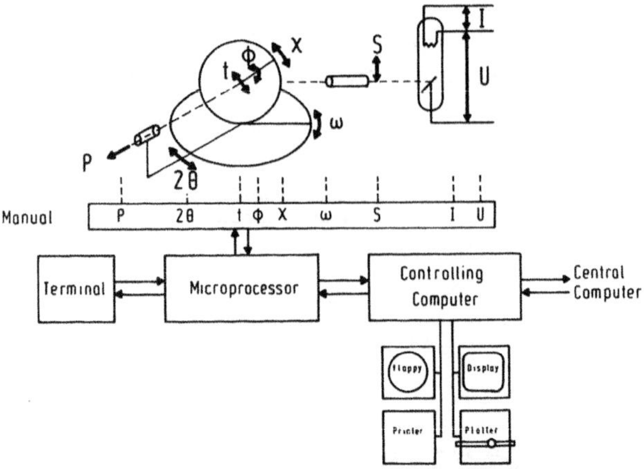

7) Modern texture goniometer
   a) schematically
   b) particular instrument

Very similar texture goniometers are being used with

— x-ray diffraction
— neutron diffraction (see e.g /10/)

Pole figure measurements can also be carried out with

— Electron diffraction

but in this case the geometrical principles used in the texture goniometer are somewhat different (see e.g. /11/).

POLE FIGURE INVERSION

As is seen in Fig.5, pole figure measurement cannot distinguish crystal orientations which are rotated about the diffraction vector s. Hence, the pole figure is an integral (generalized projection) over the orientation space.

$$P_h(y) = \frac{1}{2\pi} \int_{h\|y} f(g) d\psi \qquad \begin{array}{l}\text{Fundamental equation}\\ \text{Projection equation}\end{array} \qquad (5)$$

Nevertheless can the texture function $f(g)$ be calculated from several of its projections $P_h(y)$ (for different h) by mathematical methods which are comparable to the methods of "computer tomography" as is shown schematically in Fig.8. This procedure, called "pole figure inversion" is available in the form of routine computer programs (see e.g. /4/,/5/,/6/).

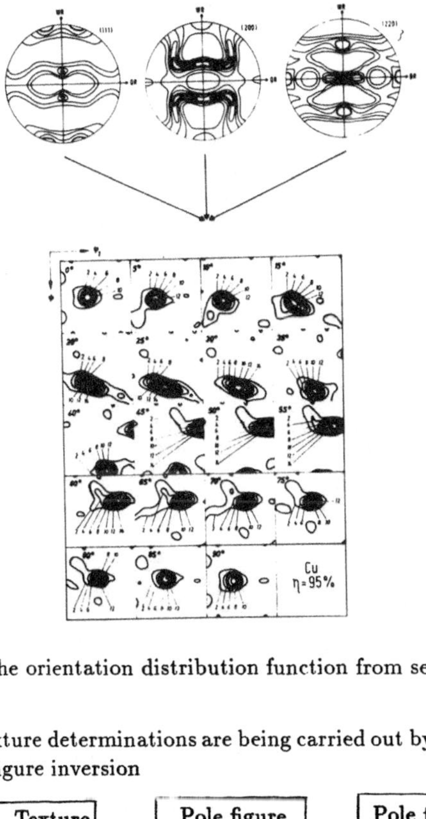

8) Calculation of the orientation distribution function from several pole figures.

The majority of texture determinations are being carried out by pole figure measurement following by pole figure inversion

Texture analysis = **Pole figure measurement** + **Pole figure inversion**

Texture Analysis - A Method of Nondestructive Characterization of Materials II

H.J. Bunge
Department of Physical Metallurgy
Technical University of Clausthal
FRG

RELATIONSHIP BETWEEN TEXTURE AND MATERIALS PROPERTIES

Materials properties in single crystals may be anisotropic i.e. they may depend on the crystal direction in which they are being measured (Young's modulus, magnetization energy and many others, see e.g. /12/). If the orientation distribution of the crystallites is not random, then also the averaged polycrystal properties may be anisotropic as is shown schematically in Fig.9. Thereby the texture is the weight function in the mean value expression.

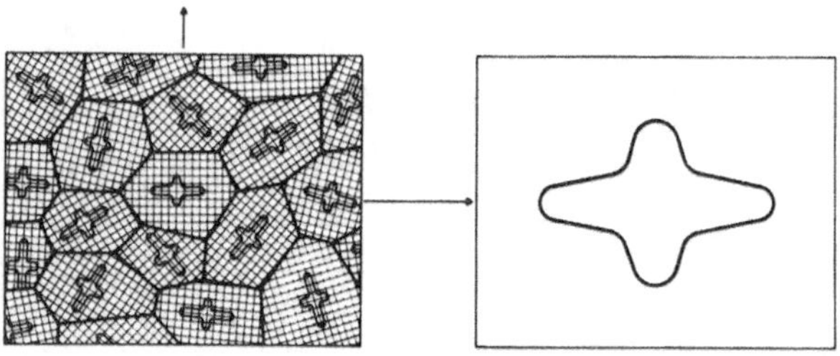

9) Averaging anisotropic crystal properties in a polycristalline material.

Polycrystal anisotropy	— TEXTURE —	Single crystal anisotropy

$$\bar{E}(y) = \oint A(h,y) \cdot E(h) \cdot dh \tag{6}$$

$A(h,y) = \frac{1}{2\pi} \int f(g) d\psi$

h = const = pole figure
y = const = inverse pole figure

Some examples of anisotropic materials properties are shown in Fig.10.

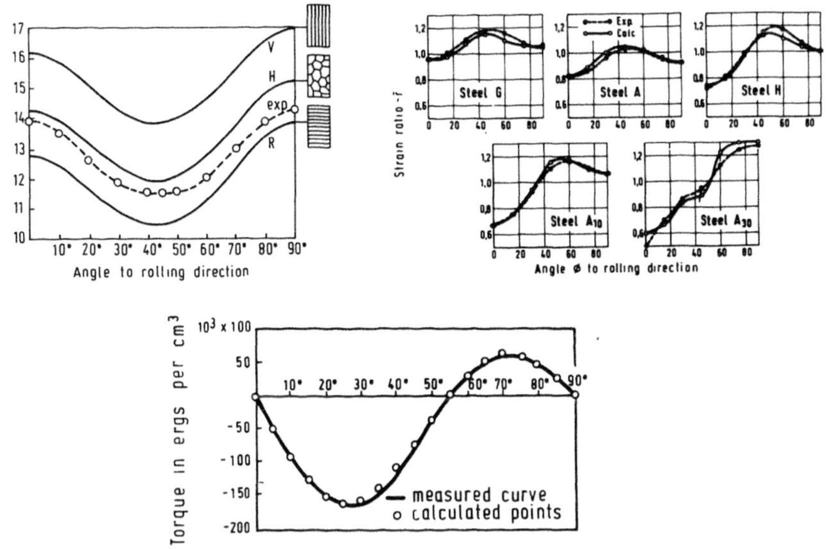

10) Anisotropic materials properties, measured and calculated from the texture.

a) Young's modulus in 90% cold rolled copper sheets.

b) Lankford parameter of several steels.

c) Magnetic torque curve in grain oriented Fe–Si transformer steel.

The texture determines the direction dependence (anisotropy) of materials properties (see e.g. /13/,/14/). It does not influence the overall value of a property (averaged over all sample directions). If we call the direction dependence the "*property profile*" then this has to be adapted to the "*application profile*" i.e. the direction dependence of required properties in the work piece (/15/)

Property profile of the material	⟷	Application profile in the workpiece

This is shown schematically in Fig.11 for the case of transformer core sheet.

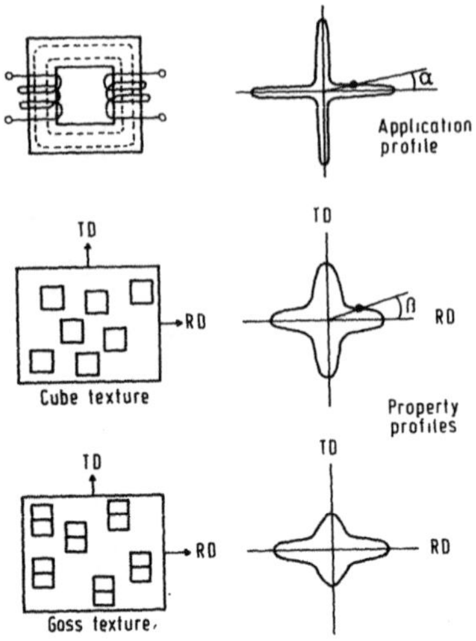

11) Adaption of the materials property profile to the application profile in transformer core sheets.

SERIES EXPANSION OF ORIENTATION FUNCTIONS

For the practical handling of orientation functions — the texture as well as the property functions — it is convenient to develop these functions into series of spherical harmonics and generalized harmonics respectively (see e.g. /4/,/5/,/6/,/16/). The functions are thus represented by the set of coefficients

	Function			Coefficient		Harmonics	
Texture	$f(g)$	=	$\sum_{\lambda=0}^{L}\sum_{\mu}\sum_{\nu}$	$C_\lambda^{\mu\nu}$	$\cdot$	$T_\lambda^{\mu\nu}(g)$	(7)
	$A(h,y)$	=	$\sum_{\lambda=0}^{L}\sum_{\mu}\sum_{\nu}\frac{4\pi}{2\lambda+1}$	$C_\lambda^{\mu\nu}$	$\cdot$	$K_\lambda^{*\mu}(h)\cdot K_\lambda^{\nu}(y)$	(8)
Property							
Single crystal	$E(h)$	=	$\sum_{\lambda=0}^{L_0}\sum_{\mu}$	e_λ^{μ}	$\cdot$	$K_\lambda^{\mu}(h)$	(9)
Poly crystal	$\bar{E}(y)$	=	$\sum_{\lambda=0}^{L_0}\sum_{\nu}$	$\bar{e}_\lambda^{\nu}$	$\cdot$	$K_\lambda^{\nu}(y)$	(10)

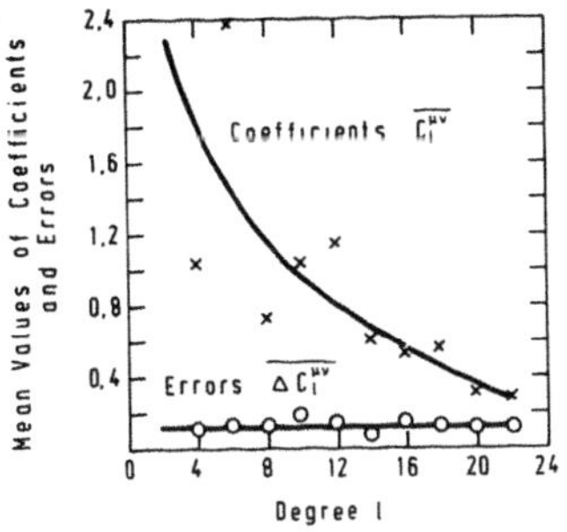

12) The mean absolute values $|C_\lambda^{\mu\nu}|$ of the texture coefficients of a 90% cold rolled copper sheet compared with the corresponding mean errors.

Substituting eq.8,9,10 in eq.6 yields a relationship between the corresponding coefficients

$$\bar{e}_\lambda^\nu = \sum_{\mu=1}^{M(\lambda)} \frac{e_\lambda^\mu \cdot C_\lambda^{\mu\nu}}{2\lambda+1} \qquad \lambda \leq [L, L_0] \tag{11}$$

In eq.11 the series expansion degree λ is restricted to values smaller than L and L_0 (otherwise either $C_\lambda^{\mu\nu}$ or e_λ^μ is zero). As is seen in Fig 12 the C-values converge to the level of their experimental uncertainties within e.g. $\lambda = 22$. The property functions $E(h)$ are either finite series with very low values of L_0 (e.g. elastic properties), or their series converges much more rapidly than that of the texture as in the case of plastic anisotropy, shown in Fig.13 /17/. Hence, only very few of the texture coefficients $C_\lambda^{\mu\nu}$ really enter eq.11

	Series degree L	L_0	Number of e_λ^μ	coefficients $\bar{e}_\lambda^\nu$	$C_\lambda^{\mu\nu}$
Texture	22				185
Property (Single crystal)					
Young's modulus		4	2		
Mag. energy		6	2		
Property (Polycrystal)					
Young's modulus		4		4	3
Mag. energy		6		8	7

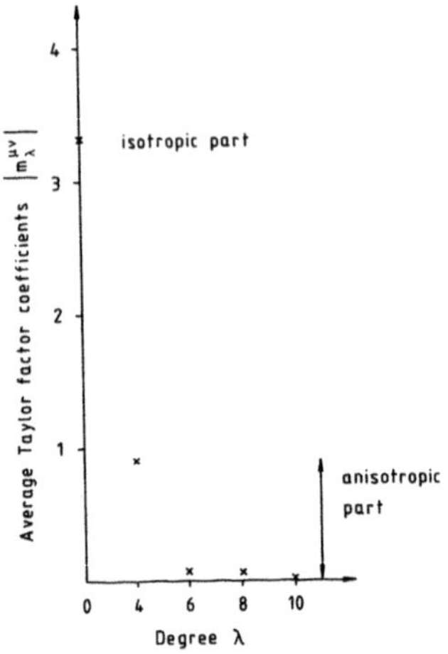

13) Mean absolute values $|m_\lambda^{\mu\nu}|$ of the series expansion of the Taylor factor describing the plastic anisotropy. (These values are comparable with ϵ_λ^μ in eq. 9,11)

PARTIAL TEXTURE ANALYSIS

Due to the low-order character L_0 of the property functions, the following conclusion can be drawn:

For the determination of physical properties of polycrystals from texture measurements, partial texture analysis with only very few coefficients $C_\lambda^{\mu\nu}$ with $\lambda \leq L_0$ is sufficient. (see /18/)

Methods of partial texture analysis

a. Anisotropy measurements

Low-order texture coefficients $C_\lambda^{\mu\nu}$ can be determined by solving eq.11 when the single crystal properties e_λ^μ are known and the polycrystal properties $\bar{e}_\lambda^\nu$ have been obtained from anisotropy measurements $\bar{E}(y)$.

$$\bar{E}(y) \longrightarrow \bar{e}_\lambda^\nu \longrightarrow C_\lambda^{\mu\nu} \qquad \lambda \leq L_0 \tag{12}$$

→ Young's modulus
→ Magnetic anisotropy

b. Fixed angle texture analyzer

Low-order texture coefficients can also be obtained from a limited set of pole figure points as is shown schematically in Fig.14. These pole figure values can be measured by a fixed angle texture analyzer in a *truely non-destructive* way (see /19/).

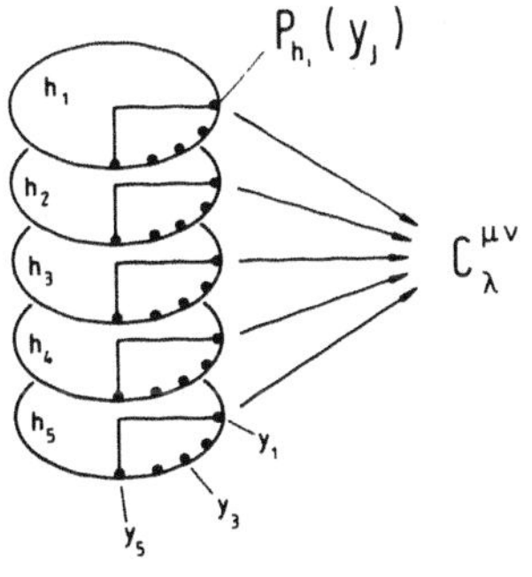

14) Principles of a fixed-angle texture analyzer. Determination of the low-order texture coefficients from a limited set of pole figure values.

TEXTURE TOPOGRAPHY

If the texture of a material is inhomogeneous then a complete texture description requires the specification of *all local textures* i.e. the textures as a function of the position $[xyz]$ in the sample (see /20/)

$$f(\{\varphi_1, \Phi, \varphi_2\}xyz) \qquad \text{Inhomogeneous texture} \tag{13}$$

This is a function of six variables which is very difficult to measure (and even its mere representation is difficult).

Partial information about the inhomogeneity of a texture can be obtained by fixing the orientational variables $\{\varphi_1, \Phi, \varphi_2\}$ and one of the spacial variables, the depth-coordinate z. Thus only the lateral coordinates xy remain. In this case the pole figure is being used instead of the ODF

$$P_h(\underbrace{\alpha\beta}_{\text{fixed}}\ xy \quad \underbrace{z}_{\text{fixed(surface)}}) \tag{14}$$

One thus obtains the pole density distribution of one pole figure point in the plane of the sample surface (see e.g. /21/)

$$P_{h,\alpha,\beta,z}(xy) \qquad \text{Texture topography} \tag{15}$$

Methods of texture topography

X-ray diffraction
  2–dimensional scanning
  photographic method (pinhole camera)
electron diffraction
  dark field image
Light optical image
  polarization microscopy

The most obvious method of texture topography is to use a conventional texture goniometer, to fix a pole figure point and to scan the sample through x and y and to measure the pole density as a function of x and y. A photographic method using one– dimensional scanning is seen in Fig.15 (/21/). Two–dimensional imaging with no further need of scanning is possible with electron diffraction. This method is nothing else than dark-field imaging with a resolving power larger than grain size. Two–dimensional imaging is also possible with the optical polarization microscope if the material is optically anisotropic or an oriented, anisotropic oxide layer can be produced on its surface (see e.g. /22/). Texture topography has been used, for instance, to visualize local inhomogeneities of the texture due to coining (after the relief of the coin had been ground off or was worn off) Fig.16a. A similar application is to re–visualize code numbers stamped into the surface of a work piece after (illegal) removing of the number Fig.16b (/21/). In this cases the texture inhomogeneities are in the millimeter range. A third example is given in Fig.16c which shows the texture inhomogeneities in two families of shear bands in the matrix of a deformed titanium sample (/22/).

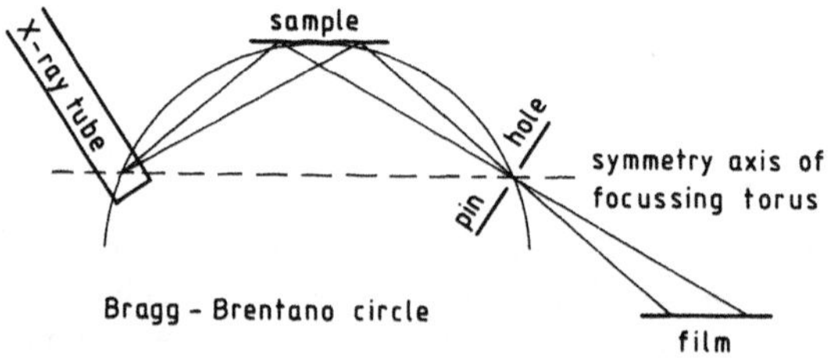

15) Principles of texture topography with a pinhole camera.

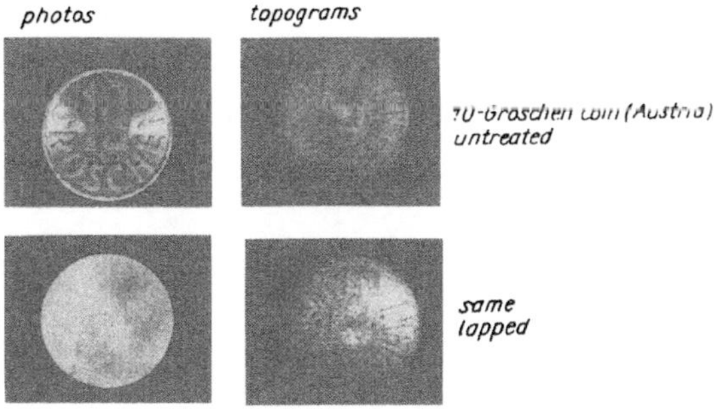

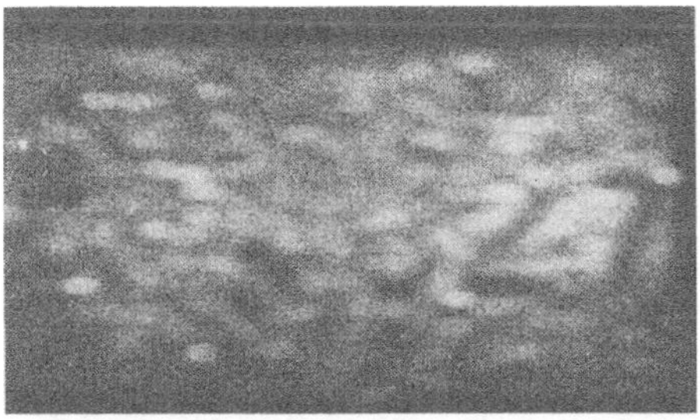

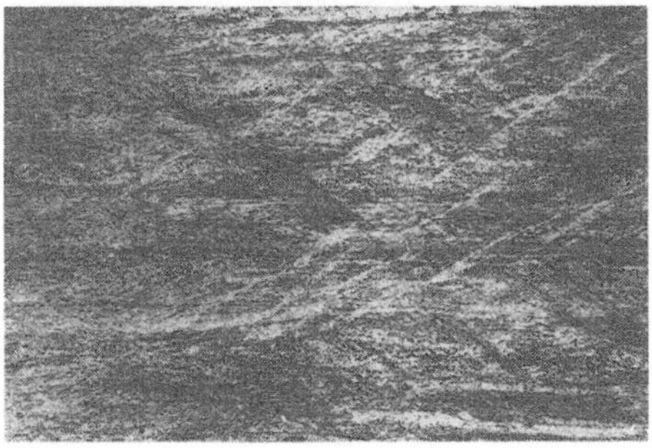

16) Texture topograms of various samples.

a) x–ray pinhole topograms of coins with the relief removed
b) x–ray pinhole topogram of an inprinted code number after removing the surface relief
c) optical topograms of shear–bands in titanium

TEXTURES IN MULTIPHASE MATERIALS

Real materials are often multiphase materials. Then a texture function is to be determined for each phase individually (see e.g. /23/)

prominent examples of multiphase materials are

—polymineralic rocks
—inhomogeneous alloys (e.g. eutectics, eutectoids)
—polyphase ceramics
—composits (e.g. cermets)

Texture measurement in these materials comprises some experimental difficulties e.g.:

—superposition of diffraction peaks
—Low intensities
—anisotropic absorption of x-rays

which must be taken care of by special experimental techniques such as for instance

—neutron diffraction
—electron diffraction
—position sensitive counter technique
—energy dispersive methods

Also special mathematical techniques are required for pole figure inversion allowing for

— multiphase pole figure superposition (see e.g. /24/)

As an example of multiphase textures, Fig.17 shows the microstructure of a highly extruded Al–Pb composite and the textures (inverse pole figures) of the two phases determined by neutron diffraction. (In this case x-ray diffraction is strongly falsified by anisotropic absorption, /25/)

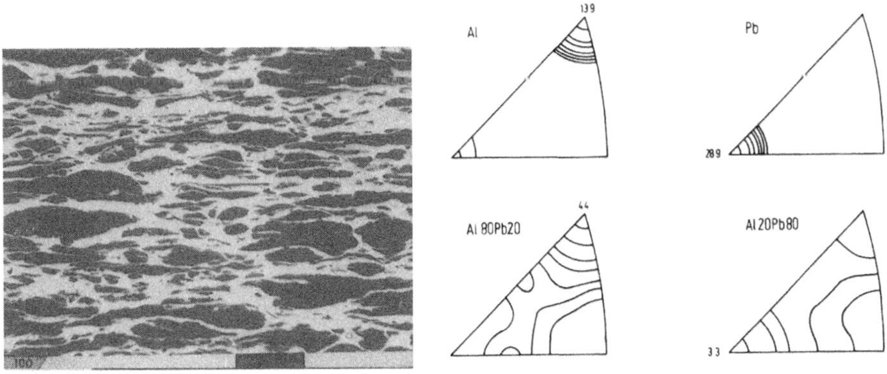

17) Textures in a highly extruded Al–Pb composite.

a) Microstructure, longitudinal section.

b) Inverse pole figure of the pure materials and of the two phases in the composite.

GENERALIZED TEXTURAL QUANTITIES

As an example out of the great number of generalized textural quantities, the grain boundary "misorientation" function or correlation function may be considered. This function is becoming more and more interesting in microminiaturized components when grain size becomes comparable with the sample dimensions. If two neighbouring grains have the orientations g_1 and g_2, Fig.18, then the grain boundary misorientation is

$$K_B^2 = \Delta g \cdot K_B^1 \qquad \text{Misorientation} \qquad (16)$$

$$\Delta g = g_2 \cdot g_{-1}$$

The misorientation function is then defined analogous to the texture in eq.2 (see e.g. /26/)

$$F(\Delta g) = \frac{dA/A}{d\Delta g} \qquad (17)$$

where dA/A is the area fraction of grain boundaries having the misorientation Δg within the limits $d\Delta g$. The grains 1 and 2 in Fig.18 may belong to the same phase (intra–phase correlation) or to different phases (inter–phase correlation). The misorientation can be determined by electron diffraction of individual crystallites. As an example Fig.19 shows the misorientation function measured in Martensite–Ferrite grain boundaries of a dual–phase steel.

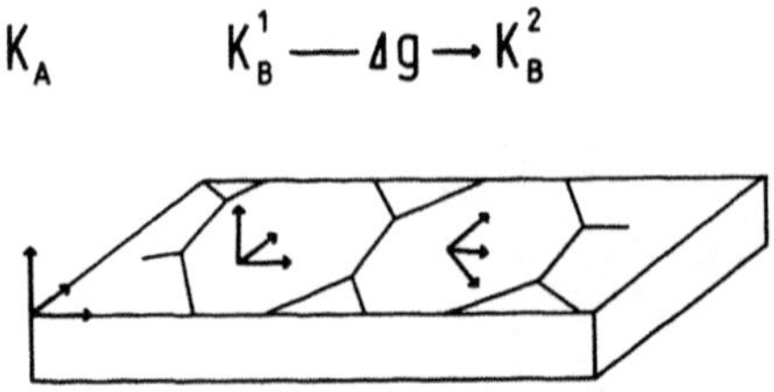

18) The misorientation of a grain boundary is defined by the two crystal coordinate systems K_B^1 and K_B^2 of neighbouring crystallites.

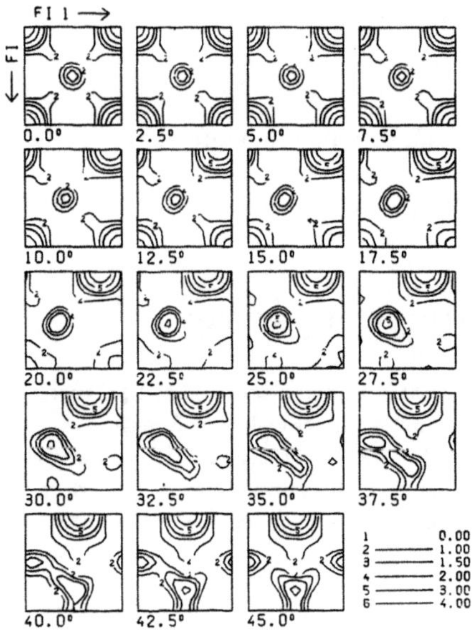

19) The misorientation function of Martensite — Ferrite in a dual—phase steel

TEXTURE AS A FEED-BACK PARAMETER

From the technological point of view, the ultimate goal of texture measurement is the production of materials having the optimum texture for a particular purpose. In an automatic production line this requires:

On-line texture measurement ⟹ available

⇓

Mathematical model of texture formation ⟹ Fundamental Research

⇓

Feed-back controll of process parameters ⟹ available

A comprehensive mathematical model of texture formation is by far nor yet available. This is due to the multifarious nature of all texture forming processes.

TEXTURE FORMING PHYSICAL PROCESSES

Cristallization from non—crystalline state
- Solidification
- Electrolytic deposition
- Sputtering
- Chemical vapour deposition
- Crystallization of glasses

Rigid rotation
- Sedimentation
- Magnetic pressing
- Forming with soft binder-phase

Plastic deformation
- Glide
- Twinning

Phase Transformation
- Martensitic
- Diffusive

Recristallisation
- Primary
- Secondary
- Grain growth

These physical processes take place in various combinations in technological processes.

TEXTURE FORMING TECHNOLOGICAL PROCESSES

Casting, welding, soldering, hot dip galvanizing
Electrolytic galvanizing, sputtering, CVD
Sedimentation, calandring, slurry forming
Rolling, wire drawing, deep drawing, forging
Extruding, machining, grinding
Sintering, heat treatment
⋮

Texture measurements are needed in principle in polycrystalline materials of any kind. Thereby *natural* and *technological* materials may be distinguished (see e.g. /27/):

MATERIALS OF INTEREST FOR TEXTURE ANALYSIS

Natural materials

- Geological materials
  - Rocks, ores, salts, soils, coal, ice
- Biological materials
  - Bones, teeth, gall-stones, shells

Technological materials

- Metals
  - Basic metals, alloys, intermetallics
- Semiconductors
  - Silicon, germanium, galliumarsenide, graphite
- Ceramics
  - Silicates, Al_2O_3, SiC, AlN, Si_3N_4 ...
  - Ceramic magnets, HT superconductors
- Polymers (partly crystalline)
  - Polyethylene, polyamid, liquid crystals
- Composites

REFERENCES

/1/Chin, H.A., Gould, R.D. (Eds.): Texture - Microstructure - Mechanical Properties.
ASM Metals Park Ohio 1984.

/2/Bunge, H.J.: Three-dimensional texture analysis. Intern. Mat. Rev. 32 (1987) 265-291.

/3/Wassermann, G., Grewen, J.: Texturen metallischer Werkstoffe. Springer Verlag Berlin 1962.

/4/Bunge, H.J.: Mathematische Methoden der Texturanalyse. Akademie-Verlag Berlin 1969.

/5/Bunge, H.J.: Texture Analysis in Materials Science. Butterworths Publ. London 1982.

/6/Bunge, H.J., Esling, C. (Eds.): Quantitative Texture Analysis. DGM Informationsgesellschaft. Oberursel 1982.

/7/Bunge, H.J., Esling, C.: Symmetries in Texture Analysis. Acta. Cryst. A41 (1985) 92-101.

/8/Bunge, H.J. (Ed.): Experimental Techniques of Texture Analysis: DGM Informationsgesellschaft Oberursel 1986.

/9/Puch, K.H., Klein, H., Bunge, H.J.: A New Computer Operated Texture Goniometer. Z. Metallkunde 75 (1984) 133-139.

/10/Brokmeier, H.G., Bunge, H.J.: Non-destructive Determination of Materials Parameters by Neutron Diffraction. This volume.

/11/Schwarzer, R.A.: Bestimmung von Polfiguren und Texturanalyse mit dem Transmissions-Elektronenmikroskop. Z. Metallkunde 73 (1982) 495-498.

/12/Nye, J.F.: Physical Properties of Crystal. The Claredon Press. Oxford 1957.

/13/Bunge, H.J.: Texture and Anisotropy. Z. Metallkunde 70 (1979) 411-418.

/14/Bunge, H.J.: Texture and Directional Properties of Materials. In: Directional Properties of Materials. Ed. Bunge, H.J. and Aernoudt, E.. DGM Informationsgesellschaft Oberursel (in print).

/15/Bunge, H.J.: Technological Applications of Texture Analysis. Z. Metallkunde 76 (1985) 457-470.

/16/Matthies, S.: Aktuelle Probleme der quatitativen Texturanalyse. ZFK Rossendorf 1982.

/17/Bunge, H.J.: Some Applications of Taylor's Theory on Polycrystal Plasticity. Kristall u. Technik 5 (1971) 145–175.

/18/Bunge, H.J.: Partial Texture Analysis. Textures and Microstructures (in print).

/19/Kopineck,H.J., Otten, H., Bunge, H.J.: On-Line Determination of Technological Characteristics of Cold and Hot-Rolled Steel Bands by a Fixed Angle Texture Analyzer. This volume.

/20/Bunge, H.J.. Inhomogeneous Textures. Z. Metallkunde 73 (1982) 483–488.

/21/Born, E., Schwarzbauer, H.: Perspective Fields of Application of Texture Topography and Further Development of its Apparatus. Kristall u. Technik 15 (1980) 837–842.

/22/Bunge, H.J., Nauer-Gerhardt, C,: Texture Topography With the Polarization Microscope. In: Experimental Techniques of Texture Analysis. Ed. Bunge, H.J.. DGM Informationsgesellschaft Oberursel 1986, 147–154.

/23/Bunge, H.J: Textures in Multiphase Alloys. Z. Metallkunde 76 (1985) 92–101.

/24/Dahms, M., Brokmeier, H.G., Seute, H., Bunge, H.J.: Quantitative Texture Analysis in Multiphase Materials With Overlapping Bragg-Reflections. Proc. 9th Risø Conference 1988.

/25/Bunge, H.J., Liu, Y.S., Hanneforth, R.: Anisotropic Absorption of X-rays in Polyphase Materials. Scripta Met. 21 (1987) 1423–1427.

/26/Bunge, H.J., Weiland, H.: Orientation Correlation in Grain and Phase Boundaries. Textures and Microstructures 7 (1988) 231-263.

/27/Wenk, H.R. (ed.): Preferred Orientation in Deformed Metals and Rocks: An Introduction to Modern Texture Analysis. Academic Press. Orlando 1985.

Characterization of Crystallographic Texture in Aluminium Can Stock by X-Ray Diffraction

C. O. RUUD[1] and D. J. SNOHA[2]

[1]Materials Research Laboratory
The Pennsylvania State University
University Park, PA, USA 16802

[2]Materials Technology Laboratory
Watertown, MA 02172-0001

Abstract

The nature and extent of preferred crystallographic orientation (texture) of grains in polycrystalline aluminum affect the formability of sheet products. The presence of texture in aluminum can stock may produce "earing" in the deep drawn cans. Rapid, nondestructive measurement of the earing quality of aluminum stock by x-ray diffraction techniques, especially in the hot rolled stage before final gage, may now be possible using a unique, position-sensitive detector.

Preliminary results are described of measurements on aluminum sheet stock of x-ray peak breadth and intensity variations as a means of differentiating qualities of texture. The results show distinct x-ray parameter differences in the samples studied, as well as a linear correlation between texture and an x-ray intensity parameter. The results provide a strong indication that the instrumentation and techniques used in this study can be optimized and refined for application to rapid, in-process, nondestructive inspection of rolled aluminum can stock.

Introduction

Nine samples of aluminum alloy can stock sheet were supplied by two manufacturers to assess the possibility of the use of rapid x-ray diffraction technology to distinguish the various texture conditions present. The samples varied in thickness from 0.128" to 0.0124", and in texture from 2.8% to 11.0%, as measured by a cup drawing "earing" test.

Using modern x-ray tubes and constant potential power supplies, a position-sensitive scintillation detector (PSSD), developed at The Pennsylvania State University, provides x-ray diffraction data collection times of less than one second -- a two order of magnitude improvement over conventional XRD instrumentation. The PSSD relies on the coherent conversion of the diffracted x-ray pattern to an optical signal (light); the amplification of this signal by electro-optical image intensification; the electronic conversion of the signal; and the transfer of the electronic signal to a computer for refinement and interpretation. This instrument has been described elsewhere [1, 2], and its application to materials characterization discussed [3].

If there is texture in the surface of the aluminum sheet, the diffracted x-ray peak intensities will vary from one psi angle to another, depending on the type and degree of texture.

Procedure

Six samples of aluminum alloy can sheet stock (samples 1-6), from one manufacturer, and three samples (x, y, and z) from another manufacturer are listed in Table 1. The (333, 511) planes and Cu K-alpha radiation were used for samples 1-6, and the (222) plane and Cr K-alpha radiation for samples x, y, and z.

Table 1
Sample Identification and Percent Texture

Sample	Thickness (Inches)	Texture	Remarks
1	0.0129	2.8%	Final gage
2	0.0124	6.9	Final gage
3	0.0128	4.0	Final gage
4	0.1270	2.6	As-rolled
5	0.1270	4.2	As-rolled
6	0.1280	4.9	As-rolled
x	0.0800	11.0	As-rolled
y	0.0800	4.0	As-rolled
z	0.0800	1.5	As-rolled

XRD Measurement on Aluminum Can Stock

Data were collected at two incident beam to specimen normal (beta) angles (18° and 30°) (yielding two psi angles each, where $\psi_1=\beta-90°+\theta$ and $\psi_2=\beta+90°-\theta$) about an axis oriented at three different phi orientations to the rolling direction of the samples (0°, 45°, and 90°), see Figure 1. The specimen-to-detector distance was allowed to vary ±1.0 mm about the ideal 45 mm; this error in R_o was corrected by the computer interfaced with the PSSD.

Data collection for each pair of psi (ψ) angles at a single beta (β) tilt required 3 seconds at 45 kv and 15 ma, with a full wave rectified power supply for Cu and Cr K-alpha radiation. A dedicated PDP 11/23 computer controlled data collection and provided data refinement.

The raw data were first corrected for electronic noise and gain of the detector; data points on either side of each peak provided a linear least-squares fit for the x-ray background correction. The calculated background then was subtracted point by point, across the collected data.

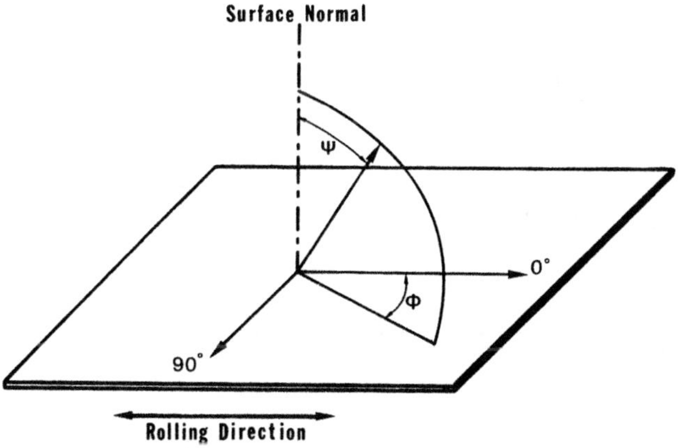

Figure 1. Measurement directions on aluminum can stock with respect to rolling direction.

Table 2

Ratio of ϕ=45° to ϕ=90° Area Functions

(Psi Angle ≈ 28°)

Sample	Area Function ϕ=45°	ϕ=90°	Area Function Ratio ϕ=45° / ϕ=90°
1	40.8	38.0	1.07
2	51.0	31.4	1.62
3	47.6	35.0	1.36
4	53.5	41.2	1.30
5	61.3	42.4	1.45
6	67.2	39.5	1.70
x	45.0	141.0	3.13
y	103.0	140.0	1.36
z	184.0	141.0	0.77

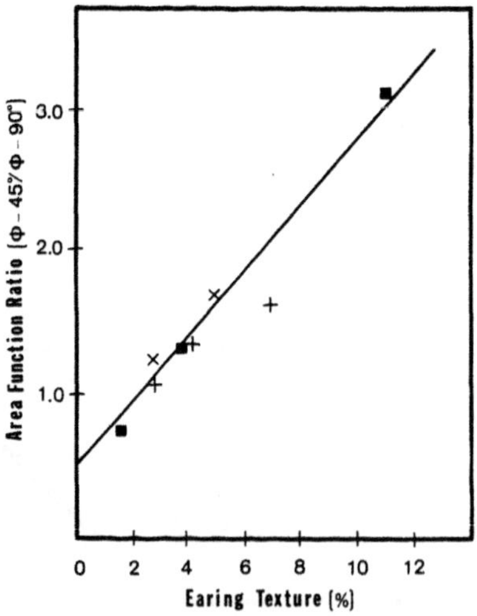

Figure 2. Plot of the area function ratio (ϕ=45°/ϕ=90°) for a psi angle of 28° ± 2° versus percent earing in aluminum can stock. Data sources are (+) from samples 1, 2, and 3; (x) from samples 4, 5, and 6; and (■) from samples x, y, and z. The regression line drawn through the points has a slope of 0.22, with a correlation coefficient of 0.958.

The intensity of the peaks above the background was measured and multiplied by the peak width at half height to obtain an intensity parameter. The height and width of the peaks from the textured samples were normalized using data from a powder sample.

The computer operations used in this study require about 60 seconds; however, this time has been reduced to about 10 seconds on the IBM PC/AT computer used on commercial versions of this instrument. Further reduction in data collection times of less than one second occur with enhancements to the IBM computer, and the use of modern x-ray tubes and constant potential x-ray power supplies.

Results and Discussion

Data were collected from diffraction patterns for the samples in the form of relative full width at half peak height (b/b_p), relative intensity (I/I_p), and the peak intensity area function percent ($b/b_p \times I/I_p \times 100$), from the two diffracted x-ray peaks at the two beta angles for Cu K-alpha radiation (samples 1-6) and Cr K-alpha radiation (samples x, y, and z). The b_p and I_p values were used to normalize the b and I values from the textured samples and were obtained from a powder specimen.

Table 2 lists the ratios of the $\phi = 45°$ to $\phi = 90°$ area functions for samples 1 through 6, and samples x, y, and z. Figure 3 is a plot of the area function ratio (a dimensionless parameter) at $\phi = 45° / \phi = 90$, for a psi angle of $28° \pm 2°$, versus the percent earing texture on the sheet surfaces for the data in Table 2. A linear regression line is drawn through the data points, with a correlation coefficient of 0.958. This is an excellent fit despite the fact that the samples came from two different manufacturers, were different gages, and were irradiated by two different radiation sources. In addition, the x-ray optics were not optimized for aluminum for the testing for either manufacturer.

Conclusions

The results for Cu K-alpha and Cr K-alpha radiation on suites of aluminum can stock samples from two manufacturers, in both hot rolled can stock precursor gages of 0.80" and 0.125" and in the final gage, show there is a good correlation between the peak intensity area function and texture. These preliminary results provide an indication that the instrumentation and techniques applied could be optimized and refined for application to rapid, in-process nondestructive inspection of rolled aluminum can stock.

References

1. C. O. Ruud, Ind. Res. and Dev. (January, 1983) pp.84-87.
2. C. O. Ruud, P. S. DiMascio, and D. J. Snoha, in Adv. in X-Ray Anal. (Plenum Press, New York, 1984), Vol. 27, pp. 273-283.
3. C. O. Ruud, in Nondestructive Methods for Material Property Determination, edited by C. O. Ruud and R. E. Green, Jr. (Plenum Press, New York, 1984), Vol 1. pp. 21-37.
4. Brakman, C. M., J. Appl. Cryst. 16 (1983), pp. 325-340.

Nondestructive Determination of Materials Parameters by Neutron Diffraction

H.-G. Brokmeier and H.J. Bunge

Institut für Metallkunde und Metallphysik Technische Universität Clausthal
Außenstelle im GKSS-Forschungszentrum Geesthacht GmbH
Postfach 1160, D-2054 Geesthacht, F.R.G.

Summary

Based on the high transmission of neutron radiation, neutron diffraction is an efficient tool for the analysis of various materials parameters of the bulk of a material in a non-destructive way. Sample sizes up to 40 mm in diameter have been used to determine the phase composition, texture and internal stresses of polycrystalline, polyphased materials. Additionally to well-known X-ray techniques which analyse the surface of a sample, neutron diffraction measurements are carried out to investigate the average behaviour of a bulk sample or the local behaviour within a compact sample. Further advantages of neutron diffraction are that small volume fractions (e.g. 0.05 Vol.% Cu in Al) can be determined. Moreover, in the case of X-ray diffraction the examination of coarse-grained materials and of complex multiphase materials is limited, and neutron diffraction has to be used. Several neutron diffractometers at different Research Centers can be used, but there are still limitations on beam time.

Introduction

Investigations of materials parameters on bulk samples in a non-destructive way require a method with a high penetration depth. In contrast to X-rays (standard radiation Cu Kα - 0.15 nm) there are some alternative radiations suitable for materials research:

- neutron radiation (Bacon [1])
- synchroton radiation (Gottstein [2])
- short wavelength X-rays such as tungsten radiation
  (W Kα - 0.02 nm) (see Brokmeier et al. [3] and Kopineck [4])
- γ-radiation

All four types of radiation have a high but different transmission and also large differences in local position resolution. Moreover, the availability of the radiations is quite different. This paper will deal only with neutron radiation. Neutron diffraction methods are closely parallel to well-known X-ray techniques.

For an analysis of a bulk sample two different kinds of studies has to be take into account:

1. A description of the average behaviour of the bulk of a material.
2. A description of a selected local area within the bulk of the sample.

Although limited by some parameters to be discussed later, neutron diffraction is suitable for both types of measurement of materials parameters. Such parameters include: the phase composition of polycrystalline, polyphased materials (phase transformation, precipitation, etc.), the texture (especially in multiphase systems and in coarse-grained materials) and internal stresses in metallic and ceramic materials (measurements of group 2).

A comparison of neutron diffraction with the well-known X-ray diffraction will be given. Additionally, neutron diffraction techniques to measure texture, phase composition and internal stresses are described and some examples illustrate the advantages of neutron diffraction. Other methods of neutron scattering in materials research are well-known especially small angle neutron scattering -SANS-, magnetic scattering and diffuse elastic neutron scattering -DENS-, but cannot be discussed in this paper.

Neutron scattering

The main difference between X-ray and neutron diffraction is the interaction between the incoming beam and the scattered atom. In the case of electromagnetic X-rays an interaction takes place with the electrons of the atoms. The neutron interacts with the atomic nucleus (nuclear scattering) and the electrons of the atom (magnetic scattering). In the following only the nuclear scattering will be considered. In contrast to X-rays the penetration depth of the neutrons in most materials is in the order of several millimeters to centimeters. Table 1 shows the penetration depth $d_{1/2}$ for some metals. The thickness $d_{1/2}$ represents a loss of 50% of the primary intensity of the incoming beam.

Tab. 1: Penetration depth $d_{1/2}$ of some metals
(X-ray and neutron scattering)

metal	$d_{1/2}$ in μm (X-rays 0.154 nm)	$d_{1/2}$ in μm (n-rays 0.150 nm)
Fe	$3.0 * 10^1$	$5.0 * 10^3$
Al	$5.3 * 10^2$	$6.3 * 10^4$
Cu	$1.5 * 10^2$	$8.0 * 10^3$
Pb	$2.0 * 10^1$	$2.3 * 10^4$

Only some elements or isotopes which have a high absorption or a high incoherent scattering (Cadmium, ^{10}B, ^{6}Li, Gadolinium, Hydrogen and some others) are difficult for neutron diffraction. Based on the higher transmission by a factor 10^2 - 10^3, neutron diffraction allows an increase of sample volume to 10^4 - 10^5 of that for X-ray diffraction. Some new techniques and applications are obtained to analyse materials parameters, see the following chapters.

Another advantage of the nuclear scattering of the neutrons is based on the geometrically small enlargement of the atomic nucleus. The scattering length b is independent of the scattering angle θ while f_x the X-ray scattering length decreases with increasing scattering angle. In practice neutron powder diagrams show more Bragg-reflections than X-ray powder diagrams from the same material. Bragg-reflections with a high hkl valuc (420) can be measured as well as lower ones (200). Hence, a high enough number of Bragg-reflections is obtained for example, for a quantitative texture analysis - calculation of the orientation distribution function ODF -. Especially in investigations of multiphase systems or of materials with a low crystal symmetry this will be a great advantage.

On the other hand an increase in the number of measured Bragg-reflections can involve a partial or total overlapping of different Bragg-reflections. New methods of evaluation are necessary to interpret the line-rich powder digrams (Jansen et al. [5], Filhol et al. [6], Dahms et al. [7]). Table 2 shows the number of Bragg-reflections calculated with the data of the JCPDS - files (Joint Commitee of Powder Diffraction Standard) for some common materials. Investigations on intermetallic compounds and composites respectively involve measurements of line-rich spectra which are rather complicated compared to metals but are attracting increasing interest.

Tab.2: Number of Bragg-reflections for some materials
(λ = 0.1 nm und 2θ = $0 - 80°$)

material	reflections	JCPDS
Fe	4	6-0696
Al	6	4-0787
SiC	12	29-1131
Ti	13	5-0682
Ti_3Al	24	16-0867
$CuAl_2$	29	25-0012
Al/Al_20_3 composite	37	4-0787 10-0173

Moreover, the neutron scattering length b is independent of the atomic number (number of electrons). In contrast to X-rays no increase of the scattering length b with increasing atomic number is found. Light elements such as carbon, oxygen or nitrogen have scattering lengths of the same order as heavy elements for example tin or tungsten (see table 3). Therefore identification of free carbon (graphite), carbides or oxides is much easier even in low volume fractions. Neighbouring elements and different isotopes can have different scattering lengths, so that superstructures and order - disorder phenomena can be determined.

Phase analysis by neutron diffraction

A knowledge of the phase composition of a multiphased material is essential to the interpretation of investigations describing the mechanical and physical behaviour of the material. Especially in materials processing such as sintering , annealing etc. changes in

Tab.3: Scattering lengths of neutrons (b) and X-rays (f)

element	b_{n-rays} (10^{-12} cm)	f_{x-rays} $(\sin\theta)/\lambda$ =0.0 (10^{-12} cm)	f_{x-rays} $(\sin\theta)/\lambda$ =0.5 (10^{-12} cm)
C	0.66	1.69	0.48
O	0.58	2.25	0.62
Fe	0.95	7.30	3.10
Al	0.35	3.65	1.55
Ti	-0.34	6.20	2.70
Sn	0.61	13.90	7.10
W	0.48	20.80	10.9

composition, fabrice and properties have to be considered. Similar to X-ray diffraction techniques neutron diffraction can be used to analyse powdered samples as well as solid materials in a non-destructive way. A typical neutron diffractometer for phase analysis is shown in figure 1. D1B is a standard equipment for powder diffraction at the high flux reactor (HFR) of the Institute Laue Langevin (ILL) in Grenoble [8]. A linear position sensitive detector covers 80° in 2θ and allows the simultaneous measurement of the whole 2θ-diagram. Moreover, ancillary equipment can be used: an Eulerian cradle (four circle diffractometer), a furnace (up to 2500° C), a cryostat (1.7 - 300 K) or an electromagnet (to 1.8 Tesla).

Investigations on a series of Al-Cu fibre composites with small volume fractions of Cu show that even small volume fractions of Cu up to 0.05 Vol.% can be determined. The neutron diffraction measurements were performed at D1B using a cylindrical sample 8 mm in high and 8 mm in diameter. Details of these experiments and a comparison with similar X-ray diffraction experiments will be given elsewhere [9].

Al-Cu composites are only stable during cold processing. In the system Al-Cu several intermetallic compounds are known. Hence, a detailed investigation of the system Al-Cu requires quantitative data on phase compositions. In X-ray diffraction several techniques are known [10] for determining volume fractions, which are also usable in neutron diffraction. However, in quantitative phase analysis, the presence of preferred orientation often leads to wrong results. Therefore a method has been developed to use a complete pole figure measurement to correct for the texture influence [11]. Experimental results of neutron diffraction measurements on different fibre composites (sample volume $\approx$ 0.4 cm^3) and on compacted graphite-siliconcarbide cylinders (sample volume $\approx$ 6.3 cm^3) have given sufficient agreements. These measurements were performed on TEX-1 [12] the neutron diffractometer at GKSS Research Center in Geesthacht (fibre composites) and at D1S [13] the four circle diffractometer at the Hahn-Meitner-Institut in Berlin (graphite-siliconcarbide).

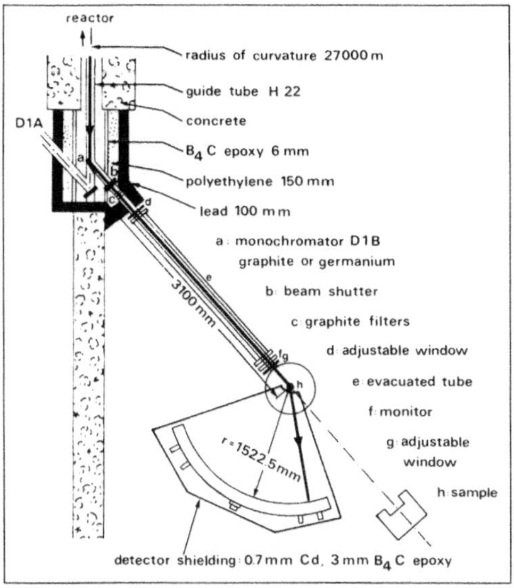

Fig.1: Schematic view of D1B a neutron podwer diffractometer at ILL in Grenoble

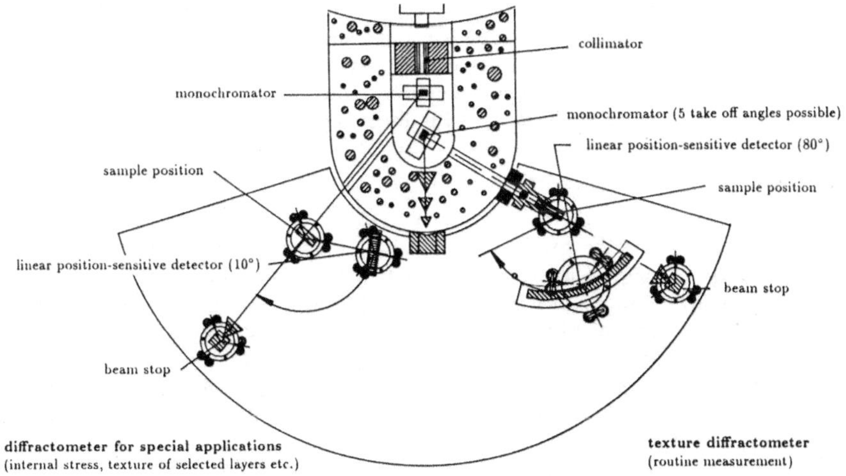

Texture diffractometers at beam tube 5

Fig.2 : Schematic view of TEX-2 the texture diffractometers at GKSS

Texture analysis by neutron diffraction

The first texture measurement using neutron diffraction was carried out by Brockhouse in 1953 [14]. Since that time neutron diffraction has attracted increasing interest in special applications. Numerous publications describe the advantages of neutron texture measurements see for example Szpunar [15] or Welch [16]. These advantages are based on the physics of neutron scattering as described before and lead to new fields in texture research:

- determination of bulk textures of compact samples (spherical sample method) [17]
- determination of texture and microstructure of the same sample
- texture determination of coarse-grained materials [18]
- rather simple sample preparation (roughness of the surface) [19]
- high accuracy of the neutron texture determination [19]
- local texture determination in compact samples ($\approx$ 8 mm^3 areas in a larger sample)
- texture determination of a second layer in a complex sample in a non-destructive way (independent of oxide coatings, no etching necessary)
- texture determination in multiphase systems without problems of anisotropic absorption [20]
- availability of position sensitive detectors

The standard texture measurements by neutron diffraction use the spherical sample methode (Tobisch and Bunge [16]) but different scanning routines (equal area, equal angular). Based on the high transmission of neutrons and the large cross sections of the primary beam, spherical samples up to 40 mm in diameter have been analysed. The reflection takes place in the whole sample volume and an average texture of the whole sample is obtained. Thereby the sample geometry can be an ideal sphere, a cylinder or even a cube. This allows the compaction of a sample to increase the sample volume performed for the texture analysis of Al-Pb fibre composites [21]. In contrast to X-ray measurements the complete pole figure ($\varphi = 0° - 360°; \chi = 0° - 90°$) will be measured.

Figure 2 shows the new texture diffractometers at the 5 MW reactor FRG-1 at GKSS Research Center in Geesthacht. In addition to the a powder diffratometer an Eulerian cradle is necessary for pole figure measurements. The right equipment is the texture diffractometer for routine measurements. A high freedom in wavelength (change of monochromator or change of the monochromator take off angle) and the linear position sensitive detector allow the texture investigation of a wide range of different materials (pure metals,, composites, intermetallic compounds and other complicated multiphase systems). A large number of pole figures can be measured simultaneous, which is necessary in routine experiments for time-saving. The left instrument (in plan) will be used for special applications. A measurement which requires a high accuracy needs a good collimated beam and is rather time-consuming, so that a second diffractometer will be used. Moreover, the 2θ range is much higher necessary for high resolution work.

An investigation of texture inhomogeneities or the texture of a local area in a compact specimen require special beam handling. By masking the primary beam as well as the scattered beam with a neutron absorber (Cadmium or Gadolinium foils) the selection of a local area within a sample becomes possible. Such an experiment was performed by Choi [22] for the examination of a copper cone sample and by Brokmeier [18] for looking at local textures in a rock salt cylinder.

Neutrons have numerous advantages in contrast to X-rays but some essential disadvantages:

- low intensities of the primary beam
  (long time for a measurement)
- radiation source
  (a reactor or a spallation source is necesssary)
- limited number of four circle diffractometers
  (beam time, in competition with physicist, chemist and biologists, priority of materials science)
- equipment of neutron diffractometers
- determination of local textures is limited
- determination of surface textures is strongly limited

Stress analysis by neutron diffraction

A texture measurement is the statistical analysis of the frequency distribution of the crystallographic grain orientations with respect to a sample coordinate system. Therefore in texture measurement - measurement of numerous pole figures - a high resolution in d-space is not necessary and a high neutron flux will be favoured for time saving. In contrast to standard texture measurements the determination of internal stresses require a neutron diffraction technique with a high resolution. A variation of the lattice parameters in the order of 10^{-3} to 10^{-4} for $\Delta d/d$ is typically and has to be determined. Hence, the resolution requirements are of the same order as in high resolution single crystal diffractometry. The specifications of such a diffractometer are different to a standard texture diffractometer:

- a high beam collimation
- a monochromator with an excellent mosaic spread $\Delta\lambda/\lambda$
- a large monochromator take off angle
- a high angular resolution
  (collimation of the scattered beam, detector diaphragm)
- measurement at large scattering angles
  (high hkl or large wavelength)

A typical high resolution powder diffractometer is the D1A at the Institut Laue Langevin (ILL) in Grenoble. Some experiments on internal stress measurements were performed at this diffractometer and have shown satisfactory results (Allen et al. [23], Priesmeyer et al. [24]. The way of masking the sample or working with diaphragm systems to select a local volume element of the sample in a non-destructive way is the same thas in local texture mesurements (see chapter before). The disadvantages of neutron stress measurements are the same thas for neutron texture measurements described before. Based on the high quality of the beam, which is necessary, relatively long measuring times have to be considered, but a method is available to analyse stresses in local areas of bulk samples.

Acknowlegements

This work was funded by the German Federal Minister for Research and Technology (BMFT) under the contract number 03BU1CLA9. The authors would like to thank the GKSS Research Center for the support in neutron diffraction at the Research Reactor FRG-1.

References

1. G.E. Bacon, Neutron Diffraction (Oxford: Clarendon Press 1975).
2. G. Gottstein, ICOTOM , ed. J.S. Kallend and G. Gottstein, (TMS-AIME 1988) 195-202.
3. H.-G. Brokmeier, B. Brehler, G. Rakuttis and A. Haase, ICOTOM, ed. J.S. Kallend and G.Gottstein, (TMS-AIME 1988),235-240.
4. H.-J. Kopineck, Experimental Techniques of Texture Analysis, ed. H.J. Bunge, (Oberursel: DGM Informationsgesellschaft 1986) 171-180.
5. E. Jansen, W. Schäfer and G. Will, Experimental Techniques of Texture Analysis, ed. H.J. Bunge, (Oberursel: DGM Informationsgesellschaft 1986) 229-240.
6. A. Filhol, J.-Y. Blanc, A. Antoniadis and J. Berruyer, ILL internal report 1988 , 88FI05T.
7. M. Dahms,H.-G. Brokmeier, H. Seute, H.J. Bunge, Mechanical and Physical Behaviour of Metallic and Ceramic Composites, ed. S.I. Andersen, H. Lilholt and O.B. Pedersen, (Risø National Laboratory 1988) 327-332.
8. Neutron research facilities at the ILL high flux reactor , Grenoble 1986
9. H.-G. Brokmeier, H. Gertel, C. Ritter, Comparison of phase analysis of Al-Cu Composites by X-ray and neutron diffraction, in preparation.
10. D.K. Smith, G.G. Johnson Jr., A. Scheible, A.M. Wims, J.L. Johnson and G. Ullmann, Powder Diffr. 2, 73 - 77, 1987.
11. H.-G. Brokmeier and H.J. Bunge, Z. Krist. 185 , 667, 1988.
12. H.-G. Brokmeier, H.J. Bunge, B. Brehler, R. Wagner, P. Wille, GKSS-Report 87/E/42.
13. H.A. Graf, HMI-B 232, 110-127, 1980.
14. B.N. Brockhouse, Can. J. Phys. 31, 339-355, 1953.
15. J. Szpunar, Atomic Energy Rev. 14, 199-261, 1976.
16. P.I. Welch, Experimental Techniques of Texture Analysis, ed. H.J. Bunge, (Oberursel: DGM Informationsgesellschaft 1986) 183-207.
17. J. Tobisch and H.J. Bunge, Texture 1, 125-127, 1972.
18. H,-G. Brokmeier, Thesis TU Clausthal, 1983.
19. H.J. Bunge , Quantitative Texture Analysis, ed. H.J. Bunge and C. Esling, (Oberursel: DGM Informationsgesellschaft 1986), 85-128.
20. H.J. Bunge, Experimental Techniques of Texture Analysis, ed. H.J. Bunge, (Oberursel: DGM Informationsgesellschaft 1986), 395-402.
21. H.-G. Brokmeier, W. Böcker and H.J. Bunge, Textures Microstr. 8/9, 429-441, 1988.
22. C.S. Choi, H.J.Prask, S.F. Trevino, J.Appl.Cryst. 12 327-331, 1979.
23. A.J. Allen , M.T. Hutchings, C.G. Windsor, C. Andreani, Advances in Physics 34, 445-473, 1984.
24. H.-G. Priesmeyer and J. Schröder, Technology Transfer Guidebook Series i.i.t.t., Fatigue and Stress, Pergamon Press Ltd. 1987 to be published.

Computer Aided Optimization of an On-line Texture Analyzer

F. Wagner*, H. Otten**, H.J. Kopineck** and H.J. Bunge***

* LM2P, University of Metz (France)

** Hoesch Stahl AG, Dortmund (FRG)

*** Institut für Metallkunde, Clausthal-Zellerfeld (FRG)

Summary

The way of calculating a partial texture from a small set of measurements obtained with an on-line texture analyzer is first recalled. By using a program conceived accordingly the influence of the several parameters on such a calculation is studied. From the so obtained results the optimization of partial texture determination is discussed and the possibility of deriving the anisotropy coefficient $r(\alpha)$ is demonstrated.

1. Introduction

Texture is an important parameter of polycrystalline materials in order to understand or calculate their anisotropic properties. The crystallographic texture is characterized by the Orientation Density Function (O.D.F.), f(g), which can be expanded on a basis of symmetrized generalized spherical

harmonic functions $T_l^{\mu\nu}(g)$ [1]:

$$f(g) = \sum_{l=0}^{l=l_{max}} \sum_{\mu\nu} C_l^{\mu\nu} \, T_l^{\mu\nu}(g) \quad (1)$$

The set of coefficients, $\{C_l^{\mu\nu}; o \leq l \leq l_{max}\}$, where l_{max} has to ensure that the convergency of the series is reached, contains then the whole texture information and consists of several hundred coefficients. In most of the cases these $C_l^{\mu\nu}$ coefficients are calculated from a set of experimental data (pole figure measurements) obtained with X-rays or neutron diffraction.
Such a texture determination has to be considered as being a destructive method of investigation because it usually requires the cutting of a piece of material and its preparation (polishing for example).
In fact the calculation of some physical properties as the Young's modulus, the magnetization energy or the anisotropy coefficient for example do not require the whole set of coefficents $C_l^{\mu\nu}$ but only the first ones (less than ten in most of the cases).
Let us call partial or low resolution texture a set of coefficients such as, $\{C_l^{\mu\nu}; o \leq l \leq L\}$ with $L \ll l_{max}$. The determination of a partial texture strongly decreases the number of necessary experimental data which then gives way to in situ measurements i.e. a non destructive method.
The first achievement in this field was an on-line texture analyzer which was designed by the Hoesch Stahl AG Compagny to control the mean anisotropy coefficient r during cold rolling or annealing processes [2]. It was based on empirical relations which can be establish between pole-figure measurements in a few points and the r value [3,4].
In this paper we will show how a more sophisticated texture analyser can lead to the determination of a partial texture and we will consider the paramaters which influence this determination. Special attention will be paid to the calculation of the anisotropy coefficient from this partial texture.

2. Experimental data

An on line texture analyser consists of a "white" X-rays source which continuously sends a beam through the sheet and a set of energy-dispersive detectors which detect the diffracted beams [5,6].
Let us assume that we have K detectors and that each one can analyze J reflexions. For a given measuring time this results in a sef of K x J diffracted intensities, I_{kj} with $1 \leq j \leq J$ and $1 \leq k \leq K$. If we denote $y_k = \{\alpha_k, \beta_k\}$ the position of a detector according to the sample coordinate system (RD, TD, ND), the I_{kj} intensity is the pole figure measurement at the point y_k of the pole figure number j. A Reasonable value for K and J is 5 : a measuring sequence results then in a set of 25 I_{kj} intensities.

If intensities, I^r_{kj}, are measured under the same conditions for a random specimen the normalized input data for the calculation of a partial texture

will be the quantities $A^{exp.}_{kj}$ defined as :

$$A^{exp.}_{kj} = I_{kj} \; / \; I^{r}_{kj} \qquad (2)$$

(It is possible to use intensities of a sample with a known texture instead of intensities of a random specimen ; this will not be considered in the following for sake of simplicity).

Because of the divergency of incident and diffracted beams, a measured intensity does not correspond to a given pole-figure point $y_k = (\alpha_k, \beta_k)$ but to a small area, called the pole-figure window w_{kj}. This intensity $\overline{I_{kj}}$ is in fact an integral over this window :

$$\overline{I_{kj}} = \int_{w_{kj}} I_{kj} \sin\alpha \, d\alpha \, d\beta \qquad (3)$$

3. Determination of a partial (low resolution) texture

Let us recall the mathematical procedure which was proposed in [7]. For a given texture, characterized by the set of coefficients $C_l^{\mu\nu}$, the normalized

intensity A^{th}_{kj} at the point y_k in the pole figure j reads :

$$A^{th}_{kj} = 1 + \sum_{l=lmin}^{lmax} \sum_{\mu=1}^{M(l)} \sum_{\nu=1}^{N(l)} (4\pi / (2l+1))\, C_l^{\mu\nu}\, k^{\mu}_l(h_j)\, k_l^{\nu}(y_k) \qquad (4)$$

where lmin , M(l) and N(l) depend on crystal and specimen symmetries and $k^{\mu}_l(h_j)$ and $k^{\nu}_l(y_k)$ are spherical harmonic functions with crystal and sample symmetry respectively [1]. The search of the unknown coefficients $C_l^{\mu\nu}$ is based on the following minimization :

$$\sum_{k=1}^{k=k} \sum_{j=1}^{j=J} \left(A_{kj}^{exp} - A_{kj}^{th} \right)^2 = \text{Min} \qquad (5)$$

By using relation (4) and restricting the search of coefficients up to the rank l = L the relation (5) becomes :

$$\sum_{k=1}^{k=k} \sum_{j=1}^{j=J} \left[A_{kj}^{exp} - 1 - \sum_{l=lmin}^{l=L} \sum_{\mu} \sum_{\nu} (4\pi / (2l+1)) C_l^{\mu\nu} k_l^{\mu}(h_j) k_l^{\nu}(y_k) \right]^2$$

$$= \text{Min} \qquad (6)$$

The differentiation with respect to a given $C_{l'}^{\mu'\nu'}$ coefficient leads to a linear equation :

$$\sum_{l=lmin}^{l=L} \sum_{\mu=1}^{M(l)} \sum_{\nu=1}^{N(l)} C_l^{\mu\nu} a_{ll'}^{\mu\mu'\nu\nu'} = b_{l'}^{\mu'\nu'} \qquad (7)$$

$$\text{with } a_{ll'}^{\mu\mu'\nu\nu'} = \sum_{k=1}^{k=K} \sum_{j=1}^{j=J} \left(4\pi / (2l+1)\right) k_{l'}^{\mu'}(h_j)\, k_l^{\mu}(h_j)\, k_{l'}^{\nu'}(y_k)\, k_l^{\nu}(y_k) \qquad (8)$$

$$\text{and } b_{l'}^{\mu'\nu'} = \sum_{k=1}^{k=K} \sum_{j=1}^{j=J} k_{l'}^{\mu'}(h_j)\, k_{l'}^{\nu'}(y_k)\, (A_{kj}^{exp} - 1) \qquad (9)$$

The set of linear equations (7) can be solved. If we denote $\beta_{ll'}{}^{\mu\mu'\nu\nu'}$ the elements of the inverse matrix of $(a_{ll'}{}^{\mu\mu'\nu\nu'})$ the solutions $C_l^{\mu\nu}$ read :

$$C_l^{\mu\nu} = \sum_{j=1}^{k=K} \sum_{j=1}^{j=J} \alpha_{l\mu\nu}^{kj} \quad [A_{kj}^{exp} - 1] \qquad (10)$$

with

$$\alpha_{l\mu\nu}^{kj} = \sum_{l'=lmin}^{l'=L} \sum_{\mu'=1}^{M(l)} \sum_{\nu'=1}^{N(l)} k_{l'}^{\mu'}(h_j)\, k_{l'}^{\nu'}(y_k)\, \beta_{ll'}{}^{\mu\mu'\nu\nu'} \qquad (11)$$

It can be seen from relation (10) that the $C_l^{\mu\nu}$ coefficients can be calculated very fast if the mathematical quantities $\alpha_{l\mu\nu}^{kj}$ are stored in a computer. Moreover a few coefficients are calculated : in the case of cubic orthorhombic symmetries there exist only 3 coefficients up to the rank L=4.
If a pole-figure window is considered the previous calculations still hold provided the $k_l^{\nu}(y_k)$ quantity is replaced by the mean value $\overline{k_l^{\nu}(y_k)}$:

$$\overline{k_l^{\nu}(y_k)} = \int_{w_{jk}} k_l^{\nu}(y)\, dy \qquad (12)$$

4. <u>Optimization of the partial texture determination</u>

A program of partial texture calculation, based on the previous relations, has been developped for cubic-orthorhombic symmetries. It allows thus to consider the influence of the several parameters : y_k (detector positions), h_j (reflexions), K and J their respective numbers, L (maximal rank of the $C_l^{\mu\nu}$ calculation) and w_{kj} (pole-figure window). In order to generate data (the normalized intensities A_{kj}^{exp}) without errors and know in advance the $C_l^{\mu\nu}$ values, we have used O.D.F.'s of annealed low carbon steels. From these O.D.F.'s it is then possible to calculate any pole-figure and select a set of A_{kj} intensities which will be used as the experimental data.

The w_{kj} pole- figure window was shown to have a very low influence on the partial texture determination and it will not be considered later.

4.1. *Influence of L*

By using the same set of data the partial texture calculation was repeated for several L values. In Figure 1 typical results are shown ; the mean relative error on the $C_l^{\mu\nu}$ coefficients for each l rank is here reported for calculations with several L values.

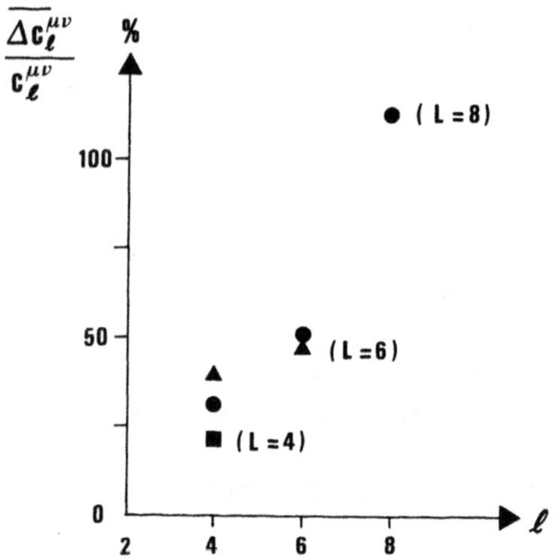

Fig.1 : Mean relative error on $C_l^{\mu\nu}$ coefficient as a function of l and L.

Two conclusions can be deduced for these textures :
- the mean relative error increases with increasing l values
- at l = 4 rank the best accuracy is obtained by using the expansion limit L = 4.

4.2. *Influence of the configuration.*

The name configuration stands for the set $\{y_k, h_j, K, J\}$ which characterizes the positions of the detectors, the used reflexions and their respective numbers. Because of the small number of the data the accuracy obtained for the $C_l^{\mu\nu}$ coefficients is influenced by the choosing of a configuration. This is illustrated in Table I : the calculated values as well as the relative errors are reported for two different configurations (the first one consists of 20 A_{kj} measurements whereas the second one consists of 25 measurements).
There is no mathematical way to define the best configuration because this depends partially on the considered texture itself. The y_k positions of the

detectors must correspond to the main features of the pole-figures. This is obtained, with our program, by "trials and errors" for a given texture class.

	true values	conf. 1 (20 pts) $C_l^{\mu\nu}$	$\Delta C_l^{\mu\nu}/C_l^{\mu\nu}$	conf. 2 (25 pts) $C_l^{\mu\nu}$	$\Delta C_l^{\mu\nu}/C_l^{\mu\nu}$
C_4^{11}	-1.602	-1.579	1%	-1.752	9%
C_4^{12}	-0.126	-0.161	28%	-0.013	90%
C_4^{13}	0.591	0.562	5%	0.550	7%

Table 1 : Influence of the con figuration on the $C_l^{\mu\nu}$ calculation.

5. Determination of the anisotropy coefficient r(α).

Among the physical properties which can be correlated with a partial texture, the anisotropy coefficient r(α) (α refers to a given direction in the sheet plane) is an important one. This correlation requires the existence of a plastic deformation model. Moreover the calculation of r(α) can be very fast if an analytical expression is available. This is the case for the Taylor model where such a relation has been proposed [8]. The anisotropy coefficient r(α) is a function of several quantities :

$$r(\alpha) = r(\alpha, C_l^{\mu\nu}, L, m_{li}^{\mu\nu}) \qquad (13)$$

where the coefficients $m_{li}^{\mu\nu}$ are linked with the Taylor model. They have been published [8] and can be easily stored in a computer at once.

Figure 2 shows the results obtained in this way by using a partial texture calculated from 25 intensities up to the rank L = 4.
The variation of r (α) is somewhat reduced in comparison with the experimental values. Nevertheless the calculated mean anisotropy coefficient ($\bar{r}$ = 1.58) is very close to the experimental one ($\bar{r}$ = 1.60)

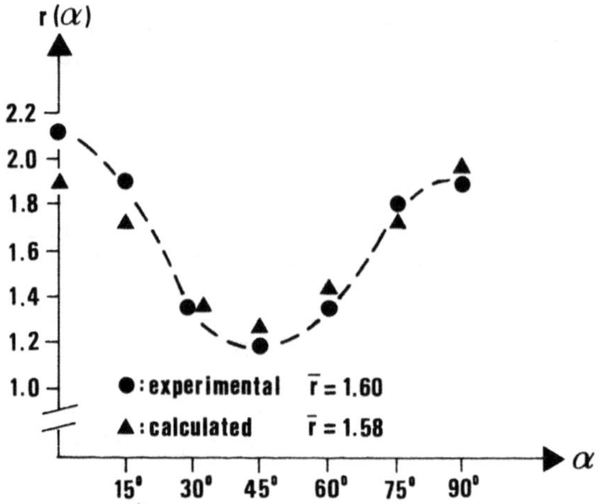

Fig. 2 : Experimental and calculated values of the anisotropy coefficient r (α)

6. Conclusion

The possibility of an on-line determination of a partial texture and then of the anisotropy coefficient is well established. As it has been demonstrated, the use of a computer program allows to define configurations which lead to a satisfactory accuracy for given texture classes.

References

1. Bunge, H.J., Texture Analysis in Materials Science, Butterworths ed. 1982.

2. Böttcher, W. and Kopineck, H.J., Stahl und Eisen, 105, (1985), 509-516.

3. Kopineck, H.J., in Experimental techniques of Texture Analysis, H.J., Bunge ed., DGM Verlag, (1986), 171-180.

4. Kopineck H.J. and Otten H., Textures and Microstructures, 7, (1987), 97-113.

5. Kopineck H.J., "Industrielle Andwendung der on-line Texturmeßtechnik", This conference.

6. Otten H., Bunge H.J. and Kopineck H.J., "Zur on-line Bestimmung technologischer Eigenschaftswerte von kalt-und warm-gewalzten Stahlbändern mittels eines Festwinkel-Textur-Analysators", This conference.

7. Bunge H.J. and Wang F., in Theoretical Methods of Texture Analysis, H.J. Bunge ed., DGM Verlag, 1987, 163-172.

8. Bunge H.J., Schulze M. and Grzesik D., Peine Salzgitter Bericht, Sonderheft 1, (1980).

Characterization of Rolling Texture by Ultrasonic Dispersion Measurement

S. Hirsekorn and E. Schneider

Fraunhofer-Institut für zerstörungsfreie Prüfverfahren,
Universität, Geb. 37, D-6600 Saarbrücken 11

Summary

Ultrasonic scattering at grain boundaries in polycrystals causes attenuation of sound waves and dispersive sound velocities. Texture, that is a weighted orientation distribution of the grains, alters the ultrasonic propagation behavior directional dependent. An ultrasonic scattering theory presented in previous papers enables to calculate the complex propagation constants and hence the texture induced birefringence of sound waves in single-phase cubic polycrystals with rolling or fibre texture from single-crystal and texture properties. Inverted, the theory can be used for texture characterization from ultrasonic measurements. Based on the theoretical results, the influence of texture on the birefringence was calculated. For these calculations, the texture coefficients of rolled steel sheets and plates were evaluated from X-ray investigations, performed by other groups. A linear correlation was found between the frequence independent and the frequency dependent part of the ultrasonic birefringence. Ultrasonic investigations using rolled steel plates confirmed this linear interdependence. A first estimation of the analytical equations for the two parts indicate that the interdependence is governed by the single-crystal elastic moduli. The only supposition concerning texture is the orthorhombic symmetry. Based on these results, ultrasonic dispersion measurements enable the characterization of texture and hence the separation of texture and stress influences on the ultrasonic birefringence. This approach has the advantage that the actual degree of texture is taken into account.

1. Introduction

Rolling textures in components of steel and Al-alloys are developed during manufacturing by deformation processes and heat treatments. Thus, the elastic behavior of the product becomes directionally dependent. Hence, the velocities of ultrasonic waves, propagating the material, change by alternations of the propagation and vibration directions. The texture influence on the ultrasonic velocities is in the range of magnitude of a few percent. This effect is used to characterize the texture and hence the deep drawability of sheets with ultrasonic techniques as shown e.g. in [1-4]. But this texture effect renders difficult the evaluation of (residual) stress states using ultrasonic velocity measurements. The stress influence on ultrasonic velocities is about one order of magnitude smaller than the texture effect.

Because of the smaller single-crystal anisotropy and the larger acousto-elastic effect in Al, both quantities in comparison with steel data, the discrimination of the texture from the stress influence in Al is not as difficult as in the case of steel samples.

Different approaches are published in order to separate the texture effect from the stress influence on the measuring quantities. They presume a "slight" texture. Experimental results, mainly found on Al-samples, demonstrate their applicability. Earlier investigations using a stainless steel bar with a large amount of preferred grain orientation (texture) showed that the ultrasonic birefringence effect increases linearly with the ultrasonic frequency [5]. Birefringence effect is caused by the difference in the propagation velocities of shear waves propagating along the same principle axis but beeing polarized perpendicular to each other. It also was found that the birefringence due to stress alone is independent on frequency. The author mentioned that a qualitative explanation of the results would be in terms of interactions of dislocations and ultrasonic vibrations.

Investigations of one of the authors using cold-rolled steel sheets confirmed a texture induced ultrasonic birefringence dispersion. But this effect is due to ultrasonic scattering in the textured polycrystals [6,7].

Ultrasonic dispersion measurements enable the characterization of rolling texture, and hence, the elimination of texture influences on birefringence in order to evaluate (residual) stress states in textured materials.

2. Texture caused ultrasonic birefringence

The texture induced birefringence effect in single-phase polycrystals is caused by the elastic anisotropy of each grain and the preferred grain orientation. Ultrasonic scattering occurs at each grain boundary and causes attenuation of sound waves and frequency dependence of ultrasonic velocities. An ultrasonic scattering theory [6,7] enables the evaluation of the complex propagation constants of ultrasonic waves in single-phase cubic polycrystals with orthorhombic texture (e.g. rolling or fibre texture). It is assumed that the changes of the elastic constants from grain to grain are small and only caused by different orientations. Boundary effects and orientation correlations between the grains are neglected. The grains are treated as spheres.

In order to calculate the ultrasonic velocities in textured polycrystals, second order perturbation theory is applied to the elasto-dynamic equation which takes into account only linear and quadratic terms of the elastic moduli. Texture enters by weighted integrations over the elastic constants and the products of them. Weighting function is the orientation distribution function in its representation as expansion with respect to generalized spherical functions. Second order perturbation theory restricts the needed expansion coefficients to those up to 8th order. In the case of cubic crystal and orthorhombic texture symmetry, there exist only 12 independent expansion coefficients up to 8th order (except that one of zeroth order which always is equal to 1), namely three of 4th, four of 6th and five of 8th order.

Under the Rayleigh condition, that means, the ultrasonic wave length is large compared to the grain size, the scattering theory yields a quadratic frequency dependence of the ultrasonic velocities. The phase velocity of shear waves, propagating and vibrating parallel to symmetry axes of the texture are given by

$$v_{ij} = d_{1,ij} - d_{2,ij} (\kappa_I a)^2 . \tag{1}$$

The subscripts i and j denote the direction of ultrasonic propagation and vibration, respectively. The wave number κ_I is the Voigt approximation in the isotropic case, and a is the grain radius.

The constants $d_{1,ij}$ and $d_{2,ij}$ contain the single-crystal elastic constants and the expansion coefficients of the orientation distribution function. They are direction dependent. A detailed description of the theoretical treatment is given in [6]. The ultrasonic birefringence caused by texture is also a quadratic function of the ultrasonic frequency. In case of propagations and polarizations parallel to symmetry axes of texture the theory yields

$$\frac{v_{ij} - v_{ik}}{v_{ik}} = C_1 + C_2 \left(\kappa_I a\right)^2 = C_1 + C_2 \left(\frac{2\pi a}{v_{TI}}\right)^2 f^2 . \tag{2}$$

v_{TI} is the Voigt approximation of the shear wave velocity in the isotropic material, and f is the ultrasonic frequency. C_1 and C_2 can be derived from the constants $d_{1,ij}$ and $d_{2,ij}$. In the following, propagation along the normal direction and hence, polarization parallel to rolling and transverse direction are considered. In this case, both constants C_1 and C_2 contain the single-crystal elastic constants and the texture coefficients of 4th order C_4^{10} and C_4^{11} as parameters. Additionally, C_2 contains the texture coefficients of 6th order C_6^{10} and C_6^{11}:

$$C_1 = C_1 (C_{11}, C_{12}, C_{44}; C_4^{10}, C_4^{11})$$
$$C_2 = C_2 (C_{11}, C_{12}, C_{44}; C_4^{10}, C_4^{11}, C_6^{10}, C_6^{11}) \ . \tag{3}$$

Concerning the texture coefficients $C_1^{\mu\nu}$ the same notation is used as in [6,7]. In other publications it is often seen that C_4^{10} is replaced by C_4^{11}, C_4^{11} by C_4^{12} and so forth. The explicit form of these relationships are given in the Appendix.

Numerical results evaluated for shear waves in cold-rolled α-brass and cold-rolled recrystallized steel samples show a linear correlation between the constants C_1 and C_2 [8,9]. Experimental investigations on cold-rolled steel specimens and additional calculations with texture coefficients from X-ray measurements confirm this interdependence as shown in Fig. 1.

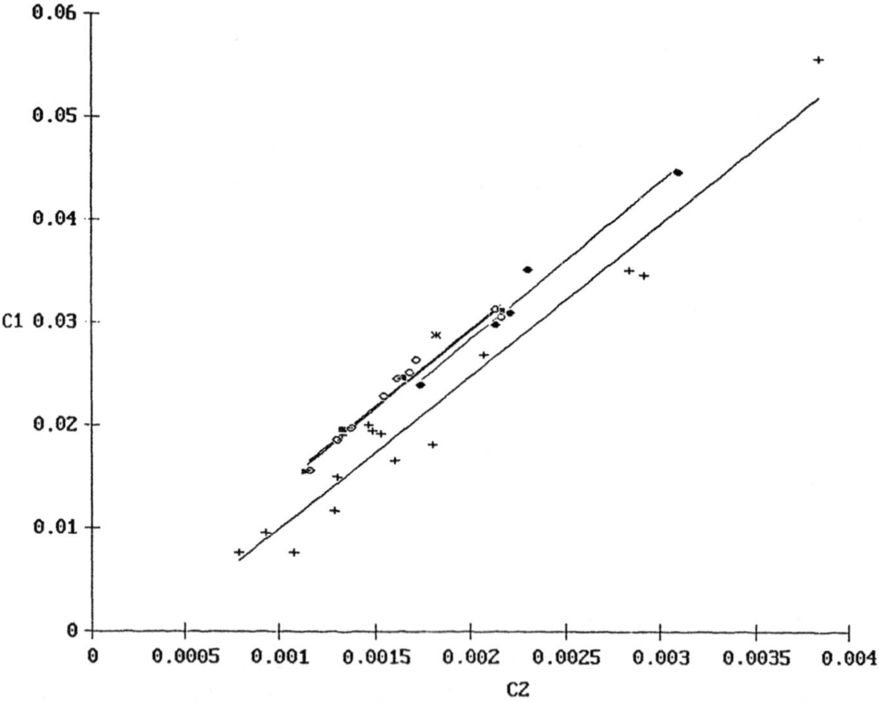

Fig. 1. Interdepence between the constant C_1 and the frequency dependent part C_2 of texture induced ultrasonic birefringence for rolled steel sheets and plates (♦ [10], * [11], + [12], ■ ultrasonic results, ◊ [13]).

3. Discussion

In Fig. 1, for a lot of different rolled steel samples the constant part C_1 of texture induced ultrasonic birefringence is plotted versus the constant C_2, which describes the frequency dependence of the birefringence in the Rayleigh region. The figure contains five essentially different data sets. Four of them follow from the numerical evaluation of the analytical results of the ultrasonic scattering theory in textured polycrystals described above. They belong to a series of five cold-rolled (30%, 50%, 70%, 84%, 86%) recrystallized low-carbon steel plates, the texture coefficients of which are determined by neutron diffraction [10], one 70% cold-rolled Al-killed steel sample [11], and nine steel sheets [13] and five steel plates [12] without definite specification from which the texture coefficients are known by X-ray measurements. From each of the five samples mentioned last, the texture coefficients at three different positions, namely at both surfaces and in the middle of the plate, are available. The constants C_1 and C_2 are calculated and plotted for each of these texture data sets. Furthermore, Fig. 1 contains the by ultrasonic measurements experimentally determined values C_1 and C_2 of four rolled steel plates from which the texture coefficients are not known.

In spite of the essentially different origin of the considered steel samples the linear dependence between C_1 and C_2 turned out to be the same within the measuring inaccuracy in each case. The regression lines of the pairs of values (C_1,C_2) for the different groups of samples and for all samples together are parallel to each other. In a first approximation, it follows from the analytical expressions for the constants C_1 and C_2, that the values of both constants are mainly dependent on the texture coefficient C_4^{11}, and the slope of the lines in Fig. 1 is dominated by the single-crystal constants.

Ultrasonic dispersion measurements enable the determination of the constant C_2 and hence, the evaluation of the texture induced part of the ultrasonic birefringence by the use of the C_1/C_2 correlation. That is an approach to separate the texture from the stress induced part of the measured birefringence. This approach has the advantage that the actual degree of texture is taken into account; the application is not limited to "slightly" textured components.

4. Appendix

The constants $d_{1,ij}$ and $d_{2,ij}$, which determine the velocities of shear waves with propagation and vibration parallel to symmetry axes of texture (i indicates the propagation and j the polarization direction) in the Rayleigh region can be taken from a previous paper [6]:

$$d_{1,ij} = v_{Tij}\left\{1+\left(\frac{A}{\langle C_{ijij}\rangle}\right)^2\frac{S_4^{ij}}{10}\left(1+\frac{2}{3}\frac{k_I^2}{\kappa_{ij}^2}\right)\right\}^{-1}, \tag{4}$$

$$d_{2,ij} = \frac{d_{1,ij}^2}{v_{Tij}}\left(\frac{A}{\langle C_{ijij}\rangle}\right)^2\frac{1}{5^2\cdot 7}\left[S_4^{ij}\left(6-\frac{4}{3}\frac{k_I^2}{\kappa_{ij}^2}+\frac{14}{3}\frac{k_I^4}{\kappa_{ij}^4}\right)+S_5^{ij}\left(3+4\frac{k_I^2}{\kappa_{ij}^2}\right)\right]\frac{\kappa_{ij}^2}{\kappa_I^2}\;; \tag{5}$$

$$\kappa_I = \frac{\omega}{v_{TI}}, \qquad \rho v_{TI}^2 = C_{44}+\frac{1}{5}A, \qquad A = C_{11}-C_{12}-2C_{44}, \tag{6}$$

$$k_I = \frac{\omega}{v_{LI}}, \qquad \rho v_{LI}^2 = C_{11}-\frac{2}{5}A, \tag{7}$$

$$\kappa_{ij} = \frac{\omega}{v_{Tij}}, \qquad \rho v_{Tij}^2 = \langle C_{ijij}\rangle = C_{44}+A\sum_{\alpha=1}^{3}\langle a_{\alpha i}^2 a_{\alpha j}^2\rangle, \tag{8}$$

$$S_4^{ij} = \sum_{\alpha=1}^{3}\langle a_{\alpha i}^2 a_{\alpha j}^2\rangle - 2\left\{\sum_{\alpha=1}^{3}\langle a_{\alpha i}^2 a_{\alpha j}^2\rangle\right\}^2, \tag{9}$$

$$S_5^{ij} = \sum_{\alpha=1}^{3}\langle a_{\alpha i}^4 a_{\alpha j}^2\rangle - \left\{\sum_{\alpha=1}^{3}\langle a_{\alpha i}^2 a_{\alpha j}^2\rangle\right\}^2. \tag{10}$$

The $a_{\alpha i}$ are elements of the rotation matrix. The angled brackets indicate a weighted averaging over all grain orientations with the orientation distribution function as weighting function. In the case of propagation along the normal direction the needed averaged values are given by

$$\sum_{\alpha=1}^{3}\langle a_{\alpha i}^2 a_{\alpha 3}^2\rangle = \frac{1}{5} - \left\{\frac{2}{15\sqrt{3\cdot 7}}C_4^{10} + (-1)^i\frac{1}{3\sqrt{3\cdot 5\cdot 7}}C_4^{11}\right\}, \tag{11}$$

$$\sum_{\alpha=1}^{3}\langle a_{\alpha i}^2 a_{\alpha 3}^4\rangle = \frac{3}{5\cdot 7} - \left\{\frac{8}{3\cdot 5\cdot 11\sqrt{3\cdot 7}}C_4^{10} + (-1)^i\frac{2}{11\sqrt{3\cdot 5\cdot 7}}C_4^{11} + \frac{2\sqrt{2}}{7\cdot 11\cdot 13}C_6^{10} + (-1)^i\frac{4}{11\cdot 13\sqrt{3\cdot 5\cdot 7}}C_6^{11}\right\}; \quad i = 1,\,2\;; \tag{12}$$

if the definition 1 = rolling, 2 = transverse, and 3 = normal direction is used. From these results texture caused ultrasonic birefringence of shear waves propagating along the normal direction can be derived analytically. One gets

$$C_1 = \frac{d_{1,31}}{d_{1,32}} - 1 , \qquad C_2 = \frac{d_{1,31}}{d_{1,32}} \left(\frac{d_{2,32}}{d_{1,32}} - \frac{d_{2,31}}{d_{1,31}} \right) . \tag{13}$$

5. Acknowledgement

The investigations were partly supported by the European Community of Carbon and Steel.

6. References

1. Spies, M. and Schneider, E.: Nondestructive Analysis of the Deep-Drawing Behavior of Rolled Sheets with Ultrasonic Techniques. These proceedings.

2. Cassier, O.; Donadille, C. and Bacroix, B.: Drawability Assessment of Steel Sheets by an Ultrasonic Method. These proceedings.

3. Clark, A.V. et al.: Characterization of Texture and Formability in Steel Sheets Using Electromagnetic Acoustic Transducers. These proceedings.

4. Li, Y.; Smith, J.F. and Thompson, R.B.: Characterization of Texture in Plates by Ultrasonic Plate Wave Velocities. These proceedings.

5. Mahadevan, P.: Effect of Frequency on Texture - Induced Ultrasonic Wave Birefringence in Metals. Nature 211 (1966) 621-622.

6. Hirsekorn, S.: The Scattering of Ultrasonic Waves in Polycrystalline Materials with Texture. J. Acoust. Soc. Am. 77 (1985) 832-843.

7. Hirsekorn, S.: Directional Dependence of Ultrasonic Propagation in Textured Polycrystals. J. Acoust. Soc. Am. 79 (1986) 1269-1279.

8. Goebbels, K. and Hirsekorn, S.: A New Ultrasonic Method for Stress Determination in Textured Materials. NDT International 17 (1984) 337-341.

9. Schneider, E.; Hirsekorn, S. and Goebbels, K.: Zerstörungsfreie Bestimmung von Volumenspannungen in Bauteilen mit Walztextur. VDI-Bericht 552 (1985) 235-264.

10. Bunge, H.J.; Schleusener, D. and Schläfer, D.: Neutron Diffraction Studies of the Recrystallization Textures in Cold-Rolled Low-Carbon Steel. Metal Science 8, (1974) 413-423.

11. Bunge, H.J.; Mathematische Methoden der Texturanalyse. Akademie-Verlag, Berlin, 1969, p. 119.

12. Ruppersberg, H.: private communication (texture coefficients of rolled steel plates determined by X-ray measurements).

13. Ruppersberg, H.: private communication (texture coefficients of rolled steel sheets determined by X-ray measurements).

Nondestructive Analysis of the Deep-drawing Behaviour of Rolling Sheets with Ultrasonic Techniques

M. SPIES and E. SCHNEIDER

Fraunhofer-Institut für zerstörungsfreie Prüfverfahren
Universität, Gebäude 37, Saarbrücken, FRG

Summary

The nondestructive characterization of the deep-drawing behavior is essential for the process control and quality assurance of rolled sheets. Complementary to X-ray diffraction techniques, ultrasonic techniques are under development to characterize the deep drawing behavior in terms of the parameters Δr and r_m. Measuring the times-of-flight of ultrasonic waves, the fourth-order expansion coefficients of the orientation distribution function are evaluated. Published empirical and theoretical results are used to link the planar anisotropy parameter Δr with the expansion coefficient C_4^{13} and the normal anisotropy parameter r_m with the coefficient C_4^{11}. In different sets of cold and hot rolled ferritic steel sheets linear correlations are found between the two deep drawing parameters and these expansion coefficients. Based on these correlations an algorithm was developed and applied to evaluate Δr- and r_m-values using ultrasonic techniques. The excellent correlation between these data and the Δr- and r_m-values given by the manufacturer of the sheets show that ultrasonic techniques hold high promise to characterize the deep-drawing behavior of sheets.

Introduction

Textures in rolled sheets are developed during manufacturing by deformation processes and heat treatments. Thus, the magnetic, elastic and plastic behavior of the product becomes directionally dependent. This can be desirable, but it can also have negative effects. In both cases a nondestructive analysis of the texture and its consequences for materials' behavior is of great interest. X-ray diffraction techniques are developed and used for that purpose /1/.

Complementary to these methods, ultrasonic techniques are under development in order to analyze the texture in the surface as well as in the bulk of rolled plates and sheets of steel and aluminum. The activities concentrate on the characterization of the deep-drawing behavior of rolled steel sheets.

1. Ultrasonic Velocities in Cubic Materials with Rolling Texture

Texture, the nonrandom distribution of the single crystals in a polycrystalline material, is described by the orientation distribution function (ODF). The ODF can mathematically be expressed as a series expansion of symmetrical generalized spherical harmonics /2/. During the rolling process of cubic materials (steels, aluminum alloys) an orthorhombic texture is developed. Because of the symmetry with respect to three orthogonal mirror planes together with the cubic crystal symmetry, the knowledge of the three fourth-order expansion coefficients $C_4^{1\nu}$ of the ODF is sufficient to describe the texture influence on the elastic properties of the material /3/. Inserting the elastic constants of a textured polycrystal /3/ into the Christoffel-Equation for orthorhombic sample symmetry /4/ the propagation velocities of ultrasonic waves in a cubic material with rolling texture are obtained:

$$\rho V_{RR}^2 = C_{11} - A\left(\frac{2}{5} - \frac{1}{70}\sqrt{\frac{7}{3}}\left(C_4^{11} - \frac{2}{5}\sqrt{5}\,C_4^{12} + \frac{1}{3}\sqrt{35}\,C_4^{13}\right)\right) \tag{1a}$$

$$\rho V_{TT}^2 = C_{11} - A\left(\frac{2}{5} - \frac{1}{70}\sqrt{\frac{7}{3}}\left(C_4^{11} + \frac{2}{5}\sqrt{5}\,C_4^{12} + \frac{1}{3}\sqrt{35}\,C_4^{13}\right)\right) \tag{1b}$$

$$\rho V_{NN}^2 = C_{11} - A\left(\frac{2}{5} - \frac{1}{70}\sqrt{\frac{7}{3}}\cdot\frac{8}{3}\,C_4^{11}\right) \tag{1c}$$

$$\rho V_{RT}^2 = C_{44} + A\left(\frac{1}{5} + \frac{1}{70}\sqrt{\frac{7}{3}}\left(\frac{1}{3}\,C_4^{11} - \frac{1}{3}\sqrt{35}\,C_4^{13}\right)\right) = \rho V_{TR}^2 \tag{1d}$$

$$\rho V_{TN}^2 = C_{44} + A\left(\frac{1}{5} - \frac{1}{70}\sqrt{\frac{7}{3}}\left(\frac{4}{3}\,C_4^{11} + \frac{2}{3}\sqrt{5}\,C_4^{12}\right)\right) = \rho V_{NT}^2 \tag{1e}$$

$$\rho V_{NR}^2 = C_{44} + A\left(\frac{1}{5} - \frac{1}{70}\sqrt{\frac{7}{3}}\left(\frac{4}{3}\,C_4^{11} - \frac{2}{5}\sqrt{5}\,C_4^{12}\right)\right) = \rho V_{RN}^2 \tag{1f}$$

ρ is the density; V_{ij} is the velocity of a wave propagating in the direction i and vibrating in the direction j. R,T,N represent the rolling, the transverse and the normal direction of the sheet. c_{11}, c_{12}, c_{44} are the single crystal elastic constants and $A = c_{11} - c_{12} - 2c_{44}$ is a measure for the single crystal anisotropy. C_4^{11}, C_4^{12} and C_4^{13} are the fourth-order expansion coefficients of the ODF.

The velocities of waves propagating in the plane of a sheet are given in several references /5,6,7/. Considering the expansion coefficient C_4^{13} up to second order and the effect of beam skewing, the velocity of the lowest order mode of a SH-wave is given by /7/:

$$\rho V_{SHo}^2(\theta) = \mu + \frac{1}{70}\sqrt{\frac{7}{3}}\,A\left(\frac{1}{3}\,C_4^{11} - \frac{1}{3}\sqrt{35}\,C_4^{13}\right) + \frac{1}{70}\sqrt{\frac{7}{3}}\,A\,\frac{8}{3}\sqrt{35}\,C_4^{13}\cos^2\theta\,\sin^2\theta$$

$$\times\left(1 - \frac{1}{70}\sqrt{\frac{7}{3}}\,\frac{16}{3}\sqrt{35}\,\frac{A}{\mu}\,C_4^{13}\left(1 - 2\cos^2\theta\right)^2\right) \tag{2}$$

where Θ is the angle between the rolling direction and the direction of wave propagation.

2. Characterization of the Deep-Drawing Behavior

In order to characterize the deep-drawing properties of rolled sheets, different quantities are used in industry; they are all based on the Lankford-parameter /8/, usually represented by R. This value is determined in a tensile test where the change of thickness d and width b of a strip is measured before (subscript o) and after (subscript ε) a plastic deformation of at least 20%. R and the quantity r, which is sometimes also called q are defined by:

$$R = \frac{\ln\left(b_o / b_\varepsilon\right)}{\ln\left(d_o / d_\varepsilon\right)} \qquad r = R/(R+1) \tag{3}$$

The quantity Δr, also called the planar anisotropy, is a measure for the direction dependence of the thickness reduction. Δr and the normal anisotropy r_m are given by:

$$\begin{aligned} \Delta r &= 0{,}5 \left(r\,(0°) - 2r\,(45°) + r\,(90°)\right) \\ r_m &= 0{,}25 \left(r\,(0°) + 2r\,(45°) + r\,(90°)\right). \end{aligned} \tag{4}$$

Δr and r_m are evaluated after the plastic deformation of strips cut at 0°, 45° and 90° to the rolling direction. It is found, that r_m is a measure for deep drawability of a sheet and Δr is found to correlate well with the effect of earing /9/.

Empirical investigations showed, that the quantities Δr and r_m correlate linearly with the Young's moduli ΔE and E_m defined analogously to Equations 4, evaluated in elastic tensile test experiments /10/.

A relationship between the fourth-order expansion coefficients and the Young's modulus E was found in theoretical investigations /3,11/: The Young's modulus measured in certain directions Θ with respect to the rolling direction can be expressed as:

$$E(\theta) = E_r + E_1 C_4^{11} + E_2 C_4^{12} \cos 2\theta + E_3 C_4^{13} \cos 4\theta \tag{5}$$

Thereby E_r is the average mean value of Young's modulus and E_1, E_2, E_3 are constants.

These results give rise to expect linear correlations between C_4^{13} and Δr on the one hand and between C_4^{11} and r_m on the other. Thus ultrasonic techniques hold promise to analyze the deep drawability by evaluating the fourth-order expansion coefficients and their correlations with the quantities Δr and r_m.

3. Determination of the Fourth-Order Expansion Coefficients

As it is seen in Equations 1 and 2, different sets of equations can be used to evaluate the three fourth-order expansion coefficients. The combination of Equations (1e) and (1f) together with $C_{11}+2C_{44} = \rho(V^2_{NR}+V^2_{NT}+V^2_{NN})$ yields /12/:

$$C_4^{11} = \frac{210}{8A}\sqrt{\frac{3}{7}}\left(2\mu - (C_{11} + 2C_{44})\frac{V^2_{NR} + V^2_{NT}}{V^2_{NR} + V^2_{NT} + V^2_{NN}}\right) \tag{6}$$

$$C_4^{12} = \frac{210}{4\sqrt{5}\cdot A}\sqrt{\frac{3}{7}}\left((C_{11} + 2C_{44})\frac{V^2_{NR} - V^2_{NT}}{V^2_{NR} + V^2_{NT} + V^2_{NN}}\right) \tag{7}$$

In these equations the sound velocities can be replaced by the times-of-flight and the pathlengths. Because the pathlengths are identical in the nominator and denominator, the times-of-flight remain as the only quantities to be measured.

Further manipulation of Equation (2), described in /12/ results in:

$$C_4^{13} = \frac{210\mu}{32\sqrt{35}\cdot A}\cdot\sqrt{\frac{3}{7}}\left(1-2\cos^2\theta\right)^{-1}\cdot\left(1-\sqrt{1-16\,\frac{1-2\cos^2\theta^2}{\cos^2\theta\sin^2\theta}\,\frac{V_{SH(\theta)}-V_{SH(0°)}}{V_{SH(0°)}}}\right) \tag{8}$$

This relationship also enables an evaluation of C_4^{13} independent of the pathlength.

It follows from Equation (2) that the velocities of SH_o-waves propagating parallel to the rolling and to the transverse directions are equal, as well as those measured at the angles θ equal to 45° and 135°. By averaging the corresponding measuring data, other influences on the quantities, e.g. residual stresses are minimized.

In order to evaluate the coefficients C_4^{11} and C_4^{12}, the times-of-flight of 20 MHz and 30 MHz shear and longitudinal waves were measured using standard ultrasonic equipments. The SH_o-wave was excited and received by EMATs held by a rigid spacer /12,13/. The accuracy of the time-of-flight measurements was $\pm$ 1 ns.

4. Correlation Between Fourth-Order Coefficients and Deep-Drawing Properties

In order to proof the expected linear correlations between the coefficient C_4^{13} and Δr and between C_4^{11} and r_m, different sets of hot- and cold rolled ferritic steel sheets were tested. The thickness of the sheets varied between 0.75 and 2.53 mm. The expansion coefficients, determined according to Equations (6) and (8), respectively, correlate linearly with the r_m- and Δr-values provided by the manufacturer of the sheets.

As representative results, Figures 1 and 2 show the correlations found for one of the investigated sets of sheets. The correlations between the expansion coefficients and the deep drawing parameters found for the other series of sheets were also very satisfying. The correlations are characterized by the linear equations $r_m = A + T \cdot C_4^{11}$ and $\Delta r = B + S \cdot C_4^{13}$.

The values found for A vary between 0.750 and 1.302, the slopes T lie between -0.288 and -0.902 for the five sets of cold rolled sheets. For the hot rolled sheets a correlation was found with A = 1.183 and T = -0.250.
The Δr-values were only given for 4 sets of cold rolled sheets. The values of B are found to vary between -0.09 and -0.251; the slopes S are between 0.889 and 1.549.

With these correlation lines between the Δr- and r_m-values and the expansion coefficients C_4^{13} and C_4^{11}, respectively, an algorithm was developed for the evaluation of the Δr(US)- and r_m(US)-values using ultrasonic time-of-flight data.

Figures 3 and 4 show the correlations between the deep drawing parameters given by the manufacturer (subscript mech) and the parameters evaluated with the ultrasonic technique (subscript US). The correlations show that ultrasonic techniques hold high promise to characterize the deep-drawing behavior of rolled sheets.

Acknowledgement

This investigation was partly supported by the European Community for Carbon and Steel. The authors would like to thank Hoesch Stahl AG, Dortmund, FRG for providing the sheets.

References

1. Kopineck, H.J.; Otten, H.; Bunge, H.J.: On-line Determination of Technological Characteristics of Cold and Hot Rolled Steelbands by a Fixed Angle Texture Analyzer. These proceedings.

2. Bunge, H.J.: Zur Darstellung allgemeiner Texturen. Z. Metallkunde 56 (1965) 872-874.

3. Bunge, H.J.: Über die elastischen Konstanten kubischer Materialien mit beliebiger Textur. Kristall und Technik 3 (1968) 431-438.

4. Auld, B.A.: Acoustic Fields and Waves in Solids. Vol. 1, New York, Wiley 1973.

5. Thompson, R.B.; Smith, J.F.; Lee, S.S.: Inference of Stress and Texture from the Angular Dependence of Ultrasonic Plate Mode Velocities. In NDE of Microstructure for Process Control 73-79, Wadley, H.N.G. (ed) ASM 1985.

6. Allen, D.R.; Langmann, R.; Sayers, C.M.: Ultrasonic SH-Wave Velocity in Textured Aluminum Plates. Ultrasonics 9 (1985) 215.

7. Sayers, C.M.; Proudfoot, G.G.: J. Mech. Phys. Solids 34 (1986) 579.

8. Lankford, W.T.; Snyder, S.C.; Bauscher, J.A.: Trans. ASM 42 (1950) 1197.

9. Vlad, C.M.: Verfahren zur Ermittlung der Anisotropie-Kennzahlen für die Beurteilung des Tiefziehverhaltens kohlenstoffarmer Feinbleche. Materialprüfung 14 (1972) 179-182.

10. Stickels, C.A.; Mould, P.R.: The Use of Young's Modulus for Predicting the Plastic Strain Ratio of Low-Carbon Steel Sheets. Metallurg. Trans. 1 (1970), 1303-1311.

11. Bunge, H.J.: Texture Analysis in Materials Science. London, Butterworth Publ. 1982.

12. Spies, M.: Diplomarbeit, Universität des Saarlandes, Saarbrücken, FRG, to be published.

13. Wilbrand, A.; Repplinger, W.; Hübschen, G.; Salzburger H.J.: EMUS Systems for Stress and Texture Evaluation by Ultrasound. These proceedings.

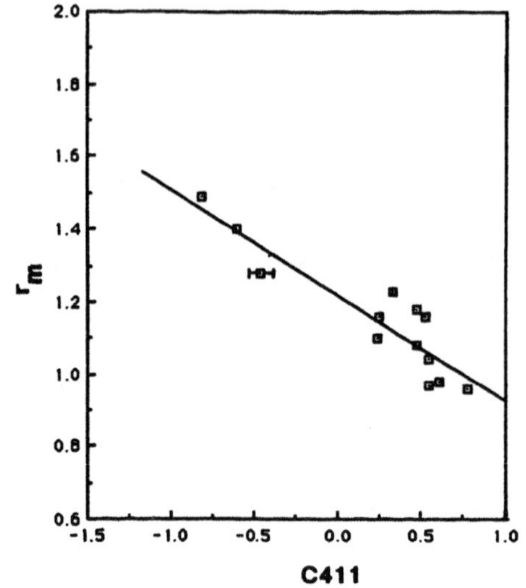

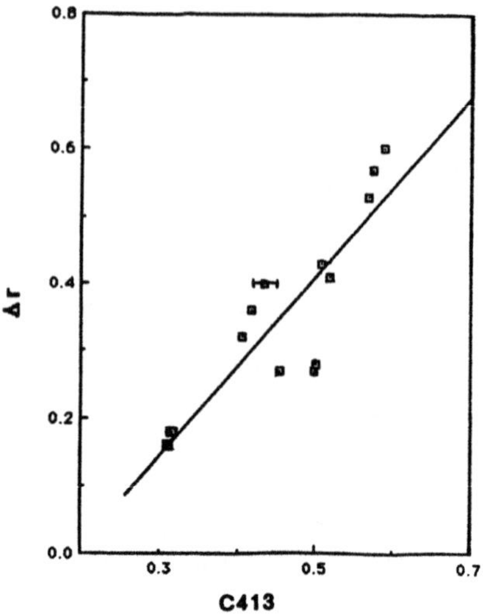

Fig. 1. Correlation between the deep drawability parameter r_m and the expansion coefficient C_4^{11}, evaluated for a set of cold rolled ferritic steel sheets.

Fig.2. Correlation between the deep drawability parameter Δr and the expansion coefficient C_4^{13}, evaluated for a set of cold rolled ferritic steel sheets.

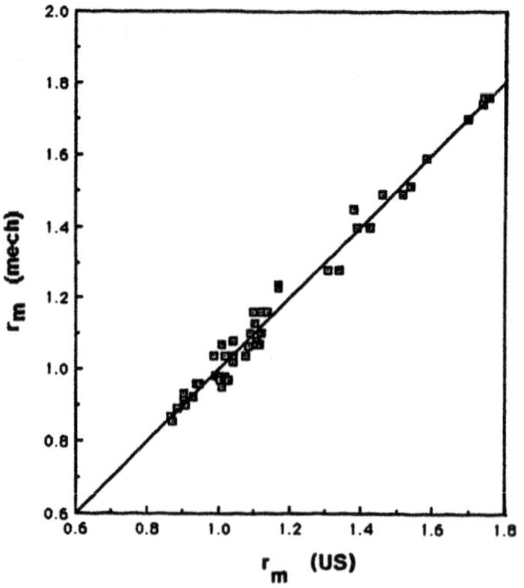

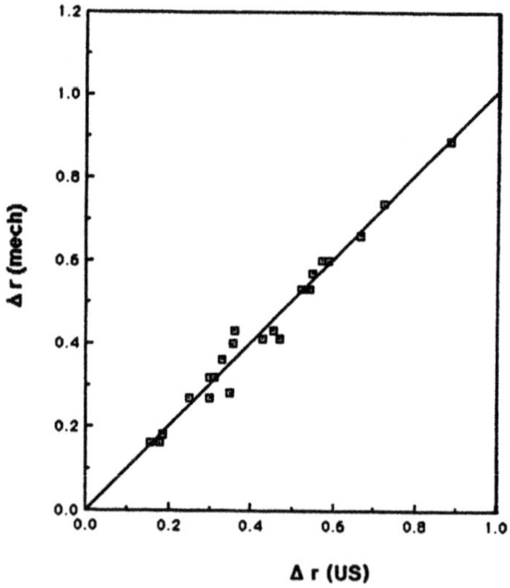

Fig. 3. Correlation between the deep drawability parameter r_m, evaluated with mechanical and ultrasonic techniques.

Fig. 4. Correlation between the deep drawability parameter Δr, evaluated with mechanical and ultrasonic techniques.

Lankford Coefficient Evaluation in Steel Sheets by an Ultrasonic Method

O. CASSIER, C. DONADILLE, B.BACROIX
CENTRE COMMUN DE RECHERCHE USINOR - SACILOR
IRSID
78105 SAINT GERMAIN EN LAYE
FRANCE

Summary

Ultrasonic (US) wave velocity measurements associated with adapted polycrystalline models of plastic flow are used to evaluate the texture and strain ratio r(α) of steel sheets. Results are compared with more classical destructive techniques (X-ray, tensile tests). A satisfactory agreement is obtained providing that US data are made consistent by the appropriate algorithm proposed.

Introduction

Texture of cold-rolled steel sheets is an important parameter which determines the formability properties of products. In steel plants, the use of continuous annealing for thin sheet production requires non-destructive testing techniques to control the metallurgical properties of the sheets. At the present time, two methods may be put forward to perform the "on-line" non-destructive evaluation of the texture : The first one uses X-ray diffraction [1]. This method is rather similar to that used in laboratory. The second one uses ultrasonic waves, the velocity of which may be expressed as a function of the expansion coefficients of the Orientation Distribution Function [2]. This last method is retained in this paper. The aim of this study is to show that ultrasonic waves are a good way of evaluating the anisotropy of cold-rolled and/or annealed sheets, as expressed in terms of the averaged strain ratios r_m and Δr. This is performed by incorporating ultrasonic data into adapted mechanical models of plastic behaviour of steel.

Theoretical background :

SAYERS [2] was the first to propose relationships between bulk wave velocities and expansion coefficients of the Orientation Distribution Function (ODF). These relationships are based on the propagation equation of a plane wave in the polycrystal, expressed in terms of the geometric parameters that describe the orientation of one individual crystallite in the sample coordinate system. This geometric function has to be weighted over the crystallites in order to take into account the texture - i.e. -the distribution of the orientations. Hence, one can introduce the ODF F(g) and its expansion on a basis of spherical harmonics as developed by BUNGE [3] and ROE [4]. Shortly, if w_{lmn} are the expansion coefficients of the ODF, according to [3] :

$$F(g) = \sum_{l=0}^{\infty} \sum_{m=-l}^{+l} \sum_{n=-l}^{+l} w_{lmn} \, Z_{lmn} \, (\theta) e^{-im\psi} \, e^{-in\phi} \qquad (1)$$

with Z_{lmn} the legendre polynomes.

The following relationships are obtained between ultrasonic velocities and the expansion coefficients : for the bulk waves propagating along the normal direction :

$$\rho V^2_{zz} = C_{11} - 2C \{1/5 - 16/35 \sqrt{2}\pi^2 \; w_{400}\} \qquad (2)$$

$$\rho V^2_{zx} = C_{44} + C\{1/5 - 16/35 \sqrt{2}\pi^2 \; (w_{400} - \sqrt{5/2} \; w_{420})\} \quad (3)$$

$$\rho V^2_{zy} = C_{44} + C\{1/5 - 16/35\sqrt{2}\pi^2 \; (w_{400} + \sqrt{5/2} \; w_{420})\} \quad (4)$$

The fourth order is the truncation order retained in these relationships. BUNGE [5] has shown that this order is sufficient to describe the elastic behaviour of materials.

In the case of plate waves, THOMPSON [6] has developed the relationships between plate waves velocities and the ODF coefficients. In this case, the long wave length limit is assumed.

From the equations (2) to (4) and from THOMPSON's relationships [6], one can easily derive the values of the three first ODF coefficients from ultrasonic velocity measurements performed on the sheet. Then, these texture coefficients may be used to introduce the anisotropy in constitutive equations of plastic behaviour of materials, in order to calculate the strain ratio $r(\alpha)$.

Two models were chosen in this study. One of them [7] consists in calculating the parameters of a HILL criterion on the basis of the texture coefficients W_{lmn}. For that, the yield surface associated to the HILL criterion is adjusted to that derived from the TAYLOR model by minimization of the difference in rate of plastic work associated to each model. Although this approach is a combination of both HILL and TAYLOR models, it is designated as the "HILL model" in the current work. In this case, $r(\alpha)$ is analytically deduced from the HILL parameters. The second model is the classical crystallographic TAYLOR model used with the "pencil glide" assumption. In this model, the r values can be directly deduced from the yield surface of the polycrystal, providing that the normality rule for the selected tensile test is satisfied [8].

This yield surface is calculated by weighting the individual yield surfaces of the grains by the ODF.

Experimental procedures :

Two sets of samples were used in the present work : four sheets in low-carbon steel of deep drawing quality and twenty sheets in 17 C_r ferritic stainless steel. Firstly, conventionnal measurements of the strain ratio were performed by tensile tests at 0°, 45° and 90° to the rolling direction, and the standard parameters r_m and Δr were derived from these tests. Secondly, texture analysis was carried out on some of the samples by X-ray diffraction. The {110}, {200} and {211} pole figures were recorded in the reflection mode and used for calculating the ODF coefficients at the expansion order l = 22. Then, these coefficients, limited to the order 4 or 8, were

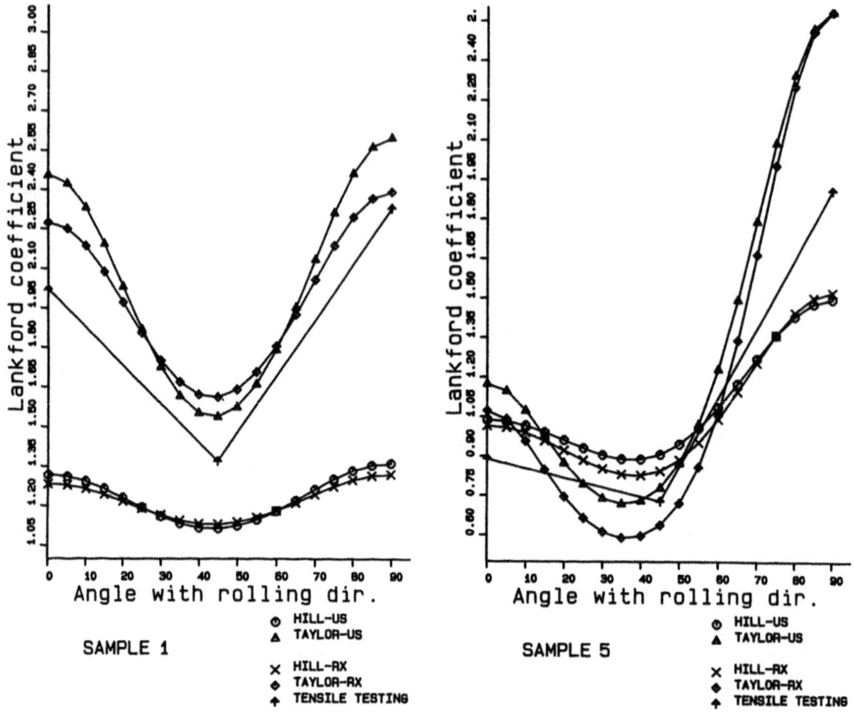

Figure 1 : Examples of evolution of the Lankford coefficient determined by ultrasonic, X-ray and tensile testing, for low-carbon steel (sample 1) ferritic stainless steel (sample 5).

underestimates them. However the aim of this paper is not to discuss the validity of the models, but rather to examine their ability to give an estimate of the standard values r_m and Δr using ultrasonic texture coefficients. In this context, it is interesting to compare the standard coefficients obtained by the ultrasonic method with those given by the X-ray method (figure 2). In fact, a good correlation between both measurement methods was obtained irrespective of the mechanical model employed.

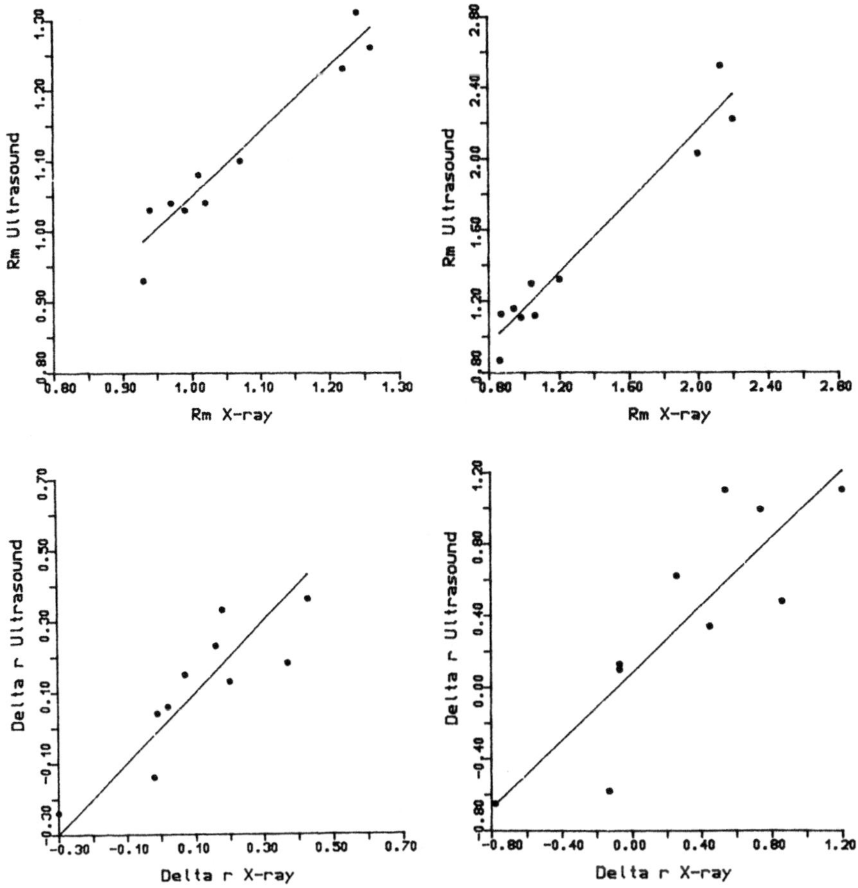

HILL TAYLOR

Figure 2 : Correlations between ultrasonic and X-ray determination of the standard values r_m and Δr. HILL and TAYLOR models.

The correlation obtained on the Δr values exhibits larger discrepancies than that obtained on the r_m values. An explanation is that the most important discrepancies between ultrasonic and X-ray techniques are observed at 45° to the rolling direction. Figure 3 shows correlations performed over all the samples between the ultrasonic method and the values given by tensile tests. The influence of the quantitative prediction of the HILL and TAYLOR models is clearly evidenced. However, in the case of r_m values lower values of deviations were obtained. This is true in the case of a model exhibiting a low-sensitivity to the anisotropy of materials (HILL). In contrast, a highly sensitive model (TAYLOR) will enhance these deviations. This is very clear for the Δr correlation. However the characteristical values of the strain ratio (r_m = 0,7 ; 1,0 ; 2,0 ; $\Delta r < 0$) are clearly pointed out. In fact, one can estimate the accuracy of the determination at ± 0,1 for the r_m values and ± 0,2 for the Δr values. The accuracy obtained with tensile tests is of the same order of magnitude. In the case of highly anisotropic steel sheets (r_m ~ 2,0) the errors on the r_m

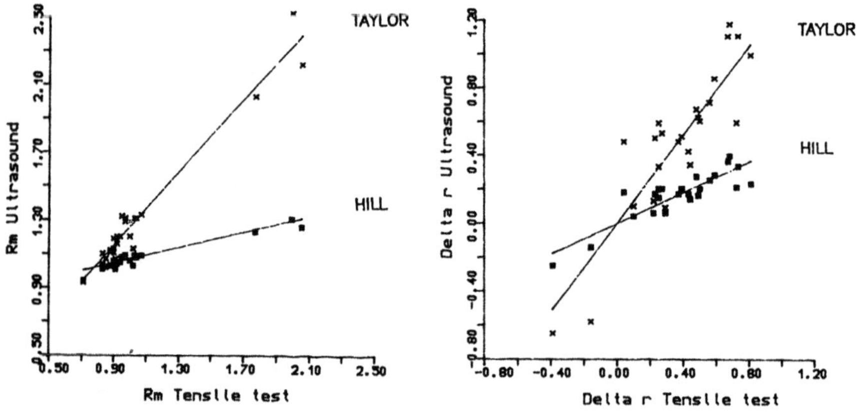

Figure 3 : Correlations between ultrasonic and tensile testing determination of the standard values r_m and Δr. HILL and TAYLOR models.

introduced into the mechanical models described above to obtain the r(α) computed from X-ray data.

The tensile as well as X-ray data were used as comparative values for evaluating the validity of the subsequent ultrasonic determination of the Lankford parameter.

Bulk wave velocities V_{zz}, V_{zx} and V_{zy} were measured along the normal direction on each sample. The SH_0 wave velocities at 0° and 45° to the rolling direction were also determined. The bulk waves were generated by means of high frequency piezoelectric transducers whereas EMATS at 0,5 MHz were used for the SH_0 waves. In order to improve the accuracy of the determination of the expansion coefficients, the measured ultrasonic velocities have been fitted using a self-consistent algorithm. In the case of SH_0 and bulk waves, two relationships that are independent of the texture can be easily derived between the propagation velocities. Since these two relationships should be simultaneously verified, it appears that the data can easily be fitted by the least square method, thus providing a significant improvement of the accuracy of the W_{400} coefficient. The accuracy of the evaluation of this last coefficient has been reported earlier to be the main problem of the ultrasonic determination of texture (especially in the case of aluminium) [9].

Results and discussion :

The expansion coefficients of the ODF determined by the ultrasonic method were used as input data in the mechanical models described above. These models allow to compute the values of the Lankford coefficient r as a function of the angle α with the rolling direction. Then, one may calculate the standard values r_m and Δr. Examples of r(α) curves obtained on low-carbon steels are given in figure 1. Results obtained with the HILL and TAYLOR models are shown. The choice of the model is of great importance in regard to the result. The TAYLOR model overestimates the variations of r while the HILL model values obtained by tensile tests are approximately ± 0,2. Consequently, the ultrasonic method has a good ability to estimate the anisotropy coefficients.

Further developments and concluding remarks :

It is shown in this study that the texture coefficients determined by ultrasounds can give a good estimate of the Lankford coefficient of steels using the TAYLOR model. Correlations obtained between the ultrasonic predicted values and those measured by tensile tests allow to have directly the right order of magnitude of r_m and Δr.

The good results obtained in this paper are promising for possible industrial applications of this technique. But the application on production lines of steel sheets requires the use of contactless transducers. The authors are developing a laboratory device using S_0 and S_{H0} waves, which would be easily transposable in industrial environment. Since the dispersive nature of the S_0 waves may be a source of uncertainties, this point will have to be verified in laboratory. The new method proposed by THOMPSON [10] to measure the W_{400} expansion coefficient using relative measurements between S_{H1} and S_0 waves will also be tested. This measurement should improve the accuracy of the determination of r_m.

Aknowledgments :

This research is partly supported by the ECSC under contract number 7210/GB/304.

References

1. Kopineck H.J.; Otten H. : Texture analyser for on-line r_m value estimation. Textures and Microstructures. Vol. 7 (1987).

2. Sayers C.M. : Ultrasonic velocities in anisotropic polycrystalline aggregates. J. Phys. D : Applied Physics 15(1982).

3. Bunge H.J. : Mathematische Methoden der Texturanalyse- Berlin Akademie Verlag (1969).

4. Roe R.J.: J. Appl. Physics 36 (1965) 2024.

5. Bunge H.J. : Über die elastischen Konstanten kubischer Materialen mit beliebiger Textur - Kristall und Technik (1968) 431-438.

6. Thompson R.B.; Lee S.S.; Smith J.F. : Relative anisotropies of plane waves and guided modes in thin orthorhombic plates. Ultrasonics 25 (1987)

7. Arminjon M. : Explicit relationships between texture coefficients and three dimensionnal yield criteria of metals - Proc. of ICOTOM 7 Holland Brakeman C.M. et al (eds) (1984) 31 - 37.

8. Bunge H.J.; Schulze, Grzesik : Peine - Salzgitter Berichte Sonderheft (1980).

9. Clark A.V.; Govada A.; Thompson R.B. ; Smith J.F.; Blessing G.V.; Delsanto P.P.; Mignonna R.B. : The use of ultrasonic for texture monitoring in aluminium alloys : a comparison of ultrasonic and neutron diffraction measurement. Ultrasonics 26 (1988).

10. Li Y.; Thompson R.B.; Smith J.F. : Use of velocities of higher order Lamb modes in the measurement of texture Proceedings of the Review of Progress in Quantitative NDE - Thompson D.O. and Chimenti D.E. (Eds.) Vol 8 (to be published).

Characterization of Textures in Plates by Ultrasonic Plate Wave Velocities

Y. Li, J. F. Smith, and R. B. Thompson

Ames Laboratory, Iowa State University
Ames, Iowa 50011, U.S.A.

Introduction

Texture (preferred grain orientation) is an important material characteristic which influences the ability of sheet metal to be formed into parts of complex shapes such as beverage cans and automobile body parts. Another consequence of texture is an anisotropy in the elastic constants and ultrasonic wave speeds. Other papers in this symposium have discussed direct relationships between the wave speeds and formability parameters [1-3]. In the present work, the status of our current understanding of the accuracy of ultrasonic characterization of texture is reviewed, and an improved procedure for determining the coefficient W_{400} is proposed.

Theory

In a polycrystal of cubic crystallites, the anisotropy of the elastic properties are fully characterized by three orientation distribution coefficients (ODC's) [4,5] which in the notation of Roe [4] are known as W_{400}, W_{420}, and W_{440}. A complete characterization of texture requires many more coefficients of higher order. However, the interest in the ultrasonic techniques is stimulated by the fact that these three coefficients play a major role in determining certain plastic anisotropy parameters which influence formability [6]. In general, many sets of three elastic wave speed measurements could be used to determine the three ODC's, W_{4MN}. However, here emphasis will be placed on measurements involving the speed of guided modes of the plate [7] as a function of propagation angle. In particular, Thompson et al. have proposed a scheme in which W_{400} and W_{440} can be deduced from measurements of the speeds of horizontally polarized shear waves, SH_o, propagating at 0° and 45° with respect to the rolling direction. W_{400}, W_{420} and W_{440} can also be deduced from the speeds of the fundamental symmetry Lamb mode, S_o, in the long wavelength limit, propagating at 0°, 45°, and 90° with respect to the rolling direction [8]. The specific formulae can be found in Refs. 7 and 8, and a detailed experimental error analysis has recently been conducted for commercial purity aluminum and copper sheets [9].

Comparison of Ultrasonic and Diffraction Predictions

Since a variety of assumptions which may not be fully satisfied in real materials go into the theory of the ultrasonic determination of the ODCs, it is useful to compare the predicted ODCs to those obtained by independent means. In a number of recent studies, either neutron or X-ray diffraction measurements have been used for this purpose, and the results of these are compared in Figs. 1-3. Most of the predictions were made from guided mode data, although the W_{420} comparison includes some ultrasonic birefringence and Rayleigh wave velocity anisotropy data.

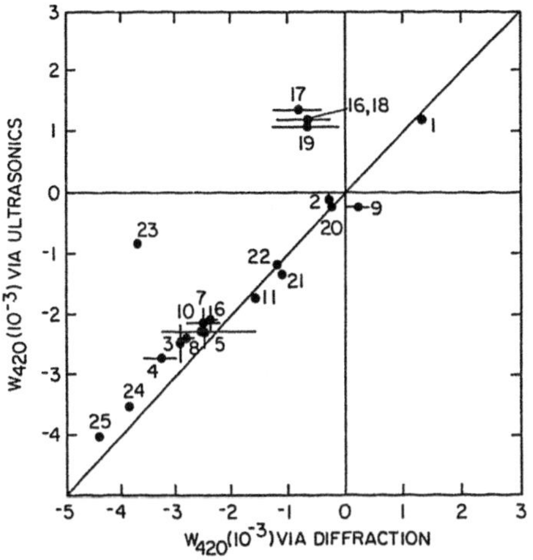

Fig. 1. Comparison of ultrasonic and diffraction predictions of W_{420}.

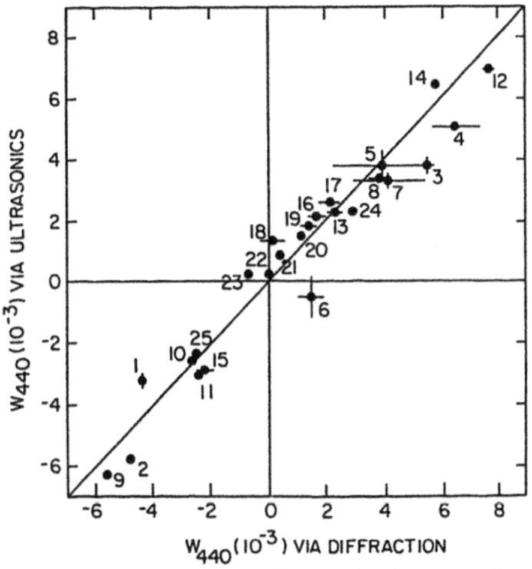

Fig. 2. Comparison of ultrasonic and diffraction predictions of W_{440}.

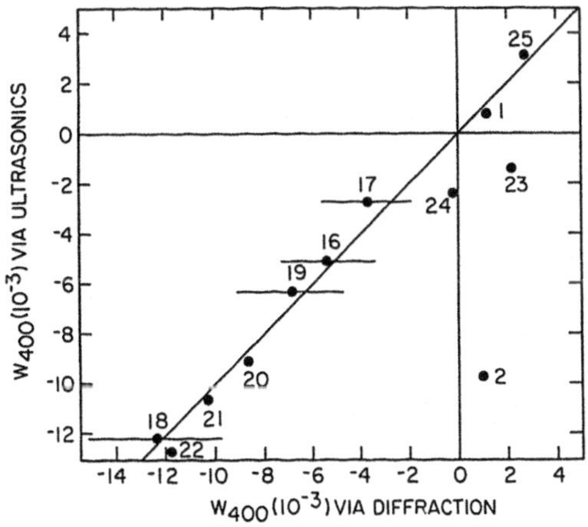

Fig. 3. Comparison of ultrasonic and diffraction predictions of W_{400}.

Figures 1 and 2 present the comparisons for W_{420} and W_{440}. The numbers of the individual data points identify the material examined as well as the original data. Where errors can be estimated from that data, they are indicated by horizontal or vertical bars. For all samples, which include aluminum, copper, and steel alloys, excellent agreement between the ultrasonic and diffraction predictions of W_{440} are observed, with errors being on the order of $|\Delta W_{440}|=10^{-3}$. Similar agreements are obtained for W_{420}, as shown in Fig. 1 with the exception of a comparison of ultrasonic to X-ray diffraction predictions for one low carbon steel sheet and comparison to neutron diffraction predictions for a set of Al-killed steel sheets (samples 16-19). Formability predictions based on both the X-ray and ultrasonic data were made for sample 23 [10], and the fact that the latter more closely coincided to experiment suggested that there may have been texture gradients which caused the surface texture, as sensed by X-ray diffraction, to not be representative of the bulk response, as sensed by ultrasonics. No reasons have been postulated for the disagreement between ultrasonic and neutron predictions in the Al-killed steel. The problem is more complex for the prediction of W_{400}. In general, rather poor results have been obtained, particularly for aluminum [11]. Recently, more encouraging results have been obtained, as shown in Fig. 3. With the exception of the aluminum data, the errors are less than $|\Delta W_{400}|= 3 \times 10^{-3}$. If one further eliminates two comparisons of steel data to X-rays, to which the aforementioned comments regarding texture gradients apply, the errors are less than $|\Delta W_{400}|=10^{-3}$. However, in some cases special techniques had to be used to stabilize the predictions against uncertainties in the assumed isotropic elastic moduli [10].

Discussion of Errors in Prediction of W_{400}

From the above studies, it is evident that significant errors can be encountered when ultrasonically predicting W_{400}, particularly for aluminum alloys. Enough experiments have not yet been performed to fully define the generality of these conclusions. However, is is possible to offer some speculations regarding possible sources of these errors.

The equations for predicting W_{400} are [9]

$$W_{400} = \frac{35\sqrt{2}\rho}{16\pi^2 C^\circ} \left[V_{SH_o}^2 (45^\circ) + V_{SH_o}^2 (0^\circ) - 2 (T/\rho)\right] \qquad (1a)$$

$$W_{400} = \frac{35\sqrt{2}\rho}{32\pi^2[3+8(P/L)+8(P/L)^2]C^\circ} [V_{S_o}^2 (0^\circ) + V_{S_o}^2 (90^\circ) \tag{1b}$$

$$+ 2V_{S_o}^2 (45^\circ) - 4 (L-P^2/L)/\rho]$$

where L, P, T are isotropic moduli, C° is an anisotropy parameter, ρ is the density, and $V(\Theta)$ is the phase velocity as a function of measurement angle with respect to the rolling directions. The subscripts refer to the guided mode type. For the S_o mode, the long wavelength limit is required, which would have the value $[(L-P^2/L)/\rho)]$ in the isotropic limit.

As in the prediction of W_{420} and W_{440}, Eqs. (1) contain C° in the denomination. The weak anisotropy of aluminum with respect to iron and copper is presumably a contributor to the poorer performance in that material since the prefactor multiplying any velocity errors is greater. An error increase of about a factor of three with respect to iron and copper would be expected on this basis.

Since the S_0 measurements are made at finite wavelength, the data must be corrected for dispersion. If one assumes that the isotropic plate theory can be used to make this correction, then

$$V_{S_o} = V_m / (V_P/V_{LIM})_{ISOTR} \tag{2}$$

where V_m is the measured phase velocity and $(V_P/V_{LIM})_{ISOTR}$ is the correction for an isotropic plate, which has the form shown in Fig. 4.

The range of validity of this correction has not yet been rigorously established. It appears clear that one would like the wavelength to be as large as possible with respect to the plate thickness, as is discussed on heuristic grounds below. In a measurement of phase velocity, one attempts to follow the motion of a phase feature, say a particular zero crossing. In a pulsed measurement on a dispersive system, this will move at a different rate than the wave packet itself, whose center moves at the group velocity. When group and phase velocities are unequal, the selected phase feature will move through the packet. It would intuitively seem reasonable to require that the phase feature stay within the packet for its entire propagation from transmitter to receiver. Suppose one selects the feature at the center of the initial packet. This then leads to the condition [8]

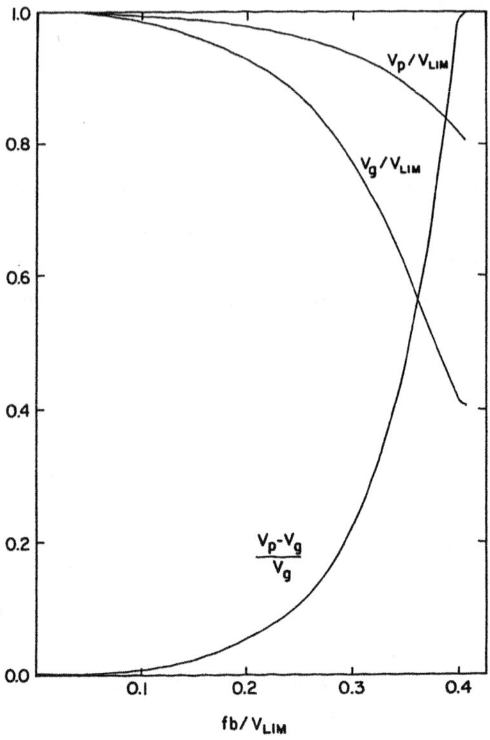

Fig. 4. Dispersion correction for V_{S_0} measurements for the phase velocity, group velocity, and wave packet spreading parameter in a plate of thickness b.

$$(V_p - V_g)/V_g < \ell/2L \qquad (3)$$

where V_p is the phase velocity, V_g is the group velocity, ℓ is the spatial length of the packet, and L is the propagation distance. In the measurements reported in Fig. 3, $\ell/2L \simeq 0.1$. For the steel, copper and some of the aluminum plates, $fb/V_{LIM} \simeq 0.15$ and Eq. (3) was satisfied. Thus, the heuristic criteria suggests that the dispersion correction should be satisfactory. However, for thicker aluminum plates studied previously and showing great error [11], $fb/V_{LIM} \simeq 0.3$ at this point. Thus, Eq. (3) does not hold and the correction may not be valid. More detailed calculations using exact theories are required to more fully examine this question.

Assuming that closer scrutiny of these ideas does in fact explain the poor results on thicker plates, the question of the better, but still unsatisfactory agreement on thinner aluminum plates such as the result shown in Fig. 3, remains. This is partly explained by the weaker anisotropy in this material and resulting greater multiplication of errors in data processing. Other factors may include errors in the theoretically computed isotropic modulii L, P and T, which do not include the effects of alloying, second phases, etc. that are present in real materials. These may play a more important role in influencing measurements of W_{400} than the other coefficients since they appear as an additive factor in brackets in Eq. (1), rather than as a simple premultiplying factor that is found in the corresponding equations for W_{420} and W_{440}.

Alternate Technique for Determining W_{400}

In order to reduce this error, one would like to predict W_{400} from relative velocity measurements. This can not be done from measurements of the angular dependence of wave speeds in the plane of the plate because W_{400} describes cross-sectional rather than in-plane variations of grain orientation. An alternate scheme that overcomes this problem has recently been described by the authors [11]. The concept is shown in Fig. 5. In summary, there is a special point for isotropic materials at which the S_o and SH_1 mode dispersion curves are tangential to one another and become known as lowest Lamé modes. At this point, each mode can be decomposed into partial shear waves propagating at 45° with respect to the plate surface. In the SH_1 mode, the waves are polarized in the plane of the plate while in the S_o mode they are polarized in the orthogonal sagittal plane. Because of this change in polarization, the presence of W_{400} will remove this tangency and cause either a mode crossing or splitting, depending on its sign, as illustrated in Fig. 6. Thus, by observing the relative behavior of these modes, one can gain specific information on W_{400} without the need for absolute measurements. Moreover, this prediction does not depend on precise knowledge of the plate thickness.

This proposed technique has been verified by exact calculations of the dispersion of anisotropic plates, and perturbation theory has been used to obtain a relationship for predicting W_{400}. The result is

$$W_{400} = \frac{7\sqrt{2}T}{40\pi^2 C^\circ k} \left[\Delta k(0^\circ) + \Delta k\,(90^\circ) + 2\Delta k(45^\circ)\right] \tag{4}$$

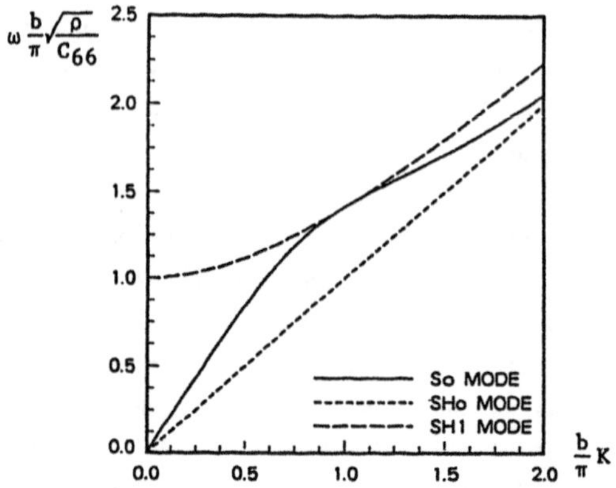

Fig. 5. Lamé point at which SH_1 and S_o modes are tangential.

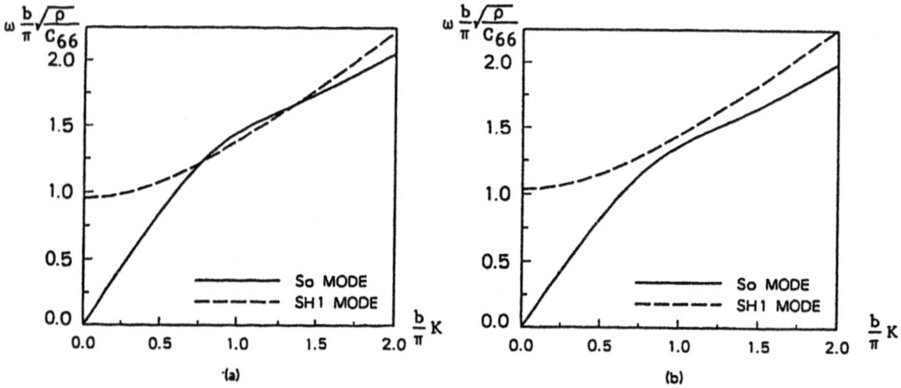

Fig. 6. Anisotropy induced crossing (a) or splitting (b) of the Lame modes.

where T is the shear modulus, k is the wavevector, and $\Delta k = \Delta k_{SH_1} - \Delta k_{S_o}$ at the frequency of mode tangency in the isotropic limit. The similarity of Eq. (4) to Eq. (1b) is evident. However, since Δk is a measure of a relative shift in dispersion curves, Eq. (4) avoids the need for absolute velocities. Numerical simulations of predictions of W_{400} based on Eq. (4) have been encouraging [11] and experimental studies are in progress.

Conclusions

Ultrasonics can be used to measure the ODC's W_{4MN} with an accuracy of $|\delta W| \sim 10^{-3}$. The predictions of W_{420} and W_{440} are based on relative measurements and seem quite reliable. Predictions of W400 require absolute velocity data. Although good results have been obtained on steel and copper, results on aluminum have been generally poor. This is not fully understood, but the weak elastic anisotropy and thickness effects appear to be important contributors. A technique to overcome this problem based on wave propagation characteristics of Lamé modes is proposed.

Acknowledgment

Ames Laboratory is operated for the U. S. Department of Energy by Iowa State University under contract no. W-7405-ENG-82. This work was supported by the Office of Basic Energy Sciences, Division of Materials Sciences.

References

1. E. Schneider and M. Spies, "Nondestructive analysis of the deep-drawing behaviour of rolled sheets by ultrasonic techniques", this proceedings.

2. O. Cassier, C. Donadille and B. Bacroix, "Lankford coefficient evaluation in textured steel sheets by an ultrasonic method", this proceedings.

3. A. V. Clark, R. B. Thompson, G. V. Blessing, R. E. Schramm, D. V. Mitrakovic, and D. Matlock, "Characterization of texture and formability in steel sheet using electromagnetic-acoustic transducers", this proceedings.

4. R.-J. Roe, J. Appl. Phys. 37, 2069 (1965).

5. H. J. Bunge, Texture Analysis in Materials Science (Butterworth, London, 1982).

6. G. J. Davies, D. J. Goodwill and J. S. Kallend, Metall. Trans, 3, 1627 (1972).

7. R. B. Thompson, S. S. Lee and J. F. Smith, Ultrasonics 25, 133 (1987).

8. S. S. Lee, J. F. Smith and R. B. Thompson, in Nondestructive Characterization of Materials, J. F. Bussiere, Ed. (Plenum Press, NY, 1988) pp. 555-562.

9. R. B. Thompson, J. F. Smith, S. S. Lee and G. C. Johnson, submitted to Metall. Trans.

10. O. Cassier, C. Donadille and B. Bacroix, Review of Progress in Quantitative Nondestructive Evaluation 8, D. O. Thompson and D. E. Chimenti, Eds. (Plenum Press, NY, in press).

11. Y. Li and R. B. Thompson, Review of Progress in Quantitative Nondestructive Evaluation 8, D. O. Thompson and D. E. Chimenti, Eds. (Plenum press, NY in press).

Microstructure, Stress State, Creep Damages

Surveillance of Material Degradation by Means of Diffraction Methods

H.-A. Crostack, W. Reimers, U. Selvadurai
Fachgebiet Qualitätskontrolle, Universität Dortmund, FRG

G. Eckold
Institut für Kristallographie, RWTH Aachen and Institut für Festkörperforschung, KFA Jülich, FRG

Summary

The plastic deformation behaviour of individual crystals embedded in the polycrystalline matrix is analysed in the Ni-base alloy IN 939. Using a single grain measuring technique based on diffraction methods the deformation inhomogeneities in the elastic deformation region are analysed as well as the changes in the reflection profiles due to the plastic deformation.

Zusammenfassung

Das Verhalten bei der plastischen Deformation einzelner Kristallite in der Matrix wird am Beispiel der Ni-Basis-Legierung IN 939 untersucht.

Hierzu wird eine auf Beugungsmethoden basierende Einkornmeßmethode eingesetzt, die sowohl die Erfassung von Deformationsinhomogenitäten von Kristallit zu Kristallit als auch die Analyse der Änderungen in den Reflexprofilen nach plastischer Verformung erlaubt.

Introduction

For the surveillance of component integrity various nondestructive testing methods are commonly used. Special problems, however, arise for the conventional testing techniques when coarse grained materials are investigated. Coarse grains are frequently found in high temperature alloys which are often used in security sensitive areas as e.g. turbin blades for aeroplants or in welding joints of reactor vessels. Thus, informative data about the changes in the structural states of the materials are needed. Especially X-ray diffraction has been shown to be a sensitive tool for the detection of plastic deformations in metallic materials /1, 2/. The plastic deformation in metallic crystals is mainly due to gliding processes of low indexed crystallographic planes in low indexed crystallographic directions. In the first stage of plastic deformation the X-ray diffraction exhibits a broadening of reflection profiles when measuring subsequently rocking curves. Also the orientation of the crystallites changes due to the external load. Although this situation allows the observation of structural changes in materials, there are still problems in the interpretation of the experimental data.

First of all this is due to the experimental techniques applied up to now. The X-ray microbeam technique in combination with a modified Debye-Scherrer detection device /3/ gives evidence for gliding processes in directions around the scattering vector, only. Gliding processes around an axis perpendicular to this direction cannot be detected. Other measuring techniques are limited to the determination of the diffraction angle /4/. Here, no direct information about the process activated in the crystallites can be obtained. Complete informations about gliding processes, orientation changes and strain status of crystallites under consideration may in principle be obtained using the Laue- technique /2/. However due to the poor accuracy of the film technique in comparison to detector techniques the data for polycrystalline sample materials could be interpreted on a qualitative basis, merely.

Also the theoretical interpretation of the mechanisms activated during plastic deformation is not unambiguous. Whereas in the free single crystal the gliding system activated is determined by Schmid's law /5/, the crystal-crystal coupling in the polycrystalline materials plays a crucial role. According to Taylor /6/ plastic deformation can take place only by the simultaneous activation of 5 gliding systems. The activated combination of the gliding systems is selected applying the principle of minimal virtual work. There is, however, a critical point in this procedure. The adaptation of the crystal to the surrounding matrix by means of gliding is calculated on the basis of macroscopical strain data. Thus variations on the microscopical scale due to the anisotropy of crystals are neglected. Recent simulation calculations /7, 8/ show that the plastic deformation behaviour is better explained by elastic-plastic coupling models.

Another access to the investigation of the plastic deformation is possible by applying the single grain measuring technique /9/ which allows the characterization of the individual crystallites embedded in their polycrystalline matrix. Here, the orientation of the crystallite under study is determined and the positions of Bragg-reflection can be calculated approximately. Careful profile measurements on individual crystallites can then be performed which yield quantitative data not only for the elastic deformation state but also for the plastic deformation state. Since these results are obtained on the microscopical scale, the elastic deformation is more precisely obtained than in the Taylor approximation. Using a four circle diffractometer, the rocking curves around arbitrary crystal directions can be analyzed.
Hence, the threedimensional deformation ellipsoid can be determined.

Systematic studies have been performed on crystals in the Ni-base alloy IN 939. The investigations give evidence that the single grain measuring technique is suited to detect even small plastic deformations. The evaluations show that the gliding process at the beginning of the plastic deformation can be forseen already on the basis of experimental data obtained in the elastic deformation region.

Measuring techniques

For the measurements of the deformation of individual crystals embedded in the polycrystalline matrix a four circle diffractometer is used.

The instrument is based on the Euler geometry. By means of three sample rotation facilities, each orientation of the crystal under study relative to the incident beam can be realised. In combination with the additional detector movement 2θ the reflecting conditions for each reflection (hkl) can be fulfilled by the appropriate sample rotations. The first step of a measurement is the selection of a particular crystallite. Using X-rays this can be done e.g. with the help of a diaphragm on the sample surface. In the case of neutrons which exhibit usually a large penetration depth, the 90° scattering technique /10/ can be applied. After having positioned the detector to the approximate 2θ value for the reflection (hkl) chosen, the spatial orientation of the Bragg-reflection is obtained by intensity measurements during systematic variations of the φ and χ sample rotations. The resulting angle configuration then defines the plane of a second Bragg-reflection of the same crystallite.The set of angles for two reflections carry already the complete information about the orientation of the crystallite with respect to the fixed laboratory system (fig. 1).

Once the orientational matrix is known, the angles corresponding to arbitrary reflections (hkl) can be determined. The reflection profiles are then measured independently in 2θ, ω and χ. The evaluation of the 2θ -profile gives the interplanar lattice spacing d_{hkl} whereas the analysis of the ω and χ profiles gives informations about the mosaic distribution of the crystal. In this context especially so called q-configuration is of interest: For a particular reflection $(h_1k_1l_1)$ the set of angles is calculated under the constraint $2\theta = \omega/2$ which can always be achieved (symmetric position). After a rotation of 90° around χ the crystal direction $[h_1k_1l_1]$ is now parallel the vertical ω -axis of the instrument. The reflecting conditions for a second reflection $(h_2k_2l_2)$ perpendicular to $[h_1k_1l_1]$ are now realised merely by means of appropriate 2θ and ω movements. By performing an ω-scan (rocking curve) on the reflection $(h_2k_2l_2)$ it is now possible to analyse the mosaic distribution of the crystal in the plane perpendicular to $[h_1k_1l_1]$. The evaluation of the rocking curves obtained on different and independent crystal directions [hkl] as rotation axes gives the mosaic distribution in form of a tensor with respect to the crystal axes. Corresponding comparative studies under external loads, e.g. in the elastic deformation region and after a plastic deformation, yield information about the mosaic enlargement of the crystal due to the plastic deformation in three dimensions.

Experiments

As sample material the Ni-base alloy IN 939 was chosen. Flat tensile specimen were prepared and installed in a tensile apparatus

which allows the application of defined uniaxial stresses. The applied stress is examined by calibrated strain gauges on the sample holder.
The neutron diffraction experiments were performed on the triple-axis spectrometer UNIDAS at the FRJ-2 reactor at the KFA Jülich. The instrument was equipped with a full circle Eulerian cradle. The experimental details are given elsewhere /10/.

The Bragg reflection measurements were performed on several grains 2. In specimen 1 the {200} and {220} reflection types were used whereas in specimen 2 the {311} reflection types were investigated. The experiments were carried out at three stres levels in the elastic deformation region (σ_{ext} = 0; 340 MPa; 510 MPa) and at one level in the plastic deformation region (σ_{ext} = 760 MPa). The usual centering routine for the Bragg-reflection yields the 2θ, ω and χ profiles. Additional rocking curves using the q-configuration were measured on grain 1, 2 and 3 of specimen 1. The corresponding profiles were taken on the {200} reflections with ⟨200⟩ directions as axes of rotation.

Evaluation and results

The data taken at the various stress levels in the elastic deformation region, were processed to yield strain and stress tensors for the individual crystals. The procedure and a part of the results are published elsewhere /10/. As the interpretation of the measurements in the plastic deformation region is concerned, it is important to note that the deformation inhomogeneity from grain to grain amounts up to 30% of the macroscopical mean deformation.
The plastic deformation of the sample material leads to significant changes in the rocking curve (fig. 1). The intensity as well as the full width half maximum of the reflections are affected.

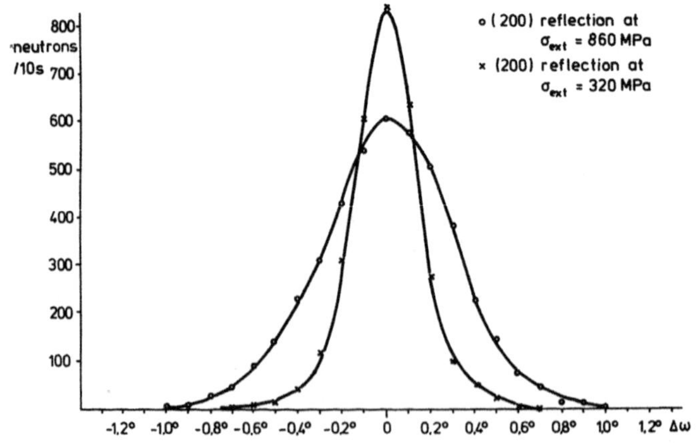

Fig. 1: Comparison of the ω-profiles of the (200)-reflection in the elastic and plastic deformation region

The enlargement of the mosaic distribution due to plastic deformation can be isotropic or anisotropic.
A further increase of the plastic deformation initiates glidings which are observed experimentally by a separation of the reflection profile.

For the evaluation of the data the broadening of the reflection profiles in has been calculated by comparing the profiles in the elastic and plastic defomation region. The resulting tensor $\underline{B}$ describes the reflection broadening with respect to the crystal axes system. Thus, the diagonal tensor components B_{ii} give a measure for the broadening of reflections when rotating the crystal around the corresponding main crystal axis (tab. 1). It was found that the off-diagonal elements vanish within the experimental accuracy. The results obtained from the least-squares refinement of the ω-reflection broadenings measured in the symmetrical mode ($\omega = 2\theta/2$) are compared to measurements in the q-configuration which gives directly the values for the crystal main axes directions (tab. 1).

Tab. 1: Broadening of the ω-profiles after plastic deformation

Broadening of the ω-profiles after plastic deformation

	tensor components B_{ii} obtained from ω-profiles measured in the symmetric position ($\omega = 2\,\theta/2$) (in degree)			full width half maximum of rocking curves measured in the q-configutation and after correction for the instrumental resolution (in degree)		
rotation around	M_{11} [100]	M_{22} [010]	M_{33} [001]	[100]	[010]	[001]
specimen 1						
grain 1	0.38±0.04	0.30±0.03	0.35±0.07	0.41±0.03	0.34±0.03	0.38±0.03
grain 2	0.16±0.08	0.40±0.08	0.21±0.22	0.23±0.03	0.43±0.04	0.24±0.03
grain 3	0.00±0.02	0.04±0.03	0.05±0.09	0.00±0.02	0.16±0.02	0.17±0.03
grain 4	0.22±0.05	0.09±0.06	1.10±0.94			
specimen 2						
grain 1	0.13±0.05	1.46±0.57	-0.58±0.52			
grain 2	0.39±0.07	0.28±0.20	0.30±0.18			
grain 3	0.46±0.14	-0.04±0.56	0.48±0.56			
grain 4	0.33±0.03	-0.09±0.35	0.37±0.35			

The comparison between the B_{ii} tensor components and the results of the q-measurements give a good agreement for grain 1 and grain 2, specimen 1. This demonstrates that the anisotropy as well as the extent of the plastic deformation can be analysed quantitatively. Moreover, the agreement between the reflection broadening values and the values of the full width half maximum for these two grains indicates that grain 1 and grain 2 had a very small mosaic distribution befor the plastic deformation. For grain 3, specimen 1, however, the q-values are larger than the values of the corresponding B_{ii}-tensor components. This indicates that already before the plastic deformation the intrinsic mosaicity of this grain was significantly larger than that of the other grains.

The increase in the full width half maximum values after plastic deformation is due to gliding processes. Following the Taylor-theory each deformation of the crystal embedded in the matrix can be realised by 5 activated gliding systems. The corresponding combination out of the 12 gliding systems in the fcc-lattice is characterised by its minimal gliding sum. Usually the calculation of the gliding sum is based on the macroscopical deformation of the sample material. In the present investigation, however, the more reliable experimental data for the microscopical deformations of single grains could be used. Thus, the inhomogenious deformation behaviour of the grains is taken into account.

The calculations were carried out for all grains of specimen 1 and 2. Since the choice of the combination of gliding-systems is not unique, the gliding sums of all energetically equivalent combinations were added. The experimental reflection profile data (tab. 1) always refer to observations in a plane perpendicular to a selected axis of rotation. The glidings, however, always take place in a special direction. Thus, for the comparison of the experimental data with the theoretically calculated glidings in the crystal, all glidings in the plane perpendicular to a main crystal axis had to be added. The corresponding gliding sums are given in tab. 2.

The comparison between the reflection broadening in the plane perpendicular to a rotation axis and the corresponding gliding sum exhibits that the anisotropy of the mosaic enlargement is (within the experimental error) in good agreement with the relative gliding sums. An exception is found for grain 1, specimen 1. Here the calculation leads to a mosaic enlargement only around two main crystal axes, wheres the experimental data prove an approximately isotropic mosaic spread. This finding may be due to the special orientation of this grain with its [100] crystal axis nearly parallel to the uniaxial stress direction. For the case of highly symmetric orientations, however, it is reported /12/ that the Taylor assumption based on 5 activated gliding systems cannot be used since here 6 or 8 gliding systems are activated.

Tab. 2: Gliding sums in the planes perpendicular to main axes of the crystal systems

Gliding sums in the planes perpendicular to main axes of the crystal systems

gliding sum	⊥ [100]	⊥ [010]	⊥ [001]
specimen 1			
grain 2	0.38	1.10	0.21
grain 3	0.01	0.94	0.44
grain 4	0.53	0.00	0.96
specimen 2			
grain 1	0.72	0.83	0.00
grain 2	0.49	0.00	1.03
grain 3	1.09	0.11	0.28
grain 4	0.47	0.21	0.63

Discussion and outlook

The experimental investigations using the single grain measuring technique demonstrate that the plastic deformation leads to significant changes in the reflection profiles. Thus, this techniques offers the potential to detect damages in the material sensitively and at an early stage of the plastic deformation damage. The experimentally observed anisotropy of the reflection broadening due to the plastic deformation was interpreted by the gliding combinations which, result from theoretical calculations based on the Taylor theory. A good agreement between theory and experiment was found for grains with general orientation. This agreement could, however, only be obtained since the inhomogenious deformations of the individual grains were taken into account, which were determined experimentally. The procedure presented here offers the possibility to predict the plastic deformation behaviour of individual grains on the basis of experiments within the elastic deformation region. Further informations about the strain-stress state of the grains within the plastic deformation region in principle can be obtained from the measurements of the interplanar lattice spacings. However, since only the elastic part of the deformation is accessible by the analysis of the diffraction angle, these data have then to be combined with further experimental informations, e.g. a detailed investigation of the strain hardening.

This work has been funded by the German Federal Minister for Research and Technology (BMFT) under the contract number 03-Cr1DOR-4.

References

/ 1/ Shriver, A., J.P. Wallace, J.J. Slade, S. Weissmann: Fatigue studies of metall crystals; Acta Metallurgica, 11, (1963), 779-789

/ 2/ Ohnami, M.: Plasticity and high temperature strength of materials; Elsevier Applied Science, London, New York, (1988), 130-137

/ 3/ Eisenblätter, J.: Forschungs- und Entwicklungsarbeiten über Früherkennung von Ermüdungsschäden in Reaktorbauteilen; Abschlußbericht zum Forschungsvorhaben RS 25, Battelle-Institut (1969)

/ 4/ Weiss, V., Y. Ohshida, A. Wu: Towards practical non-destructive fatigue damage indications; Fatigue of Engineering Materials and Structures, 1, (1979), 333-341

/ 5/ E. Schmidt, W. Boas: Kristallplastizität; Springer-Verlag, Berlin (1935)

/ 6/ Taylor, G.I.: Plastic strain in metals; J. Int. Met., 62, (1938), 307-324

/ 7/ Pedersen, O.B., T. Leffers: Modelling of plastic heterogeneity in deformation of single-phase materials; Conf. Proc. 8th Riso Int. Symp. on Metallurgy and Materials Science, (1987), 147-172

/ 8/ Leffers, T., O.B. Pedersen: Polycrystal calculations with a "universal" elastic-plastic model; Conf. Proc. 8th Riso Int. Symp. on Metallurgy and Materials Science, (1987), 401-408

/ 9/ Crostack, H.-A., W. Reimers, R. Niehus: X-ray diffraction analysis of stresses and strains in coarse grained materials; Conf. Proc. Int. Conf. on NDT, Las Vegas, 3.-8.11.1985, 622-628

/10/ Crostack, H.-A., W. Reimers: Analysis of Strain Hindering in Polycrystalline Materials; Conf. proc. 8th Riso Int. Symposium on Metallurgy and Materials Science, 7.-11.9.1987, 291-297

/11/ Eckold, G., H. Mitlacher: Dreiachsenspektrometer "UNIDAS"; FRJ2, Jülich, Jul-Spez. 141 (1982)

/12/ Bacroic, B., J.J. Jonas, F. Montheillet, A. Skalli: Grain reorientation during the plastic deformation of f.c.c. metals; Acta Metall., 34 (1986), 937-950

Grain-size Measurements by Spectral Evaluation of Line-scans in Scanning Acoustic Microscopy

A. Morsch and W. Arnold

Fraunhofer-Institut für zerstörungsfreie Prüfverfahren
Universität, Gebäude 37, D-6600 Saarbrücken 11, FRG

Abstract

The contrast mechanism of the Scanning Acoustic Microscope (SAM) allows for an excellent imaging of grain structures. We used line-scans over polycrystalline samples obtained with SAM to deduce the grain size of the material. We applied two spectral evaluation techniques: power spectrum and cepstrum analysis.

Zusammenfassung

Der Kontrastmechanismus des akustischen Reflexionsrastermikroskops (SAM) erlaubt eine ausgezeichnete Abbildung von Gefügen. Wir benutzten Linienscans über polykristalline Proben, wie man sie mit dem SAM erhält, um daraus die Korngröße abzuleiten. Wir wandten dazu zwei spektrale Auswertemethoden an: Leistungsspektrum- und Cepstrumanalyse.

1. Scanning Acoustic Microscopy (SAM)

In Scanning Acoustic Microscopy a focused ultrasonic beam produced by an acoustic lens is scanned over the sample. The sound field reflected by the sample is modulated in phase and amplitude according to the elastic properties and topography of the surface. The lens collects the returning waves and a transducer converts them back into electric signals whose intensity is displayed synchronously to the scanning of the lens on a monitor /1/.

The acoustic lens of a SAM produces surface acoustic waves on the sample, provided the opening half-angle of the lens is larger than the corresponding critical angle. Since these waves radiate energy back to the lens, they strongly influence the contrast mechanism of the microscope /2/. The surface waves are scattered by surface discontinuities such as cracks and grain boundaries, and therefore these features of the sample surface can be imaged although their size might be much smaller than the resolution limit of the lens given by its numerical aperture and the frequency employed. In our case the resolution of our instrument is 0.7 µm at 1 GHz ultrasonic frequency.

2. Imaging of grain structures with SAM

As mentioned above grain boundaries give a contrast in SAM because they scatter the surface waves produced by the lens. If the grain structure of a material with anisotropic crystallites is imaged, besides the contrast of the boundaries, a large contrast occurs because of the different orientations leading to different acoustic reflection functions of the grains /3/. Additionally, if the sample surface was etched, grain contrast arises from the topography.

3. Line-Scans

The signal V(x) received by scanning the lens along a straight line over the sample (line-scan), as shown in Fig. 1, already contains information about the grain sizes. The grain boundaries of the texture cause variations in the lens signal repeating in more or less equal distances, provided the scan velocity is constant. The evaluation of this periodicity and therefore the grain size measurement should be simplified by spectral evaluation techniques such as powerspectrum and cepstrum analysis.

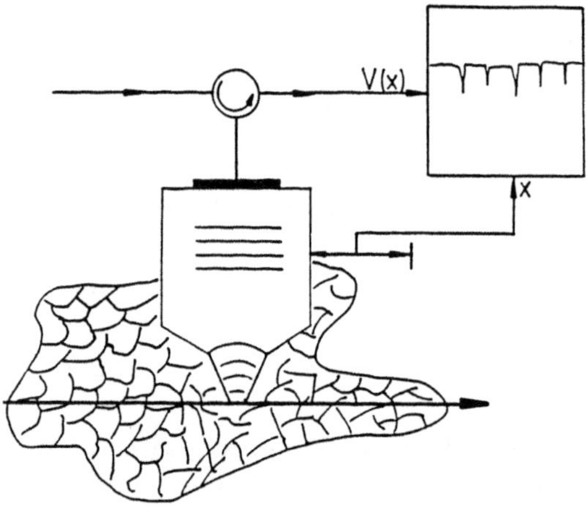

Figure 1: Principle of the measurement of a line-scan on a polycrystalline sample. The output voltage V(x) of the lens is processed.

4. Powerspectrum Evaluation

If the mean distance of the variations in the line-scan due to the grain boundaries is **a**, the frequency η_0 = **1/a** and its higher harmonics **$n\cdot\eta_0$ = 2,3,..)** dominate in the signals Fourierspectrum, meaning that its amplitude spectrum contains maxima at these frequencies /3/. These maxima can be determined directly from the video output of the microscope by an analogue spectrum analyzer. The temporal frequencies ν of the video signal are then related to the spatial frequencies by $\eta_0 = \nu/c$, where c is the scan velocity. The spectrum can also be calculated from the digitized image data with a fast discrete finite Fourier transformation algorithm (FFT) on a computer. In both cases a sequence of maxima in an approximately constant distance must be identified in the spectrum. We used the mean distance η_m of this maxima to receive a value **a** for the mean grain size from $a = 1/\eta_m$. As practice showed, the evaluation is facilitated when the power spectrum defined as the squared amplitude spectrum is calculated.

5. Cepstrum Evaluation

The autocorrelation of a signal is defined as a convolution with itself and can therefore be calculated as the inverse Fourier transformation of its power spectrum. The autocorrelation can be used to improve the S/N-ratio, and thus facilitates the evaluation of a noisy periodic signal /4/. The period of the signal is identified by strong maxima. A further improvement of the S/N ratio can be achieved by calculating the logarithm of the power spectrum before the Fourier transformation. The power spectrum of the logarithmed power spectrum is also called Cepstrum /3/. Since the Cepstrum of v(x) is a function in spatial domain, the mean grain size can be determined from the Cepstrum of the v(x) signal by calculating the mean distance of the first dominant maxima.

6. Examples

Examples for grain-size measurements on a polycrystalline silicon sample and a ferritic steel sample are shown in Figs. 2 and 3, respectively /5/. The silicon sample used had a very uniform texture with quadratic crystallites with a nearly equal side length of 25 µm. The grain structure of the steel sample was not uniform, as the grains varied in size and shape. Its mean grain size was determined metallographically to be 10 µm. Figs. 2a and 3b show the corresponding line-scans. The periodic variation of the signals due to the grains can be seen clearly. The powerspectra are shown in Figs. 2b and 3b. The powerspectrum of the more uniform silicon structure contains, in addition to a maximum at the main frequency at $\eta_0 = 46$ mm^{-1}, three higher harmonics (maxima indicated by arrows), whereas the powerspectrum of the steel texture only has a main maximum at 106 mm^{-1}. The deduced grain sizes are 22 µm and 9.4 µm, respectively, in good agreement with the metallographically determined values.

The corresponding cepstra are shown in Figs. 2c and 3c. The cepstrum determined from line scans over the silicon sample shows three well-defined maxima in a constant distance of 25.3 µm. The cepstrum of the steel sample contains two maxima (indicated by arrows) in a distance of 12.5 µm, corresponding to the grain size. This cepstrum, however, contains further sharp maxima that are not directly related to the grain size, and thus can render the grain size evaluation difficult.

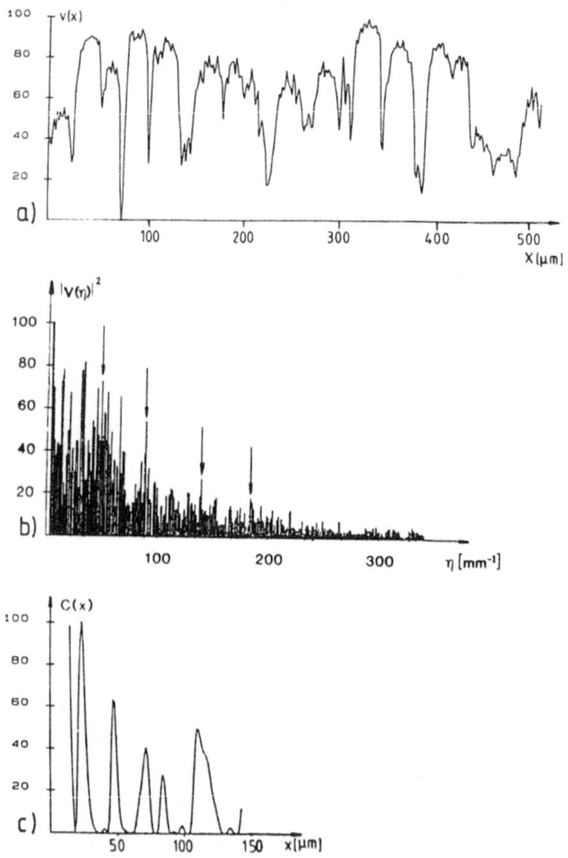

Figure 2: Grain size determination on a polycrystalline silicon sample, grain size: 25 µm
a) line-scan b) spectrum c) cepstrum

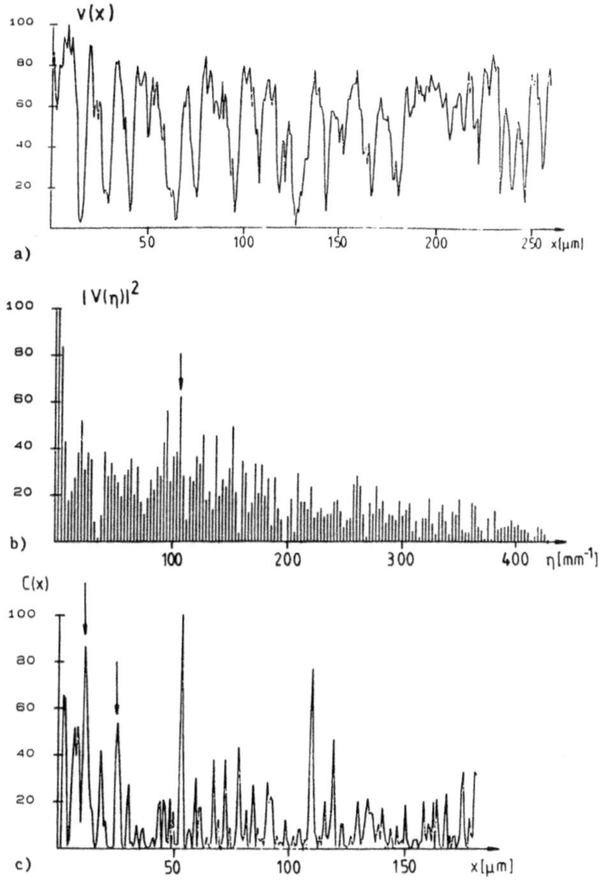

Figure 3: Grain size determination on a ferritic steel sample, grain size: 10 μm
a) line-scan b) spectrum c) cepstrum

7. Summary

We applied two spectral evaluation techniques, power spectrum and cepstrum analysis, to deduce the grain size of a polycrystalline material from line-scans with SAM. The examples we presented clearly show the practicability of the techniques. The evaluation by the powerspectrum is improved by using the cepstrum analysis. But they also show the difficulties that can arise when the grains vary strongly in size and shape or, what leads to the same results, when the sample surface, apart from the grain structure, contained many scratches or has an uneven topography. The cepstrum then contains additional maxima that do not correlate with the mean grain size but rather with these features.

8. References

1. Lemons, R.A.; Quate C.F.: Acoustic Microscopy. Phys. Acoust., ed. by Mason W.P. and Thurston R.N. 14, 1979, 1-92, Academic, New York

2. Weaver, J. M. R.; Ilett, C.; Somekh, M.G.; Briggs, G.A.D.: Acoustic Microscopy of Solid Materials. Metallography, 34, 1985, 3-35

3. Meyer, E.; Guicking, D.: Schwingungslehre. Braunschweig: Vieweg 1974.

4. Ehrenstrasser, G.: Stochastische Signale und ihre Anwendung. Heidelberg: UTB Taschenbuch 1973.

5. The measurements were performed with an instrument from the Wild-Leitz GmbH, D-6330 Wetzlar, FRG

Non-contact and Nondestructive Evaluation of Grain-size in Thin Metal Sheets

S. Faßbender, M. Kulakov*, B. Hoffmann, M. Paul, H. Peukert+, and W. Arnold

Fraunhofer-Institut für zerstörungsfreie Prüfverfahren, Universität, Geb. 37, D-6600 Saarbrücken, FRG

* Permanent Address: Institute of Radioengineering and Electronics, Academy of Science of the USSR, K. Marx Avenue 18, Moscow, USSR

+ Permanent Address: Rheinisch-Westfälischer TÜV, D-4300 Essen, FRG

SUMMARY

We use short laser pulses to generate broadband ultrasonic pulses in polycrystalline materials and interferometric techniques to detect the ultrasonic signals. The damping of the spectral components of the ultrasonic pulses can be exploited to measure the total attenuation $\alpha(f)$ as a function of frequency f. Knowing $\alpha(f)$, it is then straightforward to deduce the grain size by using the Rayleigh-approximation describing the scattering contribution to $\alpha(f)$. This technique has been applied to polycrystalline metal plates with a thickness from 1.5 to 20 mm and grain size between 20 µm and 80µm. In the case of still thinner metal sheets various technical problems arise which need a carefully engineered solution.

INTRODUCTION

The determination of the average grain size and the grain size distribution in polycrystalline materials is an important objective of nondestructive materials characterization. An ultrasonic technique to determine the grain size nondestructively is the two-frequency method using backscattering [1,2]. For thin components, however, backscattering methods become problematic to employ, because of the finite ultrasonic pulse-width reducing the axial resolution. This holds good in particular for thin metal sheets (thickness $\lesssim$ 2 mm) in which the grain size controls the yield strength [3]. We have developed a technique using broadband ultrasonic pulses generated by a pulsed laser in order to overcome this. By analyzing the spectral change of subsequent backwall-echoes, one can deduce the grain size from the ultrasonic attenuation coefficient $\alpha(f)$ by employing the theory of scattering for an ensemble of grains in the Rayleigh limit [4,5]. Different kinds of laser interferometers are used as detector for the ultrasound, making it possible to measure in a noncontact manner. Coupling problems are therefore avoided and the system can be incorporated into a production line. However, the optical components have to be protected against vibrations and rough environmental conditions. In addition the detection system must be able to work at optically rough surfaces. In this contribution the technical problems and possible solutions are discussed.

DESCRIPTION OF THE MEASURING SYSTEM

As shown schematically in Fig. 1, a laser pulse from a Q-switched ruby laser impinges onto the surface of a polycrystalline metal plate. At the rear face of the specimen the ultrasonic backwall echoes are detected using an interferometer, then digitized (sampling rate: 200 MHz) and stored, enabling further processing. A small fraction of the laser pulse is used for monitoring its beam shape and power and as a trigger signal.

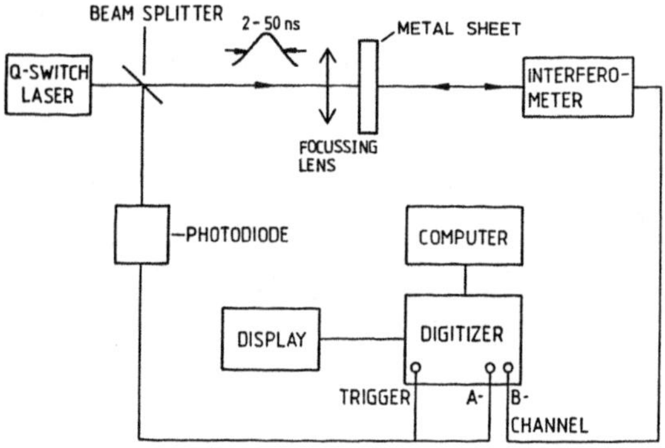

Fig. 1: Principle of measurement

Any of these backwall echoes can be selected and subjected to FFT-routines using corresponding software [6]. The ultrasonic pulses with the initial frequency components G(f) undergo attenuation between subsequent backwall reflections. Because of absorption and scattering, the total attenuation $\alpha(f)$ can be written as [2]:

$$\alpha(f) = a_1 \cdot f + S \cdot d^3 \cdot f^4 \tag{1}$$

Here, d is the average grain size, S is the scattering parameter, and a_1 is the absorption parameter. For steels, it has been found experimentally that the absorptive part is $\propto f$ [2] or sometimes $\propto f^2$ [7]. The scattering term is valid only for the Rayleigh region ($d/\lambda \leq 0.3$ with λ as ultrasonic wavelength). The frequency spectrum of any given backwall echo and the attenuation coefficient $\alpha(f)$ is given by:

$$|U_i(f)| = |G(f)| \cdot |I(f)| \cdot \exp(-\alpha(f) \cdot x_i) \tag{2a}$$

$$\alpha(f) = (\ln |U_j(f)| - \ln |U_i(f)|)/(x_i - x_j) \qquad j \neq i \qquad (2b)$$

Here, I(f) is the transfer function of the experimental set-up, U_i and U_j are the frequency components, and x_i and x_j are the path lengths of the i-th and j-th echoes, respectively. Using Eq. (2b) α(f) is deduced from the digitized data with the aid of the computer. Then Eq. (1) is fitted to the obtained α(f) curve with a_1 and d as adjustable parameters. In this way the grain size d is determined nondestructively. It should be mentioned that dispersion effects play no role up to second order [8]. As can be seen from Eq. (2b), the experimentally obtained α(f) is independent of I(f). However, the bandwidth of the ultrasound must be chosen such that the Rayleigh condition holds for the polycrystalline structure examined and that the scattering part of α(f) is sufficiently large compared to the dissipative part (see Eq. 1). The bandwidth of the generated ultrasonic pulses increases with decreasing laser pulse duration τ. From these arguments it follows that some preadjustments of the experimental parameters have to be made depending on the still unknown grain size. But this is true in a similar way for the back-scattering technique also.

EXPERIMENTAL TECHNIQUES

Laser-generated Ultrasound

The ultrasonic pulses are generated by short laser pulses obtained from a Q-switched ruby laser via the thermoelastic effect and by the recoil mechanism of ablating material due to plasma formation at higher laser intensities. The laser is operated in TEM-Multimode yielding energies up to 400 mJ. Its temporal shape is gaussian with τ ≈ 50 ns (1/e points). It can be shortened down to 2 ns enabling the examination of fine-grained materials. The temporal shape of the laser-generated ultrasound is much more complex than that of ultrasound generated piezoelectrically. Apart from longitudinal waves, transverse and surface waves are excited [9,10]. Additionally, the different modes are coupled, causing normal as well as transversal displacements during the entire interval determined by the time of flight of the longitudinal wave (P-wave) t_L and the shear wave (S-wave) t_S. The typical shape of the thermoelastically generated ultrasound (shown in Fig. 2a) is characterized by a small positive P-pulse at t_L followed by a continous decay and a step at t_S [9,11,12]. By increasing the laser intensity I_s, the shape of the ultrasonic wave changes as shown in Figs. 2b-2d. At one characteristic I_s the normal displacements nearly vanish at times later than

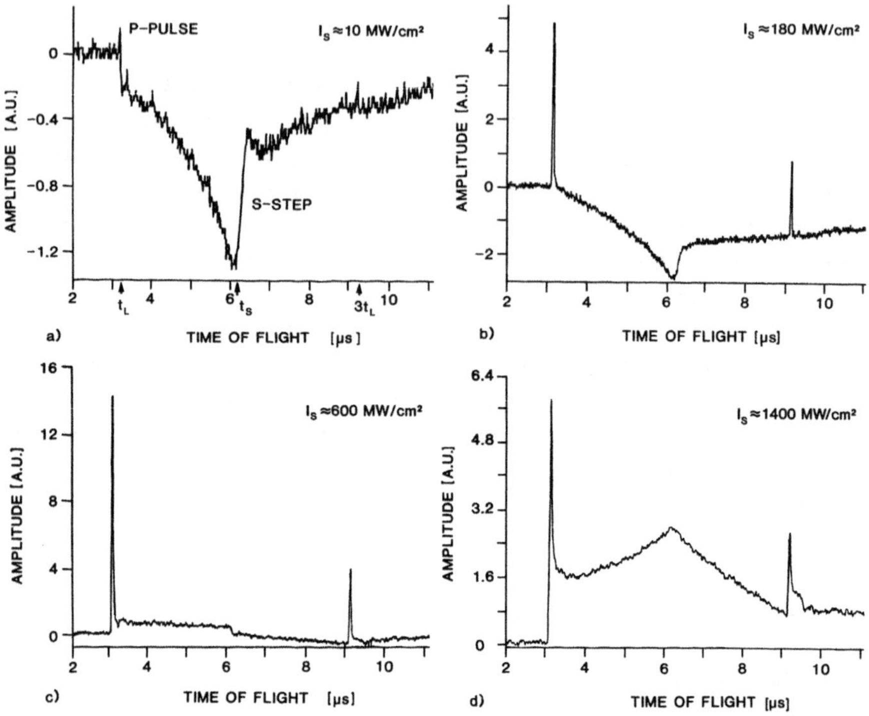

Fig. 2: Acoustic waveforms obtained at different laser intensities I_S as measured by a capacitance transducer [12] at an aluminum specimen (19.5 mm thick). a) thermoelastic region, b)-d) ablation region with increasing I_S from b) to d). The amplitude is displayed in equal but arbitrary units.

t_L and only the echoes of the P-pulse remain. For an aluminum specimen I_S = 600 MW/cm² (± 20%) and I_S = 400 MW/cm² (± 20%) for a stainless steel specimen. The uncertainty of the measurement is due to the difficulty in determining the laser-spot size because of the multimode spatial profile. The same I_S is obtained for samples of different thickness (19.5 mm, 30 mm, 50 mm), and it is independent of the laser-pulse energy. However, the amplitude of the P-pulse is proportional to the laser energy.

Detection by Heterodyne Interferometry

A heterodyne-interferometer [13] was used to detect the ultrasonic waves as shown in Fig. 3. In a Bragg cell the deflected reference beam is frequency-shifted by the driving frequency f_B. After reflection off the reference mirror and the sample under examination, respectively, both beams are once more passed through the Bragg cell, and a part of the signal beam undergoes a deflection as well as a frequency shift in the opposite direction. This part is recombined with the now unshifted part of the reference beam, and is detected by a fast Si-photodiode leading to an electrical signal with the carrier frequency of $2f_B$. The vibrations of the sample surface cause a phase modulation of the carrier signal. By demodulating the carrier with a phase-locked loop, low frequency disturbances due to vibrations and thermal unstabilities are suppressed [13,14]. In this way a signal corresponding to the ultrasonic displacement δ is obtained. A polarizer and two quarter wave plates are inserted to prevent light falling back into the laser resonator. At present a displacement of about 0.25 nm in a bandwidth of 35 MHz can be detected, resulting in a sensitivity of $4.5 \cdot 10^{-5}$ nm/$\sqrt{\text{Hz}}$. A heterodyne interferometer posesses a linear transfer function allowing the determination of the absolute value of δ, but its drawback is that its light-gathering power is small, limiting its use at rough surfaces. This might be overcome by using a technique published recently [15].

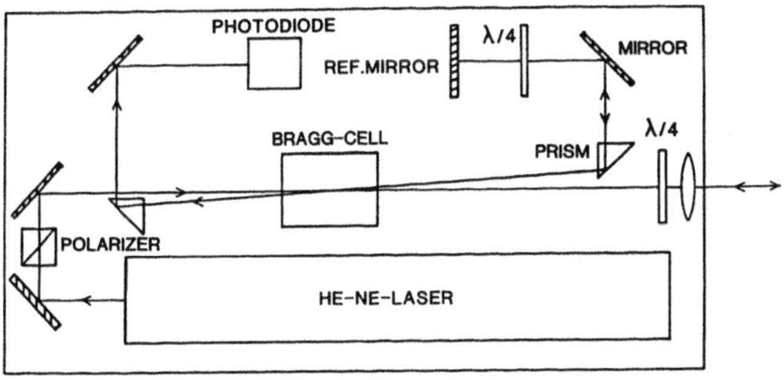

Fig. 3:
Block diagram of the heterodyne interferometer

Detection by Time-of-Flight Interferometry

Another interferometer under test was the time-of-flight interferometer shown in Fig. 4 [16]. Here, the interferometer does not make use of the surface as a mirror, but views the light scattered by it. The principle of this interferometer is that of a Mach-Zehnder type but with different arm lengths producing a time delay T between the two beams. Its light gathering power is very large, due to the fact that the backscattered wavefront interferes with itself allowing detection of ultrasound even at rough surfaces. To circumvent mechanical stabilisation of the arm lengths in our interferometer one beam is frequency-shifted with the aid of a Bragg cell, thus combining the heterodyne principle with time-of-flight interferometry. The transfer function of the time-of-flight interferometer is proportional to $\sin(\pi fT)$, resulting in a maximal sensitivity at $f = 1/(2T)$, and consequently low frequency components are heavily damped.

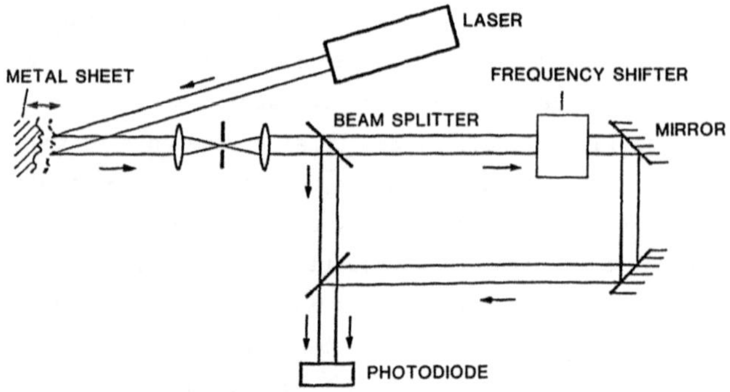

Fig. 4: Principle of the time-of-flight interferometer

EXPERIMENTAL RESULTS

Grain size measurements in different steel plates were carried out, and some experimental results together with the micrographically determined grain sizes are listed in Table 1. Usually the ultrasonically determined grain size is larger than the one metallographically determined. This is due to the fact that the ultrasonic technique averages over d^3 as can be seen from Eq. (1) and therefore overweighs larger grains. Especially in the case of a wide grain-size distribution the ultrasonic results are significantly larger than the metallographic ones.

Sample	Thickness [mm]	Grain size d_{US} [µm]	d_M [µm]
1	19.8	73	60
2	19.8	41	30
3	11.1	86	93
4	5.1	31	18
5	1.5	28	22

Tab. 1

The examination of thin sheets (0.3 - 2 mm) makes the evaluation of the ultrasonic signals much more difficult because the signal is disturbed by low-frequency oscillations saturating the demodulation electronic of the interferometer. A typical echo sequence picked up by the heterodyne interferometer is shown in Fig. 5 demonstrating this effect. Furthermore, the transfer function of the detection system is then no longer well defined. The oscillations even occur when the ultrasound is generated with the optimal laser power density as described above. The detailed origin is not yet clear to us. Finally, we should like to mention that it is possible to generate back-wall echoe sequences in very thin metal sheets if all parameters are well adjusted as can be seen from Fig. 6. In this case the signals were detected by the aid of the time of flight interferometer.

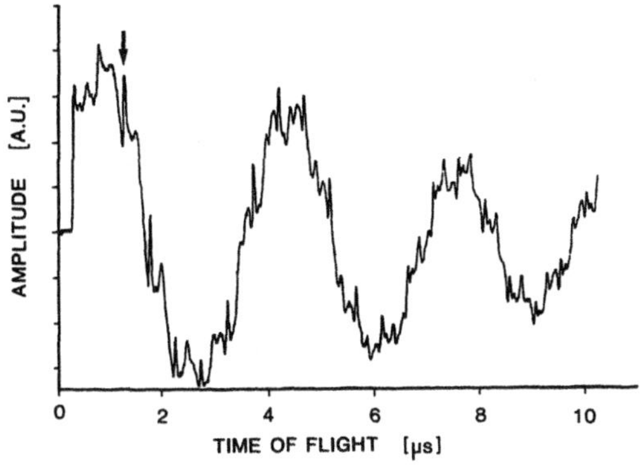

Fig. 5:
Typical output signal of the heterodyne interferometer obtained at a thin metal sheet (1.5 mm thick). Laser-pulse width τ = 25 ns. The ultrasonic pulses exploited (arrow) are superimposed on the low-frequency oscillations.

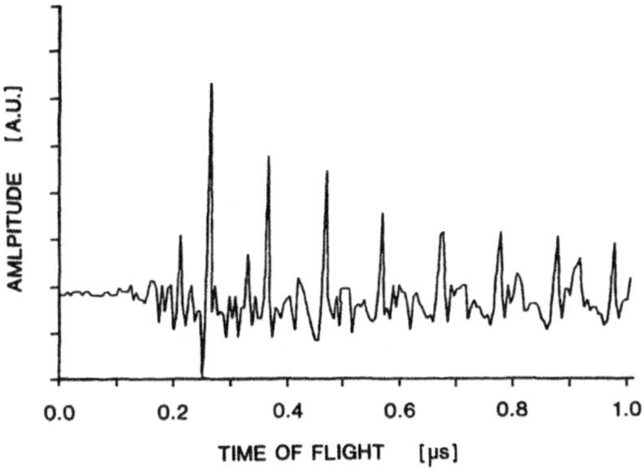

Fig. 6:
Echo sequence in a 0.3 mm thick metal sheet obtained by the time-of-flight interferometer.
Laser-pulse width τ = 5 ns.

CONCLUSIONS

In order to measure grain size in thin metal sheets in a contact free manner, we developed a system using laser generated and interferometrically detected ultrasonic pulses. Applications on sheets of different thicknesses were shown. These results encourage us to work further on this technique so that it can be employed as an on-line production control instrument.

ACKNOWLEDGEMENT

We gratefully acknowledge the financial support of the European Commission on Steel and Coal. One of the authors (M. K.) should like to thank the German Science Foundation for a grant.

REFERENCES

1. Koppelmann J.: Ultraschallmeßeinrichtung für Härtetiefenmessungen an Stahlwalzen. Materialprüfung 14, 156 (1972)

2. Goebbels K.: Structure Analysis by Scattered Ultrasonic Radiation, in "Research Techniques in NDT", edited by Sharpe R.S. (Academic, London, 1980) IV, p.87, and references contained therein

3. Meyers M.A., and Asworth E.: A Model for the Effect of Grain Size on the Yield Stress of Metals. Phil. Mag. A46, 737 (1982)

4. Lifshits I.M., and Parkomovskii G.D.: Zh. Eksp. i Theor. Fiz. 20, (1950)

5. Truell R., Elbaum Ch., and Chick B.C.: "Ultrasonic Methods in Solid State Physics", Academic Press, New York (1969)

6. Panakkal J.P., Peukert H., and Willems H.: Nondestructive Characterization of Material Properties by an Automated Ultrasonic Technique. Published in these proceedings

7. Smith R.L., Reynolds W. N., and Perring S.: The Analysis of Ultrasonic Wave Attenuation Spectra in Steels. J. Physique 44, C9-337 (1983)

8. Peukert H., Paul M., and Arnold W.: Entwicklung und Erprobung eines Prototypen zur berühungslosen Korngrößenbestimmung an dünnen Blechen mittels optisch erzeugtem Ultraschall. IzfP-Report # TW-880223, unpublished

9. Dewhurst R.J., Hutchins D.A., Palmer S.B., and Scruby C.B.: Quantitative Measurements of Laser-Generated Acoustic Waveforms. J. Appl. Phys., 53, 4064 (1982)

10. Arnold W., Betz B., and Hoffmann B.: Efficient Generation of Surface Waves by Thermoelasticity. Appl. Phys. Lett., 47, 672 (1985)

11. Schleichert U., Paul M., Hoffmann B., Langenberg K.J., and Arnold W.: Theoretical and Experimental Investigations of Broadband Thermolastically Generated Ultrasonic Pulses. "Photoac. and Phototh. Phenomena", ed. by Hess P. and Pelzl J., Springer Verlag Berlin, p. 284 (1988)

12. Schleichert U., Langenberg, K.J., Arnold W., and Faßbender S.: A Quantitative Theory of Laser-Generated Ultrasound, Proc. Int. Conf. Rev. Progr. Quant. NDE, San Diego 1988, to be published

13. Paul M.: Berührungsloser Nachweis von Ultraschall mittels Interferometrie und kapazitiven Aufnehmern. Diploma Thesis, IzfP Saarbrücken and University of Saarbrücken (1987)

14. Cretin B.: Étude et Réalisation d'un Microscope Thermoacoustique. Thèse de Docteur-Ingenieur, Université de Franche-Comté, Besançon, France (1984)

15. Paul M., Betz B., and Arnold W.: Interfermetric Detection of Ultrasound at Rough Surfaces Using Optical Phase Conjugation. Appl. Phys. Lett., 50, 1569 (1987)

16. Barker L.M., and Hollenbach R.E.: Laser Interferometer for Measuring High Velocities of Any Reflecting Surface. J. Appl. Phys., 43, 4669 (1972)

The Use of a Contactless Ultrasonic Method for Evaluating Grain Size in Steel Sheets

A. LE BRUN, J.L. LESNE
Electricité de France - D.E.R./REME
93206 Saint-Denis, France

O. CASSIER, F. GONCALVES, D. FERRIERE
IRSID
78105 Saint-Germain en Laye, France

SUMMARY

The measurement of grain size by a non-destructive ultrasonic method makes it possible to envisage monitoring changes in the characteristics of the microstructure of a metal.

Industrial applications, particularly in the nuclear and steel industries, are made possible by using contactless techniques for generating ultrasounds by laser and for receiving them by heterodyne interferometry.

Intended applications include on-line monitoring of structures while they are in use in order to evaluate their degree of ageing, and also the characterization of materials in the course of manufacture.

INTRODUCTION

In the nuclear industry, the remaining lifetime of a given installation is generally evaluated on the basis of periodic observations of the general state of its structures during routine stops for maintenance. The present trend is to develop diagnostic methods enabling the most sensitive points of a power plant to be monitored on-line, i.e. on a permanent basis. Under normal conditions of use, the materials constituting nuclear installations age by virtue of damage due to mechanical fatigue, or to thermal fatigue, or by becoming brittle under the effect of neutron flux. Non-destructive methods of investigation would make it possible to keep track of changes in the microstructure of the materials, thereby making it possible to predict the extent to which their mechanical characteristics have been degraded by the cyclical stresses induced by successively starting and stopping the plant. One of the best tools presently available for this purpose is measuring grain size ultrasonically, thereby showing up any change which may have occured in the microstructure of a given component.

In the steel industry, production methods are tending more and more towards continuous production. In particular, modern production facilities for producing sheet metal make use of continuous annealing so as to obtain a continuous throughput of metal having the desired microstructure. However, such installations require appropriate techniques for monitoring parameters which determine the application characteristics of the products. One such parameter, for example, is sheet hardness which is influenced by grain size, another is texture which affects formability properties. Both of these important parameters depend on the "quality" of the annealing, i.e. on the recrystallization of the sheet.

In both of these industries, there are obvious advantages in using non-destructive methods, and in particular ultrasonic echo techniques. Such methods are all the more advantageous in that they are capable of being adapted to the stresses of hostile industrial environments (e.g. hot products moving at speed), with such environments promoting the use of monitoring methods which are non-destructive, contactless, and remote. Photo-acoustic methods of generating and receiving ultrasounds are capable of satisfying these criteria. In addition, they have the advantage over conventional piezoelectric techniques of providing a high bandwidth frequency spectrum which improves the analysis of microstructure when using ultrasonic waves.

MEASURING GRAIN SIZE

Of the various non-destructive techniques available for monitoring grain size, numerous authors have already studied ultrasonic methods and as a result these methods offer the most highly developed theoretical basis. [1, 2].

It is now well known that the attenuation of an ultrasonic wave depends strongly on parameters related to the micro-structure of the propagation medium. This attenuation may by modelled by two mechanisms which are [1] :

- wave absorption which takes into account thermo-mechanical losses and magneto-elastic phenomena ; and
- wave scattering which is mainly due to grain size in polycrystalline materials.

Ultrasonic wave scattering verifies different relationships depending on the value of the ratio between the size of the scatterer "d" and the ultrasonic wavelength "λ". Of the various different existing scattering regions, the one which has been most thoroughly investigated experimentally in the Rayleigh scattering regime for which d/λ should lie between 0.03 and 0.3.

In this domain, the scattering coefficient α_S may be modelled by a function which takes into account the volume of the scatterer (which is assumed to be spherical) and the fourth power of the frequency of the wave :

$$a_S = S.d^3.f^4$$

where :

d is the diameter of the scatterer
f is the frequency of the ultrasonic wave ; and
S is a "dispersion" parameter which takes into account the type of wave being used (longitudinal wave or shear wave), the anisotropic properties of the crystallite and its density.

In the present case, the unknown "scatterer size" or "grain size" can thus be directly evaluated if the emission frequency is known very accurately (monochromatic emission), or else by measuring the relationship between the diffusion coefficient of the ultrasonic wave and frequency. To make this possible, the power spectrum density of the ultrasonic wave should be as wide as possible. Different frequency components of the signal can then be used for analyzing changes in the attenuation coefficient (and thus in the diffusion coefficient) as a function of frequency.

One possible method consists in calculating the spectral distribution of successive backwall echoes from which it is possible to obtain the attenuation curve of the wave as a function of frequency and thus to deduce the grain size of the material.

The method described here has been found to be highly suitable for quantitative evaluation of grain size by conventional (piezoelectric) ultrasonic techniques.

In practice, it is often difficult to use these techniques "on-line" since the reproducibility of results depends on the coupling conditions between the piezoelectric transducer and thematerial under test.

That is why there is an interest for these applications in working with unconventional contactless technologies.

EXPERIMENTAL SET-UP

A number of studies have been carried out to date to obviate the drawbacks of conventional ultrasonics.

Several authors have shown the advantages of generating ultrasound by means of lasers [3, 4, 5, 6] : a laser pulse is applied to the surface of a material under test, and the absorption of the pulse by the material gives rise to transient localized heating near to the surface. The incident energy is absorbed and converted into mechanical energy in the form of ultrasonic waves whose amplitudes vary with the incident power density.

The experimental set-up is shown in Figure 1. Ultrasound is generated in the test piece by infrared laser pulses coming from a Q-switched Nd:YAG laser (BM-Industries). Pulse duration is 10 ns with a maximum repetition rate of 30 Hz, beam diameter is 7 mm, and maximum energy is 240 mJ.

A focusing lens placed on the path of the beam serves to vary the size of the impact spot on the surface of the material, thus varying the energy density absorbed per unit area. It is thus possible to enhance a desired propagation mode (longitudinal wave or shear wave) and the most appropriate directivity as a function of the geometry of the test piece under examination.

For reception purposes, we prefer to use an interferometer rather than an EMAT transducer since, to date, the efficiency of such transducers is too low and the frequency bandwidth is too narrow for this type of application. Of the numerous optical devices for detecting ultrasound that have been described in the literature [7, 8, 9, 10], we have chosen the heterodyne interferometer which has suitable characteristics for high frequency and wideband applications, associated with high sensitivity.

The essential characteristics of our system which has been described in greater detail elsewhere [10, 11] are as follows:

polarized 2 mW He-Ne laser ;
acousto-optical modulator operating at a frequency of 70 MHz ; outlet focusing optics ; and
a fast photodetector (bandwidth 100 MHz) with very low noise preamplifier.

The usable frequency band runs from 200 kHz to 4 MHz, or 18 MHz, or 45 MHz (full bandwidth) depending on the application.

The demodulation electronics is provided with an automatic gain control loop having a dynamic range of 30 dB so as to compensate for variations in reflectivity due to the roughness of the surface of the material under test (speckle effect).

EXPERIMENTAL RESULTS

The tested sample is a 2 mm thick 304L stainless steel plate. The diameter (< 1 mm) of the generation impact source is equal to the measuring point size on the opposite side (transmission echography).

The bandwidth of the interferometer is of 0,2 to 45 MHz, and the demodulated signal is digitized with a 125 MHz sampling rate using aLECROY 9400 digital oscilloscope.

The signal is averaged over 20 acquisitions in order to improve the signal to noise ratio (figure 2), and is then transfered to the micro-computer via an IEEE parallel bus.

The data processing consists in selecting two successive backwall echoes, to compute each FFT, and then to divide the two spectra modulus (figure 3) to obtain the attenuation curve as a function of the frequency (figure 4).

The expression $s.d^3.f^4$ can then be calculated using the measured attenuation to evaluate the mean grain size "d". The grain size of the 304L sample which has been estimated to be 35 µm, is to be compared with the size of 25 µm given by the optical micrographic analysis (figure 5).

DISCUSSION AND CONCLUSION

The method presented in this paper could be a very interesting tool for industrial applications, providing that it could be applied on materials with thicknesses lying between 0,6 and 20 mm. The intended application of this method (are products with grain size between 10 and 100 µm. Several problems arise from those grain sizes and thicknesses values :

- the directivity of a laser-spot source is different from the piston model of a piezoelectric transducer [6], the attenuation observed along the ultrasonic wave propagation will be caused on a major part from the diffraction of the ultrasonic beam. This must be taken into account in the computation of the grain size : the ultrasonic wave can be modelled by a spherical wave if we assume a small spot size source compared with the ultrasonic wavelength,

- In the case of material sheets with thickness of less than 1 mm, the last results obtained in the laboratory have shown that a disturbing low frequency signal is added to the ultrasonic displacement signal. So the electronic measurement device must be modified to filter these low frequency components before the demodulating circuits, in order to keep the full dynamic measurement range for the useful signal,

- one of the limits of the experimental set-up presented in this paper is its difficulty to account for industrial material surface roughness, because of the limited laser power source of the interferometer (2 mW),

- in several applications, like pipe characterization, the emission source and the reception beam must be located on the same side of the product (reflection echography) without any interference between the generating beam and the receiving optical device.

Neverthless, the good laboratory results obtained with ultrasonic contactless methods make it possible to foresee future industrial on-line monitoring applications which could not be performed with classical instruments.

REFERENCES

[1] GOEBBELS, K., "Structure Analysis by Scattered Ultrasonic Radiation", in Research Techniques in Nondestructive Testing, Vol. IV, (sharpe ed., 1980) pp. 87-157

[2] PAPADAKIS, E.P., "Physical Acoustics and Microstructure of iron alloys". International Metals Reviews, Vol. 29, n° 1 (1984) pp. 1-24

[3] SCRUBY, C.B. ; DEWHURST R.J. ; HUTCHINS D.A. ; PALMER S.B. "Laser Generation of Ultrasound in Metals" in Research Techniques in Nondestructive Testing, Vol. V (sharpe ed., 1982) pp. 281-327

[4] HUTCHINS, D.A., "Mechanisms of Pulsed Photoacoustic Generation". Can J. Phys. (1986) 64, pp. 1247-1264

[5] LE BRUN, A. ; PONS, F. : "Non-contact Ultrasonic Testing : Applications to Metrology and Nondestructive Testing." Fourth European NDT Conference, 14-18 Sept. 1987, LONDON (G.B.)

[6] AUSSEL, J.D. ; LE BRUN, A. ; BABOUX, J.C. : "Generating acoustic waves by laser : theoretical and experimental study of the emission source". Ultrasonics, vol. 26, 1988, pp. 245-255.

[7] DRAIN, L.E. ; SPEAKE, J.H. ; MOSS, B.C. ; "Displacement and Vibration Measurement by Laser Interferometry". First European Congress on Optics Applied to Metrology - SPIE (1977) Vol. 136, pp. 52-57

[7] MONCHALIN, J.P. : "Heterodyne Interferometric Laser Probe to Measure Continuous Ultrasonic Displacements." Rev. Sci. Instrum., Vol. 56, n° 5, April 1985, pp. 543-546

[9] PAUL, M. ; BETZ, B. ; ARNOLD, W. : "Interferometric Detection of Ultrasound at Rough Surfaces Using Optical Phase Conjugation." Appl. Phys. Lett., Vol. 50, pp. 1569-1571.

[10] ROYER, D. ; DIEULESAINT, E. ; MARTIN, Y. : "Improved Version of a Polarized Beam heterodyne Interferometer" IEEE Ultrasonic Symposium 1985, pp.432-435

[11] LESNE, J.L. ; LE BRUN, A. ; ROYER, D. ; DIEULESAINT : "High Bandwidtch Laser Heterodyne Interferometer to Measure Transient Mechanical Displacements". Proceedings of the International Symposium on Industrial Opto-electronic Measurement Systems Using Coherent Light, 17-20 November 1987, Cannes, France SPIE Vol. 863, pp.13-22.

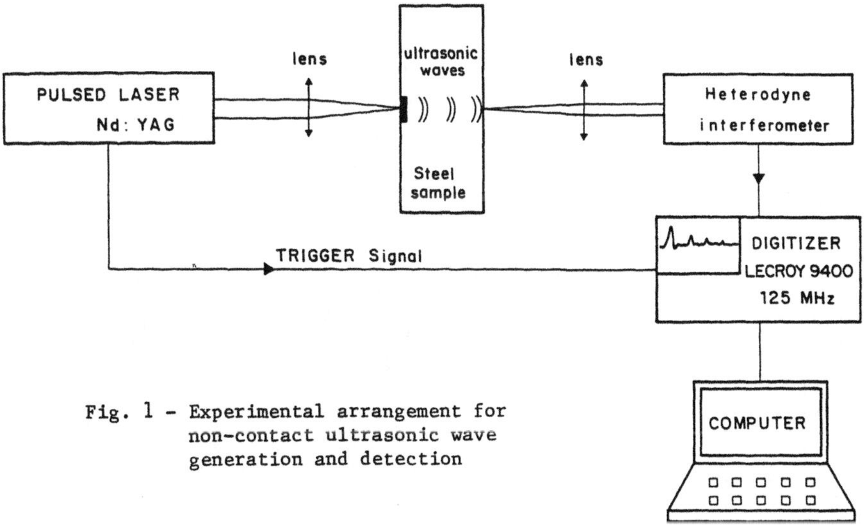

Fig. 1 - Experimental arrangement for non-contact ultrasonic wave generation and detection

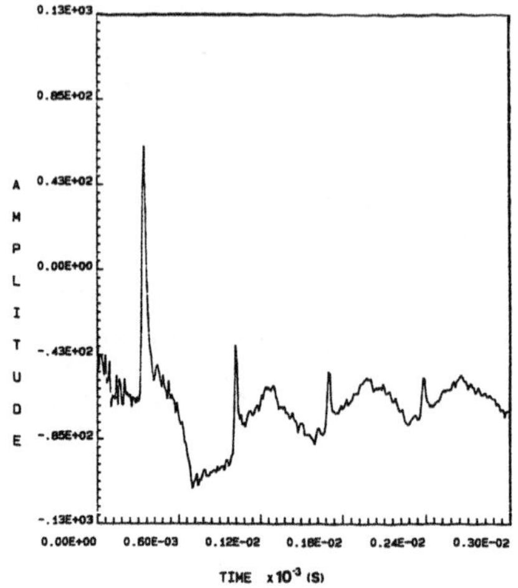

Fig. 2 - Displacement waveform measured at the epicentre on 2 mm thick 304 L steel plate

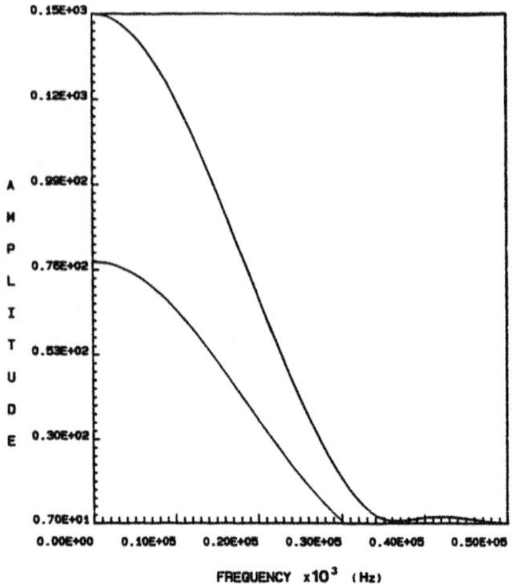

Fig. 3 - Frequency spectra of two successive echoes

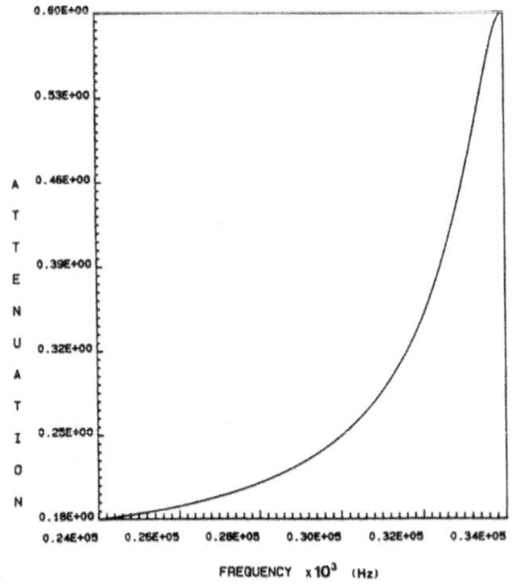

Fig. 4 - Attenuation as a function of frequency

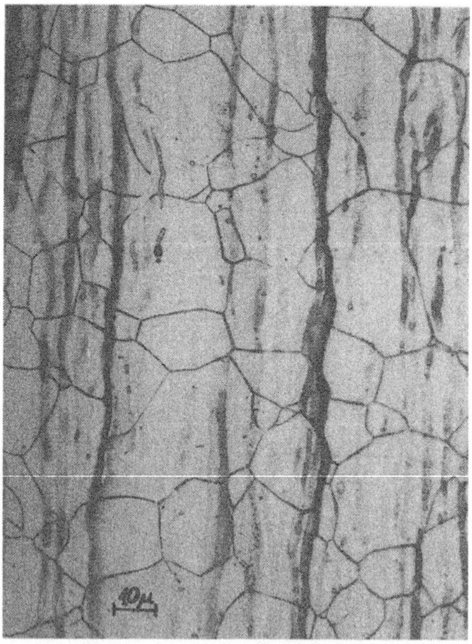

Fig. 5 - Optical micrograph of grain structure in 304 L

Analysis of the Effect of Graphite Morphology on the Elastic Properties of Cast Iron

JEAN F. BUSSIÈRE, LUC PICHÉ AND GERVAIS LECLERC

National Research Council Canada
Industrial Materials Research Institute
75 De Mortagne Blvd.
Boucherville, Québec, CANADA J4B 6Y4

Summary

Calculations of elastic constants of cast iron, consisting of an iron matrix with randomly distributed ellipsoidal graphite inclusions of different shape factors, are presented using Eshelby's tensor for a single ellipsoidal inclusion and a differential scheme to take into account finite concentrations. Good agreement with published experimental data is obtained in most cases. Varying the aspect ratio from spheres to discs causes a decrease in the elastic moduli of a factor of approximately three. Implications of this effect, related to assessments of nodularity with elastic constant or ultrasonic velocity measurements, are discussed.

Introduction

Cast irons are largely classified according to their graphite morphology, which is the dominant metallurgical factor affecting their mechanical properties. Gray irons contain interconnected graphite flakes and tend to fracture along the flake boundaries, whereas ductile (or nodular, or spherulitic graphite iron) displays much higher ductility associated with spheroidal graphite. Compacted graphite iron, on the other hand, has properties which are intermediate between gray and nodular iron and is characterized by blunt flakes which are interconnected within cells. The production of compacted graphite and nodular iron requires close metallurgical control to produce the desired morphology and mechanical properties. For this reason, nondestructive techniques to predict mechanical strength are often desirable, especially for parts which are critical for structural integrity. A number of techniques are used in industry for this purpose and have been reviewed (1,2). One of the most useful techniques is the measurement of ultrasonic velocity to determine the degree of nodularity of the graphite inclusions, which strongly affects mechanical properties. Industrial use of ultrasonic velocity for this purpose arises from empirical work which started in the 1960's (for a review, see Ref.3).

The problem of analysing the elastic properties of iron containing graphite inclusions of various shapes was initially treated by Plenard (4) in a simplified manner, which considered only the case of spherical inclusions. More recently, the problem of graphite inclusions with different aspect ratios was analysed by Papadakis et al (5) in a semi-empirical way and by Speich et al (6), who used available theories to treat the extreme cases of spheres and infinitely flat discs.

In the present paper the problem is analysed for randomly distributed graphite inclusions of spheroidal shape having an arbitrary aspect ratio. The theory is based on Eshelby's (7) tensor describing the strain caused by a single inclusion and on a differential scheme for treating the case of finite concentrations. The relation between nodularity and elastic properties is then discussed.

Theory

The theoretical analysis of the effective elastic constants of two-phase materials has been the subject of a great deal of attention. The simplest models assume either uniform strain or uniform stress throughout the composite, giving rise respectively to the Voigt and Reuss averages for the composite elastic moduli, B, E, and G. Both averages yield extreme values which can be very far apart in cases such as the present, where the elastic constants of the inclusions and the matrix can differ by a factor larger than 20. Using variational methods, Hashin and Shtrickman (8) were able to establish rigorous upper and lower values of the moduli which fall within the Voigt and Reuss averages. However, all the above models are insensitive to the detailed shape of the inclusions. Models which take into account the microscopic details of the composite are based on the calculation of the elastic field that surrounds the individual inclusion. The solution for the response of an ellipsoidal inclusion of shape factor p (ratio of the length of the axis of revolution to the length of an axis orthogonal to it) to an homogeneous static strain was given by Eshelby (4). The only assumptions contained in the model are: a) the binder and inclusion material are both elastic, isotropic and homogeneous, b) the inclusion is perfectly bonded to the matrix, c) the inclusion is isolated. The validity of the model being limited to low concentrations, schemes have been proposed to extend the theory to the case of finite volume contents. These have been briefly reviewed recently by one of the authors (9, 10). In principle, the elastic properties of composites containing randomly oriented inclusions can be obtained from those of aligned composites using a normalized integration scheme. Actually, the calculation is simplified, if the integration over all possible orientations is performed in the zero concentration limit, before allowance is made for larger values of inclusion volume contents. Such an approach was proposed by Boucher (11) in conjunction with the differential scheme (D.S.).

In what follows, we shall use this approach to analyze our data. The exercise appears reasonable in view of our previous work (10), where the D.S. was used with success on composites containing particles of different aspect ratios. Indeed, the results showed that, albeit most schemes were equivalent at low volume contents ($c \simeq 0.12$). the D.S. provided the best description of elastic moduli for larger concentrations. The theory (11) is elegant but the detailed calculations are lengthy and will not be reproduced here. It will be simply stated that, in the limit of very low volume contents, c, the bulk, B, and shear, G, moduli of the composite are expressed as:

$$(B - B_1)/(B_2 - B_1) = cF_1 \tag{1}$$

$$(G - G_1)/(G_2 - G_1) = cF_2 \tag{2}$$

where F_1 and F_2 are fourth rank tensors which are related to B_1, B_2, G_1, G_2 and p (shape factor) and characterize the random orientation. Indices 1 and 2 refer to matrix and inclusion respectively. In order to extend the results of Eqs 1 and 2 to the case of finite concentrations, the composite is progressively built up by adding new particles (D.S.). This is expressed through a system of nonlinear differential equations which must be solved numerically:

$$dB/dc = F_1 (B_2 - B)/(1 - c) \tag{3}$$

$$dG/dc = F_2 (G_2 - B)/(1 - c) \tag{4}$$

Results & Discussion

In our calculations, the matrix was assumed to have elastic constants typical of polycrystalline iron (12), $E_1 = 209$ GPa, $G_1 = 81.3$ GPa and a density of 7.9 g/cm^3. The graphite particles are nearly single crystals in the case of flake graphite, whereas in the case of nodular graphite, the growth is essentially pyrolitic with c axes of the graphite pointing radically in pyramidal regions within the nodules (5). It then seems reasonable as explained in Papadakis (5) to use c_{33} for the compressive modulus normal to the graphite spheres and c_{44} for the shear modulus, resulting in E_g = 4.41 GPa and G_g = 1.5 GPa. These values are smaller than for polycrystalline graphite by approximately a factor of two. However, since these constants are in all cases approximately twenty-five times smaller than the elastic constants of iron and correspond to approximately 10% of the volume, they have little effect on the final result. The density was taken to be 2.25 g/cm^3.

Results for the effect of shape factor on the Young's and shear moduli for the case of 11 vol % concentration of graphite are shown in figure 1. It is seen that both moduli decrease by approximately a factor of three as the shape factor is decreased from 1 (spheres) to 0 (infinitely flat discs).

In figures 2 and 3 calculated values of Young's and shear moduli are presented as a function of carbon (graphite) volume content for shape factors of 1, 0.1, 0.03 and 0. The calculations are compared to measurements taken from Speich et al (6). A description of the samples, including volume concentration of graphite, metallographically observed aspect

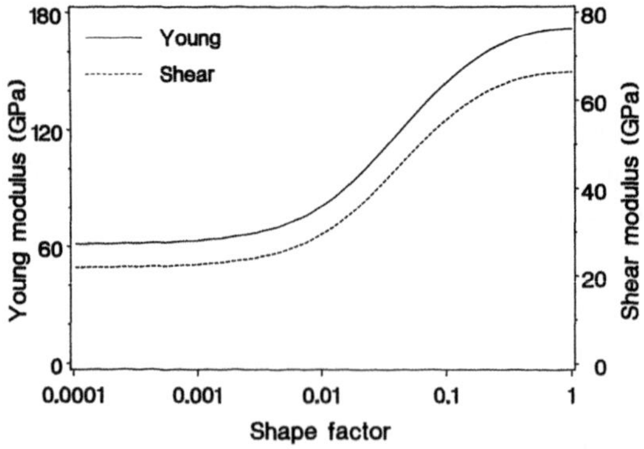

Fig. 1: Variation of Young and shear moduli of cast iron containing 11 vol. % concentration of ellipsoidal inclusions of graphite of different shape factors.

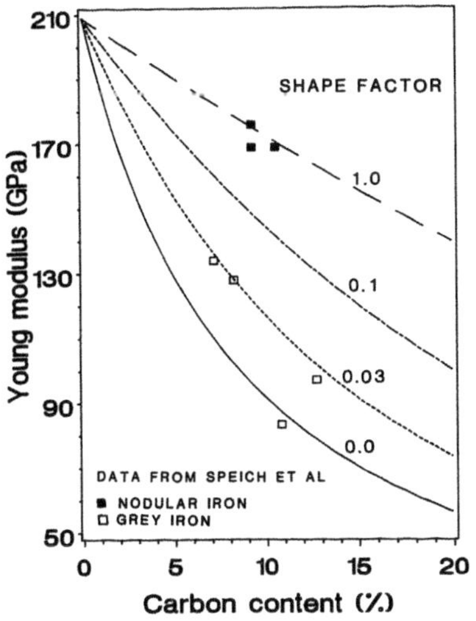

Fig. 2: Young's modulus versus graphite content for graphite of different shape factors. The data points are from ref. 6.

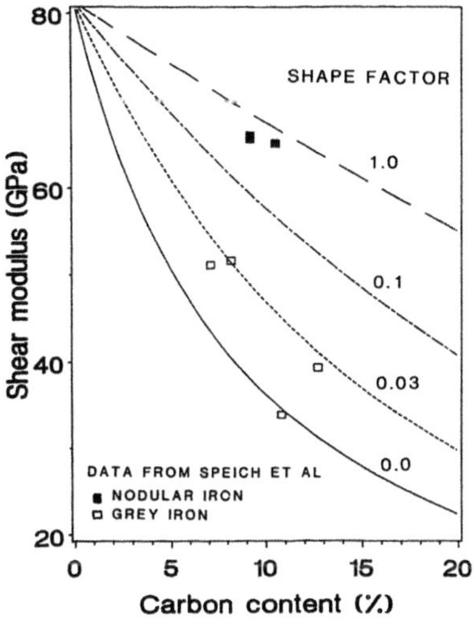

Fig. 3: Shear modulus versus graphite content for graphite of different shape factors. Data points are from ref. 6.

TABLE I

TYPE OF CAST IRON +	GRAPHITE Vol. %	AXIAL RATIO (d/t)	MEASURED MODULI+		CALCULATED MODULI	
			YOUNG+ GPa	SHEAR+ GPa	YOUNG GPa	SHEAR GPa
Gray (I)A	8.2	14.6	128	51.6	148.4	57.3
Gray (I)B	7.1	15.2	134	51.1	154.4	59.6
Gray (II)A	12.7	23.9	97.2	39.3	107.9	40.8
Gray (II)B	10.8	24.6	83.4	33.8	122.2	46.6
Nodular as cast	9.2	(1.0)	176	66.0	175.2	68.0
Nodular annealed	10.5	(1.0)	169	61.4	170.7	66.2
Nodular normalized & tempered	9.2	(1.0)	169	65.6	175.2	68.0

+ from Speich et al

ratio and measured elastic constants, are given in Table I together with calculated elastic constants. For nodular iron agreement with theory is within 5%, except for the shear modulus of the sample with 10.5% graphite which deviates by ≈ 9%. For gray irons, agreement is within 15%, except for the gray iron II B, where the deviation is ≈ 45%.

The small disagreement with theory in the case of nodular iron is probably due to the choice of the elastic constants of the matrix. For gray iron it is seen in figures 2 and 3 that better agreement is obtained for most samples with an inverse shape factor (approximately equivalent to aspect ratio) of $1/p = 33$. The use of a slightly higher inverse shape factor appears to be justified since a metallographic examination of randomly distributed spheroids of inverse shape factor $1/p$ is expected to yield a measured aspect ratio which is smaller than $1/p$. Intersection of an oblate spheroid of inverse shape factor $1/p$ by a plane going through its centre and inclined at 45° to the axis of symmetry will, for instance, yield a shape factor of $\approx \sqrt{2}/p$. One should also take into account the fact that there is a statistical distribution of shape factors, which is not considered presently. The above arguments, however, are insufficient to explain the discrepancy for gray iron II B which requires infinitely flat discs for agreement with theory (figs. 2-3). In this case, effects associated with contiguity, such as discussed by Piché and Hamel (10), may be important and are outside the scope of the present model.

We shall now analyse the problem of using elastic constant measurements, or ultrasonic velocity to determine the degree of nodularity of nodular cast iron. Nodularity is defined as the percentage concentration of graphite which is spherical. Assuming an overall graphite concentration of 11%, calculations of elastic constants and ultrasonic velocities were obtained for mixtures of spheres and prolate spheroids of different shape factors. Elastic constants were obtained by first introducing the spheres in the differential scheme and then adding the prolate spheroids. Results for the longitudinal ultrasonic velocity are shown in figure 4 for shape factors of the prolate spheroids between .003 and 0.5. It is clearly seen, as expected, that a wide range of ultrasonic velocity can be obtained for a given nodularity level. For this reason it is expected that elastic constant or ultrasonic velocity measurements will give a better indication of mechanical properties than metallographic estimates of nodularity, as is in fact observed experimentally (1,2).

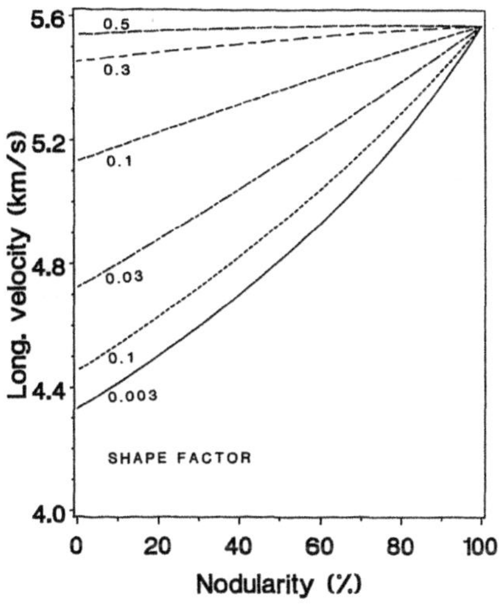

Fig. 4: Calculated values of longitudinal ultrasonic velocity as a function of nodularity for cast iron containing 11% graphite, assuming different shape factors for the non-nodular graphite.

Conclusion

Calculations of the elastic constants of cast iron were presented assuming a pure iron matrix containing randomly distributed ellipsoidal graphite inclusions. The model, which contains no adjustable parameters, is based on a differential scheme for extending Eshelby's solution to finite concentrations and was found to be in reasonable agreement with experimental data published by Speich et al, except for one gray cast iron. Limitations of the model appear to be associated with the fact that all inclusions are assumed to be independent and noninteracting. The relation between nodularity and ultrasonic velocity was found to strongly depend on the aspect ratio of the non-nodular graphite. This limits assessments of nodularity from elastic measurements but not necessarily assessments of quality as it relates to mechanical properties.

References:

1. Fuller, A.G.; Emerson, P.J.; Sergeant, G.F., Report on the effect upon mechanical properties of variation in graphite form in irons having varying amounts of ferrite and pearlite in the matrix structure and the use of nondestructive tests in the assessments of mechanical properties of such irons. Trans. American Foundrymen's Soc. 88 (1980) 21-50.

2. Sergeant, G.F.; Fuller, A.G., The effect upon mechanical properties of variation in graphite form in irons having varying amounts of carbide in the matrix structure and the use of nondestructive tests in the assessment of the mechanical properties of such irons, Trans. Amer. Foundrymen's Soc. 88 (1980) 545-574.

3. Krautkrämer, J.; Krautkrämer, H., Ultrasonic Testing of Materials, 3rd edition, Springer-Verlag 1983, 528-536.

4. Plenard, E.; The elastic behavior of cast iron, in Recent Research in Cast Iron, Merchant, H.D. ed., New York: Gordon and Breach 1968, 707-792.

5. Papadakis, E.P.; Bartosiewicz, L.; Altstetter, J.D.; Chapman, G.B., Morphological severity factor for graphite shape in cast iron and its relation to ultrasonic velocity and tensile properties. Amer. Foundrymen's Soc. 91 (1983) 721-728.

6. Speich, G.R.; Schwoeble, A.J.; Kapadia, B.M., Elastic Moduli of Gray and Nodular Cast Iron, J. Appl. Mechanics 47 (1980) 821-826.

7. Eshelby, J.D., Proc. of the Royal Soc. 241 Series A (1957) 376.

8. Hashin, Z; Shtrikman, S., J. Franklin Inst. 271 (1961) 336-341.

9. Piché, L.; Hamel, A., Ultrasonic evaluation of Filled Polymers: I. Techniques and Models for the Elastic moduli of a resin filled with iron spherical inclusions, Polymer Composites 7 (1986) 355-362.

10. Piché, L.; Hamel, A., Ultrasonic evaluation of Filled Polymers: II. Elastic moduli of a resin filled with iron inclusions of different aspect ratios, Polymer Composites 8(1987) 22-28.

11. Boucher, S., Revue M 21 (1975) 243; Revue M 22 (1976) 31.

12. Gschneider, K.A., Solid State Physics 16 (1964) 275-307.

Effect of Graphite Aspect Ratio on Cast-iron Elastic Constants

HASSEL LEDBETTER

Institute for Materials Science and Engineering
National Institute of Standards and Technology
Boulder, Colorado 80303-3328

SUBHENDU DATTA

Department of Mechanical Engineering and CIRES
University of Colorado
Boulder, Colorado 80309-0449

Using a scattered-plane-wave ensemble-average model developed for composite materials, we calculated the effective elastic constants of cast iron. We focused on the effect of graphite-particle aspect ratio on the Young modulus. Between model and observation, we found good agreement. Oblate-spheroidal graphite flakes lower the elastic stiffness much more than do spheres. To obtain good model-measurement agreement, one must use graphite's lower third-order-bound (Kröner-bound) elastic constants.

Introduction

Graphite-particle shape affects cast-iron's properties, both physical and mechanical. Experimentally, Okamoto et al. [1] studied how graphite shape affects four physical properties: electrical resistivity, thermal conductivity, Young modulus, and internal friction. Experimentally and theoretically, Löhe et al. [2] considered thoroughly how graphite shape affects the Young modulus. For special shapes — rods, spheres, discs — they calculated how the Young modulus depends on graphite volume fraction. Their calculations confirm the long-known result that spherical particles produce a smaller property change than produced by either rods or discs. Also, they confirmed that discs in the oblate limit produce a much larger property change than rods. Their model calculations omitted explicit Young-modulus dependence on particle aspect ratio, c/a. Other notable studies include those by Speich et al. [3] and by Anand [4]. Speich et al. measured the Young moduli and shear moduli of nodular and grey cast irons, where the graphite-particle shapes are approximately spheres and discs, respectively. They considered three models: Hashin [5] for spheres, Rossi [6] for discs, and Wu [7] for discs. When the inclusion is elastically softer than the matrix, Hashin's result equals the upper elastic-stiffness bound. None of these three models consider aspect ratio explicitly. To carry out similar calculations, Anand used better models: the self-consistent method [7–10] and the differential-computation method [11–14]. Anand found that all measurements lie between the Hashin-Shtrikman [15] bounds and that disc-shaped particles strongly decrease elastic stiffness with increasing volume concentration. Anand's calculations also failed to consider aspect ratio explicitly.

The present study considers theoretically how cast-iron's Young modulus depends on graphite-particle aspect ratio. We represent graphite particles as biaxial ellipsoids where the aspect ratio varies from zero (oblate-disc limit) to unity (spherical limit).

The theoretical model [16,17] uses a scattered-plane-wave ensemble-average approach to predict the effect of graphite aspect ratio on longitudinal and transverse sound velocities. The model considers three of the inclusion's geometrical properties: volume fraction, shape (sphere to disc), and orientation (random).

Figures 1 and 2 show microstructures of both nodular and grey iron, where the respective graphite-particle shapes are spheres and discs.

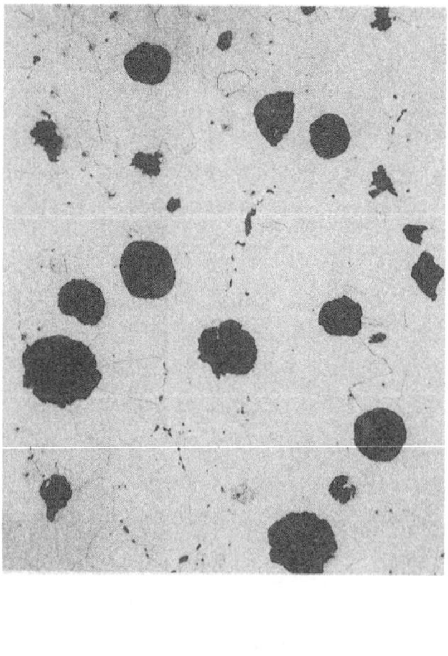

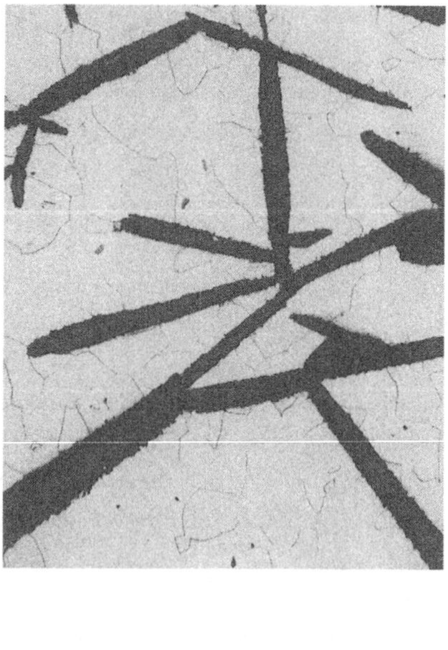

Fig. 1. (Left) Microstructure of a nodular cast iron. Spherical graphite particles are approximately 0.05 mm in diameter. Fig. 2. (Right) Microstructure of a grey cast iron. Graphite flakes are approximately 0.5 mm in diameter with an aspect ratio (c/a) of 25.

Modeling

Here, we consider what theory predicts for the sound velocities of cast iron containing variously shaped voids. Especially, can a model explain the observed strong dependence of properties on particle aspect ratio?

For a model, we adopted one used previously [16] to calculate the effective elastic properties of a reinforced composite material. The present case differs principally in that the occluded phase is elastically softer than the matrix phase. For simplicity, we assumed the inclusions to be homogeneous, isotropic, oriented randomly, distributed homogeneously, geometrically identical, and bonded firmly to the matrix. Omitted here for brevity, the model's details occur elsewhere [17].

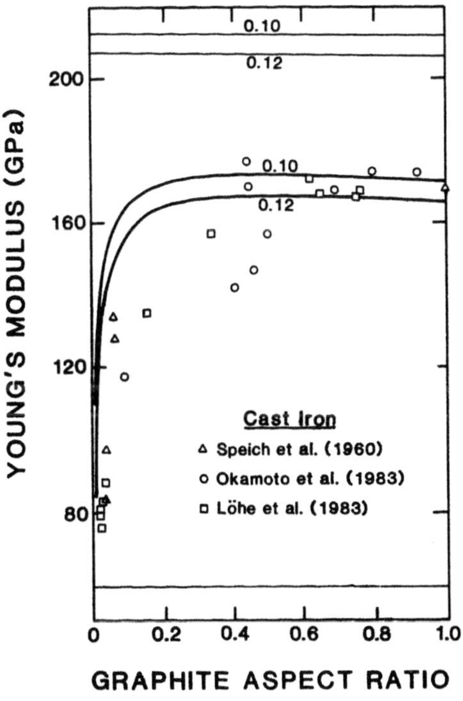

Fig. 3 Young's modulus versus graphite-particle aspect ratio. Symbols represent measurements. Curves represent model-calculation results for two volume fractions: 0.10 and 0.12. Upper, nearly horizontal, curves represent graphite's upper third-order-bound (Kröner-bound) elastic constants. Lower curves represent graphite's lower third-order bound.

Results

Figure 3 shows the study's principal result: the predicted change of Young modulus with graphite-particle aspect ratio. The figure contains experimental results from three studies: Speich et al. [3], Löhe et al., [2] and Okamoto et al. [1]. As reviewed by Löhe et al., many more experimental studies exist. But, the three shown in the figure define sufficiently the total experimental results.

The figure contains predicted curves for two volume fractions — 10 and 12 percent — a range that contains most of the studied cast irons.

The two upper, nearly horizontal, curves correspond to graphite's upper third-order elastic-constant bounds. The two lower curves correspond to the lower bounds. From monocrystal elastic constants, using equations given by Kröner and Koch [18], for graphite, Wawra et al. [19] calculated third-order elastic-constant. They found the following effective quasi-isotropic elastic constants: E = 4.17 (1.34) GPa; G = 1.41 (0.45) GPa; B = 36.56 (35.77) GPa; ν = 0.1965 (0.2225). Values outside parentheses denote upper bounds; those inside denote lower.

For the matrix phase, we took E = 206 GPa [2] and ν = 0.2880, the value for alpha iron. These two quasi-isotropic elastic constants imply G = 80.0 GPa and B = 162 GPa.

Other predicted results — shear modulus, bulk modulus, and Poisson ratio — will appear elsewhere.

Discussion

Corresponding to observation, our model predicts a strong dependence of Young modulus, E, on aspect ratio, c/a. Near the spherical limit (c/a = 1), E varies slowly with r. Near the oblate-disc limit (c/a = 0), E varies rapidly with r. This case provides another example that physical properties depend more on an occluded disc-shaped phase than on an occluded sphere-shaped phase.

A surprising result is that graphite's lower-bound quasi-isotropic effective elastic constants fit observation so well. Löhe et al. [2] found the same result. Kröner [20] emphasized that the various-order bounds correspond to different geometrical properties of the polycrystalline aggregate. (Here, we imagine an aggregate of graphite monocrystals in a matrix of ferrite.) Zeroeth-order bounds (the extreme monocrystal shear moduli) imply that we know nothing concerning the material's texture, particle size, particle shape, particle-boundary shape, or distributions of the geometrical features. First-order bounds (Voigt and Reuss) imply absence of texture: randomly oriented crystal axes. Second-order bounds (Hashin-Shtrikman) imply noncorrelated elastic moduli (or crystal axes) with either particle shape or particle size. Third-order (Kröner) bounds imply no special distribution of grain shapes. Infinite-order bounds imply perfect disorder: absence of any special distributions related either to elastic constants or to particle geometry. No one yet calculated fourth-order bounds for graphite. From results shown in Fig. 3, we guess that graphite in cast iron falls outside (below) the fourth-order bound. Thus, graphite particles in cast iron depart from perfect disorder. This order appears at the fourth-order bound and may correspond to particle-boundary geometry.

Results in Fig. 3 show that all the graphite-upper-bound cast-iron elastic constants are nearly aspect-ratio independent. The larger changes near $c/a = 0$ may represent a calculation artifact caused by an unrealistic particle-distribution function as one approaches the oblate limit for finite volume fractions. Closer to $c/a = 1$, the calculations avoid this limitation. The graphite-upper-bound case yields unphysically high elastic-stiffness predictions, too slow varying, and turning up instead of down as c/a approaches zero. The extraordinarily high elastic anisotropy of monocrystalline graphite leads to noncoincident third-order bounds, which provide a good test of elastic-property-prediction models.

We fail to understand why the predicted young modulus fails more slowly than the observed young modulus as c/a decreases from unity to zero. Perhaps, the low-aspect-ratio cast irons fail to meet the model's requirements: macroscopic homogeneity, isotropy, and disorder.

Conclusions

From this study, there emerged three conclusions:

1. A scattered-plane-wave ensemble-average model predicts the dependence of cast-iron's elastic constants on graphite-particle aspect ratio. Oblate particles soften the material much more than do spherical particles.

2. Concerning graphite's effective quasi-isotropic elastic constants in cast iron, the model shows that the appropriate choice is the lower third-order (Kröner) bound.

3. Graphite's upper third-order bounds give results contrary to observation.

Acknowledgment
Results of this study emerged from computer programs written by Dr. R.D. Kriz of NIST and from computational help from Ming Lei, visiting scientist from the Institute of Metal Research (Shenyang). Dr. G.R. Speich, formerly at U.S. Steel, now at Illinois Institute of Technology, contributed the photomicrographs.

References

1. T. Okamoto, A. Kagawa, K. Kamei, and H. Matsumoto, Imono 55, No. 2 (1983).

2. D. Löhe, O. Vöhringer, and E. Macherauch, Z. Metallk. 74, 265 (1983).

3. G.R. Speich, A.J. Schwoeble, and B.M. Kapala, J. Appl. Mech. 47, 821 (1980).

4. L. Anand, Scr. Metall. 16, 173 (1982).

5. Z. Hashin, J. Appl. Mech. 29, 43 (1962).

6. R.C. Rossi, J. Amer. Ceram. Soc. 51, 433 (1968).

7. T.T. Wu, Int. J. Sol. Struct. 2, 1 (1966).

8. R. Hill, J. Mech. Phys. Sol. 13, 213 (1965).

9. B. Budiansky, J. Mech. Phys. Sol. 13, 223 (1965).

10. L.J. Walpole, J. Mech. Phys. Sol. 17, 235 (1969).

11. R. Roscoe, Rheol. Acta 12, 404 (1973).

12. S. Boucher, Rev. Tijdschrift 22, 31 (1975).

13. P. McLaughlin, Int. J. Eng. Sci. 15, 237 (1977).

14. M.P. Cleary, I.W. Chen, and S.M. Lee, ASCE J. Eng. Mech. 106, 861 (1980).

15. Z. Hashin and S. Shtrikman, J. Mech. Phys. Sol. 11, 127 (1963).

16. H.M. Ledbetter, S.K. Datta, and R.D. Kriz, Acta Metall. 32, 2225 (1984).

17. H.M. Ledbetter and S.K. Datta, J. Acoust. Soc. Amer. 79, 239 (1986).

18. E. Krönor and H. Koch, Sol. Mech. Arch. 1, 183 (1976).

19. H.H. Wawra, B.K.D. Gairola, and E. Kröner, Z. Metallk. 73, 69 (1982).

20. E. Kröner, personal communication, Boulder, Colorado (1981).

Ultrasonic Determination of Materials Characteristics

R. L. SMITH and T. E. DIXON

National NDT Centre
Harwell Laboratory

Summary

It is increasingly important to ensure that not only are materials defect free but also that they possess the physical properties suitable for their intended use. Parameters such as yield strength and fracture toughness have a complex dependence on the microstructure of the material, its state of stress (applied and residual) and environment. There are many highly sophisticated destructive tests and instruments that can give the materials design engineer guidance on the qualities needed of a particular material. However it would be of immense value if some or all of the parameters could be verified in a nondestructive way on the components themselves. Ultrasonics is an attractive tool to use to achieve these goals and our work has concentrated on the use of ultrasonic attenuation spectral analysis to obtain information on grain size and dislocation content in metals.

We have also recently been investigating another approach for the measurement of dislocation content based on the decay of incoherent ultrasonic energy.

Recent results obtained from these approaches are discussed along with a description of some of the instrumentation developed for the application of these techniques.

Introduction

The National NDT Centre at the Harwell Laboratory has been developing a number of ultrasonic techniques for the nondestructive characterisation of materials. These include accurate velocity measurements for texture and residual stress analysis [1] and attenuation techniques for microstructural features such as grain size and dislocation content [2]. This paper will concentrate on the latter work where we are developing ultrasonic attenuation techniques for the determination of microstructural characteristics which can influence the fracture properties of a material. The work is based on the assumption that the attenuation of ultrasonic waves is governed by two processes, scattering from elastic discontinuities such as grain boundaries and absorption from damping factors such as dislocations. Hence we can write,

$$\alpha(f) = \alpha_a(f) + \alpha_s(f) \tag{1}$$

where $\alpha_a(f)$ is the absorption contribution and $\alpha_s(f)$ the scattering contribution to the total attenuation α. f is the frequency as, in general, these coefficients are frequency dependant and it is the frequency dependence of these mechanisms that has been used to characterise the processes. Two techniques are discussed in this paper, ultrasonic attenuation spectral analysis and ultrasonic reverberation measurements.

Ultrasonic Attenuation Spectral Analysis

This technique has been extensively investigated by the National NDT Centre both experimentally and theoretically and the results have been recently summarized [2]. The technique has been based on the original work of Vary [3] but crucially relies on the theoretical interpretation of the measurements proposed by Reynolds and Smith [4]. This analysis is summarized in Fig.1.

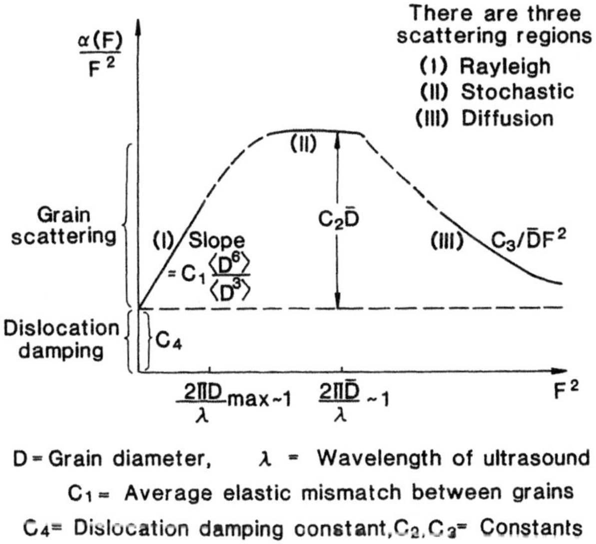

Fig. 1 Ultrasonic Attenuation Spectral Analysis

Ultrasonic attenuation measurements in polycrystalline metals are dependent on the mean grain size and, just as importantly, the grain size distribution and these effects have been demonstrated in a number of metals and alloys [2] including those of industrial importance [5,6]. More recent developments have concentrated on the design and implementation of instrumentation systems to enable the application of the spectral analysis techniques to real inspection situations. Fig.2. illustrates the compact microprocessor controlled instrumentation developed to implement the inspection, based on the Harwell Sweepscan system. In this case the backwall echo from the component is digitized and the frequency content of the signal measured using FFT routine built into the instrument. The signal spectra are than compared to those previously recorded and the changes quantified. These changes are then interpreted by comparison with calibration spectra. This system has been used on an existing inspection line to monitor grain size uniformity.

Ultrasonic attenuation spectral analysis has also been used to monitor dislocation distributions in weakly scattering materials such as aluminium [7]. Fig.3 illustrates the results obtained where the dislocation content can be monitored by the displacement of the overall spectra (Fig.1). In Fig.3 the effect of low temperature annealing on strained aluminium is recorded whereas Fig.4 demonstrates the effect of room temperature self annealing in strained aluminium. The long time required to relax the aluminium sample to its original state possess some measurement problems but a control system is currently under development to enable accurate measurements to be made over a period of many days. By comparison of relaxation times at different temperature it should be possible to measure the activation energy of the relaxation process. This could give an insight into the detailed ultrasonic absorption mechanisms in aluminium and aluminium alloys.

Ultrasonic Reverberation

A promising technique to measure the dislocation density directly is ultrasonic reverberation [8]. This is a technique not based on the attenuation of coherent waveforms but on the attenuation of a diffuse ultrasonic field made up of incoherent waveforms, measured over a long period of time. The assumption is made that the diffuse ultrasonic field is

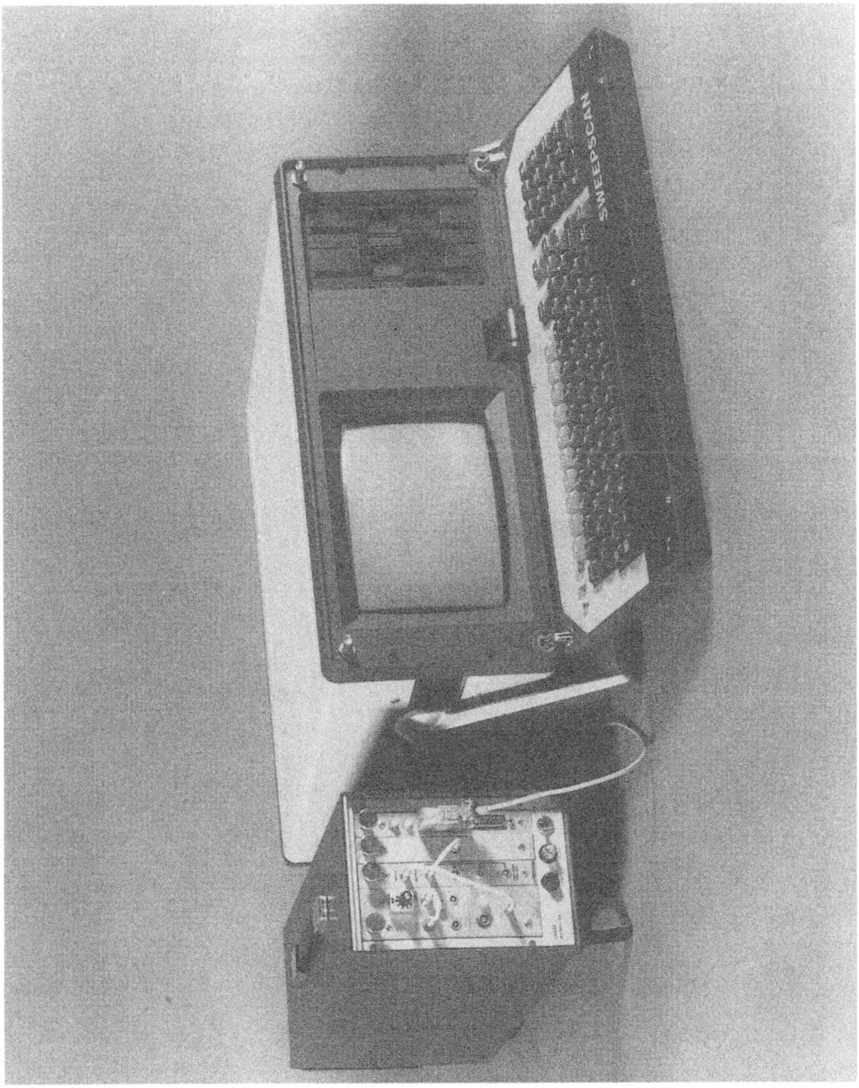

Fig.2 Portable Equipment for the Implementation of ultrasonic attenuation spectral analysis on-line

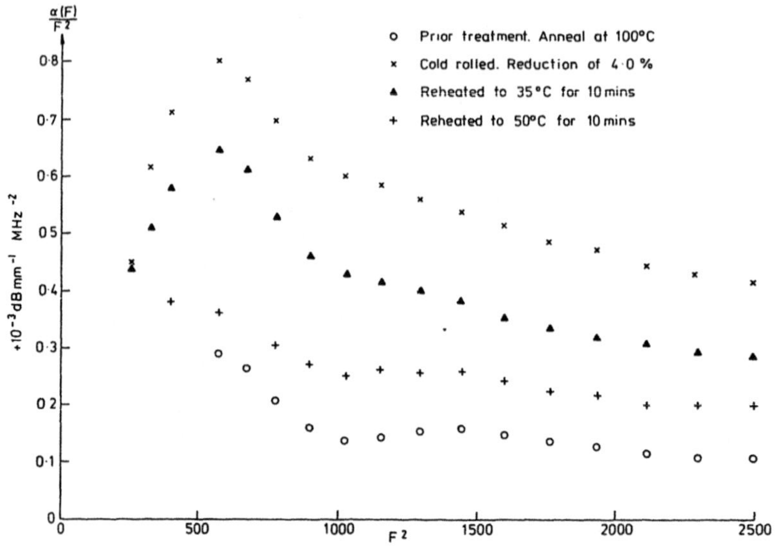

Fig.3 Ultrasonic attenuation spectra in strained and annealed pure aluminium

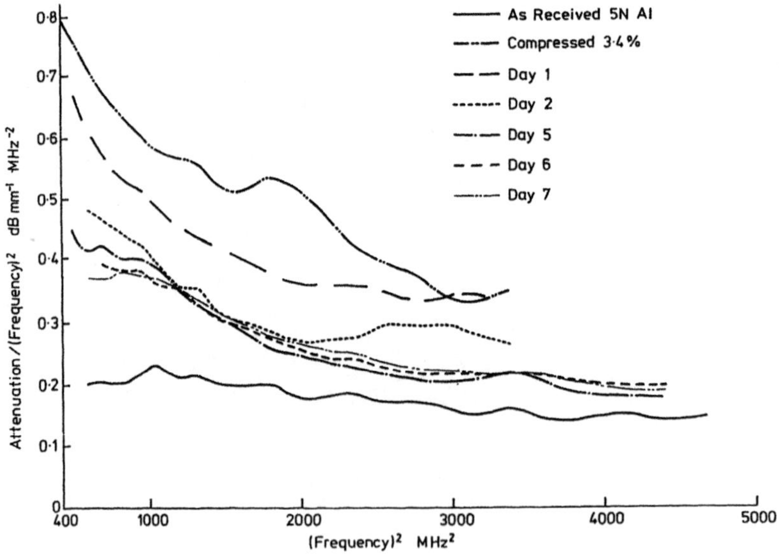

Fig.4 Relaxation of strained aluminium as a function of time

attenuated soley by absorption mechanisms. Experimental evidence for the sensitivity of this technique was reported by Willems [9] and our work is aimed to verify these measurements and develop practical measurement systems. Our initial work is described in reference [10] and some of the results shown in Fig.5. The samples were of Inconel 800H suitably treated to produce different grain sizes and dislocation densities. The input pulse consisted of 20 cycles of sine-wave oscillations at 5MHz to achieve homogeneity of ultrasound in the samples. The coarse grain structure scatters the ultrasonic waves and breaks up the coherent echoes into incoherent and diffuse ultrasonic energy which reverberates within the sample. The reverberation decay follows an approximately exponential form and a logarithmic amplifier was used on the received signal to enable a linear fit to be applied to the gradient. The signal was then rectified and collected by a Tektronix 7854 digitising oscilloscope. To obtain the gradient, a linear regression routine was incorporated into a standard signal acquisition

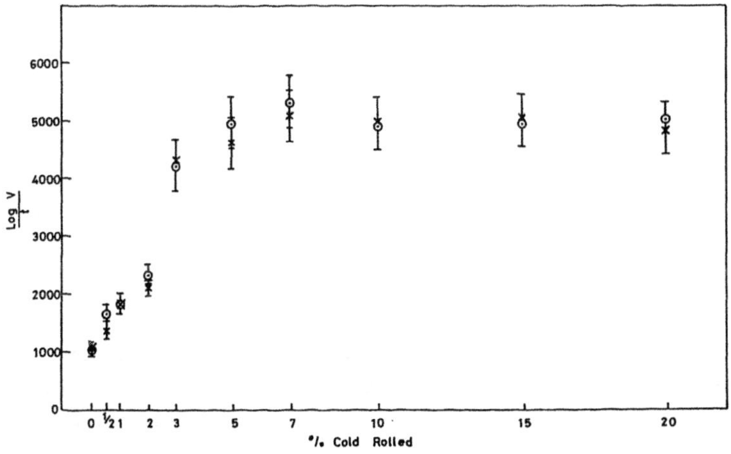

Fig. 5 Ultrasonic reverberation measurements of samples of two grain sizes (175 μm and 225 μm) as a function of % cold reduction.

programme on a Tektronix 4052 microcomputer. This acquired the signal from the oscilloscope and plotted it, fitting a straight line to the decay gradient and giving the gradient figure in log (Volts)/secs.

Some of the ultrasonic reverberation decay gradients measure on the samples are given in Fig.5 and are plotted against the amount of cold rolling i.e. dislocation density. Ageing of the sample is shown to double the reverberation decay gradient of series. This is the expected response from knowledge of the metallurgical conditions of the samples. From metallurgical data on similar material the dislocation densities are expected to vary from 10^7 to 10^{11} cm^{-2}.

The reverberation decay gradients for samples of different grain size were also investigated. There was no apparent difference in decay gradients for the grain sizes used (175μm and 225 μm) as illustrated in Fig. 5.

The error bars plotted in Figure 5 represent three times the standard error of the linear regression fit, i.e. 99% of points fall within these limits. They also accommodate the reproducibility of the gradient values due to the effects of the ultrasonic couplant.

Conclusions

Ultrasonic attenuation techniques have proved to be a useful tool for the investigation of the microstructural characteristic of metals. The techniques can also be developed to produce instruments which can carry out inspections on real components and one such instrument has been illustrated which monitors the uniformity of gain size. However the interaction of ultrasonic waves with the microstructure is a very complex process and must be carefully quantified for the particular material if the ultrasonic measurements are to give unambiguous results. This has been clearly demonstrated by our work on the measurement of dislocation densities where the absorption and scattering components have to be separated. In the case of aluminium this is easy to achieve because of the weak grain scattering. However a newer technique, that of ultrasonic reverberation, has shown its potential for the measurement of dislocation effects in coarse grained steels. This technique is still in the early stages of development and further work will aim to obtain more quantitative values of the dislocation density as well as examining the effects of sample volume and shape.

References

1. Pritchard, S. E.: The use of ultrasonics for residual stress analysis NDT International 20(1)(1987) 57-60.
2. Smith, R. L.: Ultrasonic materials characterization NDT International 20(1)(1987) 43-48.
3. Vary, A.: Ultrasonic measurement of material properties in Research Techniques in Nondestructive Testing IV ed Sharpe, R. S. Academic Press, New York 1980 159-204.
4. Reynolds, W. N. and Smith, R.L. : Ultrasonic wave attenuation spectra in steels. J. Phys. D. Appl. Phys. 17 (1984) 109-116.
5. Smith, R.L.: The application of ultrasonic attenuation spectral analysis to materials characterisation. Report AERE-R11885 HMSO London (1985).
6. Smith, R. L. and Scudder, L. P.: Characterising the microstructure of industrial steels by ultrasonic attenuation spectral analysis. Ultrasonic International 87 Conf. Proc. (1987) 832-837.
7. Perring, S.; Smith, R. L.; Hudson, B. and Reynolds, W. N.: The role of dislocations in ultrasonic attenuation. Philosophical Magazine A 52(5) (1985) 721-727.
8. Weaver, R. L.: Indications of Materials Character from the Behaviour of Diffuse Ultrasonic Fields. Symp. Nondestructive Characterisation of Materials, Montreal, July (1986).
9. Willems, H.: A New Method for the Measurement of Ultrasonic Absorption in Polycrystalline Materials, Review of Progress in QNDE, San Diego (1986).
10. Dixon, T. E. and Smith, R. L.: Ultrasonic determination of materials characteristics - assessment of dislocation content by ultrasonic reverberation measurements. Annual British Conference on NDT. (1988) to be published.

Real Time Cure Monitoring and Control of Composite Fabrication Using Nondestructive Dynamic Mechanical Methods

B. Z. Jang, H. B. Hsieh, and M. D. Shelby

Composites Research Labs
Materials Engineering Program, 201 Ross Hall
Auburn University, AL 36849
(205) 826-4575

SUMMARY

The techniques of mechanical impedance analysis (MIA) have been developed for the characterization of the dynamic mechanical properties of materials. The MIA concept is based on the vibrational analysis of the motion response of a material or structure to a controlled excitation. The frequency response function (FRF), can be obtained from the ratio of the fourier transforms of excitation and response signals. The FRF spectra contain the information needed to calculate the dynamic mechanical properties of a material, such as storage modulus (E'), loss modulus (E''), and loss tangent (tan δ). This MIA method is therefore useful for polymer viscoelasticity study. The spectra obtained during curing of thermosetting resins or composites can be used in real time to calculate the dynamic mechanical properties as a function of cure time and temperature. Optimal cure cycles of a new resin can be determined by using the MIA technique to characterize the phase transformation of the material. The MIA technique is found to be a sensitive and direct means for in-process monitoring of the overall cure states of composite structures of various geometries. The signals obtained can be integrated in a closed-loop feedback control system for composite fabrication.

INTRODUCTION

The mechanical properties of a composite is, to a great extent, dictated by how well a resin is cured. Many techniques have been applied for quality assurance of resins and prepreg materials [1]. These include dynamic mechanical analysis (DMA), differential scanning calorimetry (DSC), high performance liquid chromatography (HPLC), and fourier transform infrared spectrometer (FTIR). These techniques can be employed to study the phase transitions in a thermosetting system and, therefore, are useful for cure cycle design and optimization of a new resin. However, because of the geometry constraints, they can not be readily utilized for real-time in-process cure monitoring of composite fabrication. Very few direct measurement tools can be successfully used to monitor the mechanical properties as a function of the degree of curing in an industrial composite manufacturing environment. The most promising techniques include: (a) microdielectric analysis [2-4], (b) acoustic emission (AE) [7,8], (1c) ultrasonic and acoustic wave measurements [5,6,9-12], (d) fluorescence [13-17], and (e) dynamic or vibration-based analysis [1]. A vibration-based mechanical impedance analysis (MIA) technique has been used as a NDE tool for material inspection and quality control. A new application of this MIA technique was proposed by Jang et al. [1] for monitoring the cure behavior of polymers and composites. Preliminary results of using this MIA technique for composite cure monitoring have been reported [18-20]. The purpose of the present paper is to report on the recent progress of our investigation concerning monitoring and control of composite fabrication using the method of MIA.

THE BASIC PRINCIPLES OF MIA

The mechanical impedance analysis technique basically involves a quantitative measurement of the effect of a vibratory force on a structure. The motion and the force signals in the time domain are transformed to the frequency domain by using a fast fourier transform (FFT) spectrum analyzer. The ratio of motion to force or its reciprocal in the frequency domain is termed "Frequency Response Function (FRF)", which is a parameter that implicitly contains the dynamic mechanical properties of a material [1].

For a rectangular beam subjected to a vibrational motion with both ends free (free-free vibration), the complex modulus and the loss tangent of the material can be estimated from the impedance spectrum [18-20]. The storage modulus (E') can be obtained from the resonant frequency squared while the loss tangent from the shape and size of a resonant peak by the techniques of half-power points or curve-fitting [1,18-20]. This sample geometry and the boundary conditions utilized are useful for determining the viscoelastic behavior of a polymer or composite since the dynamic modulus and loss tangent can be recorded and plotted as a function of the test temperature and frequency. In this respect, the MIA method will provide similar information as obtained by using other thermomechanical techniques. The same MIA method with a rectangular beam geometry can also be employed to determine the best cure cycles and the phase transformation behavior of a composite material sample (e.g. epoxy-carbon prepreg) that is capable of supporting its own weight during vibration. Other alternative boundary conditions, such as end-loaded, free-clamped, and clamped-clamped, can also be adopted whenever deemed more convenient.

A multilayer beam model in general and a two-layer beam model in particular can be utilized to determine the dynamic properties of a resin or a composite material, when the material is too soft to support its own weight or has a damping factor too high to exhibit resolvable resonance peaks. The two-layer beam model consists of a stiff structural layer to support the uncured liquid resin or soft composite prepreg. Classical dynamics theory concerning the vibration of a two-layer beam has shown that the absolute values of the storage modulus E' and the loss tangent of the material can be estimated from appropriate equations by measuring the dynamic properties of the substrate alone and those of the two-layer beam [21-35].

Mechanical impedance analysis of a moving element immersed in a liquid resin can be used to obtain the complex shear modulus of the resin. A resonant shearing motion can be established at the interface between the viscous or rubberlike resin and the moving mass as random oscillation signals are applied to this moving element. Ferry et al. [22] proposed an expression for the mechanical impedance of the moving element. By measuring the absolute values of peak displacement (X) and peak force (F) at resonance, the loss shear modulus can be estimated. The loss tangent of the material can be obtained from the damping factor, which again can be measured by the half-power bandwidth technique [1]. This sample geometry and boundary conditions can be applied, for instance, to the case of press curing of thermosetting resins and composites. An extension rod (a probe) connected to the impedance head can be inserted into a hole where the properties of the overflowed resin can be monitored. Alternatively, a small probe can be inserted directly into the composite structure for cure state sensing in an autoclave or press curing environment. The vibration response of the whole composite assembly, including either the vacuum bag or the mold, can also be followed in real time. The dynamic mechanical properties of a material in a more complex structure can be obtained from the finite element analysis or the modal analysis technique.

EXPERIMENTAL

1. MIA Technique Setup

Figure 1 shows one of the available apparatus for viscoelasticity study and cure monitoring of composites by the MIA technique. The excitation and the response signals were detected by the impedance head, which was connected to the composite structure being studied. These signals were amplified and fed into a fast fourier transform (FFT) spectrum analyzer. The frequency response function (FRF) was then obtained from the ratio of the fourier transforms of excitation and response signals. These FRF spectra were continually recorded and analyzed while the specimen was being excited by an electro-magnetic shaker, which was driven by a power amplifier fed with an input signal of random frequencies. The force level is easily controlled through the power-amplifier knob. The random excitation signal is generated by the noise source of the FFT analyzer. When the material phase transformation proceeds, both the resonant frequencies and the damping characteristics will change as a function of time. In a viscoelasticity study, these characteristics will also vary as a function of temperature. As the excitation and response signals are fed into the FFT analyzer, the FRF spectra over a desired frequency span can be continually displayed by a HP-9133 computer in real time. A HP-3565S data acquisition system is programmed to collect and store spectrum output from the analyzer for further calculations and analysis. A representative FRF spectrum, in terms of inertance values, will show several resonant peaks for a continuous, elastic material. From the FRF spectrum, the resonant frequency and damping ratio can be calculated by a SMS Modal 3.0 software package, which is compatible with the data acquisition system software. Zoom and curve fitting techniques can also be applied to increase the spectrum resolution. With the existing system, ensemble averaging and Hanning window function can be performed to remove the extraneous noise, nonlinearities, and distortion effects.

2. Materials Utilized and Sample Configuration

The dynamic mechanical properties of a 50/50 blend of polycarbonate and polyethylene (PC/PE) sample were measured over a temperature range of -40 to +160 C. Three different boundary conditions were tried on rectangular beams: loaded at the center with both ends free, loaded at one end with the other end free, and center-loaded with both ends clamped. Several small rectangular samples were prepared from strips of fiber-epoxy prepreg tape for cure monitoring study. The sample is covered with aluminum foil to prevent leakage of epoxy resin. The conditions of free-free ends and center-loading are preferred in this case. Alternatively, vibration of a constrained two-layer composite beam can be performed to follow the cure state of a soft composite or liquid resin that can not support its own weight. For instance, a layer of resin can be coated on a steel strip with known dynamic properties. To determine the sensitivity of the MIA probe in press curing, the impedance head is connected to a small extension rod that is inserted into the press mold through a hole that can act as a resin flashing channel. This extension rod serves as a probe to measure the shear force and motion between the rod and resin. The dynamic response of the epoxy resin was then followed during a ramping temperature stage and a subsequent isothermal stage.

RESULTS AND DISCUSSION

1. Dynamic Mechanical Characterization

The dynamic mechanical properties of a PC/PE blend, as measured by the MIA technique using the end-loading condition, are shown in Figs. 2a and 2b. The values of resonant frequency squared, which are proportional to the magnitudes of storage modulus, are plotted in Fig. 2a over a range of test temperatures. The corresponding loss tangent values are plotted in Fig. 2b. Several transitions can be identified from these curves. At the higher temperature end, for instance, one transition occurs at about 128 C while a second transition at 150 C, corresponding to the fusion of PE and the glass transition of PC, respectively. The storage modulus data of the same MIA sample but loaded under a free-free boundary condition show similar transition or relaxation behavior. Similar results were also obtained by using a dynamic mechanical thermal analyzer (PL-DMTA) under a dual cantilever beam clamped-clamped condition. These results provide partial evidence to verify the sensitivity of the MIA method when applied for the determination of the dynamic mechanical properties of materials.

2. FRF Spectra During Composite Curing

Representative inertance spectra continually measured over a broad frequency range during the curing processes of the E-glass/epoxy composite are shown in Fig. 3. These inertance spectra show the temperature effect on the frequency response functions before the desired isothermal temperature (140°C) was achieved. At room temperature prior to the heating process, the uncured resin has a moderate viscosity. Therefore, a resonant peak can be seen in the FRF spectrum (first diagram in Fig. 3). As the test temperature increases and when appreciable chemical reaction has not yet occurred, the viscosity of the uncured resin would decrease, as would the shear storage modulus. This is reflected by a decrease in the natural frequency and an increase in the damping coefficient (second diagram in Fig. 3). The peak becomes less sharp, which is characteristic of a less viscous liquid. Significant changes in the spectrum for the subsequent isothermal curing stage can be clearly identified as polymerization and crosslinking start and continue. The peaks become sharper and occur at higher frequencies. A sharp resonant peak can be observed in the 6th diagram of Fig. 3, when the laminate is completely cured and the temperature cooled down. The results of applying the MIA technique in cure monitoring of the E-glass/epoxy laminates are summarized in Fig. 4. In-situ adjustment of curing parameters can be carried out by comparing the dynamic mechanical properties detected from the MIA sensor with the theoretical values predicted by the approach of cure model analysis.

3. Monitoring of Press Curing Using The Methods of MIA

Press molding represents one of the more commonly used methods in composite fabrication. Inertance spectra for the curing of the Araldite-507/HY-956 matrix have been obtained by applying the shear resonance with a moving rod [24]. Although the vibration created by the operation of a hydraulic press sometimes can complicate the resolution of the FRF spectra, relative changes in the spectrum can still be observed. These changes are due to the reaction and solidification of the thermosetting matrix. The noise effect from the press operation could cause severe scattering of the response signals only at the final stage of composite fabrication when the curing of resin is nearly completed. Autoclave curing is a more commonly used process for the industry-scale composite fabrication. To simulate this process, a graphite fabric reinforced Araldite-507/HY-956/ CTBN thermoset system has been cured on a rectangular supporting plate with a peel layer covered on the surface of the material. The results will be presented elsewhere [24].

4. Alternative Ways of Presenting The FRF Data

Several different ways can be used to present a FRF spectrum. An inertance spectrum can be broken down into real part, imaginary part, and phase angle, each as a function of frequency. The imaginary part can also be plotted against the real part. The dynamic properties can then be deduced from these diagrams [24].

CONCLUSIONS

The MIA technique has been successfully employed to directly measure the storage and loss components of the dynamic response of a viscous liquid or a viscoelastic solid. This technique is highly sensitive for characterizing both the molecular relaxation behavior of a polymer solid and the cure states of a thermosetting resin and composite. This technique can be used for cure cycle design and optimization for any developmental or new thermosetting resin. The same technique can also be used as an alternative to other commercially available instruments for quality assurance of the in-coming materials (resins and prepregs). When applied to the fabrication of composite structures, this technique can serve as a real-time in-process dynamic mechanical sensor for the cure state characterization. This technique is expected to contribute to ensuring the manufacturing reliability and reproducibility of composites. This method is not subject to any constraints on the test specimen geometry and dimensions. Although not addressed in the present investigation, the same technique can also be applied for assessing the damage or integrity of the composite structure during the service life [30]. It can therefore be concluded that the MIA technique is a highly versatile tool for the determination of the dynamic mechanical properties of a polymer or composite as a function of the material physical and chemical state.

ACKNOWLEDGMENTS

We are grateful for the financial support provided by the Manufacturing Engineering Program of NSF.

REFERENCES

1. B. Z. Jang and G. H. Zhu, J. Appl. Polymer Sci. 31, 1986, p.2627.
2. W. M. Sanford and R. L. McCullough, Proc. of Am. Soc. for Composite Conference, 1987, pp.21-30.
3. R. D. Hoffman, J. J. Godfrey, D. E. Kranbuehl, L. Weller and M.

Hoff, 41st Ann. Conf., RP/Compo. Inst., Soc. of Plastics Ind., Inc., Jan. 27-31, 1981.
4. S. D. Senturia, 28th National SAMPE Symp., April 12, 1983, p.851.
5. R. T. Harrold and Z. N. Sanjana, Polymer Engr. and Sci., 1986, 26, no. 5, pp.367-372.
6. Eamor M. Woo and James C. Seferis, ANTEC '86, SPE, p.375.
7. Y. L. Hinton, R. J. Shuford and W. W. Houghton, Proc. of the Crit. Review; Tech. for the Charac. of Compo. Mater., May 82, p.25.
8. J. R. Mitchell, Physical Acoustic Corp. 1987, TR-103-60D-7/87.
9. Harrold and Z. N. Sanjana, 31st Inter. SAMPE Symp., 4/86, p.1713.
10. E. J. Juegel and H. T. Hahn, Advances in Modeling of Composites Precesses, pp.129-136.
11. W. P. Winfee and F. R. Parker, Review Prog.in Quanti. NDE 85,p.1.
12. S. I. Rokhlin, J. Acoust. Soc. Am. 76(9), June 1986, pp.1786-1793.
13. R. L. Levy and D. P. Ames, Organic Coatings and Appl. Polymer Sci. Preprints, ACS, 48(1983) pp. 116-120.
14. R. O. Loutfy, Macromolecules, 14(1981) pp. 270-275.
15. A. Bur, F. W. Wang, and R. Lowry, ANTEC '88, SPE., p.1107.
16. A. Stroeks, M. Shmorhun, A. M. Jamieson, and R. Simha, Polymer, 29 (1988) pp. 467-470.
17. F. W. Wang, R. E. Lowry, and B. M. Faconi, Polymer 27 (1986) 1529.
18. B. Z. Jang, M. D. Shelby, H. B. Hsieh and T. L. Lin, 19th Tech. Conf., SAMPE, Crystal City, Virginia, Oct. 13-15, 1987.
19. B. Z. Jang, H. B. Hsieh and M. D. Shelby, 14th Conf. on Production Research and Technology, SME-NSF, Ann Arbor, Michigan, 10/87.
20. B. Z. Jang, H. B. Hsieh and M. D. Shelby, the 46th Annual Tech. Conf of SPE, Atlanta, Georgia, April 18-21, 1988.
21. T. J. Dudek, Journal of Composite Material, 4, 1970, pp.74-89.
22. J. D. Ferry, Viscoelastic Properties of Polymers, John Wiley and Sons, Inc., 1970, NY, pp.125-157.
23. A. F. Lewis, M. J. Doyle and J. K. Gillham, Polymer Engineering and Science, vol. 19, Aug. 1979 no. 10, pp.683-686.
24. B. Z. Jang et al., submitted to Composite Sci. and Technology.

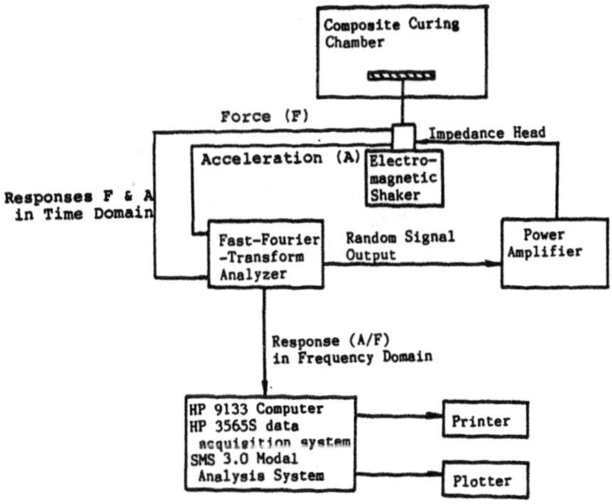

Fig. 1: The Block Diagram of a MIA Cure Monitoring System.

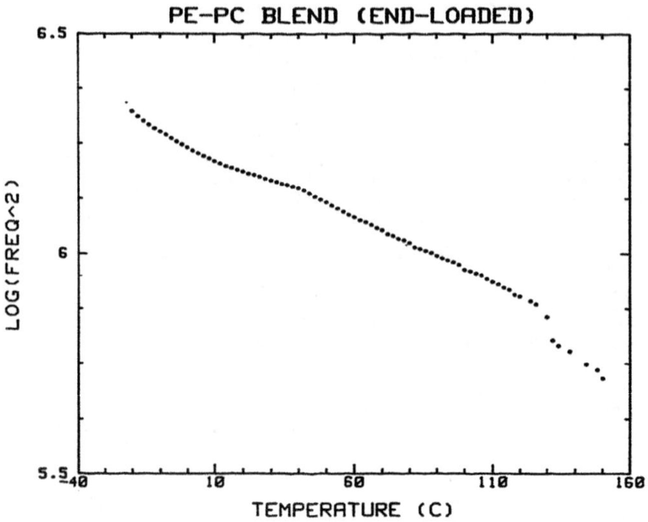

Fig.2a. Resonant frequency squared, proportional to the storage modulus, plotted versus test temperature.

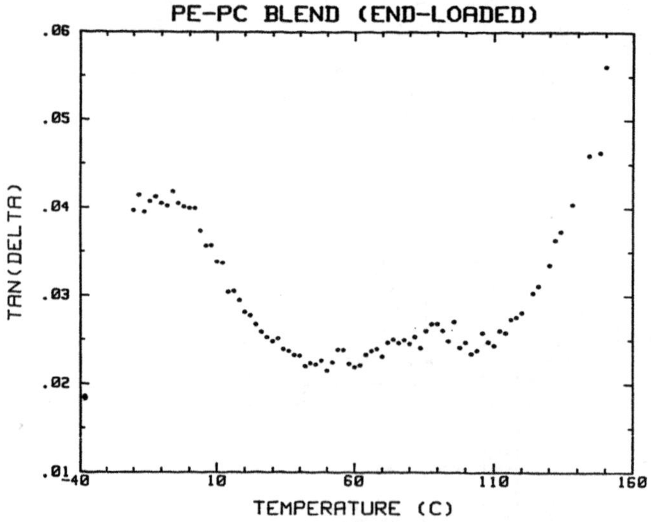

Fig.2b. Corresponding loss tangent versus temperature of a PC/PE blend.

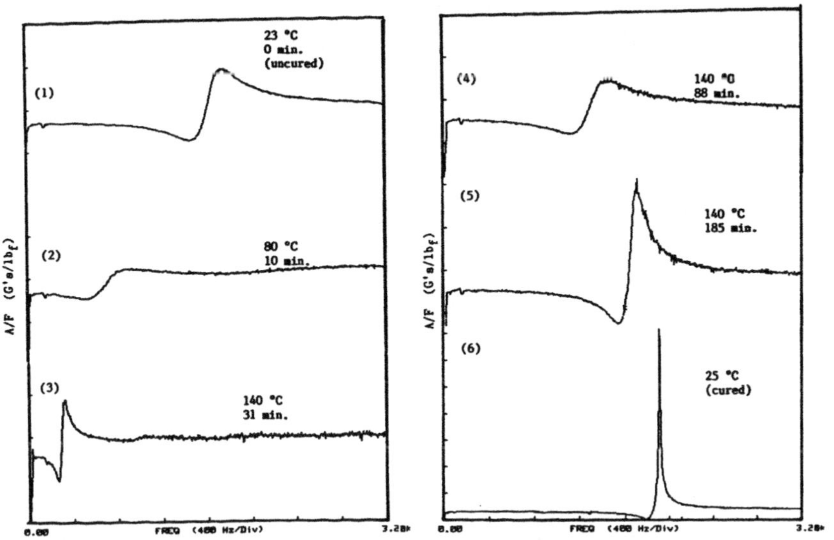

Fig. 3: Inertance Spectra Obtained at Various Stages of E-glass/Epoxy Laminate Curing.

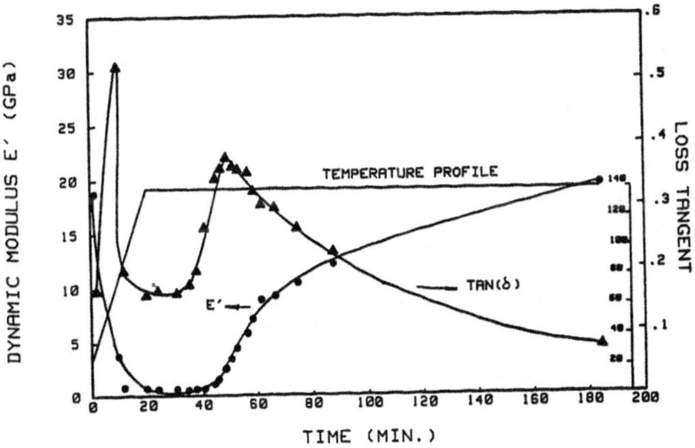

Fig. 4: Dynamic Mechanical Properties versus Time during the Curing of an E-glass/Epoxy Laminate.

Characterization of the Elastic Behaviour of Nanocrystalline Materials by Scanning Acoustic Microscopy

A. Morsch, D. Korn*, H. Gleiter*, M. Hoppe**, and W. Arnold

Fraunhofer-Institut für zerstörungsfreie Prüfverfahren,
Universität, Gebäude 37, D-6600 Saarbrücken 11, FRG

* Abteilung für Werkstoffwissenschaften und Technologie,
Universität, Gebäude 2, D-6600 Saarbrücken 11, FRG

** Wild-Leitz GmbH, D-6330 Wetzlar, FRG

Abstract

The V(z)-measurement technique in Scanning Acoustic Microscopy (SAM) was used for an attempt to characterize the elastic behavior of nanocrystalline materials that are polycrystals whose grain sizes comprise a range from 1 to 100 nm. From the periodicity of the V(z)-curves the velocities of surface waves can be deduced. We present the results of such measurements on nanocrystalline Pd-samples which were performed with the Ernst-Leitz SAM at 1.0 and 1.6 GHz. We interprete the observed periodicity as caused by a skimming compressional wave. The deduced phase velocity is about 30% lower than the velocity of compressional bulk waves as determined by time of flight measurements.

Zusammenfassung

Die V(z)-Meßtechnik mit dem akustischen Reflexionsrastermikroskop (SAM) wurde benützt, um das elastische Verhalten nanokristalliner Materialien - das sind Polykristalle, deren Korngröße von 1 bis 100 nm reicht- zu charakterisieren. Aus der Periodizität der V(z)-Kurven kann die Geschwindigkeit von Oberflächenwellen abgeleitet werden. Wir stellen die Resultate aus solchen Messungen an nanokristallinen Pd-Proben, die mit einem SAM der Fa. Leitz bei den Frequenzen 1.0 und 1.6 GHz durchgeführt wurden, dar. Wir interpretieren die beobachtete Periodizität als durch oberflächennahe Kriechwellen verursacht. Die daraus abgeleitete Phasengeschwindigkeit ist etwa 30% niedriger als die Geschwindigkeit der longitudinalen Volumenwelle, wie man sie mit Laufzeitmessungen bestimmt.

1. Nanocrystalline Materials

Nanocrystalline materials are a new kind of polycrystalline solids whose grain sizes comprise a range from 1 to 100 nm. The volume portions of the grain boundaries in these materials are 20 to 50% of the total volume. X-ray scattering experiments revealed that the structure of the grain boundaries is similar to that of a gas /1/. In order to learn more about such an uncommon structure its elastic behavior is investigated. As the first samples were very thin (< 200 μm) and brittle, common methods for the elastic characterization such as time of flight measurements of acoustic pulses turned out to be unpracticable. Therefore, we used a Scanning Acoustic Microscope (SAM) for an attempt to measure the elastic properties of the nanocrystalline samples.

2. Scanning Acoustic Microscopy

In Scanning Acoustic Microscopy a focused ultrasonic beam produced by an acoustic lens is scanned over the sample. The sound field reflected by the sample is modulated in phase and amplitude, according to the topography and elastic properties of the surface. The lens collects the returning waves and the transducer converts them into electric signals whose intensity is displayed synchronously to the scanning of the lens on a monitor /2/.

Additionally, the acoustic lens of a SAM produces surface acoustic waves on the sample, provided the opening half-angle of the lens is larger than the corresponding critical angle. Since these waves radiate energy back to the lens, they strongly influence the contrast mechanism of the microscope /2,3/.

3. V(z)-Measurements

In a nonscanning mode the SAM can be used for material characterization. The output voltage V(z) of the lens as a function of the distance z of the lens to the sample displays periodicities which depend on the elastic properties of the sample. Due to this the V(z)-curves are also called Acoustic Material Signatures (AMS). The periodicities are produced by the interference of the leaky wave modes excited by the rays in the lens under oblique incidence and the central rays. The phase velocities of these modes travelling parallel to the sample surface can be related to the periods Δz /3,4/ by:

$$c_{sw} = \frac{c_{sw}}{\sin\Theta_c}, \text{ where } \Theta_c = \arccos\left(1 - \frac{c}{2\,\nu\,\Delta z}\right) \qquad (1)$$

Here, c is the velocity in the coupling medium, and ν is the ultrasonic frequency.

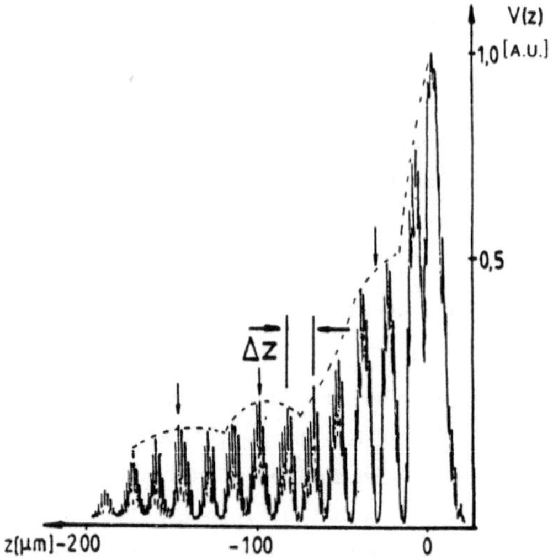

Figure 1: V(z)-curve of an Al-sample. The well-defined periodicity results from a Rayleigh-wave. The compressional surface skimming wave only causes a faint modulation (dotted line). The frequency employed was 400 MHz.

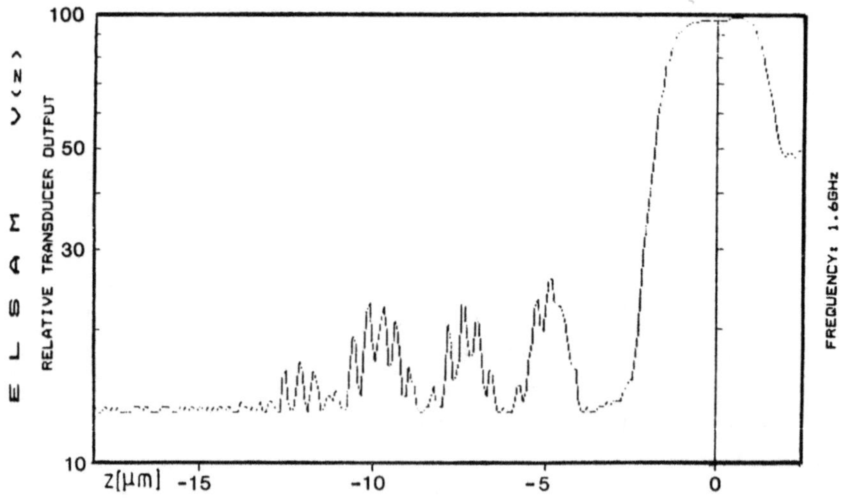

Figure 2: V(z)-curve of an Arsenic Trisulfide sample. Since the Rayleigh-wave velocity in this material is smaller than the longitudinal wave velocity in water, Rayleigh-waves are not excited and well-defined maxima result from compressional surface skimming waves.

The opening half-angle of the lens has to be larger than the critical angle Θ_c for the exitation of the corresponding leaky wave mode. On an isotropic surface that can be considered as semiinfinite, two leaky wave modes can exist: leaky Rayleigh-waves and leaky skimming compressional waves /5/.

Fig. 1 shows the V(z)-curve obtained on an Al-sample at an ultrasonic frequency of 400 MHz. The mean distance of the peaks is 15.0 µm yielding a surface wave velocity of 3.1 km/s in good agreement with the Rayleigh-wave velocity of 2.95 km/s given in literature. The skimming compressional wave only produces a faint modulation of the main peaks (dotted line). Fig. 2 shows the V(z)-curve measured on a As_2S_3-sample at an ultrasonic frequency of 1.6 GHz. Since the Rayleigh-wave velocity of this material is lower than the velocity of sound in water, the Rayleigh-mode can not be excited by the lens. In this case the well-expressed periodicity of the curve yields a surface wave velocity of 2.6 km/s corresponding to the velocity of the compressional volume wave in this material deduced by a time of flight measurement.

These examples illustrate the fact that the periodicity produced by the Rayleigh-wave mode is much more pronounced than that produced by the compressional mode. If, however, the Rayleigh-mode is supressed the periodicity produced by the compressional wave becomes clearly visible and can be used to deduce its velocity.

4. V(z)-measurements on nanocrystalline Pd-samples

V(z)-measurements were performed on nanocrystalline Pd-samples at the ultrasonic frequencies 1.0 and 1.6 GHz. The opening half-angle of the lens was 50°. High-resolution images of the sample surfaces show that their surface is rather inhomogeneous. In order to realize the condition for an undisturbed propagation of the surface waves and thus for proper velocity measurements, we had to choose places for the measurements where the surface was as homogeneous as possible. Figures 3a and 3b show two of the V(z)-curves obtained at 1.6 GHz. Fig. 3a demonstrates that well-defined maxima and minima can be obtained in the V(z)-curve. The second maxima of the V(z)-curve in Fig. 3b, however, may be distorted due to an interaction of the surface waves with the inhomgeneties of the surface.

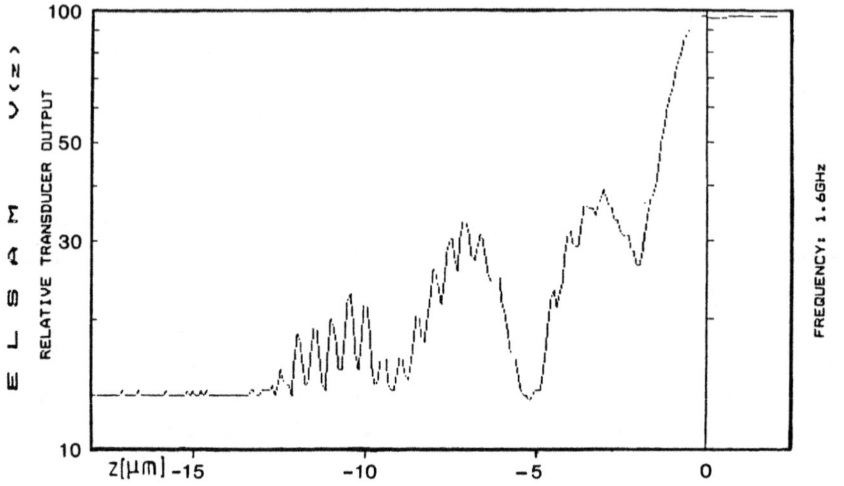

Figure 3a: V(z)-curve of a nanocrystalline Pd-sample showing a well-defined periodicity.

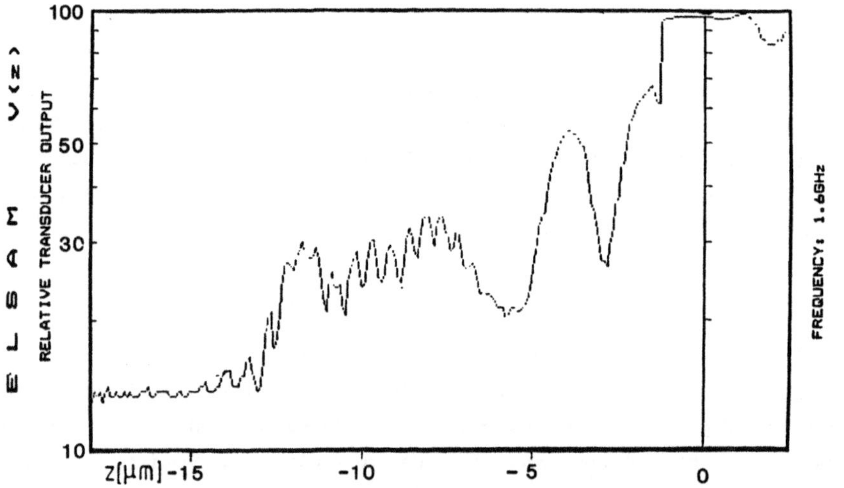

Figure 3b: V(z)-curve of a nanocrystalline Pd-sample showing distorted maxima due to the interaction of the surface waves with the inhomogeneties of the sample surface.

5. Results

Evaluating the mean distance of the maxima of V(z) curves measured at different sites on the same sample proved that this period is reproducible within a range not exceeding 15 % and that the deduced velocities are within a range of about 8 %. Therefore, we conclude that the interpretation of the periodicity as caused by one and the same surface wave mode is still valid despite the inhomogeneous surface. The mean surface wave velocities obtained at different samples with different grain sizes (8-15 nm) varied from 2.7 to 3.4 km/s. Another result is that the velocities deduced from measurements at an ultrasonic frequency of 1.0 GHz are systematically lower than those obtained at 1.6 GHz.

6. Comparison with results from time of flight measurements and conclusions

From time of flight measurements of broadband longitudinal and transversal ultrasonic pulses on two nanocrystalline samples with parallel surfaces the longitudinal, the transversal and the Rayleigh-wave velocities have been deduced. The measurements yielded v_L = 4 km/s and v_T = 1.8 km/s and hence a Poisson ratio of 0.37. From this values the Rayleigh velocity could be calculated to be 1.7 km/s. It turned out that the critical angle for Rayleigh-wave excitation is greater than 50°. This means that with the lens used for our V(z)-measurements no Rayleigh-waves are excited, and hence we interprete the deduced velocities as the velocities of surface skimming longitudinal waves. Comparing the values for the longitudinal volume wave (from the time of flight measurements) with that of the corresponding surface wave (from V(z)), we note that the velocity of the latter is about 30 % lower. Whether this deviation is a result of the inhomogeneous surface or whether it is caused by the structure of the nanocrystals itself is a matter still under investigation.

7. Acknowledgement

Two of us (A. M. and W. A.) should like to thank E. Matthaei and H. Vetters from the "Institut für Werkstofftechnik" in Bremen for the use of their SAM-instrument with which the V(z)-curve in Fig. 1 was obtained.

8. References

1. Birringer, R.; Gleiter, H.; Marquardt, P.; Klein H.P.: Nanocristalline Materials- An Approach to a Novel Solid Structure with Gas-like Disorder? Phys. Lett., 102a, 1984, 365-369.

2. see for example Briggs, A.G.D.: An Introduction to Scanning Acoustic Microscopy. Oxford: University Press, 1985

3. Lemons, R.A.; Quate, C.F.: Acoustic Microscopy. Phys. Acoust., ed. by Mason W. P. and Thurston R.N., 14, 1979, 1-92, Academic, New York.

4. Weglein, R.D.: A Model for Predicting Acoustic Material Signatures. Appl. Phys. Lett., 34, 1979, 179-181.

5. Bertoni, H.L.: Ray-Optical Model Evaluation of V(z) in the Reflection Acoustic Microscope. IEEE Trans., SU-31, 1984, 105-116.

6. Kushibiki, J.; Chubachi, N.: Material Characterization by Line-Focus-Beam Acoustic Microscope. IEEE Trans., SU-32, 1985, 189-212.

Variation of the Ultrasonic Propagation Velocity Due to Creep Strain in Austenitic Steel

F. LAKESTANI* and P. RIMOLDI**
*NDE Laboratories, CEC, JRC Ispra, Italy.
**AGIP Centro di Medicina, Italy.

ABSTRACT

- Ultrasonic velocity measurements were performed on austenitic stainless steel test-pieces, with two different compositions, after creep straining up to rupture.

- The applied methodology considered the variation of the ultrasonic velocity, for longitudinal and shear waves, in two directions of propagation : parallel and orthogonal to the axis of the stress; the measurements were performed on samples taken from the test-pieces at different distances from the ruptured section.

The results showed that :

- Generally speaking all the velocities decrease by some percents near the ruptured section.

- The changes of the velocities for waves with particles vibration perpendicular to the stress axis are 2-3 times smaller than for the waves with particles vibration parallel to the stress axis. This might be due to the presence of grain boundary microcracks perpendicular to the stress axis.

For shear waves which propagate orthogonally to the axis of the stress, the creep creates an important variation in anisotropy (defined as the variation of the velocity between shear waves polarized parallel and orthogonally to the planes of the microcracks), when compared to the initial anisotropy in the corresponding virgin material.

This methodology might be applied to the in service inspection for life prediction of components. It has the advantages to be rather independent of the transducers and of the instrumentation characteristics and is not needing supplementary references, due to the use of the relative measurement between the two polarizations of the shear wave.

Introduction

- The life time prediction in structural materials submitted to creep damage may be improved by using ultrasonic measurement.

- Actually when we consider non-destructive techniques, only the metallographic replica taken in a few places give an information on the superficial state of damage. Besides this superficial aspect, the technique can only be applied reasonably to limited areas of the component [1].

- The microporosities created in materials by creep straining, influence the density and elastic constants; their slight variations may be detected by sensitive methods of ultrasonic velocities measurements.

- Several authors [2], [3], [4] describe some theoretical and experimental aspects of this interaction between microporosities and ultrasonic propagation velocity.

[4] shows that the effect of creep on the velocity of the compressional waves depend on their direction of propagation. In our work we prove the general aspect of this property : in the case of uniaxial creep straining, the ultrasonic velocity variation depends on the direction of the particles vibration. So, independently of the wave polarisation (longitudinal or shear), the effect on the velocities is stronger when the particles vibration is parallel to the stress axis.

Experimental work

. The tests were performed on 2 types of austenitic stainless steel : AMCR 33 and AISI 316 L. The chemical composition of these material is given in Table 1.

. The samples were cut in test-pieces submitted to creep straining up to rupture. The condition of creep straining are given in Table 2. The dimensions of the samples and their location in the test pieces are shown on Figures 1 and 2. The flat surfaces of the samples were rectified.
Precise measurements of the thickness with tolerance of +/- 0,002 mm were performed where the ultrasonic tests had to be realized.

. The time of flight measurements are realised in echo mode configuration using two 0° contact transducers. The first one is a compressional wave transducer with a crystal diameter of 0.25 inch. and a crystal frequency of 10 MHz; the second one is a shear wave transducer with a crystal diameter of 0.25 inch. and a crystal frequency of 5 MHz. For each test piece, the measurements of the velocities are referred to those of the sample cut in the head of the test-piece in order to eliminate the effect of the heat treatment during creep.

The relative measurements are performed in the following way: each echo in the samples is compared with the equivalent one in the reference sample; by this way, as all the samples have the same geometry, the effect of the beam divergence, the reflection on the side walls and the initial attenuation of ultrasound in the material is eliminated. The two signals have nearly the same shape and the delay between them is only due to velocity and thickness variation. This relative measurement is feasible using a digital oscilloscope with a time base monitored by quartz.
However, some change remain in the shape of the signals; this is due to different factors :

- modification of the ultrasonic attenuation by the creep;
- variation of the coupling conditions and of the relative position of the transducers on the samples;
- superposition of parasitic signals due to spurious mode generated by the transducer and to mode conversion at the boundary of the sample;
- grain and microporosities noises.

We tried to minimize the effect of some of these factors using

- a system of compressed air for having constant coupling conditions,
- a good couplant (honey) and micrometric monitoring for the transducer positioning;
- an adequate choice on the shape of the excitation signal in order to increase the signal to noise ratio.

Because of the overall effect of these phenomena and the precision on the thickness measurement, the uncertainty on the relative value of the velocities is 0.1%.

Results

General behaviour :

All the ultrasonic velocities decrease by some percents near the ruptured section but for waves with the particles vibration parallel to the stress axis, this decrease is 2-3 times higher than for waves with the particles vibration orthogonal to the stress axis (Fig. 2).

Typical results

For the considered waves, the direction of propagation and the direction of the particles vibration are shown on Figure 3. The values of the velocity variation are average values obtained on 4 test-pieces for AMCR 33 and on 5 test-pieces for AISI 316L, the velocity variations are between the rupture section and the head of the test pieces.

Conclusions

1. Generally speaking all the velocities decrease by some percents near the ruptured section.

2. The uniaxial creep straining produces grain boundary microcracks having their larger dimensions preferentially perpendicular to the stress axis. This may be the reason why the effect on the velocities is 2-3 times higher when the direction of particles vibration is parallel to the stress axis.

3. The uniaxial creep modifies the material anisotropy. For the shear waves which propagate orthogonally to the stress axis, this anisotropy may be defined as the difference between the velocity Vp of the wave with polarisation parallel to the stress axis and the velocity Vn of the wave with polarisation orthogonal to the stress axis. Typical variation of this anisotropy is shown on Figure 4. This measurement is merely realized turning the 0° shear wave transducer of 90° on itself. With this method, the measurement on a reference specimen is not more necessary and the thickness measurement is eliminated too.

4. The experimental uncertainty is about 0.1%.

5. This procedure might be applied to determine the actual creep damage. However, for its validation, supplementary experiments are necessary.

References

[1] Willems, H.; Bendick, W.; Weber, H., Proc. of the 2nd Internat. Symp. on Nondestructive Characterization of Materials, Plenum Publishing Corporation, 1987, 451-460.

[2] Sayers, C.M.; Smith, R.L.; Ultrasonics, Sept. 1982, 201-205.

[3] Ledbetter, H.M.; Fields, R.J.; Datta, S.K., Acta Metallurgica; Vol. 35, No. 9, 1987, pp. 2393-2398.

[4] Willems, H., Proc. of the International Conference "Life Assessment and Extension", The Hague, Netherlands, June 1988, pp. 86-91.

AMCR 33 :

Mn 17.3%	Cr 10.1%	Si 0.55%	Ni <0.1%	V 0.015%
Mo <0.065,	Cu <0.06,	Al <0.005,	C 0.1,	N 0.2%
P 0.02 ,	S 0.008,	Pb < 1ppm	B 25ppm	

AISI 316L :

Ni 12.33%	Cr 17.44%	Mn 1.81%	C 0.024%	Cu 0.2%
Mo 2.3%	Si 0.46%	Co 0.17%	S 0.001%	
Ta 0.01%	N 0.06%	B 8ppm	P 0.026%	

Table 1 : Chemical composition in weight -%

AMCR 33 4 test-pieces

T[°C]	500	650	650	700
Load [MPa]	330	100	150	70
Life time duration [h]	>10000	2850	3350	1400

AISI 316L 5 test-pieces

T[°C]	550	550	550	550	550
Load[MPa]	300	320	340	360	380
Life time duration [h]	2760	1470	1120	700	620

Table 2 : Creep straining conditions

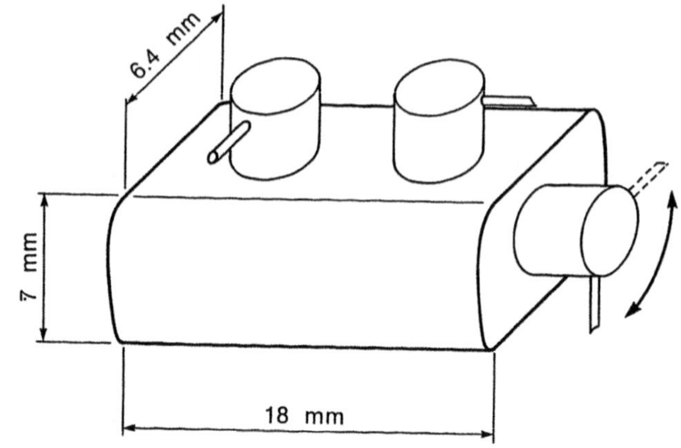

Fig. 1. Geometry of the samples and position of the transducers.

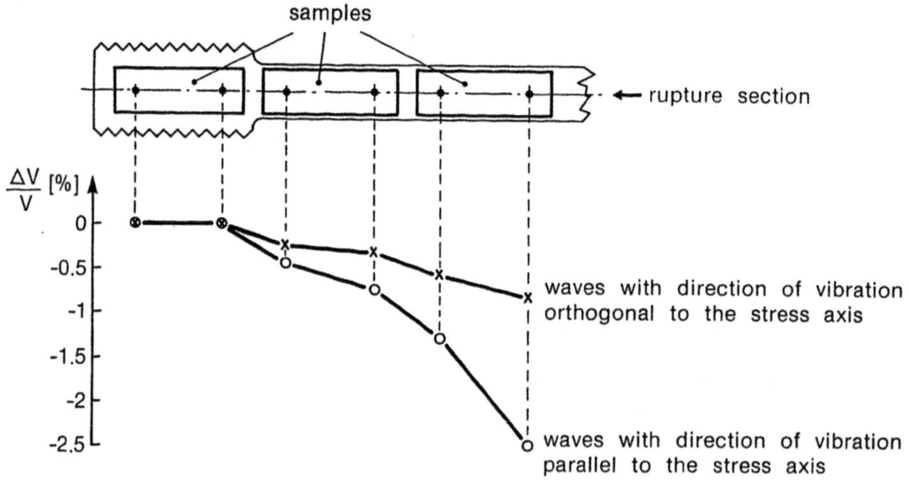

Fig. 2. General behaviour of the velocities.

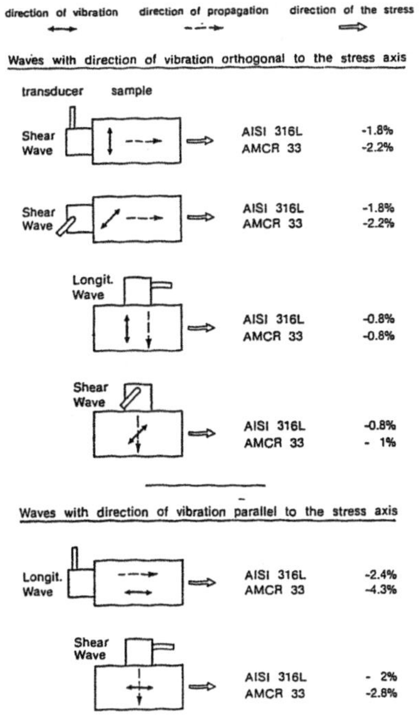

Fig. 3. Average values of the velocity variation between ruptured section and the head of the test pieces.

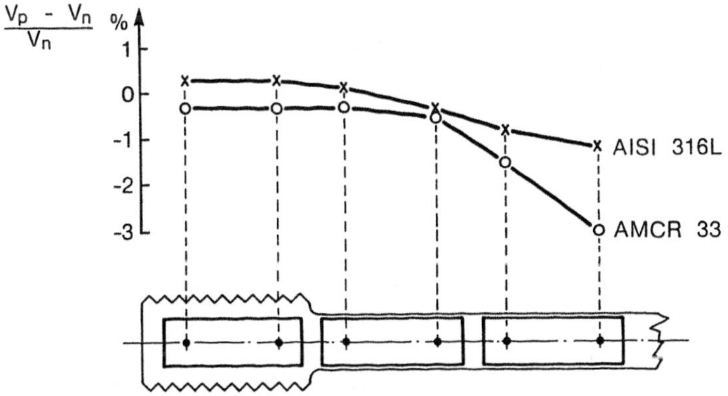

Fig. 4. Typical variation of the anisotropy; shear waves propagating orthogonally to the stress axis.

Application of Ultrasonic Methods for the Characterization of ZrO_2 Pellets

S.Ekinci*, A.N.Bilge**

* Çekmece Nuclear Research and Training Center, P.K.1, Havaalani-Istanbul, TURKEY
** Istanbul Technical University, Institute for Nuclear Energy, Maslak-Istanbul, TURKEY

Summary

This work describes the characterization of ZrO_2 pellets simulated for UO_2 by using ultrasonic methods. By sound velocity and sound attenuation measurements, density, porosity concentration, homogenity and elastic constants (E- and G-modulus and Poisson ratio) have been determined.

The sound velocities and the sound attenuation coefficients have been measured by a conventional and a computer aided system and correlated with the total porosity concentrations of the pellets. The results of both systems have been compared.

Zusammenfassung

Diese Arbeit beschreibt die Charakterisierung von für UO_2 simulierten ZrO_2-Pellets durch Ultraschallverfahren. Schallgeschwindigkeits- und Schallschwächungsmessungen dienten zur Bestimmung von Dichte, Porositätskonzentration, Verteilung der Porosität und elastischen Konstanten (E- und G-Modul und Poissonsche Zahl).

Die Schallgeschwindigkeiten und die Schallschwächungskoeffizienten wurden mit einem herkömmlichen und einem rechnergestützten System gemessen und mit den gesamten Porositätskonzentrationen der Pellets korreliert. Die Ergebnisse der beiden Systeme wurden verglichen.

Introduction

Density, porosity, grain structure and elastic constants of UO_2 ceramic pellets are important parameters influencing the behaviour of nuclear fuels with respect to in-service performance /1/. Ultrasonic methods can be applied for the determination of these parameters.

In an isotrope and homogeneous medium, the sound velocity and the sound attenuation are influenced by the porosity /2/. They change with the frequency (dispersion) /3/ and the porosity concentration /4/.

The sound attenuation is mainly caused by the scattering of the ultrasonic waves on the grain boundaries and the pores.

In the Rayleigh-region, the scattering is proportional to the fourth power of the frequency /5/.

In the works carried out at Çekmece Nuclear Research and Training Center (Istanbul), the longitudinal sound velocity (V_L) has been measured on UO_2 pellets by using a sound velocity measuring instrument (CL 204 and CLF 5, Krautkraemer) and correlated with the density as the percent of the theoretical density (% T.D.), Fig. 1.

The further works have been carried out at Fraunhofer-Institute, (IzfP) by simulating UO_2 pellets with ZrO_2 pellets. As standard, ZrO_2 pellets have been prepared from ZrO_2 powder stabilized with 6% ($^w/_w$) CaO by using different parameters. The real densities and the total porosity concentrations of the pellets have been determined according to DIN 51056 norm /6/. During the correlations, only the total porosity concentrations have been considered, because the density as the percent of the theoretical density (% T.D.) and the total porosity concentration (P_T %) make up to 100%.

The sound velocity and attenuation measurements have been carried out by using a conventional and a computer aided system. The results of both systems have been compared.

Sound Velocity and Sound Attenuation Measurements

Measurements by a Conventional System

For the measuring of the sound velocity in the porous ZrO_2 ceramic, a conventional system consisting of an ultrasonic instrument (USIP 11, Krautkraemer) and a double-channel osciloscope (Rohde und Schwarz) has been equipped. The time of flight has been measured by using pulse-echo overlapping method and direct contact technique and the velocity determined. Fig. 2 shows the relation between the total porosity concentration and the longitudinal sound velocity measured at 5 MHz (KB-A, Krautkraemer). Fig. 3 shows the relation between the total porosity concentration and the transversal sound velocity (V_T) measured at 5 MHz (V 115, Panametrics).

A linear correlation between the sound velocity and the total porosity concentration has been made. According to /4/, this relation is not exactly linear, but this nonlinearity is considerable only in an interval larger than 10 %. Since the porosity interval of the pellets sintered at 1800°C was 3%, a linearity can be assumed within this range.

The changing of the sound velocity by means of the porosity concentration has led to investigate the porosity distribution of the pellets by scanning their surface with a suitable miniature probe (V_{116}, Panametrics). Fig. 4 shows the changing of the longitudinal sound velocity along the measured axis on the surface of the pellets sintered at 1400°C.

E-module (Young's module), G-module (Shear-module) and Poisson ratio have been determined by using the longitudinal and transversal velocities measured at 5 MHz, the densities and the related formulas given in /7/.

To investigate the dispersion, the phase velocity of the longitudinal waves has been measured by changing the frequency at a system consisting of a pulse generator (Hewlett Packard), a double-channel osciloscope (Rohde und Schwarz) and a probe (V 111, Panametrics). The group velocity and the correlated phase velocity have been obtained according to /8/, Fig. 5.

The attenuation of the longitudinal waves has been measured with a conventional ultrasonic instrument (USIP 11) by pulse-echo method and direct contact technique. At 3,5 MHz (V 111, Panametrics) the correlation between the sound attenuation coefficient (α_L) and the total porosity concentration was not good, since the effects related to the geometry were dominant at this frequency. At 7,5 MHz (KB-A, Krautkraemer) a better correlation has been obtained, Fig. 6. The middle frequencies of the probes used have been determined by the frequency spectrum.

Measurements by a Computer Aided System

As comparison to the conventional system the sound velocities and the sound attenuation coefficients have been measured by a computer aided system developed by H.Willems *, Fig. 7. In this system, the sound velocity and the sound attenuation coefficient could be determined by using a special programme, Fig. 8. The measurements have been carried out at 7,5 MHz. Fig. 9 and 10 show the measuring results.

Conclusion

In this work, a simulation of ZrO_2 ceramic has been accomplished as a base of UO_2 ceramic. The results of the computer aided system have proved relatively better correlations than that of the conventional system. The total porosity concentration or the density could be determined with an accuracy of 1%.

These methods can also be applied to UO_2 pellets by selecting suitable measuring parameters.

* H.Willems, IzfP-Saarbrücken, FRG.

References

1. W.Dörr et al., "Bestimmung der Dichte, offenen Porosität, Porengrößenverteilung und spezifischen Oberfläche von UO_2-Tabletten", Journal of Nucl. Mat. 81 (1979), 135-141.

2. R.B.Thompson et al., "Relative effects of porosity and grain size on ultrasonic wave propagation in iron compacts", Review of Progress in Quantitative NDE, 5B, 1643-1653 Plenum Press, NY, 1986.

3. C.M.Sayers, "Ultrasonic velocity dispersion in porous materials", J.Phys. D: Appl. Phys. 14 (1981), 413-420.

4. C.M.Sayers and R.L.Smith, "The propagation of ultrasound in porous media", Ultrasonics, 20 (1982), 201-205.

5. E.P.Papadakis, "Revised grain scattering formulas and tables", J.of Acous. Soc. of Am. 37 (1965) No. 4, 703-710.

6. "Bestimmung der Wasseraufnahme und der offenen Porosität", Deutsche Norm DIN 51056, August 1985.

7. J.Krautkraemer und H.Krautkraemer, "Werkstoffprüfung mit Ultraschall", 5.Auflage, Springer Verlag, Berlin 1986.

8. S.Hirsekorn, "Streuung von ebenen Ultraschallwellen an kugelförmigen isotropen Einschlüssen in einem isotropen Medium unter Berücksichtigung der Mehrfachstreuung", IzfP-Bericht Nr. 790218-TW, 1979.

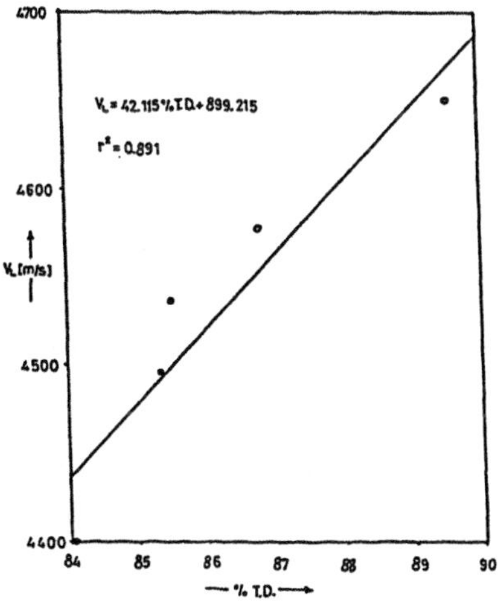

Fig. 1 Relation between V_L and % T.D. of UO_2 pellets

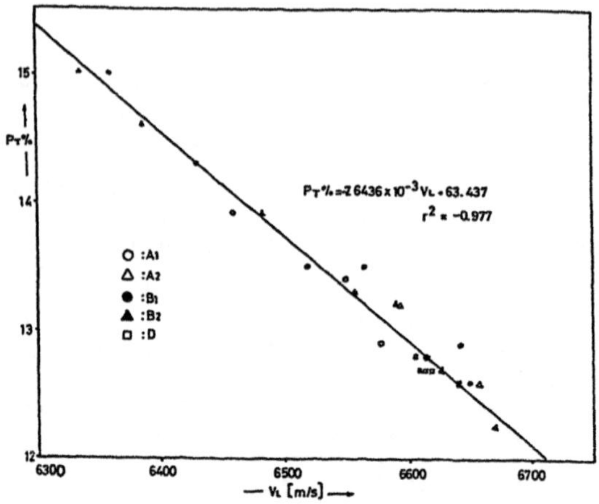

Fig. 2 Relation between V_L and P_T % at 5 MHz

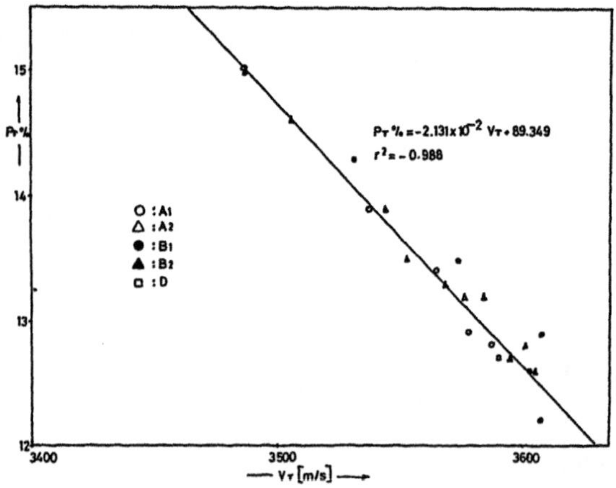

Fig. 3 Relation between V_T and P_T % at 5 MHz

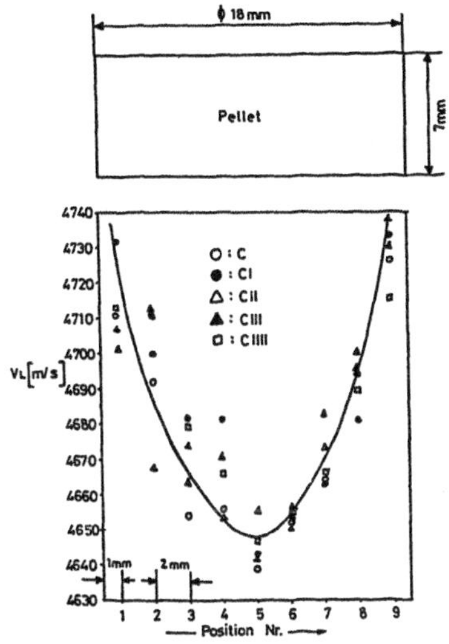

Fig. 4 Changing of the longitudinal sound velocity on the pellets sintered at 1400°C

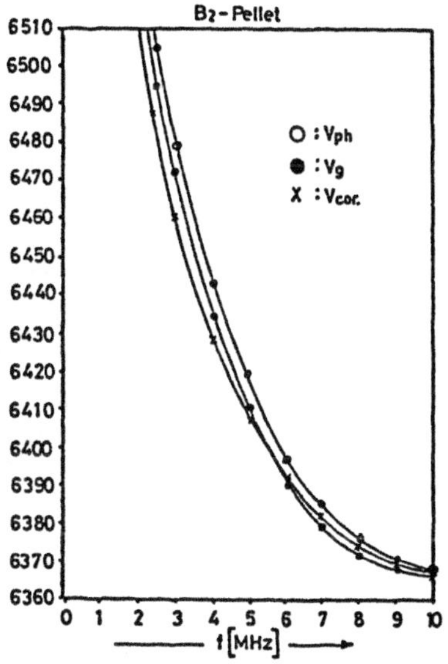

Fig. 5 Phase velocity, group velocity and correlated phase velocity at frequency variation

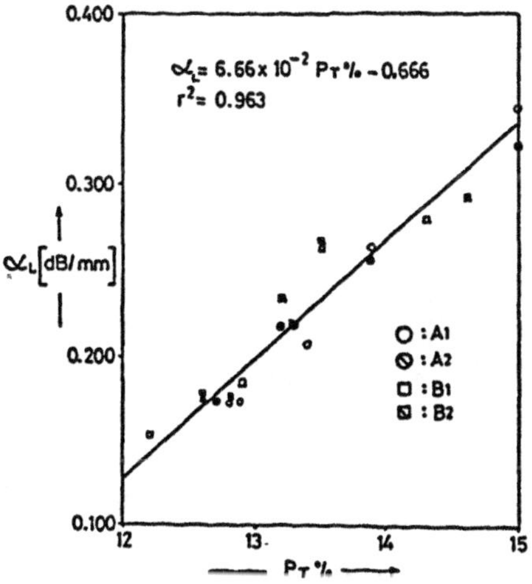

Fig. 6 Relation between α_L and P_T % at 7,5 MHz

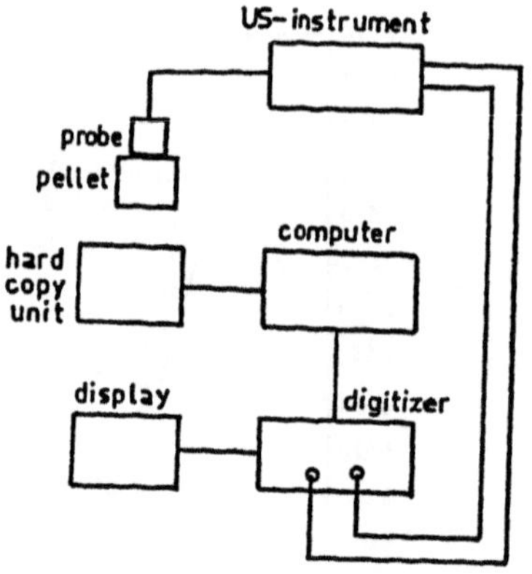

Fig. 7 Block-diagram of computer aided system

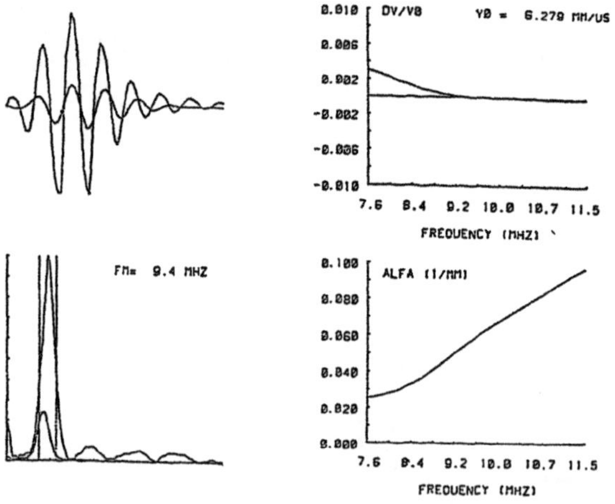

Fig. 8 Computer output of computer aided system

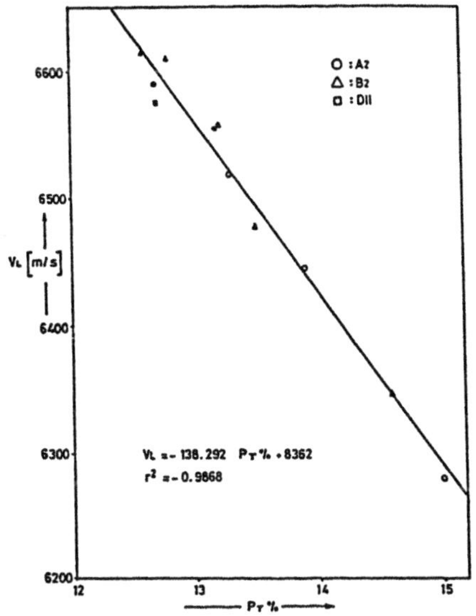

Fig. 9 Relation between V_L and P_T % at 7,5 MHz

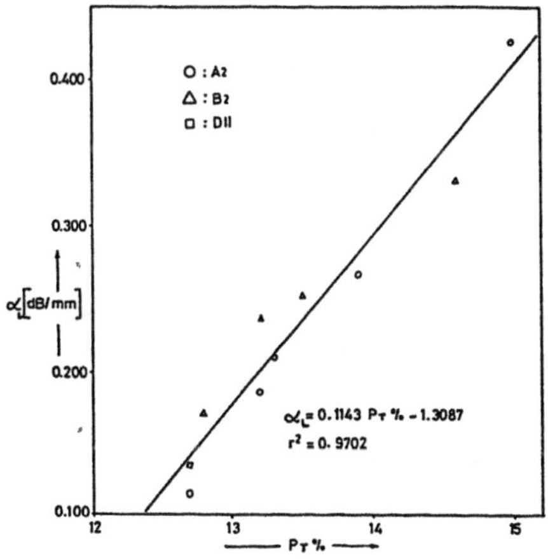

Fig. 10 Relation between α_L and P_T % at 7,5 MHz

Simultaneous Residual Stress and Retained Austenite Measurement by X-Ray Diffraction

C. O. RUUD[1], G. H. PENNINGTON[2], E. M. BRAUSS[3], and S. D. WEEDMAN[1]

[1]Materials Research Laboratory
The Pennsylvania State University
University Park, PA USA 16802

[2]Denver X-Ray Instruments, Inc.
Sixth Ave. and Forty-fifth St.
Altoona, PA USA 16602

[3]Proto Manufacturing
2175 Solar Cr.
Old Castle, Ontario NOR 1LO
Canada

Abstract

A new fiber optic based XRD instrument is described that is capable of measuring simultaneously both residual stresses and phase composition present in steels. The hardening of steel requires that the material first be heated to a high enough temperature to produce a face-centered cubic polycrystalline solid solution called austenite. The material is then rapidly quenched to form a hard, metastable body-centered tetragonal solid solution called martensite. In practice, the transformation is incomplete, leaving some austenite at room temperature. Because the presence of austenite is often detrimental to the quality of the steel, it is of considerable interest to be able to quantitatively determine the amount of the austenite phase present (the retained austenite), as well as the amount of residual stress induced in the martensite.

This paper describes a new three detector XRD instrument that is capable of making simultaneous measurement of both residual stresses and retained austenite in steels at the same location on the sample in a few seconds.

Introduction

X-ray diffraction techniques have been used extensively to measure both residual stress and phase composition in steel parts. However, the procedures for these measurements are time comsuming and must be performed sequentially with any x-ray instrument, except the one that is described here. The advantages that this new instrument offers include very short measurement times and the measurement of materials properties in exactly the same location at the same time.

Background

The position-sensitive scintillation detector (PSSD) developed at The Pennsylvania State University provides unprecedented x-ray diffraction measurement speed consistent with excellent accuracy (Fig. 1). The PSSD relies on the coherent conversion of the diffracted x-ray pattern to an optical signal (light); the conduction of this light over several linear centimeters of fiber optic bundles; the amplification of this signal by electro-optical image intensification; the electronic conversion of the signal; and the transfer of the electronic signal to a computer for refinement and interpretation. The PSSD uses two independent detector surfaces to collect data from two positions on the Debye ring simultaneously, thus providing a unique capability of precision stress measurement by the single exposure technique (SET) [1]. The instrument has been described more thoroughly elsewhere [2,3] and its application to materials characterization discussed [4].

Recently, the PSSD has been modified by the addition of a third detector (Fig. 1), for the additional task of measuring retained austenite. For residual stress measurements, the diffracted peak positions from two orientations of grains, designated with respect to the psi(ψ) angle, are measured. For the retained austenite determination, the integrated areas of the austenite peak from detector 3, is compared to the martensite peaks from detectors 1 and 2 (see Fig. 1).

With the PSSD, data collection times of less than one second are possible using modern x-ray tubes and constant potential power supplies -- a two order of magnitude improvement over conventional XRD instrumentation. A compact version of this instrument has been developed that is capable of making

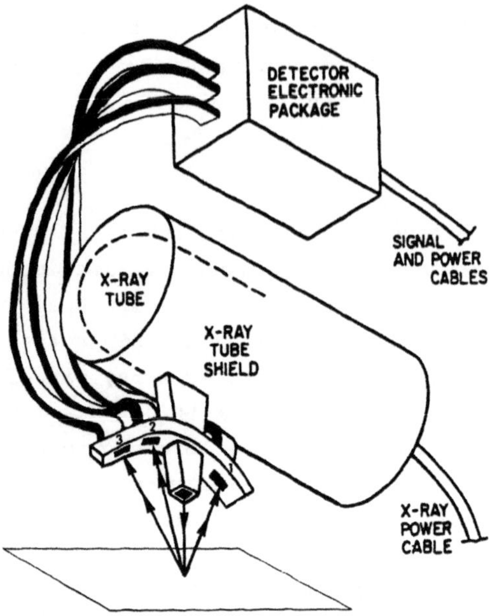

Figure 1. Schematic of the position-sensitive scintillation detector (PSSD), showing the new modification of an added detector (3) for retained austenite determination.

measurements on the inside of a pipe as small as 100 mm inside diameter [4]. Also, the geometry of its x-ray optics allow the stress readings in confined areas such as gear teeth, pipe and turbine vane bases.

Residual stress

When a metallic crystalline material is stressed, the elastic strains in the material are manifested in the crystal lattices of its grains. The stress applied externally or residual within the material, if below its yield strength, is taken up by uniform interatomic macrostrain that is spread over several tens of grains.

The SET method of stress calculation is based on the fact that a single incident x-ray beam is diffracted at a constant angle such that a cone of diffracted radiation is formed. A plane perpendicular to the cone axis intercepts the cone as a circle when the specimen is unstressed, and as an ellipse, if stressed [5]. Deviation from a circle, then, is a measure of that stress. To read this deviation, detectors are placed 180° apart, on that ellipse. Until the development of the PSSD, the SET method had been restricted to the use of film camera devices [1]. The variables used in the SET method are defined in Figure 2 with respect to the PSSD geometry.

The stress measurement is made with:

$$\sigma = \frac{E}{1+\nu} \frac{S_2 - S_1}{4R_0 \sin^2 \theta_0 \sin 2B} \tag{1}$$

where R_0 is the detector-to-specimen distance (see Fig. 2), is the angle between the specimen surface normal and the incident x-ray beam, θ_0 is the diffraction angle of a sample of unstressed material, (S_2-S_1), as shown in figure 1, is a measure of the distortion of the cone of diffracted radiation, E is the elastic modulus, and ν is Poisson's ratio.

For the x-ray method to be used as an absolute measure of the residual stress, the d-spacing of planes of the same Miller indices must be measured for at least two different orientations with the metal surface, ψ_1 and ψ_2. Absolute measurement means that no previous or subsequent measurement of the metal piece need have been obtained in a zero-stress condition for comparison.

Retained Austenite

Austenite is an interstitial solid solution of carbon in γ-iron (FCC). Below 723°C it normally decomposes into ferrite, a solid solution of carbon in α-iron (BCC), and cementite (Fe_3C). During quenching these two products may not have time to form, with the result that an unstable body-centered

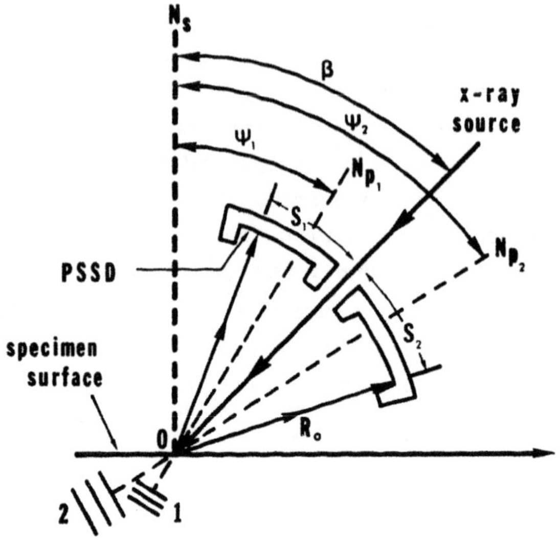

Figure 2. Schematic of the single-exposure technique (SET) showing the incident x-ray beam at an angle, β, to the specimen surface normal (N_s) and two orientations of grains with sets of interatomic planes and normals, N_{p1} and N_{p2} at angles of ψ_1 and ψ_2 to N_s.

tetragonal product, martensite forms, usually with some residual untransformed austenite. Because of the significant effects of this retained austenite on the properties of the steel, it is important to be able to measure quantitatively the amount present.

The method used to calculate the amount of retained austenite was originally proposed by Averbach and Cohen [6], and further discussed in Cullity [5]. The fundamental equation that defines the intensity of radiation diffracted by any single phase powder can be rewritten in the general form [5]:

$$I = \frac{K_2 R}{2\mu} \tag{2}$$

where K_2 is a constant, and R will vary with θ, with (hkl), and with the kind of substance. This expression can be rewritten for austenite (γ) and martensite (α) as:

$$I_\gamma = \frac{K_2 R_\gamma C_\gamma}{2\mu_m} \quad (3)$$

$$I_\alpha = \frac{K_2 R_\alpha C_\alpha}{2\mu_m} \quad (4)$$

where μ_m = the linear x-ray absorption coefficient of the steel alloy, c_γ is the concentration of austenite, and c_α is the concentration of martensite. These two equations can be divided to yield:

$$\frac{I_\gamma}{I_\alpha} = \frac{R_\gamma C_\gamma}{R_\alpha C_\alpha} \quad (5)$$

Now, the intensities can be measured by the PSSD, and R can be determined experimentally with the use of known standards that are available from the United States Bureau of Standards. Given that $c_\gamma + c_\alpha = 1$, the individual concentrations of austenite and martensite can then be determined. Therefore, the integrated area of the austenite peak, derived from detector 3, can be compared to the areas of the martensite peaks from detectors 1 and 2 to determine the ratio of austenite to martensite.

Conclusions

Recent modification of the PSSD of the addition of a third detector allows the simultaneous measurement of both residual stresses and retained austenite in steels from the same volume of material. The advantages of this new instrument are that it offers very short measurement times and of these properties in the same location on the sample.

References

1. SAE, Residual Stress Measurement by X-Ray Diffraction -- J784a, (Soc. of Auto. Eng., Warrendale, PA, 1971).

2. C. O. Ruud, Ind. Res. and Dev. pp. 84-87 (January, 1983).

3. C. O. Ruud, P. S. DiMascio, and D. J. Snoha, in Adv. in X-ray Anal. (Plenum Press, New York, 1984), Vol. 27, pp. 273-283.

4. C. O. Ruud, in Nondestructive Methods for Material Property Determination, edited by C. O. Ruud and R. E. Green, Jr. (Plenum Press, New York, 1984), Vol. 1, pp. 32-37.

5. B. D. Cullity, Elements of X-Ray Diffraction, Second Edition (Addison-Wesley, Reading, MA, 1978).

6. B. L. Averbach and M. Cohen, Trans. AIME 176 401 (1948).

Acoustic Characteristics of Porous Materials in Simple and Complex State of Stresses

Z. Pawlowski

Institute for Basic Technological Reasearch
of the Polish Academy of Science, Warsaw, Poland

SUMMARY

Coal is a combustible, porous rock. The knowledge of its chemico-physical properties is needed when predicting the danger of sudden outbursts during coal strata exploitation. The paper presents the results of acoustic response of coal samples when subjected to axial compression with superimposed hydrostatical pressure in the range between 25 MPa and 300 MPa. There were measured the time-of-flight of ultrasonic waves, 1 MHz frequency, and the level of the pulse transmitted through the sample. From six characteristics one can obtain mainly the changes in ultrasonic wave velocity against stress or strain which are presented and discussed.

INTRODUCTION

In coal mining there exists a need to recognize physico-chemical properties of rocks, especially of coal which is a porous, combustible rock of organic origin. These properties serve as the basis for predicting the danger occuring during exploitation. To the most important dangers there belong sudden outburst and gas explosions. The knowledge of physico-chemical properties allows one also to determine the workability conditions of coal stratum and the load carrying capacity of coal piles. The factors which greatly influence the coal behaviour are: * the state of stresses, * the gas content, mainly CO_2, and * the ability of gas penetration through the strata.

To explore the coal characteristics a research project on "rocks as multiphase medium" has been started. One of the project parts is devoted to the acoustic characteristics of rocks. The research project CPBP 03.06 is coordinated by Strata Research Institute of the Polish Academy of Sciences.

ACOUSTIC CHARACTERISTICS

The following acoustic characteristics have been investigated:

* stress-strain at various levels of three-axiality
* ultrasonic wave velocity changes against stress and/ or strain
* attenuation of ultrasonic waves against stress and/ or strain
* changes of ultrasonic wave velocity during sorption and desorption of CO_2
* acoustic emission at straining, as well as at sorption, desorption and gas flow

This paper presents only a part of investigations on the change of ultrasonic wave velocity when straining dry samples, i.e., not saturated with liquid, at various levels of three-axiality.

SAMPLES AND INSTRUMENTATION

Due to the fact that the strength of coal samples taken from the stratum is very low, it was imposssible to cut out samples needed for conducting the experiments. Therefore, special cylindrical samples, about 40 mm long and 22 mm in diameter, were prepared from coal powder. Thus the samples created a coal agglomerate which can be machined easily to the shape wanted.

The straining of the samples was carried out in a loading arrangement GTA [1,2] . It allows one to realize the hydrostatical compression up to 400 MPa and to superimpose additional axial compression up to 1500 kN. Further possibities of the apparatus are: * the testing at elevated temperatures up to 400 °C, testing samples having various water or gas saturation inducing the increase of the pore pressure in the sample. The apparatus is also furnished with a system for saturating the samples with gas under pressure up to 0.4 GPa. There exists also the possibility of inducing in the loaded specimen the filtration of noncorrosive gases or liquids with controlled pressure along the sample.

Figure 1 illustrates how the loading of the sample is carried out. The sample, positioned in the pressure chamber, is first

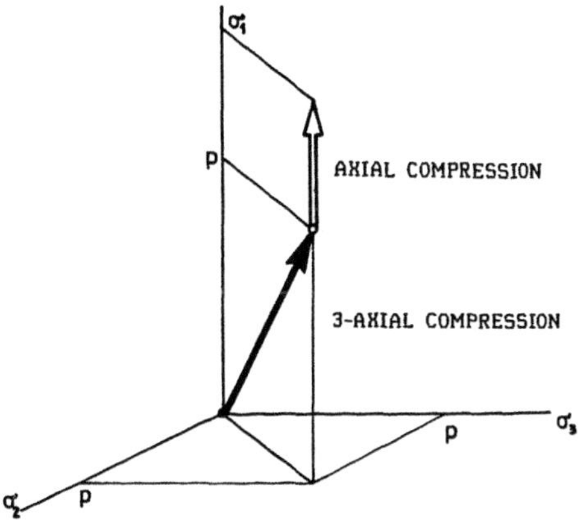

Fig. 1. Loading path of the sample

subjected to three-axial compression and then the additional load is being applied up to the final fracture. During the experiments, which have been carried out, all samples were loaded hydrostatically to 300 MPa and then, after decreasing the hydrostatical pressure to one of the levels corresponding to 25, 50, 75, 100, 150, 200, 250 MPa, the sample was loaded additionally with axial compression. The mechanical response, i.e., the stress-strain characteristic is shown in the centre of Figure 2. After subtracting the hydrostatical pressure σ_0 from the total axial stress and the deformation ε_0 from the strain we get the characteristics shown in the right-hand part of Figure 2.

The loading path, shown in Figure 2, runs as follows: i) increasing the hydrostatical pressure up to 300 MPa, ii) diminishing this pressure to the level wanted, and iii) applying the axial compression stress.

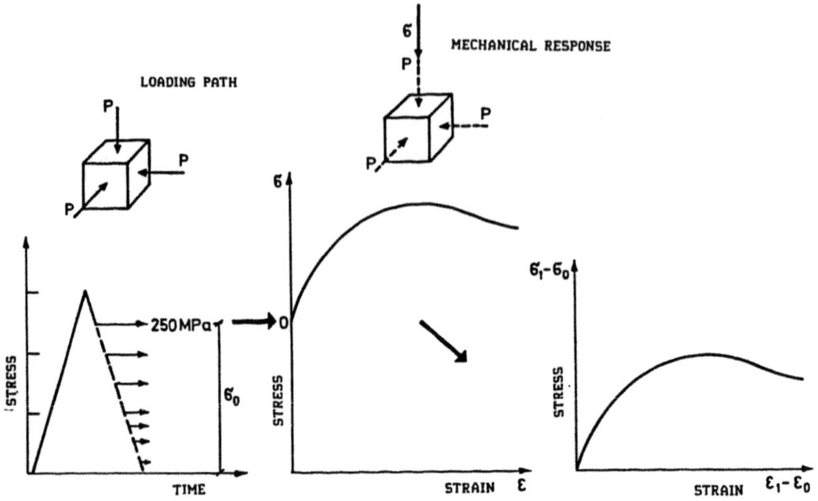

Fig. 2. Loading path and mechanical response of coal samples subjected to hydrostatical pressure and additional axial compression

CHARACTERISTICS STRESS-STRAIN

The stress-strain dependence was recorded for superimposed hydrostatical pressure at 8 various levels, i.e., for 25, 50, 75, 100, 150, 200, 250 and 300 MPa. Figure 3 shows 5 characteristics for 25, 50, 100, 200 and 300 MPa. They indicate the increasing value of stress and strain with the superimposed hydrostatical pressure increase. The strain reaches considerable values, up to 40%, before sample fracture occurs. There were observed distinct changes in fracture modes, from slant fractures at 0 MPa, 25 MPa, to cone fracture associated with considerable bulging of the sample.

Due to the fact that the diameter of the sample increases when the sample is compressed, the real stress was calculated by assuming, as a first approximation, constant specimen volume at every straining rate. The deviation from this assumption may probably appear but on this stage of investigations we do not dispose of results which can indicate what are volume changes at various levels of straining. It should be noted that the test with the sample subjected to 300 MPa hydrostatic pressure was interrupted before reaching 40% contraction, therefore, the characteristic was not completed in the range of large deformation values.

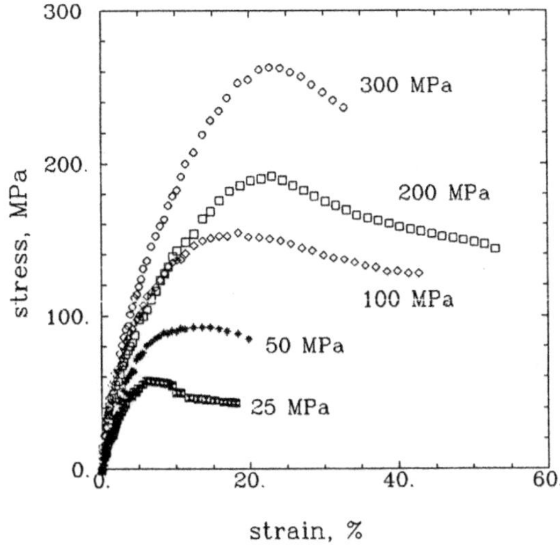

Fig. 3. Stress-strain relations for samples tested at various levels of hydrostatic pressure applied

CHARACTERISTICS ULTRASONIC WAVE VELOCITY CHANGE AGAINST STRESS

Figure 4 shows the changes observed in the velocity of ultrasonic waves with increasing stress. Almost linear behaviour was observed for 25 MPa hydrostatical pressure. However, with increasing value of hydrostatic pressure, the sensitivity to wave velocity change decreases with increasing superimposed hydrostatic pressure. For larger values of hydrostatic pressure the sensitivity to velocity changes decreases with the larger superimposed hydrostatical pressure, where, after a short range of insensitivity to stress changes, the change in velocity starts to go in the opposite direction.

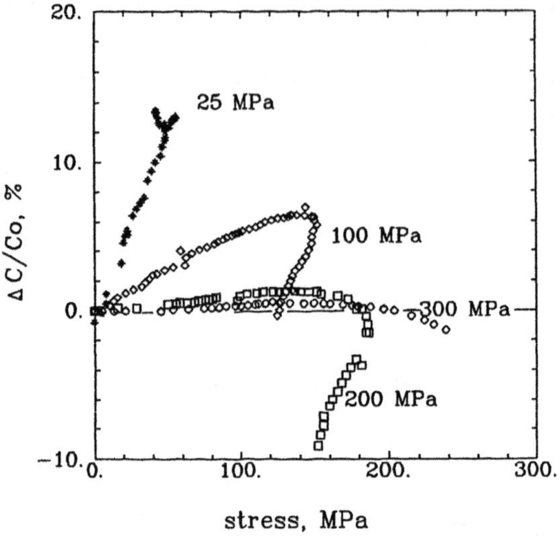

Fig. 4. Change of ultrasonic wave velocity as a function of axial compression for various values of superimposed hydrostatic pressure

This means that starting from the point of maximum velocity increase (reaching, e.g., about 7% for 100 MPa), the velocity begins to decrease. It can be seen clearly from the characteristics for 100 MPa and 200 MPa hydrostatic pressure that the range of "insensitivity and change of sign" of the elasto-acoustic coefficient corresponds to the range being close to the maximum value of stress.

CHARACTERISTIC ULTRASONIC WAVE VELOCITY CHANGE AGAINST STRAIN

Figure 5 presents four characteristics of ultrasonic wave velocity change against strain. Two trends are observed. First, the sensitivity to wave velocity change decreases markedly when the hydrostatical pressure increases. For large values, e.g. 200 MPa and 300 MPa in the strain range between 0% and 10%, the elasto-acoustic coefficient is very small. It means that almost insensitivity to strain changes appears. The second tendancy is the change of sign of the elasto-acoustic coefficient.

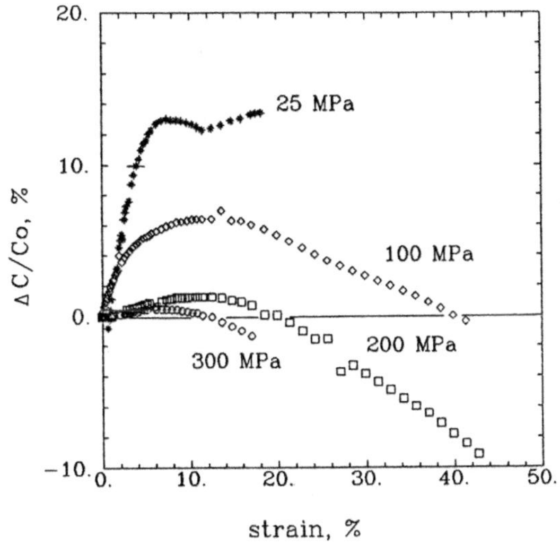

Fig. 5. Changes of ultrasonic wave velocity as a function of strain for various values of superimposed hydrostatical pressure

Both phenomena can be interpreted when we take account of the internal structure changes at increasing levels of strains. Figure 6 contains three characteristics taken for 250 MPa of hydrostatical pressure. There are the characteristics: * wave velocity change against strain, * time-of-flight against strain and the stress-strain curve. One notices that the initial increase in wave velocity corresponds to the elastic range of deformation. Up to about 90% of σ_{max} the changes in ultrasonic wave velocity are very small, while the transition to negative values of the elasto-acoustic coefficient starts before the stress reaches its maximum value. The trend described above could be explained on the basis of T.O.F.-strain characteristic.

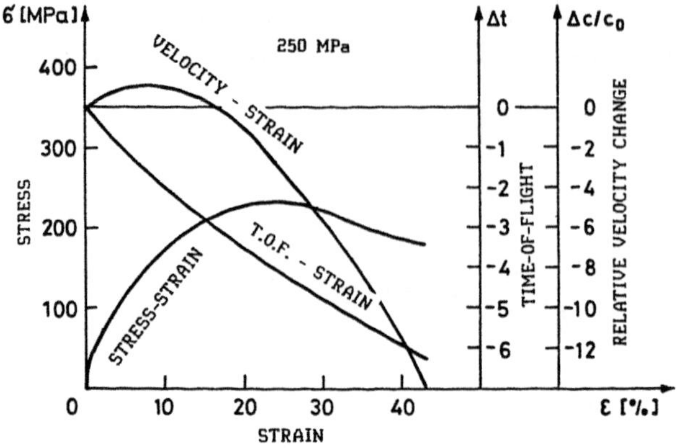

Fig. 6. Change of wave velocity, stress and time-of-flight as a function of strain

CONCLUSIONS

It was shown that the characteristic for wave velocity change is not linear. The values of this coefficient changes from positive to negative. It appears also that the changes in material structure, leading to final fracture, start long before the maximum value of compressive stress is attained.

REFERENCES

1. Dlugosz M., Gustkiewicz J. Wysocki, A., Apparatus for investigation of rocks in threeaxial state of stresses. Part I Characteristics of apparatus and of the investigated method, Archivum Gornictwa, Tom 26, 1981, Zestzyt 1, 17-22
2. Dlugosz M., Gustkiewicz J., Wysocki A., Apparatus for investigations of rocks in threeaxial state of stresses. Part II, Some investigations concerning certain rocks, ibid. pp. 29-41
3. Gutkiewicz P., Cislowski W., Pawlowski Z., Set-up for measuring the time of flight of ultrasonic waves using sampling converter, /in Polish/, 12th National Conference on Non-destructive Testing, Rydzyna 1983, pp 139-144

Crack Inspection of Railroad Wheel Treads by EMATs

R. E. SCHRAMM, P. J. SHULL,
A. V. CLARK, JR., and D. V. MITRAKOVIĆ*

National Institute of Standards and Technology
Fracture and Deformation Division
Boulder, Colorado 80303 U. S. A.

Abstract
Railroad safety depends on many factors. The integrity of the wheels on rolling stock is one that is subject to nondestructive evaluation. For some years, ultrasonic testing has been applied to the detection of cracks in wheel treads, with particular attention to automatic, in-rail, roll-by methods. We have begun constructing a system aimed at using relatively low frequency Rayleigh waves generated by electromagnetic-acoustic transducers (EMATs). The current design uses a permanent magnet to maintain a compact structure and minimize the size of the pocket machined into the rail. Measurements thus far indicate a responsiveness, even to small flaws. With the development of a signal processing and analysis system, field tests should soon be possible.

Introduction

Cracks in railroad wheels may result from high stresses due to dynamic or static loads and residual stresses generated by such events as heating during braking. These flaws generally originate in the tread surface or flange and can lead to catastrophic wheel failure resulting in considerable equipment damage and possible derailment. This threat to personnel safety and the potential costs in time and money are inducements to search for an effective, automated method for nondestructive examination that will identify damaged wheels needing replacement.

The current method of examination is visual observation. In the early 1970's, an ultrasonic method was introduced [1-4]. This involves an in-rail system of piezoelectric transducers that generate Rayleigh waves, electronics for signal generation and processing, and a method for tagging suspect wheels. The goal is to examine each wheel of a train as it rolls by a checkpoint in a railyard. Two of these systems are currently in operation in this country. A companion portable hand-held

---

* Guest scientist on leave from the University of Belgrade, Belgrade, Yugoslavia.

system subsequently verifies any flaw indications. The Fraunhofer Institute (IzfP) in Saarbrucken, Federal Republic of Germany, has been developing another ultrasonic system along this line [5], but using electromagnetic-acoustic transducers (EMATs). IzfP has placed a prototype into operation.

Our current research in this area [6-8] is the development of an ultrasonic system using EMATs designed to work on American-style wheels and rails. The objective is to check every wheel on a train, also in a roll-by mode. Like the earlier systems, ours uses Rayleigh waves that travel around the wheel tread. We are using EMATs for two main reasons: 1. No ultrasonic path in the transducer itself and low sensitivity to mode-converted signals mean a simplified, low-noise signal that is relatively easy to interpret with high reliability, even in an automated system; 2. Their noncontact nature eliminates the need for ultrasonic couplants, i.e., no water sprays or liquid-filled boots. Two extensive reviews of EMAT designs and applications have recently been published [9, 10].

Equipment

In the electronic system, a function generator provides the rf signal for a gated MOSFET power amplifier which drives the transmitter. The toneburst consists of 5-10 cycles at 500 kHz with a pulse repetition frequency (PRF) of 60 Hz. The receiver preamplifier is a very low noise design. Electronic impedance matching is very important for both transmitter and receiver to ensure maximum efficiency.

Our EMAT device (Fig. 1a) is meant for insertion into a recess cut out of the rail. Therefore, one goal in the transducer design is to keep the configuration as compact as possible to minimize both the amount of required machining in the rail and the loss of weight-bearing surface. Toward this end, we use a single permanent magnet (Nd-Fe-B for maximum field strength) 32 mm X 26 mm X 52 mm with the magnetization direction along the 32 mm direction and oriented normal to the tread. For pitch-catch operation, there are separate transmitter and receiver coils. These are meanderlines with a periodicity of 6 mm to generate a Rayleigh wave with this wavelength at 500 kHz (velocity = 3 km/s). This design generates and receives bidirectional Rayleigh waves traveling normal to the coil legs.

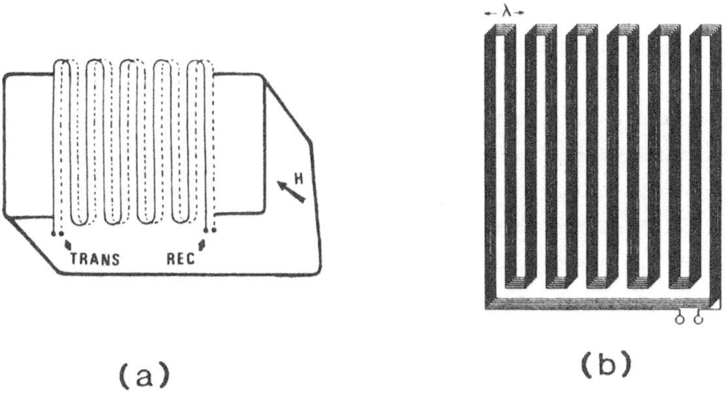

Fig. 1. Rayleigh-wave EMAT construction.
(a) Two meanderline coils, shifted to prevent shielding, placed atop a magnet with field H. A foil eddy current shield lays over the magnet and a thin, compliant foam sheet is under the coils.
(b) Details of the wirewound meanderline.

Rather than stacking the transmitter and receiver coils directly atop one another, we shifted them by a quarter period so they do not shield each other. To prevent signal complications from ultrasound production in the magnet, we cover it with a thin foil of copper or aluminum to minimize eddy current generation in the Nd-Fe-B. Between this shield and the coils are a few millimeters of compliant material (currently, polymer foam); this layer compresses under the wheel's weight and their flexible substrate allows the meanderlines to conform to the curvature and taper of the wheel tread. This minimizes liftoff for maximum efficiency and signal/noise.

The present coil design contains eight cycles or loops in the meanderline. The receiver coil is AWG 36 enamel-coated wire that has been wound through the pattern six times on acetate-based adhesive tape (Fig. 1b). This multiplicity adds to the sensitivity since the repetitions are series-connected. The flexible tape allows the coil to conform to the wheel's curvature.

We've constructed the transmitter coil in two forms. The first is a printed circuit on polymer film. The conductor is about 1 mm wide and 0.025 mm thick. In this form, the impedance is very low (about 1.3 Ω including a current limiting resistor). Our power amplifier can deliver about 140 A of rf current into this coil. The second form is identical to the wire-wound receiver, and the much higher impedance limits the drive current to about 30 A.

The greater the current flow through the transmitter, the greater is the eddy current induced in the specimen. While the printed circuit coil permits the maximum current, the wirewound coil passes the current through the EMAT aperture multiple times and the current density (A/mm^2) induced in the specimen is roughly the same, i.e., the size of our ultrasonic signal is nearly identical with either form of transmitter. Using the lower current coil will increase the longevity of the power amplifier, and will allow an increase in the pulse repetition frequency (PRF) to as much as 250 Hz, should this prove desirable later. At present, the limit on the PRF is due to two factors in the power amplifier: ohmic heating of the MOSFET output drivers and the time required to recharge the power supply filtering capacitors.

Experimental Results

We have recently constructed a short (4 m) section of track in which we mounted the transducer (Fig. 2). Rolling an actual wheel set (two wheels mounted on an axle) over the device much more closely simulates field conditions than our initial static measurements [7]. The goal is to debug several parts of the system: mechanical mounting, compliant coil backings, geometric variations in the size and position of magnet and coils, triggering (if necessary), and signal collection and storage. Since the wheels have a few centimeters of side-to-side play in normal operation, this rolling test will also show us the importance of the wheel's transverse position with respect to the transducer.

For these initial measurements, we had two wheel sets, well used but nominally uncracked. The wheels were rolled onto the transducer and held stationary or rolled over the EMAT site at a few kilometers per hour. In both wheels of one set, this arrangement produces a signal which was still detectable after 16 round trips or about 42 m of travel. In both wheels of the other set, the signal traveled around the tread only six times before being attenuated into the noise level. Metallurgical variables due to such factors as manufacture, wear, etc., have a very large influence on the wheel's acoustic response. Any signal analysis will have to account for this effect.

The critical flaw depth is about 6 mm, so our interest is in distinguishing between cracks of greater depth (no longer safe) and those more shallow (safe for continued use). Initially, however, we saw-cut a very shallow circular flaw into the center of the tread along a wheel radius. This flaw depth into the wheel was almost 2 mm at the maximum; the length was oriented along the tread width and 12 mm long at the surface.

To determine how far a wheel can move across the transducers and still return a useful signal, we rolled the wheel in small steps across the EMATs and measured the signal strength as the initial contact point moved across the coils. This indicated that the effective EMAT aperture within which the wheel must be located to both transmit and receive our test signal is about 5-8 cm long. This aperture is actually slightly longer than the coils because the wheel radius is relatively so much larger, and the flexible coils conform to the tread curvature. While it would be desirable to introduce multiple pulses into the wheel in order to improve statistics for a higher signal-to-noise ratio, this may not be feasible since the Rayleigh wave travels several times around the circumference. To avoid processing confusion, it will be necessary to delay any new pulse until the prior one has sufficiently decayed. Since each round trip takes nearly one full millisecond, this means a delay interval of about 10-16 ms. To get even two test pulses into the wheel (one at the beginning and end of a 5 cm window) means the train cannot travel faster than 5 cm/10 ms or about 18 km/h. This would likely be unacceptably slow for a field system, so we shall probably have to settle for a single transmitter pulse. A likely scenario is to use a trigger device to start the pulse as the wheel enters the window. Recording just the data arriving before the return of second round trip signal (see below) takes nearly 2 ms, so the maximum train speed could be about 5 cm/2 ms, or 90 km/h. There are other factors (signal processing time, timing accuracy of pulse trigger, etc.) that will likely limit the allowable speed to somewhere under this value.

There are possibilities for triggering mechanisms to signal the proper moment during a wheel's transit to pulse the EMAT. For these initial tests, we placed a simple membrane switch between the coils and the magnet. The compliant foam transmitted the wheel pressure and switch closure triggered both the high current pulser and a digital oscilloscope to capture the transient ultrasonic signal from the single pulse. At the low speeds possible on our short test track, this system has been both reliable and consistent with our current spatial window.

The transducers are bidirectional and generate Rayleigh waves traveling clockwise and counterclockwise around the railroad wheel tread. Figure 3 indicates schematically the signal paths to a flaw and the resulting scope trace. Signals A and C are echos that have traveled about 22% and 78% of the way around the wheel circumference to the 2 mm deep flaw and then back. Signals B and D are first and second round trip signals.

Fig. 2. Current test configuration.
(a) Standard wheel set rolled along a short length of track.
(b) EMATs mounted in rail. A protective film covers the coils and they are represented schematically in this photo.

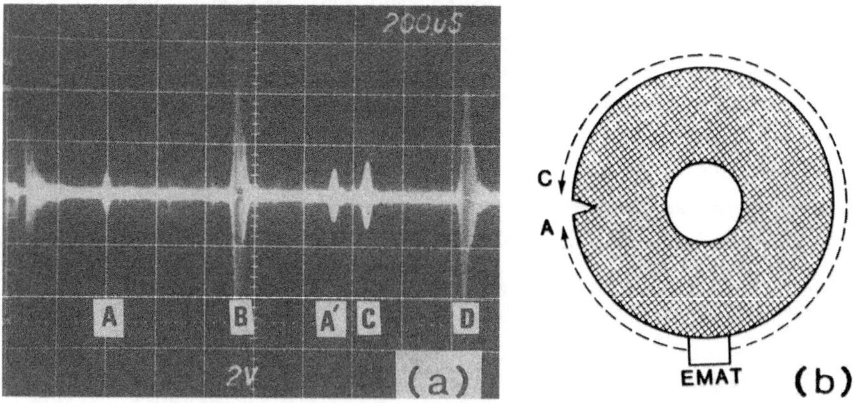

Fig. 3. Flaw echoes.
(a) Typical oscilloscope signal for 2 mm deep cut. A and C are short and long path flaw echoes. B and D are first and second round trip signals. A' is the short path echo after a complete round trip (A + B).

Signal A' is the short path echo after the initial pulse has already traveled once around the entire circumference. Frequently some splitting occurs in these signals and this likely has at least two sources. Each round trip signal is a combination of counter-rotating signals. The path length is quite long (about 2.6 m) and some phase incoherence between these two becomes likely over this distance before they again combine on arriving back at the transducer. Also, the Rayleigh wave has some tendency to spread out to the wheel flange where it has a somewhat longer path than along the tread. The amount of ultrasonic energy in a wheel depends on the closeness of EMAT coupling and the wheel condition, e.g., grain size and residual stress state. Consequently, we are presently using the amplitude of the first round trip signal as a normalizing factor; the amplitude ratios A/B and C/B are our current flaw depth indicators.

The arrival time of signal B remains constant, of course, while A and C move closer to B as the EMAT-flaw distance increases until they all coincide when the flaw is exactly opposite the transducer. Thus, there are two zones on the circumference where meaningful measurements are not possible. Both of these are about 24 cm long; one is centered at the transducer and the other is exactly opposite. The first is due to recovery of the receiver amplifier following the transmitter pulse; the second is due to the merging of the round-trip and flaw signals when their acoustic path lengths become identical. As a result, approximately 80% of the tread can be inspected with each pass. A second pair of EMATs located 0.5-1 m down track would assure 100% inspection.

Future Work

While our present EMAT design is producing excellent signals, there are several possible improvements we will be investigating. Among these are the use of a physically smaller magnet to reduce the size of the pocket machined into the rail, alternate materials for the compliant layer under the coils, and variations in the design and physical structure of the coils. Other wheel-detection mechanisms to trigger the current pulse are possible.

Acknowledgments

We are grateful for discussions with H. J. Salzburger of IzfP as well as John Cowan and Richard Conway of The Fax Corporation. This work was sponsored by the U.S. Department of Transportation, Federal Railroad

Administration, Washington, D.C. We also thank the Transportation Test Center, American Association of Railroads, Pueblo Colorado, and, in particular, Britto Rajkumar and Dominic DiBrito for providing test wheels.

This paper was also submitted to the Review of Progress in Quantitative Nondestructive Evaluation, vol. 8, edited by D. O. Thompson and D. E. Chimenti for publication by Plenum Press, N.Y., N.Y., U.S.A.

References

1. Inspect Wheels "In Motion," Progressive Railroading, Jan., 1974.

2. Spotting the Defects in Railroad Wheels, Business Week, Sept. 28, 1974, p. 44.

3. Cracked-Wheel Detector "Could Save Millions," Railway Systems Control, May, 1975.

4. U.S. Patent 3,978,712: Sept. 7, 1976, John Vincent Cowan, Gerald De G. Cowan, and John Gerald Cowan.

5. Salzburger, H. J.; Repplinger, W.: Ultrasonics International 83, conference organizers Z. Novak and S. L. Bailey, Kent, Great Britain: Butterworth & Co. Ltd., 1983, pp. 8-12.

6. Clark, A. V.; Schramm, R. E.; Fukuoka, H.; Mitraković, D. V.: Ultrasonic Characterization of Residual Stress and Flaws in Cast Steel Railroad Wheels, Proceedings, IEEE 1987 Ultrasonics Symposium, Vol. 2, edited by B. R. McAvoy, New York: Institute of Electrical and Electronic Engineers, 1987, pp. 1079-1082.

7. Schramm, R. E.; Clark, Jr., A. V.; Mitraković, D. V.; Shull, P. J.: Flaw Detection in Railroad Wheels Using Rayleigh-Wave EMATs, Review of Progress in Quantitative Nondestructive Evaluation, edited by D. O. Thompson and D. E. Chimenti, New York: Plenum Press, 1988, Vol. 7B, pp. 1661-1668.

8. Schramm, R. E.; Shull, P. J.; Clark, Jr., A. V.; Mitraković, D. V.: EMAT Examination for Cracks in Railroad Wheel Treads, submitted to Proceedings: Nondestructive Testing and Evaluation for Manufacturing and Construction, Aug. 9-12, 1988, Washington, DC: Hemisphere Publishing Corp.

9. Maxfield, B. W.; Kuramoto, A.; Hulbert, J. K.: Materials Evaluation 45 (1987) 1166-1183.

10. Alers, G. A.; Burns, L. R.: Materials Evaluation 45 (1987) 1184-1194.

Computer Aided Threedimensional Deformation Analysis Using Holographic Interferometry

K. Kussmaul, A. Ettemeyer

Staatliche Materialprüfungsanstalt (MPA) Universität Stuttgart, FRG
Rottenkolber Holo-System GmbH, Kirchheim/Munich, FRG

Summary

An evaluation system for holographic interferograms is presented providing the possibility of calculating threedimensional deformations, respectively strains. Using the "mirror method" for the acquisition of three consecutive holograms and an image processing system with phase-shift fringe interpolation computer aided deformation analysis is derived.

Introduction

At the Staatliche Materialprüfungsanstalt (MPA), University of Stuttgart, FRG, sometimes arise problems of measurement which can hardly or not at all be solved. For instance, when determining extensive displacements and strains in fracture mechanics tests in order to enable an adaption of finite element calculations on the existing proportions. The holographic interferometry is especially suitable for this task because it allows concurrently precise local and integral non-contact displacement measurements.

The conservative holographic systems have not allowed up to now quantitative statements concerning the spatial deformation behaviour of an object. For this reason, the holographic system at MPA was extended to assure that quantitative and threedimensional measurements and strain calculations could be carried out by the holographic interferometry. The system is made up of two parts, the holographic set-up and the evaluation system. The complete development was sponsored by the Deutsche Forschungsgemeinschaft within the scope of a research project.

Fundamentals

The holographic measuring principle is based on the comparison of two spatial states of an object. For this purpose two spatial pictures of the object to be investigated are recorded into a hologram, one picture in the undeformed condition and the other in the deformed condition. The simultaneous reconstruc-

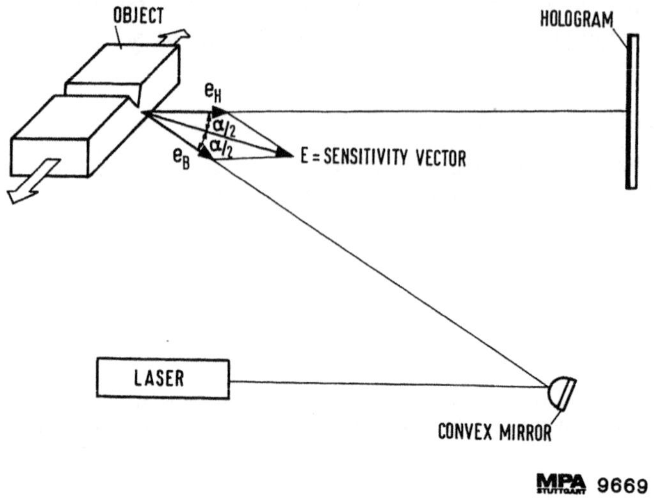

Figure 1. Definition of the sensitivity direction

tion of both states shows the differences through interference lines. These interference lines, however, represent only a displacement component in the so-called sensitivity direction which is defined by the holographic set-up, Figure 1, and lies in the half angle between illumination and observation direction. The holographic set-up is insensitive to displacements, which will not be recorded perpendicular to this sensitivity direction /1/.

On the other hand, the component of displacement in sensitivity direction is recorded by interference fringes resulting from an alternation of the optical path from the holographic set-up to the object. The specific interference fringe order over the various points of the object surface shows the displacement components in sensitivity direction in fractions of the light wave length.

It is easy to delineate the fact mathematically. The measured displacement component is the projection of the displacement vector $\vec{d}$ on the sensitivity unit vector $\vec{e}$

$$\vec{e} \cdot \vec{d} = \lambda \cdot N \tag{1}$$

λ = light wave length

N = interference order at the observed point.

It is obvious that three sensitivity directions have to be generated for three-dimensional measurements. This can be attained by various methods, for instance, by observing the object with a mirror, Figure 2. Thus, simultaneous measurement from different directions is possible /2/. The spatial displacement vector can be computed for each point from the three interference arrangements in the three sensitivity directions. This results in an equation system with three equations (3 sensitivity directions) and 3 unkown displacement components (in x, y, z-direction). This system can be solved, e.g. with Gauss elimination.

Mathematical terms are:

$$\vec{e}_1 \cdot \vec{d} = \lambda \cdot N_1$$
$$\vec{e}_2 \cdot \vec{d} = \lambda \cdot N_2 \qquad (2)$$
$$\vec{e}_3 \cdot \vec{d} = \lambda \cdot N_3$$

$\vec{d}$ = wanted displacement vector
N = interference order

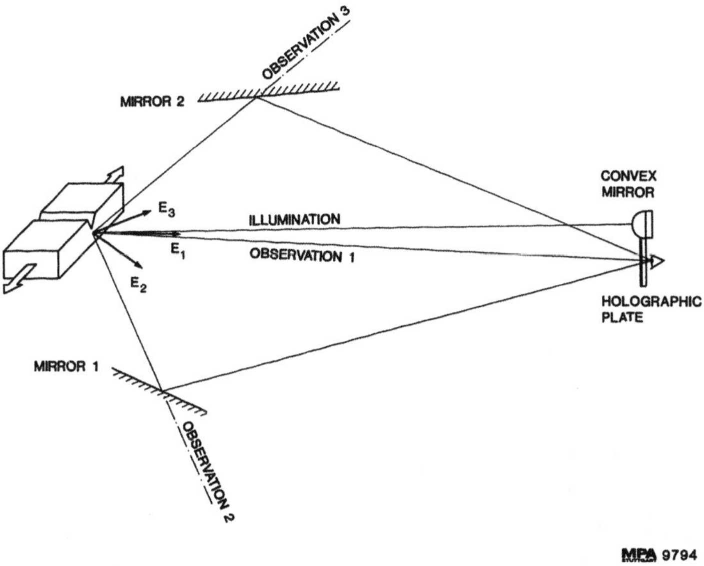

Figure 2. Holographic set-up with mirrors: 1 illumination, 3 observation directions

Holographic set-up

As mentioned in the previous chapter, it is possible to use a holographic set-up with mirrors to generate three sensitivity directions, Figure 2. This set-up, which has one illumination direction and three observation directions, is disadvantageous, because in practical operation the 3 sensitivity directions include a maximum angle of 30^0 to 40^0. Better conditions are obtained by working with 3 illumination and 3 observation directions. It is simultaneously illuminated and observed with two mirrors, Figure 3. However, in this case the various illumination waves disturb each other and form disturbing interference patterns in the hologram (Moiré-effect). Therefore, a set-up has been developed to avoid these interferences, Figure 4.

A standard holographic set-up is established. A two-reference-beam-module can be inserted /3/ in the reference beam for post evaluation according to the phase-shift method. The reference and object beams are put into a so-called "3-d-holography-module" prior to directing the reference beam to the hologram surface and the object beam to the beam expanding mirror. There they are split up into 3 single beams, and the path of each reference/object beam pair is extended by 2 m, respectively 4 m against the others. Consequently, the coherence condition for the not corresponding reference and object beams is no longer met and thus unwanted interferences will no longer occur. The system is fixed on an ordinary holography table and it is possible to complete the general set-up with the corresponding beam pairs. The three views of the object are then recorded by the three reference beams at various spots of the hologram plate, so that they will not superimpose each other during reconstruction.

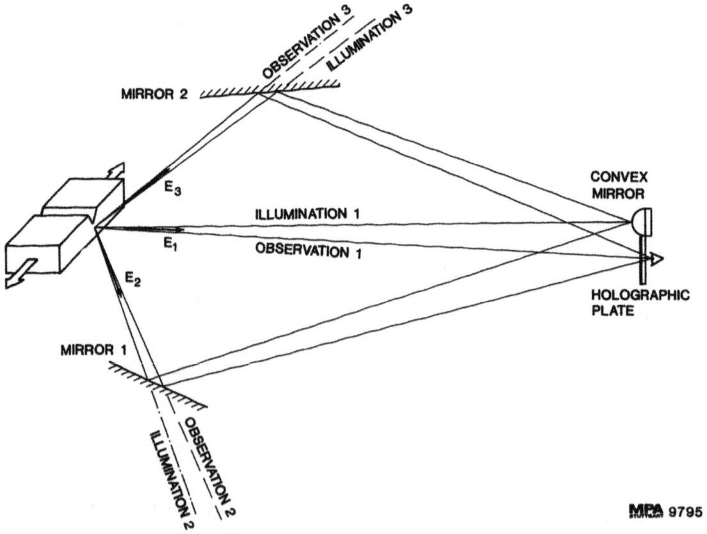

Figure 3. Improved holographic set-up: 3 illumination, 3 observation directions.

Evaluation system

The evaluation of the holographic interferograms is performed computer aided. Figure 5 shows the complete evaluation system with reconstruction set-up and computer. The evaluation facility is completely decoupled from the holographic set-up, so that holograms can be made and evaluated at the same time.

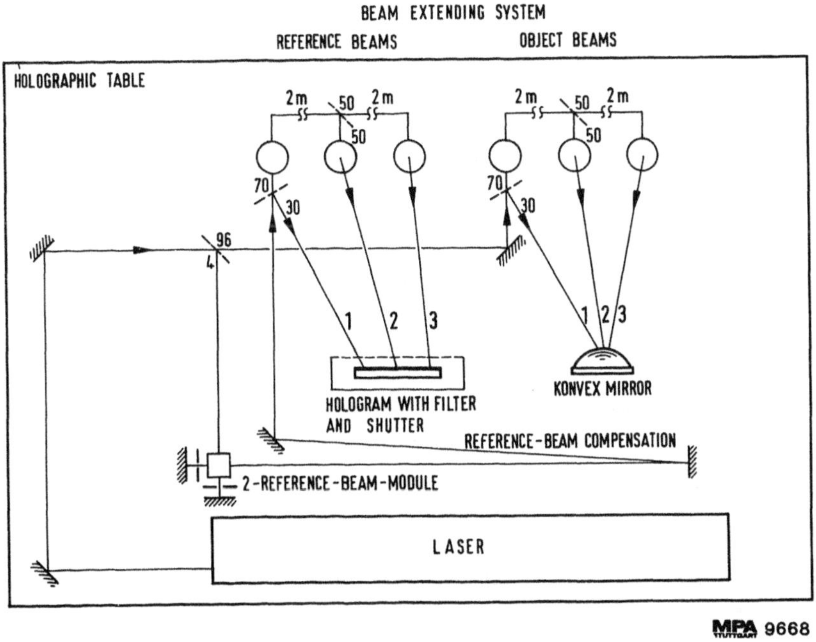

Figure 4. Holographic set-up to avoid Moiré effects.

The interference fringe order is determined by the phase-shift method which shall not further be considered here. The holograms are read with a TV camera into the computer which consists of a pdp $11/23^+$ processor with 20 Mbyte hard disk and 256 kbyte core memory. The image memory comprises 512 kbyte; this means that it is possible to store simultaneously 4 pictures with 256 x 256 points à 8 bit in the image memory and 3 additional pictures of the same size in the core memory. A picture with 512 x 512 points or 4 pictures with 256 x 256 points can concurrently be represented on the colour monitor.

The image processing system INCOS/E (Messrs. Signum, Munich) is mounted in the computer. It is supported by the operating system RT-11 and offers all essential basic image processing routines. The modular structure of the system allows the completion by user programmes which can be programmed in PASCAL and be linked to the picture processing programme.

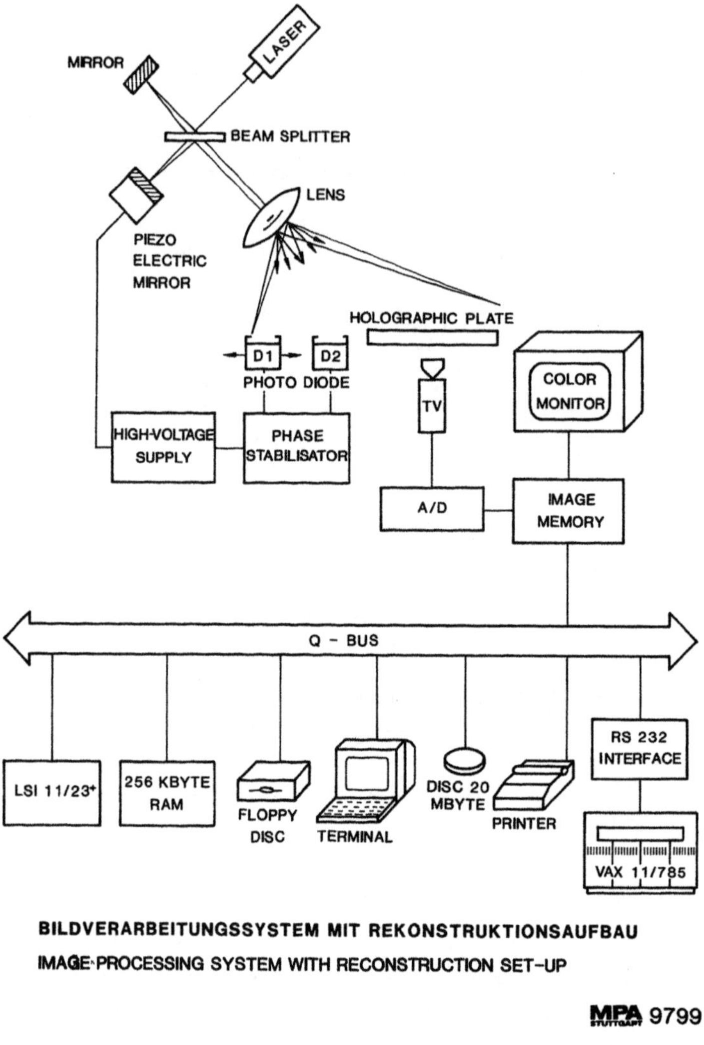

Figure 5. Evaluation system for holographic interferograms.

A completion for determining the interference order according to the phase-shift method was already implemented on purchase of this system. MPA Stuttgart has completed the system with the threedimensional evaluation of holographic interferograms.

The process of the threedimensional hologram evaluation is represented in the following. The complete flow chart is shown in Figure 6.

Three holograms with various sensitivity directions are recorded from an object and read into a computer by means of a TV-camera. The various views are accordingly distorted by the three different observation directions. In order to rectify them later so that the corresponding object points on the three views will also be positioned on the same image points, the cursor will determine the positions of three on the surface of the object defined points. These data will be written together with the geometrical data of the holographic set-up into a data file which is represented in Figure 6 by the box "geometry of the set up".

HOLOGRAPHIC DISLOCATION-AND STRAIN-CALCULATION

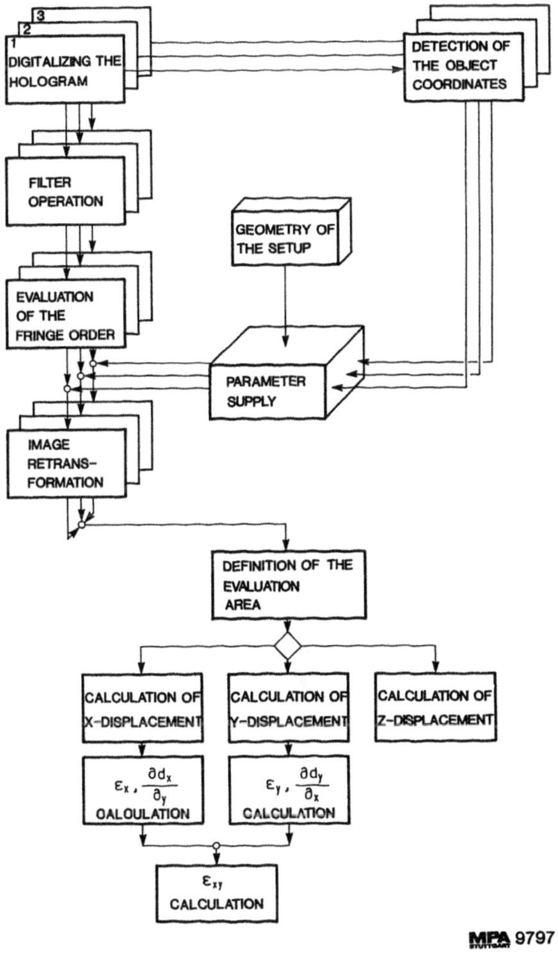

Figure 6. Flow chart of the evaluation.

First of all, the interference arrangements in the three views are calculated with the phase-shift method, and then the above mentioned image rectification takes place. Three rectified images have resulted containing the interference orders in the three sensitivity directions. They are represented in pseudo colours on the monitor and are standardized at 250 values.

The next step ist that the calculation of the displacement components in x-, y- and z-direction are calculated from these three result images according to the above delineated equations.

It has to be considered that the sensitivity direction has to be calculated again for each point from the geometry of the holographic set-up, because its direction over the object changes a little and may have a relatively considerable effect on the result. The result is again represented in pseudo colours and standardized at 250 values.

The strains are received from the displacements by differentiation. For this purpose pieces of smoothing lines are put on the displacement functions and always shifted further one by one point, Figure 7. This is necessary to minimize inadmissible errors during the differentiation caused by the unavoidable noise of the displacement curve.

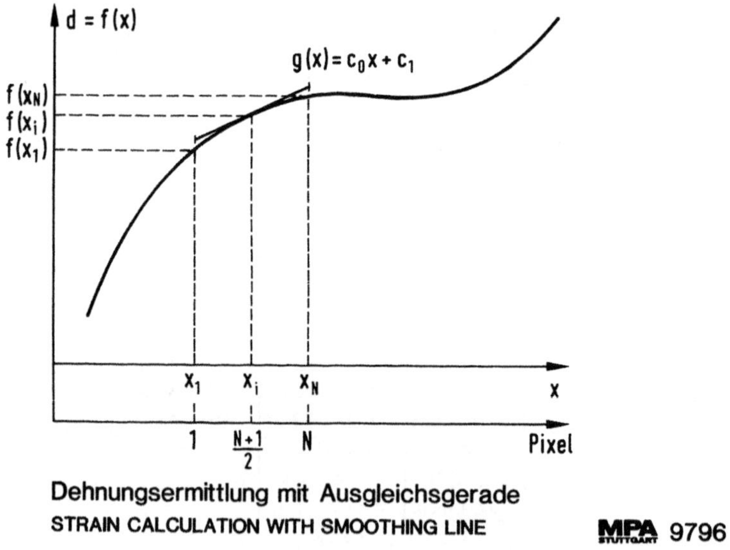

Figure 7. Numerical differentiation with smoothing lines.

Possibilities and limits of the system

The possibility was shown how to calculate with the delineated evaluation system spatial displacements and strain fields. Highest accuracy can be obtained by the fact that the holographic set-up can be chosen so that angles between the sensitivity directions obtain 50° - 90°. The essential limit is in the local solution of the object surface, i.e. in the main memory capcity. If, for instance, digitalization would be made with 1024 x 1024 points instead of 256 x 256 points, errors would be reduced to a minimum when rectifying the views.

A further reason for the limitation of the accuracy of the system introduced here is also due to the main memory capacity of the computer: All intermediate results have to be standardized to 256 values with 8 bit solution per point. This limits greatly the evaluation accuracy for the time being. The representative calculation time per image can be shortened due to larger computer capacity and faster computers. There should be the possibility of accelerating the evaluation by the factor 10.

Conclusion

The introduced system for the purpose of threedimensional hologram evaluation represents a prototype which in our opinion will step up the use of holographic interferometry in measuring and testing techniques. With this system, the currently rather qualitative use of holography could be extended to quantitative measurements.

References

/1/ Wernicke, G; Osten, W.: Holografische Interferometrie, Physik Verlag, Weinheim, 1982.

/2/ Schönebeck, G.: Eine allgemeine holografische Methode zur Bestimmung räumlicher Verschiebungen. Dissertation TU München, 1979.

/3/ Breuckmann, B.: Rechnergestützte Auswertung holografischer Interferogramme. Report to the "Holografie Seminar", MPA Stuttgart, 1985.

Practical Application of Ultrasonic Stress Analysis on Thin Walled Components

Praktische Anwendung der Ultraschall-Spannungsanalyse auf dünnwandige Bauteile

H: Jörgens, H. Wiessiolek

Rheinisch-Westfälischer Technischer Überwachungsverein
Essen, Germany

Summary

Three variations of the completely nondestructive ultrasonic stress analysis method can be derived from the fundamental equations. Accompanied by the results of a test demonstrating the effect, the investigation of an burried storage tank and the results of the measurement of environmental additional loads of a pipeline are described.

Zusammenfassung

Die Grundlagen der Ultraschall-Spannungsanalyse werden kurz dargestellt und drei unterschiedliche Varianten des völlig zerstörungsfrei arbeitenden Verfahrens angegeben. Neben den Ergebnissen eines Demonstrationsversuchs werden als weitere Beispiele qualitative Untersuchungen an einem unterirdischen Lagerbehälter und quantitative Messungen an einer Rohrfernleitung unter äußerer Zusatzbeanspruchung dargestellt.

The physical base of ultrasonic stress analysis is the dependence of the sound velocity on the state of mechanical strains and stresses in an component by the second and third order elastic constants. For strains transformed to principle axis in a rectangular coordinate system the three following fundamental equations can be deducted for waves propagating along one of these axis /1/:

$$1)\quad \rho \cdot v_{11}^2 = \lambda + 2\mu + (2l + \lambda) \cdot \xi + (4m + 4l + 10\mu) \cdot \varepsilon_1$$

$$2)\quad \rho \cdot v_{12}^2 = \mu + (\lambda + m) \cdot \xi + 4\mu \cdot \varepsilon_1 + 2\mu \cdot \varepsilon_2 - 0.5 \cdot \varepsilon_3$$

$$3)\quad \rho \cdot v_{13}^2 = \mu + (\lambda + m) \cdot \xi + 4\mu \cdot \varepsilon_1 + 2\mu \cdot \varepsilon_3 - 0.5 \cdot \varepsilon_2$$

λ and μ are the second, l, m and n the third order or Murnaghan constants; $\xi = \varepsilon_1 + \varepsilon_2 + \varepsilon_3$ is the sum of the main strains.

The first one is related to a longitudinal wave, the other two equations to two shear waves with planes of polarization perpendicular to each other. To compute strains from the measured velocities the elastic constants of higher order have to be known. As there are only few values reported in the literature, own measurements are necessary, except the load of the component can be varied in a quantitative manner.

As usually the length of the sound path is not known exactly, so you have to compare your time of flight measurement with another one performed at a reference specimen of the same material in the same metallurgical but stress free condition, if longitudinal waves are used according to equation 1); nevertheless there are some possibilities of practical application without the exact knowledge of the Murnaghan constants and without a reference specimen, if only the local or chronological change of stresses is the matter of interest, as in the case of burried gas containers described later on. From the equations (2) and (3) another equation can be derived for the dependence of the difference of time of flight of two orthogonally polarized shear waves on the difference of the principal strains in the plane parallel to the surface. In this case the path of the shear waves is perpendicular to the surface and the time between two bottom echos is measured. Additionally this method is useful to determine the principal stress directions because the form of the echos changes typically turning the probe around its axis. Sometimes a combination of longitudinal and shear waves can be used resulting in the computation of an absolute value for one principal stress.

The measuring device consists of parts listed at table 1. Principally it is possibel to grant exception to a computer which is used for controlling the measurement as well as for evaluation of the results. But the reliability and the order of accuracy can be improved with the number of measurements; therefore especially a portable computer is suitable for this purpose. Probes, pulser and receiver are the same as used in conventional ultrasonic technique. The time of flight measurement can be performed manually; first experiences are made using an automatical device /2/. The whole equipment can be carried in the trunk of a car. Because of this feature measurements outside the laboratory at components in the field are possible; but environmental influences as temperature and dust cannot be neglected.

It is a special token of ultrasonic strain analysis that it is a pure nondestructive method. A further advantage is the great number of possible modifications; this means that for a special problem a well adopted procedure can be found easily. As disadvantage especially the great dependence of the sound velocity on the structure and the texture of the material have to be mentioned.

To demonstrate the effect we placed probes at a common gas cylinder so that the change in time of flight caused by the change of the internal pressure could be measured. The readings are plotted at table 2.

At burried containers for liquid natural gas, the stress state near internal stiffener rings was to be evaluated, especially the changes caused by filling and exhausting. Therefore a shaft was digged from the surface of the soil to the top of the cylindrical containers. The time of flight of ultrasonic pulses running in axial as well as in circumferential direction of the containers was measured at different locations near the stiffeners before and after filling, using longitudinal waves and an arrangement of two probes as transmitter and receiver. The readings werde compared to values measured at a location which was estimated to be approximately free of stresses. The most important result was that we found remarkable local changes of the stress within a region of few inches ; the amount of stress change reached more than half of the yield stress. After the containers were filled, the spots of maximum load migrated some inches away; that meant that the cyclic stress-loading of the container could not be neglected with regard to their time of life.

The third example is a 20"-pipeline subjected to increasing addtional loads, which from time to time are relaxed by cutting and inserting or removing a short piece of a pipe. During such an event we carried out a sequence of measurements. In this case the pipeline was free of environmental stresses and stressed by internal pressure only immediately after cutting and rewelding. So a quantitative evaluation of additional load acting before the relaxation was possible despite the fact, that no reference specimen was available for calibration. The time of flight values measured under the different load conditions (see table 3a) are plotted at table 3b.

References:

/1/ E. Schneider and K. Goebbels
Zerstörungsfreie Bestimmung von Eigenspannungen
Paper published in
E. Macherauch and V. Hauk
Eigenspannungen (Vol.2), Oberursel 1983

/2/ E. Schneider and R. Herzer
Automatisierte Spannungsanalyse mit Ultraschall
(Automated Analysis of Stress States Using Ultrasonic Techniques)
Paper presented at 13. MPA-Seminar 1987 at Stuttgart

Ultrasonic Stress Analysis
Measuring Device

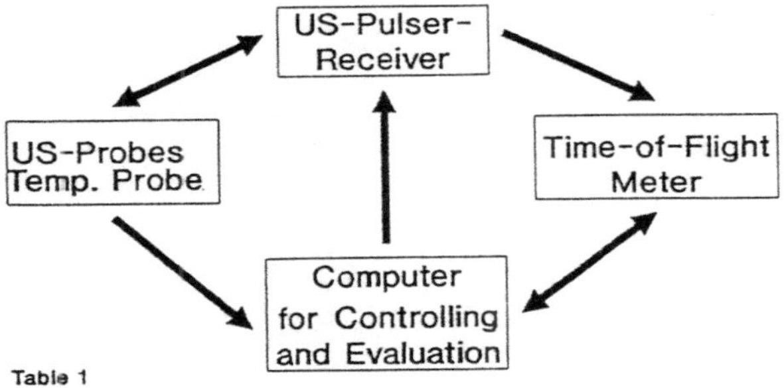

Table 1

Pressure Gas Cylinder
Ultrasonic Pressure Meter

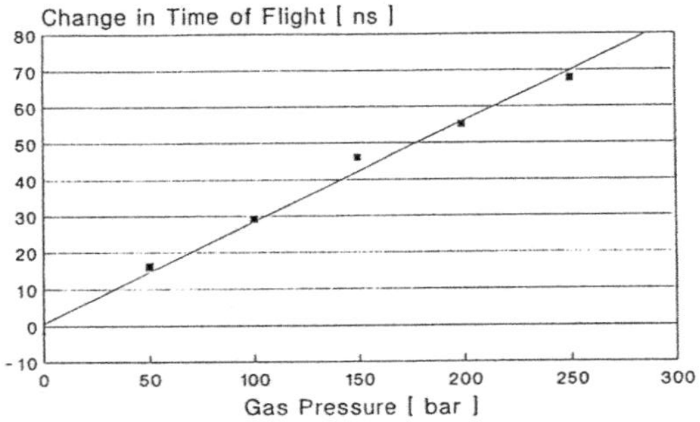

Skimming Longitudinal Wave Technique

Table 2

Ultrasonic Strain Analysis of a Pipeline
Change of Stress Strain Conditions

Step	Date	Load Condition
1	before Cutting	Internal Service Pressure and Additional Load
2	before Cutting	Addit. Load only
3	after Cutting	Free of Service and Add. Loads
4	after Cutting	Internal Pressure (Hydraulic Test)

Table 3a

Ultrasonic Strain Analysis of a Pipeline
before and after Mechanical Relaxation

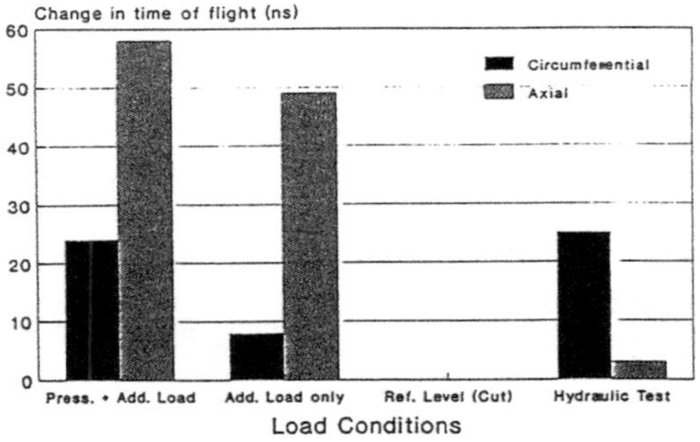

Skimming Longitudinal Wave Technique

Table 3b

Stress Field in a Cold-Rolled Nickel Plate Deduced from Diffraction Experiments Performed with Synchrotron Radiation at Varied Penetration Depths

H. Ruppersberg and M. Eckhardt

FB Ingenieurwissenschaften, Universität des Saarlandes, D-6600 Saarbrücken
W-Germany

Abstract

$\varepsilon(\sin^2\psi)$ curves measured for six lattice planes of a cold-rolled nickel plate using synchrotron radiation with the penetration depth varied between 1.6 and 10.8 µm deviate more or less strongly from the usual linear behaviour. An approximate interpretation, taking texture into account, is possible only by assuming an internal stress field which varies in a non-linear way with increasing distance from the surface. A completely satisfactory description was not possible. At least the (400)-curve behaves anomalously.

Introduction

For a cold-rolled nickel plate we [1] studied hardness, texture, width of the diffraction peaks and the variation of these properties on annealing. With X-rays [2] and synchrotron radiation [1] we performed X-ray stress measurements. Bragg peaks from six different lattice planes were investigated with and without applying external forces, and for the (420) reflection the penetration depth of the radiation was changed by a factor 6 by varying the wavelength through the K-absorption edge of nickel.

The texture is not strongly pronounced. Nevertheless it has a distinct influence on the results of the stress measurements as we [3] showed recently for the Voigt and the Reuss cases. For some of the lattice planes investigated the $\varepsilon(\sin^2\psi)$ curves remained nearly linear even under load, in spite of the texture. But the apparent X-ray elastic constants (XEC) turned out to be changed by about 10 % with respect to the values of an isotropic sample. While we succeeded in relating to texture the non-linear variation of the $\varepsilon(\sin^2\psi)$ curves which we observed for several lattice planes on changing the applied stress, it was not possible to find an analogous interpretation for the non-linear behaviour based on the internal stress field. The most obvious example is the behaviour of the (420) curves obtained with penetration depths of 1.6 and 10.8 µm, respectively which are given in figures 1 and 2 and which

look completely different. An interpretation is possible only by assuming gradients of the internal stress field. We [4] proposed a simple formalism for evaluating linear variations of the stress components $\sigma_{11}(z)$ and $\sigma_{22}(z)$ in the z-direction perpendicular to the surface of the specimen, taking texture into account. A least-square fit from the $\varepsilon(\sin^2\psi)$ curves observed for the two penetration depths yields the following equations:
$\sigma_{11}(z) = -400 + 40z$ [N/mm^2] and $\sigma_{22}(z) = 80 - 15z$ [N/mm^2], z is given in µm. Texture is of little influence, but we had the impression that second order derivatives $d^2\sigma_{ii}/dz^2$ should be taken into account for a better description of the experimental data.
In this paper we try to evaluate the whole set of 7 different $\varepsilon(\sin^2\psi)$ curves.

Formalism

We discuss a single-phase homogeneous and chemically uniform polycrystalline specimen which has a texture close to orthorhombic with respect to the coordinates x(1), y(2) and z(3). The surface of the plate is perpendicular to the z-axis. $\varepsilon_{ij}(x,y,z)$ and $\sigma_{ij}(x,y,z)$ (with i,j = 1,2,3) are the components of the local strain tensor and stress tensor, respectively. $\varepsilon_{\varphi\psi}(x,y,z,h)$ is the local longitudinal strain in the direction p of the normal on the reflecting plane. The latter is characterized by the Miller indices (h). ψ is the angle between p and the z-axis. The p-z plane intersects the x-y plane along a line which forms the angle φ with the x-axis. From the angular position Θ of the centre of gravity of a Bragg peak one obtains

$$\varepsilon_{\varphi\psi}(h) = \int_{V(h)} \varepsilon_{\varphi\psi}(x,y,z,h)\, e^{-z/\tau}\, dV \Big/ \int_{V(h)} e^{-z/\tau}\, dV \qquad (1)$$

The integrals are taken over the volume V(h) of the reflecting grains. τ is the absorption factor. For experiments in ψ-mode [5] it is given by $\tau = \tau_o \cdot \cos\psi = (\sin\Theta/2\mu) \cdot \cos\psi$. μ is the absorption coefficient.
Each local $\varepsilon_{ij}(x,y,z)$ varies proportionally to the macroscopic stress σ_{kl} (k,l = 1,2,3), and for τ large enough for surface anisotropy to be neglected, the following equation was given by Brakman [6] and by Barral and Sprauel [7]:

$$\varepsilon_{\varphi\psi}(h) = F_{ij}(\varphi\psi,\, R^{RV}_{33ij}(h)) \cdot \sigma^{RV}_{ij} \qquad (2)$$

for which the Einstein summation convention applies. $R_{ijkl}(h)$ are the generalized XECs which may be calculated from the orientation distribution function for both the Reuss (R) and the Voigt (V) cases. σ_{ij}^{RV} are the macroscopic stresses (external or internal) which, corresponding to R or V, yield the strain given on the lefthand side of Equ. 2.

We now suppose that the local strain and stress components are uniformly distributed only for x-y planes, in which they fluctuate about the mean values $\varepsilon_{ij}(z,h)$ and $\sigma_{ij}(z)$, respectively. Equ. 1 may be written

$$\varepsilon_{\varphi\psi}(h) \;=\; \int \varepsilon_{\varphi\psi}(z,h)\, e^{-z/\tau}\, dz / \int e^{-z/\tau}\, dz \tag{3}$$

with

$$\varepsilon_{\varphi\psi}(z,h) = \int\int_{x,y,V(h)} \varepsilon_{\varphi\psi}(x,y,z,h)\, dx\, dy \,/ \int\int_{V(h)} dx\, dy \tag{4}$$

Equ. 2 yields:

$$\varepsilon_{\varphi\psi}(z,h) = F_{ij}(\varphi,\psi,\, R_{33ij}^{RV}(z,h)) \cdot \sigma_{ij}^{RV}(z) \tag{5}$$

with

$$\sigma_{ij}(z) = \int_x \int_y \sigma_{ij}(x,y,z)\, dx\, dy \,/ \int dx\, dy \tag{6}$$

Equilibrium conditions require $\sigma_{3i}(z=0) = 0$ in the absence of external forces on the x-y (z=0) surface, and

$$\begin{aligned} \partial\sigma_{13}/\partial x + \partial\sigma_{23}/\partial y + \partial\sigma_{33}/\partial z &= 0 \\ \partial\sigma_{12}/\partial x + \partial\sigma_{22}/\partial y + \partial\sigma_{32}/\partial z &= 0 \\ \partial\sigma_{11}/\partial x + \partial\sigma_{21}/\partial y + \partial\sigma_{31}/\partial z &= 0 \end{aligned} \tag{7}$$

in the absence of body forces. For vanishing gradients with respect to x and y one obtains $(\partial\sigma_{3i}/\partial z) = 0$ throughout in the specimen. Finite values of $\partial\sigma_{3i}/\partial z$ will occur locally in polycrystalline plates with anisotropic elastic behaviour of the individual grains. However, because $\int\int(\partial\sigma_{ij}/\partial x)\, dx\, dy$ and $\int\int(d\sigma_{ij}/\partial y)\, dx\, dy$ will vanish for areas of the specimen sufficiently far away from the edges of the rolled plate the corresponding integral values of $(\partial\sigma_{3i}(z)/\partial z)$ should vanish as well, which together with the condition $\sigma_{3i}(0)=0$ means that $\sigma_{i3}(z)=0$ everywhere in this part of the specimen. In the present paper we will discuss only the case $\varphi=0$ and σ_{12} has no influence on $\varepsilon_{o\psi}(h)$.

Thus we consider only the fields of the $\sigma_{11}(z)$ and $\sigma_{22}(z)$ stress components. For "compatible" deformations there are also restrictions concerning the second derivatives of ε_{ij}, which yield $\partial^2\varepsilon_{22} / \partial z^2 = \partial^2\varepsilon_{11}/\partial z^2 = \partial^2\varepsilon_{12}/\partial z^2 = 0$ for vanishing gradients of ε_{ij} with respect to x and y. This means that from the second derivatives of $\varepsilon_{ij}(z)$ only the $\partial^2\varepsilon_{3i}(z)/\partial z^2$ terms have to be considered. We make no use of this restriction because it is not clear to us whether the observed strains are really compatible. We suppose further that texture (i.e. $R_{33ij}(z,h)$, equ. 5), which in the present case is only of minor importance, does not depend on z. Equ. 5 is evaluated for the Hill case, i.e. $\sigma_{ij}(z) = (\sigma^V_{ij} + \sigma^R_{ij})/2$, and its combination with equ. 3 may be given in a more convenient form:

$$\begin{aligned}\varepsilon_{o\psi}(h) = &\left(\tfrac{1}{2}S_2(\psi)\sin^2\psi + S_1(\psi)\right)\int\sigma_{11}(z)e^{-z/\tau}dz/\int e^{-z/\tau}dz \\ &+ S_1(\psi)\int\sigma_{22}(z)e^{-z/\tau}dz/\int e^{-z/\tau}dz\end{aligned} \tag{8}$$

The $S_i(\psi)$ values are the XEC, corrected for texture (Hill case). The specimen is thick enough for the integrals to be calculated within the limits ranging from $z = 0$ to $z = \infty$. The equation resulting from (8), if linear variation of $\sigma_{ii}(z)$ with z is assumed, was given in a previous publication [4]. In addition we will discuss a more general ansatz proposed by Hauk and Krug [8]:

$$\sigma_{ii}(z) = (A_i + B_i\, z + C_i\, z^2)\, \exp(-D_i\, z) \tag{9}$$

which yields for the ψ-mode experiments:

$$\begin{aligned}&\int\sigma_{ii}(z)e^{-z/\tau}dz/\int e^{-z/\tau}dz = \\ &= \left[A_i + (B_i + 2A_iD_i)\tau_o\cos\psi + (2C_i + A_iD_i + B_iD_i)\tau_o^2\cos^2\psi\right]/(1 + D_i\tau_o\cos\psi)^3\end{aligned} \tag{10}$$

(10) inserted into (8) yields an equation which allows us to calculate the eight parameters describing $\sigma_{ii}(z)$ from a set of experimental $\varepsilon_{o\psi}(\sin^2\psi,h)$ curves obtained for different τ_o values.
For $\sin^2\psi = 1$ one obtains $\varepsilon_{090}(h) = S_2/2 \cdot \sigma_{11}(0) + S_1(\sigma_{11}(0) + \sigma_{22}(0))$ for both the linear $\sigma_{ii}(z)$ and the damped quadratic ansatz, and this should generally be true. For evaluating surface stresses this equation is however of little help because $\varepsilon_{o\psi}(\sin^2\psi)$ turns out to be perpendicular at $\psi = 90°$. This fact underlines the importance of extending the experiments to ψ angles as large as possible, which has long since been postulated by Hauk and coworkers [8].

Results and Discussion

In figs. 1 and 2 the $\varepsilon_{o\psi}$ versus $\sin^2\psi$ curves are given in the order of increasing τ_o values. The incomplete (331) curve was determined by X-ray diffraction. The other results were obtained with synchrotron radiation using the Two-Axis-Diffractometer at HASYLAB. Details of the experimental set-up are given elsewhere [3] together with the orientation distribution function which was used to calculate the $F_{ij}(\psi)$ coefficients of equ. (5).
For the present discussion we neglect ψ-splitting which is strongest for the smallest penetration depth.The deduction of $\sigma_{ii}(z)$ from the $\varepsilon(\sin^2\psi)$ curves turned out to be a tedious puzzle. The reason is obvious: a given stress field $\sigma_{ii}(z)$ yields a unique solution for $\varepsilon_{\varphi\psi}$, but a single experimentally obtained $\varepsilon_{\varphi\psi}(\sin^2\psi)$ curve may result from very different $\sigma_{ii}(z)$ distributions, especially because the data for the large ψ-angles are missing. For inverting the Laplace transform which is obtained from the combined equations (3) and (5) one needs an analytical ansatz for $\sigma_{ii}(z)$ and absurd results may be obtained if an insufficient ansatz is used together with an insufficient amount of experimental information. To give an example, we tried a least-square fit for the (420) curve given in fig. 2, based on a quadratic expansion of $\sigma_{ii}(z)$, and obtained $\sigma_{11}(z)$ [N/mm^2] = -46000 - 1200 z + 40 z^2 and $\sigma_{22}(z)$ = -150000 - 7000 z + 2400 z^2, z given in µm. This is complete nonsense but fits the experimental data almost perfectly!

The linear expression for $\sigma_{ii}(z)$ given in the introduction, and which is shown at the bottom of fig. 2 as dashed lines, matches quite well the two (420) curves from which it was calculated. It even gives correctly the shallow maximum at $\sin^2\psi$ = .15 in fig. 2. Also the (400) and (331) curves are satisfactorily described by this plot. It fails however for the (220), (311) and (222) reflections.
Inspection of the whole set of $\varepsilon_{o\psi}$ curves shows an impressive variation on going from τ_o = 1.6 to 3.4 µm and much less change beyond this penetration depth. This behaviour indicates that drastic changes of the stress field occur within the first micrometers from the surface. We therefore tried a quadratic ansatz but finally obtained better results using equations (9) and (10) proposed by Hauk and Krug [8]. We calculated the coefficients A_i to D_i by minimizing the absolute values of the deviations between the experimental $\varepsilon_{o\psi}$ values and the recalculated data. With 0.14 the mean absolute deviation of the (400) reflection turned out to be much larger than the one of the other lattice planes, which was 0.06 and smaller. The error of the latter

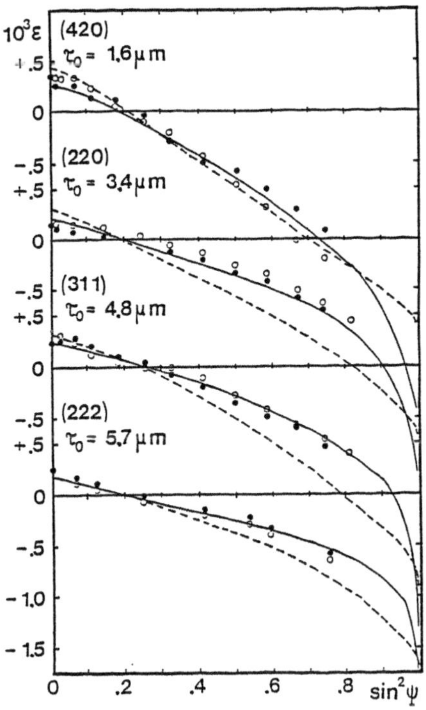

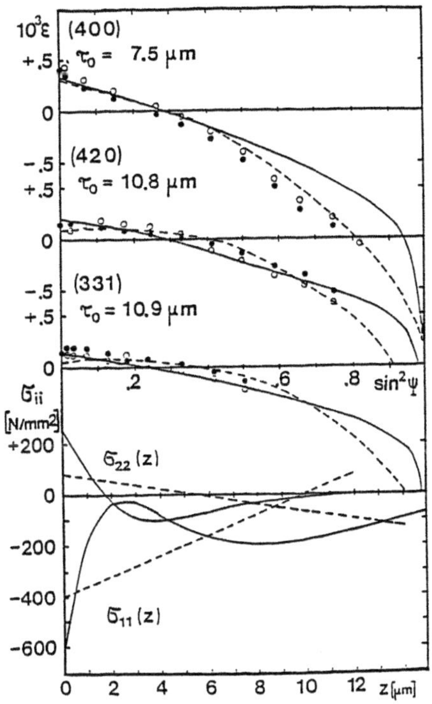

Fig. 1. $\varepsilon_{o\psi}(\sin^2\psi)$

Fig. 2. upper part: $\varepsilon_{o\psi}(\sin^2\psi)$
lower part: $\sigma_{ii}(z)$

Experimental values circles: $\psi>0$, points: $\psi<0$

Calculated curves dashed lines: linear $\sigma_{ii}(z)$

full drawn curves: damped quadratic $\sigma_{ii}(z)$

became distinctly smaller, 0.04 on the average, if the (400) values were disregarded and in this way were obtained the results given in the figures as full drawn lines. The error of the (400) data was evidently increased and became 0.19. Thé only plausible reasons we can imagine for the anomalous behaviour of the (400) reflection is a failure in our case of the damped quadratic ansatz, special σ^{II} distributions related to the plastic deformation or gradients in x and y direction which then allow for finite $\sigma_{i3}(z)$ values. The last mentioned case was recently studied by Hauk and Krug [9] and these authors proposed a $\sigma_{i3}(z)$ ansatz which is different from the damped squared relation (9). Also for the other curves some unsatisfactory details remain: it was for example not possible to improve the curvature of the (420) curve, fig. 2, without increasing too strongly the deviations for the (220) reflection and the above mentioned maximum which was almost perfectly given by the linear plot could not be reproduced.

The minimum with respect to variations of the A_i to D_i parameters of the absolute deviations mentioned above is very shallow. For the ψ-range investigated the $\varepsilon_{o\psi}(h)$ curves are quite insensitive with respect to variations of $\sigma_{11}(z<1\mu m)$ and values of $\sigma_{11}(o)$ between the yield strength and -300 N/mm^2 seem compatible with our $\varepsilon_{o\psi}$ curves. A linear plot of the $\varepsilon(\sin^2\psi)$ curve obtained for the (420) reflection at the smallest penetration depth studied, yields σ_{11} = -300 N/mm^2. It seems logical assume that in our case this is indeed an upper limit of $\sigma_{11}(0)$. This statement would have been impossible without the experiment done at a penetration depth of only 1.6 µm. An extension of the ψ-range to $\sin^2\psi$ values of at least 0.95 would have been necessary to obtain the same information from the other curves.

The $\sigma_{22}(z)$ values are only of minor importance for the general behaviour of the recalculated $\varepsilon_{o\psi}$ curves, and the corresponding plot of $\sigma_{22}(z)$ in fig. 2 is just supposed to give the general trends. These trends are however important for special details of the calculated $\varepsilon_{o\psi}$ values and the fit is significantly improved by proper choice of $\sigma_{22}(z)$.

More systematic theoretical work and additional experiments at φ = 45°and 90° are necessary to obtain more satisfactory results.

Acknowledgements

We thank the Deutsche Forschungsgemeinschaft and HASYLAB for supporting this work and we are indebted to the team of the Two Axis Diffractometer for their valuable help. We thank Prof. Berveiller (Metz) for helpful discussions.

References

1. Eckhardt, M; Ruppersberg, H.: in "Residual Stresses in Science and Technology", eds. Macherauch, E.; Hauk, V.: DGM Oberursel (1987) 377.

2. Ruppersberg, H.; Schmidt, W.; Weiland, H.: Z. Metallkde. 71 (1980) 475.

3. Eckhardt, M.; Ruppersberg, H.: Textures and Microstructures (1988) in print.

4. Eckhardt, M.; Ruppersberg, H.: Z. Metallkde. (1988) in print.

5. Wolfstieg, U.: HTM 31 (1976) 19.

6. Brakman, C.M.: J. Appl. Cryst. 16 (1983) 325.

7. Barral, M.; Sprauel, J.M.: in "Eigenspannungen", eds. Macherauch, E.; Hauk, V., DGM Oberursel, Vol. 2 (1983) 31.

8. Hauk, V.; Krug, W.K.: same as 1., page 303.

9. Hauk, V.; Krug, W.K.: HTM 43 (1988) 164.

A Hybrid Computer for Phase Images Visualization and Correlation Based Recognition

Paolo SIROTTI Paolo DEMANINS

Dipartimento di Elettrotecnica Elettronica Informatica
University of Trieste, Italy

Summary

We examine some possibilities of an optical-digital hybrid computer for the processing of phase images constituted of surfaces or transparent thin materials. Different modalities of operation are possible, depending on the stage of the process in which it is more convenient to turn from optical to digital operation and on the objective of the process. Two subjects are treated: first a phase image is visualized employing a rotating derivative filter, secondly we survey some applications of a correlation based recognition method to surface textures and to mechanical or thermal stresses in thin transparent materials.

Es werden einige Möglichkeiten für den Einsatz eines optisch-numerischen (hybriden) Rechners bei der Prüfung von aus Oberflächen oder durchsichtigem dünnen Material bestehenden Phasenbildern untersucht. Verschiedene Vorgangsweisen sind möglich, je nach dem Ziel des Verfahrens und dem Stadium, in welchen es am besten erscheint, vom optischen zum digitalen Betrieb überzugehen. Zwei Themen werden behandelt: zunächst wird ein Phasenbild vermittels eines derivativen Rotationsfilters visualisiert, sodann werden einige Anwendungen der korrelationsgestützten Erkennungsmethode zur Bestimmung von Oberflächentexturen und mechanischen oder thermischen Spannungen bei transparenten dünnen Materialien durchgegangen.

Introduction

In coherent optical processing input images are often considered as real. However cathegories of images exist that can be actually considered as input images for a coherent optical processing system and at the same time can be represented as pure imaginary or complex functions. Thin planar transparent objects whose refractive properties are negligible and whose refractive index is a function of two spatial coordinates are pure phase images. Similarly let us consider surfaces that reflect, do not scatter, as a plane mirror in which only the phase and the amplitude (but not the direction) of each incident ray is altered by a point function $f(x,y) = |f(x,y)| \exp j\varphi(x,y)$. When illuminated by a uniform plane wave these surfaces originate images for which optical processing is possible. Transparent objects or reflected images (phase images), in so far as they do not overcome the limits of Fourier optics [1] , can be optically Fourier transformed, filtered or correlated.

Some possibilities of an optical-digital (hybrid) computer (fig.1) for phase images processing are examinated. The optical part of the system consists of two He-Ne lasers: one of them is to process transparent images, the other to process reflected images. A convergent lens performs the Fourier transform of the transmitted or reflected light. This transform can be filtered and reconstructed in the same optical part. The digital part consists of a TV camera, which acquires the Fourier transform or the reconstructed image, of a PC IBM AT with mathematical coprocessor, frame grabber, graphical board, optical disk and of two high resolution monitors.

Different modalities of operation are possible depending on the stage of the process in which it is more convenient to turn from optical to digital operation and on the objective of the process: this may be a filtering to give a modified copy of input image or a correlation based recognition.

In the following two processes are considered. First the hybrid system is employed to visualize phase images. In particular we illustrate a technique based on a special class of amplitude filters, the rotating filters [2], [3]. Secondly we survey some applications of a correlation based recognition method to surface textures and to mechanical or thermal stresses in thin transparent materials.

Phase images visualization using derivative rotating filters

Adopting for the phase image $f(x,y)$ a monodimensional notation, the intensity of its derivative, that is the image that can be seen or revealed if the original image is subjected to a derivation is:

$$\left|\frac{d}{dx}|f(x)|\right|^2 + \left|\frac{d}{dx}\varphi(x)\right|^2 |f(x)|^2 .$$

This expression shows that the derivation makes visible the phase information of the image. If $|f(x,y)| = 1$, that is the image is a pure phase image, the derivation makes the square of the derivative of $\varphi(x,y)$ appear.

An optical derivative filter can be obtained with the technique of rotating filters. These filters are opaque sectors rotating around the optical axis on the Fourier plane. If $T(r)$, (r, ϑ polar coordinates), is the circularly symmetrical transmittance to be realized, the profile of the sector results:

$$\vartheta(r) = 2\pi T(r) .$$

A derivative filter on the Fourier plane is a filter whose transparence linearly increases with the spatial frequency. Under obvious boundary conditions the profile of the rotating derivative filter results:

$$\vartheta(r) = 2\pi r/R ,$$

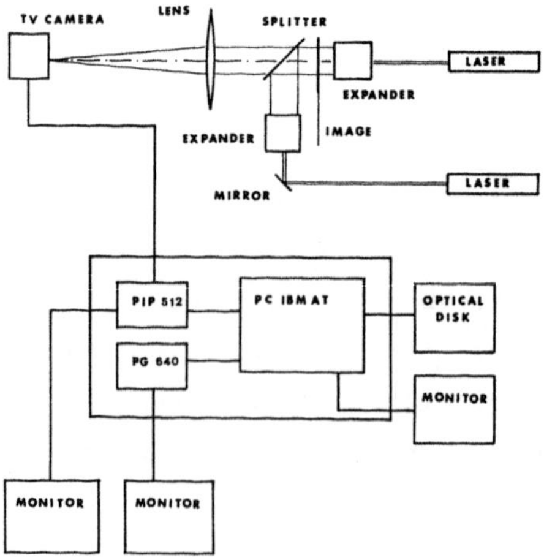

Fig.1. Optical digital computer

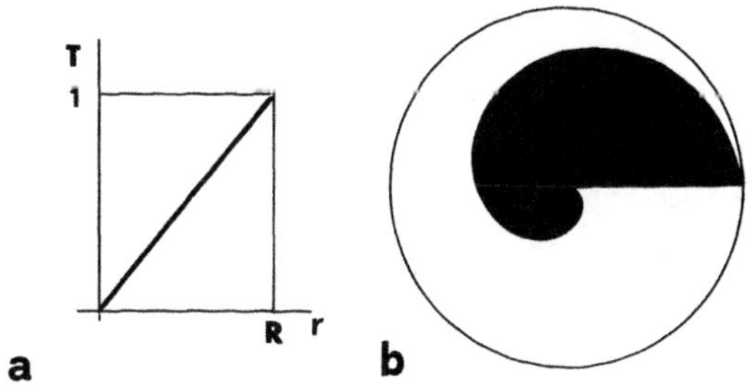

Fig.2. a) Transmittance of the derivative filter;
b) Profile of the rotating derivative filter.

where R is the radius of the filter bandwidth. Fig. 2a represents the transmittance T(r), while the profile of the filter is given in fig. 2b.

Fig. 3a shows a plastic sample filtered by the derivative filter. The sample (polymethylmetacrylate) is 2 mm thick, about 3x3 mm wide, with a hole (0.5 mm Ø), and a resistive thread in it. The figure is the static initial situation, produced by a previous heating of the thread. Fig. 3b,c are photographs of the sample taken at time intervals of 10s, after a current of 1.5A has began to flow along the thread. When the current is interrupted the deformation proves to be reversible, returning to the situation of fig. 3a. Without filtering the sample appears as in fig. 3d, that has been taken after the heating sequence.

Similar results can be obtained using more conventional filters, for instance a $\lambda/4$ phase shifter acting on the zero order frequency. Some trials indicate that the derivative filter is probably less sensitive than the phase shifting filter to low frequency phase variations. On the other side it can be easily centered: its alignement can be verified on an amplitude test image. Similar to a high pass filter (but more effective in the tests performed), it visualizes the phase as a bright image on dark field.

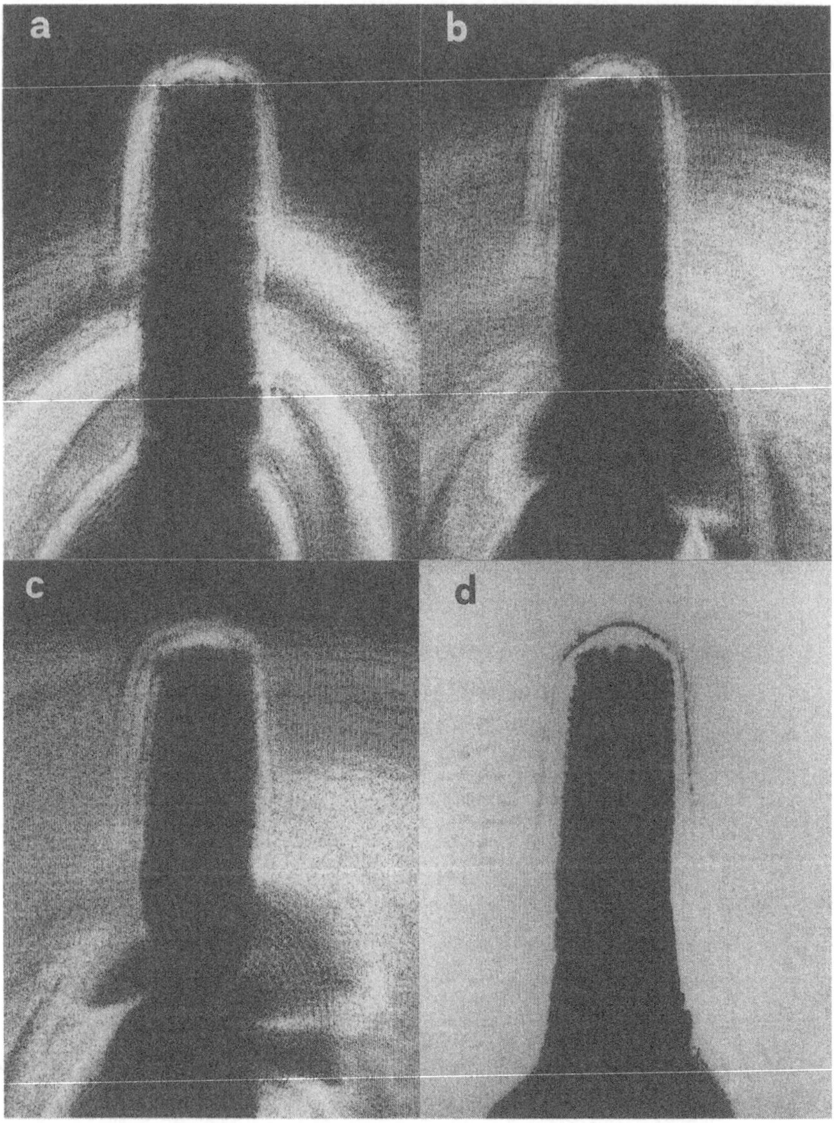

Fig.3. a) Plastic sample filtered with a derivative filter;
b), c) Same sample thermally stressed by a current of 1.5 A flowing along the thread inside the hole, 10s and 20s after the beginning of the heating;
d) Original

Correlation of phase images

A correlation based method for the recognition of phase images has already been tested in several situations. In [4] the presence of mechanical stress was recognized in plastic samples. In [5] patterns of time varying thermal stresses were recognized, while the possibility of the method to follow, working in reflection,the evolution of surface textures was examined in [6]. The method works correlating an unknown image with a set of reference images and evaluating the resulting crosscorrelation functions. Unknown and reference images are set at the input plane of the optical system. Their joint Fourier transform is quadratically recorded and again transformed (optically or digitally), giving a correlation pattern in which the crosscorrelations of all pairs of input images can be identified. A possible disposition of input images, that avoids the overlapping of output correlation terms and reduces the errors of the system, is reported in fig. 4a. Fig. 4b is the corresponding output pattern.

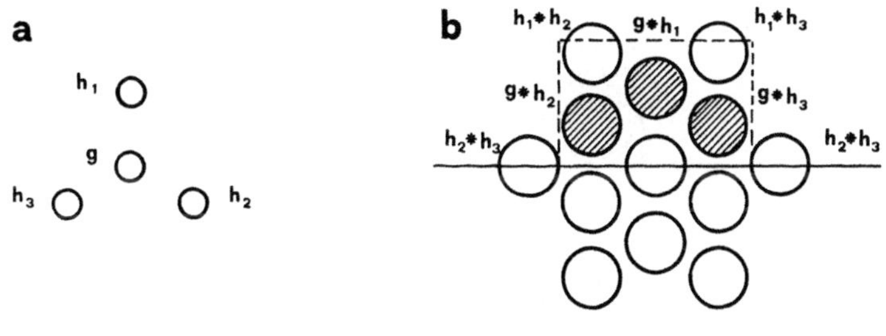

Fig.4. a) Possible input disposition for the correlation of 4 images;
b) Corresponding output plane

The output correlation pattern is symmetrical, so only a half can be considered.

The capability of the method of monitoring the istantaneous similarity of input images has already been proved and will not be discussed here.

The evaluation of correlation areas can be greatly aided using a suitable software. Fig. 5a,b,produced on the monitor of the graphic board and successively on a plotter,shows one of the steps digitally analyzed of a sequence taken to follow, working in reflection, the evaporation of a solvent spread on microscope cover glasses.

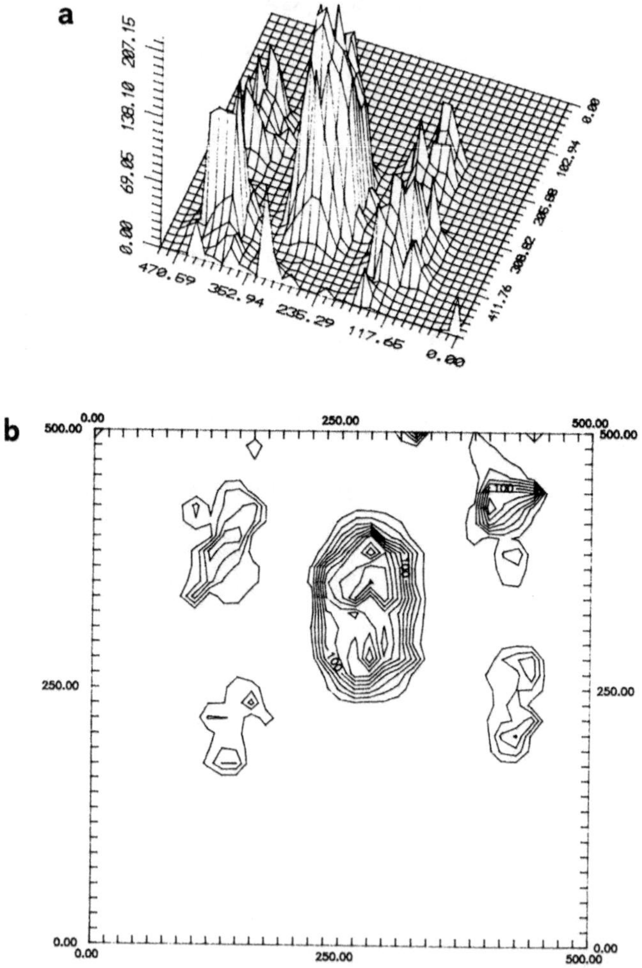

Fig.5. a),b) Orthographic projection and contour map. Situation of high correlation $g*h_1$. The central zone (sum of all autocorrelations of input images) has been partially obscured by a stop

References

1. Yu, F.T.S.: Introduction to diffraction, information processing and holography. MIT Press 1973.

2. Mottola, R.; Sirotti, P.: Filtri di ampiezza rotanti per l'elaborazione ottica delle immagini. Alta Frequenza, No. 3, 1978.

3. Rizzatto, G.; Sirotti, P.: Optical processing of linear array ultrasonic images. Acoustical Imaging. Vol. 12, 225-233. Ash, E.A.; Hill, C.R. (eds.) Plenum Publishing Corp. 1982.

4. Sirotti, P.: Optical joint Fourier transform correlation for phase objects recognition. Nondestructive Characterization of Materials II. Bussiere, J.F.; Monchalin, J.P.; Ruund, C. O.; Green, R.E., Jr. (eds.) Plenum Publishing Corp. 1987.

5. Sirotti, P.: Coherent optical recognition of time varying thermally stressed phase images. Proc. of the 16th Symp. on Nondestructive Evaluation. Gray, C.D.; Matzkanin, G.A. (eds.) 1987.

6. Sirotti, P.: Coherent optical analysis, filtering and correlation of reflected images. In press.

Three Dimensional Surface Representations of Linear Elastic Anisotropy in Cubic Single Crystals

Edward J. Tucholski and Robert E. Green, Jr.

Center for Nondestructive Evaluation and Department of Materials Science and Engineering, The Johns Hopkins University, Baltimore, Maryland U.S.A. 21218

Summary

The magnitude of the velocities of the three possible modes of elastic wave propagation are calculated for single crystal chromium, aluminum, and iron at intervals over all possible propagation directions and plotted to form wave velocity surfaces. These materials were chosen because they vary widely in isotropy factor as defined by Kittel [1]. Additionally, the energy flow vectors associated with each mode are found, allowing the angle between the energy flow vector and the propagation direction to be calculated. Energy flux deviation angles are then displayed on three dimensional polar plots to form three dimensional surfaces. The relative elastic anisotropy of these single crystals is compared graphically and quantitatively using isotropy factors.

The maximum longitudinal mode energy flux deviation angles are calculated for a number of cubic crystals. It is shown that near room temperature, maximum longitudinal mode energy flux deviation is linearly related to the isotropy factor. Also the angle with respect to the <100> direction at which this maximum longitudinal mode flux deviation occurs decreases exponentially with the isotropy factor.

Introduction

Various ultrasonic techniques are used for nondestructive characterization of real materials. Underlying most of these techniques is the assumption that the material in question is perfectly isotropic. Unfortunately, this is usually not the case. Wood and synthetic composite materials are obvious examples of materials that are anisotropic because of their macroscopic structure. In metals, the degree of anisotropy is sometimes more elusive due to surface finishes which mask the anisotropy below.

In many cases, the macroscopic anisotropy of a composite material is very similar and in fact mathematically identical to that found in a perfect single crystal of a particular symmetry [2]. For this reason, representing a material by elastic stiffness moduli is very appropriate. With the advent of digital computers, many [2,3,4] have performed calculations to predict such parameters as elastic moduli, wave phase velocity, and energy flux deviation angle. Plots displaying this data were typically two dimensional slices of the material. In order to present a complicated material, enough slices had to be presented to adequately display the symmetry. Two dimensional models present limited information about three dimensional phenomena in the same way that scalars and vectors fail to adequately describe physical quantities requiring tensors.

Recent advances have been made in the field of computer graphics [5], particularly in the areas of hidden line algorithms and shading routines. The present work exploits these advances to construct three dimensional plots of surfaces which at a single glance depict the manner and degree of the anisotropy of a material. Computer Aided Design (CAD) systems allow the flexibility to rotate and superimpose these surfaces in order to study the particular material properties in question. From both an educational and design point of view, these capabilities are extremely useful.

Phase Velocity

The theory explaining the propagation of linear elastic waves propagating in anisotropic solids has been worked out by many authors [3,4]. Applying Hooke's law and the definition of strain to an unbounded, continuous medium in the absence of body forces results in a general wave equation. Assuming the solution is a traveling plane wave moving in a direction normal to the plane with propagation direction cosines, $l_m = (l,m,n)$, and with particle displacement direction cosines, $\alpha_i = (\alpha,\beta,\gamma)$, results in the following relation:

$$\left(c_{ijkl}\, l_j\, l_l \;-\; \rho v^2\, \delta_{ij}\right) \alpha_k \;=\; 0. \qquad (1)$$

Eqn. (1) is often called Christoffel's Equation referring to the 1877 paper by Christoffel [6]. Knowing only the second order elastic stiffness moduli, c_{ijkl}, the density of the material, ρ (assumed homogeneous), and a particular set of propagation direction cosines, the magnitude of three phase velocities ($v = \omega / k$) in that propagation direction can be obtained by solving for the eigenvalues of Eqn. (1). Substitution of each eigenvalue into Eqn. (1) allows the calculation of the eigenvector direction cosines of the particle displacements, α_i.

The wave propagation mode whose eigenvector is primarily in the direction of wave propagation is known as quasi-longitudinal, while the two wave propagation modes whose eigenvectors are primarily in the directions perpendicular to the direction of wave propagation are known as quasi-transverse.

Three dimension polar plots of these three velocities form what Musgrave [3], defined to be a velocity surface of three sheets, i.e. the locus of the termination of a radius vector drawn through the origin of length proportional to the phase velocity in the direction of the vector. Using elastic constants [7] and material densities [8] from standard tables, Fig. 1 displays the wave velocity surfaces for chromium, aluminum, and iron. Calculations of the wave velocity solutions were done on a digital computer using the method of Jacobi [9]. Plotting was then done on a VAX-11/780 in a CAD environment.

The shapes of the three surfaces for chromium are typical of those found for cubic elements with an isotropy factor, I, as defined by Kittel,

$$I = \frac{2\, c_{44}}{c_{11} - c_{12}} \tag{2}$$

which is less than unity. The wave velocity surfaces for aluminum and iron are typical of the shapes found for materials with isotropy factors greater than unity and with low and high relative anisotropy, respectively. The resemblance between the general

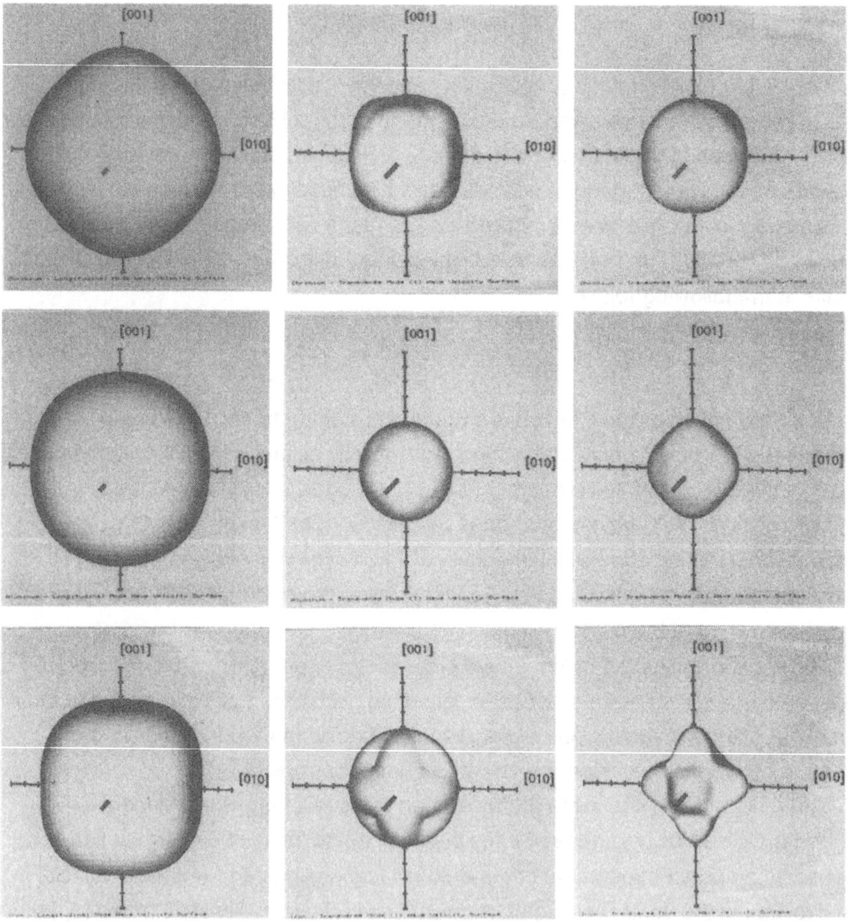

Fig. 1. Quasi-longitudinal, fast quasi-transverse, and slow quasi-transverse wave velocity surfaces for (a) chromium (row 1), (b) aluminum (row 2), and (c) iron (row 3). (1 division = 1000 m/sec)

shape of these surfaces, and the effective Young and shear modulus surfaces plotted for cubic crystals by Schmid and Boas [10], is unmistakable.

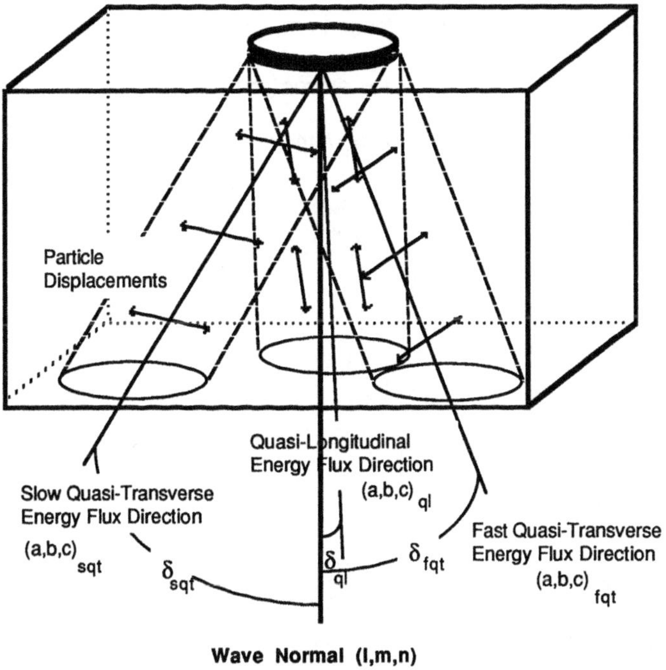

Fig. 2. Schematic diagram illustrating quasi-longitudinal and quasi-transverse wave propagation in a crystal along a non-pure mode axis.

Energy Flux Deviation

To determine the direction of maximum energy flux in these refracted waves, we use the energy flow vector derived by Love [11].

$$E_j = - \sigma_{ij} \dot{u}_i \tag{3}$$

Using Hooke's Law and the definition of strain allows the stress tensor to be found in terms of particle displacements for a cubic crystal. Substituting an assumed plane wave solution, and its first time derivative yields the time varying energy flow vector in terms of propagation direction cosines and particle displacement direction cosines. Taking the time average over one period gives the time averaged energy flow vector components from which the direction cosines of the energy flow can be found [4]. The direction of energy flow is not necessarily the same as the propagation direction (l,m,n) which is perpendicular to the plane wavefront.

The angle between the propagation direction and energy flow for a particular propagation mode can then be found from the following relationship,

$$\delta = \arccos(al + bm + cn) \quad (4)$$

where a,b,c are the direction cosines of the energy flow for a particular propagation

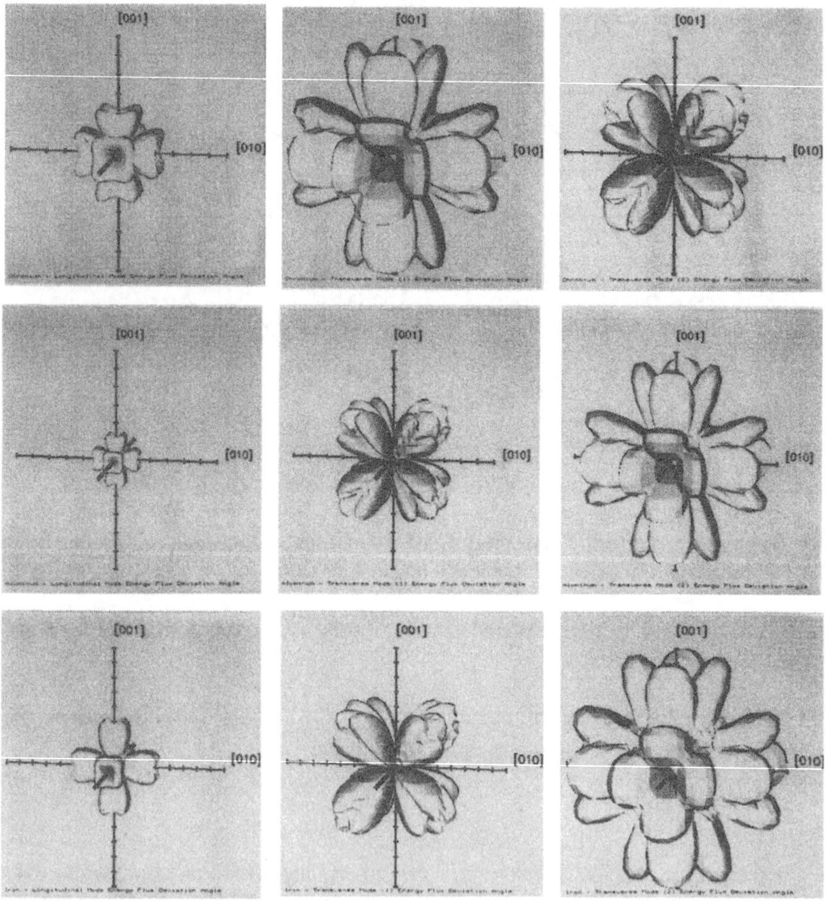

Fig. 3. Quasi-longitudinal, fast quasi-transverse, and slow quasi-transverse energy flux deviation angle surfaces for (a) chromium (row 1, 1 div. = 3°), (b) aluminum (row 2, 1 div. = 2°), and (c) iron (row 3, 1 div. = 5°).

mode. All directions at an angle, δ, around a particular propagation direction are not the possible directions of energy flow, since energy will only travel in one particular direction given by Eqn. (3). In a manner similar to the wave velocity surfaces, energy flux deviation surfaces are calculated and plotted as Fig. 3 for chromium, aluminum and iron.

Discussion

By displaying parameters calculated for anisotropic materials in three dimensions, certain observations concerning the general shapes of these parameters and the

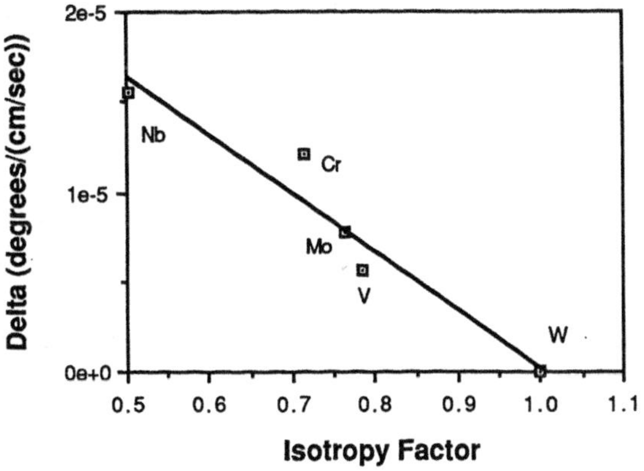

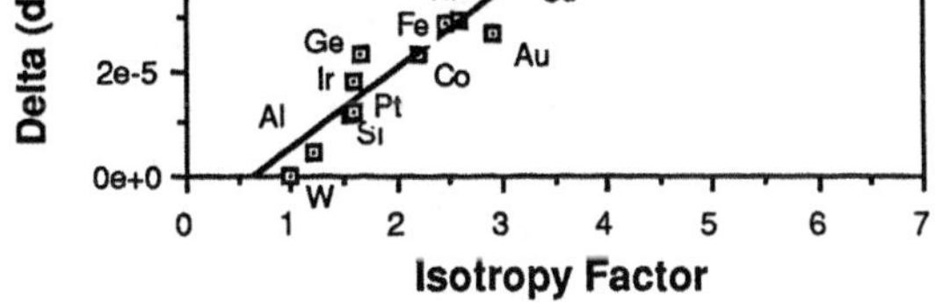

Fig. 4. Velocity normalized maximum longitudinal mode energy flux deviation angle for materials with (a) isotropy factor < 1 (upper), (b) isotropy factor > 1 (lower).

relationship to the material isotropy factor can be made. Fig. 1 displays the wave velocity surfaces for chromium (I = 0.714), aluminum (I = 1.23), and iron (I = 2.46). By arranging these surfaces in order of increasing isotropy factor, the effect of the material anisotropy is easily observed. As is perhaps intuitive, the topological features of the surface become more pronounced as the anisotropy of the crystal becomes greater. In terms of the isotropy factor, this is equivalent to a smaller isotropy factor (approaching zero) for those materials with isotropy factors less than unity and a larger isotropy factor (approaching infinity) for those with isotropy factors greater than unity.

The quasi-longitudinal mode energy flux deviation angle varies as a function of the material anisotropy as noted from the scales of the plots in Fig. 3. In other words, the maximum energy flux deviation angle, which always lies in a {110} plane for cubic crystals generally increases with the degree of anisotropy. More exactly, if this energy flux deviation angle is normalized by the velocity in the direction of maximum energy flux, the relationship to isotropy factor is quite linear for cubic crystals at room temperature as shown in Fig.4.

Additionally, viewing energy flux deviation angle surfaces in order of increasing isotropy factor reveals that the angle, θ, at which this maximum energy flux deviation angle strays from the <100> direction generally increases as isotropy factor (not degree of anisotropy) increases. This relationship has a somewhat exponential fit as shown in Fig. 5.

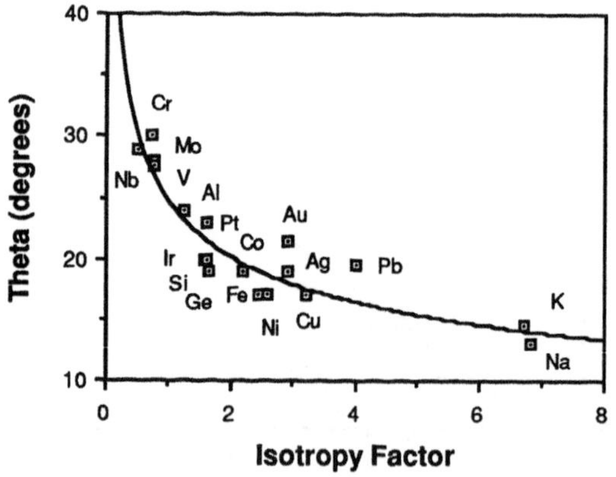

Fig. 5. Propagation direction resulting in maximum energy flux deviation angle with respect to the nearest <100> direction and in a {110} plane.

Conclusion

Using surface representations, this paper has described the anisotropic nature of elasticity in a number of solid materials. Too often in the nondestructive evaluation community, the attempt is made to describe linear elasticity with only two coefficients as if the material in question was isotropic. In fact the required number of coefficients may be as great as twenty one. In the case of single crystal tungsten, molybdenum, or aluminum the assumption of elastic isotropy may be good enough. In the case of single crystal iron (or nickel, silver, gold, copper, zinc, magnesium, etc.) the assumption may lead to significant experimental errors. Locations of material cracks, for example, may be incorrectly determined by ignoring energy flux deviation. The same principle applies for polycrystals with significant texture or composite materials with anisotropy induced by the method of formation or manufacture.

Acknowledgement

The authors are deeply indebted to the staff at the Computer Aided Design/ Interactive Graphics group at the United States Naval Academy. The assistance of these fine programers was invaluable to the completion of this work. Additional thanks is also due to the staff of the Photo Lab and Educational Resources Center at the Academy for their assistance in preparing the manuscript.

References

1. Charles Kittel, *Introduction to Solid State Physics*, (Wiley, New York, 1953), page 52. (This definition is deleted in later editions.)
2. Ronald D. Kriz and Hassel M. Ledbetter, "Elastic Representation Surfaces of Unidirectional Graphite/Epoxy Composites," In *Recent Advances in Composites in the United States and Japan: ASTM Special Technical Testing Publication 864*, Edited by J. R. Vinson and M. Taya, (American Society for Testing and Materials, Philadelphia, Pennsylvania, 1985).
3. M.J.P. Musgrave,*Crystal Acoustics*, (Holden-Day, San Francisco, 1970).
4. Robert E. Green, Jr., Ultrasonic Investigation of Mechanical Properties, in *Treatise on Materials Science and Technology, vol. 3*, edited by Herbert Herman, (Academic Press, New York, 1973).
5. David F. Rogers, *Procedural Elements for Computer Graphics,* (McGraw Hill, New York, 1985).
6. E. B. Christoffel, "Ueber die Fortpflanzung von Stössen durch elastische feste Körper," *Annali Di Matematica Pura Ed Applicata,* Series II, Vol 8, 193-243 (1877).
7. R. F. S. Hearmon, *Landolt-Börnstein: Numerical Data and Functional Relationships in Science and Technology, Volume 11,* edited by K.H. Hellwege, (Springer-Verlag, Berlin, 1979).
8. Robert C. Weast, *Handbook of Chemistry and Physics*, (The Chemical Rubber Company, Cleveland, Ohio, 1970).
9. Forman S. Acton, *Mathematical Methods that Work*, (Harper and Row, New York, 1970).
10. E. Schmid and W. Boas, *Plasticity of Crystals*, (Chapman and Hall, London, 1968).
11. A. E. H. Love, *Treatise on the Mathematical Theory of Elasticity*, Fourth Edition, Cambridge University Press, Cambridge, 1927).

Determination of Shot Peened Surface States Using the Magnetic Barkhausen Noise Method

V.HAUK, P.HÖLLER*, R.OUDELHOVEN, W.A.THEINER*,

Institut für Werkstoffkunde, Rheinisch-Westfälische Technische Hochschule, Aachen,

*Fraunhofer-Institut für zerstörungsfreie Prüfverfahren (IzfP), Saarbrücken

Summary

The experiments have been performed on different shot peened sheets of Armco iron and the steel C60. The residual stress (RS) distributions have been determined by the X-ray method and by the bending arrow etching method. From these results, the residual stress distributions of first, second (and third) kind have been evaluated. The nondestructive testing has been carried out using the magnetic Barkhausen noise method. From these Barkhausen noise events which are rectified and recorded over the tangential field strength, different stress and microstructure sensitive quantities have been derived. The most important parameter for these testing procedures is the analyzing frequency. This investigation has shown that shot peened surface states can be characterized by these micromagnetic quantities: especially the homogeneity of shot peened-, the work-hardened profile- and the quantitative (after calibration) residual surface-stress state.

Zusammenfassung

Die Untersuchungen beziehen sich auf unterschiedlich gestrahlte Bleche der Stähle Armco Eisen und C60. Der Eigenspannungs(ES)-Verlauf wurde über der Randtiefe röntgenographisch und nach dem Biegepfeilabätzverfahren bestimmt. Aus diesen Messungen werden die Eigenspannungstiefenverläufe (I.II.(III.) ES) berechnet. Als mikromagnetische Meßgröße wird das magnetische Barkhausenrauschen genutzt. Aus den über der tangentialen H-Feldstärke aufgenommenen und gleichgerichteten Barkhausenrauschereignissen werden unterschiedliche, spannungs- und gefügeempfindliche Meßgrößen abgeleitet. Wichtigste Prüfvariable ist die Analysierfrequenz, bei der das magnetische Barkhausenrauschen aufgenommen wird. Die Ergebnisse zeigen, daß mit mikromagnetischen Meßgrößen die gestrahlten Randschichtzustände richtungsabhängig und über der Tiefe charakterisiert werden können. Unter Nutzung dieser mikromagnetischen Meßgrößen kann der gestrahlte Zustand hinsichtlich seiner Homogenität richtungsabhängig und über dem Ort überprüft werden, ferner kann die kaltverfestigte Randschichttiefe und der Oberflächeneigenspannungszustand nach einer Eineichung in quantitativer Weise bestimmt werden.

Specimens

The steel sheets which were used for shot peening had a dimension of about 300 mm length and 200 mm width. The thickness was about 2 mm. The specimens of Armco iron and C60 in the as delivered condition had averaged Vickers hardness and yield point values as given in table 1. One half of these sheets have been shot peened on both sides /1/ with shot peening parameters as indicated in table 2. By using different pressure, mass flow and distance from the nozzle to work piece, different work hardening profiles were achieved in the surface; further different residual stress distributions were maintained by the variation of the above parameters. The microhardness result given in fig. 1 shows that there is no work-hardening and softening effect on the surface region of the examined C60 sheet. Depending on the local measuring point, these microhardness values show a varation of about fifty.

	HV 0.05	Rp 0.2 [N/mm²]
Armco Iron	150	250
C60	300	580

Tab.1: material parameter

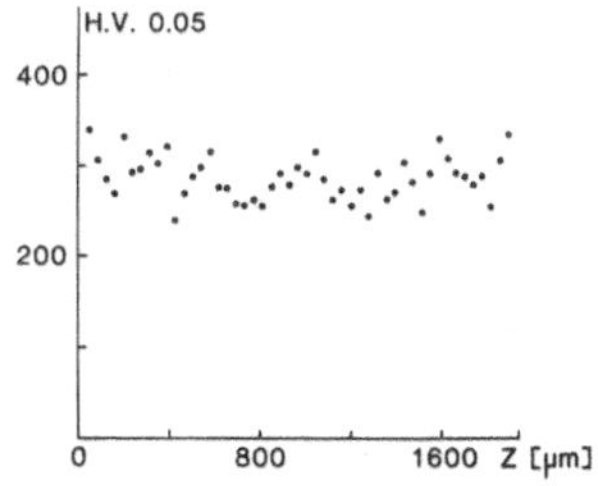

Fig. 1: Microhardness.
Material: C 60. Specimen: C1R

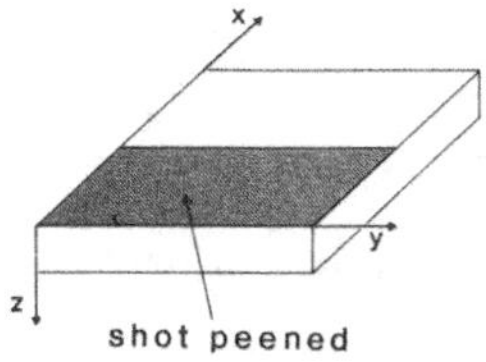

Fig.2: Specimen-shot peened area

Results determined by X-ray measurements

The as delivered and the shot peened state are further characterized by X-ray measurements and mechanical tests. The diffractometer type and the X-ray parameters which have been used for the residual stress determination are given in table 3. The amount of residual stresses is evaluated using the well known $\sin^2\psi$ formula /2/. The surface stress states have been determined before and after shot peening (table 4). One can see that the as delivered states show residual compressive stresses between 0 and about -140 N/mm². (The direction σ_{11} represents the direction x as indicated in fig. 2, σ_{22} the direction y and σ_{33} the direction z). Because these sheets did not show any texture, changes in the micromagnetic results are given by stress anisotropy as discussed later in the section dealing with magnetic measurements.

material	specimen	media	p (bar)	m (kg/min)	HD (mm)	α (°)	VD (mm/min)	ΔS (mm)
Armco	A1R	glass MGL	1	2	100	0	3000	10
	A2R		0.5			0		
	A3R		1			60		
C60	C1R	ø0.4-0.8mm	1			0		
	C2R		0.5			0		
	C3R		1			60		
	C4R	glass MGL ø0.23-0.3 mm	1		200	0		
	C5R		1		100	0		
	C6R		0.36		100	0		
	C7R		0.36		200	0		

p-pressure
m-mass flow
HD-distance from nozzle to workpiece
α-impact angle
VD-nozzle velocity
ΔS-distance of traces

Tab.2: Shot peening parameters

type of diffractometer	Ω
radiation	Cr-K_α
{hkl},2 Θ	{211},156°
penetration depth $\tau_{0.3}$	4.4 μm
radiated area at ψ = 0°	20 x 2 mm^2
$\frac{1}{2}\ s_2$	$5.76 \cdot 10^{-6}\ mm^2/N$
s_1	$-1.25 \cdot 10^{-6}\ mm^2/N$

Tab.3: X-ray parameters

Besides these stresses, the half width (HW) resp. FWHM values of these sheets have also been determined. After shot peening, the half width values increase by about 0.45 degrees, whereas the residual stress state becomes quasi isotropic (table 4). One can also see from the amount of the residual stress values that the chosen shot peening parameters did not change the magnitude of the compressive surface stress level in a drastic way. These results demonstrate that the RS level at the surface is not sufficient to characterize shot peening states /3/. To avoid these unambiguous values, one has to determine depth profiles as given in fig.

3. One can see from these results that the RS depth profile is changing in a systematic way with decreasing pressure and shot peening intensity.

material	probe	as delivered: HW degree	as delivered: $\sigma_{11}-\sigma_{33}$ N/mm²	as delivered: $\sigma_{22}-\sigma_{33}$ N/mm²	shot peened: HW degree	shot peened: $\sigma_{11}-\sigma_{33}$ N/mm²	shot peened: $\sigma_{22}-\sigma_{33}$ N/mm²
Armco	A1R 1	0.92	-43±8	-31±6	1.50	-235±11	-260±11
	A1R 2	0.96	-45±9	-42±8	1.46	-265±14	-282±15
	A2R 1	0.92	-33±7	-38±7	1.33	-283±14	-307±16
	A2R 2	1.0	-59±8	-54±7	1.33	-266±15	-294±15
	A3R 1	1.0	-7±5	-14±8	1.33	-258±11	-231±13
	A3R 2	0.92	-32±7	-39±8	1.33	-288±12	-243±14
C60	C1R 1	1.17	-138±13	-110±9	1.58	-390±10	-402±12
	C1R 2	1.17	2±8	-64±10	1.60	-350±9	-351±11
	C2R 1	1.17	-57±16	-94±9	1.60	-409±8	-400±9
	C2R 2	1.17	3±8	-60±9	1.52	-377±9	-363±8
	C3R 1	1.27	-70±15	-125±9	1.77	-391±14	-336±21
	C3R 2	1.25	-74±15	-101±9	1.76	-306±10	-291±21
	C4R 1		-40±5			-386±17	
	C4R 2						
	C5R 1		-40±5			-382±9	
	C5R 2						
	C6R 1		-31±12			-326±17	
	C6R 2						
	C7R 1		-31±12		1.64	-354±6	-363±10
	C7R 2						

Tab.4: Half-width (HW) and RS I-values, evaluated from X-ray parameters.

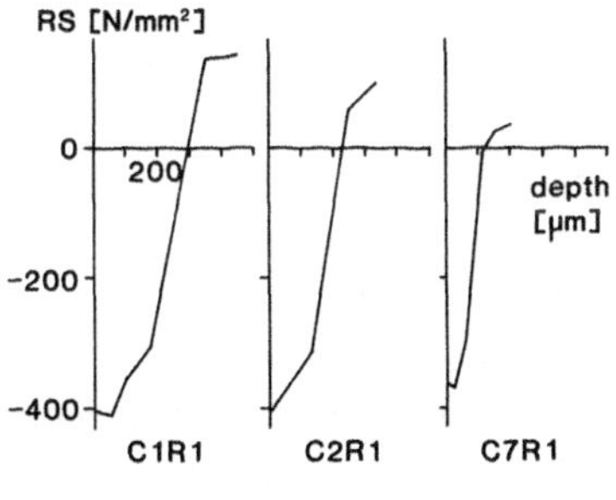

Fig. 3: RS-depth profiles for different shot peened states (see tab.2)

Fig.4 shows the (RS I+II, RS I, RS II and HW)-depth profiles for one shot peened state CIR (see table 2). The indicated macrostresses (RS I) have been measured by the bending arrow etching method, whereas the RS I+II levels are determined by the X-ray method. One can see from these results that this shot peened state is characterized by the great compressive RS I stresses at the surface and that the direction dependent stress distribution is quite homogeneous. If one subtracts the mechanically determined macro-stresses from the X-ray results, one can get the contribution of residual stresses of second kind as outlined in /4/. These results demonstrate that the RS II depth profile can have a quite different stress distribution as the macroscopic mechanical stress profile. These results must be checked and recognized in future for discussing micromagnetic interactions with stress fields.

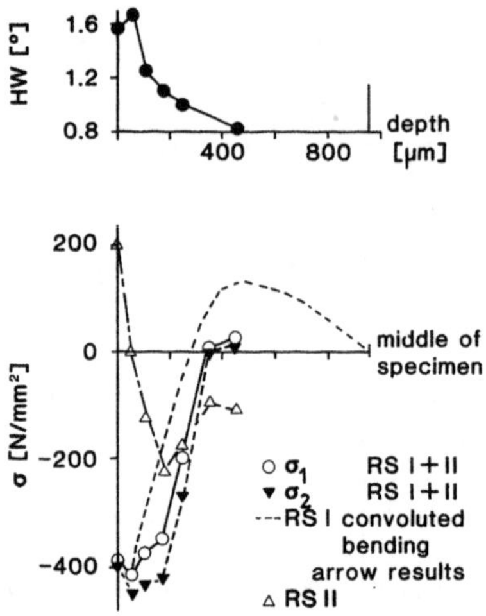

Fig. 4: Residual stresses (RS I,II) and half width values (HW) for the shot peened state C1R (see table 2). Material: C60.

Results by magnetic measurements

To characterize shot peened states by nondestructive methods only a few articles have been reported /5,6/. In this paper the magnetic Barkhausen noise method is used to characterize the homogeneity of shot peened states and the work-hardening profile depth. The procedure of performing residual stress measurements in a quantitative way are discussed in detail elsewhere /7/. For the homogeneity tests the 3MA-equipment and for the profile measurements laboratory setups have been used.
The measuring quanitities which have been used are the coercivity H_{cM} and the Barkhausen noise amplitude M_{max}. Qualitatively the coercivity value increases with increasing dislocation densities and increasing compressive macro stresses, whereas the Barkhausen noise amplitude M_{max} shows an inverse behaviour in contrast to the coercivity. The measuring trace in fig.5 along shot peened and non shot peened states shows this characteri-

stic behaviour in a very clear way. In the interface between work-hardened and as delivered states the H_{cM}, M_{max}-quantities are also influenced by macrostresses.

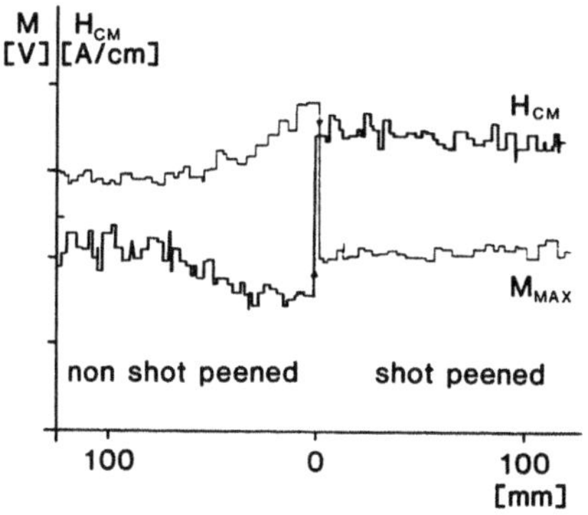

Fig.5: Barkhausen noise; homogeneity of sheet states. Material: C60

In fig. 6 the magnetic pole fig.s are given for the as delivered and for the shot peened states. Since the as delivered sheets did not show any texture in the X-ray results the given anisotropy of the coercivity and the Barkhausen noise amplitude must be explained by macroscopic stress fields. After the shot peening procedure (fig. 6b) the sheet state becomes isotropic. This can be derived from the constant M_{max} and H_{cM}-angle dependences.

To characterize the work-hardening depth profile the most important variable is the analyzing frequency f_A of the received Barkhausen noise events. One characteristic result is given in fig. 7. The penetration depth is calculated with constant permeability and electrical conductivity values; the analyzing frequency f_A is the middle frequency of the used band pass filter. The magnetic profile curves are showing that the as delivered state is characterized by constant coercivity values of about 6 A/cm. In contrast to this constant level, the shot peened sheet shows magnetic H_{cM}-values of about 8 A/cm at the surface. By this measuring procedure work-hardening profiles can be characterized in a simple quantitative nondestructive way. This can be proved by systematic investigations of different work-hardened surface states as demonstrated in fig. 7b. These H_{cM}-depth curves compare well with X-ray results (fig. 3). In fig. 8 the coercivity H_{cM} is plotted over the X-ray half width value (both values correspond to penetration depth of about 67 µm). One can see that shot peening increases the half width values by about 0.45° whereas the coercivity value increases by 2.4 A/cm. One can conclude from these results that the workhardening process of the ferritic phase is the main reason for the increase in both measuring quantitives HW and H_{cM}.

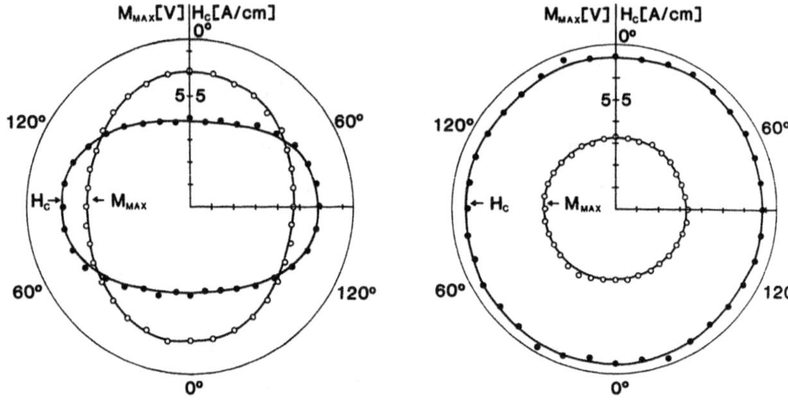

Fig. 6a,b. Barkhausen noise; "pole figures". Material: C60, left as delivered, right-shot peened;

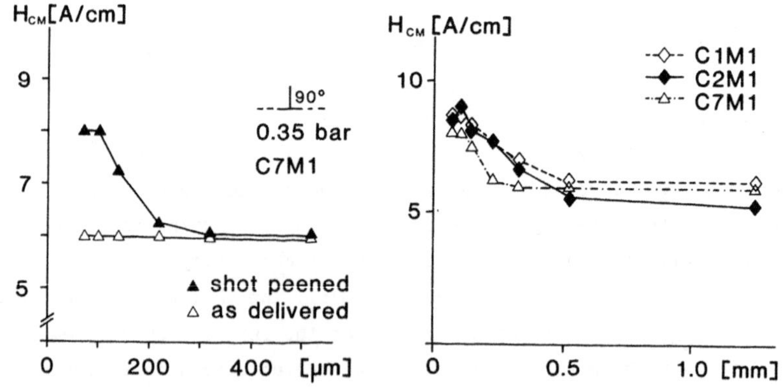

Fig. 7a,b: Magnetic Barkhausen noise: Work hardening depth profile.

Acknowledgement: This contribution is based on work performed with the support of the Deutsche Forschungsgemeinschaft (DFG).

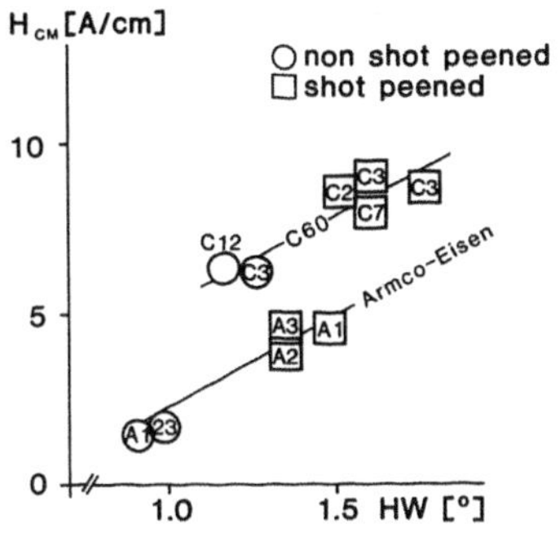

Fig. 8: Coercivity H_{cM} over half width HW for different shot peened states.

References:

1. Kopp, R., Ball, H.: RWTH-Aachen 1987.

2. Faninger, G., Hauk, V., Macherauch, E., Wolfstieg, U.: Empfehlungen zur praktischen Anwendung der Methode der röntgenographischen Spannungsermittlung (bei Eisenwerkstoffen). HTM 31, 109-111, 1976.
   Hauk, V., Macherauch, E.: Die zweckmäßige Durchführung röntgenographischer Spannungsermittlung (RSE), in Eigenspannungen und Lastspannungen, Moderne Ermittlung-Ergebnisse-Bewertung. Herausgeg. von V.Hauk und E.Macherauch, Carl Hanser Verlag München Wien (1982) 1-19;
   A useful guide for X-ray stress evaluation (XSE), Adv. X-ray Anal. 27 (1984) 81-99;
   The actual state of X-ray stress analysis. in: Residual Stresses in Science and Technology. Edited by E.Macherauch and V.Hauk, DGM Informationsges. Verlag Oberursel 1 (1987) 243-255.

3. Ruud, C.O., Snoha, D.J.: Residual stress mapping of gears and bearings, in ASM's 1987 Conference on Residual Stress in Design, Process and Materials Selection, Ed: W.B. Young.

4. Theiner, W.A., Höller, P., Hauk, V., Oudelhoven, R.W.H.: 1.Zwischenbericht (1.3.86-31.7.87), DFG-Vorhaben Ho 716/8-1.

5. Filinov, V., Shaternikov, V.: Testing of Shot Blasting Regimes and Metal Product Surface Hardening Parameters by Barkhausen Effect Method. Shot peening. Science/Technology/Application. Edited by H.Wohlfahrt, R.Kopp, O.Vöhringer. DGM-Verlag 1987, 407-414

6. Tietz, H.D., Schott, G., Christoph, R., Warch, H.: Zerstörungsfreie Charakteristik der Oberflächenverfestigung durch das Barkhausenrauschen. Neue Hütte, 33. Jahrgang, Heft 1, 12-13, Januar 1988.

7. Schneider, E., Altpeter, I., Theiner, W.: Nondestructive determination of residual and applied stress by micromagnetic and ultrasonic methods. Nondestructive Methods for Material Property Determination. (Eds.: Ruud, C.O., Green, R.E.) Plenum Press, N.Y., 115-122, 1984.

Development of Creep Damages on Heatresistant Ferritic Steel

H. Weber

1. Introduction

When high-temperature ferritic steels are subjected to creep stress, plastic creep procedures take place in the material which limit the service life of a component. The damage probability therefore generally increases parallel to the operating time /1/. The service life itself, respectively the point in time when damage occurs, can however only be predicted with great uncertainty, as the input data for life time analysis are not usually available in the degree of precision required for the purpose. Consequently, the longer the operating time lasts, the greater the significance attached to inhouse monitoring procedures including repetitive tests by means of non destructive testing methods /2/.

The aim of the repetitive tests is to determine deterioration caused by operating conditions. This report occupy with creep damage under laboratory conditions and in components, and examines the questing as to which requirements must be fulfilled by non-destructive test methods to ensure reliable works monitoring.

2. Development of creep deterioration

Norton's creep law reveals the extent to which creep rate depends on the essential influencing variables such as strength of the material, temperature and stress (fig. 1).

The end of service life has been reached when strain ϵ resulting from creep rate $\dot{\epsilon}$ reaches fracture elongation A_u.

A characteristic feature of an exposure in the creep range is a permanently effective stress $\sigma_e = \sigma_o - \sigma_i$, which remains greater than zero because when high temperatures prevail, the material does not harden in parallel with the extent of outer stress /3/. This can be observed in the case in ferritic materials above approx. 450 degress C.

$\dot{\epsilon} \sim (\sigma_o - \sigma_i)^n \cdot e^{\frac{Q}{R\cdot T}}$	Creep law
$\epsilon = \epsilon \cdot t$	Creep strain
$A_u = \dot{\epsilon} \cdot t_R$	Rupture strain
σ_i = Internal stress A_u = Rupture strain	Material characteristics
σ_o = Mechanical stress T = Service temperature	Service conditions
$\dot{\epsilon}$ = Creep rate ϵ = Creep strain	Variables
$n > 4$ Norton Exponent Q = Activation energy	Constants

<u>Figure 1</u>: Material creep characteristics

Up until the rupture time, the material passes through various creep stages, as shown in <u>fig. 2</u> in the form of a time-elongation curve. Further microstructural reactions take place parallel to the plastic creep deformations on account of the high temperatures. The first and second creep stages are determined by dislocation reactions such as hardening and recovery togehter with phase reactions towards a more thermal, stable microstructural state; the latter reactions are widely held to be reversible. The third creep stage is characterized by the addition of irreversible mircostructural deterioration caused by micropores and microcracks /4/.

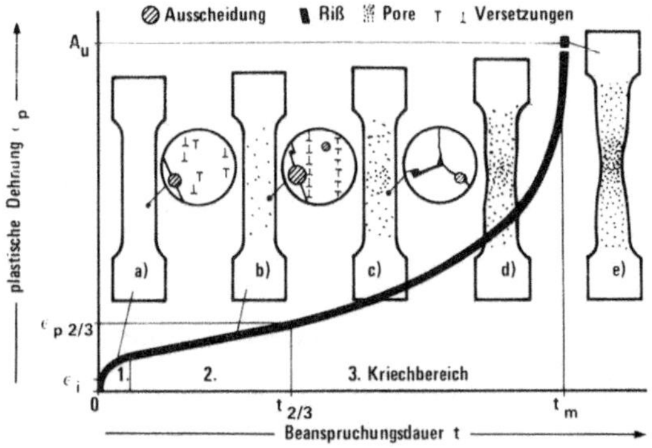

Figure 2: a) Primary creep : Dislocation density increases
b) Secondary creep: Development and resolution of dislocations at equilibrium
c) Tertiary creep : Material damages due to: formation of micropores formation of microcracks creep crack growth

The increasing creep rate to be observed in the third creep stage is generally caused by increasing stress as a result of a reduced cross section with micropores and local constructions. In addition, further material reactions such as microstructural changes or even recrystallization can begin to take effect. When greater stresses exist, the stress increasing influenced caused by local reduction in area prevails; when stress are less great, 3rd stage samples have revealed micropores before the formation of local reduction in area.

This creep stage also shows that the creep mechanism changes from a transcrystalline process whenn greater stress exist to an intercrystalline process at lower stress /5,6/. The following study refers only to the effects of intercrystalline creep strain on the development of damage.

The damage forms and degress of deterioration contained in <u>fig. 3</u> form the basic standard for the remaining remarks. The damage grades 1 to 4 identify damage forms with micropores; damage grade 5 identifies microcrack formation and damage grades 6 and 7 identify creep crack propagation.

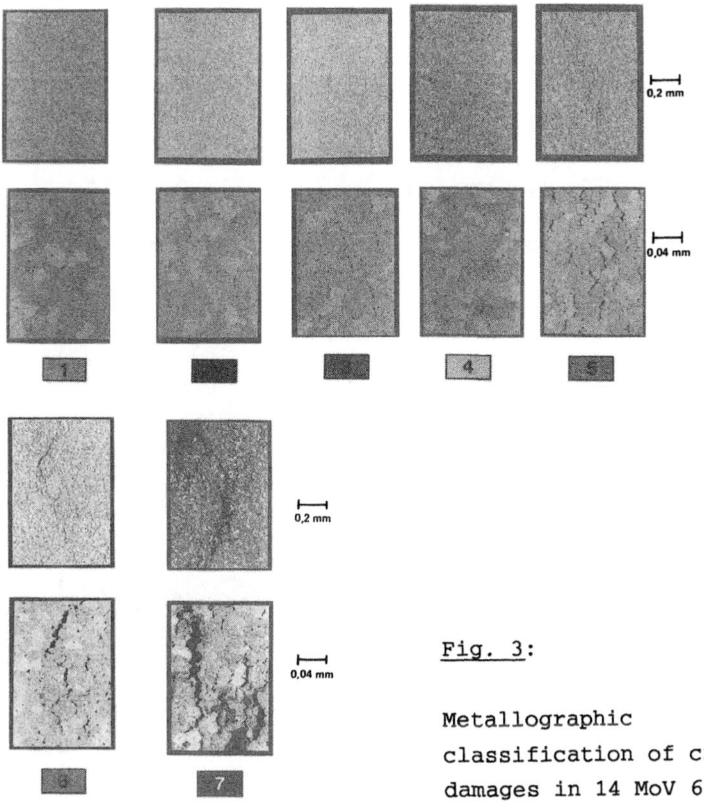

<u>Fig. 3</u>:

Metallographic classification of creep damages in 14 MoV 6 3

The micropores in damage grades 1 to 4 are only a few µm in size and cannot usually be detected in customary non-destructive testing. Creep deterioration cannot generally proven until after creep crack propagation.

3. Description of long period elongation behaviour

Fig. 4 shows some time-elongation curves of standard ferritic steels as per DIN 17 175 after approximately the same periods have elapsed. The end of the linear creep stage is reached after approx 50% of the total time period. The progress of the time-elongation curve in the third creep stage is influenced by fracture elongation. As revealed by the reduction in area, the fracture elongations differ in extent in the individual material. Subsequent metallographic investigations have revealed that the extent depends on the fracture mechanisms. The relatively extensive fracture elongation in 10 CrMo 9 10 is caused by a transcrystalline fracture mechanism, and the slight elongation in 14 MoV 6 3 by an intercrystalline fracture mechanism.

2. 10 CrMo 9 10, 550°C, 80 N/mm²
4. X 20 CrMoV 12, 550°C, 160 N/mm²
1. 13 CrMo 44, 555°C, 80 N/mm²
3. 14 MoV 6 3, 530°C, 150 N/mm²

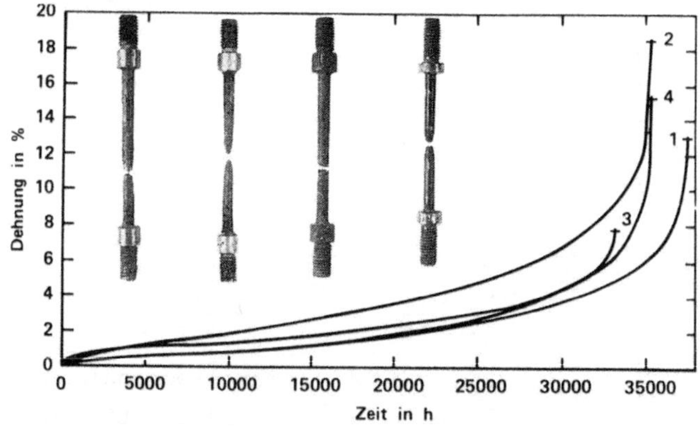

Figure 4: Time-strain curves and broken creep samples

It is however possible to combine these differing creep curves in a standard presentation by taking elongation after half the fracture time $\epsilon_{0,5\ t_B}$ as standardizing variable in addition to the fracture time, instead of the fracture elongation, and by anticipating the phase reactions by suitable pretreatment; e.g. by long period service exposure. The experimental curve progression shown in fig. 4 can serve as basis for further analysis of creep behaviour using theoretical models.

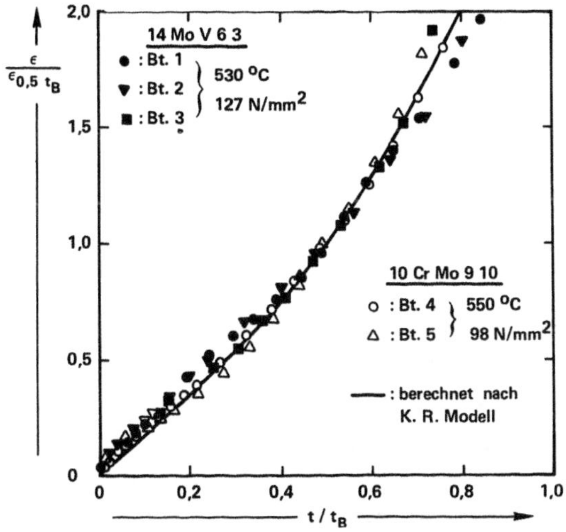

Figure 5: Standard creep curves of 14 MoV 6 3 and 10 CrMo 9 10 after service condition

The so-called damage model as per Kachanov and Robotnov (K.R. model) has been selected for further study /7,8/. This model adds a damage parameter w to the simple form of Norton's creep law:

$$\dot{\epsilon} = \frac{A \cdot \sigma^n}{(1-w)^n} \quad (1)$$

$$\dot{w} = \frac{B \cdot \sigma^q}{(1-w)^\eta} \quad (2)$$

A = Constant

$0 \leq w \leq 1$

The integration of the differential equation results in:

$$\epsilon/\epsilon_B = 1 - (1-t/t_B)^{\alpha} \tag{3}$$

$$W = 1 - (1-t/t_B)^{\frac{1-\alpha}{n}} \tag{4}$$

with ϵ_B fracture elongation
t_B fracture time and
$\alpha = 1 - \frac{n}{1+\eta} = 0,29$

The ratio t/t_B is a measure of materials exhaustion. Introduction of the standard parameter $\epsilon_{0,5\ t_B}$ results in the relationship

$$\epsilon/\epsilon_{0,5\ t_B} = \frac{1 - (1-t/t_B)^{\alpha}}{1 - 0,5^{\alpha}} \tag{5}$$

The curve in fig. 5 shows the test results of two materials with widely differing behaviour.

The K.R. model is based simply on an abstract model of material damage. The possibilities of this model for describing real damage by micropores are to be examined, such as can be observed in material such as 14 MoV 6 3 under creep conditions. In this case it possible to interpret the damage parameter w as that part of the grain boundary affected by micropores. A deterioration grade is defined in accordance with the K.R. model, resulting from the stress bearing grain boundary area A and the porous part A_h:

$$f_h = \frac{A_h}{A} = \frac{A - Ae}{A} \tag{6}$$

$$0 \leq f_h \leq 1$$

Ae = effectively stress bearing grain boundary area

Creep damage leads to the formation of pores as described in model in fig. 6.

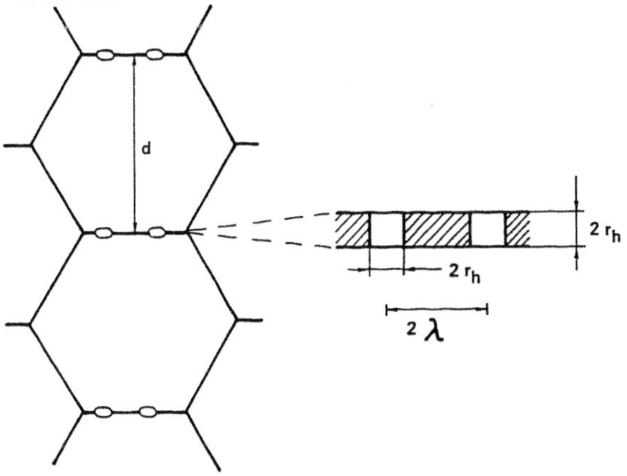

Figure 6: Creep damage due to formation of pores

This model presumes that the micropores consist of small cylinders with 2 r_h diameter and height. The measured length of a sample l can then be divided into cells with the diameter $d + 2r_h$. On this basis, it is possible to derive a connection between density change $\frac{\Delta\rho}{\rho_o}$ through micropores and the degree of deterioration:

$$\frac{\Delta\rho}{\rho_o} = \frac{V_h}{V_o} \tag{7}$$

V_h = Volume of pores

$$V_h = \frac{l}{d + 2r_h} \times A_h \cdot 2r_h$$

$$V_o = V_{ges} - V_h = l \cdot A - V_h$$

$$f_h = \frac{\frac{\Delta\rho}{\rho_o}\left(\frac{d}{2r_h} + 1\right)}{\frac{\Delta\rho}{\rho_o} - 1} \tag{8}$$

Creep rate is increased by the reduction in the cross section area caused by the micropores. In this way it is also possible to determine the degree of damage from the creep tests. On the presumption that Norton's creep law applies, tests with a constant load L resulted in the following:

$$\dot{\epsilon} = C \cdot \left(\frac{L}{A_e}\right)^n \tag{9}$$

with $Ae = A - A_h$

For $t = 0$ ist $A = Ao$ and $A_h = 0$:

$$\dot{\epsilon}_{(t=0)} = \dot{\epsilon}_s = C \left(\frac{L}{Ao}\right)^n \tag{10}$$

With $A/Ao = e^{-\epsilon}$, 6, 9 and 10 result in

$$f_h = 1 - \left(\frac{\dot{\epsilon}}{\dot{\epsilon}_s}\right)^{1/n} \cdot e^{\epsilon} \tag{11}$$

The models developed here have been checked using samples and components of material 14 MoV 6 3. Density change was checked using creep samples which were removed after certain strains. Fig. 7 shows the connection between the creep curve and density change. The micropores were shown using a

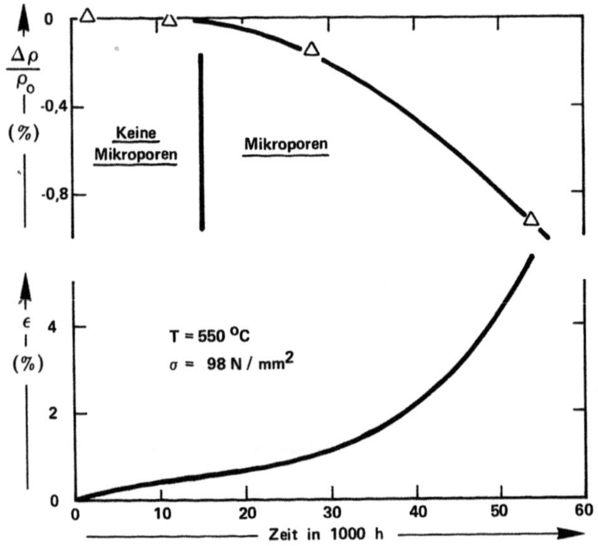

Figure 7:

light-optical method. The fact emerges that the light-optically revealed micropores mark the beginning of tertiary creep. Density shows marked drop in the third creep stage.

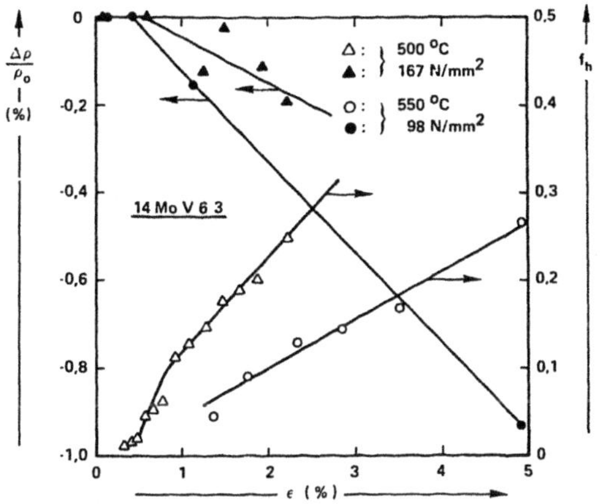

Figure 8: Dependence of damage parameters and density changes upon creep elongation

Fig. 8 shows the measured density values and the degrees of damage calculated from the creep curves according to equation 11 depending on strain. Fig. 9 shows a summarized presentation of the degrees of damage calculated from

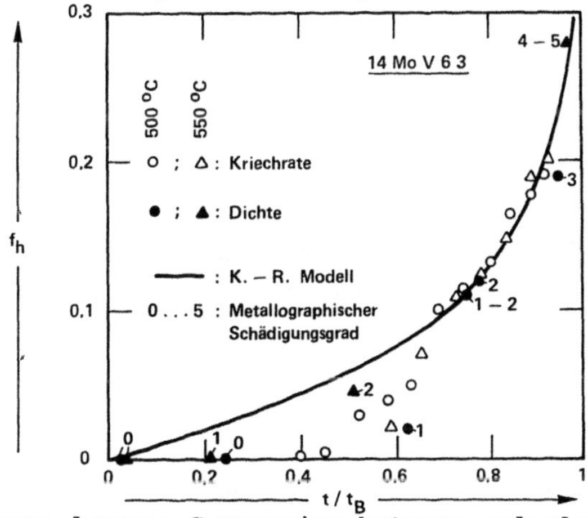

Figure 9: Creep damage: Comparsion between calculations according to the model and experimental results from density measurements and creep rate

density with equation 8 and creep rate with equation 11 in dependence on exhaustion t/t_B. The results of the metallographical investigation identify the state of damage as per forms 3 and 4. Above creep exhaustion of approx. 60%, the results are uniformly located on the curve calculated an the K.R. model. Below 50% no damage is measured, so that the model cannot be used in this range. The points of the calculated curve are closer together in the intermediate zone. Analysis reveals clearly that it would be purposeful to differentiate between material in an undamaged and damaged condition.

Material exhaustion in the third creep stage develop in samples homogeneously subjected to single axis stress under laboratory conditions to a great extent by the formation of micropores, which can currently only be revealed in a non-destructive method by means of metallographical investigations. Microcracks are formed towards the end of the last section of the third creep stage and the rupture is caused by creep crack propagation.

4. Summary and conclusions

The progession of damage under creep conditions is influenced to a great extent by the state of stress. Tests under homogeneous stress show that material exhaustion is caused to approximately 100% by the formation of micropores. The degree of deterioration can be calculated according to the K.R. model from creep curves and density measurements. The micropores influence both the physical and mechanical properties of the material.

Customary non-destructive test methods cannot be considered as a safe, preventive measure in damage progression of this nature, as these methods are not capable of recognizing such faults. At present, metallographic testing by means of microstructural impressions has to be recommended. For this reason the development of a new test method is considered necessary.

Literary references

/1/ Bendick, W. und H. Weber:
Beurteilung von Bauteilen nach langer Zeitstandbeanspruchung
Teil 1: VGB KRAFTWERKSTECHNIK 66 (1986), H. 1, S. 63- 77
Teil 2: VGB KRAFTWERKSTECHNIK 66 (1986), H. 2, S. 170-177

/2/ TRD 508:
Zusätzliche Prüfungen an Bauteilen, berechnet mit zeitabhängigen Festigkeitswerten

/3/ Ilschner, B.:
Hochtemperaturplastizität.
Springer-Verlag, Berlin, Heidelberg, New York (1973)

/4/ Brühl, F., G. Kalwa und H. Weber:
Bruchverhalten von Bauteilen unter Zeitstandbeanspruchung
VGB KRAFTWERKSTECHNIK 65, H. 11, Nov. 1985, S. 1059-1068

/5/ Ashby, M.F., C. Ghandi und D.M.R. Tablin:
Acta Met. 27 (1979), pp. 699-729

/6/ Garfalo, G.:
Fundamentals of Creep and Creep Ruputure in Metals.
Macmillan Series in Materials Science 1966

/7/ Kachanov, L.M.:
Ruputure time under creep conditions.
Festschrift MUSKHELISVILI (Problems of continuum mechanics), Soc. Ind. and Appl. Math., Philadelphia 1961, S. 306-310

/8/ Rabotnov, G.N.:
Some problems of the theory of creep.
Moskau 1948, Übersetzung: NACA TM 1353, 1953

Nondestructive Metallurgical Investigations for the Evaluation of Turbines

N. KASIK

Asea Brown Boveri (ABB)

Zusammenfassung

Bei der Revision thermisch belasteter Bauteile im Kraftwerk bleibt die Abdrucktechnik zusammen mit der Härteprüfung die wertvollste Hilfe für die Ermittlung der Werkstoffschädigung im Betrieb.
Die Veränderung des Werkstoffzustandes in Abhängigkeit von Zeit, Temperatur und Beanspruchung wurde an Flachproben mit Messchneiden ermittelt. Diese Probenform ermöglicht einerseits eine genaue Erfassung der Zeitstanddaten, andererseits eine periodische Gefügekontrolle mittels Oberflächenabdrücken.
Auf diese Weise wurde eine Beziehung zwischen Kriechgeschwindigkeit, Gefügedaten und dem Lebensdauerversuch für die CrMoV- und 12%Cr-Stähle aufgestellt.
Den beiden Werkstoffen gemeisam ist die Tatsache, dass Hohlräume erst knapp von deren Lebensende auftreten. Aus diesem Grunde wurden andere, zum Teil reversible Gefügeveränderungen, wie z.B. das Wachstum der inter- oder intrakristallin ausgeschiedenen Karbide mit den Kriech- und Härtekurven verknüpft.

Introduction

The life extension of fossil power plants demands techniques for the accurate determination of time and service dependent material properties.
A number of conventional nondestructive tests including X-ray diffraction, dye penetrant, ultrasonic and eddy current techniques have been successfully used for determination of flaws with a diameter larger than 1 mm. In recent years attempts have been made to apply these and other methods, based on the change of physical properties caused by material degradation, such as softening, embrittlement etc, for the prediction of remaining life. The practical use of these methods is still restricted due to serious problems such as the difficulty in quantifying results. Some nondestructive methods detect only the first stage of the creep curve (coercive force, electrical resitivity [1], the others the end of the third

stage of the creep curve by revealing microcracks, cracks or a larger density of cavities - US-velocity, eddy currents [2-5]. At present, surface replication [6-8] accompanied by hardness checks [9] are in use to reassess the material properties after time in service.

Creep tests

Constant-load creep tests on flat specimens with knife edges (fig. 4) have been performed and the microstructural changes related to the corresponding stress, time and temperature. The shape of the specimens enabled the investigation of the microstructure by means of surface replication. This method has been chosen for two reasons. Firstly for the periodic examination of the material condition with few specimens, secondly to obtain more experience in interpretation of material degradation from the inverted SEM-image.

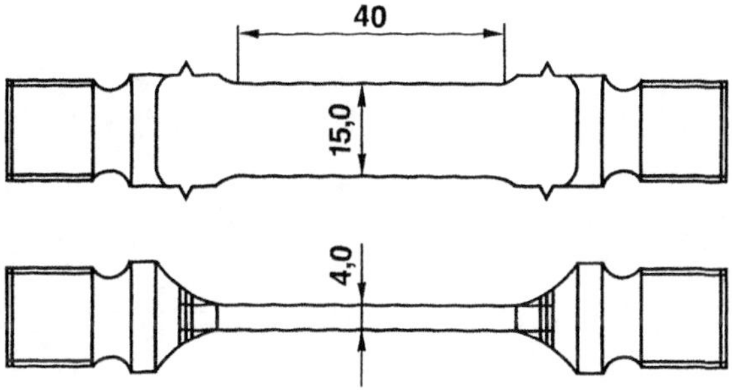

Fig. 1: Creep test specimen

The rupture time has been aimed for 3000 and 10'000 hours respectively. After periods of 500 hours in the short-term tests and 2000 hours in the long-term tests the specimens have been removed from the creep machines. After grinding, polishing and etching of one side alternately the hardness has been measured and the replica taken.

The following steels have been tested:

2 1/4 CrlMo	(10CrMo9 10)
CrMoV	(21CrMoV 5 11)
12%Cr	(X20CrMoV 12 1)

In the 2 1/4 Cr1Mo-steel for steam pipes cavitation occurs already in the secondary creep stage and the remaining life can be related to the density of cavities, as described in literature [10-12].
As the CrMoV- and 12%-steels cavitate only towards the end of the tertiary creep stage investigations described in this paper have been focussed on these "difficult" steels.

12%Cr-Steel

Creep specimens of 12%Cr-steel were tested under 200 MPa at 550°C and under 100 MPa at 600°C. At 550°C cavities and microcracks could be observed only after rupture. In figure 2 the microstructures of the broken specimens exposed at 550°C and 600°C for 3000 hours can be compared.

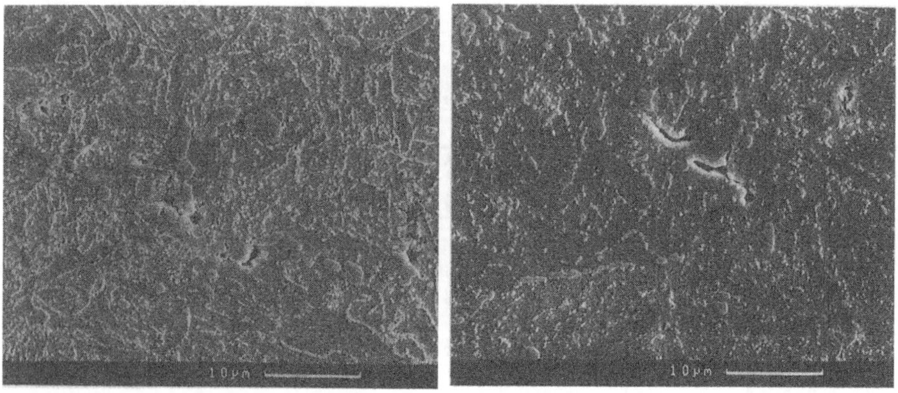

a) b)

Fig. 2: 12% Cr-Steel after stress rupture
a) at 550°C, 200 MPa, 3000 h
b) at 600°C, 100 MPa, 3000 h

The microstructure of the as received materials consisted of tempered martensite with a small ammounts of δ-ferrite. The different carbides in the decomposed martensite were identified as M_3C, M_2C, M_6C, $M_{23}C_6$ and M_7C_3 [13-16]. During the creep process all these carbides transform to only one type $M_{23}C_6$, which can be found along former austenite grain boundaries as well as along the boundaries between α'- and δ-phase while smaller carbide particles border the laths of the former martensite. Thus coagulation and growth of all carbide particles can serve as a parameter to correlate with the advanced creep damage. The size of carbide particles in different periods of interrupted creep tests have been measured in SEM-photographs using the mini-magiscan analyser from Joyce-Loebl. The fit between the results gained on replicas and on cross sections was satisfactory. The increase of the length, breadth and area of carbides has been expressed using histogramms (fig. 3) or mean values.

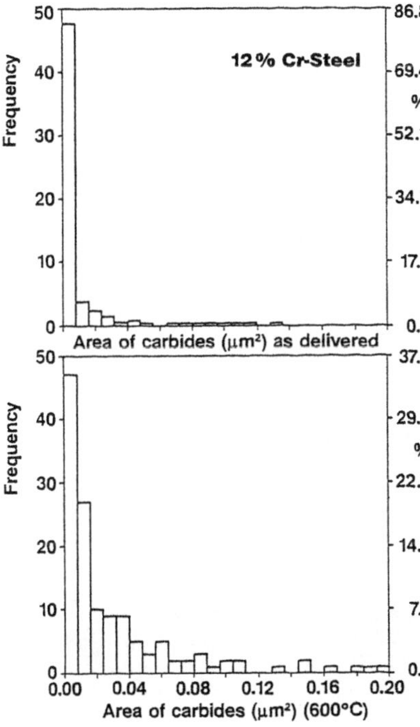

Fig.3:Area of carbides

a)as delivered

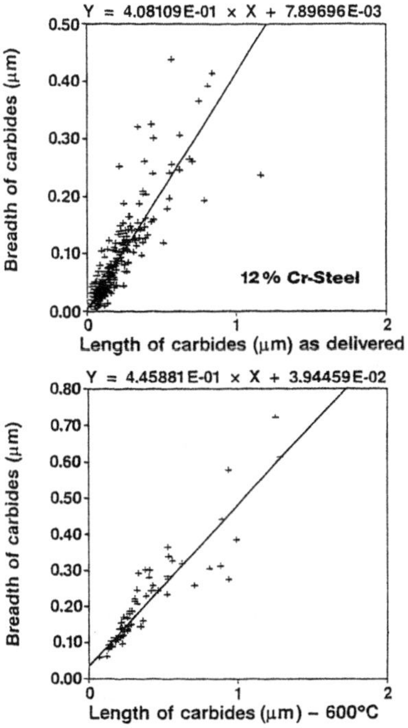

Fig.4:Correlation between the breadth and length of carbides

b)after 3000h at 600°C, 100MPa

The correlation between the length and breadth of carbides after exposure at 600°C and before does not reveal any significant change of shape (fig. 4). These findings are in agreement with recent results reported in the literature [17,18].

previous investigations [19] the microstructure at the rotor surface, consisting almost entirely of bainite, cavitates towards the end of the lifetime. Therefore cavitation cannot be a reliable indicator of material degradation in the early stages.

Another problem arising in the CrMoV-steels is the embrittlement after long-term exposure at temperatures over 500°C. Above all cross-sections thicker than 100 mm are only unsufficiently hardened, so that in the core of a forging considerable amounts of ferrite can be observed (compare figure 6a and 6b). Consequently the impact thoughness in the core is much lower than in the rim.

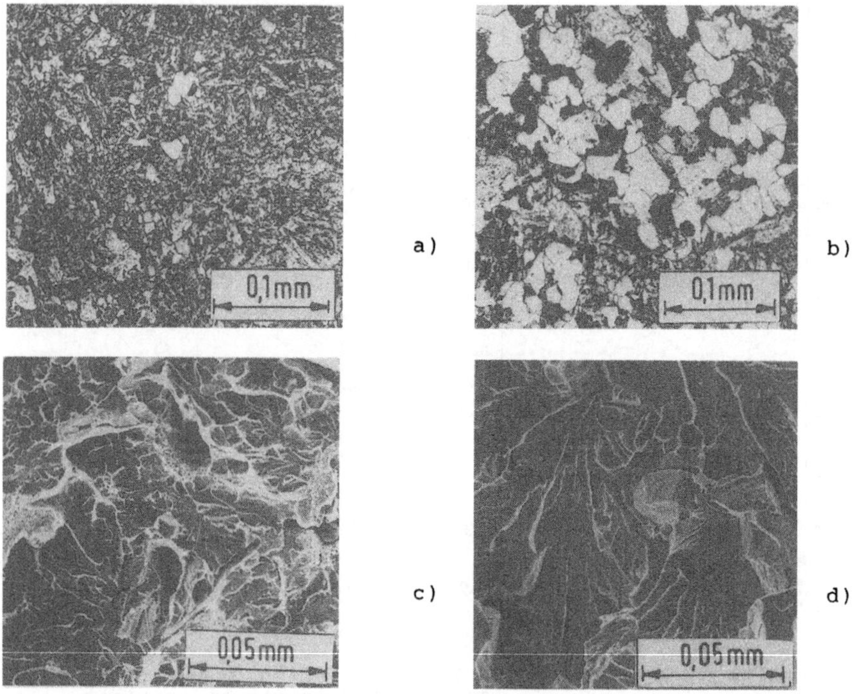

Fig. 6: Microstructure and fracture surface of HP-rotor after thermal embrittlement a) c) - rim b) d) - core

Whereas in the high pressure rotors the temperature in both zones is about the same, the stress in the core is higher;

In figure 8 the carbide size and creep rupture stress are plotted against Larson-Miller parameter. Supposing that stress and either time or temperature are known, from the carbide size the remaining life (difference between two LMP-values in curve 8b) can be estimated. In addition rupture data of two flat specimens have been marked (points I, II in figure 8). Both are near the 99,9% lower limit curve.

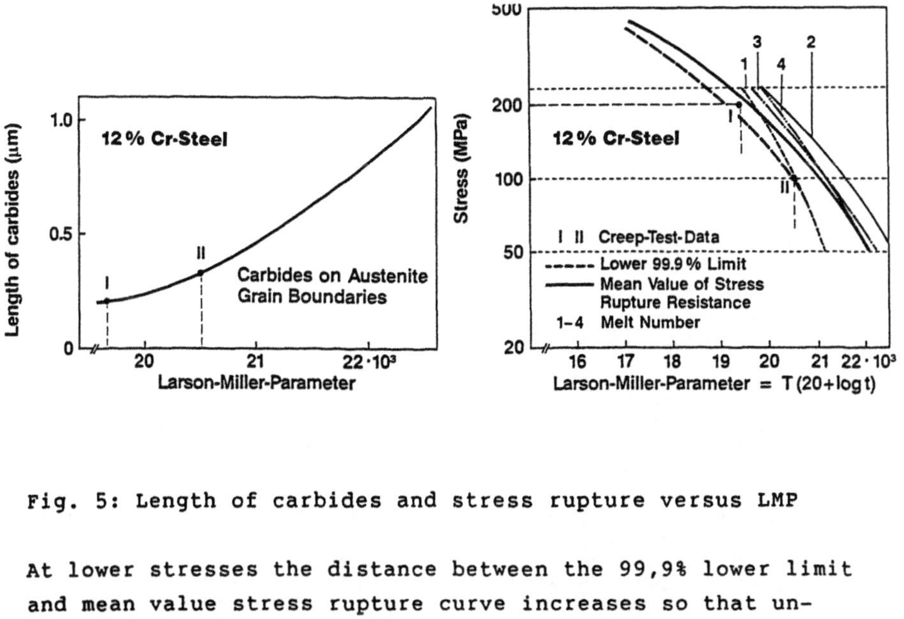

Fig. 5: Length of carbides and stress rupture versus LMP

At lower stresses the distance between the 99,9% lower limit and mean value stress rupture curve increases so that uncertainity of lifetime prediction increases. Former investigations on long-term creep specimens had shown that the lifetime of the coarse grained melts 2-4 under 50 MPa was about 100'000 hours longer than for the fine grained melt 1. Other factors of importance are impurities as well as tempering time and temperature. If the estimation of remaining life should not be too conservative then all these factors have to be taken into account.

CrMoV-Steel

In the CrMoV-steels the moment of the first cavitation depends on the grain size and on the phase composition. After our there is a much higher danger of brittle cracking than of creep damage. In one HP-rotor taken out of service after 110000 hours the toughnes decrease at the rim is from 48 to 15 J and in core is from 15 to 7 J. Material from both zones of the cold end of the rotor has been used for flat specimens to test the influence of the different phase composition on the creep behaviour.

The characteristic features of temper embrittlement are the precipitation of molybdenum enriched carbides M_2C in ferrite as well as larger deposites of carbides $M_{23}C_6$ at grain boundaries (figure 7). Phosphorus segregation at grain boundaries which is supposed by the Japanese to cause embrittlement and detected by electrochemical method [20] could not be confirmed neither by means of microprobe analysis nor of Auger spectroscopy. The fracture surface of impact test specimens at room temperature was not intergranular, as observed by Goto, but transgranular (fig.6c,d).

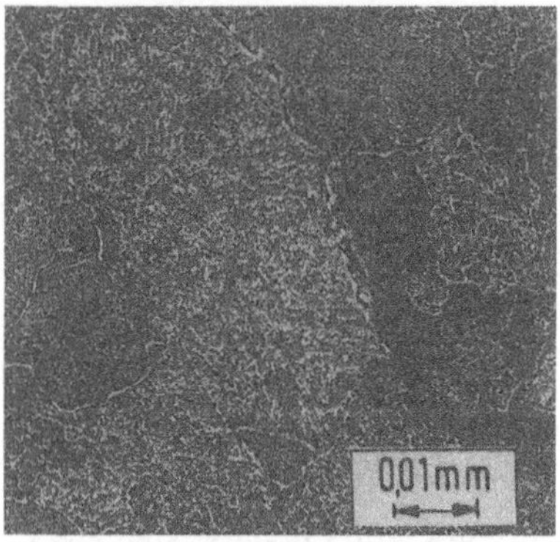

a) Cross-section

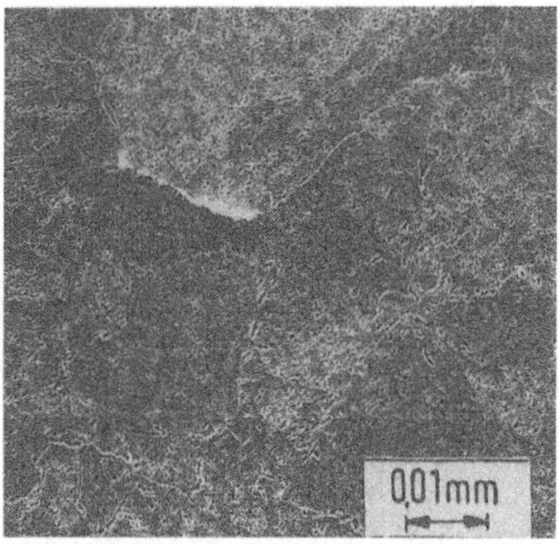

b) Replication

Fig. 7: CrMoV-Steel after 10'000 h at 550°C, 150 MPa

The characteristic feature of creep damage is the growth of carbides at the grain boundary and in the bainite. However at temperatures under 550°C the growth rate of carbides is very small, so that the hardness measurement remains the most useful method. On the contrary of the temper embrittlement the hardness decrease in the crept specimens is very significant [21]. In figure 8 hardness and creep rate are plotted against the lifetime fraction. Independent of the stress level there is only little scatter in both up-to the life consumption of about 80 per cent, provided the hardness in the initial state is the same.

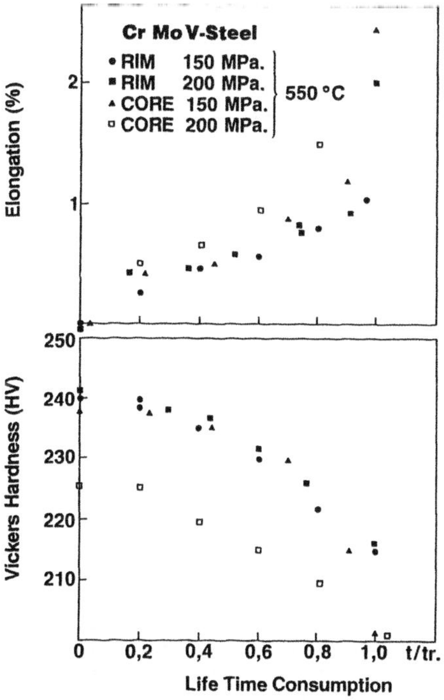

Fig. 8:Creep elongation and hardness versus lifetime consumption in CrMoV-Steel

Concluding remarks

Surface replication and hardness measurement are still the best testing methods used in the overhaul of older power plant components.
With the help of surface replication, quantitative analysis of the microstructure not only concerning the grain size, phase distribution and density of cavities but also the growth rate and distribution of the finest carbides can be performed.
The fit between the analysis results obtained with polished cross-sections and replicas was surprisingly good. The creep behaviour of the 12% Cr- and CrMoV-steels is untypical in that they both cavitate only shortly before rupture. Thus the growth rate of carbide $M_{23}C_6$ is the most suitable sign of creep

degradation of 12% Cr-steels. As for CrMoV-steels the degradation of material in service consists partly of the decrease of the creep rupture stress and partly of temper embrittlement.

The former is caused by coarsing of carbides in bainite, the latter by the appearance of particles $M_{23}C_6$ at the grain boundaries and by the precipitation of the fine M_2C carbides from the ferritic phase. The creep damage is accompanied by a hardness decrease even when there is neglidgible carbide coarsening. When analysing the surface of HP-rotors, the more highly loaded inner part of low toughness, which is unaccessible for surface replication, should be taken into account. For this reason the findings at the rim have to be considered very conservatively.

References

1. Szombatfalvy, A.: Wiss. Zeitschr.O.v. Guerike Magdeburg 10, 1972
2. Clark, W.G.: Metala, M.S.: EPRI Conf. 1986
3. Thompson, R.B.: Journ.of Applied Mechanics, 59, 1983
4. Willems, H., Bendick, W., Weber, H.: 2nd Int.Symp.on Nondestructive Characterisation of Materials, Montreal, Canada, July 21-23, 1986
5. Willems, H., Degischer, H.P.: DGZFP-Tagung, Lindau 1987
6. Neubauer, B.: Nuclear Technology Vol. 66, 1984
7. Arnswald, W., Blum, R., Neubauer, B., Poulsen, E.: VGB Kraftwerkstechnik 59, 1979
8. Blum, R., et.al., Summary Report on a Round Robin Testing Programme, 1987
9. Equotip by PROCEQ, Zürich, Switzerland
10. Bendick, W., Weber, H.: VGB Kraftwerkstechnik 66, 1986
11. Hagn, L., Schüller, H.J.: VGB-Sondertagung, Essen, 1984
12. Kussmaul, K., Maile, K., Zili, Li.: VGB Kraftwerkstechnik 9, 1986
13. Irwine, K.J., Crowe, D.J., Pickering, F.B.: J. Iron Steel Inst. 195, 1960
14. Hede, A., Aronsson, B.: J. Iron Steel Inst. 207, 1969
15. Koutsky, J.,et.al.: Arch. Eisenhüttenwesen 45, 1974
16. Sandström, R. Karlsson, S, Modin, S.: Report Swedish Institute for Metals Research, Stockholm 1981
17. Eggeler, G., Nilsvang, N., Ilschner, B.: Steel Research 1 1987
18. Nilsvang, N., Eggeler, G.: Pract. Met. 24, 1987
19. Lempp, W., Kasik, N., Feller, U.: (see 4.)
20. Shoji, T., Takahashi, H.: EPRI Conference, Life Extension and Assessment of Fossil Plants, 1986
21. Goto, T, et al.: (see 4)

Early Recognition of Creep Damages by Means of Nondestructive Test Methods

H.-A. Crostack, V. Beckmann, W. Bischoff, R. Niehus
Fachgebiet Qualitätskontrolle, University of Dortmund, FRG

Summary

Nondestructive testing methods for the early detection of creep damage are reported upon. Correlations between structural parameters and received test signals are discussed in detail. Further, the limiting conditions for testing which apply to the transfer of experimental data to components are considered in detail. The nondestructive test methods used are eddy-current methods, ultrasonic testing, and a thermoelectric measuring method.

Zusammenfassung

Es wird über den Einsatz zerstörungsfrei arbeitender Prüftechniken beim Nachweis von Zeitstandschäden berichtet. Hierzu werden die Wechselwirkungen zwischen dem Gefügeaufbau und den Prüfanzeigen eingehend diskutiert. Darüber hinaus wird auf die Randbedingungen bei der Prüfung hinsichtlich der Übertragbarkeit der Ergebnisse auf Bauteile detailliert eingegangen. Als Prüfverfahren wurden die magnetische Prüfung, die Ultraschallprüftechnik und ein thermoelektrisches Meßverfahren eingesetzt.

1. Problem

Under thermal and mechanical load, the structure of steels changes with time. Creep damage occurring in the primary and secondary region of the creep curve is for the most part reversible. This means that heat treatment of the damaged regions can lengthen the lifetime of the component. The early recognition of creep damage thus requires the availability of test methods which provide information on the current state of damage within the structure. The early recognition of damage primarily comprises the region of primary and secondary creep, i.e. structural changes are detected which produce few pores even after longer creep times. Moreover, according to TRD501/1/, only plastic elongations $< 2\%$ are of interest. The non-destructive early recognition of damage takes advantage of the fact that the acoustic, magnetic and for example electric material properties change in dependence upon the current structure, and are consequently physically measurable. It must be taken into account that owing to creep fatigue several damage mechanisms can be simultaneously effective so that measured effects may compensate each other. Early damage recognition therefore requires several mutually independent measuring techniques. The choice of testing techniques further depend largely on the material to be investigated. The basic procedure is illustrated in Fig. 1. Fundamental measurements require on the one hand that different creep-damaged materials be investigated in order to detect the physical parameters; on the other hand, defined specimen material is of importance, i.e. uniform basic structure and constant specimen dimensions. Otherwise, only restricted transfer of the fundamental results to components is possible. This approach requires that specimens be subjected to defined creep load and that the structural changes occurring be measured at short time intervals by non-destructive testing methods and then be compared with metallographic results. To obtain results within a reasonable time, short period creep tests were performed rather than long period creep tests. For this, the test stress, but not the test temperature, was increased relative to the operating stress. Structure formation and precipitation behaviour of steels are temperature-dependent, as illustrated in the time-temperature diagrams.

2. Thermoelectric testing

This method makes use of the thermoelectric voltage in the µV range (Seebeck Effect) which exists between two contact points of different temperature within a homogeneous conductor circuit. Fig. 2 includes the results for the nickel-bases alloy Inconel 617 at different test stresses and temperatures. It was measured on cylinders which were worked down from fractured circular specimens. At temperatures of T = 550°C and 650°C the thermal differential voltage (reference is made to specimen 77) increases with the load time and permits a separation of different damage conditions. When the accuracy of thermoelectric contact voltage measurements (+/-3%) is taken into account, the specimens 84 and 85 can no longer be clearly separated. At T = 550°C the increase of thermoelectric voltage can be correlated directly with the increase in the content of the $M_{23}C_6$ carbides which are preferentially precipitated at the grain boundaries after longer times, Fig. 3. At T = 650°C, the difference in $M_{23}C_6$ carbide content of the specimens is no longer so apparent. Specimen 83 already reaches a content which increases only slightly under further load. Furthermore, TiN and the γ'-phase occur at this temperature. In contrast to the foregoing results, at T = 850°C chromium carbides are preferentially formed on crystallographic planes in the grain and not at the grain boundaries; and a γ'-phase formation cannot be observed in contrast to T = 650°C. No voltage differences are relative to reference specimen 77 measured with this constellation.

3. Magnetic testing

The investigations of early damage recognition were performed on temperature-resistant structural boiler steel 13CrMo44. The creep damages were set at 450°C and at an increased test stress (σ= 120 Nmm^{-2}). Fundamental investigations for the determination of structural changes by means of ND methods require that the measuring signals are not influenced by perturbations resulting from geometric variations or inhomogeneous primary structure of the specimens. For this reason, flat specimens were used which, according to TRD 508/1/, do not exhibit significant local constrictions as a result of load in the interesting 2% plastic elongation range, Fig. 4. "Zero measurements" were performed on all specimens by means of the CS pulsed eddy-current technique /2/ to exclude specimens with significantly deviant eddy-current characteristics resulting from different structure. Measurements were performed with passing coils by means of the difference method (reference specimen 13/100). A transmitter signal whose spectrum exhibits constant amplitude and phase in the frequency range of 50 Hz to 150 Hz was pre-defined. The specimens were finally selected by forming the cross-correlation between the response signal of any chosen specimen with the response signals of the remaining specimens. Specimens were then chosen which are very similar in correlation coefficient (>0,99) and time delay (+/-8). This zero test allowed the selection of specimens with approximately equal starting conditions, thus reducing the influence of different initial conditions on the later measured effect for creep-loaded specimens.
The eddy-current measurements on damaged specimens were performed in analogy to the zero measurements. In addition, at the appropriate measuring time points, already-measured undamaged specimens were once again investigated in order to detect the scatter of the results within the different measuring cycles, Tab. 1. The redundant zero tests, Tab. 1a, resulted in nearly identical values for the correlation coefficients (specimens 111, 112, 114, 115, 117, 121, 122, 140, 143 and 144) with an average of <1% difference and time delay values which vary less than the 3rd. significant figure. As concerns the damaged specimens, the measuring effects are more distinct, Tab. 1b, within part contradictory tendencies being ascertained for the dependence on the load time of the correlation coefficients and time-displacement measurements. This is illustrated again in Fig.5 for the case of the amplitude maxima differences, where the measuring position along the longitudinal specimen axis has been additionally taken into account. The greater the dissimilarity of a component to the reference specimen (13/100) (which is equivalent to a deviant structure or a deviant geometry at the specimen ends), the larger

the difference amplitude. For damaged specimens (*), these differences occur not only in the specimen centre but are also more prominent towards the ends of the specimens than in the case of undamaged components. This means for the degree of damage present that the indications resulting from damage and from deviating geometry are amplified in the same direction.
In analogy to the CS pulsed eddy-current testing, monofrequency comparison measurements were performed at 100 Hz. The magnitudes of the voltage amplitudes (vector components) within the complex voltage plane were in this case evaluated along the longitudinal axis of the specimen, these values being compensated against specimen 13/100, Fig. 6. The coil voltage during the monofrequent meansurement is likewise dependent upon the creep load.
The metallographic findings on the specimens of steel 13CrMo44 showed that structural changes such as increase of grain coarseness or carbide precipitations had not yet taken place in the present plastic elongation range. Glide processes are instead involved. This means that the measurements are determined by internal stress distribution, and hence by the dislocation structure. At later times, the residual stress decreases, which explains the fall in the measuring indications.

4. Ultrasonic testing

A further method was the ultrasonic testing technique, using very narrow-band CS pulses as transmitter signals (bandwidth 1.4 MHz for 40dB decrease). In contrast to conventional broad-band excitation, the centre frequency and band width can be pre-defined here. Demodulation effects with the resulting distortion of the pulse shape of the received signal are largely avoided. Of particular advantage is the coherent length of the signal that permits interferometric measurement /3/.
As in the case of holographic data collection, measurements are performed by means of the pulse-echo method with the immersion technique. During evaluation, the transmitter pulse reflected from the rear face of the specimen is modulated by two reference signals which possess the same centre frequency as the transmitted signal but which are phase-displaced to each other by 90°. The real and imaginary components thus generated allow determination of the phase angle. The accuracy attained depends on the dynamics of the measuring system and lies in the parts per thousand range for the present test problem. The specimen dimensions must be known exactly. Fig. 8 presents the result of an interferometric measurement. The reference phase of the undamaged material condition is constant along the specimen length. The phase shift, however, exhibits significant differences which depend upon the load time. As concerns the results so far obtained, the difference phase angle increases with the load time.

5. Conclusions

Structural changes resulting from load can successfully be detected by means of non-destructive testing methods. This is particularly true for load times whereby long period creep pores have not yet occurred within the steel. The examples of eddy-current and ultrasonic testing demonstrated that the individual methods result in part in contradictory indications of the measured values relative to load time. This means that one non-destructive method alone does not suffice for the detection of creep damage. Thermoelectric difference-voltage measurements have shown that this method allows quantitative assignment of the defects at one test temperature only. The application of these results to other test-temperature ranges would result in incorrect declaration of the extent of damage.

The investigations were funded by the Deutsche Forschungsgemeinschaft. We wish to thank the DFG at this point.

Literature

/1/ TRD 508 (Technische Regeln für Dampfkessel): Zusätzliche Prüfungen an Bauteilen, berechnet mit zeitabhängigen Festigkeitskennwerten, Okt. 1978, Heymanns Verlag, Köln, Beuth-Verlag, Berlin

/2/ Crostack, H.-A., J. Nehring, W. Oppermann: Einsatz von Korrelationsimpulsen (CS-Technik) zur Wirbelstromprüfung, Materialprüfung 26 (1984) 1/2, S. 16-20

/3/ Crostack, H.-A., W. Roye: Profile and deformation measurements with high axial resolution using the acoustic holography, Nuclear Engineering and Design 102 (1987), p. 313-317, North Holland, Amsterdam

Fig.1: Procedure of damage prediction

Fig.2: Thermoelectrical voltages of different creep damage conditions

Fig.3: Metallographies of In 617 after different creep test conditions

Fig.4: Specimen geometry for creep tests with respect to nondestructive testing methods

Fig.5: Spatially dependent of maximum amplitude as a function of creep damage, CS-pulsed eddy current technique

Fig.6: Results in eddy current testing of damaged specimen 13/126 at different loading times; complex voltage plane

Fig.7: Ultrasonic interferometry; measurement of phase difference as a function of creep damage

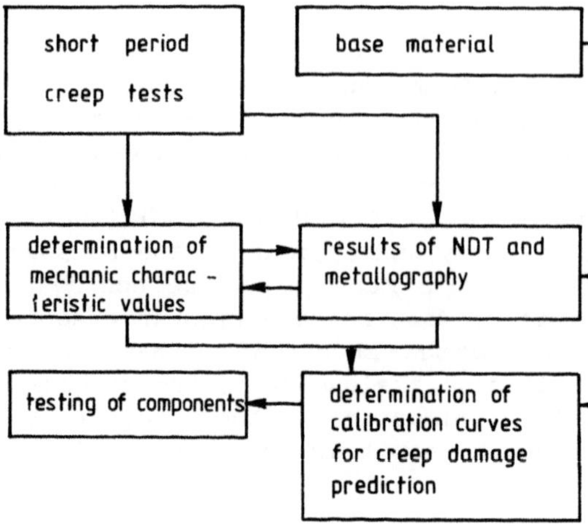

Fig. 1

Inconel alloy 617

σ [N mm⁻²]	486	463	399	177	409	341	284	178	98	73	47
t_B [h]	343	905	2104	31163	24	82	198	18953	163	1713	10004*
T [°C]	550	550	550	550	650	650	650	650	850	850	850
specimen	77	78	79	82	83	84	85	86	89	90	94

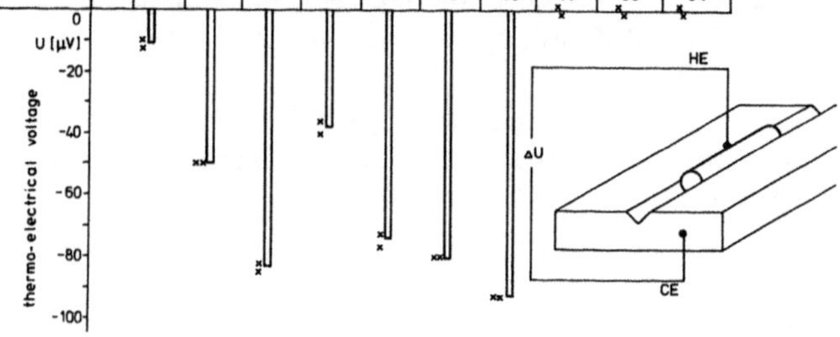

Fig. 2

Fig. 3

a) t_B = 343 h

σ = 486 N mm⁻²

T = 550 °C

b) t_B = 31163 h

σ = 177 N mm⁻²

T = 550 °C

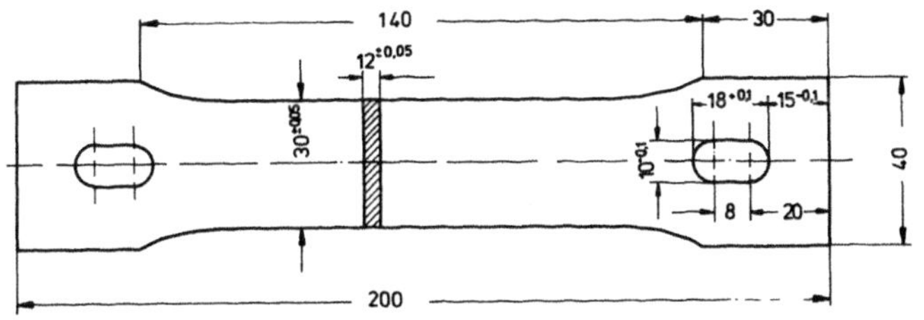

Fig. 4

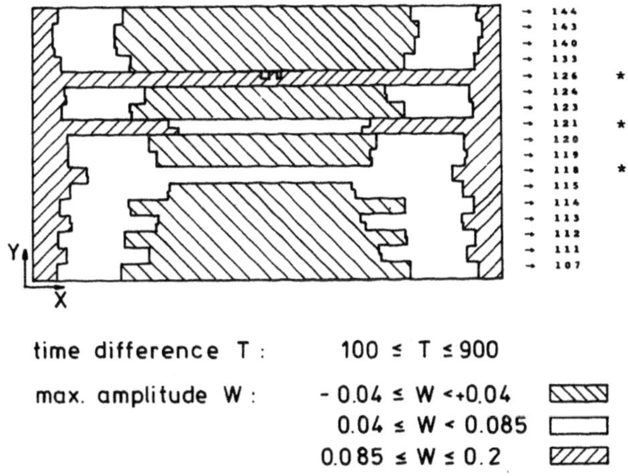

Fig. 5

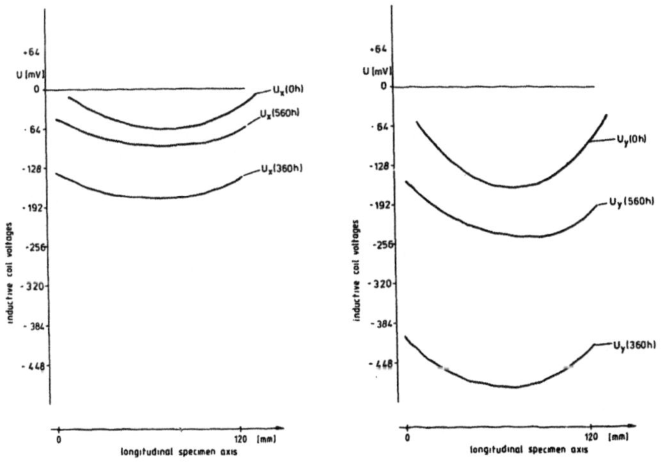

Fig. 6

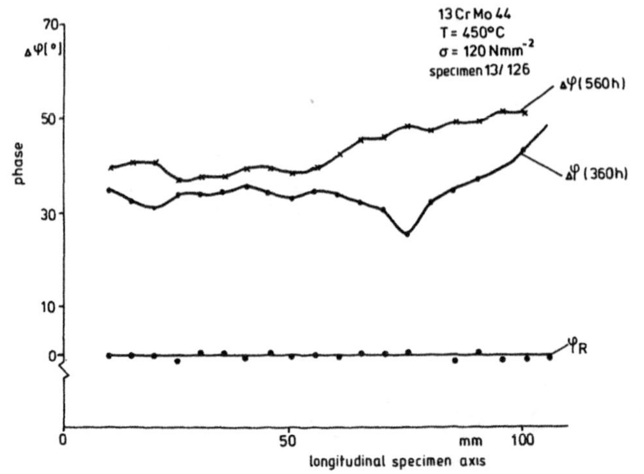

Fig. 7

Specimen 13/...	1st test		2nd test	
	W |10⁻⁴|	T |steps|	W |10⁻⁴|	T |steps|
111	9877	506	9900	503
112	9919	513	9990	512
114	9915	511	9943	509
115	9933	502	9884	501
117	9854	518	9980	517
121	9974	508	9916	507
122	9932	518	9963	518
124	9919	505	9926	504
140	9975	514	9977	517
143	9965	516	9969	518
144	9975	511	9984	512

Tab.1a: Scattering of values measured by pulsed eddy current method, undamaged material; time between measurements greater three months

Specimen 13/...	1st test		measurements after short period - creep tests			
	W |10⁻⁴|	T |steps|	W |10⁻⁴|	T |steps|	t |h|	ε_{pl} |°/oo|
110	9946	508	9701	504	168	1,75
132	9970	510	9342	517	261	1,79
118	9939	508	9860	494	360	0,00
126	9906	506	9739	483	360	0,86
126	9906	506	9525	517	560	1,77
121	9974	508	9739	483	650	1,98

T: delay; W: correlation coefficient; t: loading time; ε_{pl}.: plastic strain

Tab.1b: Values measured by pulse eddy current method of undamaged material and after different loading times

Electromagnetics

The Origin of Nondestructive Determination of Characteristic Material Parameters Using Electromagnetic Methods

Friedrich Förster, Reutlingen

Conversion of characteristic material parameters to measurable variables using suitable sensors is a field of activity to which I have devoted 57 years of my life. In 1931, I developed a method of converting the growth rate of plants to an acoustic frequency variation at the University of Göttingen. This made it possible to "hear the grass grow" in the true sense of the word. My doctoral thesis which I completed in 1932 dealt with what was then a new ultrasonic method for converting the velocity of sound in fluids to electrical measured values, thus permitting the effect of molecular association on the velocity of sound to be measured.

In 1935 I joined the Kaiser-Wilhelm-Institut of Metallurgical Research in Stuttgart which had then just been founded (and which is now the Max-Planck-Institut). As early as the first few weeks, I realized that the capabilities of electronics, a field which was still entirely new at the time and a discipline in which I, as a physicist, was particularly interested, were still largely unknown in this institute which had just been opened. By contrast, when this research institute was started, there was an extraordinarily great need for new measuring methods to determine structural parameters and their gradients as a function of time and temperature etc.

Owing to my great love of music, I first applied myself to the sound of metallic materials, i.e. the elastic and anelastic properties, and in 1935 I developed a method of measuring the modulus of elasticity and attenuation as a function of temperature and time. For this purpose, it was necessary for me to develop magnetic pickups and sensors since suitable ones were not available commercially at the time. This acoustic measuring method which led to the adoption of the particularly important characteristic material parameter of attenuation in the field of metallurgical research is now used widely in the fields of metallurgical, ceramic, glass, plastics and wood research, in addition to the corresponding industries.

As early as 1936, I used the acoustic sensor which I developed in 1935 for the initial experimental investigations and recordings of acoustic emission in metallographic processes such as martensite conversion, the conversion of zirconium and the twinning of mechanically stressed bismuth monocrystalline wires. In the following year, 1937, the acoustic sensor which I developed in 1935 was used again for the far-removed field of magnetic measurements and surveys. For this purpose, the ferromagnetic body under investigation was connected to the acoustic sensor by a miniature quartz tube to a very slight extent, e.g. with a weight of a fraction of a gram. The test piece which was located in a small electric furnace was magnetized by means of a DC field coil outside the furnace so that it was characterized by a magnetic moment. This magnetic moment was, at the same time, influenced by the gradient of an alternating magnetic field generated outside the furnace. This subjected the test piece to an alternating force which the acoustic sensor converted to an AC voltage. After amplification, the magnetic moment of the test piece was indicated by a measuring instrument. The measuring instrument could be used to trace physical metallographic processes, such as processes of conversion etc. as a function of temperature and time, through to determining the temperature of the Curie point. In the 30s, this acoustic sensor which was originally developed for totally different purposes and which was patented in the 1937 for magnetic measurements, was used for a whole series of theses.

Even these initial sensors which I developed in 1935 for acoustic measurements taught me something which has been confirmed time and time throughout the long course of my professional career, namely that developing a sensor generally leads to applications which far exceed the initial development objective of the sensor in question.

As early as 1936 to 1937, the manifold problems which face a metallurgical research institute made me realize that electromagnetic methods embodied extraordinary potential capabilities for research and industry. Owing to my lack of experience at the time, I naively assumed that it would be possible to open up the capabilities of using magnetic methods for non-destructive determination of characteristic material parameters simply by conducting series of experimental tests. I asked myself what was the influence of electrical conductivity, magnetic permeability and the dimensions (diameter, tube wall thickness and plating etc.) of metallic test pieces located in the electromagnetic field of a test coil on the impedance of the test coil if an alternating current of frequency f is passed through this coil. I tried to answer this question by conducting extensive series of experimental tests. But this turned out to be a mistake since there is an infinite number of possible test piece parameter and test frequency combinations. Owing to these unsatisfactory, experimental results, I resolved to thoroughly and painstakingly study the entire field of non-destructive determination of characteristic material parameters using electromagnetic methods in order to derive from this the laws of similarity which permit the many and varied influences of the specified parameters to be reduced to one single standardized curve. I was certainly aware that theoretically establishing the effect of influencing parameters of regularly cylindrical bars, tubes, level surfaces and spheres on the electromagnetic data of a through-type coil and the influencing parameters of test pieces and foils on the electromagnetic data of an approach-type coil or a transmission coil array would be no easy task.

For years, I was confronted by many unsolved problems since the extensive experimental results which I had obtained previously could not be incorporated significantly in a generally applicable law before establishing the theoretical background, i.e. the laws of similarity.

However, when the theoretical fundamentals of eddy-current testing were elaborated in 1940, the great understanding of the inner relationships which this provided made it possible to elaborate the endeavored simplification. The infinite number of possible material property combinations: electrical conductivity, magnetic permeability and geometrical dimensions and shapes of the test piece with the frequency of the test instrument could be represented with a normal curve for specific test piece shapes. At the same time, it was possible to theoretically derive the result that variations in the electrical conductivity, permeability and dimensions of a test piece must occur in different directions in the impedance plane of the test coil. This finding was a tremendous challenge involving displaying the impedance plane of the test coil which had then been defined only in theory or displaying the equivalent complex voltage plane of the test coil on the screen of a cathode ray tube in order to draw conclusions in respect of the cause in the test piece on the basis of shifts in the impedance value of the test coil. The method for quantitative display of the impedance plane of the test coil on the screen of a CRT was developed in December 1940.

Nowadays, displaying the impedance plane or the equivalent complex voltage plane of the test coil on the screen of a CRT is a foregone conclusion world-wide. This is why, today, no one could comprehend the excitement and emotion which I felt when I pushed a bar made of an aluminium alloy with a slight local diameter reduction produced by emery paper and local punctiform heating, i.e. a reduction in the electrical conductivity, through the test coil during my first attempt to display the test coil impedance plane on a fluorescent screen. Before my very eyes, as the section of the bar with reduced diameter passed through the coil, the bright luminous spot on the screen moved in a direction different to the direction in which it moved when the point with the conductivity change passed through the coil, and this was exactly as I had predicated in my theory which I had just concluded. This phenomenon which excited me so much, i.e. that the luminous spot on the screen moved in different directions, just as my theory had predicted, made me realize immediately that this two-dimensional display of the test coil value would become the standard method of eddy-current testing in the future. And today, 48 years later, this eddy current impedance plane analysis method actually is used throughout the world.

Between the early 40s and early 50s, the impedance values or the complex voltage values of the test coil were calculated for the following test pieces:

1.) Regularly cylindrical bars made of non-ferromagnetic material for various test coil space factors.
2.) Regularly cylindrical bars made of ferromagnetic material for various permeabilities.
3.) Thin-walled tubes made of non-ferromagnetic material.
4.) Non-ferromagnetic tubes with various wall thicknesses.
5.) Non-ferromagnetic bars or thick-walled tubes with an outer plating comprising a non-ferromagnetic metal and characterized by differing electrical conductivity.
6.) Non-ferromagnetic thick-walled tubes, with the test coil inside the tube.
7.) Ferromagnetic spheres with a high magnetic permeability in a short test coil.
8.) Ferromagnetic spheres or short, regularly cylindrical bodies with various magnetic permeabilities in a short test coil.
9.) Non-ferromagnetic spheres in a short test coil.
10.) Flat sheet material or foil with primary or secondary coil on opposite sides (transmission method).

So-called sensitivity diagrams were calculated for the above test pieces and test coil arrays with the purpose of quantitative evaluation of eddy-current measurements and selecting an optimum frequency for specific tasks involving non-destructive determination of characteristic material parameters. These diagrams show the amplitude and phase angle on the screen when the parameters of property for the various test pieces, such as electrical conductivity, magnetic permeability, diameter and wall thickness, change by 1 %, as a function of the ratio of the eddy current test instrument's test frequency to the fundamental characteristic parameter of the eddy-current theory, namely the limit frequency of the test piece. The appropriate formulae are available for the limit frequencies of the above-listed test pieces with the relevant dimensions and physical properties, termed "Foerster limit frequency" in American literature. All above-mentioned impedance diagrams and the sensitivity diagrams just mentioned have been calculated, tabulated and recorded for the ratios frequency to limit frequency. For this purpose, approximation equations for regularly cylindrical bodies, for spheres, for thin-walled tubes and for thin, flat sheet materials have been derived for low and high values of these ratios. It is only the ratios of frequency to limit frequency which make it possible to apply the laws of similarity of the eddy-current technique for any parameters electrical conductivity and magnetic permeability in addition to dimensions of a test piece for any frequencies of the coil field to these impedance and sensitivity diagrams. This quantitative explanation of the theoretical fundamentals of the eddy-current method which I have but touched upon above and a major part of which is described in the US Non-destructive Handbook Electromagnetic Testing - Editions I and II turns this approximate indicator method into a non-destructive measurement method. And it was just this aim which I set myself exactly half a century ago. I have not discussed eddy-current crack testing because it is not the subject of this congress. But may I, at this point, mention that extensive mercury model tests on which flaws were simulated by suitable plastic bodies have permitted the complex test voltage effects to be determined for a broad range of test frequency to limit frequency ratios for a large number of flaw configurations and flaw locations. The laws of similarity of the eddy-current theory also now permit us to predict the influences of material flaws on the complex voltage of the test coil, whereby such influences cannot be calculated strictly mathematically.

However, during my research into the experimental and theoretical fundamentals of the eddy-current method in the late 30s, I was disturbed by the fact that, over and above crack testing, it was possible to non-destructively determine only two material parameters - electrical conductivity and magnetic permeability. I must explain that the metallurgical research institute for which I was working urgently required a method of determining numerous physical and technological characteristic material parameters occurring in conjunction with metallurgical processes non-destructively and, for thermal investigations, wherever possible also without physical contact. The main emphasis was on magnetic measurements for answering questions related to metallography. Ferromagnetic investigations, e.g. the method of punctiform, ballistic recording of an hysteresis loop which was conventional at the time, were extremely time-consuming. When, in 1939, the first CRTs with a screen diameter of 18 cm became available, I devoted myself to the task of developing an instrument for stationary display of the hysteresis loop on a fluorescent screen. In 1938, on the occasion of the Institute's annual conference, it was possible to demonstrate this instrument in numerous example applications. In addition to quantitative measurement of the coercive force, remenance, saturation, initial permeability, Rayleigh's constant and magnetostriction as a function of the temperature up to the Curie point, the differential hysteresis loop method provided a sensitivity for non-destructive determination of various characteristic material parameters which extended way beyond the previous magnetic measurement methods (e.g. by a factor of 1000). The differential hysetersis loop method consists in arranging two identical test pieces, such as wires or strips etc. which are magnetized in a primary field, in two back-to-back-connected secondary coils so that only the difference in the magnetic properties between the two test pieces is displayed on the screen. If one of the two identical test pieces is modified, either thermally or mechanically, the magnetic effect of such a modification can be amplified, e.g. by a factor of 1000, as compared with the sensitivity of the loop display itself. In addition to being used for normal series of measurements, this instrument was used immediately for establishing the true physical limit, i.e. the start of an irreversible hysteresis loop change of a ferromagnetic wire made of iron or nickel as a function of the tempering temperature of these wires. At a tempering temperature of 900 °C, this physical limit was one power of ten below the values determined mechanically until then. As early as 1940, the German "Zeitschrift für Metallkunde" (metallographic periodical) received two extensive papers elaborated with the aid of this new differential hysteresis loop method, entitled "Magnetic Investigations of Internal Stresses". The following characteristic material parameters were investigated with this new instrument which permitted the magnetic properties of ferromagnetic wires with diameters of as low as 50 μm and at field strengths of $H_{max} = 10^{-3}$ to $H_{max} = 2 \times 10^{3}$ Oe to be determined:

1.) Quantitative build-up of internal second-order stresses by stretching in the plastic region on ferrous and nickel wires until the wires break.
2.) Quantitative tracing of the process of reduction of internal stresses with increasing tempering temperature.
3.) Determining the physical yield point when the first irreversible change in remanence occurs as a function of the tempering temperature.
4.) Development of first-order internal stresses when burring a wire.

5.) Influence of second-order stretching stresses on the first-order stress state of burred wire.
6.) Influence of the tempering temperature on the first-order stress state of a burred wire.

The extreme sensitivity for the vertical magnetization axis on the screen of the ferrograph clearly showed the Barkhausen effect on the screen trace. For this reason, investigations were then conducted into what characteristic material parameters could be derived from the Barkhausen effect. Consequently, a paper entitled "On the question of magnetic fold-over processes in iron and nickel" was concluded in October 1940. This paper dealt with the following investigations:

1.) The influence of tempering temperature on the mean range magnitude of Barkhausen effects on cold-burred carbonyl iron wire. The same influence was investigated on carbonyl nickel wire. This established the role of first and second-order stresses on the mean range magnitude.
2.) The influence of second-order stresses when stretching cabronyl iron wires on the mean range magnitude of the Barkhausen effect.
3.) Initial oscillographic traces of the Barkhausen effect as a function of time, carried out in 1939, provided important insights into the interrelationships between the range volumes of the Barkhausen effect and the material properties.
4.) Investigation of the reduction of a large Barkhausen effect complex by stretching the wire with increasing initial field strength.
5.) Simultaneous recording of the longitudinal and circular components of the same Barkausen effect with a dual-trace oscilloscope. This clearly establish the mechanism of the Matteucci effect.

To conclude, I shall discuss a sensor for magnetic DC fields, which resulted from the above magnetic investigations and which, in the same way as the eddy current impedance plane analysis method, is now in use throughout the world as a Förster probe, ferro-probe, flux gate detector or second harmonic detector. This magnetic sensor which, when it was developed in 1939, was the most sensitive device for measuring magnetic DC fields, with high indicating stability for fields with a strength of even less than one hundred thousandth of the earth's magnetic field formed the basis for a whole series of methods and instruments for non-destructive determination of characteristic material parameters. I shall name but a few by way of example:

1.) Instruments for determining the coercive force even on very small test pieces, and which today incorporate an automated measuring sequence.
2.) Instruments for susceptibility measurement in the low range on non-magnetic steels, in addition to tracing the stability with mechanical stressing of austentic steels.
3.) Residual field test instruments for sorting mass-produced parts on the basis of hardness, tempering temperature, carbon content and case depth.
4.) Instruments for measuring the rolling mill texture on ferromagnetic sheets with the aid of the point-pole method.
5.) Instruments for measuring magnetic leakage flux with the normal leakage-flux method or with the magnetography method described by the author in 1950. Förster micro-probes with dimensions 1 x 1 x 1 mm are used for this purpose.
6.) Detection instruments with Förster differential probes for quantitative measurement of relative characteristic material parameters of unknown sought objects concealed either underground or on or below the seabed. These characteristic material parameters relate to: magnitude of the magnetic moment, axial direction of the magnetic moment in space, depth of the sought object and point on the surface perpendicularly above the focus of the sought object's center of magnetic moment. Detection devices equipped with Förster differential probes have

been used since the war to locate millions of unexploded bombs and mines, thus permitting them to be disposed of.

7.) Finally, these magnetic field sensors surveyed the magnetic field and, thus, the magnetic moment of the planet Venus, as the start of the series of space projects as early as 1962 by the American "Mariner 2" space probe. Here we have a planet as the test piece and the planet's magnetic moment as the characteristic material parameter.

Progress in the Micromagnetic Multiparameter Microstructure and Stress Analysis (3MA)

G. Dobmann, W.A. Theiner, R. Becker

Fraunhofer-Institut für zerstörungsfreie Prüfverfahren, IzfP
Universität des Saarlandes, Geb. 37
D-6600 Saarbrücken, FRG

Summary

Material characteristics such as hardness, yield-strength, ductility etc. can be controlled and designed for a given material in certain limits. For steels manufacturing parameters like the chemical composition, block- or continuous-casting, hot-rolling, forging, cold-rolling, heat-treatments, and machining influence the microstructure and residual stress state which in combination with additional load conditions during service (mechanical static/dynamic, thermal, corrosive) determine the material properties and behaviour. In mechanical- and technological tests (standard tensile, notch-impact, etc.) the service loads are simulated mainly under maximum load conditions in order to characterize the material fitness for service. 3 MA-techniques and their inspection quantities (Barkhausen-noise, incremental permeability, eddy-current impedance spectroscopy, etc.) are also influenced by microstructure and stress in steels. The emphasis in the following contribution will be on the presentation of these facts with the goal to show correlation between material characteristics and 3 MA-quantities.

The Multiparameter Approach

Fig. 1 explains the fact of the multi-dependence of the material characteristics. On the influence of microstructure and residual stress on the material behaviour .Theoretical and empirical models are reported [1, 2]. In the same way we find a multidependence of the ndt-quantities, i.e. the 3 MA-quantities as function of the microstucture and residual stresses. Most of the theoretical models existing [3] are restricted to single crystals and Bloch-wall interactions with lattice imperfections. More or less they can be used for a qualitative description of the observed magnetic properties. The general aim for 3 MA is to solve the inverse problem, to characterize and analyze the microstucture and residual stress states by the use of 3 MA- quantities. In order to perform such an analysis, i.e. to find a solution of the inverse problem, a set of measuring quantities has to be selected. This means that quantities which react with weighted sensitivities on the different microstructure and stress parameters have to be used. This selection of independent quantities is successful because 3 MA allows to use reversible and irreversible magnetization processes which contribute independent information. Furthermore the different types of Bloch-walls (in iron-based materials 180° and 90°) are the sensors in the material which are more or less stress sensitive; the 90°-Bloch-walls interact directly with stresses.

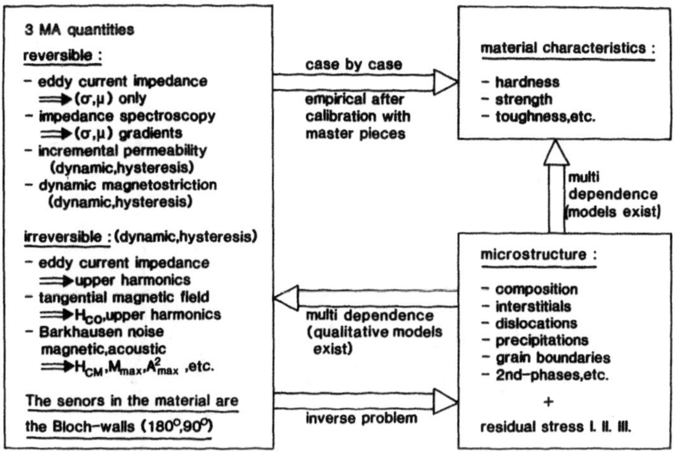

Fig. 1. Multidependence of Material Characteristics and NDT-Quantities as Functions of the Microstructure and Stress

In a case by case solution correlations between the material characteristics and the 3 MA-quantities can be derived. Nevertheless these solutions cannot be generalized like a physical law. Correlations obtained after an empirical calibration at well defined master pieces are restricted on the knowledge base used for multi-regression, i.e. an extrapolation is not allowed. In the contribution two typical 3 MA-applications are discussed.

Stress and Hardness Measurements at Turbine Blades

W.A. Theiner [4] had initially reported in Hershey on this nd-inspection task. Turbine blades (X20Cr13, ASTM A276Gr403) of the last stage row in low-pressure turbine sections must operate in regions of high moisture. This can result in water droplet erosion of the leading edge. To increase the resistance of the blades to this type of erosion, a flame hardening process is used by Siemens-KWU, Mülheim. The process, if not properly applied increases the potential to stress corrosion cracking. Sufficiently high tensile stress must be present for corrosion cracking. The residual stress state near the leading edge is mainly determined by the hardening process. The critical residual surface stresses must therefore be detected by ndt. For the past six years, the company uses a 3MA-device designed by the IzfP for the inservice inspection. Additionally a second device is beeing used for the inspection after hardening in the fabrication for the past two years [5].

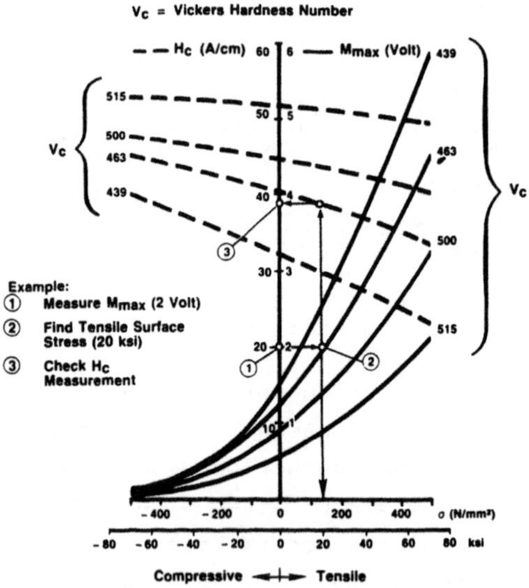

Fig. 2. Nomogram for Stress (σ) and Hardness (V_c) Prediction, 3MA-Quantities (Hc, M_{max}); Turbine Blades X20Cr13^c

Fig. 2 shows the nomogram which is used for the interpretation. For different microstructures, characterized by the Vickers hardness number and representing the different hardened states, the coercivity and the maximum Barkhausen-noise amplitude have been measured as function of compressive and tensile stress, both quantities are derived from a Barkhausen-noise measurement. There exist a bi-unique transformation between the pair (σ_{mech}, HV_{10}) and the pair (M_{max}, H_{cM}). For calibration σ_{mech} is measured in a bending experiment by strain gages or by x-ray-diffraction.

Whereas Siemens-KWU uses an iterative procedure to find the (σ_{mech}, HV_{10})-pair in the nomogram, IzfP has applied a multi-regression algorithm. In a R + D-project for the German machinery industry turbines of a second manufacturer (steel X21CrMoV12 1) have been characterized by 3MA. I. Altpeter has used the quantities coercivity H_{cM} and M_{max}, maximum of the Barkhausen-noise but only up to a (incomplete) 2-D-polynominal approximation in the form [6].

$$HV_{30}\ nd = f_1\ (H_{cM},\ Max * H_{cM}) \quad (1)$$

$$\sigma_{mech}\ nd = f_2\ (H_{cM},\ M_{max} * H_{cM}) \quad (2).$$

Fig. 3 and 4 show the results. In Fig. 3 the residual stress distribution as function of the distance from the leading edge after x-ray measurement is compared with the 3MA-model (2), Fig. 4 represents the results for the Vickers hardness HV_{30} (1). Standard deviations $\pm$ 47 MPa and $\pm$ 32 hardness numbers have been obtained. The reason for these large values is that the two techniques sense different material sheet thickness (x-ray ~ 30 μm, 3 MA ~ 200 μm). The result can be improved, when for calibration masterpieces (different hardness states) are investigated in bending tests, using strain ganges. Then comparable material volumes are analyzed.

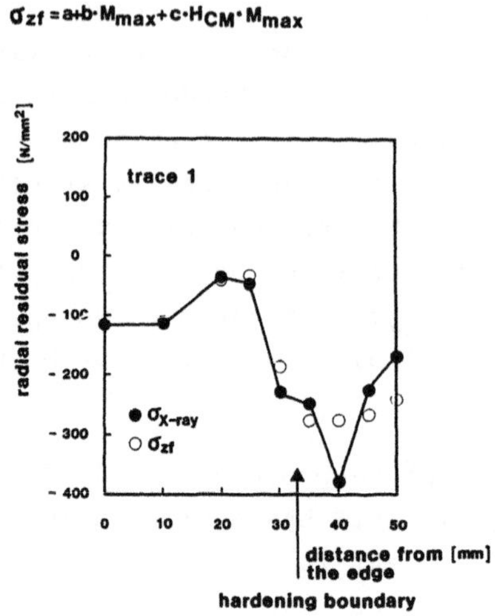

Fig. 3. Residual Stress Distribution at the Surface of a hardened Turbine Blade, Correlation of X-Ray Results with 3MA-Results (X21CrMoV12 1)

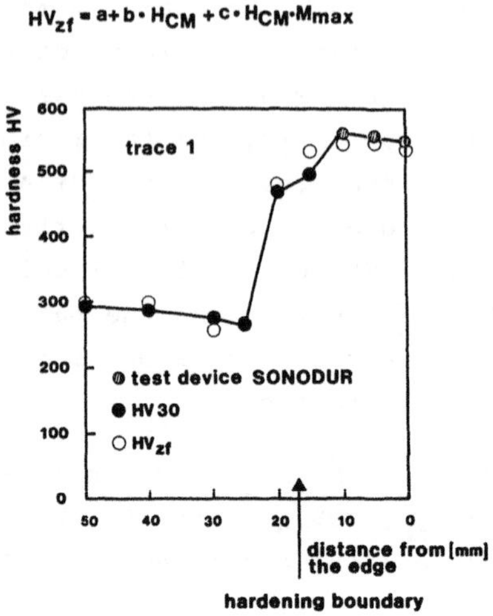

Fig. 4. Hardness Distribution at the Surface of a hardened Turbine Blade, Correlation of HV_{30} Results with 3MA-Results (X21CrMoV12 1

Improved (σ, μ)-Measurement

In [7] we have reported on an eddy current approach for the material characterization of steels. Applying numerical modelling a bi-unique transformation is found which can translate the real- and imaginary-part of the eddy current coil impedance into a pair of (σ, μ)-values (σ electrical conductivity, μ-initial magnetic permeability). Pick-up-coils are needed for inspection tasks requiring high spatial resolution. Air pick-up-coils which can be modeled analytically unfortunately show a good sensitivity together with the potential to separate σ-and μ-effects only in the low μ-range ($\mu < 10$) and therefore they are insufficient for ferromagnetic materials ($\mu < 200$). In order to overcome the problem ferritic-cup-core-coils have been modeled numerically using 2D FE-software [0].

Fig. 5 shows the coil geometry and the FE-mesh used for the discrete numerical problem. The progress here is that now (σ, μ)-calibration curves can be modeled for a wider range of frequencies (100 Hz $\leq f \leq$ 500 kHz) and in addition that a frequency can be selected to provide optimal resolution in (σ, μ).

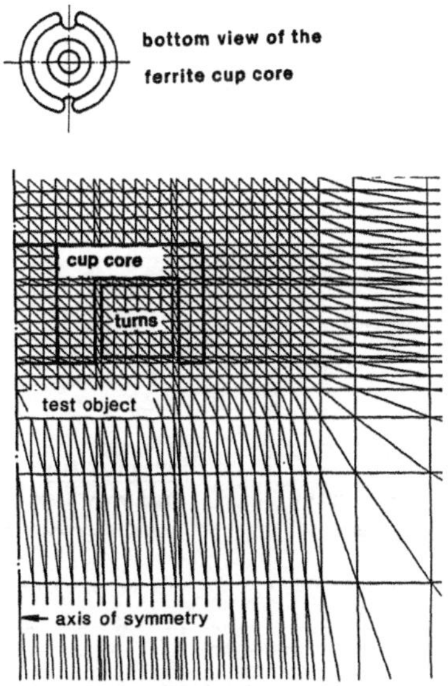

Fig. 5. Ferrite Cup Coil Geometry and Finite Elemente Mesh for the discrete numerical problem

Fig. 6 documents the results for a series of frequencies. Furthermore the approach was extended with a new feature. Because the well known phase-selection cannot be used to suppress the lift-off of the probe, the lift-off is measured by a second coil which is integrated in the first and excited by a current of 1 MHz. Fig. 7 shows the blockdiagram of the new designed (σ, μ)-meter in which an amplitude correction-algorithm is used in order to eliminate the lift-off-effect.

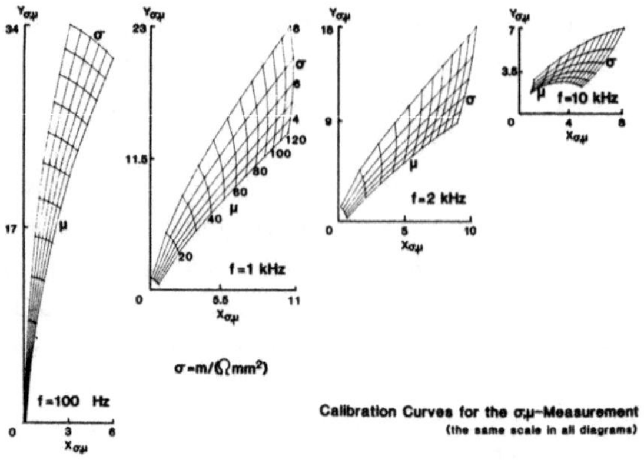

Fig. 6.

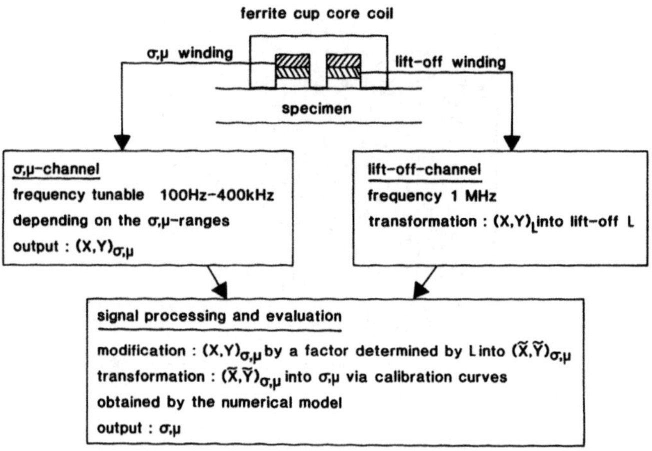

Functional Diagramm of σ,μ-measurement

Fig. 7.

Research and development was performed under contract of the German Ministry for Research and Technology.

References

1. Embury, J.D.: Strengthening by Dislocation Substructures. Strengthening Methods in Crystals. London: Appl.Sci.Publ. LTD 1971

2. Kelly, A., Nicholson, R.B.: The Strength of Martensite. ibid

3. Kneller, E., Ferromagnetismus. Berlin, Heidelberg, New York: Springer Verlag 1962

4. Schneider, E., Altpeter, I., Theiner, W.A., Ruud, C.O., Green, Jr., R.R. (eds.): Nondestructive Methods for Material Property Determination. New York, London: Plenum Press 1984

5. Stücker, E., Gartner, G.: Quality Control of Last Stage Blades with Flame Hardened Leading Edges. Jt. ASME/IEEE Power Generation Conference, Miami Beach, Florida: October 4-8, 1987

6. Altpeter, I., Theiner, W.A.: Weiterentwicklung von mikromagnetischen Verfahren zur Bestimmung von Einsatzhärtungstiefen bzw. Einhärtetiefen und Eigenspannungen an Werkstückoberflächen. Abschlußbericht zum Forschungsvorhaben Nr. 700 86, Arbeitskreis Materialspannungen des Forschungskuratoriums Maschinenbau des VDMA. Saarbrücken: IzfP-Eigenverlag 1987

7. Dobmann, G., Becker, R., Rodner, Ch.: Quantitative Eddy Current Varients for 3 MA. Review of Progress in Quantitative Nondestructive Evaluation. Thompson, D.O., Chimenti, D.E. (eds.). New York, London: Plenum Press 1988

8. Palanisamy, R.: Finite Element Modeling of Eddy Current Nondestructive Testing Phenomena. Fort Collins: Dissertation Colorado State University 1980

Detection of Stress in Steels from Differential Magnetic Susceptibility

D. C. JILES* AND P. GARIKEPATI+

Center for NDE
Iowa State University
Ames, Iowa 50011
USA

*Department of Materials Science and Engineering
+Department of Electrical Engineering

Abstract

Following the publication of the theory of ferromagnetic hysteresis considerable advances have been made in understanding the mechanisms involved in the magnetization process in ferromagnets. One of the areas in which an extension of the theory has been made is the effect of applied stress on magnetization. This has important consequences for NDE of stress via magnetic measurements since for the first time it is possible to give a quantitative prediction of changes in the magnetization curve with stress. This paper discusses the development of the theory and shows the successful application of the model to the case of a steel specimen under tensile or compressive stress.

Introduction and Background

Recent work on the theory of ferromagnetic hysteresis [1] has resulted in a differential equation describing the variation of magnetization with magnetic field in the case of uniformly impeded domain wall motion. The model equation, which has been used for materials as diverse as pipeline steels [2] and magnetic components in electronic circuits [3], is characterized by five independent parameters which depend on the material properties.

The essential idea behind the model is that the rate of change of magnetization M with field H is dependent upon the displacement of magnetization from an ideal equilibrium value known as the anhysteretic magnetization M_{an}. The anhysteretic magnetization curve is a single valued function of the magnetic field, although it is of course dependent upon external factors such as temperature and stress.

Anhysteretic Magnetization

The dependence of the normal DC magnetization curve on stress is extremely complex. Experimental work was conducted on this more than forty years ago, but until recently our understanding of the mechanisms behind the observed behavior has remained as crude as it was in the 1930's when Becker and Doring [4] considered the problem.

The behavior of the anhysteretic magnetization curve under stress of a typical sample of high strength constructional steel [2] is much simpler, particularly at low fields where the magnetostriction of iron is positive. In this region the application of a compressive stress reduced the magnetization, whereas a tensile stress increased the magnetization. At higher field strengths the behavior was more complex due to the change in sign of the magnetostriction λ, which was a function of both stress and field.

In the most general case discussed [1] no assumption about the general form of the anhysteretic magnetization curve was made. This ensured that the hysteresis equations remained totally general. In practice, in order to make use of the model, some assumptions need to be made about the form of the curve. In earlier papers we have suggested a modified Langevin function can be used to represent the anhysteretic. This treatment is identical in form to that used by Bean and Livingston [5] to derive a magnetization curve for a superparamagnet.

$$\frac{M_{an}(H)}{M_s} = \coth\left(\frac{H+\alpha M}{a}\right) - \left(\frac{a}{H+\alpha M}\right) \qquad (1).$$

Extension of the Model: Incorporation of Stress

The Gibbs free energy for a magnetic material under a stress σ is given by

$$G = U - T.S + \frac{3}{2}\sigma\lambda \qquad (2)$$

where U is the internal energy, T is the temperature and S is the entropy. The effective magnetic field H_e can be defined in terms of the Helmholtz free energy, which in S.I. units is,

$$H_e = \frac{1}{\mu_o}\left(\frac{\partial A}{\partial M}\right)_T \tag{3}$$

It is known that the Helmholtz and Gibbs free energies are related by

$$A = G + \mu_o H M \tag{4}$$

and furthermore, the internal energy in this case can be empressed in terms of the magnetization, since, the magnetostatic energy is given by

$$U = \frac{\alpha}{2\mu_o} M^2 \tag{5}$$

where the coefficient α represents a combination of the interdomain coupling and any demagnetiziing effects, both of which lead to energies dependent on M^2. Therefore the effective field is [6]

$$H_e = \frac{1}{\mu_o}\left(\frac{\partial A}{\partial M}\right)_T = H + \alpha M + \frac{3}{2}\frac{\sigma}{\mu_o}\left(\frac{\partial \lambda}{\partial M}\right)_T \tag{6}$$

This can be written in the form

$$H_e = H + \alpha M + H_\sigma \tag{7}$$

where H_σ is the magnetic field that is equivalent to the stress

This field can now be included in the expression for the anhysteretic magnetization,

$$\frac{M_{an}(H,\sigma)}{M_s} = \coth\left(\frac{H + \alpha M + H_\sigma}{a}\right) - \left(\frac{a}{H + \alpha M + H_\sigma}\right) \tag{8}$$

The behavior of the anhysteretic magnetization at the origin, $\chi'_{an}(H=o)$ can be found from this equation,

$$\frac{M_{an}(H=o,\sigma)}{M_s} = \frac{H + \alpha M + H_\sigma}{3a} \tag{9}$$

from a simple series expansion of the hyperbolic contangent.

If the relation between magnetostriction λ and magnetization M is expressed as

$$\lambda = bM^2 \tag{10}$$

$$\frac{\partial \lambda}{\partial M} = 2bM \tag{11}$$

Substituting this into the expression for the anhysteretic magnetization above, and differentiating leads to

$$\frac{dM_{an}(H=o,\sigma)}{dH} = \frac{M_s}{3a - \left(\alpha + \frac{3b\sigma}{\mu_o}\right)M_s} = \chi'_{an}(H=o,\sigma) \tag{12}$$

and from this it is easily shown that

$$\frac{3b\sigma}{\mu_o} = \frac{1}{\chi'_{an}(\sigma=o)} - \frac{1}{\chi'_{an}(\sigma=\sigma)} \tag{13}$$

This equation, which uses only the simplifying assumption that magnetostriction is quadratic in M, predicts a very simple relationship between the anhysteretic differential susceptibility at the origin and the stress.

Results

Measurements have been taken on a range of specimens in order to examine the validity of the theoretical prediction. In this paper we present results on AISI 4130 steels measured under both compressive and tensile stresses.

The variation of magnetostriction with magnetization is shown in Fig. 1. As can be seen the magnetostriction does depend on the square of the magnetization in steel, and this holds true for magnetizations up to 1.3 x 10^6 A/m (magnetic inductions of 16 k. Gauss or 1. Tesla). The coefficient of proportionality b in equation (10) above has the value 2.4 x 10^{-18} $(A.m^{-1})^{-2}$.

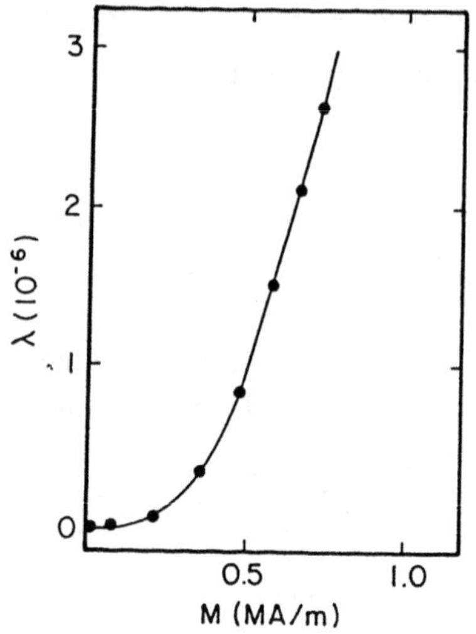

Fig. 1. Variation of the magnetostriction λ with magnetization M for AISI 4130 steel.

From the measurement of points along the anhysteretic magnetization curve at several different levels of stress the variation of χ'_{an} with stress was determined and this is shown in Fig. 2. as the open circles. The variation of χ'_{an} with stress based on equation (13) above yielded the theoretical curve shown in Fig. 2. It is clear from this figure that the agreement between theory and experiment is good.

Significantly the theory and experiment both show an asymmetry in the dependence of $\chi'_{an}(H=o)$ under tensile and compressive stress. The variation of χ'_{an} with stress is much stronger under tensile stress than under compressive.

Transforming this graph to reciprocal susceptibility $\frac{1}{\chi'_{an}(H=o,\sigma)}$ against stress yields a straight line as shown in Fig. 3, where once again the experimental data appear as the open circles. The theoretical curve is given by the equation,

$$\frac{1}{\chi'_{an}(H=o,\sigma)} = 2.95 x 10^{-3} - (5.73 x 10^{-12})\sigma \tag{14}$$

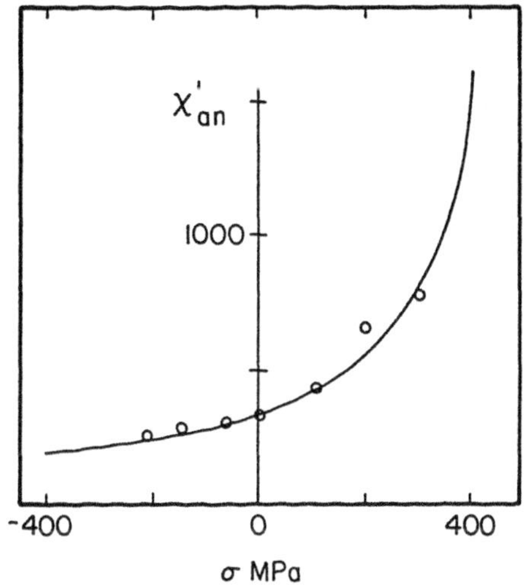

Fig. 2. Anhysteretic susceptibility at the origin $\chi'_{an}(H=o)$ as a function of stress. (circles are experimental results, the curve is the theoretical prediction).

Conclusions

This paper presents the development of a new theory of the effects of stress on magnetization in ferromagnetic materials such as steels, and experimental measurements taken to test the validity of the theory.

The variation of the anhysteretic susceptibility at the origin of the M, H plane is predicted to be dependent upon the reciprocal of the stress, with the coefficient of proportionality depending on the magnetostrictive behavior of the material.

The usefulness of the anhysteretic susceptibility in the evaluation of stress has been hinted at in earlier papers [2], however in this paper we show for the first time how quantitative estimates of stress can be obtained from such a measurement.

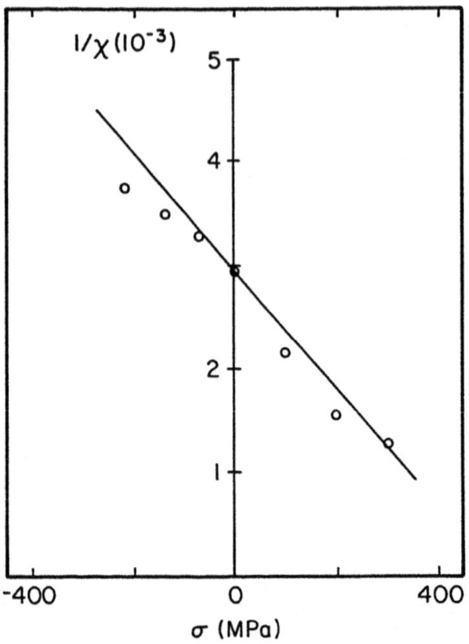

Fig. 3. Reciprocal anhysteretic susceptibility at the origin as a function of stress (circles are experimental results, the straight line is the theoretical prediction).

The theoretical work therefore gives not only an important indication of how the materials behave under stress but also indicates a new method of stress evaluation that can be applied to iron and steel components.

Acknowledgement

This work was sponsored by the Center for NDE at Iowa State University and was performed at the Ames Laboratory. Ames Laboratory is operated for the U.S. Department of Energy by Iowa State University under Contract No. W-7405-ENG-82.

References

1. D. C. Jiles and D. L. Atherton, J. Mag. Mag. Mater. 61, 48, (1986).
2. D. C. Jiles and D. L. Atherton, J. Phys. D. 17, 1265, (1984).
3. P. Tuinenga, "A guide to circuit simulation and analysis using PSpice" (Prentice Hall 1988).
4. R. Becker and W. Doering "Ferromagnetisms" Springer-Verlag, Berlin 1939.
5. C. P. Bean and J. D. Livingston, J. Appl. Phys. 30(s), 120, (1959).
6. M. J. Sablik, G. L. Burkhardt, H. Kwun and D. C. Jiles, J. Appl. Phys. 63, 3930, (1988).

Nondestructive Characterization of Austempered Ductile Irons

Seung S. Lee and Sekyung Lee

Korea Standards Research Institute
P.O. Box 3, Taedok Science Town, Taejon, Chungnam 300-31
Republic of Korea

Introduction

Recently developed austempered ductile irons have good mechanical properties. Comparing the austempered ductile irons with the normal ductile irons the austempered ductile irons have about two times or higher tensile strength than the normal ductile irons at the same level of elongation as shown in Fig.1 [1]. Austempered ductile irons are made from ductile cast irons by austempering heat-treatment. Ideally, during austempering, the ductile cast irons are cooled rapidly from the austenitizing temperature to the temperature range for bainite formation, avoiding the zone of pearlite formation. However, wide spread use of austempered ductile irons have not yet occurred. The main problem is in the difficulties of identifying and controlling the austempering heat-treatment parameters to achieve optimum properties consistently [1]. Those heat-treatment parameters are austenitizing temperature, austempering temperature, and austempering holding time. Austempering reaction of ductile cast irons is usually divided into two stages. First stage is toughening stage. Austenite decomposes into bainitic ferrite and carbon riched austenite. These two phase mixture of bainitic ferrite and austentie is the reason for the outstanding mechanical properties of austempered ductile irons. However, extended holding at the austempering temperature causes a severe reduction in ductility [2]. The reduction in toughness is caused by retained austenite decomposition into additional ferrite and carbide during the second stage of austempering [3]. The carbide precipitation is the reason for the reduced toughness. Therefore, there is a range of austempering time and temperature for the optimum mechanical properties of the cast irons [1]. Even if the cast irons are austempered in the processing range determined by mechanical test we are not sure about the qualities of the austempered ductile irons because casting and heat-treatment process are so complex. Therefore, if there is a nondestructive quality assurance technique of austempered ductile irons, it is helpful to assure the quality of the castings by examining all the castings. Identifying the possible ways of nondestructive quality assurance techniques of austempered ductile irons is the object of this study.

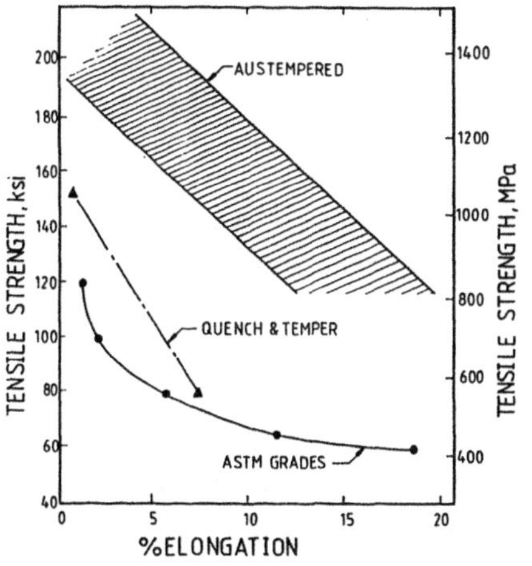

Fig.1 Comparison of mechanical properties for austempered ductile irons with standard ductile iron grade.

Sample preparation

Samples were prepared in the following way. Commercial ductile cast iron Y-blocks were used for austempering. There were two kinds of ductile cast iron Y-blocks as shown in Table 1. Both of them were JIS FCD 45 and manufactured at the same company. The composition and ferrite percent were slightly different since each was manufactured by different batch. Number 1 ductile cast irons were austempered with heat-treatment of 250 °C and 350 °C ; number 2 ductile cast irons were austempered with heat-treatment of 200 °C, 300 °C and 400 °C. Austempering holding times were same in both cast irons such as 10, 20, 30, 60, 120 minutes at each austempering temperature. Austenitizing treatments were also same in both cast irons.

Table 1. Sample preparation.

		# 1	# 2
Original ductile cast iron before heat-treatment	Composition(%)	C 3.75, Si 2.60, Mn 0.25, P 0.04, S 0.017	C 3.85, Si 2.65, Mn 0.32, P 0.03, S 0.015
	Ferrite content (%)	70	60
Austempering condition	Austenitizing treatment	900 °C, 1.5 hr	900 °C, 1.5 hr
	Austempering temperature	250 °C, 350 °C	200 °C, 300 °C, 400 °C
	Austempering time	10,20,30,60,120 mins	10,20,30,60,120 mins

Before austempering heat-treatment, the Y-blocks were cut into impact test specimens (unnotched subsize Charphy impact test specimens with a size of 8 x 10 x 55 mm^3) and nondestructive test specimens (rectangular shape with a size of 20 x 40 x 60 mm^3). The nondestructive test specimens were used as hardness test specimens after nondestructive tests. The test specimens were heat-treated at each austempering temperature and holding time as a batch of three impact test specimens and one nondestructive specimen.

Mechanical test results and microstructure observation

Fig.2. and Fig.3 show the results of Charphy impact test and Brinell hardness test respectively. The values of the impact energy increase or do not change as a function of austempering time at all austempering temperatures except 400 °C. The reason why the values of the impact energy drop after 30 minutes at 400 °C specimens could be due to the carbide precipitation from the second stage austempering reaction. Microstructures of the test specimens at different austempering temperatures are shown in Fig.4. The microstructure of 200 °C specimen looks like martensite. The martensite structure is the reason of the low impact energy and high hardness of 200 °C specimen. As austempering temperatures increase the bainitic ferrites shown by the dark regions in the microstructures become coarse and distinct. The carbide precipitation of 400 °C specimen cannot be resolved at these 2000 X magnification. The results of impact test, hardness test, and microstructural analysis match each other pretty well.

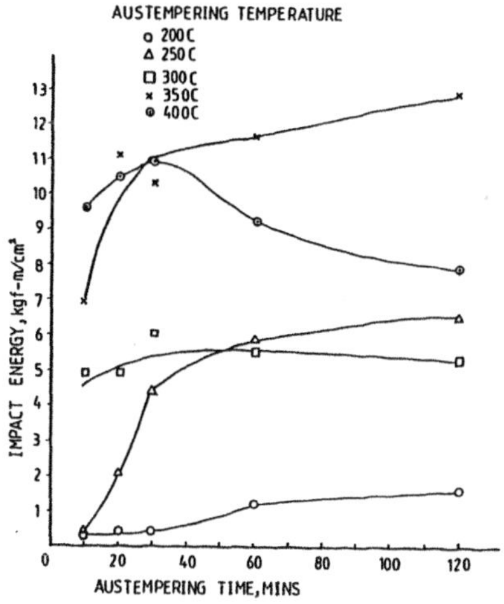

Fig.2 Influence of austempering time on the impact energy of ductile cast iron austenitized at 900 °C and austempered at 200 °C, 250 °C, 300 °C, 350 °C, and 400 °C ; as cast state # 1 has 9.5 kgf-m/cm² and # 2 has 10.0 kgf-m/cm².

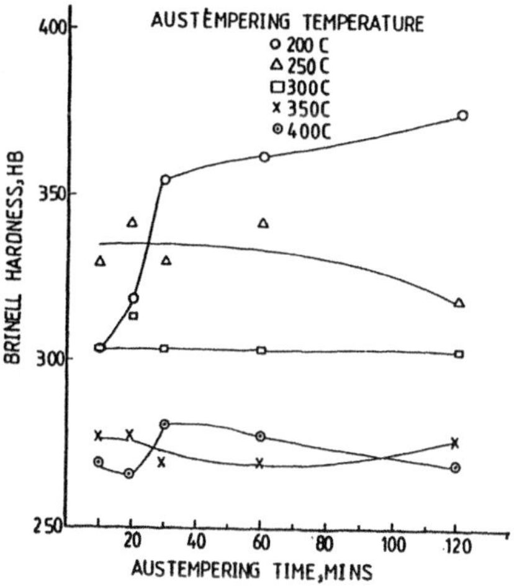

Fig.3 Influence of austempering time on the Brinell hardness of ductile cast iron austenitized at 900 °C and austempered at 200 °C, 250 °C, 300 °C, and 400 °C ; as cast state # 1 has 166 HB and # 2 has 163 HB.

Ultrasonic velocity measurement and electrical resistivity measurement.

In the nondestructive characterization of the austempered ductile irons, the factors which affect the measurement parameters are graphite and matrix structures such as bainitic ferrite, carbon enriched austenite, martensite, and carbide portion. Because austempering heat-treatment does little affect the graphite the nondestructive measurement parameters are mainly affected by the change of the matrix structure during austempering. Ultrasonic velocities were measured by Pulse Echo Overlop method [4]. Fig.5 shows the ultrasonic velocities as a function of austempering time at different temperatures. (A) is the result of the number 1 ductile cast iron and (B) is the result of the number 2 ductile cast iron. Consider the ultrasonic velocity measurement results and the impact test results. For the originally same ductile cast irons before heat-treatment, ultrasonic velocities decrease and impact energies increase generally as austempering temperatures increase at every fixed austempering time. The trends of ultrasonic velocities changes as a function of austempering time are different between

a) 200 °C

b) 250 °C

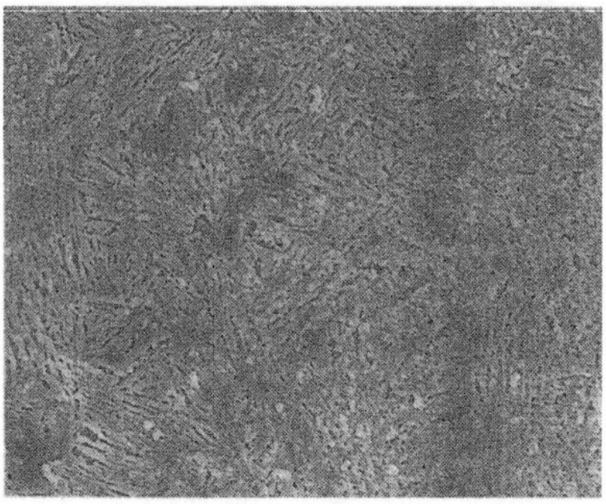

c) 300 °C

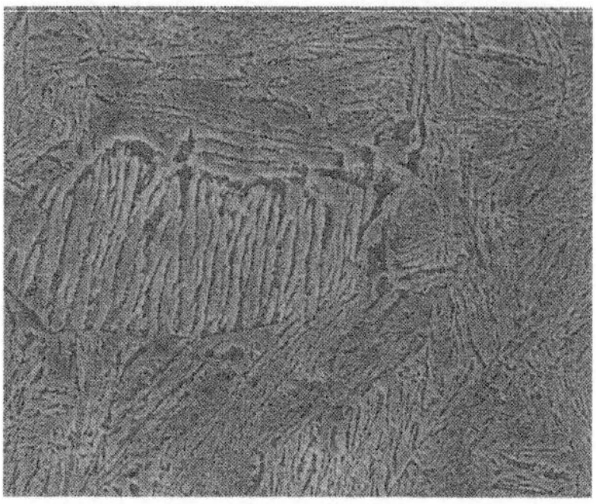

d) 350 °C

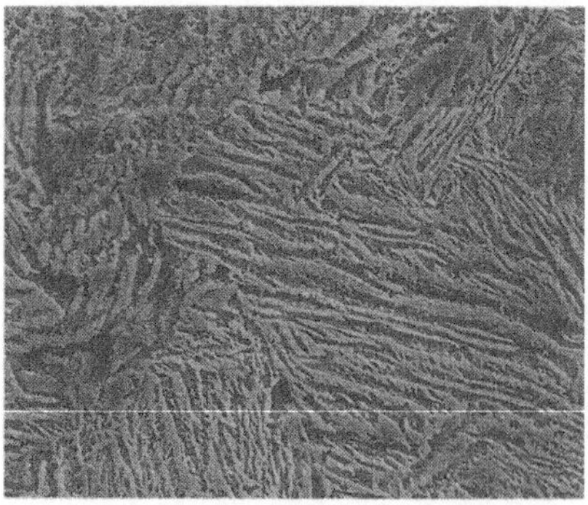

e) 400 °C

Fig.4 Microstructures of specimens austempered at various temperatures ; austempering times are same as 120 minutes at all the samples ; revealed by the SEM (magnification 2000 X)

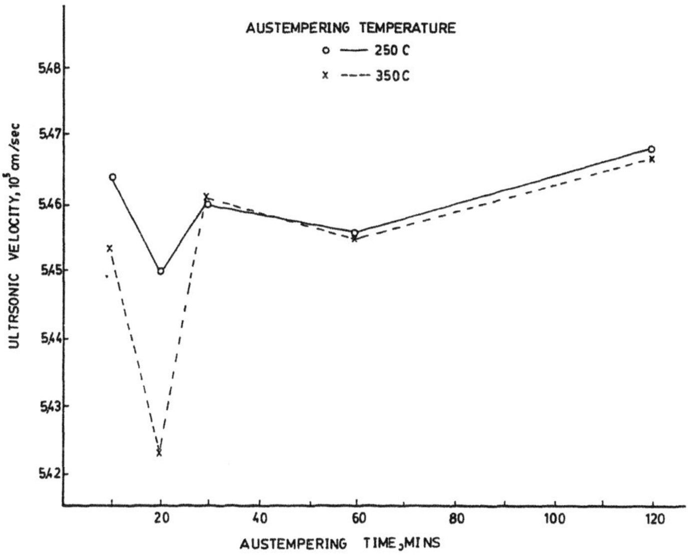

(a)

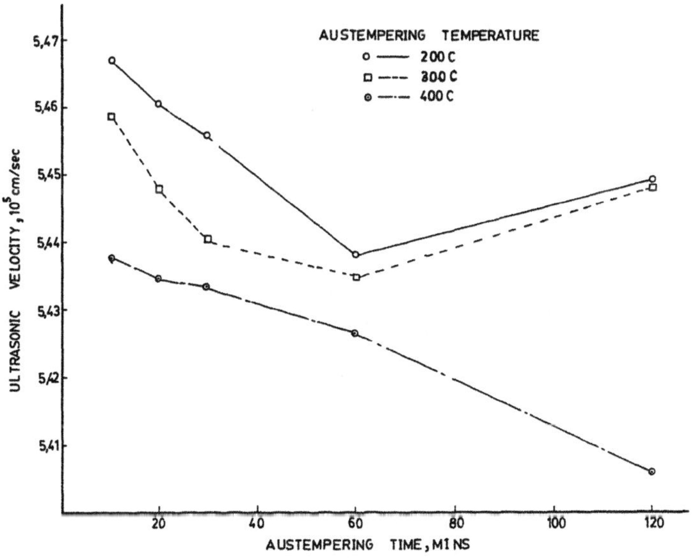

(b)

Fig.5 Influence of austempering time on the ultrasonic velocity
(a) austempered at 250 °C and 350 °C
(b) austempered at 200 °C, 300 °C, and 400 °C

Fig.5 (a) and Fig.5 (b). The changes of ultrasonic velocities at 200 °C, 300 °C, 400 °C show the same trends ; the changes of ultrasonic velocities at 250 °C, and 350 °C show the same trends. It means that the austempering reactions depend on the structures of the original ductile cast irons before austempering. Because ultrasonic velocities depend on the structures of the original ductile cast irons it looks hard to use ultrasonic velocity measurement as the quality assurance tool for the austempered ductile irons. However, the fact that Fig. 5 (a) and Fig.5 (b) show the distinct different trends might be used for the study of austempering reaction mechanism. More research will be needed in this field. Electrical resistivities were measured by D.C. potential drop method [5]. Fig.6 shows the results. In the figure, the trends of electrical resistivities changes of the austempered ductile

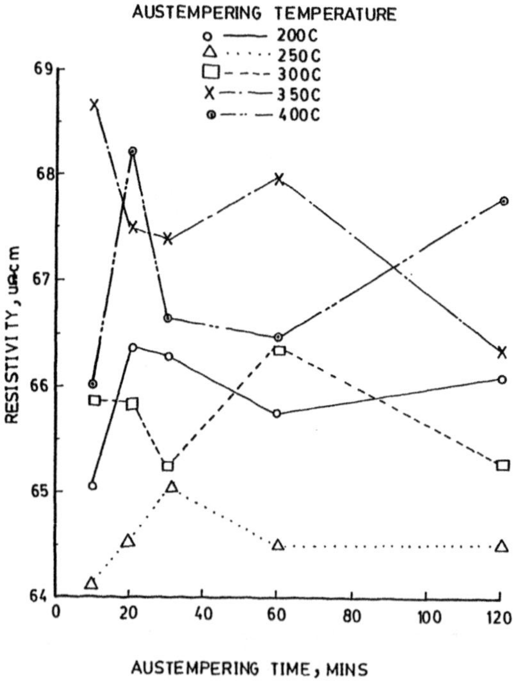

Fig.6 Influence of austempering time on the resistivity of ductile cast iron austenitized at 900°C and austempered at 200 °C, 250°C, 300°C, 350 °C, and 400 °C.

irons as a function of austempering time do not depend on the structures of the original ductile cast irons. The fact is good for the characterization of austempered ductile irons by electrical resistivity measurement. As austempering temperatures increase t the values of electrical resistivity increase generally except 200 °C samples. Because 200 °C is near martensite starting temperature austempering temperature of 200 °C is not usually applied in practice. In fact, the microstructures of 200 °C samples were martensite structure. Considering the electrical resistivity measurement results and impact test results, the values of impact energy generally increase as the values of electrical resistivity increase.

Process window

It is possible to set up process windows of the austempered ductile irons in terms of austempering time and temperature by mechanical test and nondestructive measurement. Process window is the region where we can obtain the austempered ductile irons having optimum properties by austempering heat-treatment. Fig.7 shows the process window determined by impact test for obtaining higher toughness than 9 kgf·m/cm². Fig.8 shows the process window determined by electrical resistivity measurement for obtaining higher resistivity than 66.5 μΩ-cm. Comparing the process window determined by the two methods, the regions are almost identical. It means that there is a possibility to use the electrical resistivity measurement as a quality assurance technique for the austempered ductile irons.

Conclusion

From the above results, we conclude the following. The ultrasonic velocities depend on the structures of the original ductile cast irons before austempering heat-treatment. However, the ultrasonic velocity measurement could be used for the study of austempering reaction mechanism. The electrical resistivities do not depend on the structures of the original ductile cast irons before austempering heat-treatment, and there is a possibility to use the electrical resistivity measurement as a quality assurance technique for the austempered ductile irons.

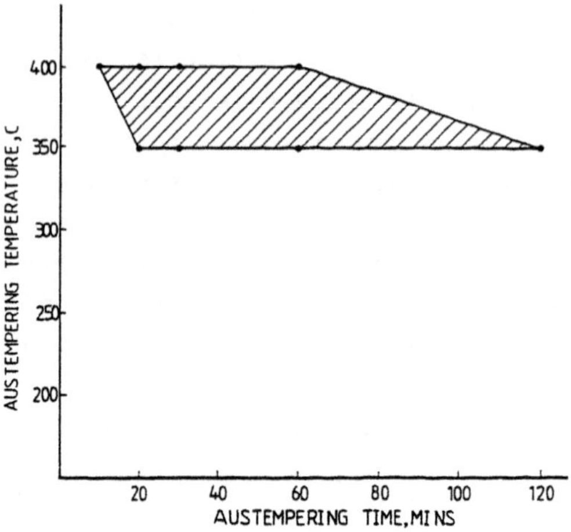

Fig.7 A processing window, in terms of austempering time and temperature for obtaining higher toughness than 9 kgf·m/cm².

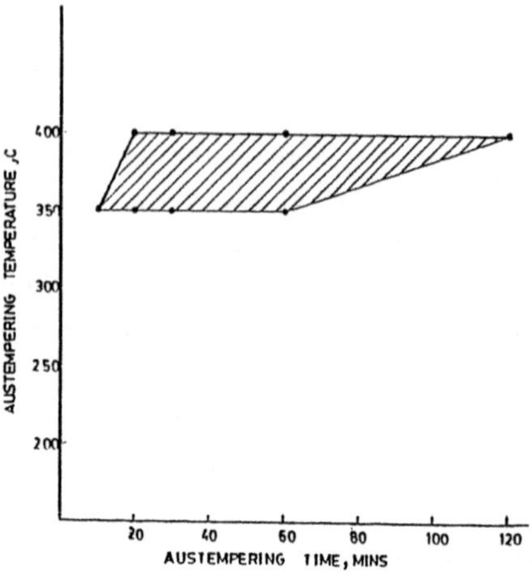

Fig.8 A processing window, in terms of austempering time and temperature for obtaining higher resistivity than 66.5 μΩ-cm.

References

1. Janowak, J.F. ; Gundlach, R.B. : Development of a ductile iron for commercial austempering. AFS Transactions 83-54 (1983) 377-388.

2. Voigt, R.C. : Microstructural analysis of austempered ductile cast iron using the scanning electron microscope. AFS Transactions 83-89 (1983) 253-262.

3. Gundlach, R.B. ; Janowak, J.F. : Austempered ductile iron combines strength with toughness and ductility. Metal Progress July (1985) 19-26.

4. Lee, S ; Lee, S.S. : Quality control and evaluation methods of cast irons. KSRI-88-64-IR.

5. Cohen, R.L. ; West, K.W. : Characterization of metals and alloys by electrical resistivity measurements. Materials Evaluation 41 (1983) 1074-1077.

Theory of Eddy Current Characterization of Magnetic Conductors

R. E. BEISSNER

Southwest Research Institute
San Antonio, Texas

Summary

Eddy current nondestructive testing is widely used for measurements of the electrical conductivity of materials with known magnetic permeability and, to a lesser extent, for measurement of the permeability of materials with known conductivity. In either application, probe liftoff must be known for correct interpretation of the data. This can present difficulties if, for example, the surface is coated with a material of unknown thickness. The procedure described here makes use of multifrequency eddy current data and an iterative, least squares algorithm for fitting model predictions to these data to determine either the conductivity and liftoff or the permeability and liftoff. Numerical results are presented which demonstrate convergence of the method using simulated multifrequency data with and without errors due to noise.

Introduction

The impedance of an eddy current probe in the presence of an electrical conductor depends on the conductivity and magnetic permeability of the material as well as on geometrical factors such as the probe coil dimensions, probe liftoff, and size and shape of the conductor. If geometrical factors are held constant, and if the conductivity or permeability is also constant within a given population of material samples, then the variation in impedance from sample to sample provides a measure of the variation of the unknown material property, either permeability or conductivity, within the population. With proper calibration on a set of samples with known properties, probe impedance data can be directly converted to materials property data, thus providing a noncontacting, nondestructive method for monitoring such properties [1].

Difficulties arise when two or more of the factors that determine probe impedance are unknown. This can happen, for example, with a ferromagnetic conductor where both the conductivity and permeability may vary from sample to sample in some unknown way. It can also happen that a geometrical factor such as liftoff is unknown, perhaps due to a coating of unknown thickness on the surface of the sample. In such cases, a single impedance measurement is insufficient to determine the unknown parameters; one needs more than one measurement taken under different conditions to obtain useful results.

The work described in this paper is a theoretical exploration of the use of multifrequency eddy current data to determine two unknown parameters in the measurement of conductivity or permeability. The investigation is based on a simple model of probe impedance which is described in the next section. This model is used to examine the uniqueness of frequency-dependent impedance-plane loci as a function of conductivity, permeability, and liftoff. The results suggest an iterative, least squares approach to determining either the conductivity and liftoff or the permeability and liftoff from multifrequency data. The paper concludes with a numerical demonstration of the feasibility of the proposed approach and suggestions for further applications.

Probe Impedance Model

The impedance of an eddy current probe in the presence of a conductor can be expressed in terms of the electric field $\vec{E}$ and the magnetic field $\vec{H}$ on the conductor surface as follows [2]:

$$z = \frac{1}{I^2} \int_S [\vec{E}_0 \times \vec{H} - \vec{E} \times \vec{H}_0] \cdot \vec{n} dS, \tag{1}$$

where z is the impedance change relative to the probe impedance in free space, I is the current in the coil, $\vec{n}$ is the outward normal to the conductor surface, and $\vec{E}_o$, $\vec{H}_o$ are the probe fields in free space. For a plane nonmagnetic conductor and a probe coil of arbitrary shape and orientation, the surface fields $\vec{E}$ and $\vec{H}$ can be calculated from the formulas given in Ref. 3. The generalization of this theory to include magnetic conductors with constant permeability is easily obtained by a similar calculation. The result is

$$\vec{E} = -i\omega \nabla \times (\Psi \hat{Z}), \tag{2}$$

$$\vec{H} = \frac{1}{\mu_0} \nabla \times \nabla \times (\Psi \hat{Z}), \tag{3}$$

with

$$\Psi = -\frac{i\mu_r}{\pi} \int \frac{\hat{Z} \cdot \vec{k} \times \vec{a}_0}{k(\mu_r k + \lambda)} e^{i\vec{k}\cdot\vec{\rho}} d^2 k, \tag{4}$$

$$\lambda = \sqrt{x^2 - i\omega\mu_0\mu_r\sigma}, \tag{5}$$

where $\hat{Z}$ is the unit vector normal to the surface, $\vec{\rho}$ is the position vector in the plane surface, μ_r is the relative permeability, σ is the conductivity, and $\vec{a}_o$ is the Fourier transform of the vector potential in the absence of the conductor, as in Ref. 3. The probe geometry enters through $\vec{a}_o$, which can be calculated for any coil configuration from the Biot-Savart formula.

For the present purpose, it suffices to consider the special case of a circular coil of infinitesimal cross section with the coil axis normal to the plane surface. In this case Eq. (1) reduces to [4,5]

$$z = i\omega\mu_0 r_0 \int_0^\infty J_1^2(x) \frac{\mu_r x - \Gamma}{\mu_r x + \Gamma} e^{-2\frac{z}{r_0}x} dx \tag{6}$$

with

$$\Gamma = \sqrt{x^2 + 2i(r_0/\delta)^2} \tag{7}$$

where r_o is the coil radius, δ is the skin depth, and Z_o is the liftoff. Eqs. (6) and (7) contain the essential features that relate material and probe properties to the impedance, and were used for all of the numerical work reported here.

Effects of Permeability, Conductivity, and Liftoff

It would appear, from Eqs. (6) and (7), that the impedance can be considered a function of four independent variables: liftoff; coil radius; permeability; and the product of frequency, permeability, and conductivity. This being the case, it should be possible to design a set of three or more measurements that would allow one to solve for permeability, conductivity, and liftoff for given values of the coil radius. It turns out, however, that for frequencies and coil sizes normally used in eddy current testing, where the coil radius is much greater than the skin depth, the impedance is sensitive only to the ratio of the conductivity to the permeability [6]. The eddy current data alone are not sufficient to determine both of these material properties.

With multifrequency measurements, it is possible to determine either (1) the liftoff and permeability, if the conductivity is known, or (2) the liftoff and conductivity, if the permeability is known. The separate dependence of impedance on permeability and liftoff, with conductivity held constant, is illustrated in Fig. 1. This is a plot of impedance loci generated by varying the excitation frequency. The labels on the curves are values of the relative permeability, and the three groups of curves correspond to three different liftoff values. Although all curves have similar shape, and there is considerable overlap at lower frequencies in the upper left part of the figure, arc lengths are

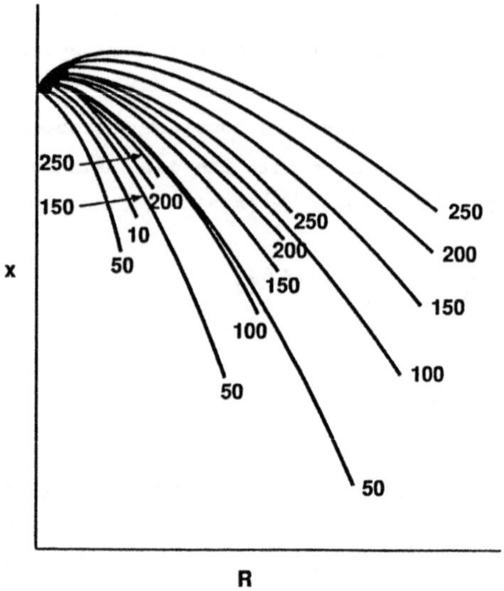

Fig. 1. Impedance loci as a function of frequency. Labels on the curves are values of the relative permeability, and the three groups of curves correspond to three values of liftoff.

strongly liftoff dependent regardless of the value of the permeability. A similar grouping of curves according to liftoff is obtained when the permeability is held constant and conductivity is allowed to vary.

The loci shown in Fig. 1 suggest that it should be possible to determine two unknowns, conductivity or permeability and liftoff, from a set of two or more impedance measurements at different frequencies. One way to do this is to determine the values of the unknowns which, when substituted in Eq. (6), give the best least squares fit of the calculated impedance curve to the experimental data. An iterative procedure for realizing such a least squares fit is described in the next section.

Iteration Method

If Z_i is the measured complex impedance at frequency i, and z_i is the calculated impedance for trial values of the unknown parameters, then the values of the residuals

$$f_i = |Z_i - z_i|^2 \tag{8}$$

determine how well the calculated and measured data agree in the least square sense. The problem, then, is to find the values of the unknown that minimize the f_i.

The Newton-Raphson method [7] is a well-known technique for solving such optimization problems by successive approximations. For the ease when the permeability and liftoff are the unknowns, the relevant equation is

$$f_i + \frac{df_i}{d\mu_r}(\mu_r^{j+1} - \mu_r^j) + \frac{df_i}{dZ_0}(Z_0^{j+1} - Z_0^j) = 0, \tag{9}$$

where the superscript j denotes the jth estimate of each unknown and f_i and its derivatives are evaluated at the values of permeability and liftoff determined in the jth iteration. If there are N frequencies ($N \geq 2$), then the Newton-Raphson formula provides N equations for the two unknowns μ_r^{j+1} and Z_0^{j+1}, which can be computed by the least squares method. The procedure is then repeated, using new values of the permeability and liftoff, until successive approximations agree to within some prescribed tolerance.

The procedure is, therefore, as follows: choose starting values of the two unknowns, evaluate f_i and its derivatives at the starting values of the unknowns for each of N frequencies, substitute in Eq. (9), solve for new estimates of the unknowns, then repeat the process to determine a third set of estimates, etc. Convergence of the result to the correct values of the unknowns depends on how well the measured impedance data can be fit by calculated values and, to some extent, on the choice of starting values of the unknowns. The numerical examples described in the next section were intended as tests of the convergence of the method.

Numerical Examples

The first series of tests was performed using Eq. (6) to generate impedance data at five frequencies from 100 to 500 kHz, with the coil radius 6.35 mm, liftoff 0.635 mm, relative permeability 125, and conductivity 10-percent IACS. The five complex impedances thus obtained were used in place of the experimental impedance Z_i to test the sensitivity of convergence to the initial estimate of permeability and liftoff.

Figs. 2 and 3 show estimates of the permeability and liftoff, respectively, as a function of iteration number. The initial estimates used here were 250 for the permeability and 1.27 mm for the liftoff, both of which are twice the correct values. As can be seen from the figures, both the permeability and liftoff estimates converge smoothly to the correct values in fewer than ten iterations. This is typical of the convergence behavior for a large number of similar tests, some of which made use of initial estimates that were in error by as much as a

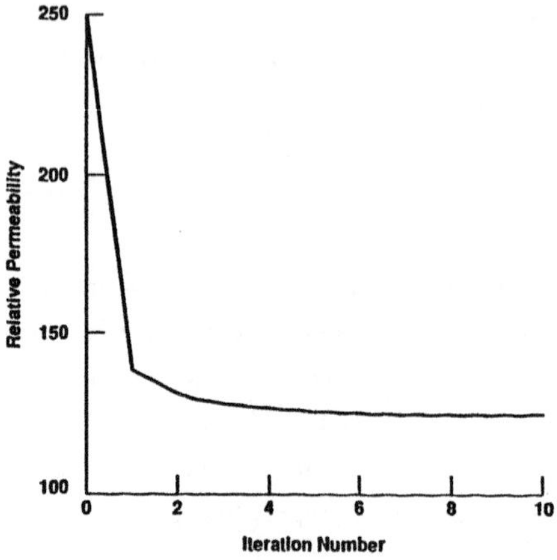

Fig. 2. Convergence of the relative permeability as a function of iteration number

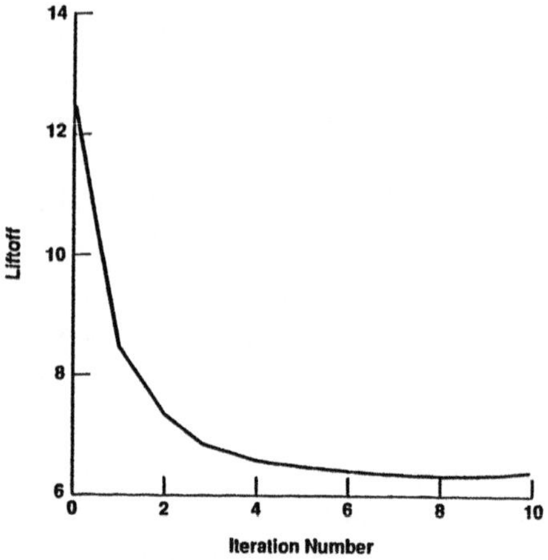

Fig. 3. Liftoff convergence as a function of iteration number

factor of ten. These results indicate that convergence is quite insensitive to the initial estimates, provided that the input impedance data can be fit by predictions of the model.

The next question addressed concerns convergence behavior in the presence of noise in the experimental data. To investigate this effect, the impedance data used as input were altered as shown in Fig. 4 to simulate rather large experimental error. The resulting convergence curves for the permeability and liftoff are shown in Figs. 5 and 6 for initial estimates of 500 and 0.0635 for the permeability and liftoff, respectively.

It is evident from Figs. 5 and 6 that the experimental error simulated in Fig. 4 has led to much slower convergence than was the case when exact input data were used (e.g., Figs. 2 and 3). Also, it is evident that convergence is now non-uniform, i.e., that successive estimates do not always result in better predictions. On the other hand, considering the large deviation from the ideal impedance curve (Fig. 4), which is probably much worse than would be observed in practice, convergence behavior is actually quite good and provides reasonable estimates of both the permeability and liftoff in about ten iterations.

Again, the results shown in Figs. 4 through 6 are only one of a number of examples of the effects of noise on estimates of the permeability and liftoff. In addition, several calculations similar to those shown here were performed for

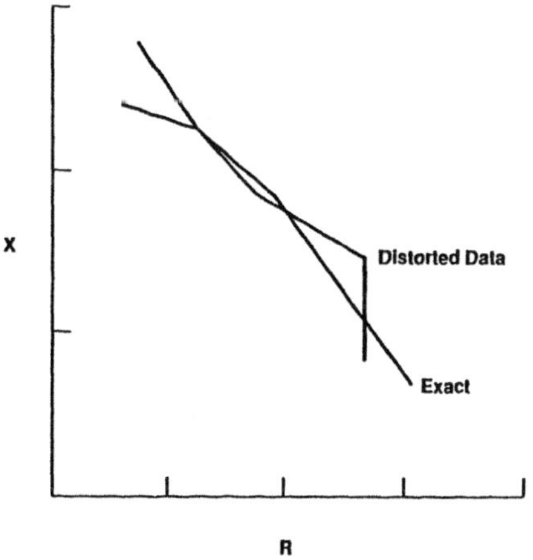

Fig. 4. Exact impedance curve as a function of frequency and the distorted curve used as input for the convergence test shown in Figs. 5 and 6

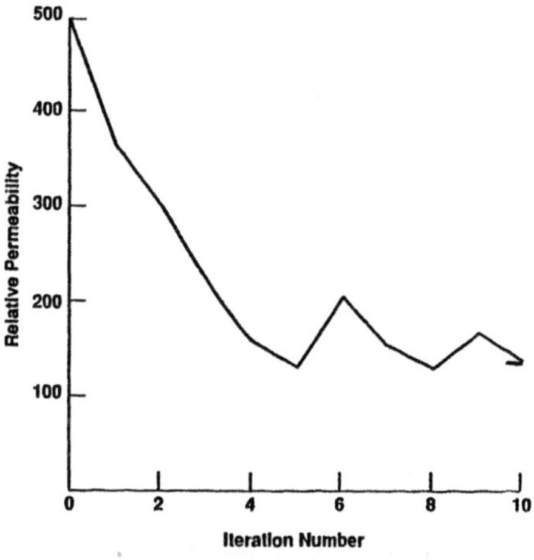

Fig. 5. Convergence of the relative permeability estimate based on the distorted data shown in Fig. 4. The exact value is indicated by the tic mark on the right side of the figure.

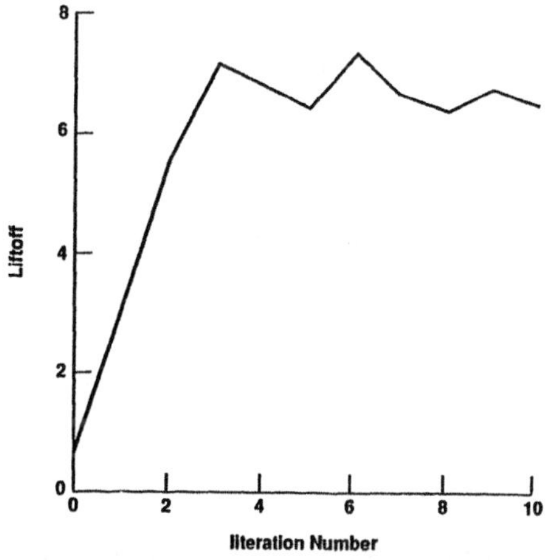

Fig. 6. Convergence of the liftoff estimate under the same conditions as in Fig. 5

different numbers of frequencies in the data set, and for the complementary case where the permeability was assumed known and conductivity and liftoff were estimated. The overall conclusion is that the method is insensitive to initial estimates of the unknowns and is reasonably tolerant of errors due to experimental noise, with the results shown in Figs. 4 through 6 being an extreme case of very large noise effects.

Conclusion

The proposed iteration method seems to provide a simple and robust technique for estimating either the conductivity and liftoff or the permeability and liftoff from multifrequency eddy current data. It is applicable to magnetic as well as nonmagnetic conductors provided that either the conductivity or the permeability is known from auxiliary measurements.

While the numerical results presented here pertain to a very simple probe configuration, Eqs. (1) through (5) provide the theoretical basis for extending the calculations to more complex probes of arbitrary orientation. Also, the iterative method need not be restricted to the plane sample geometry assumed here. In particular, the analytical solutions developed by Dodd and Deeds [8] seem to be well suited to extensions of the method to other sample geometries such as plates of finite thickness or cylindrical shells.

References

1. McIntire, P. (ed.): Introduction to electromagnetic testing: Section 1, Part 1 and Part 2. Nondestructive testing handbook: electromagnetic testing. Columbus, Ohio: American Society for Nondestructive Testing, 1986, 1-20.

2. Auld, B.A.: Theoretical characterization and comparison of resonant-probe microwave eddy-current testing with conventional low-frequency eddy-current methods. Eddy current characterization of materials and structures. Birbaum, G., and Free, G. (eds.). Philadelphia, Pennsylvania: American Society for Testing and Materials, 1981.

3. Beissner, R.E.; Sablik, M.J.: Theory of eddy currents induced by a nonsymmetric coil above a conducting half-space. J. Appl. Phys. 56 (1984) 448-454.

4. Zaman, A.J.M.,; Long, S.A.; Gardner, C.G.: The impedance of a single-turn coil near a conducting half space. J. NDE. 1 (1980) 183-189.

5. Burke, S.K.: A perturbation method for calculating coil impedance in eddy-current testing. J. Phys. D: Appl. Phys. 18 (1985) 1745-1760.

6. Mayos, M.; Segalini, S.: Different methods to evaluate electromagnetic parameters on steels. Paper presented at the Third National Seminar on nondestructive evaluation of ferromagnetic materials, Western Atlas Co., Houston, Texas, 1988.

7. Korn, G.A.; Korn, T.M.: Mathematical handbook for scientists and engineers, New York: McGraw Hill (1968) 728-729.

8. Dodd, C.V.; Deeds, W.E.: Analytical solutions to eddy-current probe-coil problems. J. Appl. Phys. 39 (1968) 2829-2838.

Nondestructive Approach to Characterizing the Strength and Structure of Cast Iron

W. MORGNER and J. GOMEZ

"Otto von Guericke" Technical University
Magdeburg, G.D.R.

Summary

Results of multivariate correlation analysis are presented demonstrating that the strength of grey cast iron can be determined and, moreover, the structure evaluated with a comparatively high accuracy. An account is given of how methods for nondestructive materials characterization should be purposefully developed on the grounds of structure-to-property relationships.

1. Principles of the development of nondestructive testing methods

The majority of cast iron grades such as lamellar iron, nodular iron, or malleable cast iron are characteristic in that they have a matrix conforming to the metastable iron-cementite diagram and, as per the stable iron-carbon diagram, contain graphite in varied quantities and configurations as a secondary structural component.

In developing suitable nondestructive inspection methods adapted to characterize essential mechanical properties, it may be assumed that, depending on previous treatment and/or length of service, every material condition features a real structure which can be characterized submicroscopically by (i) its lattice type, (ii) lattice atoms arranged as a function of chemical composition, (iii) existing vacancies, dislocations and other lattice defects, as well as microscopically by phases present in the structure and their configurations, grain boundaries, but also by the presence of internal stresses. While accounting for the material's quality features this real structure, even though not initially regarded by the production engineer, also determines its physical properties.

In fact, all nondestructive methods are just based on these physical properties, i.e. macroscopic such as mechanical, magnetic, electrical, thermal and thermoelectric, as well as microscopic properties and effects such as diffraction of X-rays, Barkhausen noise, or nuclear magnetic resonance.

In the search for a suitable nondestructive quality testing method, the main interest is always focussed on which effect quality variations associated with structural alterations have on the change in physical properties. Typically, in cast iron materials such quality variations may originate from improper graphite formation or from a wrong ferrite-to-pearlite ratio.

At this point it is convenient to remember the classification of macroscopic physical properties previously common in the science of metals, i.e. into structure-sensitive and structure-insensitive, or structure-sensitive and phase-sensitive properties [1]. Even though, in view of the state of the art, the present high-sensitivity instrumentation is liable to reflect any physical property under the action of structural defects (dislocations, lattice strains, foreign atoms) as structure-sensitive, this philosophy may still be well adopted in the nondestructive characterization of cast iron.

Considering cast iron, the modulus of elasticity, the electrical conductivity and the absolute differential thermoelectric voltage have proved to be little sensitive to the above-mentioned structural defects. All the better do these properties respond to changes in the quantity and configuration of graphite. The absolute differential thermoelectric voltage indicates variations in silicon and manganese contents. Permeability, remanence and coercive field strength are sensitive to variations in the ferrite-to-pearlite ratio.

In some cases a functional relationship can be established between physical and mechanical variables or other quality parameters. Typically this is possible when the modulus of elasticity or the thickness of components is determined via the time of flight in a nondestructive approach.

The classical physical measuring methods, however, are not directly suited for nondestructive testing. Hence, indirect methods must be resorted to almost throughout to establish a nondestructive test approach. Measuring the ultrasonic velocity is excellently adapted to determine the modulus of elasticity. To select the most suitable magnetic property one wants to know whether only surface effects are to be measured, or if a mean value for a larger volume is of interest. The latter is true in determining the strength of cast iron as the skin is not typical of the total cross-section. Therefore, measuring the coercive field strength by means of yoke-type coercive-force meters should be adopted. This method is also advantageous in that it shows a low lift-off sensitivity since, unlike the permeability on which eddy current methods are based, it is not influenced when the demagnetizing factor N is greater than unity.

2. Nondestructive strength determination on GGL lamellar iron

There are two major reasons why nondestructive strength determinations on lamellar iron are of particular interest. On the one hand, as distinct from steel, nodular iron or malleable cast iron, the tensile strength of lamellar iron only poorly correlates with the hardness and, unlike for steel, it cannot be determined directly on components. On the other hand, the tension bar specimen cast separately for a heat is hardly relevant to a particular casting (or the entire heat) if there are major wall thickness differences in the component(s).

Therefore, an interest was taken at an early time in determining the strength of lamellar iron by nondestructive methods [2-13]. Attempts have been made to correlate the hardness, acoustic attenuation, ultrasonic velocity, the coercive field strength and other magnetic properties with the strength of cast iron. In al-

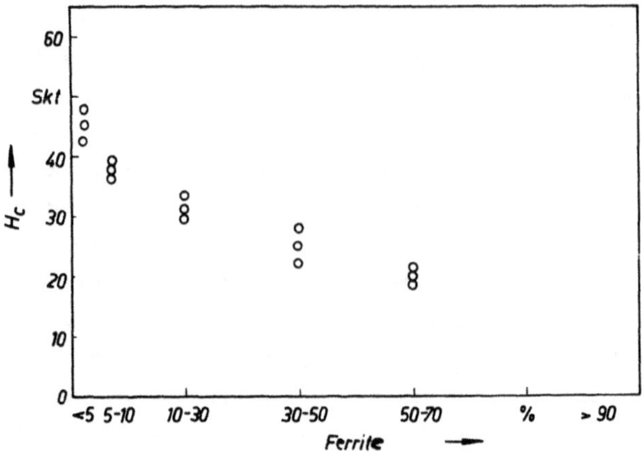

Fig. 1. Coercive field strength vs. ferrite-to-pearlite ratio (annealed specimens)

most all these cases, however, only one of the afore-mentioned properties at a time has been studied as a function of strength. It is merely the advantageous multiplicative linking between modulus of elasticity and hardness that has been emphasized [5].

Referring to the structure-to-property relationships mentioned earlier in this paper, it is an obvious approach to link more than one structure-to-property relationship for cast iron as the time of flight of an ultrasonic pulse is prolonged mainly by the graphite flakes, whereas the magnetic properties are essentially influenced by the matrix.

Specimen materials available were grades of lamellar iron with an almost identical ferrite-to-pearlite ratio but different graphite configurations. The strength levels reached ranged from 150 to 370 MPa. These materials were subjected to heat treatment to obtain another specimen assortment featuring an almost identical graphite configuration but varied ferrite-to-pearlite ratios. The parameters measured were the tensile strength R_m,

Brinell hardness HB, ultrasonic velocity c_l, coercive field strength H_c, acoustic attenuation α, and the noise integral F (area under the structure noise signal on the monitor of an ultrasonic instrument). As expected the first, i.e. non-annealed, series of specimens with identical ferrite-to-pearlite ratio showed no differences in the coercive field strength, whereas the well-known, good correlation was seen between the strength and the ultrasonic velocity. Similarly, the correlation between strength and the product of ultrasonic velocity and hardness was even better.

Inversely the other, i.e. annealed, series of specimens with identical graphite configuration exhibited an almost constant ultrasonic velocity, while the coercive field strength decreased proportionally because of the varied ferrite content (Fig. 1). To be correct, the coercive field strength measured by means of a yoke-type coercive-force meter is given as scale divisions as per the indication of the demagnetizing current. Calibration was later achieved through comparative measurements using a calibrated coercive-force meter on rod-shaped specimens of the same material. Table 1 reflects the coercive field strength data converted to A/cm, while also showing the correlation coefficients of the relationships investigated with simple regression, multiplicative linking of parameters studied, and multiple regression analysis. The latter particularly well reflects the extent to which ultrasonic velocity measurement, hardness measurement, coercive field strength measurement and attenuation measurement have a bearing on the overall result of nondestructive strength determinations, viz.

$$c_l : HB : H_c : 1/\alpha = 22 : 7 : 2 : 1 .$$

Judging from the above, measuring the ultrasonic velocity in conjunction with a hardness measurement would appear to be sufficient. Still, when adopting this or that method a number of boundary conditions have to be allowed for which must be known to permit proper interpretation of the results.

Table 1. Results of regression analysis in nondestructive strength testing

	$R_m = f\ (x)$	r	S [MPa]	
A	$R_m = 0.195\ c_1 - 658.8$	0.84	19.2	I
	$R_m = -860\ \alpha + 448.8$	0.81	21.1	
	$R_m = 1.49\ HB - 67$	0.79	21.7	
	$R_m = 9.56 \cdot 10^{-3}\ F + 116.6$	0.73	25.0	
	$R_m = 343.87\ S_c + 582.3$	0.52	31.1	
A	$R_m = 0.301 \cdot 10^{-3} \cdot c_1 \cdot HB - 48.42$	0.89	16.3	II
	$R_m = -144 \cdot 10^{3} \cdot \alpha/HB + 408.12$	0.88	17.0	
	$R_m = -771.27 \cdot \alpha \cdot S_c + 418.46$	0.83	19.9	
	$R_m = 36 \cdot 10^{-6} \cdot F \cdot HB + 140.70$	0.79	22.0	
A	$R_m = 0.093\ c_1 + 0.535\ HB + 13.604\ 1/\alpha + 1.786\ F \cdot 10^{3} + 30.8 \cdot 1/S_c - 417.16$	0.94	13.0	III
	$R_m = 0.834\ HB + 0.137\ c_1 - 569.54$	0.91	14.8	
B	$R_m = 178\ c_1 - 633$	0.70	15.0	I
	$R_m = 15.14\ H_c + 132$	0.85	11.2	
	$R_m = 3.17 \cdot 10^{-3}\ c_1 \cdot H_c + 151.3$	0.89	9.8	II
	$R_m = 0.09\ c_1 + 0.634\ HB + 4.08\ H_c + 3.98\ 1/\alpha - 329$	0.96	6.6	III

R_m - Tensile strength in MPa; c_1 - Sound velocity in m/s;
α - Attenuation in dB/mm; F - Selective noise integral (measured on the area under the backscatttered signal in a computer-assisted approach;
S_c - Degree of saturation; HB - Brinell hardness; H_c - Coercive field strength in A/cm; S - Residual scatter;
r - Correlation coefficient
I - Simple regression; II - Multiplicative linking;
III - Multiple regression
A - Smelted in different ways; non-annealed. B - Annealed

There is an inherent disadvantage of the hardness measurement in that it is not a "noninvasive" method. Furthermore, it sometimes "detects" surface properties (skin, chill) which are not typical of the entire casting. It is a point in favour of additionally measuring the coercive field strength that the latter is the only parameter which is directly proportional to the ferrite-to-pearlite ratio. The specific value of this is evident in assessing the machinability when cast iron is to be machined on automatic machine tools or by means of robots. It should be taken into account, however, that the coercive field strength is responsive to batch differences in chemical composition even if they have a minor influence on strength. The ultrasonic velocity measurement requires accurate thickness gauging, and the attenuation measurement is comparatively inaccurate in view of the fact that the coupling conditions are not readily reproducible.

Comparison of the data reported by various workers for the error of nondestructive testing methods revealed that the figures have ranged from 10 to 30 percent. Ascribing this solely to the methods of nondestructive inspection would be unfair as the strength testing of cast iron too is subject to a large scatter. Typically, comparative investigations along the circumference of the cylinder liner of a Diesel engine, i.e. almost uniform structure and uniform chemical composition, showed that the coefficient of variation in measuring the ultrasonic time of flight was as small as ± 0.5%, whereas that noted in classical strength determination on tensile specimens cut from the solid was ± 7% [14].

In general it should be warned not to define hardness or strength tests on cast iron as the sole criterion of quality monitoring, especially since failure of castings may be attributable to other causes such as faulty designs, internal casting stresses, stresses thermally induced in service, thermal and mechanical fatigue, as well as inner flaws.

Hence, characterizing cast iron should be aimed at contributing towards more comprehensive materials characterization rather than just desiring to substitute any other approach for hardness testing or strength testing.

References

1. Morgner, W.: Material sensors in non-destructive testing. Proc. 4th ECNDT, Vol. 3, pp. 1874-1884. London: Pergamon Press 1987.

2. Ziegler, R.; Gerstner, R.: Die Schallgeschwindigkeit als kennzeichnende Größe für die Beurteilung von Gußeisen. Giesserei 45 (1958) 8, 185-192.

3. Gerstner, R.: Die Ultraschallprüfung von Gußeisen. Materialprüfung 3 (1961) 6, 213-217.

4. Frielinghaus. R.; Koppelmann, J.; Goosens, M.: Ultraschalluntersuchungen an Gußeisen mit Lamellengraphit. Gießerei, Techn. wiss. Beihefte 16 (1964) 2, 95-110. 5. Felix, W. A.: Nondestructive "tensile testing" of cast iron. Metal Progress, Febr. 1963, 92-95.

6. Bierwirth, G.: Qualitätskontrolle von Gußwerkstoffen mit Ultraschall. Berg- u. Hüttenm. Monatsh. 105 (1960) 3/4, 76-83.

7. Kipka, S.; Pursian, G.: Beitrag zur Bestimmung der Zugfestigkeit bei grauem Gußeisen durch Ultraschall. Gießereitechnik 13 (1967) 12, 373-376.

8. Pawlowski, Z.: Zerstörungsfreie Prüfung von Gußerzeugnissen. Bewertung der Gußeisenfestigkeit mittels US-Messungen. Gießereitechnik 28 (1982) 3, 81-85.

9. Mironenko, V.V.: Nondestructive strength testing on castings of grey cast iron (in Russian). Liteynoye proizvodstvo (1970) 39-41.

10. Sarko, A.K.: Current status and prospects of the development of acoustic techniques to test the strength properties of engineering materials (Outline) (in Russian). Defektoskopiya 19 (1985) 5, 72-87.

11. Weis, W; Lampic, M.; Ortha, K.: Untersuchungen über die Prüfbarkeit von Gußeisen mit Lamellengraphit mittels Ultraschall. Gießereiforschung 27 (1975) 1, 1-11.

12. Vetiska, A.: Zerstörungsfreie Ermittlung mechanischer Kenngrößen an Gußstücken. Gießereitechnik 17 (1971) 9, 213-217.

13. Skrbek, B.; Weiss, J.: Schnelle zerstörungsfreie Kontrolle von verwickelten Eisengußstücken mittels der Methode des punktförmigen Feldes. Proc. 3rd ECNDT, Vol. 4, pp. 126-138. Firenze 1984.

14. Richter, H.U.: Personal communication, Magdeburg 1987.

Nondestructive Measurement of Mechanical Properties of Steel Plates

Isamu Komine
Katsuyuki Nishifuji

Electronics Research Center, NKK Corporation,
1-1 Minami-watarida, Kawasaki-ku, Kawasaki 210, JAPAN

Introduction

To assure quality of steel plates, various kinds of material testing have been used. The classical testing for strength, toughness and other properties by applying load requires machined test pieces. So, it causes loss of material, and requires labor and time.

There is a strong demand for nondestructive measurement of mechanical properties. Many nondestructive material testing methods for various properties have been proposed, but no nondestructive measurement of mechanical properties has been developed.

Meanwhile, great progress has been achieved in the field of computerization of steel manufacturing processes. So, all the manufacturing data of materials: chemical contents, thickness, width, rolling temperature etc., could be obtained.

Under these recent progress, on-line nondestructive measurement of mechanical properties of steel plates using magnetic characteristics and manufacturing data was studied.

Incremental permeability

In magnetic characteristics of steel, coercive force and incremental permeability have been studied for nondestructive testing. Especially, incremental permeability has been proved to be in close relationship with microstructure of steel which can be considered as an important factor of mechanical properties of steel. Incremental permeability of steel plates is measured by a probe coil under a magnetic field.

The manufacturing process of steel plates of Keihin Works,

N.K.K., is completely computerized, and all the manufacturing data could be obtained from the process computer system.
So, a measuring system of incremental permeability having a high temperature-durable probe was developed, and a nondestructive material testing method of steel plates using incremental permeability and manufacturing data was studied.

Measurement on cut samples of steel plates

Fig. 1 shows the measurement of the incremental permeability $\mu\Delta$ by superimposing an alternating magnetic field with a small amplitude but a rather high frequency on the hysteresis loop. This superposition is performed using a probe coil.

Since the incremental permeability may be closely related to micromagnetic quantities (Magnetic Barkhausen Signals), the variation of the incremental permeability on the hysteresis loop was used for this study.
Fig. 2 shows the instrument with integrated yoke and probe.

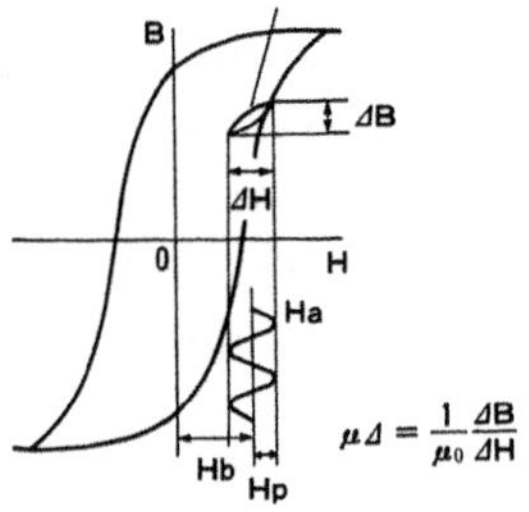

Fig. 1
Incremental Permeability $\mu\Delta$

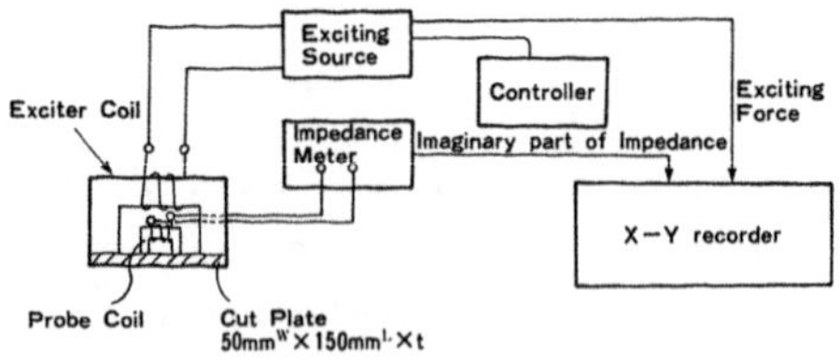

Fig. 2 Blockdiagram of Measurement for Incremental Permeability

Fig. 3 shows a principle of measurement by self-induction.

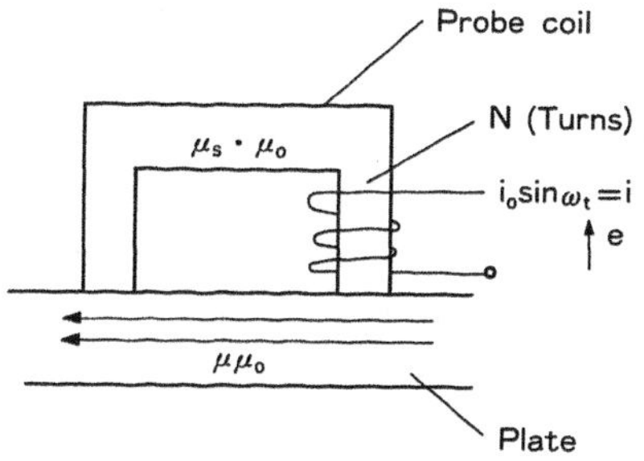

Fig. 3 Principle of Measurement

The following assumptions were made.

1) The sensor is at close contact with the plate.
2) The permeability of the sensor, μo μs, is constant during measurement.

Applying sensing current, i = io sin ωt, the impedance of the probe coil, z = r + jωL was measured. Here, r is the real part of z, ωL is the imaginary part of z, ω is the angular velocity and L is the inductance. The number of turns of the probe coil, N, magnetic flux, ϕ, inductance, L, and sensing current, i, are related by equations (1) and (2).

$$N\frac{d\phi}{dt} = L\frac{di}{dt} = L\, io \cos \omega t \tag{1}$$

$$\left(\frac{\ell s}{\mu o\ \mu s\ As} + \frac{\ell}{\mu\Delta\ \mu o\ A}\right)\phi = Ni = Nio \sin \omega t \tag{2}$$

where, μo is constant, μs is the permeability of the probe core. *As* is the effective area of cross section of magnetic path of probe, ℓs is the effective length of magnetic path is probe, μΔ is incremental permeability of the plate, and *A* and ℓ are the effective area of the cross section and the effective length of the magnetic path in the plates, respectively.

Therefore, L is derived from equation (3).

$$L = L\,(\mu\Delta) = \frac{N^2}{\dfrac{\ell s}{\mu o\ \mu s\ As} + \dfrac{\ell}{\mu o\ \mu\Delta\ A}} \tag{3}$$

Equation (3) shows that L is a function of the incremental permeability, $\mu\Delta$. Angular velocity, ω, is fixed.
In order not to influence the manufacturing process, and to obtain magnetic characteristics from deeper part of plates, sweep time of the hysteresis loop and sensing frequency were set to 12 seconds and 70 Hz respectively.

Fig. 4 shows an example of measurement of a cut plate. Magnetic Index (MI) is the variation of the imaginary part of the impedance.

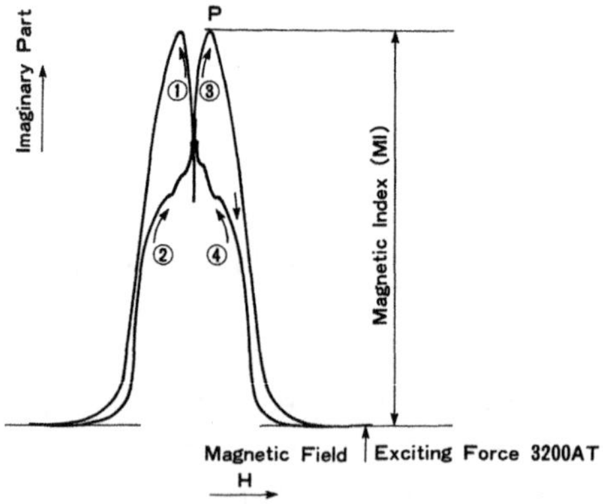

Fig. 4 An Example of Measurement on a Cut Plate
(11.69 mmt x 50 mmW x 150 mmℓ)

Experimental result of cut plates

Fig. 5 shows relationship between MI and mechanical properties.

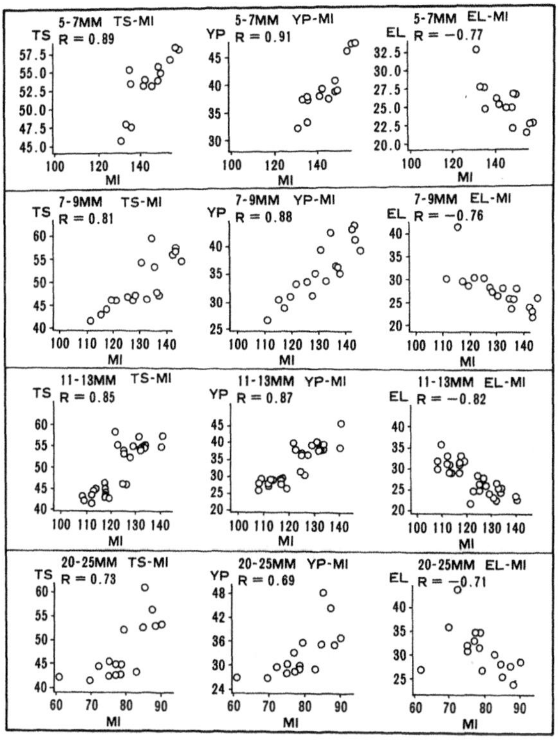

Fig. 5 Relationship between Mechanical Properties and Magnetic Index (MI) for Cut Plates

TS : Tensile Strength (kgf/mm^2)

YP : Yield Point (kgf/mm^2)

EL : Elongation (%)

R : Correlation Coefficient

It shows that correlation coefficients between MI and tensile strength are from 0.7 to 0.9 in the thickness range of 5 mm to 25 mm.

Outline of online measurement of MI.

Measuring instrument

Table 1 shows specification of the system.

Table 1 Specification of Online Test System

Terms	Specification
Material	Hot Plates
Thickness	5~40mm
Temperature	5~400°C (Max.)
Measuring Cycle Time	25sec.

Measuring head was designed to closely contact the plate and to be durable to plate temperatures up to 400°C. This online instrument was constructed at marking position where cutting lines are marked on the steel plate by a marking machine.
MI was obtained during marking operation and connected to the manufacturing data given by the process computer.
Fig. 6 shows the measuring head in operation.

Fig. 6 Measuring Head in Operation

Experimental result of online test

Table 2 shows a result of linear regression using MI and manufacturing data.

Table 2 Result of Linear Regression for Online Test

Thickness 18.0 - 22.9 mm, N=55

Online Data / **Result of Regression**

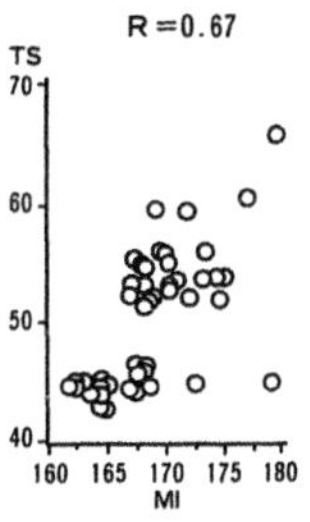

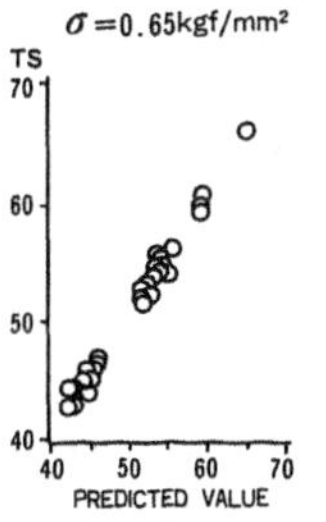

VARIABLE	PARAMETER ESTIMATE
INTERCEP	35.05
C	0.03608
SI	0.2255
MN	0.04872
P	0.08493
S	0.02293
CU	0.05664
NI	−0.06244
CR	−0.001834
MO	0.98360
V	−0.005911
TI	0.09297
NB	0.2827
TAL	−0.08983
SAL 1	−0.02828
THICK	0.0004090
WIDTH	0.00005961
TPF	−0.0.618
MI	0.05152
ITEMP	0.006451

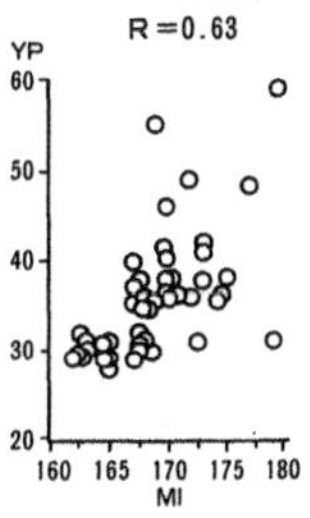

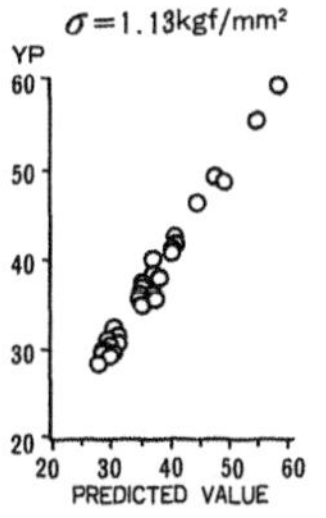

VARIABLE	PARAMETER ESTIMATE
INTERCEP	45.16
C	−0.02344
SI	0.2828
MN	0.01386
P	0.06103
S	0.09707
CU	0.07356
NI	−0.1216
CR	−0.003752
MO	1.3306
V	−0.03197
TI	0.1807
NB	0.4385
TAL	−0.1049
SAL 1	−0.01608
THICK	0.0005245
WIDTH	0.0002789
TPF	−0.03241
MI	0.05269
ITEMP	−0.001020

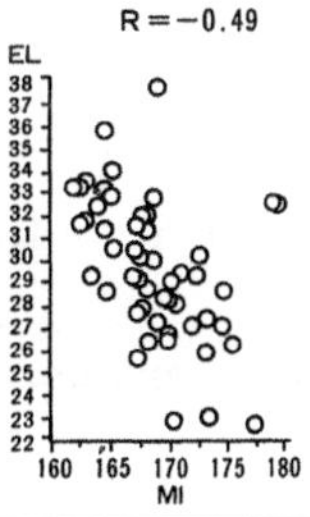

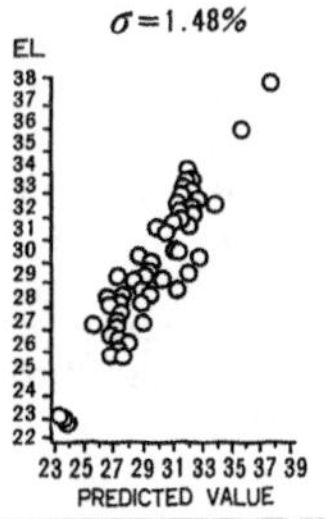

VARIABLE	PARAMETER ESTIMATE
INTERCEP	17.49
C	0.04289
SI	−0.1107
MN	−0.02450
P	−0.05457
S	−0.1334
CU	−0.4443
NI	0.7779
CR	0.01755
MO	−6.457
V	0.2992
TI	0.2221
NB	−0.1763
TAL	0.01160
SAL 1	0.03916
THICK	0.0004538
WIDTH	0.00005661
TPF	0.02821
MI	−0.08659
ITEMP	0.0003062

Table 3 shows the summary of linear regression for all data. Estimate errors (standard deviation) of TS, YP, and El are 0.6 - 1.8 kgf/mm^2, 0.9 - 3.0 kgf/mm^2 and 1.0 - 2.7% respectively.

Table 3 Linear Regression for All Data

Data Group	Thickness (mm)	Number	Tensile Strength		Yield Point		Elongation	
			R	σ kgf/mm²	R	σ kgf/mm²	R	σ %
1	5.0~6.9	32	0.55	0.60	0.53	1.39	−0.52	2.66
2	7.0~8.9	22	0.51	1.62	0.45	1.83	−0.36	1.93
3	9.0~11.9	49	0.42	1.81	0.45	3.00	−0.34	1.79
4	12.0~14.9	42	0.76	0.79	0.78	1.52	−0.36	1.89
5	15.0~17.9	28	0.73	1.00	0.41	1.83	0.1	1.85
6	18.0~22.9	55	0.67	0.65	0.63	1.13	−0.49	1.48
7	23.0~27.9	27	0.72	0.52	0.60	1.07	−0.69	0.96
8	28.0~32.9	17	0.59	0.11	0.64	0.85	−0.47	1.89
9	33.0~46.0	24	0.62	0.66	0.70	0.97	−0.53	1.52

R : Correlation Coefficient between mechanical properties and Magnetic Index
σ : Standard Deviation of Regression

Summary

Nondestructive material testing for steel plates using MI (variation of incremental permeability) and manufacturing data was studied.
It was proved that there was a close correlation between mechanical properties and MI on cut plates: correlation coefficients of tensile strength and MI were from 0.7 to 0.9. Experimental result of online test at Keihin Works proved that the estimate errors of tensile strength, yield point, and elongation are 0.6 to 1.8 kgf/mm^2, 0.9 to 3.0 kgf/mm^2 and 1.0 to 2.7% respectively.
Next theme of this study is to gather more data and to obtain good equations that estimate mechanical properties for nondestructive material testing.

References

P. Höller, "NON-DESTRUCTIVE ANALYSIS OF STRUCTURE AND STRESSES BY ULTRASONIC AND MICROMAGNETIC METHODS", Nondestructive characterization of Materials II (Published 1987), PP211 - 225.

Simultaneous Electromagnetic Determination of Various Material Characteristics

K. Grotz and B. Lutz

Institut Dr. Förster, Reutlingen, F.R.G.

Summary

With the aid of statistical methods it is possible to determine quantitatively material characteristics, e.g. case-hardening depth, from magneto inductive measurements. In many practical cases, however, there exist simultaneously varying influences like surface hardness, alloy composition or geometry. Such further influences can only be eliminated in a multi-dimensional measurement.

A multi-dimensional regression algorithm is introduced, by which it is possible to discriminate disturbing influences and simultaneously determine various independent characteristics combining the result of several measuring parameter settings (e.g. different frequencies or harmonics). Integrating this algorithm in a magneto inductive test equipment allows for the first time nondestructive testing of such material characteristics during production process.

Zusammenfassung

Mit Hilfe statistischer Auswerteverfahren ist es möglich, aus magnetinduktiven Meßgrößen Werkstoffkenngrößen, wie z.B. Einhärtetiefe, quantitativ zu bestimmen. Vielfach existieren jedoch weitere gleichzeitig variierende Einflußgrößen wie etwa Oberflächenhärte, Legierungszusammensetzung oder auch Geometrie. Diese können nur in einer mehrdimensionalen Messung eliminiert werden.

Es wird ein mehrdimensionales Regressionsverfahren vorgestellt, mittels dessen, durch Kombination mehrerer Meßgrößen (z.B. verschiedene Prüffrequenzen, Oberwellen), unerwünschte Einflußgrößen unterdrückt und gleichzeitig mehrere unabhängige Kenngrößen bestimmt werden können. Die Integration dieses Auswerteverfahrens in ein magnetinduktives Prüfgerät erlaubt erstmals eine zerstörungsfreie Prüfung solcher Werkstoffparameter während des Produktionsprozesses.

Introduction

The magneto inductive method can be used to characterize materials nondestructive. The advantage of this method, which in principle is based on the influence of a test piece on the inductivity of a sensor coil, is its suitability for fast automated applications.

However, the relation between the magneto inductive signal (MIS) and a characteristic technological quantity, like surface hardness, is not quantitatively known. The situation is sketched in fig. 1. At the bottom there are various influences, e.g. heat treatment, alloy composition, geometry etc. On one side the characteristic quantities depend upon these parameters and also the MIS is influenced. Most of these dependencies are not quantitatively known. Moreover, often by far not all influences are known. Only empirical relations can be established, relying on a calibration procedure.

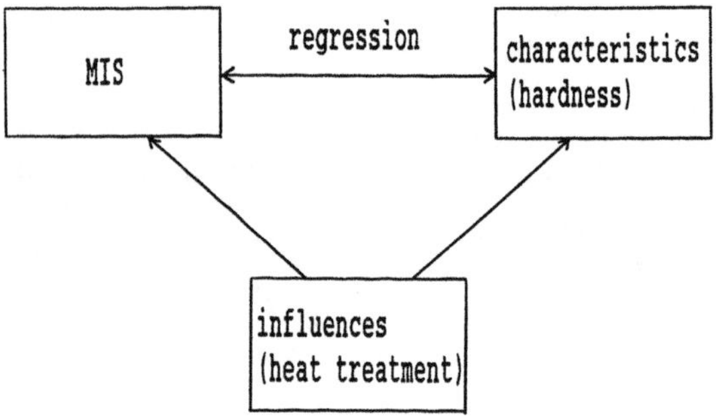

Fig. 1: Scheme of the interrelation between primar influences (bottom), material characteristics (right top) and magneto inductive signals (left top). The various dependences are not quantitatively known.

This can be accomplished by statistical means. If the outcome of the measurement has to be a quantitative value for the desired characteristic, a regression method must be used. It is the aim of this paper to demonstrate the power of a multi-dimensional regression algorithm. Let us start with describing the principle of a simple one-dimensional regression.

Theoretical considerations

A one-dimensional linear regression approximates the relation between a measured quantity x (e.g. real part of MIS) and a characteristic quantity y by a straight line as shown in fig. 2.

$$\hat{y} = a_0 + a_1 \cdot x \qquad (1)$$

The coefficients a_0 and a_1 are determined by the condition of minimal quadratic deviation of known calibration points from the regression line. For every x the relation (1) defines a so-called estimate $\hat{y}$, which means a prediction for y. The quality of this estimate can be judged from the correlation coefficient k_0:

$$k_0^2 = \frac{[\Sigma(x_i-\bar{x})(y_i-\bar{y})]^2}{\Sigma(x_i-\bar{x})^2\Sigma(y_i-\bar{y})^2} \qquad (2)$$

In the case of only few calibration points k_0 however depends upon the number of calibration points. By means of an appropriate correction function this dependency can be eliminated yielding the corrected correlation coefficient k:

$$k = c(k_0, fg) \qquad (3)$$

In eq. (3) fg means the number of statistical freedoms.

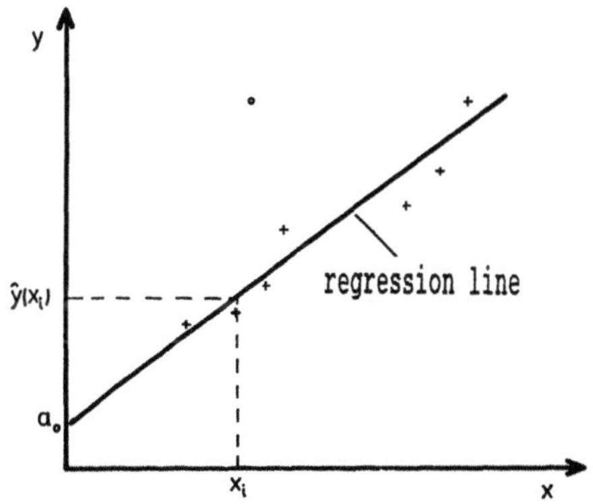

Fig. 2: Principle of a one-dimensional linear regression. The relation between the measured quantity x and the quantity y, which is to be determined, is approximated by a regression line. The closer the calibration points (crosses) are to this line, the better the correlation.

The simple ansatz eq. (1) normally doesn't work describing material characteristics based on the magneto inductive method. The reason is the fact already mentioned, that more than one parameter influences the measurement. These further influences can arise from probably unknown disturbing parameters as well as can be connected with further characteristic parameters. An example will be shown. In any case such further influences may have an effect on the primarily measured signal x, producing data points (circle in fig. 2) away from the regression line, making the regression useless. Relation (1) states a one by one relation between x and $\hat{y}$, but one would need the freedom of varying x without varying $\hat{y}$.

This possibility is given in a multi-dimensional analysis. This fact is well-known for a two-dimensional analysis in the complex plane of the MIS. As shown in fig. 3 one finds a characteristic direction for a variation of the desired quantity in this plane. If there exists a disturbing factor, the corresponding direction in general will be different. The disturbing influence can therefore be eliminated by analysing the signal perpendicular to this latter direction. This corresponds to a two-dimensional regression function:

$$\hat{y} = a_0 + a_1 x_{Re} + a_2 x_{Im} \tag{4}$$

The analysing direction in fig. 3 is determined by the ratio $a_1 : a_2$. For small angles φ the sensitivity however goes down like $\sin \varphi$.

By performing a two-dimensional regression in the complex MIS plane it is possible to eliminate one disturbing factor. For practical applications this often is not sufficient. As mentioned already, there may be several influences. This condition especially arises when using hand-held probes and positioning is not very accurate.

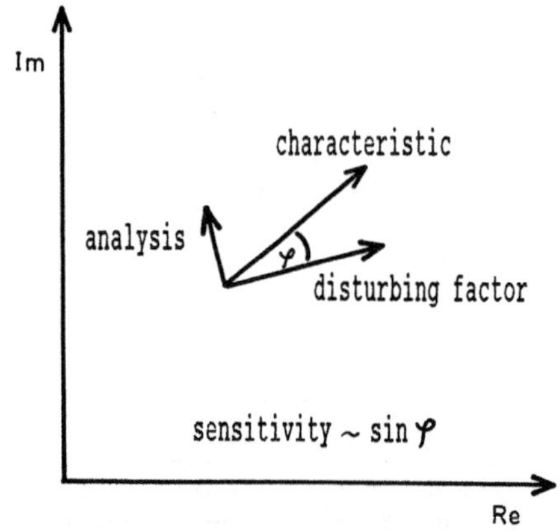

Fig. 3: The complex plane of the magneto inductive signal. A variation of one specific characteristic quantity corresponds to a variation of the signal in a characteristic direction. A disturbing influence yields a second direction. Analysing the signal perpendicular to this latter direction eliminates the disturbance.

In such cases it is necessary to apply a n-dimensional regression. This can be done by working with more than one test parameter setting. For example taking in addition to the carrier frequency (f1) the 3rd harmonic (f3) it is possible to perform a 4-dimensional regression calculation. Another possibility is to work with different carrier frequencies. The important point is to have nonequivalent information from the different parameter settings chosen. Theoretically it is possible to eliminate n-1 disturbing quantities when working with n measuring quantities. The second very interesting application of such a multi-dimensional regression follows, if the need exists to determine more than one characteristic of the same test piece. This can be done simultaneously analysing only one set of MIS data. The situation mathematically is described by a set of linear equations:

$$\hat{y}_1 = a_{10} + a_{11}x_1 + \ldots + a_{1n}x_n$$

$$\hat{y}_2 = a_{20} + a_{21}x_1 + \ldots + a_{2n}x_n \qquad (5)$$

$$\vdots \qquad \vdots$$

In respect to a first characteristic y_1 a second characteristic y_2 has the same effect as a disturbing influence. Via a second regression equation y_2 is extracted and y_1 eliminated using the same values x_1 to x_n. This system of linear equations therefore allows to determine several characteristics simultaneously. In principle besides the determination of k characteristics there can be eliminated further n-k disturbing influences.

Limitations of the described procedure arise from computing power and also from practical aspects: The time consuming part of the algorithm is the calculation of the regression coefficients in the calibration phase. Up to dimension 16 this is done in less than one minute on MC68000 processor but computing time grows rapidly for even higher dimensions. In the test mode the computing time is negligible compared to measuring time, which however also grows with the number of parameter settings used. Another limiting factor is the number of available calibration parts. A n = 8 regression does not work well for less than 15 - 20 calibration points. Moreover, the analysis becomes more sensitive for unprecise calibration values, the higher the dimensionality. Therefore, in most practical applications more than n = 8 - 10 will not be reasonable.

Measurements

The discussed multi-dimensional regression algorithm we have implemented in a magneto inductive test instrument. In the following we want to demonstrate the method by an example. We show results obtained with this instrument testing gudgeon pins. The test pieces where characterized by three quantities, namely surface hardness, case-hardening depth and core tensile strength. There exist six groups of test pieces with the characteristics given in table 1.

Identification	CHD mm	SH HV 10	CTS N/mm^2
1	0.15	630	930
2	0.45	630	930
3	0.30	830	1260
4	0.45	780	1300
5	0.25	890	1370
6	0.50	850	1370

Table 1: The characteristics of the six different groups of gudgeon pins tested with the MAGNATEST S equipment. The three different characteristics are mixed and do not correlate with each other.

Recognize the fact that the three characteristics do not correlate with each other and, therefore, really can be considered as independent quantities.

We start with the result of a one-dimensional regression, which is shown in fig. 4a. The abscissa of the diagram gives the calibration values for the case-hardening depth of each test piece. The ordinate shows the estimates resulting from a one-dimensional regression for these test pieces. As input was used the imaginary part of the MIS at 128 Hz test frequency. This frequency was selected to yield the best correlation. However the achieved correlation is only k = 29%, which means nearly no correlation. The estimated values are useless. For perfect correlation all data points should be lying on the regression line.

Performing a two-dimensional regression with real and imaginary part at 128 Hz the result looks already much better as can be seen in fig. 4b. We have k = 88%. But still there is a splitting of the groups with identification 2 and 4, both having case-hardening depth 0.45 mm but different case hardness. This splitting is expected from the theoretical point of view, because by two-dimensional analysis only one influence can be eliminated but in our example there exist at least two influences, disturbing the result.

In fig. 4c the result of a 4-dimensional regression is shown. The parameter settings used were 128 Hz carrier frequency (f1) together with the 3rd harmonic (f3). Now the correlation coefficient is raised to 97%. An even better result is obtained from an 8-dimensional regression with parameter settings 128 Hz f1, 128 Hz f3, 32 Hz f1 and 32 Hz f3. From fig. 4d it is seen, that the precision of the estimated values for the case-hardening depth is better than 5/100mm.

In fig. 5 and 6 we show the equivalent results to fig. 4d for the other two characteristics using exactly the same data as input. The precision of the surface hardness prediction is better than 20 HV10 units.

In practical applications the test instrument offers the results of the analysis in form of numerical tables to the user. Examples can be seen in fig. 7a+b. Besides this automatic action can be taken in the case of exceeding a predefined tolerance.

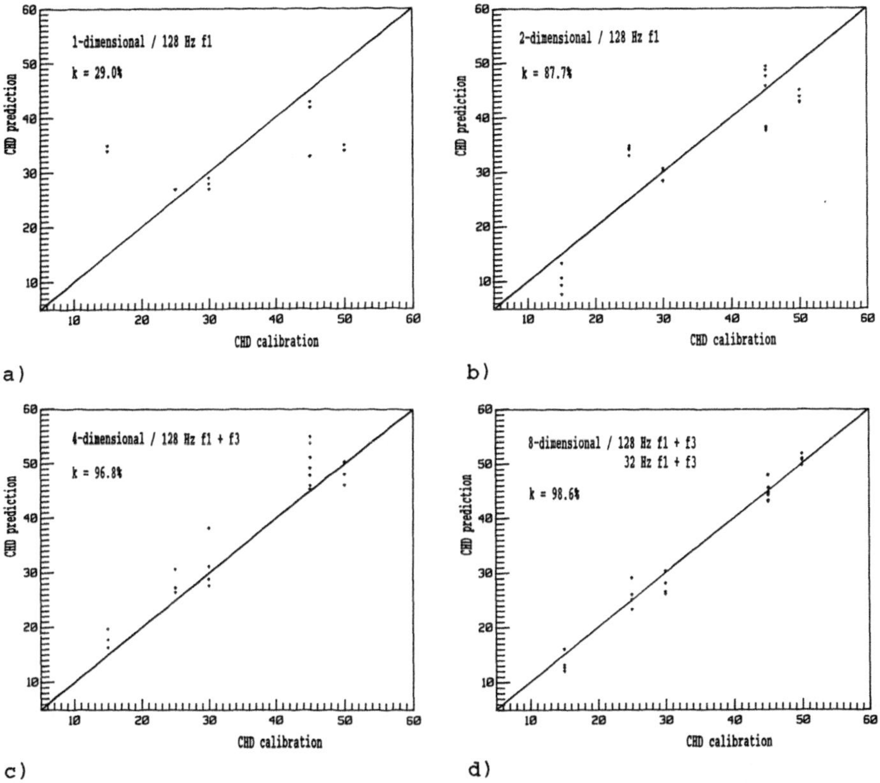

Fig. 4: Results of regression analysis for case-hardening depth. The diagrams show the predictions of regression calculations (CHD prediction) versus the real calibration values (CHD calibration) for each test piece. For perfect correlation the data points should ly on the straight line. The various parameter settings were selected to yield best correlation.

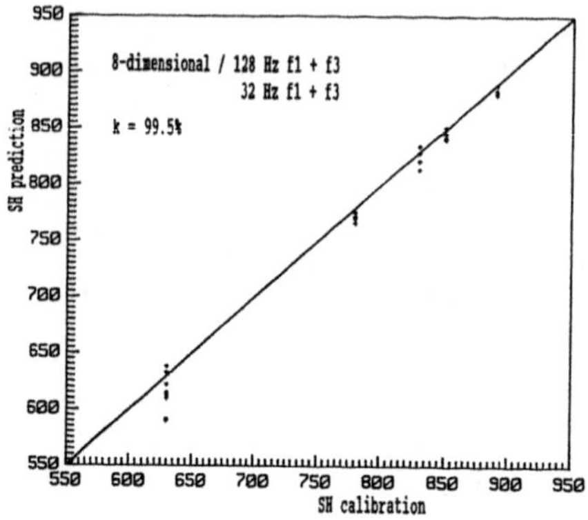

Fig. 5: Same as fig. 4d for surface hardness, using the same MIS data.

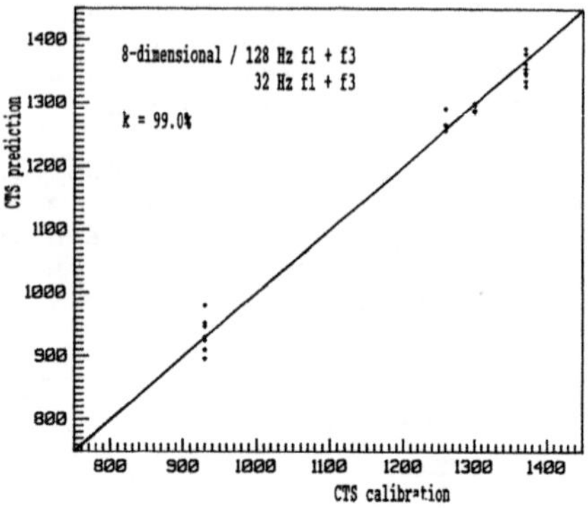

Fig. 6: Same as fig.5 for core tensile strength

Piece	Number	Parameter set 1 2 3 4 5 6 7 8	Class	Value	P	S
PT	12	A A (3 4)	AH	653.9	1	1
PT	12	A A (1 2)	AH	18.4	1	1
PT	11	A A (3 4)	AH	662.8	1	1
PT	11	A A (1 2)	AH	47.4	1	1
PT	10	A A (3 4)	AH	887.2	1	1
PT	10	A A (1 2)	AH	26.6	1	1
PT	9	A A (3 4)	AH	850.3	1	1
PT	9	A A (1 2)	AH	29.2	1	1
PT	8	A A (3 4)	AH	897.0	1	1
PT	8	A A (1 2)	AH	28.3	1	1
PT	7	A A (3 4)	AH	770.6	1	1
PT	7	A A (1 2)	AH	44.5	1	1
PT	6	A A (3 4)	AH	845.3	1	1
PT	6	A A (1 2)	AH	26.9	1	1
PT	5	A A (3 4)	AH	783.4	1	1
PT	5	A A (1 2)	AH	44.6	1	1
PT	4	A A (3 4)	AH	640.2	1	1
PT	4	A A (1 2)	AH	45.0	1	1
PT	3	A A (3 4)	AH	653.4	1	1
PT	3	A A (1 2)	AH	20.5	1	1
PT	2	A A (3 4)	AH	848.9	1	1
PT	2	A A (1 2)	AH	49.1	1	1

Fig. 7: Result table offered by the MAGNTEST S. Part of a table of determined characteristics is given.

Conclusion

We summarize the advantages of a multi-dimensional analysis. Performing a n-dimensional regression up to n-1 influences can be eliminated. It is also possible to determine simultaneously several independent characteristics and in addition to eliminate disturbing factors. The limits of the method, arising mainly from computing time restriction and from the availability of calibration pieces, are not serious. We have demonstrated in an example that the three quantities case hardening depth, surface hardness and core tensile strength can be simultaneously determined. The simultaneous determination of various further quantities together with one characteristic could be used to obtain information about the reason of variations of this characteristic. We have also demonstrated that more independent measuring information yields better correlation. In the case of analysing harmonics together with the carrier frequency it is not necessary to switch between different frequencies during measurements, which allows for very short test cycle times.

Nondestructive Testing of Forged Components Using CS-pulsed Eddy-current Technique

H.-A. Crostack*, W. Bischoff*, J. Nehring**
*Universität Dortmund, Fachgebiet Qualitätskontrolle, **GeWerTec, Dortmund

Zusammenfassung

Die Prüfung von Sicherheitsbauteilen gewinnt bei der Produktion metallischer Halbzeuge speziell im Bereich des Automobilbaus in jüngster Zeit zunehmend an Bedeutung.

Kostengründe und die Notwendigkeit der Gewichtsersparnis lassen den Materialeinsatz bei der Dimensionierung derartiger Bauteile immer stärker schrumpfen. In gleichem Maße wachsen auch die Sicherheitsanforderungen an diese Komponenten und damit die Einhaltung bestimmter Grenzen für den Werkstoffzustand. Für den Bereich der Schmiedeindustrie resultiert hieraus, daß angefangen vom Wareneingang über den Fertigungsprozeß bis hin zum Warenausgang der gesamte Herstellungsablauf des geschmiedeten Bauteils genau überwacht werden muß.

Hierzu eignet sich die Wirbelstromprüfung, die nicht nur empfindlich auf Legierungs- und Gefügeänderungen des Werkstücks reagiert, sondern darüber hinaus neben einer schnellen und berührungslosen Prüfung von Bauteilen auch ein hohes Maß an Automatisierung erlaubt.

Bei der Gefüge- und Verwechslungsprüfung von Schmiedestücken mit dem CS-Impulswirbelstromverfahren läßt sich gegenüber der monofrequenten Wirbelstromprüfung eine erhöhte Trennschärfe durch die gleichzeitige Aufnahme und Verknüpfung vieler unabhängiger Kennwerte erreichen.

Durch den konsequenten Einsatz sehr vieler Prüffrequenzen sowie durch eine gleichzeitige, gezielte Auslegung des Prüfimpulses kann hierbei eine eindeutige Trennung der Wärmebehandlungen schmiederoh, schmiedeperlitisch, normalgeglüht, vergütet und gehärtet an den Geometrien Knüppel, Lenkhebel und Pleuel aus dem Werkstoff 41Cr4 und C45 erreicht werden.
Die hohe Trennsicherheit wird durch den Einsatz der sendeseitigen Kreuzkorrelationstechnik bei vermindertem Kalibrieraufwand - im Gegensatz zur konventionellen Technik muß hier lediglich der Soll-Zustand vorliegen - durch Auswertung der aus der Korrelation ermittelten Kennwerte (Ähnlichkeit W und Zeitversatz T) erzielt, so daß eine zerstörungsfreie Bauteilprüfung mit hoher Prüfsicherheit bei gleichzeitig kurzen Prüfzeiten realisiert werden kann.

1. Introduction

The testing of safety components has recently become more and more important for the production of metallic semifinished material, particularly in the realm of automobile construction.

Cost considerations and the necessity of saving weight mean that such components are dimensioned so as to include ever less material.

Simultaneously, the safety requirements on these components increase and so do the requirements to fulfil particular tolerances for the material condition. This means for the forge industry that the entire forging production sequence must be monitored, from the receipt of goods via the manufacturing process to the delivery of goods.
Hence, a testing method is required that subjects components to a 100 % test (i.e. testing of all items) and which can guarantee their safety.

This potential is offered by eddy-current testing or by magnet-inductive determination of material exchanges under which name this test method at low test frequencies is also known. Not only does this method react to alloy and structural changes in the component but, alongside fast and non-contact component testing, the method also allows a high degree of automation which, in addition to the check of component receipt and final control, allows further applications in the realm of product development and continuous component inspections.

2. Monofrequency structural testing and determination of material exchanges

Eddy-current testing for the classification and characterisation of material states employs the principle of electromagnetic induction. A monofrequency alternating current is passed through a coil, and the component to be tested is introduced into the alternating magnetic field thus produced. The component itself, as well as all material properties affecting the permeability μ and the conductivity σ, lead, via the excitation of eddy-currents according to Lenz's rule, to a change of the induced magnetic field and hence to a measurable impedance change in the coil. The properties of the component on hand can then be deduced from the measurement of the coil impedance /1,2/.

It must be noted here that the penetration depth of the eddy-currents depends strongly on material and frequency. Fig. 1 illustrates the dependence of penetration on frequency for various permeabilities and conductivities. This means that low test frequencies are necessary for all tests by which material properties and flaws at greater component depths are to be evaluated (e.g. control of heat treatments, determination of hardening depths, etc.), and that higher test frequencies are required for the evaluation of surface-near component effects (e.g. edge zone properties such as partial decarburization, weak spot, etc.).

The measuring principle is further restricted by the fact that measurements of the component properties to be investigated can be overlaid by additional interfering factors which result for example from geometric and filling-factor changes of the test coils.
For a clear test declaration, it is therefore absolutely necessary that the disturbing influences are either eliminated or are at least held constant during testing. The risk of ambiguity in a test result is shown in Fig. 2. The complex voltage plane /1/ is shown with four different test problems and dimensions in normalized form (a to d).

If the test frequency for two different components is unfavourably chosen, it conceivable that a permeability change from μ_r = 10 to μ_r = 5 in the one component to be tested could be compensated by an additional diameter variation (filling-factor modification) from η = 0.5 to η = 1 in the second component, thus leading to an identical measured voltage (point I of the curves b and c).

The possible ambiguity of the measurement leads to the requirement to precede every measurement by the determination of a correlation between the instrument reading (electric/magnetic characteristic value) and all material conditions which could possibly occur before and during the production. This determination is achieved by appropriate choice of the test frequency and coil excitation in order to determine the optimum test-parameter setting (instrument calibration) for each separation. The required calibration measures are shown in the following test problem taken from a serial testing.

Tests were performed on safety components from automobile construction (connecting rods, bars and lever rods) made of the materials 41Cr4 and C45 which had been subjected to the following heat-treatment conditions: undefined cooling rate (1), pearlitic structure (2), normalized (3), conventionally tempered (4), tempered by forging temperature (5) and hardened (6). To achieve the statistical reliability of the measurements, 20 specimens from each heat treatment were available, Fig. 3.

The monofrequency measuring results are presented in the complex voltage plane with the choice of relative units, and are restricted to the material 41Cr4 and to the geometry of the lever rod as an example for the complete measurement series.

In order to obtain reliable statistical information on the selectivity, the measured values are in addition overlaid with a two-dimensional normalized distribution approximation. The tests carried out on the specimen sets each containing 20 components indicate the applicability of the normalized distribution in this range, although its validity for the number of specimens present and the type of measurement has not yet been rigorously demonstrated.
The evaluation of the measured results indicates a strong dependence of the separability of different heat treatments upon the test frequency and coil excitation.

Fig. 4 makes this clear for the separation of heat treatments at the constant test frequency of 3 Hz with excitations of 220 A/m and 3500 A/m (upper part of the picture) and heat treatments at an additionally changed test frequency of 100 Hz with an excition of 350 A/m (lower part of the picture). Only a combination of 3 Hz and 3500 A/m results in a clear separation of the hardened condition (6) from all other heat treatments.

In addition to a strong dependence on frequency and excitation, the choice of optimum test data is significantly dependent on the test geometry of the component. Fig. 5 documents this by means of clusters for the three investigated component geometries of 41Cr4 material (bar, connecting rod and lever rod) at a constant test frequency of 100 Hz and an excitation of 350 A/m. A modification of the component geometry thus always requires renewed optimization of the original test parameters.

3. CS-pulsed eddy-current testing

In order to increase the test reliability, all influencing factors resulting from the combination of many frequencies and excitations must be taken into account within a purposeful multi-frequency test. The expenditure of time and instrumentation increases at the same time in such a way that its practical use is severely restricted.

However, there are advantages if several test-parameter combinations can simultaneously be taken into account in one single measurement and thus be used for component separation /3,4/.

The CS-pulsed eddy-current technique allows a spectral range of defined amplitude and phase content to be pre-determined, thus allowing a free choice of pulse form. The system response to this short test pulse allows the complete impedance curve within the pre-defined spectral range to be recorded by one single test shot.

The defined structure of the spectral combination offers new perspectives for the use of the eddy-current technique in rapid and certain on-line detection of certain material conditions, beginning with the matching of the transfer properties between test instrument and component and ending with the optimization of the test pulse by means of the inclusion of appropriate signal-processing algorithms.

Under consideration of the coil transfer function, the broad-band test pulse is pre-determined, Fig. 6 (above left), and the response to the reference condition ("desired condition" of a heat treatment) is recorded as the receiving signal, Fig. 6 (centre). A new problem-matched transmitting pulse, Fig. 6 (above right), can then be formed from the crosscorrelation between the test pulse and the response signal which containes all frequency-dependent characteristics of the desired condition.

If the test is performed with this problem-matched transmitting pulse, the similarity to the desired condition can be determined directly from the received signal by evaluating the amplitude maximum (similarity W) and the point of maximum (time delay T), Fig. 6 (below right).

The testing of a component identical to the reference part results in a transmitter-side correlation, Fig. 6 (below right), which is the so-called autocorrelation with a normalized amplitude W = 1 and a time delay of $T_0 = 0$.

The use of the transmitter-side crosscorrelation allows time optimization of the method to the extent that, after the single determination of the test pulse and calibration of the test instrument, serial on-line testing of the further components can take place.

The number and the type of the castings used for calibration is of decisive importance for test reliability. The type of sampling should represent the fundamental entirety as well as possible; a larger sample is more representative.

Since many independent characteristics are already taken into account and combined for the determination of the desired condition, the high selectivity of the multi-frequency technique allows outliers to be recognized early and excluded during calibration.

This is shown in Fig. 7 by means of the scatter of components within the individual heat treatments upon lever rods of material 41Cr4. The number of specimens has been plotted as ordinate, and as abscissa have been plotted the amplitude values W in the range $0.7 \leq W \leq 1$, Fig. 7 (left), as well as the time delay values T in the range $-60 \leq T \leq +60$, Fig. 7 (right), for each individual heat treatment.

It becomes apparent that, with the exception of individual outliers within the normalized lever rods (3) marked with an arrow which were already excluded during calibration, the majority of the measured values show autocorrelation values (W 1, T 0).

In contrast to this, the separation of an individual heat treatment from all others yields normalized crosscorrelation functions with smaller and in some cases negative similarities with changed time delay values.

The separation behaviour of a complete heat-treatment charge is demonstrated in Fig. 8 for the example of a pearlite-forged lever rod of material 41Cr4. The extreme values of the similarity values for all 20 specimens are shown in Fig. 8 (above), and their time delay values as the interval, Fig. 8 (below). Owing to the low scattering within a heat-treatment group and a strong amplitude and time delay change for the separation of the pearlite-forged condition from all others, high selectivity can be achieved. A similar separation behavior can principally be observed for all other reference heat treatments.

4. Testing of connecting rods in a forging factory

The possibilities of selectivity between different heat treatments which are available if using defined pulse excitation are demonstrated by means of a problem from the testing practice of a forging factory.

This problem concerns the automated, computer-controlled material and structure control of tempered automobile connecting rods of material 41Cr4, Fig. 9.

In addition to the reliable elimination of improperly heat-treated connecting rods (e.g. without heat treatment, pearlitic structure, normalized, hardened), the high selectivity of a problem-matched transmitting pulse also resulted in a classification of the tempered connecting rods produced (I.O. parts) into different strength classes.

Fig. 9 (below) shows this in terms of the measured values for the strength class 900-1000 N/mm2 as well as for the connecting rods of strength class 800-850 N/mm2 excluded during a serial measurement.

For the measurement a frequency band of 10-100 Hz was used, which with a test time of approx. 1/10 s enables problem-free integration of the CS-pulsed eddy-current technique within the production sequence.

Alongside test sequence control by mutual data exchange with the robot system, the measuring system CS-TEST 1016 used additionally transmits all measured values to a central CAQ system so that comprehensive quality control of these components can be achieved.

5. Summary

Structure tests and determination of material exchanges for forged components by means of the CS-pulsed eddy-current method result in a better selectivity than obtained in monofrequency tests because of simultaneous recording and combination of many independent characteristic values.

The approprate use of many test frequencies as well as a simultaneous controlled definition of the test pulse lead to a clear separation of the heat treatments undefined cooling rate, pearlitic structure, normalized, tempered and hardened on the geometries of bars, lever rods and connecting rods of materials 41Cr4 and C45.
High selectivity is obtained by use of the crosscorrelation method within the transmitting signal with reduced calibration expenditure - in contrast to the conventional technique the desired condition only must be known -, by evaluating the characteristic values computed from the correlation (similarity W and time delay T) so that non-destructive component testing can be realized with high test reliability combined with short test times.

Literature

/1/ F. Förster, K. Stambke — Theoretische und experimentelle Grundlagen der zerstörungsfreien Werkstoffprüfung mit Wirbelstromverfahren
III. Verfahren mit Durchlaufspule zur quantitativen zerstörungsfreien Werkstoffprüfung
Zeitschrift für Metallkunde 45 (1954) 4, S. 166-179

/2/ E. A. Becker, M. Vogt — Wirkungsweise und Anwendungsmöglichkeiten eines elektromagnetischen Gefüge- und Verwechslungsprüfgerätes
Materialprüfung 15 (1975) 5, S. 182-185

/3/ J. Nehring — Untersuchungen zum Einsatz problemangepaßter Prüfimpulse (CS-Technik) bei der Wirbelstromprüfung
VDI-Fortschrittsberichte, Reihe 5
Nr. 132 (1987)

/4/ H.-A. Crostack, W. Bischoff, J. Nehring — Zerstörungsfreie Prüfung von Schmiedeteilen
Industrieanzeiger, Schmiedetechnische Mitteilungen
Heft 41 S. 32 - 35, Heft 55/56 S. 18-30 (1988)

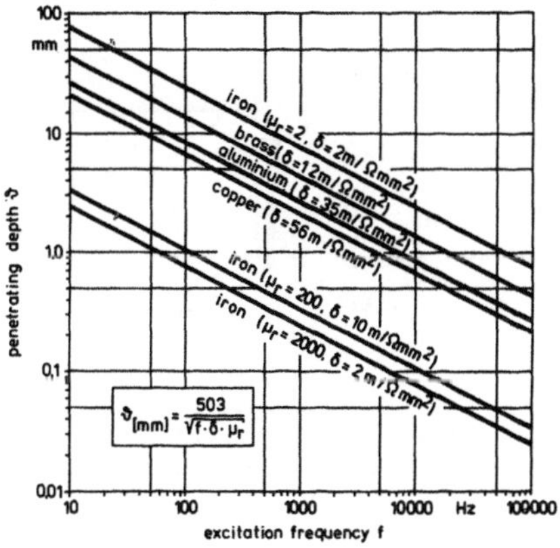

Fig.1:
penetration depth versus excitation frequency f

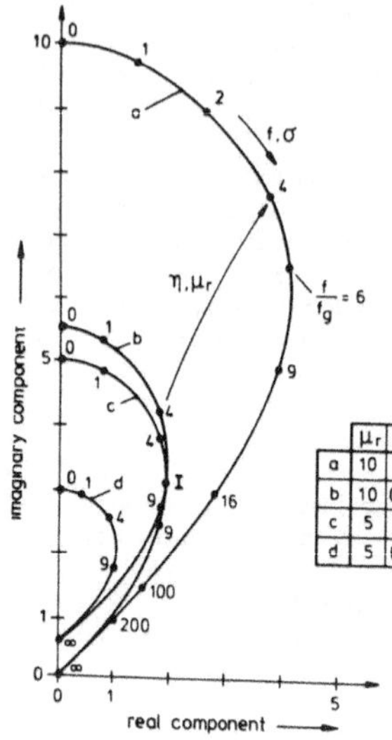

Fig. 2:
complex voltage plane at several test situation in dependence on test frequency

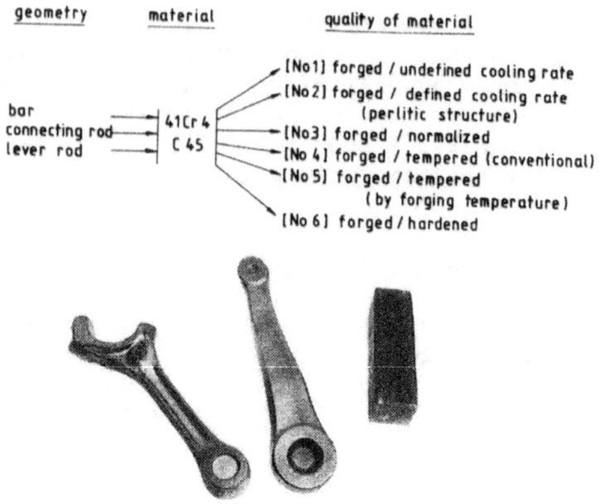

Fig. 3:
test specimens

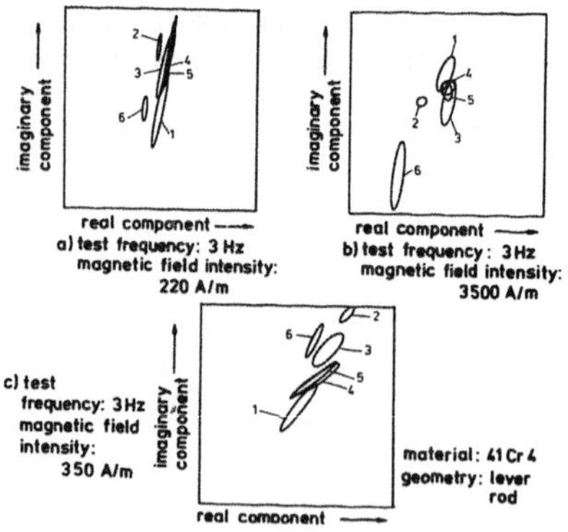

Fig. 4:
statistical straggling range of heat treatments at several frequencies and magnetic field intensities

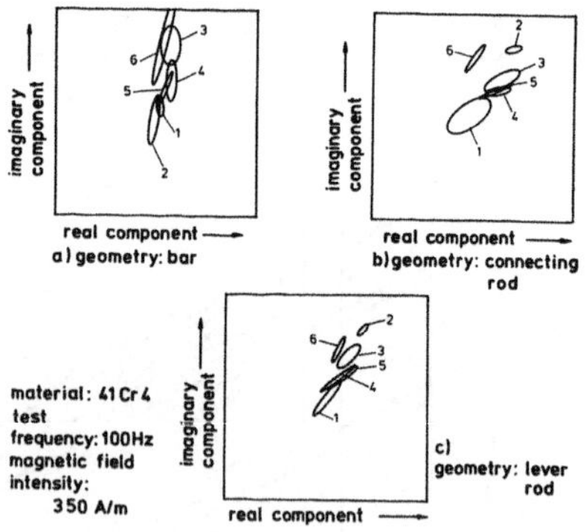

Fig. 5:
statistical scattering range of heat treatments at several test geometries

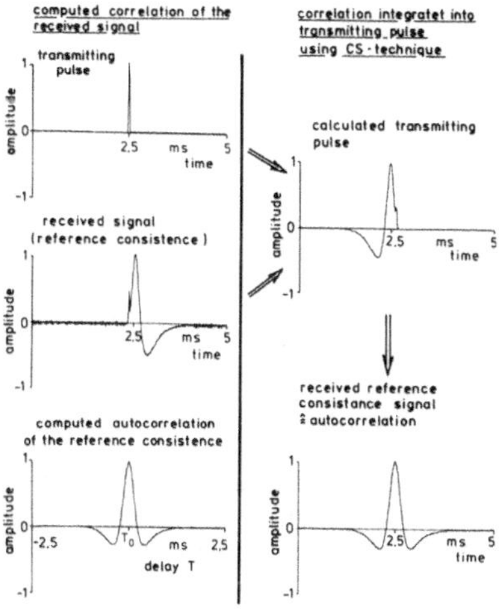

Fig. 6:
crosscorrelation using
CS-technique

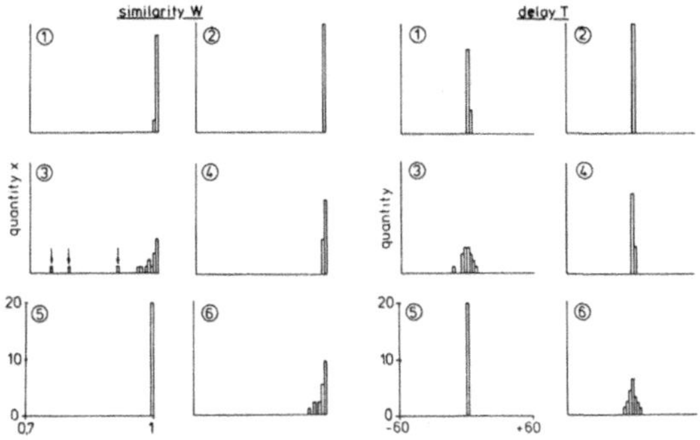

Fig. 7:
abundance distribution (W,T)
within heat treatment
material: 41Cr4,
geometry: lever rod

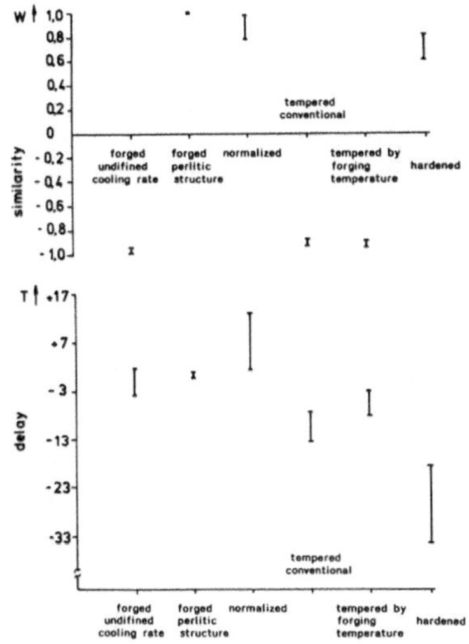

Fig. 8:
extrem values of similarity W
and of time delay T
reference: lever rod, 41Cr4

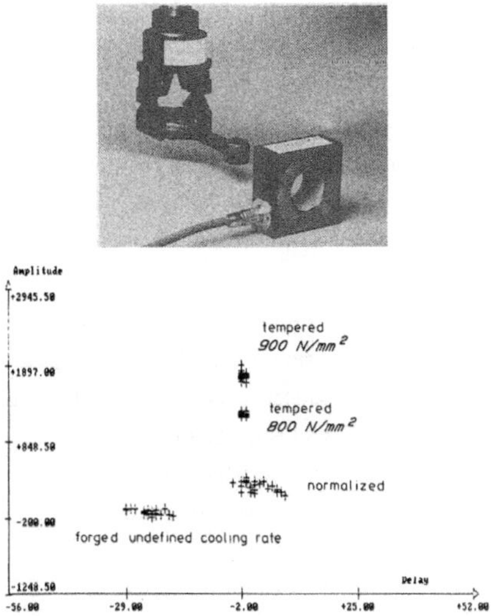

Fig. 9:
series test of forged safty
components measuring equipment:
CS-TEST 1016

Applications of Capacitive Array Sensors to Nondestructive Evaluation

P.J. SHULL and A.V. CLARK

Fracture and Deformation Division
National Bureau of Standards
Boulder, CO

B.A. Auld

E.L. Ginzton Laboratory
Stanford University
Stanford, CA

Summary

The capacitive array sensor is a versatile and promising device for nondestructive evaluation of dielectric material. Responding to the complex dielectric constant of the interrogated material, the device is sensitive to detection of surface and subsurface features in dielectric materials and to surface features in conductive materials.

We describe here the work of an on going project at NIST (formerly the National Bureau of Standards) on the capacitive array sensor [1,2]. This work consists of the characterization and use of the sensor in the two basic configurations - absolute and differential. We also report the results of comparison of liftoff measurements to an existing theoretical model.

The probe in the differential configuration was characterized on a series of well defined surface and subsurface notches in a dielectric material and surface notches in a conductor. Also characterized were the effects of liftoff, dielectric constant, sensitivity, shielding, and lead configuration. In the absolute configuration the characterization was primarily the effects of liftoff and dielectric constant.

We investigated several possible applications. Among them were: a) detection of surface and subsurface flaws and features in insulators such as cracks in ceramics, b) cure monitoring of polymer composites, c) determination of porosity and thickness of thermal barrier coatings, and d) dielectric constant monitoring of ceramics during sintering.

Design

The basic element of the capacitive probe is a parallel plate capacitor whose electrodes have been unfolded to lie in the same plane. When a potential difference is impressed between the electrodes, an electric field is produced which in this configuration is elliptical and nonuniform, figure 1. This electric field produces a displacement current that flows between the electrodes. With the introduction of a sample material into the lower half-space of the electric field, the amount of displacement current that flows between the electrodes is either attenuated (conductors) or amplified (dielectric).

The probe's sensing ability can be varied by changing the number and geometry of the electrodes and the size of the sensor. The field penetration can be increased by adding electrodes, figure 1a. With this increase in field penetration the probe trades sensitivity near the surface for increased interrogation depth.

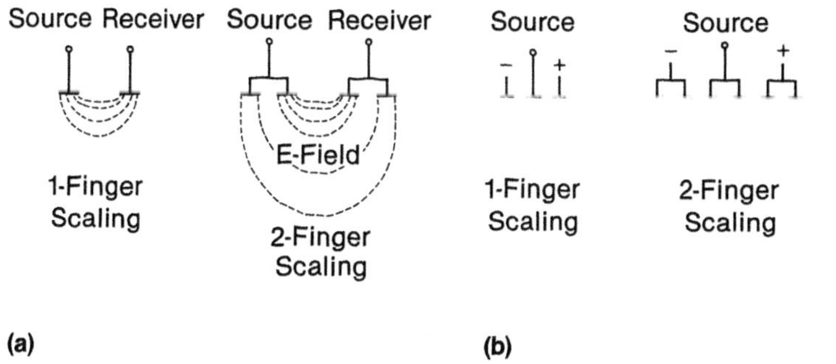

Figure 1. Capacitive array electrode configurations.
a) The array in absolute mode with single finger and double finger spacing.
b) The array in differential mode with single finger and double finger spacing.

Sensitivity to liftoff, a common problem with electromagnetic sensors, can be reduced by operating the probe in a differential geometry, figure 1b. The probe is used in this mode to detect variations in the sample that are small in size compared with the probe's sensing area. Examples of such variation are edges or steps and flaws such as voids or cracks.

We designed and characterized three different probes - one in the differential mode and two in the absolute mode. The differential probe had a single finger configuration with electrodes dimensions of 2.5 x 0.4 mm with 0.2 mm electrode spacing. In this particular design, the leads were in the same plane as the electrodes. To reduce capacitive coupling to the workpiece, leads were shielded with aluminum foil.

We characterized two absolute probes -shielded and unshielded [3]. Excluding the shielding, both probes were nominally the same. Both probes had eight electrodes, each of which were connected to a switch that allowed the spatial frequency to be altered. The unshielded probe was designed to determine if shielding was necessary to maintain a reasonable signal to noise ratio. The unshielded probe also reduced the capacitive coupling to the probe case.

Experiment

The experiments are separated into those made in the differential mode and those made in the absolute mode. Two basic experimental setups, described below, were used to make these differential and absolute measurements. In both cases the probe was mounted on a three-axis computer controlled scanner with the probe face oriented downward and parallel to the horizontal plane.

For the differential mode, the capacitive probe, which includes the sensor and an amplifier, was connected to a gain-phase meter, which measured the ratio of output voltage to the input voltage. This transfer function (Vout/Vin) was measured at frequencies between 20 and 100 kHz. The differential measurements were made by scanning the probe across the workpiece in discrete increments whose size was dependent on the probe and flaw size, typically 0.1 mm. The computer recorded the output, Vout/Vin, of the gain-phase meter and plotted the magnitude and phase of the signal as a function of scan position.

For the liftoff experiments in the absolute mode, we made measurements of the probe impedance as a function of liftoff. The probe which was mounted on the scanner was set at an initial liftoff from the specimen - typically 78 μm. After loading the initial liftoff data into the computer, the admittance in both magnitude and phase were determined and stored with the corresponding liftoff data. The computer then increases the liftoff by raising the probe head a predetermined amount. The admittance at this new liftoff is now measured and stored with the corresponding liftoff value. This process is continued until a saturation value of the admittance is obtained, which corresponds to the probe admittance in air.

Results and Discussion

Figure 2a shows the response to a series of notches in PMMA, $\epsilon_r = 3.7$, ranging in depth from 0.03 mm to 0.5 mm. As the probe was scanned across the surface of the workpiece, the response increased with increasing notch depth. This trend continued until the signal saturated at a depth of about 0.25 mm. Flaw signals were asymmetric and the position of the peaks and valleys did not

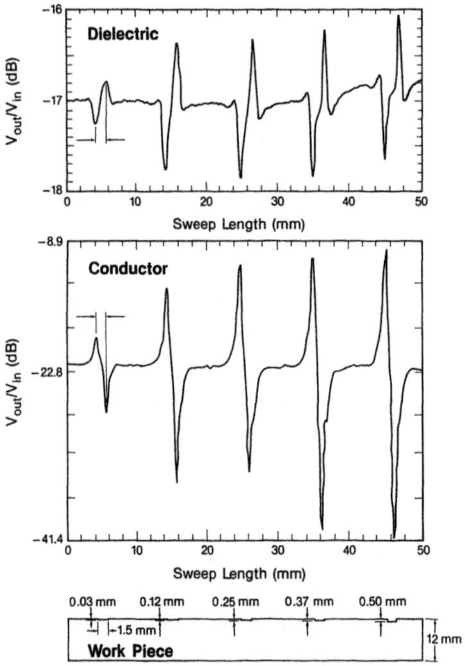

Figure 2. Capacitive probe response to a series of surface notches in a) dielectric material, b) $\epsilon_r = 3.7$, and a conductor.

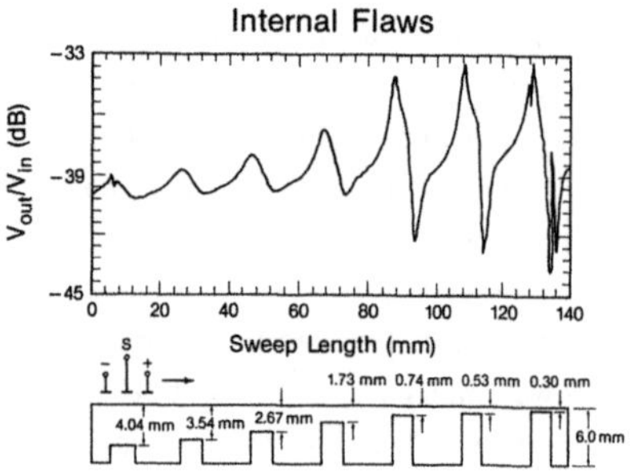

Figure 3. Capacitive probe response to a series of sub-surface or buried flaws.

coincide with leading and trailing edges of the notches. The signal asymmetry is caused by a slight residual imbalance in the probe amplifier. When the probe is scanned across a notch the finite width of the electrodes will cause the peak (valley) of the response signal to be offset by approximately one half of the effective width of the center electrode(s).

To determine the effective depth of interrogation, a series of notches 6.0 mm wide were made from the bottom of a plate of PMMA. The depth from the top surface to the buried notches range from 0.3 to 3.0 mm. At first we expected this range to be more than adequate to yield the depth limitation of the particular probe and material used, but we found it was necessary to add an additional notch at a depth of 4.0 mm. Figure 3 shows that even at this depth the notch is clearly discernable.

The electric field does not penetrate the surface of a good conductor, and therefore the capacitive probe can only detect surface features on metals. Figure 2b shows the response to a series of 1.5 mm wide notches in Al that ranged in depth from 0.03 mm to 0.5 mm. The clear signal response to a 30 μm step indicates the extreme sensitivity of this probe to surface features.

A theoretical model that predicts the probe response to liftoff is useful for many noncontacting applications. Typically in noncontacting applications not only are the material properties unknown but the liftoff is also an unknown. To complicate the problem further, both the material properties and the liftoff may change during the measurement. An example is the monitoring of a ceramic during the sintering process. While the material is being fired, both the dielectric constant and the size of the material are changing. Using the Gimple/Auld model [3] and the multiplexing capability of the sensor, we propose a technique to first determine the liftoff and then the dielectric constant of the workpiece.

The current theory for the linear array capacitive probe shown in equation 1 was developed by Gimple and Auld [3].

$$\frac{\Delta Y}{A} = \frac{a\epsilon_0}{\Gamma_r e^{2ad} - 1} \quad ; \quad \Gamma_r = \frac{\epsilon_r + 1}{\epsilon_r - 1} \qquad \text{EQN 1}$$

Where $\Delta Y/A$ is the change in admittance between the value with the workpiece present and the value in air normalized to the active area of the probe. d, ϵ_0, ϵ_r, a, Γ_r are respectively the liftoff distance, the free space permeability, the relative permeability of the workpiece, the spatial frequency of the probe electrodes, and the relative reflection coefficient. The theory assumes a sinusoidal potential (spatial frequency, a) prescribed in a plane at distance, d, above a dielectric half-space.

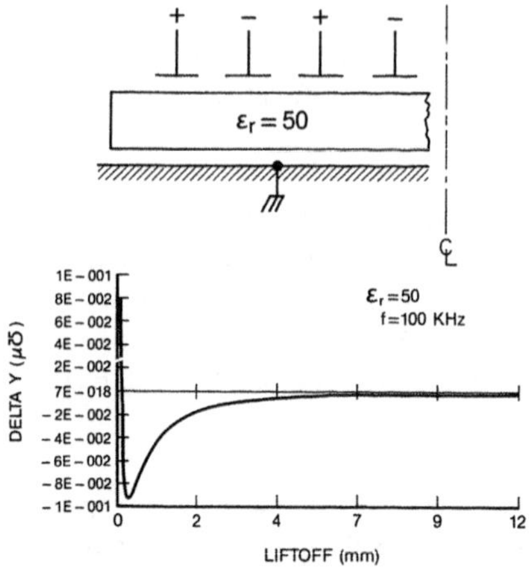

Figure 4. ▲Y vs. liftoff for the shielded probe. The workpiece was 8 mm thick and $\epsilon_r = 50$.

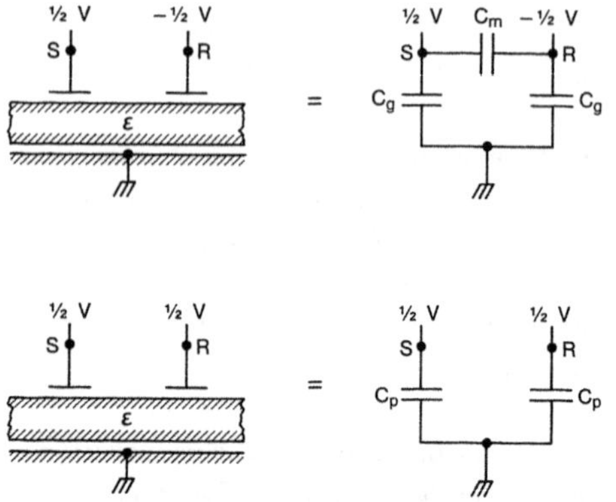

Figure 5. Superposition and lumped element models of the a) nonzero spatial frequency case, b) zero spatial frequency case.

If in equation 1 we vary the liftoff ,d, while holding the other variables constant, the equation would predict a monotonic decrease from some initial value. As indicated in figure 4, the experimental ▲Y versus liftoff curve initially decreases rapidly followed by a recovery - very different from the theoretical prediction.

To understand the recovery effect we used a superposition model of the sensor and the network analyzer used to make the admittance measurements. The probe is modeled as the sum of zero and nonzero spatial frequency components equalling the actual system as shown in Figure 5. The zero spatial frequency component is necessary because the network analyzer causes the receiver electrode be at ground potential. In addition we developed a lumped element circuit model of each of these two superposition components.

The nonzero spatial component, figure 5a, has a potential of +1/2 V on the source electrodes and -1/2 V on the receiver electrodes. In this balanced system there is coupling capacitance between source and receiver (Csr), between the source and ground (Csg), and the receiver and ground (Crg). It is important to note that the direction of the displacement current is out of the source and into the receiver electrode. If Csg is equal to Crg then the current that flows from the source to ground is equal to the current that flow from the ground to the receiver. Under this condition - Csg = Crg all the displacement current that flows out of the source electrode arrives at the receiver electrode.

As shown in figure 5b, the potentials on the electrodes for the zero spatial frequency are equal. This lumped element model suggests that there is no coupling between the source and receiver because they are at equal potentials. In this case the direction of the displacement current is away from or out of both the source and the receiver electrodes. Because the only measured current in the system is in the receiver electrode, the current that arrives at the ground from the source in the zero spatial frequency case is not measured. In addition, current flows from the receiver to ground which is opposite in sign to the current flow in the nonzero spatial frequency case.

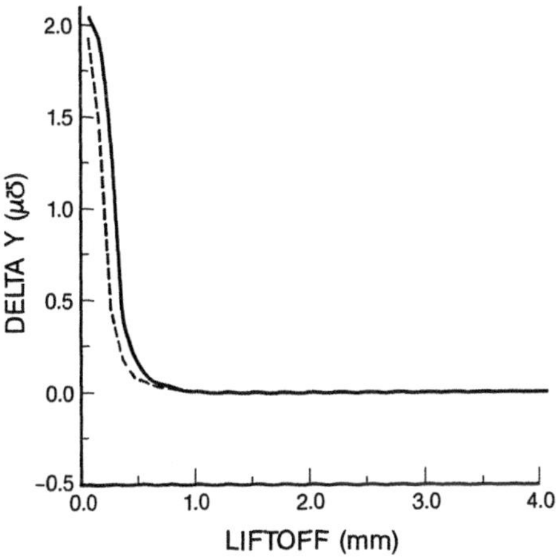

Figure 6. ▲Y vs. liftoff comparison of the experimental results of the unshielded probe to equation 1.

Adding the two superposition elements produces a potential +V on the source electrode and a zero potential with reference to ground on the receiver, the condition of the physical system. When the currents are added it is noted that the current from the source to ground for the zero spatial frequency case is not measured. Furthermore, the current that flows from the receiver to the ground for this component subtracts from the overall current. The Gimple/Auld theory requires that all of the current that leaves the source electrode arrive at the receiver electrode. To simulate this condition, we forced all of the current to pass through the receiver by connecting the ground plane that was below the specimen to the receiver. Because both the receiver and the ground plane were at the same potential this did not alter the environmental interaction. The effect was that any displacement current that arrived at the ground would be routed to the receiver. The results indicate that this technique indeed reduces the recovery but does not eliminate it.

We suspected the remaining recovery was associated with parasitic coupling to the probe case. Because we were measuring relatively small quantities we had assumed that the probe connections and leads had to be shielded. In an attempt to reduce the coupling, we built a second probe with no probe case. The results of the liftoff measurements for this probe, figure 6, show very good agreement with the theoretical model. The theoretical curve was fitted to the experimental data of the initial liftoff which is equivalent to determination of the effective area of the probe. The slight difference in the curves is probably a result of the uncertainty in the initial liftoff measurement.

APPLICATIONS

We investigated a variety of applications for the capacitive probe as an NDE sensor. Figure 7 shows the response to a through crack in a 3.0 mm thick alumina plate; the crack was approximately 40 μm wide and much longer than the extent of the probe. We observed the characteristic bipolar flaw signal for a notch,with excellent signal-to-noise ratio. The width of the flaw signal is governed by the size of the probe, not the crack width.

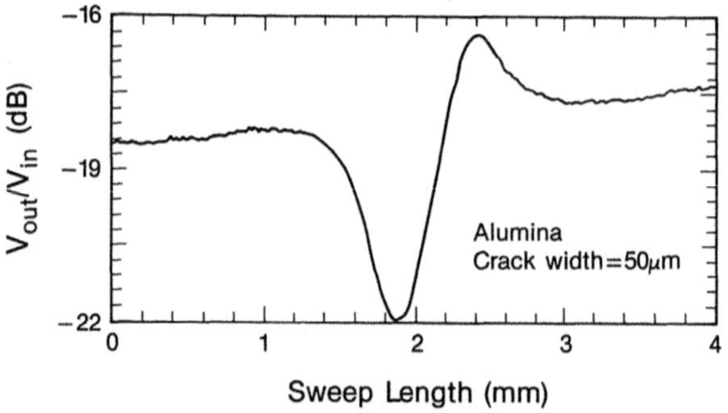

Figure 7. Response to a through crack in alumina approximately 40 μm.

A more realistic crack, approximately 4.0 μm wide, was made in plate glass. The probe response qualitatively agrees with the response to the alumina crack and had about the same signal-to-noise ratio.

Another potential application of the capacitive sensor is noninvasive cure monitoring in composites. We performed proof of concept experiments to show the ability to detect changes in the dielectric constant in conductive composites. In our experiments we used graphite/epoxy plates (6 x 150 x 100 mm) with uniaxial fibers. The probe was inverted with the electrodes facing upward and the composite panel was placed directly on the face of the probe. A curing epoxy was then placed on the opposite side of the panel. The probe impedance was monitored as a function of time with different temporal and spatial frequencies. Measurements were made with the probe's electric field oriented both parallel and perpendicular to the graphite fibers.

Typical results are shown in figure 8 for 100 kHz and with the probe's electric field parallel to the fibers. During the curing process, there is about 1% maximum change in Z, the probe impedance magnitude and a change of approximately 0.03 á in the phase angle. The measurement was repeated with the field oriented perpendicular to the fiber direction. We found that this change in orientation caused no significant change in our measured results.

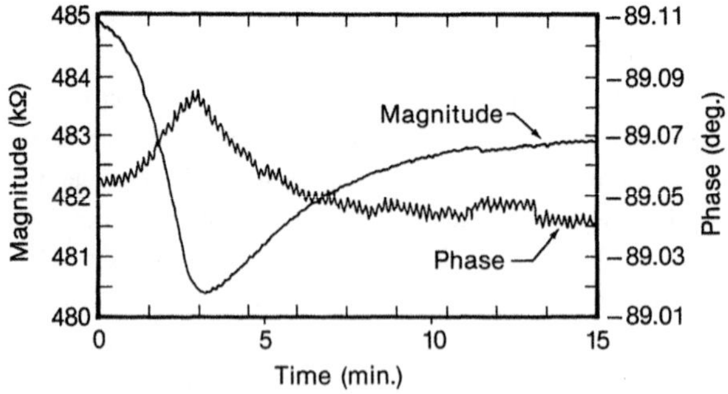

Figure 8. The impedance in magnitude and phase of the capacitive probe in proximity to a curing epoxy vs. time. A graphite/epoxy panel (6 mm thick) was placed between the probe and the curing epoxy to determine if the electric field could penetrate a partially conductive composite.

The effects of conductivity are exemplified by changing the frequency of operation. When we contrast the maximum change in Z at 10 to 100 kHz we find that the value at the lower frequency is nearly double that of the higher frequency. We also note that the change in the phase angle is about an order of magnitude greater at the lower frequency. This shows, as expected, that the conductivity effects increase with increasing frequency.

Remark

This paper is a brief of the references 1, 2. For more detailed information consult these references.

REFERENCES

1. P.J. Shull, J.C. Moulder, P.R. Heyliger, M. Gimple, and B.A. Auld, in Review of Progress in Quantitative NDE, V. 7A, D.O. Thompson and D.E. Chimenti, eds. (Plenum, NY, 1988), p. 517.

2. P.J. Shull, A.V. Clark, and B.A. Auld, in Review of Progress of Quantitative NDE, to be published.

3. M. Gimple, Capacitive Arrays for Robotic Sensing, Ph.D. Dissertation, Stanford University, 1987.

Modelling of the Electromagnetic Field Diffracted by an Inhomogeneity in Metal: A First Step in Magnetic Imaging

R. ZORGATI, A. BERNARD and F. PONS,

Retour d'Expérience, Mesures, Essais. Direction des Etudes et Recherches (EDF)
25, Allée Privée - Carrefour Pleyel
93206 Saint-Denis Cedex, France.

B. DUCHENE, D. LESSELIER and W. TABBARA

Equipe Electromagnétisme. Laboratoire des Signaux et Systèmes (CNRS-ESE)
Plateau de Moulon
91192 Gif-sur-Yvette, France.

Summary
Non-destructive multifrequency imaging of anomalous conducting regions in a non-magnetic metallic half-space is carried out from the fields due to the eddy-currents induced in the anomaly by a low-frequency current source located above the half-space. First, the behavior of the anomalous electric and magnetic fields is investigated as a function of the electrical and geometrical parameters of the anomaly and of the current source. Both synthetic -exact and Born's approximate- and experimental data are used. Second, an imaging procedure based on diffraction tomography is introduced. Preliminary results obtained for small anomalies are then shown in order to study the resolution of the imaging technique.

Introduction

In many cases of interest, variations of conductivity or of permeability in metallic structures which should be otherwise homogeneous indicate the presence of anomalies or defects. In particular, it is known that local stress or corrosion cause variations of permeability, whereas cracks or notches modify the conductivity. We are only interested herein to probing variations of conductivity within a homogeneous block of metal; to do so, we observe and process via a diffraction tomography procedure the variations of the anomalous magnetic field we observe just above the block surface when a known low-frequency current source (a solenoid) is placed near the anomaly. Diffraction tomography has been shown to be the basis of useful non-destructive testing procedures of buried targets in both microwaves and ultrasonics [1-2], and we would like to explore herein the viability of such an approach at much lower frequencies using the eddy-currents. First, we investigate the variations of the electric and magnetic fields associated to a given anomaly by varying location, shape, and conductivity of the anomaly, and frequency (in the 1-300 kHz range) and location of the source. Both synthetic (exact and Born's approximate) and experimental data are used. Then, we attempt to image the eddy-currents within the anomaly in diverse configurations using diffraction tomography. (At the present time, such images are only calculated from synthetic fields.)

These last years, only a few authors seem to have used eddy-current data in order to image anomalies in conducting materials. Among others are refs. [3-5]. In [3], the author studies the reconstruction of the conductivity and permeability profiles of a rope placed along the axis of a solenoid from the variations of the input impedance of the solenoid. In [4], the authors propose using a least-square technique to retrieve the variations of conductivity and apply it to the mapping of a flaw in the wall of a cylindric tube. The conductivity varies into the radial and azimuthal directions only and the first-order Born's approximation is assumed to be valid (this is a 2-D linear inverse problem). Holographic techniques have also been developed for the first time to our knowledge in eddy-current imaging in [5]. (Note that a unified description of the solutions of the linear inverse problem, i.e., within the frame of the Born approximation or Physical Optics, is now available [6].)

The anomalous fields: formulation and measurements

In the following, we consider linear isotropic non-magnetic materials. An anomaly is a domain where the conductivity differs from the conductivity of the homogeneous metal in which this anomaly is present. The geometry of the problem is depicted in Fig. 1 below.

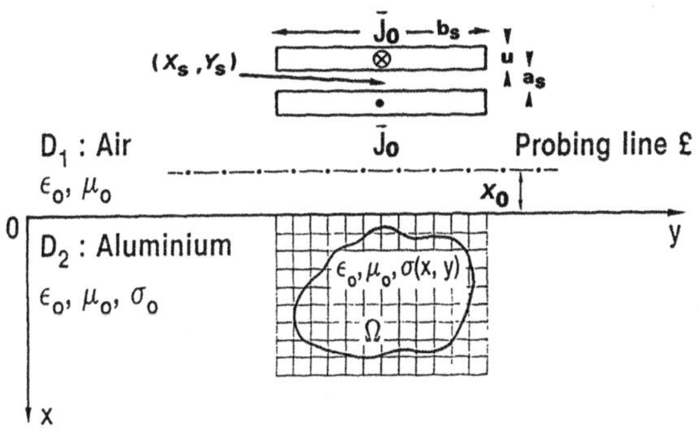

Fig.1: Geometry of the problem

The space is divided into two half-spaces with interface Γ. The upper one D_1 is air, with permittivity ε_0 and permeability μ_0, the lower one D_2 is a homogeneous conductor with same permittivity and permeability and with conductivity σ_0. D_2 contains a cylindrical anomaly with axis parallel to $\underline{z}$ and whose conductivity σ differs from σ_0 and vary with position $\underline{r}$=(x,y) in a domain of finite cross-section Ω. A given current source (with $e^{-j\omega t}$ time-dependence) $\underline{J}_0 = J_0(\underline{r})\underline{z}$ is located in D_1 and the anomalous fields are observed on the probing line L above and parallel to the interface at height x_0. The electric field is axial, $\underline{E} = E_z \underline{z}$, and the magnetic field deduced from $\underline{E}$ by Faraday's law is transverse, $\underline{H} = H_x \underline{x} + H_y \underline{y}$, for reasons of symmetry. (Both components of the magnetic field, the normal one H_x and the tangential one H_y, are measured using a microcoil suitably oriented and mechanically displaced.)

An exact scalar integral formulation of the electric field in the whole space as a function of the eddy-currents induced in the anomaly Ω can be derived from the Helmholtz equation satisfied by the field by using Green's theorem. Such a derivation is similar to those presented earlier [1-2], which we refer the reader to. We let the propagation constants be k_1, k_2, k_Ω in D_1, D_2,

Ω ($k_1^2=\omega^2\varepsilon_0\mu_0$, $k_2^2=j\omega\mu_0\sigma_0$, $k_\Omega^2(\underline{r})=j\omega\mu_0\sigma(\underline{r})+\omega^2\varepsilon_0\mu_0$). We have

$$E(\underline{r})=E_0(\underline{r})+\int_\Omega G_{m,2}(\underline{r},\underline{r}')\,J(\underline{r}')\,d^2\underline{r}' \qquad \underline{r}=(x,y)\in D_m,\ m=1,2,\text{ and } \underline{r}'=(x',y')\in\Omega \qquad (1)$$

where $J(\underline{r}')=(k_\Omega^2(\underline{r}')-k_2^2)E(\underline{r}')$ represents the induced sources and $E_0(\underline{r})$ the field with the anomaly absent (the "incident" field). $G_{m,2}$ is the Green function, i.e., the field observed at point $\underline{r}$ in D_m, m=1, 2, when a line-source $\delta\underline{z}$ is placed at $\underline{r}'$ in D_2; its plane-wave expansion is known and its values can be accurately computed when needed by FFT algorithms. If we assume that the field in the anomaly is equal to the field with the anomaly absent, we get the first-order Born approximation of the anomalous field E_a in D_1:

$$E_a(\underline{r})=\int_\Omega G_{1,2}(\underline{r},\underline{r}')[k_\Omega^2(\underline{r}')-k_2^2]E_0(\underline{r}')\,d^2\underline{r}' \qquad \underline{r}=(x,y)\in D_1 \text{ and } \underline{r}'=(x',y')\in\Omega \qquad (2)$$

(the total field is equal to E_0+E_a). Otherwise, the computation of the anomalous field E_a on the probing line is made in two steps. First, we compute the electric field in Ω by applying a method of moments with point-matching; the resulting linear system derived from (1) is either solved directly using a Gauss-Jordan algorithm, or iteratively using a conjugate-gradient algorithm. Second, we compute the anomalous field by integration over Ω. (The magnetic field is then estimated by central finite-difference.)

In the experiments, the current source is a solenoid oriented along $\underline{y}$ (current I_0, N turns, length b_s, diameter a_s, center (x_s,y_s)). As is usual, this source is modeled (see Fig. 1) by two rectangular sheets (thickness u) of axial currents with constant but opposite values ($\pm NI_0/ub_s$). In that case, exact closed-form formulations of E_0 and of the corresponding magnetic field $\underline{H}_0$ are obtained in D_1 and D_2. The experimental set-up is shown in Fig. 2.

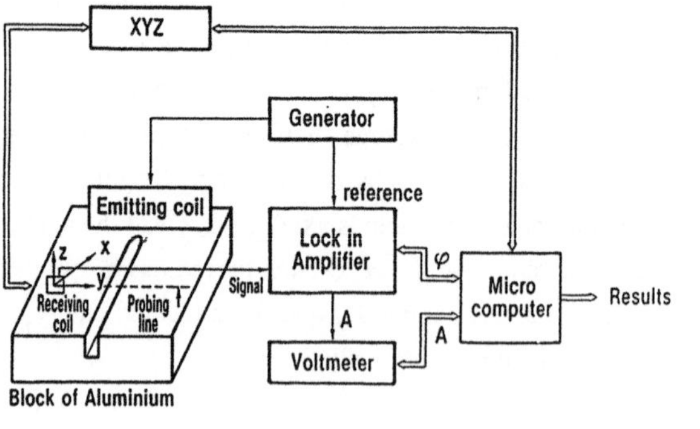

Fig.2: Layout of the experimental setup

Experiments described here have been performed on empty (σ=0) notches of various depth and width at the surface of the metal (an aluminium alloy AU4G where σ_0=1.96 10^7 S/m), but others are in progress on conducting anomalies (regions filled with different alloys) .

The eddy-current imaging procedure

Imaging the anomaly (in effect, the eddy-currents induced within) is made via a diffraction tomography procedure which was developed for microwave and ultrasonic imaging of buried targets by the *Equipe Electromagnétisme*. Complete analyses are found in [1-2]. Here we only give a short summary of the procedure we employ to get images of the anomaly. We assume that the current source is such that we can assimilate the incident field to a plane wave, i.e., that we have $E_0(\underline{r})=e^{jk_1x}+Re^{-jk_1x}$ ($x\leq 0$) and $=Te^{jk_1x}$ ($x\geq 0$), where R and T are the Fresnel reflection and transmission coefficients. We introduce a spatial Fourier transformation which associates a function $\hat{f}$ of the wave vector $\underline{K}$ to a function f of the vector of position $\underline{r}$ (they both depend also on the temporal frequency ω): $\hat{f}(\underline{K})=\int f(\underline{r})e^{-j\underline{K}.\underline{r}}\,dV(\underline{r})$ with integration over the entire space ($\mathbb{R}^2$, or $\mathbb{R}^1$ when f depends on one variable only). We define accordingly the 1-D and 2-D Fourier transforms $\hat{E}_a(x_0, K_y)$ and $\hat{J}_n(\underline{K})$ of the anomalous electric field on the probing line $E_a(x_0,y)$ and of the eddy-currents normalized with respect to the incident field $J(\underline{r})/E_0(\underline{r})$. Using the plane wave expansion of the Green function, and on condition that k_1 and k_2 are real-valued, we obtain after some calculations the relationship in $\underline{K}$-space:

$$\hat{J}_n(\mu, \nu) = -j\, T^{-1}\, (\beta_1+\beta_2)\, e^{j\beta_1 x_0}\, \hat{E}_a(x_0, \beta) \tag{3}$$

$$\mu=-k_2-\beta_2,\quad \nu=\alpha \quad \text{and} \quad \beta_i=\sqrt{(k_i^2-\alpha^2)},\ i=1, 2,\ \text{with } |\alpha| < \text{Inf}\,(k_1, k_2).$$

When k_1 and k_2 are complex-valued as it is obviously the case in this study, we have to replace them in (3) by their real parts in order to get a meaningful (but approximate) relationship between induced eddy-current and anomalous field Fourier transforms. For each frequency ω we obtain spectral values $\hat{J}_n(\underline{K})$ on a circular arc in the spectral plane. By varying the frequency ω, we cover a limited area of the spectral plane. We consider that we are dealing with the same function whatever be the frequency, function which we still denote as $\hat{J}_n$, and the Fourier inversion theorem allows us to go back to $\underline{r}$-space and to obtain $J_n(\underline{r})$, i.e., the multifrequency image of the anomaly. In short we get this image by coherent superposition in $\underline{K}$-space. (We display the modulus of J_n in the examples herein.) Fourier transformations are implemented using FFT algorithms from a finite number of regularly spaced samples on the probing line obtained for a finite number of frequencies. A nearest-neighbor interpolation algorithm is used to get $\hat{J}_n$ on the required cartesian mesh.

The numerical and experimental results

Fig. 3 (a, b) shows exact and experimental values of the normal and tangential components of the magnetic field $\underline{H}$ (displayed versus position of the probe) at 10 kHz in the case of a notch 1.5 mm wide and 3 mm deep centered at $y=y_s=170$ mm (with $x_0=-1.45$ mm). Fig. 4 shows the frequency dependence of the normal component of the measured field for the same anomaly. Fig. 5 and 6 show exact and Born's approximate amplitudes and phases of the anomalous magnetic field at 1 and 50 kHz, respectively, for a smaller notch (1 mm^2, with $x_0=-0.5$ mm). From these figures and other results obtained during our investigations, we conclude the following. (i): The experimental and synthetic values of the fields are in good agreement on condition that the discretization of this anomaly is conveniently chosen. Convergent and stable numerical results are obtained when it is divided into square cells with side $\Delta \leq \delta_p/7$, where δ_p is the depth penetration $\sqrt{(2/\omega\sigma_0\mu_0)}$. This agreement is better close to than far from the anomaly, which might be due to the discrepancy between the solenoidal source and its model. (ii): The tangential component of the magnetic field is much more sensitive to the presence of the anomaly than the normal component. (iii): The Born approximation provides results which closely agree to the exact ones at low frequency (a few kHz) and for small anomalies. At higher frequencies or for larger anomalies, Born's approximate results largely differ (this is especially true for the phase). (iv): As is obvious, the higher is the frequency, the larger are the variations of the field near the anomaly.

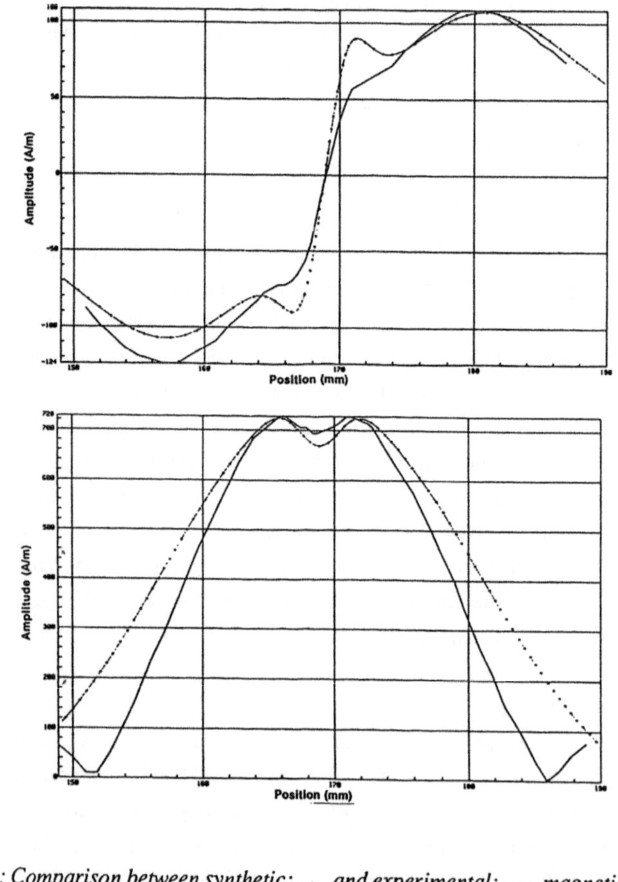

a)

b)

Fig. 3: Comparison between synthetic: —•— and experimental: —— magnetic fields. Normal component H_x: (a) and tangential component H_y : (b), observed at 10 kHz for the 1.5 X 3 mm² notch (see text).

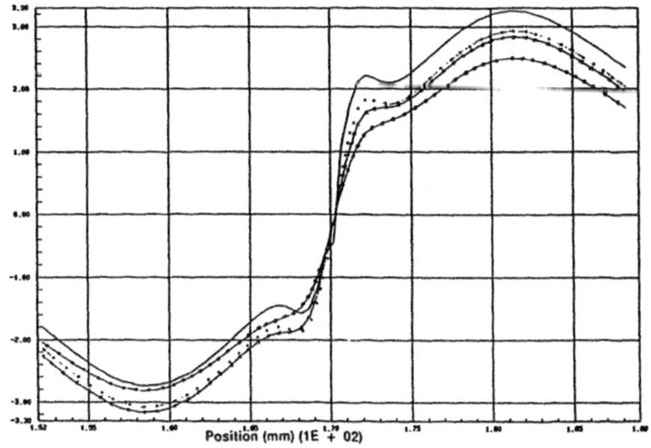

Fig.4: Frequency dependence of the magnetic field: normal component measured in the configuration of Fig. 3 at 10: —o—, *20:* —x—, *50:* —•—, *100:* —— *kHz.*

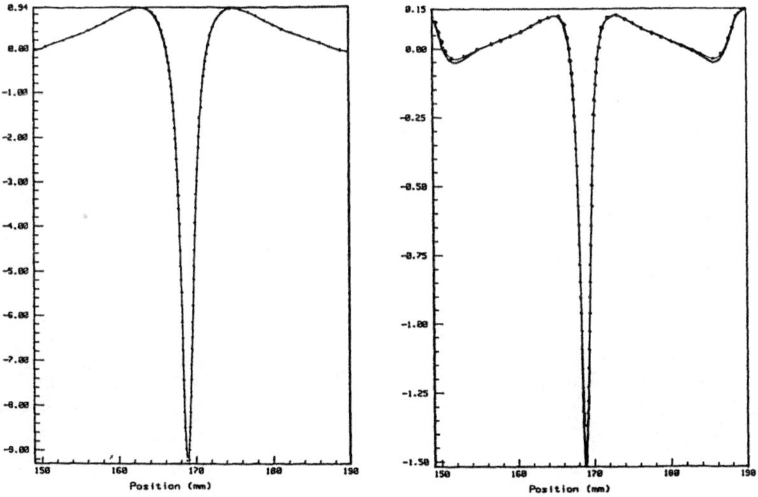

Fig. 5: Comparison between exact: —— *and Born's approximate:* —•— *synthetic fields for the small 1 x 1 mm² notch (see text). Amplitude in A/m (left) and phase in degrees (right) of the anomalous field at 1 kHz .*

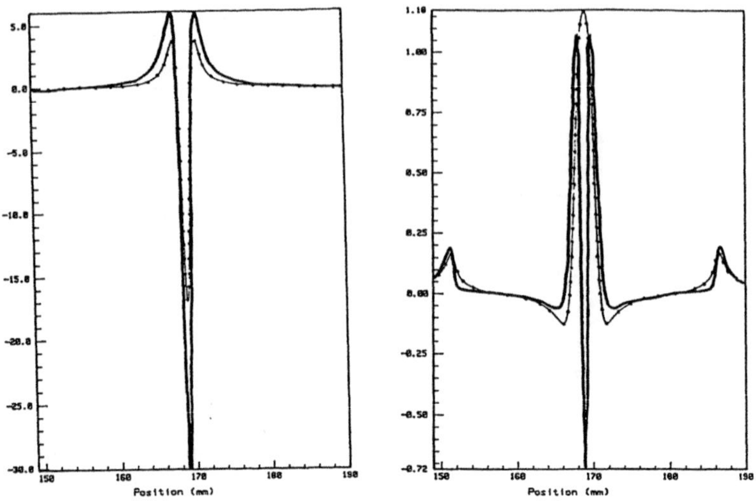

Fig. 6: Comparison between exact: ___ and Born's approximate: —•— synthetic fields for the small 1 X1 mm² notch (see text). Amplitude in A/m (left) and phase in degrees (right) of the anomalous field at 50 kHz .

Figs. 7 and 8 are multifrequency images of point-like anomalies calculated using the exact field observed at five different frequencies (100-150-...-300 kHz). Fig. 7 shows the image of one squared (empty) notch with side 0.04 mm long (i.e., the impulse response).

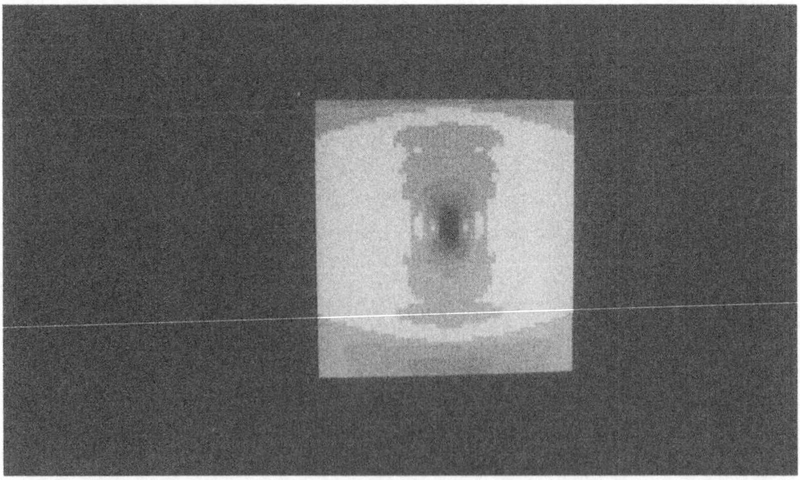

Fig. 7: The impulse response of the imaging procedure: multifrequency image of a point-like anomaly (the side of the square is 12.8 mm long).

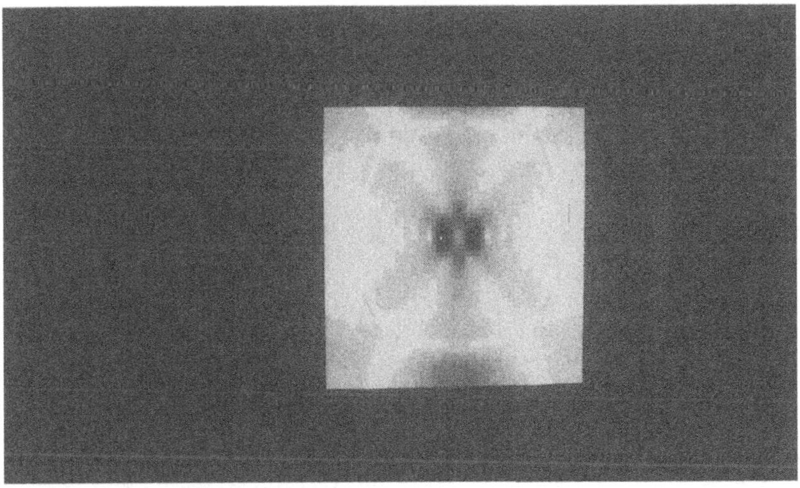

Fig. 8: The resolution of the imaging procedure: multifrequency image of two point-like anomalies separated by 1.4 mm. (same scale than in Fig. 7).

Fig. 8 shows the image of two similar squared notches separated by 1.4 mm. From the above and other results, we conclude the following. (i): The impulse response has transverse and longitudinal half-height widths equal to about λ and 2λ, respectively, where λ ($=2\pi\delta_p$) is the minimal wave-length in the metal (which corresponds to the higher operating frequency in the imaging procedure). (ii): We cannot discriminate between anomalies closer than λ from one another.

References

[1] Chommeloux L. *et al.*, "Electromagnetic modeling for microwave imaging of cylindrical buried inhomogeneities", IEEE Trans. Microwave Theory Tech., MTT-34, 10, 1064-1076, 1986.

[2] Duchêne B. *et al.*, "Acoustical imaging of 2D fluid targets buried in half-space: a diffraction tomography approach", IEEE Trans. Ultrason. Ferroelec. Freq. Control, UFCC-34, 5, 540-549, 1987.

[3] Seznec R., "Problème inverse des courants de Foucault. Application au contrôle non-destructif des cables", Rev. CETHEDEC, 77, 21-29, 1983.

[4] Sabbagh H. A. and Sabbagh L. D., "An eddy-current model for three-dimensional inversion", IEEE Trans. Magn., MAG-22, 4, 282-291, 1986.

[5] Hildebrand B. P. *et al.*, "Holographic principles applied to low-frequency electromagnetic imaging in conductors", in IEEE Proc. 10th Int. Optical Computing Conf., 59-66, 1983.

[6] Langenberg K., in *Basic Methods of Tomography and Inverse Problems*, P. Sabatier Ed., Adam Hilger, Bristol, Part II, 127-467, 1987.

Comparative Micromagnetic and Mössbauer Spectroscopic Depth Profile Analysis of Laserhardened Steel X210Cr12

R. KERN*, W.A. THEINER*, P. SCHAAF**, U. GONSER**

* Fraunhofer-Institut für zerstörungsfreie Prüfverfahren, IzfP
D-6600 Saarbrücken, FRG

** Universität des Saarlandes, Fachrichtung Werkstoffphysik und Werkstofftechnologie, D-6600 Saarbrücken

Summary

Nondestructive magnetic testing by magnetic Barkhausen noise and incremental permeability is applied to study the transformation of microstructure of a commercial X210Cr12 steel submitted to laser irradiation. Magnetic quantities are derived to determine the variations of the martensite and retained austenite content observed up to 2 mm in depth to the laser treated surface. Conversion x-ray-Mössbauer measurements in backscattering technique (CXMS) are performed to get an exact phase analysis, including information about phase formation. The magnetic and Mössbauer results are compared in connection with Vickers hardness and metallographic data.

Zusammenfassung

In lasergehärteten Schichten auftretende Gefügeveränderungen können mit dem magnetischen Barkhausenrauschen und der Überlagerungspermeabilität empfindlich nachgewiesen werden. Daraus abgeleitete zerstörungsfreie (zf) Meßgrößen ermöglichen eine Ermittlung der Einhärtungstiefe, des Härteprofils und bei der hier untersuchten Stahlgüte X210Cr12 eine Aussage über den Restaustenitanteil im laserbehandelten Gefüge. Mössbauerspektroskopische Rückstreumessungen (CXMS) an bis zu diskreten Tiefen abgetragenen Proben, die mit unterschiedlicher Laserleistung bestrahlt wurden, ermöglichen eine genaue qualitative und quantitative Bestimmung der einzelnen Phasenanteile. Die magnetischen und Mössbauer-Ergebnisse werden in Verbindung mit Vickershärte und metallographischen Daten miteinander verglichen.

Introduction

In order to improve the static and dynamic mechanical properties of steel components, laser irradiation was found to be a powerful and promising technique. There are several applications including hardening, alloying and welding. In this report we investigated the phase transformations which occur at the surface and at various depths of a commercially available X210Cr12 cold forming tool steel, submitted to single-pass laser irradiation with different laser power levels. Two different methods are applied: By means of a micromagnetic technique a determination of laser hardening parameters such as the hardening depth, the surface hardness and the retained austenite can be obtained [1]. Additionally Mössbauer spectroscopy performed in the backscattering mode by detecting the reemitted x-rays after absorption (CXMS) is used to determine exactly the phase fractions in the laser treated surface zones [2].

Experimental

The laser treatment was carried out by MAN Technologie GmbH, Munich, with a Cw CO_2-laser on a commercially available X210Cr12 cold forming tool steel. The X210Cr12 steel contains 2.1 wt % C and 12 wt % Cr. Samples were submitted to a single pass square 16x16 mm laser beam, using several incident powers: 2700 W for specimen 1, 2850 W for specimen 2 and 2950 W for specimen 3.

Vickers hardness measurements (HV 0.5), also performed by MAN, are shown in fig. 1. For sample 1 the hardness starts to increase at about 0.4 mm depth below surface and peaks at about 1.1 mm. For samples 2 and 3 a similar hardness increase is observed but shifted to lower depths. As displayed in

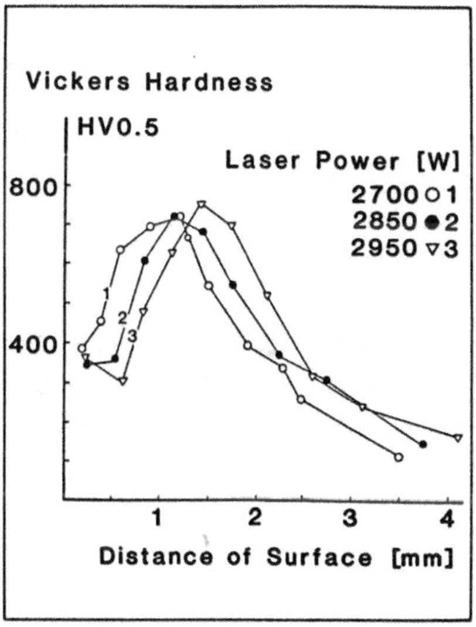

Fig. 1: Vickers hardness (HV0.5) versus depth

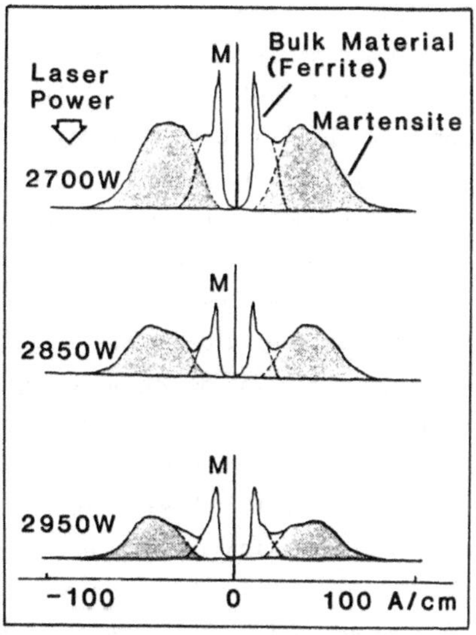

Fig. 2 Magnetic Barkhausen noise versus applied magnetic field for X210Cr12, laser treated with different power levels

fig. 1, the mechanically soft surface region is extended with increasing laser power level whereas the extension of the hardness maximum remains almost the same.
The investigations in this report were performed at the laser treated surface and additionally for specimen 1 (2700 W) at increasing depths down to 2 mm, obtained after removing layers mechanically and by electropolishing.

Results and discussion

The magnetic Barkhausen noise M(H) and the incremental permeability $\mu_{\Delta}(H)$, both measured during the dynamic magnetization of ferromagnetic materials, have been successfully applied to microstructure and stress analysis [1,3]. The Barkhausen noise signals received by an inductive sensor at the specimen surface are amplified and rectified and usually plotted versus the strength of the alternating magnetic field (frequency f_E = 50 mHz). The main activity of the Barkhausen noise M(H) is normally observed around the coercivity H_c of the magnetic hysteresis curve B(H). The coercivity can also be deduced from the magnetic field strength corresponding to the maximum of the Barkhausen noise amplitude.

In fig. 2 magnetic Barkhausen noise curves M(H) are displayed for three specimens of X210Cr12 steel, laser treated with power levels 2700 W (1), 2850 W (2) and 2950 W (3). As demonstrated the three M(H) curves can be composed by two contributions of different phase fractions. The outer peak at higher magnetic field strength corresponds to the laser-hardened martensite phase whereas the inner peak at low magnetic field strength originates from the magnetically and mechanically soft ferritic bulk material. With increasing power level of the incident laser beam the whole M(H) curve gets smaller in amplitude that means the contributions of both phases are decreasing. This can be explained by the nonmagnetic retained austenite phase near to the surface covering the martensitic and ferritic phases below it. Therefore the Barkhausen noise signals (micro eddy currents) originating from the magnetic phases are damped to an extent depending on the retained austenite fraction and on the thickness of this surface layer.

Fig. 3 demonstrates the magnetic Barkhausen noise M(H) for the 2700 W specimen, recorded at different depths after removing surface layers mechanically and by electropolishing. Using a high analyzing frequency of 100-315 kHz there is only a small magnetic penetration depth (~ 100 µm) or interac-

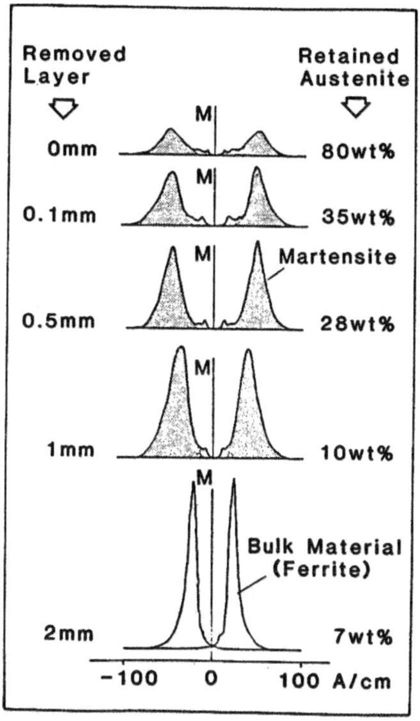

Fig. 3: Magnetic Barkhausen noise for the 2700 W specimen at different depths

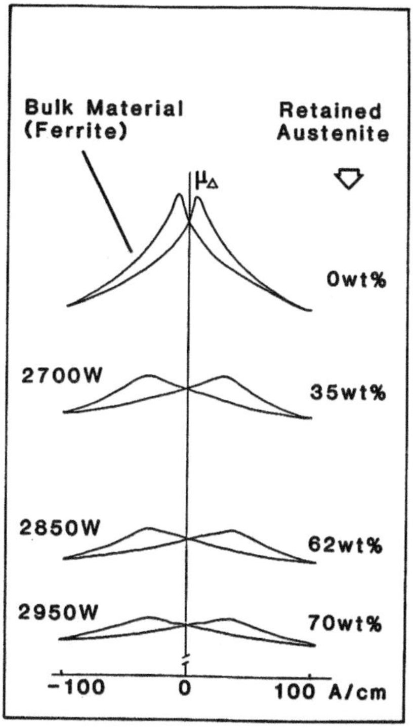

Fig. 4: Incremental permeability for X210Cr12, untreated and laser treated with different laser powers

tion volume to receive Barkhausen noise signals due to the eddy current damping. As shown in fig. 3 the magnetic Barkhausen noise amplitude measured near to the laser treated surface is small. After removing surface layers the martensite contribution (chaded regions) continuously increases with depth up to 1 mm. At a depth of 2 mm ferritic bulk material dominates, because the martensitic phase is almost completely removed.

The incremental permeability is another appropriate measuring quantity to evaluate the retained austenite content in surface hardened layers. This is demonstrated in fig. 4 for the different laser-hardened X210Cr12 specimens which are compared with the $\mu_\Delta(H)$ curve of the untreated base material. It can be seen that with decreasing laser power the retained austenite content and extension also decreases and consequently the μ_Δ-values are approaching to the μ_Δ-values of the ferritic base material containing no retained austenite. If one can calibrate the μ_Δ-axis e.g. by Mößbauer spectroscopy the content of the retained austenite can be determined magnetically.

The Mössbauer investigation was performed at room temperature. The spectra were recorded in the backscattering mode by detecting the reemitted 6.4 keV x-ray. Details of the experimental set up are explained in ref. [4]. Mössbauer-spectra taken for two velocity ranges are displayed in fig. 5 for the 2700 W sample. The spectrum corresponding to the untreated surface is well fitted by means of three magnetically split subspectra and two central paramagnetic doublets. The relative abundancies of the three sextets are in

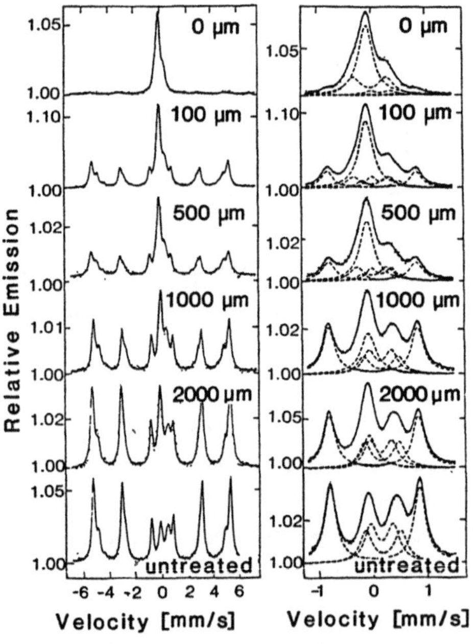

Fig. 5: Mössbauer spectra for the 2700 W treated specimen at different depths (removed layers)

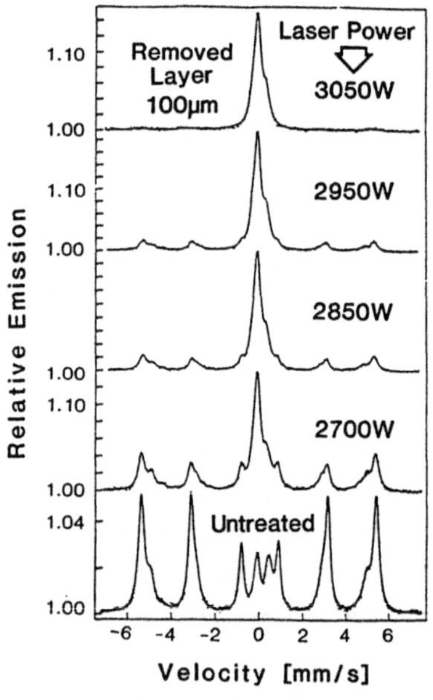

Fig. 6: Mössbauer spectra for X210Cr12, laser treated with different powers

agreement with a random distribution of chromium atoms in the first and second neighbourhood of a given iron atom. The two quadrupole split doublets must be ascribed to the inequivalent iron sites of a chromium-iron-carbide M_7C_3. From the chemical composition and a computed chromium content of 2 wt % in the ferrite, the formula $(Cr_{0.5}\ Fe_{0.5})_7C_3$ is derived. The spectra recorded at the laser treated rough surface exhibits a dominant paramagnetic component superimposed on a ferromagnetic one. The improved resolution of the low velocity range spectra leads to a fit with four subspectra for this central component: i) a single line plus a quadrupole doublet, the hyperfine parameters of which are consistent with those of retained austenite with interstitial carbon [5]; two quadrupole doublets with the same hyperfine parameters as those encountered in the untreated sample and therefore ascribed to undissolved M_7C_3 carbides. With increasing depths the amount of the magnetically split subspectra increases. They can be attributed to martensite with low carbon content. The results of the Mössbauer phase analysis for the 2700 W sample are displayed in fig. 7.

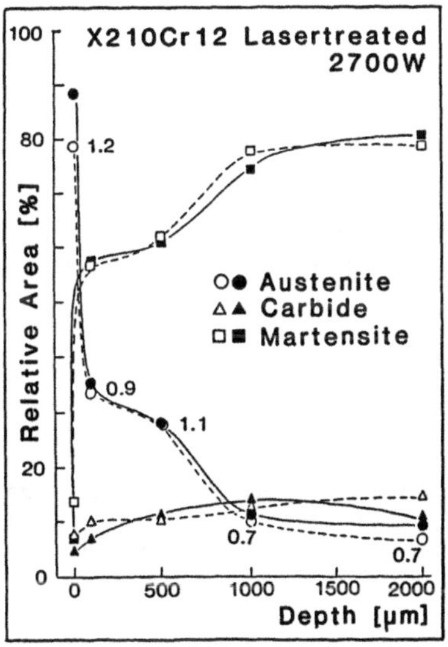

Fig. 7: phase fractions versus sample depth for the 2700 W treated specimen. Filled symbol for high-velocity and open symbols for low-velocity range spectra

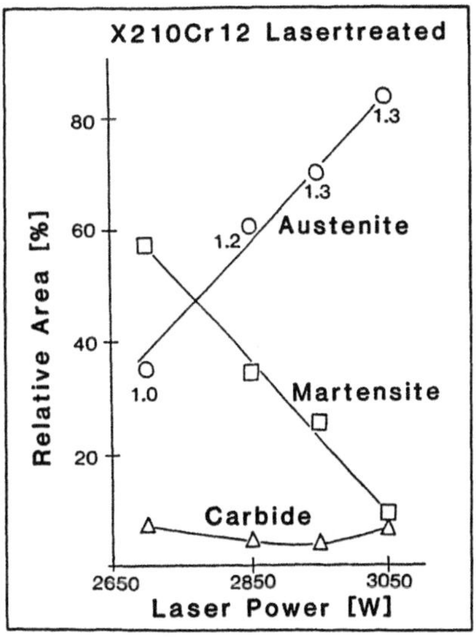

Fig. 8: phase fractions of laser treated X210Cr12 versus laser power (removed layer 100 µm)

In fig. 6 the Mössbauer spectra recorded at a depth of 100 µm below the lasertreated surface of the different specimens (different laser power levels) are combined and compared with the untreated base material. As can be seen, the amount of the ferromagnetic subspectra continuously increases with decreasing laser power. As displayed in fig. 8 the phase analysis yields an approximately linear increase of the retained austenite fraction with increasing laser power in a depth of 100 µm, whereas contrarily the amount of martensite increases with decreasing laser power. This result is in agreement with the micromagnetic data discussed above. The variations of the martensite and retained austenite content observed between 100 and 2000 µm explain the micromagnetic results displayed in figs. 2-4. The content of retained austenite in the surface near zone and its extension are responsible for the damping of the micromagnetic measuring signals arising from the martensite and ferrite phases.

Conclusions

Two completely different techniques, micromagnetic inspection by magnetic Barkhausen noise and incremental permeability and conversion x-ray Mössbauer spectroscopy in backscattering technique (CXMS) are combined to investigate the phase transformations which occur at the surface and at various depths of a commercial X210Cr12 cold forming tool steel, laser irradiated with different power levels. The depth profile determined by means of both Mössbauer and magnetic analysis are in accordance with the metallographic and Vickers hardness data. The mössbauerspectroscopic determination of the retained austenite fraction can be used to calibrate the micromagnetic measuring quantities.

Acknowledgement

This contribution is based on work performed with the support of the German Ministry for Research and Technology. We gratefully thank Dr. W. Amende from MAN Technologie GmbH, Munich, for supplying the laser irradiated samples with metallographic and hardness data.

References

1. R. Kern, W.A. Theiner: Nondestructive magnetic testing of laser hardening parameters in: Laser Treatment of Materials, Ed. Barry L. Mordike, DGM Inf. Verlag (1986) 427-434.

2. P. Schaaf, Ph. Bauer, U. Gonser: Mössbauer measurements in backscattering technique (CXMS) of laser irradiated cold forming tool steel (X210Cr12), submitted to Hyperfine Interactions.

3. W.A. Theiner, I. Altpeter: Determination of Residual stresses using micromagnetic parameters in: New Proceedures in Nondestructive Testing. Ed. P. Höller, Springer Verlag Berlin, Heidelberg 1983.

4. L. Blaes, H.G. Wagner, U. Gonser, J. Welsch, J. Sutor: Hyperfine Interactions 29 (1986) 1571.

5. M. Ron in Applications of Mössbauer-Spectroscopy II, Ed. R.L. Cohen, Academic Press, New York, (1980) 329.

Characterization of Cementite in Steel and White Cast Iron by Micromagnetic Nondestructive Methods

I.ALTPETER, R.KERN, P.HÖLLER

Fraunhofer-Institut für zerstörungsfreie Prüfverfahren (IzfP), Saarbrücken

Summary

The influence of different cementite contents and cementite modifications on micromagnetic measuring quantities of steels with ferritic-perlitic, martensitic annealed and martensitic soft annealed microstructure states and white cast iron are analyzed. For this, micromagnetic measuring quantities derived from the magnetic Barkhausen noise are used. By heating up to temperature values above the Curie temperature of cementite, characteristic changes of the measuring quantities are shown. Cementite produces actively its own magnetic Barkhausen noise, whereas it also influences the magnetic Barkhausen noise of the iron matrix as a foreign body and by its stress fields of the second kind. Both influences act in a different way as well in steel as in white cast iron. In steel, the stress induced effect is the dominating one; in white cast iron, the active contribution to Barkhausen noise dominates.

Zusammenfassung

Der Einfluß unterschiedlicher Zementitgehalte und Zementitausbildungsformen auf mikromagnetische Meßgrößen wird an ferritisch-perlitischen, martensitisch angelassenen und martensitisch weichgeglühten Gefügezuständen sowie an weißem Gußeisen analysiert. Hierzu werden aus dem magnetischen Barkhausenrauschen abgeleitete Meßgrößen genutzt. Durch Aufheizen über die Curie-Temperatur von Zementit werden charakteristische zementitspezifische Änderungen der Meßgrößen aufgezeigt. Zementit trägt einerseits aktiv zum magnetischen Barkhausenrauschen bei und beeinflußt andererseits als Fremdkörper sowie über seine Spannungshöfe 2-ter Art das magnetische Barkhausenrauschen der Eisenmatrix. Beide Beiträge sind in Stahl und weißem Gußeisen unterschiedlich ausgebildet. In Stahl ist der Spannungseffekt, in weißem Gußeisen der aktive Rauschbeitrag der Zementitphase dominant.

Introduction

Cementite precipitations contribute to an increase of the yield point. According to some theories of strengthening methods [1], there are several influences increase in strength like precipitation shape, -density, -phase content. The nondestructive determination of the yield point requires the knowledge of the quantities named above, which are normally determined by microscopic methods. To determine characteristic material parameters like hardness, strength etc. by micromagnetic methods, it is necessary to determine the cementite content in a nondestructive way. All conventional methods for the determination of cementite content like Mössbauer-spectroscopy [2], the quantitative x-ray phase analysis [3], the extraction method [4] and the image analysis of transmission electron or optical micrographs [5] are very expensive, time consuming and destructive. In addition to this, the knowledge of the influence to cementite on micromagnetic measuring quantities is useful for the improvement of the microstructure characterization with nondestructive testing methods. It is also useful in sorting steels and white cast iron.

Description of equipment

Cylindrical specimens were magnetized in an electromagnet excited by a bipolar power-supply. The excitation frequency of the external magnetic field was 50 mHz and was supplied by a function generator. The magnetic Barkhausen noise [6] was measured with a differential coil device. The hysteresis curves were measured with a hystrometer. The specimens were heated with a heating coil.

Measured results/discussion

The influence of cementite on some measuring quantities derived from the magnetic Barkhausen noise is investigated on steels with different microstructure states (perlitic microstructure states with globular (P_G) and lamellar cementite modification (P_L), martensitic annealed (M+A) and martensitic soft annealed (M+W) microstructure states just as white cast iron. The microstructure states are different concerning their cementite modification, cementite size and cementite distribution. The different microstructure states with a cementite size between 0.28µm and 3.4µm demonstrate characteristic changes of the hysteresis curves and a different Barkhausen noise activity as a function of their cementite content.

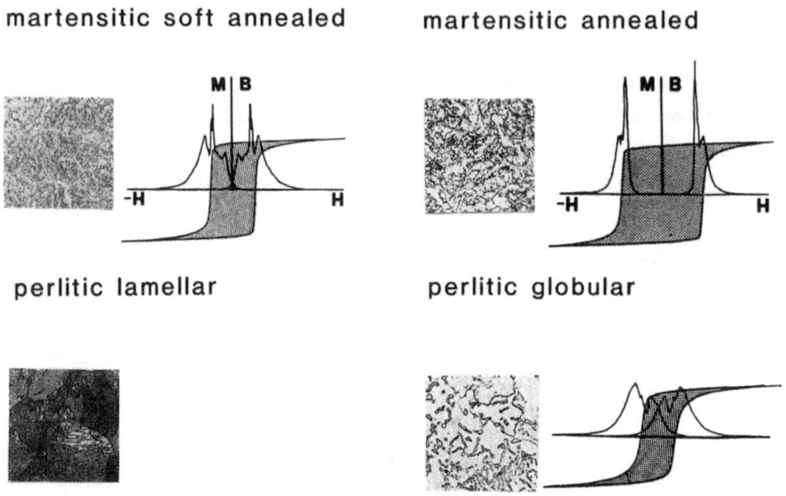

Fig. 1: Magnetic Barkhausen noise curves and hysteresis curves of four different microstructure states

In figure 1 the hysteresis curves as well as the Barkhausen noise curves of four microstructure states for the steel D45 (7 wt.%Fe_3C) are presented. As the figure demonstrates, the hysteresis curves are remarkably different in their coercivity values and in the H-field region of the Bloch-wall-nucleation just as in the area spanned by the hysteresis curves (shaded in fig. 1). The Barkhausen noise curves are different in their structure (single-peak, double-peak and triple-peak) and their maximum amplitude-values. The different structure of the Barkhausen noise curves for several microstructure states can be explained by the stress fields in the region near the cementite precipitation. These stress fields originate in the different thermal expansion coefficient of the ferrite and cementite phase [7]. These stress fields of second kind have an influence on the magnetic domain structure, which characterizes the iron matrix.

According to Laszlo [8] the maximum tensile stress is found in the interface "precipitation/matrix". The stress decreases following a $1/R^3$ function for spherical precipitation and a $1/R^2$ function for cylindrical precipitation. For identical dispersion states, there are long-range stress fields for lamellar cementite and short-range for globular cementite. The long-range stress fields lead to an interaction between residual stress fields of the cementite precipitations and therefore also to a stronger interaction with residual stress fields of 90°-Bloch-walls in the iron matrix. The result is a stronger hindrance to the 90°-Bloch-wall-motion. Subsequently the Barkhausen events will be shifted from the knee region to the region near the coercivity H_c, where the 180°-BW are active. This is the reason for the single peak in the M(H)-curves of perlitic lamellar microstructure steel and white cast iron. By heating up a two phased material to temperatures above the cementite Curie temperature, it can be shown, that cementite contributes actively and directly to the magnetic Barkhausen noise. Figure 2 shows the ratio $M_{max}(T)/M_{max}(RT)$ (T: temperature, RT: room temperature) for soft iron and compact cementite sample. This ratio decreases 20% for soft iron and 100% for cementite in the temperature intervall (40°C $\leq$ T $\leq$ 240°C).

The Curie temperature of the compact cementite sample is reduced to -168°C by a manganese content of 3% (investigations of Jellinghaus [9]). With increasing temperature the order of the magnetic moments decreases, this means, that the spontaneous magnetization decreases in the same way. At the Curie temperature, the ferromagnetic domain structure completely disappears and there is no spontaneous magnetization. The transition from ferromagnetic to paramagnetic state is therefore accompanied by a decrease of magnetic domain structure and consequently the Barkhausen noise intensity decreases. The white cast iron-samples show a temperature dependence in magnetic Barkhausen noise which can be compared with that of the compact cementite sample. The maximum value of the Barkhausen noise amplitude decreases with increasing temperature and decreasing cementite content (fig. 3). This behaviour evidently demonstrates that cementite contributes to the Barkhausen noise as a second ferromagnetic phase. The active noise contribution of the cementite phase is greater in white cast iron than in steel, as a consequence of the greater volume fraction. In steel there exist different $M_{max}(T)$-curves dependent on the microstructure state.

The $M_{max}(T)$-values of P_L-structures and white cast iron are similar, the smaller effects with P_L-structures originate in the lower content of cementite. For microstructure states with a globular cementite shape, it is found that all $M_{max}(T)$ curves have a maximum between room temperature and Curie temperature. According to Laszlo [8] residual tensile stress near globular cementite particles increases with temperature until Curie temperature and then decreases again. This is the result of the expansion anomaly of cementite [7].

For a determination of the cementite content it is necessary to use the 3MA-analysis (micromagnetic multiparameter microstructure and stress analysis). Each magnetic measuring quantity depends at least on several structure and/or stress influences. To find a solution one has to combine several micromagnetic measuring quanitities. It must be assured that the information obtained by the different magnetic measuring quantities are independent of each other. For the determination of cementite we prefer measuring quantities derived from the magnetic Barkhausen noise because of their high sensitivity to changes of the microstructure.

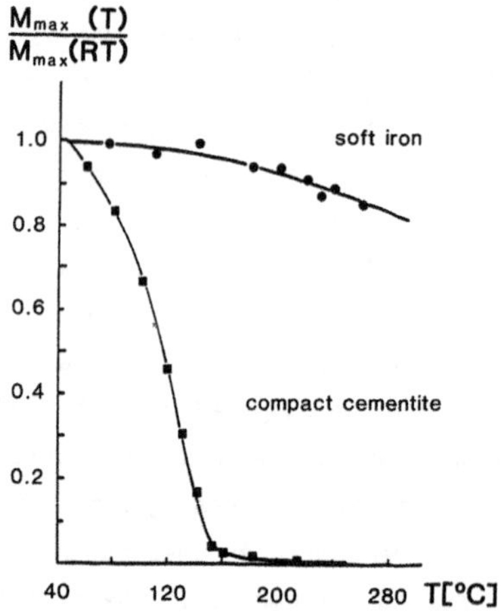

Fig. 2: Magnetic Barkhausen noise amplitude as function of temperature soft iron - cementite sintered

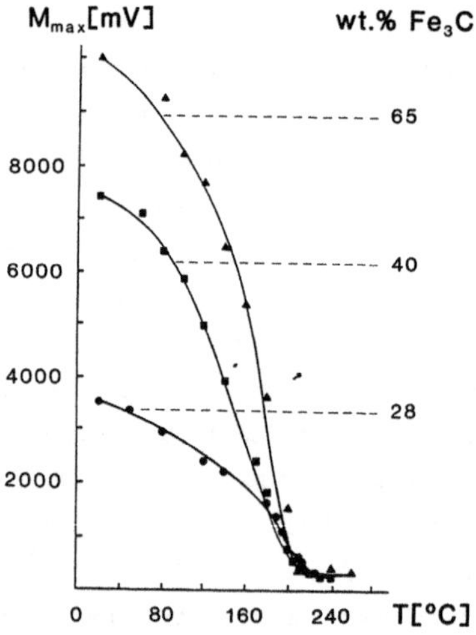

Fig. 3: Magnetic Barkhausen noise amplitude as function of temperature - white cast iron

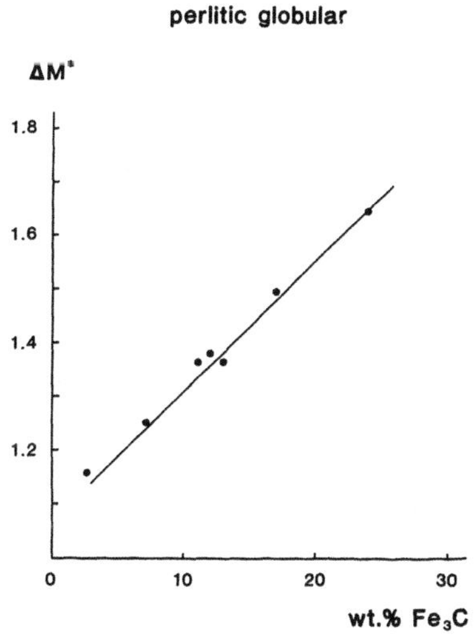

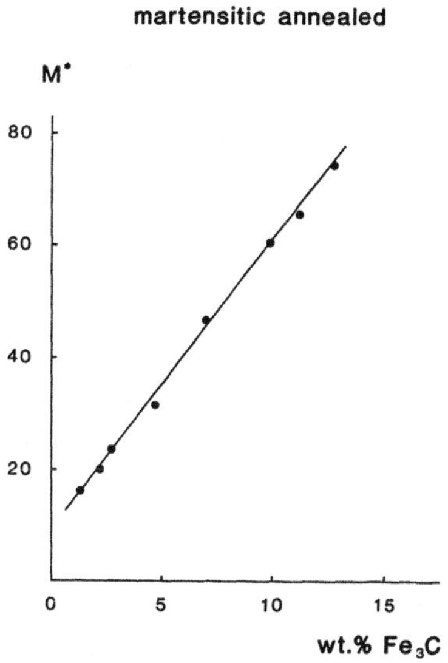

Fig. 4: Measuring quantity ΔM^* as function of cementite content

Fig. 5: Measuring quantity M^* as function of cementite content

The determination of the cementite content requires information on the microstructure state of the steel. For a known steel quality, the cementite morphology can be determined from the shape of the M(H)-curve. If the steel quality is unknown, the coercivity can be derived from the M(H)-curve and used to characterize the microstructure state.

Each microstructure has its own calibration curve to determine the content of cementite. The measuring quantities for the states under inspection are:

- The width of the M(H)-curve at 25% of M_{max} at RT (room temperature) for the P_L-microstructure states.
- The width of the M(H)-curve at 25% of M_{max} at RT divided by the same quantity at 200°C, ΔM^*, for the P_G-microstructure states (Fig. 4).
- For the structure states M+W:

$$\frac{M_{max}\,(RT) - M_{max}\,(T=T_{Curie})}{M_{max}\,(RT)}$$

- For the structure states M+A (Fig. 5):

$$M^* = \frac{M_{max}(RT) - M_{max}(200°C)}{M_{max}(RT)}$$

These measuring quantities are the most sensitive ones. Moreover they can be gathered in a rather simple way under practical conditions.

Conclusion

The aim of these investigations was to determine the influence of cementite on micromagnetic measuring quantities for steel and white cast iron. Specimens consisting of ferrite and cementite were heated beyond the Curie temperature of cementite in order to measure changes of magnetic properties caused by this phase transition. Structures with different diameter, shape and distribution of cementite show characteristic variations of Barkhausen noise curves as function of the content of cementite. The multiple peak structure of the Barkhausen noise curve is correlated with diameter and shape of the cementite particles. The variations of the peak maxima M_{max} in the Barkhausen curves as function of the temperature can be explained by the decrease of active noise contribution by the cementite phase and by a change of residual stresses. Measuring quantities derived from M_{max}(T)-curves can be used to determine the content of cementite, if the microstructure state is known.

Acknowledgement

This work has been carried out under the research promotion of the Deutsche Forschungsgemeinschaft.

References

1. Brown, L.M., Ham, R.K: Dislocation-Particle Interactions. in: Strengthening Methods in Crystals. Hrsg.: Kelly, A., Nicholson, R.B. Appl.Sci.Publ. LTD London 1971.

2. M. Ron: Iron-carbon and iron-nitrogen systems. in: Applications of Mößbauer Spectroscopy. edited by Cohen, S.329. 1980.

3. Faber, F., Hartmann, U.: Quantitative röntgenographische Phasenanalyse am System α-Fe/γ-Fe/Fe_3C. Härterei-Techn.Mitt. 28. 1973.

4. Schrader, A: Die Gefügeuntersuchung von Eisen und Stahl mit dem Elektronenmikroskop. Schweizer Archiv (1960) 1963.

5. Puskeppel, A., Harsdorff, M.: Anwendung der quantitativen Bildanalyse bei der Auswertung elektronenmikroskopischer Hellfeld- und Dunkelfeldaufnahmen. in: IMANCO, Fortschritte der quantitativen Bildanalyse 20, 1975.

6. Theiner, W.A., Altpeter, I.: Determination of Residual Stresses Using Micromagnetic Parameters. in: New Procedures in Nondestructive Testing (Proceedings). Editor P. Höller. Springer-Verlag Berlin, Heidelberg, 1983.

7. Jellinghaus, W.: Massiver Zementit, Archiv für das Eisenhüttenwesen. 37.Jahrgang, Heft 2, S. 187, 1966.

8. Laszlo, F., Nolle, H.: J.Mech.Phys.Solids, 7, 193-208, 1958/59.

9. Jellinghaus, W.: Massiver Zementit, Archiv für das Eisenhüttenwesen, Heft 2, S.183 ff, Febr. 1966.

In-situ Ferrite Content Measurement of Duplex Steel Structures in the Chemical Industry. Practical Applications of the Alternating Field, Magnetoinductive Method

Dr. W. Staib,

Helmut Fischer GMBH + Co., Sindelfingen, W. Germany

Dipl. Ing. H. Künzel,

Lambrecht, W. Germany.

Abstract:
The ferrite content of austenitic and duplex steels has a great deal of influence on their properties. By using the FERITSCOPE® to determine magnetic permeability, ferrite contents between 0 - 80% can be non-destructively measured. This paper discusses the measurement principle, the instrument design and the results of practical measurements made on duplex steel structures (Steel No. 1.4462) such as flue-gas scrubbing plant. By utilising the instrument capabilities to the full, significant savings in time and money can be realized.

Introduction

Tanks, pipelines, valves and fittings in the chemical, energy and oil industries are often exposed to high temperatures, high pressures and aggressive environments. These conditions demand corrosion and acid resistant steels which exhibit high temperature strength e.g.:

- normal constructional steel with an austenitic chrome steel welded cladding.
- austenitic stainless steel
- duplex steel.

For austenitic steels, with typical ferrite contents of 0.5 to 12%, the measurement of residual ferrite within the cladding or weld often gives important insight into future mechanical and corrosion resistant behaviour. Too little ferrite encourages susceptibilty to hot cracking, too much reduces the mechanical strength and corrosion resistance.

For duplex steels,with ferrite contents of 40 - 60% (recently 25 - 75%), too little ferrite in a weld is an indicator for future stress cracking.

In both cases special welding material has to be used to ensure that the specified ferrite contents are maintained within the welded zone. The FERITSCOPE® also provides a means of checking that the correct material has been used, even when the weldment is complete.

Measurement Principle

The small handheld probe of the FERITSCOPE® provides the means to quick and exact, non-destructive, in-situ determination of ferrite content in welds or claddings. The instrument works on the magnetic inductive principle, whereby the ferrite content is obtained from the magnetic permeability. The measurement principle is shown in Fig. 1 for a single pole probe; two-pole probes can also be used.

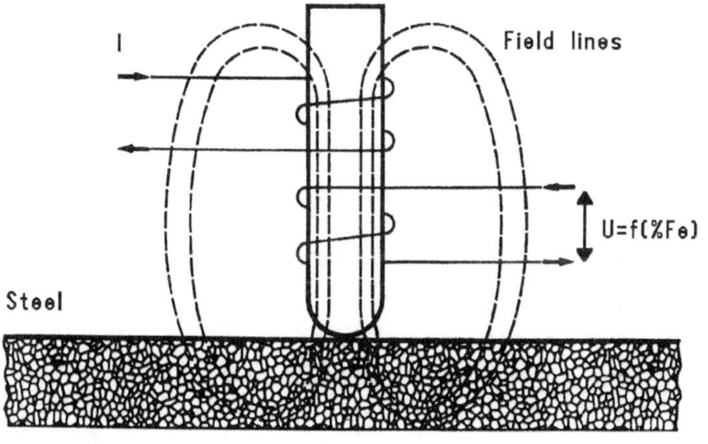

FIG. 1: Schematic diagram of measuring principle

A low frequency alternating current, I, flows through the coil of the probe, generating an alternating magnetic field in the ferrite-containing substrate. The interaction between field and coil induces an alternating voltage, U, the magnitude of which is a function of the ferrite content of the test piece. This alternating voltage is the measuring effect.

Master calibration

In view of the fact that, depending on previous mechanical working, ferrite in austenitic steel can appear in many forms, no universally valid relationship between measuring effect and %Ferrite can be derived which exploits to the full the accuracy attainable by the method for a specific measurement problem. Instead the instrument relies upon an empirical, internally stored, master calibration curve via which the measured signal is converted into %Ferrite. The curve is based on a large variety of ferrite specimens drawn from leading laboratories which have been measured by standard methods, see Fig. 2.

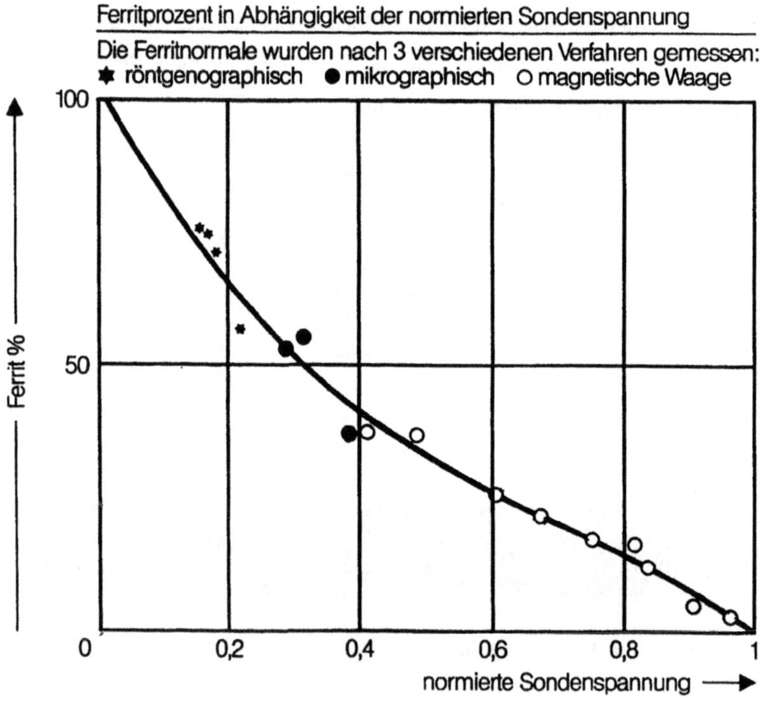

Fig. 2: Master calibration curve

Corrective calibration

Several external factors can affect measurements made by the magnetic method. These are primarily:

- the magnetic properties (morphology) of the ferrite,
- specimen curvature,
- specimen thickness and
- edge effects.

The instrument is tuned to current ferrite morphology and specimen geometry by means of a corrective calibration. User standards of known ferrite content, with geometries and ferrite form corresponding to the objects to be tested, are measured during the corrective calibration procedure. A modified conversion characteristic is then derived from the master calibration. The instrument now measures the problem at hand reliably and accurately.

The master characteristic can be regenerated at any time by calibrating with equivalent standards, which can be supplied with the instrument. These comprise soft iron discs covered by plastic foils of different thicknesses which produce equivalent ferrite readings. The standards, which are available for both %Ferrite and WRC Numbers, also provide an excellent basis for comparative measurements. Where the variation in ferrite content rather than an accurate absolute value is of interest, the instrument can immediately be used without calibration (the corrective calibrations are stored in non-volatile memories).

Instrument design

The FERITSCOPE® is available as a bench top instrument (M11 see Fig. 3) or as a handheld instrument for in-situ measurements (MP3 see Fig. 4). Both instruments are microprocessor-based and possess non-volatile memories for 10 x 500 readings or 4 x 1000 readings respectively plus the associated calibration parameters. Both have measuring ranges from 0 - 80 %Fe or 0 - 110 WRC. The advantages can be summarized as follows:

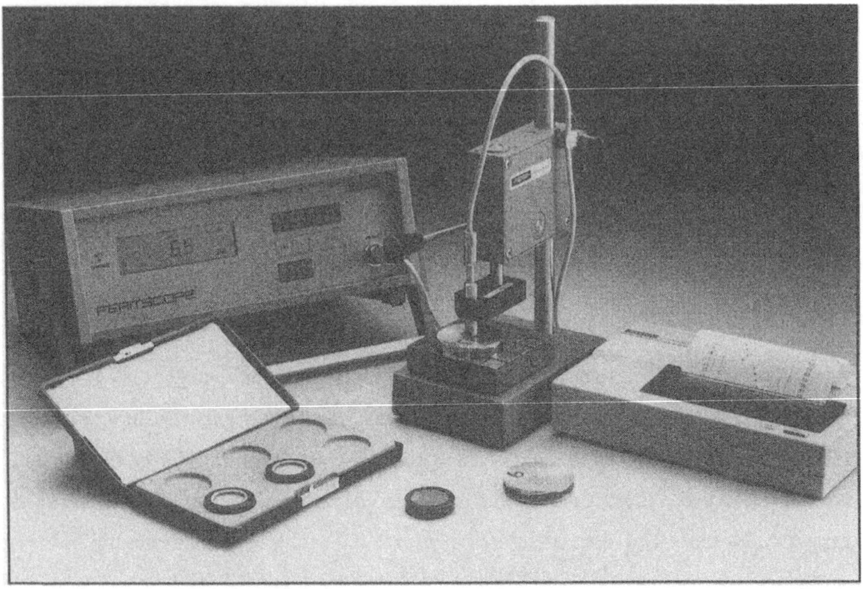

Fig. 3: FERITSCOPE M11 with printer and stand.

Fig. 4: FERITSCOPE MP3 handheld instrument.

- The instruments are intelligent. The stored working characteristics contain in condensed form the experience of various international laboratories. By taking customer-specific reference standards to extend the measuring range or to increase accuracy in a particular subrange the instrument can be easily and quickly adapted to technological innovations or to specific applications.
- The influence of material properties on the measurement is completely or at least for the most parts eliminated. This is particularly true for fluctuations in conductivity for claddings and for oxide layers.
- The FERITSCOPE® is light, portable and suitable for measurements on-site. It has ten applications in which the measurements taken at the object - presorted were necessary - can be stored for logging at a later time. For data transmission to a central computer, an (optional) RS-232 serial interface is provided.
- The instrument is equipped with a evaluation program which displays or prints the minimum, maximum and mean-value as well as standard deviation at the touch of a button. In addition measurements can be graphically presented in the form of a histograms. Tolerance limits can also be set: the MP3 provides an acoustic signal on violation of the set limits, the M11 a log in addition.

Fig. 5: Flue gas scrubber at Marl, BASF

Practical measurements with the FERITSCOPE®

As mentioned in the introduction, ferrite measurements fall roughly into two catagories:

- measurement of contents between 0.5 and 12% (also confirmation of zero content) in austenitic claddings and weldments such as tanks, reactors and pipelines,
- measurement of contents between 25 and 75% in duplex steels as used in heat exchangers, pipelines, scrubbers, marine drive shafts and desalination plant.

The measurement of ferrite is essential in both cases if optimum mechanical and anti-corrosion properties are to be obtained, in particular in welded areas. For austenitic steels too little ferrite increases susceptibility to hot cracking by favoring the formation of low melting point eutectics with Nb, Si, P and S at grain boundaries. Too much ferrite encourages corrosion by the reduction of chromium and nickel and leads to unfavorable deformation properties, characterised by fissures along the ferrite lines.

Duplex steels, which are ferrite-containing austenitic steels with Mo, Mn, Si and N additions and only traces of C, provide better mechanical properties than fully austenitic steels. In addition they offer good resisitance against pitting and stress-crack corrosion in chloride-bearing and generally acidic environments. These properties are, however, attained only when the ferrite content, which is also dependent upon heat treatment, lies between 15 and 75%.

Figs. 5 - 7 show plant components recently investigated by one of the authors. Fig. 5 is a flue gas desulphurizing scrubber installed at the BASF Marl power plant. The internal components of the absorber tower on left were manufactured from duplex steel and Hastalloy C 276, the bypass filter and components from duplex steel No. 1.4462.

Fig. 6 shows the inlet (Manufacturer: Widmann and Sohn, Mannheim), made of 1.4462 duplex steel with a wall thickness of 8 mm. The flange, of the same material, was welded on by TIG and electrical hand welding techniques. As part of the final acceptance procedure, the components were checked for ferrite content using the FERITSCOPE® M11 and GAB1.3 probe. The acceptable tolerance range was 28 to 56 %ferrite. Random determination of the ferrite content by ambulant metallography confirmed the magnetic measurements.

Pipes for the plant (Manufacturer: Stahl, Geräte- und Apparatbau, Viernheim) are shown in Fig. 7. These were seamless, made of duplex steel No. 1.4462, with a wall thickness of 22 mm. The flanges were welded on using a 1.4462 welding rod. The ferrite content of the solution treated pipes averaged 50%: a content 50% ± 25% was specified for the welded zone. Measurements were again made with the FERITSCOPE® M11 and checked by ambulant metallography. Agreement was with a ± 10% ferrite content tolerance. The magnetic method averaged a reading every 5 - 10 seconds, bringing a considerable saving in time, effort and expense when compared to ambulant metallography at approx. 20 minutes per measurement.

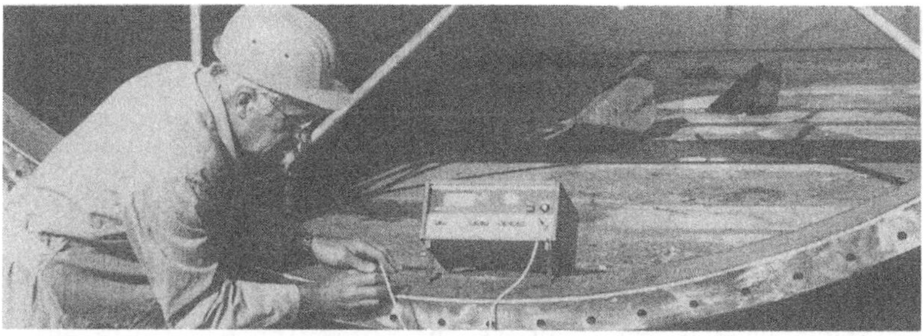

Fig. 6: Ferrite measurements on welded flange.

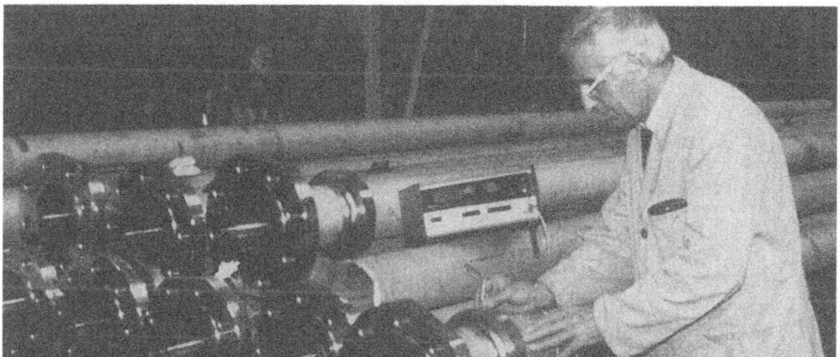

Fig. 7: Checking ferrite content on seamless pipes.

Changes in Magnetic and Mechanical Properties and Microstructure During Annealing of the Stainless Soft Martensitic Steel X5 CrNi 13 4 (1.4313)

G. Maußner, A. Seibold

Siemens AG UB KWU
Erlangen

Summary

The quenched and tempered steel X 5 CrNi 13 4 (1.4313) belongs to the family of soft martensitic stainless steels used for components in pumps, valves and turbines.

The changes in the magnetic and mechanical properties and in the microstructure after quenching were investigated by tempering series of specimens in the temperature range between 450 and 650 °C on two melts containing different amounts of nitrogen.

Three kinds of changes in the as-quenched microstructure are caused by tempering: precipitation of $M_{23}C_6$ carbides, matrix recovery (decrease in dislocation density) and formation of stable austenite by diffusion controlled decomposition of the matrix.

The magnetic properties respond very sensitively to these changes of the microstructure in the temperature range between 500 and 600 °C where the formation of stable austenite predominates.

The interaction between microstructure, mechanical and magnetic properties is complex and cannot be reduced to simple correlations.

The results show that much more detailed information on the microstructure is to be expected from the measurement of other magnetic/electrical properties (e.g. Barkhausen noise, magnetostriction, etc. ...).

1. Introduction

The material 1.4313 belongs to the group of soft martensitic stainless steels. It offers a combination of good mechanical properties, resistance to uniform and selective corrosion and dimensional stability during machining; it is suitable for the full range of nondestructive examinations and it can be welded to materials of similar composition.

This steel is widely used, e.g. for components in pumps, valves and turbines, thanks to this special combination of properties /1 - 3/.

This steel derives its macroscopic properties from a tempered martensitic microstructure built up by means of quenching and tempering (quenching: 950 - 1000 °C/air; tempering: 500 - 600 °C/h) /4/.

Microstructural characteristics can be varied extensively by varying the temperature at which the steel is tempered. This process gives rise to different mechanical and magnetic properties.

The objective of the study was to determine and quantify the microproperties of the material (microstructure) and to correlate these with its magnetic and macroscopic properties (strength, toughness, hardness).

2. Specimen Material and Test Methods

These correlations were investigated after quenching (at 1000 °C 1 h/air) by tempering series of specimens in the temperature range between 450 and 650 °C with holding times at temperature for 1 - 10 h.

Comparative tests were carried out on specimen material taken from hot-rolled rounds (dia. 90 mm) from two melts with different nitrogen concentrations.

Chemical Composition of the Melt (% by weight)

	C	Si	Mn
Melt A	0.027	0.44	0.73
Melt B	0.023	0.50	0.72
	P	S	Cr
Melt A	0.018	0.002	13.06
Melt B	0.018	0.002	12.90
	Mo	Ni	N
Melt A	0.55	4.14	0.0114
Melt B	0.55	4.18	0.032

The specimens were analyzed to determine their microstructural characteristics and their mechanical and magnetic properties.

3 Results

3.1 Microstructural Analysis by means of Optical Microscopy

Metallographic microsections were taken to analyze the microstructural condition of the specimens.

There are two methods of highlighting microstructural characteristics:

- Colour-etching using the Lichtenegger technique (to reveal differences in the chrome concentration of the microstructural constituents)
- Etching using V_2A etchant

The colour etching technique shows up differences in the structure and chemical composition (chromium content) of the fine-lamellar secondary grain microstructure (Figs. 1,2).

Etching with V_2A as etching agent is mainly used to reveal the degree of carbide precipitation which varies as a function of tempering temperature (Figs. 3 - 5).

Examination of specimens under an optical microscope cannot provide quantitative information on the microstructural characteristics affecting the material's mechanical and magnetic properties.

3.2 Microstructural Examinations by means of Transmission Electron Microscopy (TEM)

Analytical transmission electron microscopy is the only method of analyzing the microstructural characteristics of the steel.

The structure of the fine-lamellar matrix was examined using electrolytically thinned metal foil specimens.

The platelet martensitic phase structure with a platelet width of about 0.2 µm (Fig. 6) is still extensively present at a tempering temperature of 450 °C.

At higher tempering temperatures from about 520 - 540 °C most of the microstructure of the matrix has recovered.

The matrix now consists of adjacent lamellar ferrite and stable austenite /5, 6/ containing a higher concentration of nickel and chromium (Fig. 7).

Figure 10 shows the dislocation density, the stereologically determined volume fraction of tempered austenite and the nickel equivalent (according to Schaeffler /7/) in the tempered austenite as a function of the tempering temperature and time.

The volume fraction of stable austenite is shown in the lower portion of Fig. 10. A 10 h holding time at temperature as low as 500 °C produces about 4 % of the total volume and its maximum volume fraction of around 10 % is obtained at about 560 °C.

The nickel equivalents (nickel equivalent = % Ni + 0.5 % Mn) in the tempered austenite of the 10 h tempering series are shown in the central portion of Fig. 10 /7/.

Extraction replicas are of advantage for evaluating the extent of precipitation, especially carbides.

With the exception of a few oxide or sulphide inclusions the matrix is free from precipitation after quenching.

Very finely distributed $M_{23}C_6$ carbides with a size characteristic for the tempering temperature/time start to precipitate at low tempering temperatures (Fig. 8, 9).

The mean carbide diameter increases with tempering temperature and time from about 50 nm to 200 nm (Fig. 11, lower portion).

The chromium concentration in the carbides likewise increases with tempering temperature. The $M_{23}C_6$ carbides contain 60 - 80 % Cr, 4 - 6 % Mo and Fe (remainder), Fig. 10.

3.3 Mechanical Properties

To determine the mechanical properties of the steel at room temperature, HV 10 hardness measurements on metallographic microsections, tensile tests to determine tensile strength, elongation and reduction of area and charpy-V impact tests were performed.

Hardness in the regime of 380 - 400 HV are obtained for the material in its initial air-quenched condition. Figs. 11 and 12 plot hardness performance as a function of tempering temperature for holding times at temperature of 1 h and 10 h respectively.

Specimens from both melts reveal a tensile strength of about 1,100 N/mm² after quenching (Figs. 14, 15). Beyond about 520 °C a tempering time of 10 h produces a reduction in mechanical strength. A marginal increase in mechanical strength is recorded beyond about 610 °C (re-formation of martensite!).

High values in excess of 100 J are generally obtained for absorbed energy in impact tests at room temperature (mean value from three specimens) for all tempering temperatures. The absorbed energy curve for both melts contains a trough at 505 °C and then increases very rapidly, reaching a high plateau of about 200 J and about 170 J for melts A and B respectively even at low tempering temperatures.

3.4 Magnetic Properties

The magnetic hysteresis curve was plotted for specimens measuring 10 x 10 x 50 mm up to a max. field intensity of 150,000 A/m.

Coercive force Hc and saturation permeability B (at H max.) were derived from this.

Saturation magnetizability B decreases with increasing tempering temperature (time). Conversely, coercive force Hc increases, attaining an intermediate peak at temperatures at which the secondary hardness peak also occurs (Figs, 16, 17).

Saturation magnetizability of the melt with the higher nitrogen concentration starts to decrease at low tempering temperatures. This correlates with the tendency of melt B to form stable austenite more rapidly on account of its higher nitrogen content.

TEM using Lorentz microscopy techniques was used to observe the appearance of the magnetic domains.

The size of the domains in microstructures tempered at the lower temperature, is of the same order as that of the initial austenitic grain (Fig. 18).

From about 500 °C the domains start to shrink to about the size of the lamellar secondary grain (or stable austenite) as shown in Fig. 19.

4. Summary and Conclusions

During tempering three principal types of changes occur in the microstructure of material 1.4313:

- Precipitation of type $M_{23}C_6$ carbides, whereby carbide morphology and the chemical composition of the microstructure are characteristic of the tempering temperature and the nitrogen content of the melts.

- Recovery of the matrix appearing in the microstructure in the form of a reduction in dislocation density or as polygonization.

- Formation of stable austenite. Tempered austenite derives its stability from the diffusion-controlled decomposition of the matrix. This

  stabilization effect becomes less and less pronounced on further entering in the $\alpha + \gamma$ - Fe dual - phase regime with the result that new martensite is formed beyond about 610 °C/10 h.

Recovery of the matrix (self diffusion inhibited) and carbide precipitation are delayed in the melt with the higher nitrogen concentration.

Carbide precipitation in the temperature range between 500 and 550 °C results in scarcely observable relative secondary hardening maxima.

The formation of stable austenite is accelerated in the nitrogen containing melt.

The magnetic properties react very sensitively to changes in the microstructure during tempering in the temperature range between 500 and 625 °C. Saturation permeability primarily correlates with the fraction of stable austenite and decreases with increasing austenite fraction.

Coercive force increases with tempering temperature. There is presumably a correlation between the intermediate peak and the secondary hardening maximum.

The coercive force approximately correlates inversely to the hardness of the material.

Although the correlations between the different microstructural parameters and mechanical and magnetic properties are complex, the formation of stable austenite seems to be the major factor affecting at least the magnetic properties of this steel.
Since the magnetic properties are clearly influenced by the chemical composition of the melt and the microstructural condition obtained after tempering, they can be analyzed to provide information on the material's microstructure and macroscopic properties.

These results show that more detailed information on the microstructure is to be expected from the measurement of other magnetic/electric properties (e.g. Barkhausen noise, magnetorestriction etc.).

5 References

/1/ P. Brezina
"Martensitische CrNi-Stähle mit niedrigem C-Gehalt",
Härterei Technische Mitteilung, 28 (1983) H. 5

/2/ J. Niederau
"Entwicklungsstand nichtrostender weichmartensitischer Stähle unter besonderer Berücksichtigung des Stahles X 5 CrNi 13 4"
Stahl und Eisen, 98 (1978) H. 8, S. 385 - 392

/3/ W. Gysel, E. Gerber, A. Trautwein
"G-X 5 CrNi 13 4: Neuentwicklung auf der Basis von 20 Jahren Erfahrung"
Konstruieren und Gießen, 7 (1982) Nr. 2

/4/ VdTÜV-Werkstoffblatt 394
Aufgabe von 08.84, Maximilian Verlag, Herford

/5/ W. Heimann, F.-H. Strom
"Eigenschaften und Gefüge einiger weichmartensitischer nichtrostender Chrom-Nickel-(Molybdän)-Stähle"
Thyssen Edelstahl
Technische Berichte, 8 (1982) H. 2, S. 115 bis 125

/6/ G. Maußner
"Gefügebau und Eigenschaften des nicht rostenden, weichmartensitischen Stahles 1.4313"
Tagungsband "Mikrostruktur and Gebrauchseigenschaften" der 12. Sitzung des Arbeitskreises Rastermikroskopie in der Materialprüfung des DVM (1987) SS. 173 - 193

/7/ A. L. Schaeffler
"Metal Progress", November 1949, S. 680, 680 B

6 List of Figures

Fig. 5: Microstructure after annealing at 610 °C/10 h, melt A, V_2A etchant

Fig. 6: Microstructure after annealing at 450 °C/10 h, melt A, TEM-thinfoil specimen

Fig. 7: Microstructure after annealing at 540 °C/10 h, melt A, TEM-thinfoil specimen

Fig. 8: Carbides precipitated during annealing at 520 °C/1 h, melt A, TEM extraction replica

Fig. 9: Carbides precipitated by annealing at 650 °C/1 h, melt A, TEM extraction replica

Fig. 10: Dislocation density, volume fraction of stable austenite and nickel equivalent according to Schaeffler in the austenite versus the annealing temperature

Fig. 11: Average diameter and chromium concentration of the $M_{23}C_6$-Carbides versus the annealing temperature

Fig. 12: Hardness as a function of the annealing temperature, melt A

Fig. 13: Hardness as a function of the annealing temperature, melt B

Fig. 14: Tensile strength (Rm), yield strength ($R_{p1.0}$), fracture elongation (A_5), reduction of area (Z) and impact toughness (Av) as a function of annealing temperature, melt A

Fig. 15: Tensile strength (Rm), yield strength ($R_{p1.0}$), fracture elongation (A_5), reduction of area (Z) and impact toughness (Av) as a function of annealing temperature, melt B

Fig. 16: Magnetic flux density and coercive force as a function of the annealing temperature, time on temperature = 10 h, melt A

Fig. 17: Magnetic flux density and coercive force as a function of the annealing temperature, time on temperature = 10 h, melt B

Fig. 18: Magnetic domains in the microstructure after annealing at 450 °C/10 h, melt A

Fig. 19: Magnetic domains in the microstructure after annealing at 610 °C/10 h, melt A

Fig. 1: Microstructure after annealing at 505°C/10 h, melt B, colour etched

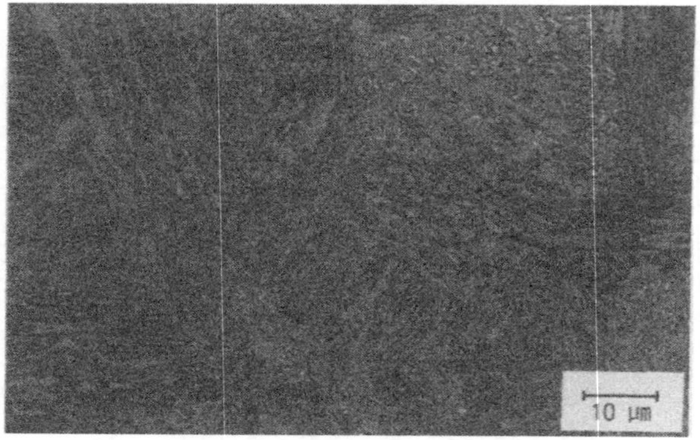

Fig. 2: Microstructure after annealing at 610°C/10 h, melt B, colour etched

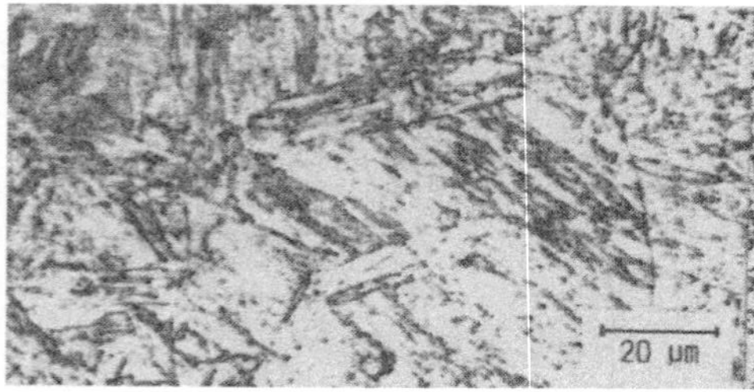

Fig. 3: 505°C/10 h

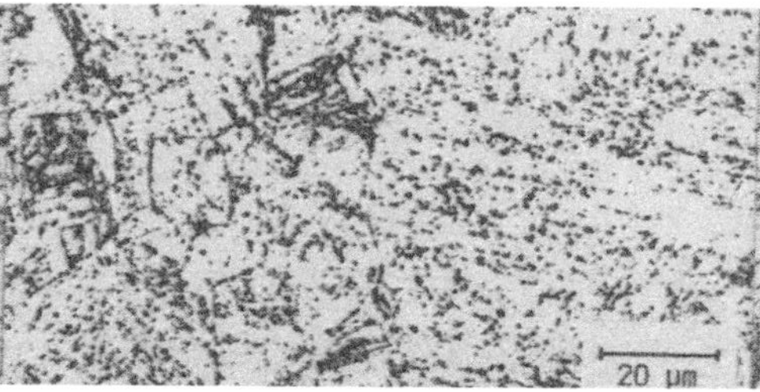

Fig. 4: 540°C/10 h

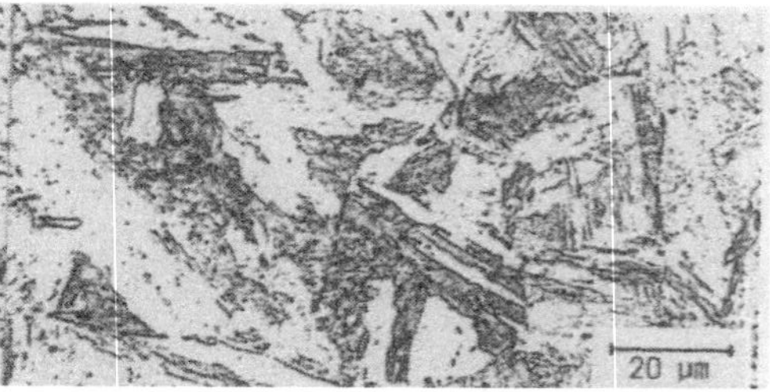

Fig. 5: 610°C/10 h

Fig. 3 - 5: Microstructure after annealing, melt A, V_2A etchant

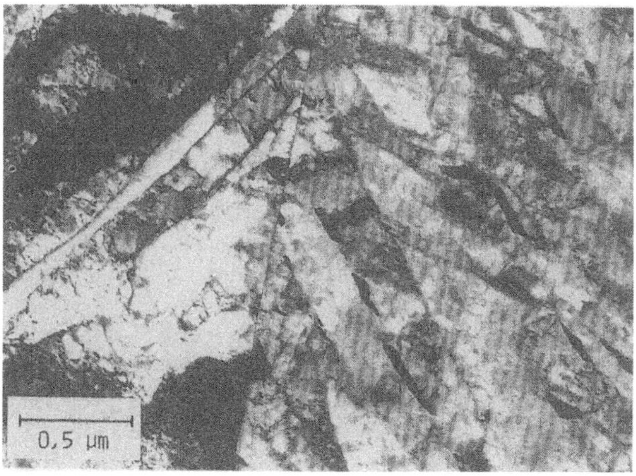

Fig. 6: Microstructure after annealing at 450 °C/10 h, melt A, TEM-thinfoil specimen

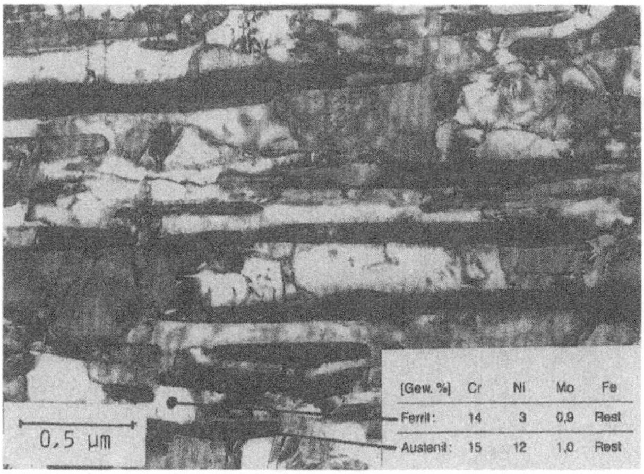

Fig. 7: Microstructure after annealing at 540 °C/10 h, melt A, TEM-thinfoil specimen

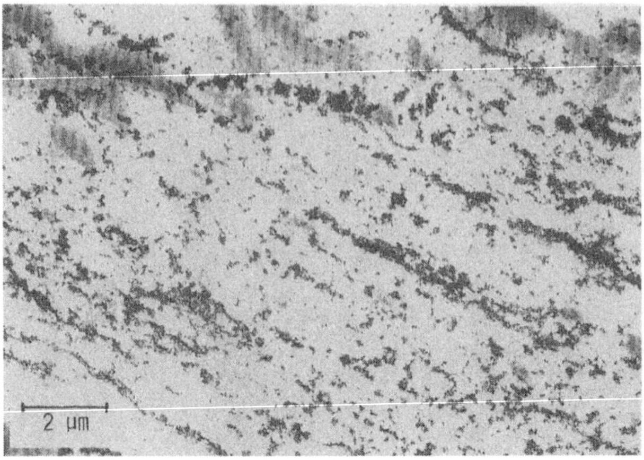

Fig. 8: Carbides precipitated during annealing at 520 °C/1 h, melt A, TEM extraction replica

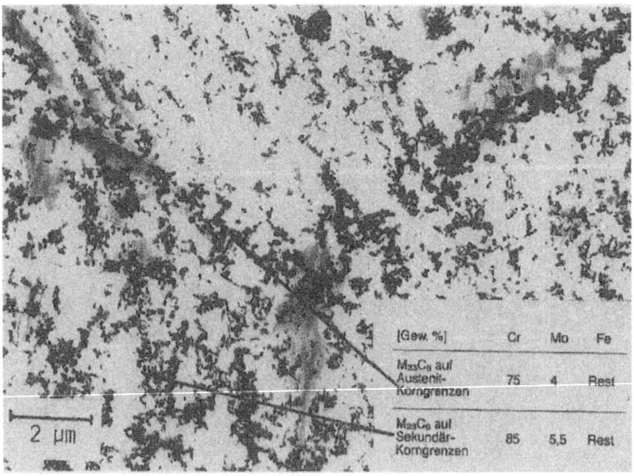

Fig. 9: Carbides precipitated by annealing at 650 °C/1 h, melt A, TEM extraction replica

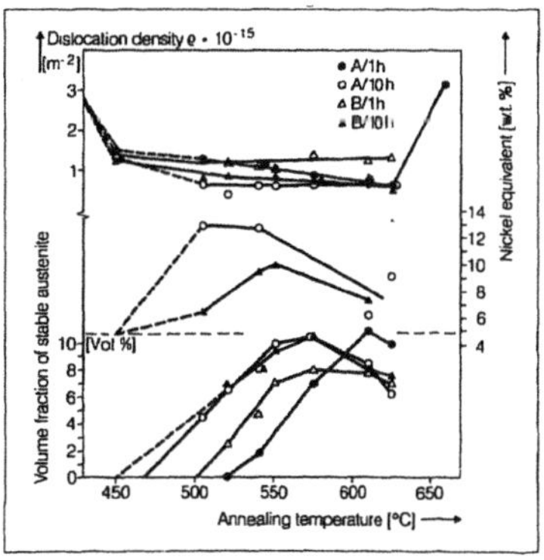

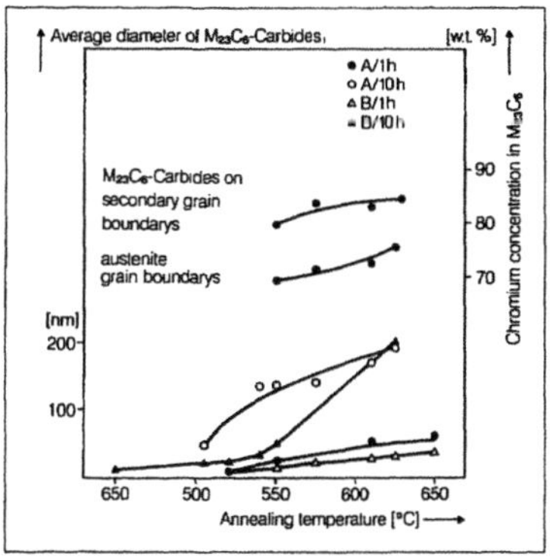

Fig. 10: Dislocation density, volume fraction of stable austenite and nickel equivalent according to Schaeffler in the austenite versus the annealing temperature

Fig. 11: Average diameter and chromium concentration of the $M_{23}C_6$-Carbides versus the annealing temperature

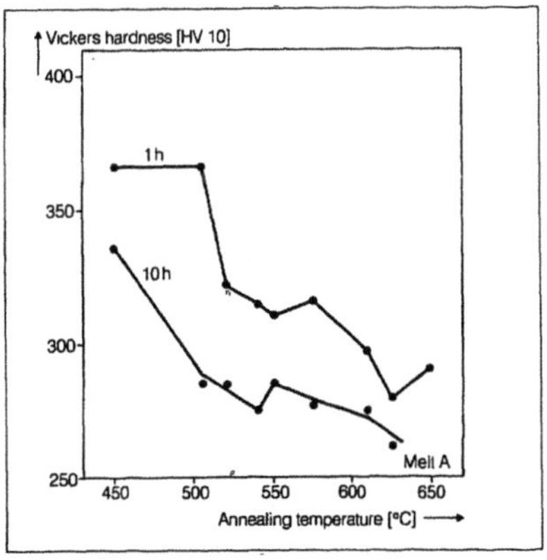

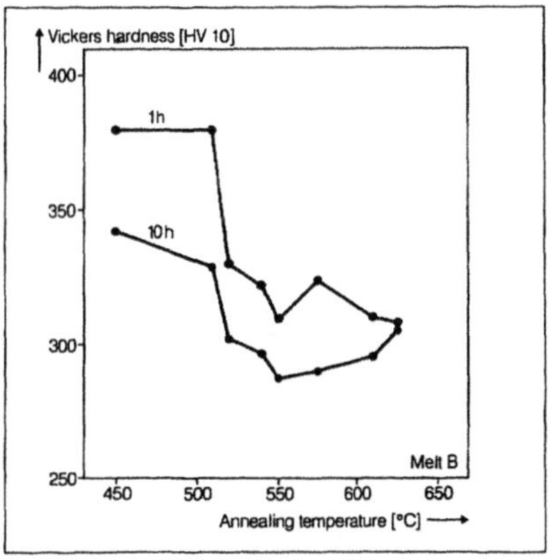

Fig. 12: Hardness as a function of the annealing temperature, melt A

Fig. 13: Hardness as a function of the annealing temperature, melt B

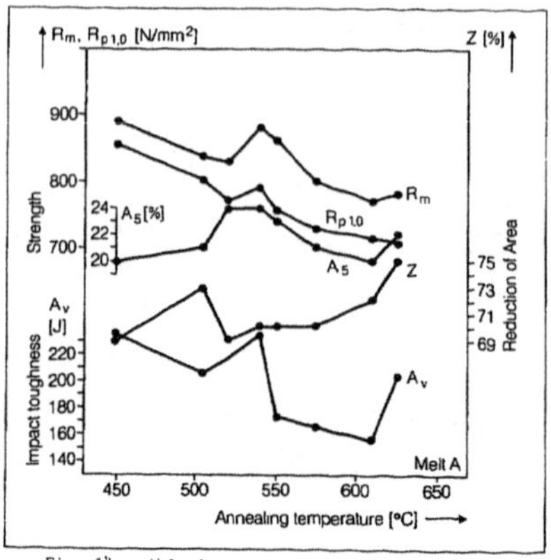

Fig. 14: Melt A

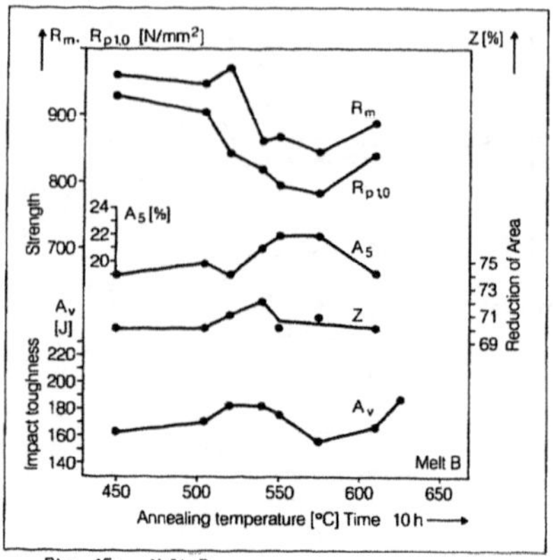

Fig. 15: Melt B

Fig. 14: Tensile strength (Rm), yield strength ($R_{pl.0}$), fracture elongation (A_5), reduction of area (Z) and impact toughness (Av) as a function of annealing temperature, melt A

Fig. 15: Tensile strength (Rm), yield strength ($R_{pl.0}$), fracture elongation (A_5), reduction of area (Z) and impact toughness (Av) as a function of annealing temperature, melt B

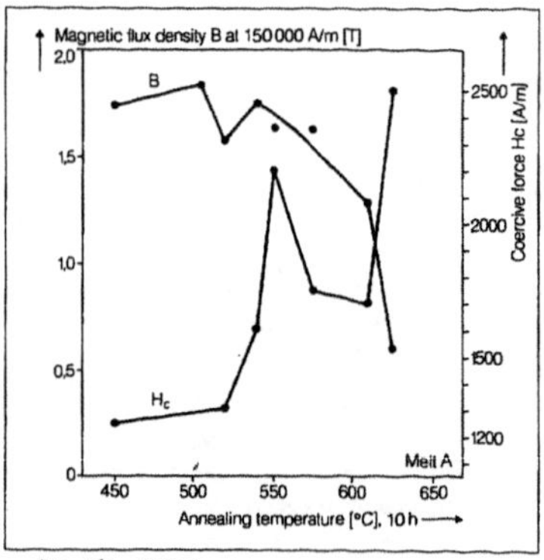

Fig. 16: Melt A

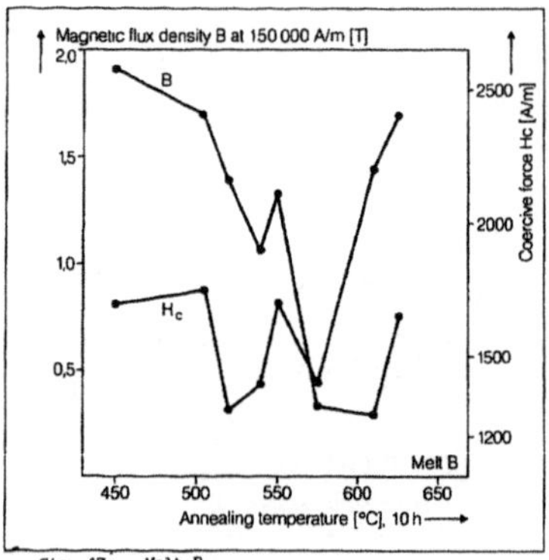

Fig. 17: Melt B

Fig. 16: Magnetic flux density and coercive force as a function of the annealing temperature, time on temperature = 10 h, melt A

Fig. 17: Magnetic flux density and coercive force as a function of the annealing temperature, time on temperature = 10 h, melt B

Fig. 18: Magnetic domains in the microstructure after annealing at 450 °C/10 h, melt A

FIg. 19: Magnetic domains in the microstructure after annealing at 610 °C/10 h, melt A

Magnetic Tangential Field-Strength-Inspection, a Further NDT-Tool for 3MA

G.DOBMANN*, H.PITSCH**

* Fraunhofer-Institut für zerstörungsfreie Prüfverfahren (IzfP), Saarbrücken
** Robert Bosch GmbH, Stuttgart

Introduction

For the real-time quality control in the production process the signification of ndt-methods is more and more increasing. In order to solve these problems within short cycle times with micromagnetic techniques (magnetic Barkhausen noise, incremental permeability) it is necessary to use excitation frequencies for alternating magnetization in the range greater than 10 Hz. According to these restrictions the fore mentioned inspection techniques are performed by the 3MA-analyzer /2/. In this range of exciting frequencies the analyzing depth of these methods is restricted to surface near material zones (ca. 300-400μm). The up to now existing 3MA-draft is expanded by two new measured quantities in the following. These quantities are derived by upper harmonic analysis in the time signal of the tangential field strength during alternating magnetization. In this case the analyzing depth is only restricted by the penetration depth of the exciting magnetic field.

Possibilities to solve inspection problems by the fore mentioned ndt-techniques are demonstrated by several examples.

Examples of application:
- Steel grading and separation of microstructure variations
- Characterization of tempering states
- Measurement of case hardening depth (Eht)
- Measurement of hardness (HV10)

(3MA: Micromagnetic Multiparameter Microstructure- and Stress-Analysis)

Physical background

The physical basics of the measured quantities derived by magnetic Barkhausen noise- (maximum value: M_{max}; coercivity: H_{cM}), incremental permeability- (maximum value: $\mu_{\Delta max}$; coercivity: $H_{c\mu}$) measurement are described in /1/. These measurements are performed by the 3MA-analyzer /2/; see Fig. 1a. Fig. 1b shows the yoke with integrated sensors (coil: measurement of micromagnetic quantities; Hall generator: measurement of tangential field strength).

For the evaluation of the coercivity using micromagnetic measurements the parallel measurement of the tangential magnetic field strength is necessary. This is done due to a commercial Hall generator /3/. The inductance of a yoke depends on the material of the ferromagnetic core and the magnetic properties of the specimen under test (magnetic circuit). Furthermore the permeability of ferromagnetic materials depends on magnetic field strength (influence of ferromagnetic hysteresis).

As discussed in /4/, during alternating magnetization a non-sinusoidal magnetic field strength within the specimen is obtained. Because of the influence of hysteresis, only the odd upper harmonics are generated beside the exciting frequency /4/. Some applications of upper harmonic analysis in ndt are described in /5,6,8/. These effects can be observed and used also in the time signal of the tangential field strength.

In the following two measured quantities are introduced which are derived from the upper harmonic analysis of this signal. The measurement was performed by Fourier analysis. In nearly all cases of application for the reconstruction of the time signal of the tangential field strength within sufficient accuracy only the complex Fourier coefficients up to the seventh order have to be respected. The amplitudes of higher harmonics are in the resolution range of a 10 bit-ADC.

The experimental results show that the following two measured quantities are reliable for non-destructive testing and can be used for the characterization of different material states:

- The amplitudes of the upper harmonics. According to /7/ the distortion factor K is derived as an inspection quantity.

  Definition:

$$K := \sqrt{\frac{|\tilde{A}_3|^2 + |\tilde{A}_5|^2 + |\tilde{A}_7|^2}{|\tilde{A}_1|^2}} \cdot 100\ \%$$

with:

$\tilde{A}_1$: complex Fourier coefficient of exciting frequency

$\tilde{A}_3, \ldots, \tilde{A}_7$: complex Fourier coefficients of the corresponding harmonic

- The evaluation of the time signal obtained by the phase-true superposition of the upper harmonics allows the determination of the coercivity. The first zero of this time signal after the remanence gives the trigger to measure the coercivity H_{co} which is the value of the tangential field strength at this point; see Fig. 2.

By the analytical discussion of the hysteresis characteristics i.e. the behaviour of the differential permeability, the additional inflection point observed in the tangential magnetic field time signal is understood. The field value at this inflection point is the coercivity of the hysteresis. There exists a strong correlation between this real H_c-value and the H_{co}-value named above, whereas the determination of H_{co} can be performed in a simple way.

Fig.3 documents the correlation between the defined coercivity H_{co} and the coercivity determined by hysteresis measurement H_{cB-H} (f_E: excitation frequency; H_{max}: maximum value of tangential field strength; r: correlation coefficient after linear regression).

In Fig.4 the distortion factor K is correlated with the maximum of electrical voltage which is induced in a coil encircling the specimen. It is obvious that the quantities K and H_{co} determined in picking up technique, are well correlated with quantities determined in circumferential technique by hysteresis measurements.

Experimental results and discussion

For all experiments the excitation frequency f_E=50Hz was used. The maximum value of tangential field strength (H_{max}) was adjusted to the special problem of inspection.

- The grading of steels and different microstructure states is demonstrated in Fig. 5.

Examined steels and heat treatments ([...]):
(Specified according to DIN 17014)
C15 [N, GKZ-N]
C45 [N, GKZ-N]
C100 [N, GKZ-N]

Fig.5 demonstrates the ability to separate steels and microstructure states, analyzing nomograms K versus H_{co} (reproducibility of the measurement: K ± 0.1%; H_{co} ± 0.1 A/cm).
Further examples are described in /9/.

- Fig.6 shows the dependence of K respectively H_{co} as function of the tempering temperature of hardened steel 100 Cr 6.
  Especially the high resolution within the range up to 300°C is significant and important for practical applications.
  Further examples are described in /9/.

- The measurement of case hardening depth (Eht) was carried out at the steel 16 MnCr 5.
  Fig.7a shows the value of distortion factor K with respect to the metallographic determined case hardening depth.
  Fig.7b shows the coercivity H_{co} as a function of case hardening depth. Besides two measuring points, this function is unique. The specimens according to these two points are characterized by a coarse-grained microstructure (metallographic result). Generally the coercivity depends on grain size like $H_c \sim 1/\sqrt{d}$ (d: grain diameter).
  Restricted on the given steel grade, the calibration curves in Fig. 7a,7b allow at first the unique determination of the case hardening depth and in addition to this, the variation of undesired microstructures are recognized.

- The result of a hardness measurement is shown in Fig.8. The samples were ferromagnetic steel plates with different hardness values (HV10). These plates should be separated in classes according to hardness. Experimental conditions are optimized so that especially the class 500 - 600 HV10 can be characterized significantly.

Summary

The described experimental results underline the reliability and performance of the defined inspection quantities coercivity H_{co} and distortion factor K as a further ndt-tool for 3MA.

The upper harmonics analysis offers the possibility to expand the analyzing depth of magnetic measurements up to the range of the penetration depth of the exciting magnetic field (typically some millimeters).

Whereas the micromagnetic quantities derived from magnetic Barkhausen noise and incremental permeability for magnetizing frequencies greater than 10 Hz are restricted on analyzing depth of some hundred microns, the new 3MA modules H_{co}, K are a completion to deeper analyzed surface near sheets.

References

1. Theiner, W.A., Altpeter, I.: Determination of Residual Stresses Using Micromagnetic Parameters. New Procedures on Nondestructive Testing; Springer Verlag; Berlin, Heidelberg; 1983

2. Theiner, W.A., et al.: Stress measurements on components with nondestructive ferromagnetic methods. Residual Stresses in Science and Technology, S.167-174; DGM Informationsgesellschaft mbH; Oberursel, 1987

3. Siemens Datenblatt: SV 200 C

4. Koch, K.M.; Jellinghaus, W.: Einführung in die Physik der magnetischen Werkstoffe. Verlag Franz Deuticke; Wien; 1957

5. Ershov, R.E., et al.: Nondestructive Testing of hardness of grey iron casts by upper harmonics. Materialprüfung, 24, Nr.10, S.363-366, 1982

6. Ershov, R.E., et al.: Testing cast iron and steel by the upper harmonics mágnetic method with simultaneous heating. NDT International, Vol.18, No.4, S.206-208, August 1985

7. Christian: Magnettontechnik; Franzis Verlag; München; 1969

8. Institut Dr.Förster. Bedienungsanleitung Magnatest S

9. Pitsch, H., Jonck, R., Höller, P.: 3MA-Verfahrensansätze zur zerstörungsfreien Klassierung von Stählen. DGZfP-Jahrestagung, Siegen, 09.-11.Mai 1988, S.190-209

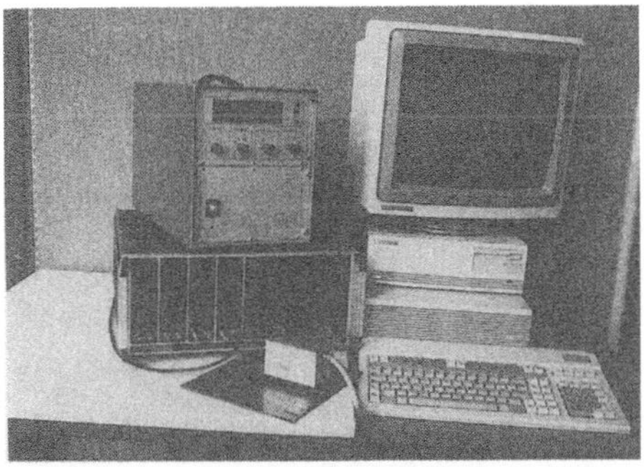

Fig. 1a: 3MA-Analyzer

Fig. 1b: Yoke with integrated sensors

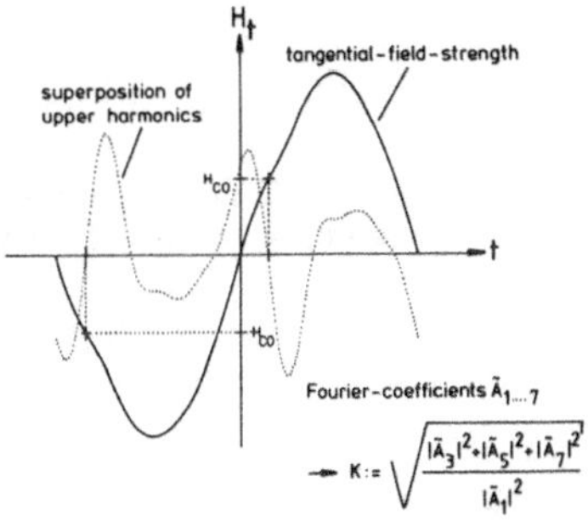

Fig. 2: Deduction of the measured quantities

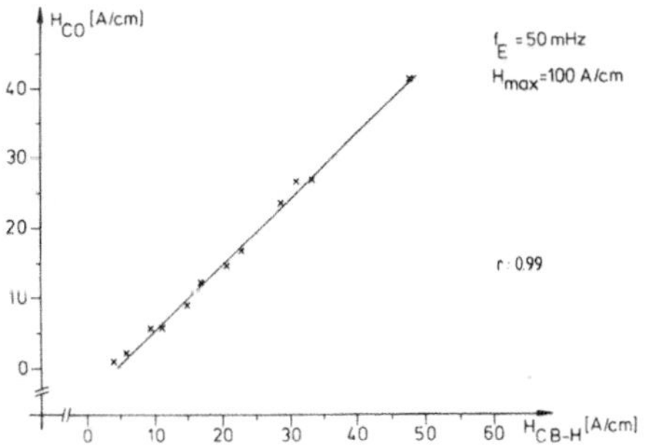

Fig. 3: Correlation between H_{co} and H_{cB-H}

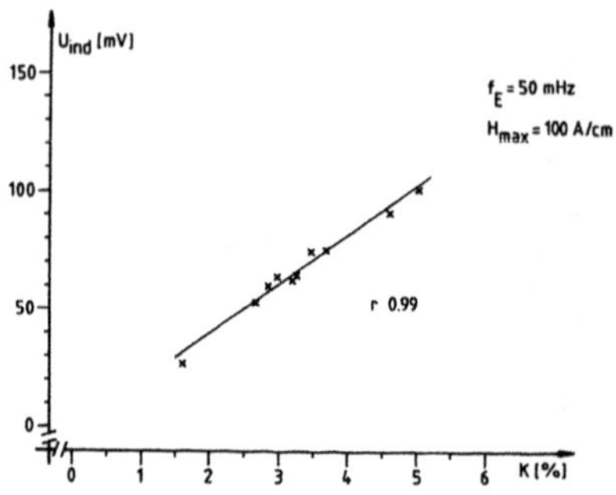

Fig.4: Correlation between induced voltage and K

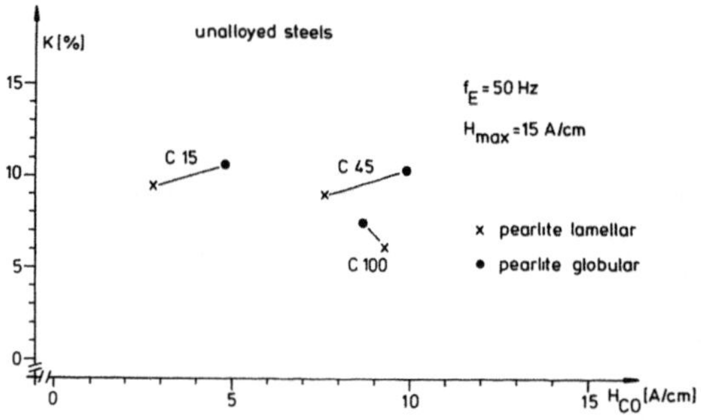

Fig.5: Steel grading, unalloyed steels

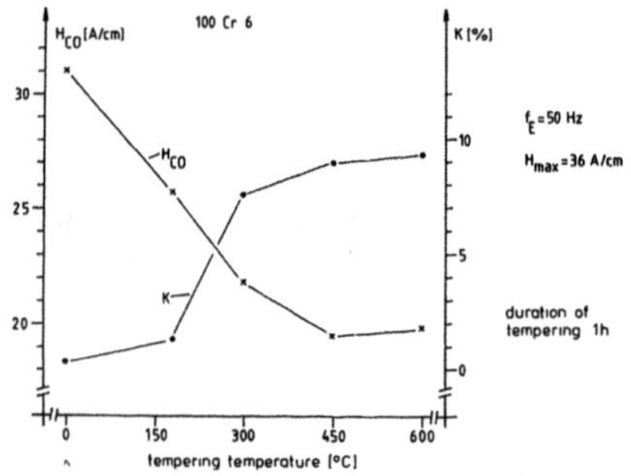

Fig.6: Characterization of tempering states, 100 Cr 6

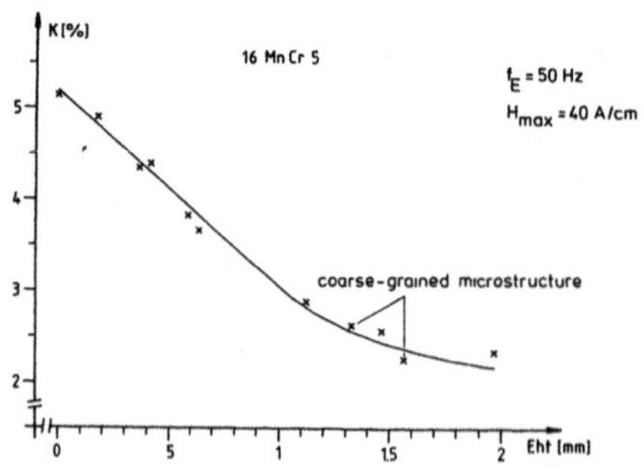

Fig.7a: Distortion factor K as a function of case hardening depth

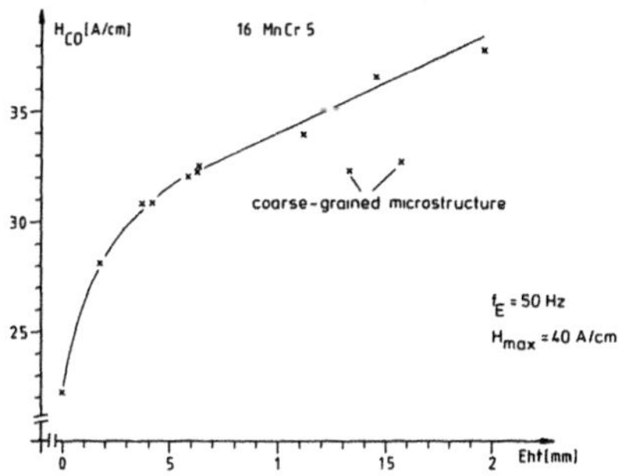

Fig.7b: Coercivity H_{co} as a function of case hardening depth

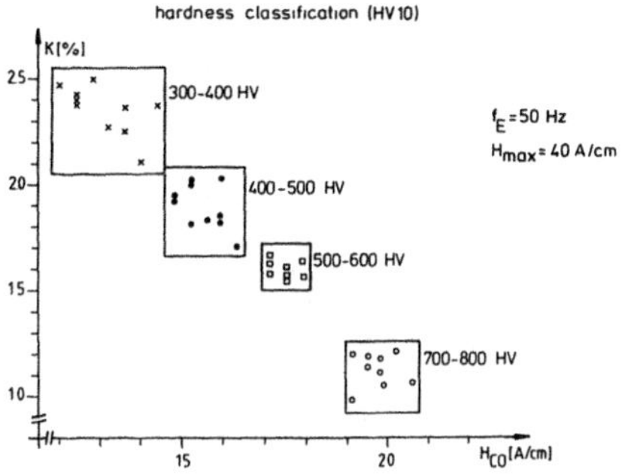

Fig.8: Hardness measurement

A Modulus for the Evaluations of the Dynamic Magnetostriction as a Measured Quantity of the 3MA Method

R. Koch

NUKEM GmbH, Alzenau

P. Höller

IzfP, Saarbrücken

Summary

The principle function of an apparatus for the evaluations of the dynamic magnetostriction as a measured quantity of the 3MA method is discussed. An ultrasonic wave is generated by an electromagnetic acoustic transducer due to a magnetostrictive interaction mechanism. This wave is received by an electromagnetic acoustic transducer due to an inductive interaction mechanism. Amplitude and phase of the ultrasonic pulse are evaluated depending on the magnetic field strength of the transmitter, keeping constant the magnetic field of the receiver. Measurement results like the influence of stress or the influence of texture on amplitude and phase curves of the dynamic magnetostriction are demonstrated.

Zusammenfassung

Es wird ein Gerätekonzept vorgestellt, mit dem die dynamische Magnetostriktion zur Charakterisierung von Struktur- und Spannungszuständen genutzt werden kann. Mittels elektromagnetischer Ultraschallwandler wird über einen magnetostriktiven Wechselwirkungsmechanismus eine Ultraschallwelle angeregt, die in einiger Entfernung vom Ort der Sendespule über einen induktiven Wechselwirkungsmechanismus empfangen wird. Amplitude und Phase der angeregten Ultraschallwelle werden in Abhängigkeit von der Magnetfeldstärke am Ort der Sendespule bei konstanter Feldstärke am Ort der Empfangsspule aufgenommen und ausgewertet. Anhand von Meßergebnissen wird die Struktur- und Spannungsabhängigkeit der Amplituden- und Phasenkurven der dynamischen Magnetostriktion verdeutlicht.

Physical Backround

Beside of the coercivity, the Barkhausen-noise (magnetic or acoustic) and the incremental permeability, there exists also the dynamic magnetostriction usable as a measured quantity of the 3MA (Micro Magnetic Multiparameter Analyses) method.

The magnetostriction itself shows a strong dependence on structure and stress, but there is no easy way to measure magnetostriction. Like the magnetostriction, the curves of the dynamic magnetrostriction - the superposition of a quasistatic and an alternating magnetic field - are influenced by structure and stress. With an electromagnetic acoustic transducer there exists a possibility to use the dynamic magnetostriction in setting up technique free of coupling medium. The quasistatic magnetic field is generated with an electromagnet and the alternating field with a coil of wire carrying a dynamic current. Amplitude and phase of an ultrasonic signal generated in this kind are directly correlated to the incremental magnetostriction.

During studies of the magnetostrictive generation of ultrasonic surface and guided waves, R. B. Thompson (1) first used the efficiency of this generation for the detection of strain. Because he worked with an EMAT meander coil transmitter and a receiver which was a standard piezoelectric wedge, he had to use coupling medium with all negative influences on the signal amplitude. Also, in case of specimen thickness changes, he had to change the wave length of the meander coil.
W. Theiner (2) used the dynamic magnetostriction for the measurement of stress by an apparatus using an EMAT meander coil transmitter and receiver under the same yoke. So he could work free of coupling medium, but he was dependent on the shape of the specimen because he needed an edge echo as received ultrasonic signal.

With the apparatus presented in this paper, it is possible to use the dynamic magnetostriction as a measured quantity without these disadvantages.

Principle function of the apparatus

Figure 1 shows the recording unit of the 3MA dynamic magnetostriction modulus. An adjustable quasistatic magnetic field (frequency 50 Hz) parallel to the surface of the specimen is generated by an electromagnet. The geometry of the yoke is designed in a way that the outside leakage field is as low as possible. The magnetic field strength is measured by a hall probe. A line coil EMAT transmitter is used. The direction of the dynamic current through the coil is perpendicular to the applied magnetic field. The trigger for the initial pulse (frequency 300 kHz) is set in the turning point of the hysteresis loop. The ultrasonic pulse, generated magnetostrictively in this way, is received by an EMAT meander coil transducer working due to an inductive mechanism. A permanent magnet in the receiver is responsable for a high magnetic field perpendicular to the surface of the specimen. The distance between transmitter and receiver is fixed.

For a high resolution on the surface of the specimen, it is necessary to use a small line coil transmitter. Depending on the specimen thickness, this broad band transmitter generates guided waves as well as surface waves and transverse and longitudinal waves in different angle directions. So a narrow band meander coil receiver is chosen to filter the surface wave or a guided wave. The bandwith of the transmitter and the receiver is set in such a manner that it is possible to generate and receive a guided wave nearly independent from the specimen thickness, only causing a small shift of the centre frequency. So the penetration depth (about 0.1 mm for a 300 kHz inital pulse frequency) is constant.

Amplitude and phase of the received ultrasonic pulse are evaluated depending on the magnetic field strength of the transmitter, keeping constant the magnetic field of the receiver. Figure 2 shows the blockdiagram of the modulus. Using this system,it is possible to evaluate the dynamic magnetrostriction in settiñg up technique free of coupling medium with a high surface resolution and independent from the specimen thickness.

Measurement of amplitude and phase curves

Amplitude and phase curves are automaticly measured under the control of a microprocessor.

Before the start of a measurement, the system has to be calibrated with a specimen of the desired thickness. For this calibration, first a high magnetic field (200 A/cm) is set where the electromagnetic ultrasonic generation still works magnetostrictive. For higher fields there is an increasing influence from Lorentz forces. The A-scan of the signal received for this field strength is digitised (figure 3). Start and end of the time window for different modes (Rayleigh or Lamb waves) are automatically set by the software. This is necessary because the time of flight for the guided waves is changed with the probe thickness. Then the centre frequency of the signal received in the time window is calculated by FFT. This is done to eliminate the influence of from dispersion of guided waves. Amplitude and phase of the received signal have to be evaluated for the centre frequency because, in the region of lower magnetic fields, the spectrum of the magnetostrictive ultrasound generation is changed by irreversible magnetic processes. Figure 4 shows also the calculated spectrum. For the rayleigh wave, there is no influence of the specimen thickness on the spectrum, so the centre frequency of the received signal is the same as for the initial pulse.

After the calibration, it is possible to measure amplitude and phase curves. The magnetic field is increased step-by-step from 0 to 200 A/cm. The signal, received in the time window, is digitisated and, for each field strength, amplitude and phase are calculated. Figure 4 shows the amplitude and phase curves for a steel type St 37, 100 mm thick, belonging to the A-scan and spectrum in figure 3.

Measurement results

For the demonstration of the efficiency of the apparatus, some measurements have been done to show the influence of stress and texture on the magnetostrictive amplitude and phase curves.

A point in the amplitude curve, very sensitive for small tensile and for all compression stresses, is the magnetic field strength position of the relative minimum. For measuring the residual stress in the heat-affected zone of a weld of a steel type 22NiMoCr3.7, 4 mm thick, a calibration curve was fitted up with a specimen of the same steel type and a thickness of 10 mm. Therefore the change of the magnetic field position of the relative minimum dependent on an applied stress is registered.

The upper part of figure 5 shows the measured values of the magnetic field strength position of the relative amplitude minimum versus the distance to the centre of the weld. The orientation of the transmitter is parallel to the weld, so that the residual stress perpendicular to the weld is measured. The lower part of figure 5 shows the calibrated stress values of the dynamic magnetrostriction and the results of X-ray measurement. The difference in the curves may be explained by the different penetration depth (X-ray about 0.02 mm, dyn. magnetrostriction about 0.1 mm) and the different spatial resolution on the specimen surface for both methods.

The texture measurement is done by turning the recording unit of the modulus over the specimen surface around the centre point of the transmitter coil. The specimen is a cold rolled strip of the steel type St 37 with a thickness of 0.77 mm. In figure 6 the phase for a magnetic field strength of 200 A/cm is drawn. Assuming that there is a small tensile stress on the surface in the region of the penetration depth of this method (results of measurements with applied stresses on the specimen), it is possible to correlate this "pole figure" of the dynamic magnetostriction to the texture. For high magnetic fields and tensile stress the influence of structure is greater than the influence of stress. Outgoing of the direction dependence of the magnetostriction for a stress free monocrystalline iron it is possible to postulate, that there is a great shift in phase between the (100) and (110) direction for higher magnetic field strengths. Because, for the examined specimen, the $(11\overline{2})$ direction is parallel and the $(1\overline{1}0)$ is perpendicular to the rolling direction (results of X-ray measurements), the symmetry of the "pole figure" can be explained and correlated to the texture.

But for an exact separation of the influence of stress and texture and a correlation specially to the r_m-values of the X-ray measurements, more experiments are necessary. Also the correlation to other micro magnetic parameters of the 3MA method have to be investigated in the future.

adjustable quasistatic magnetic field parallel to the surface of the specimen

high static magnetic field perpendicular to the surface of the specimen

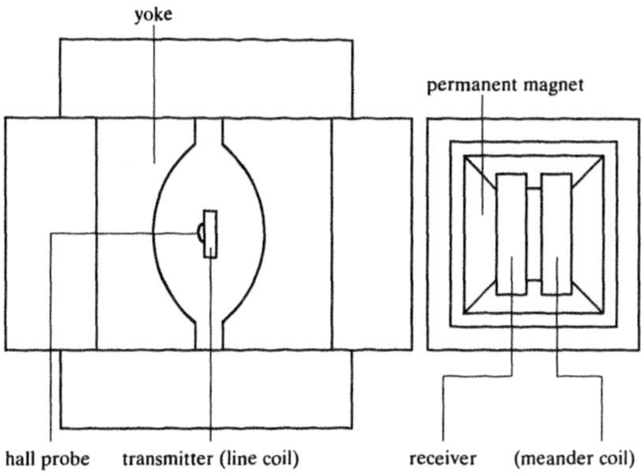

Fig. 1 recording unit of the 3MA dynamic magnetostriction modulus

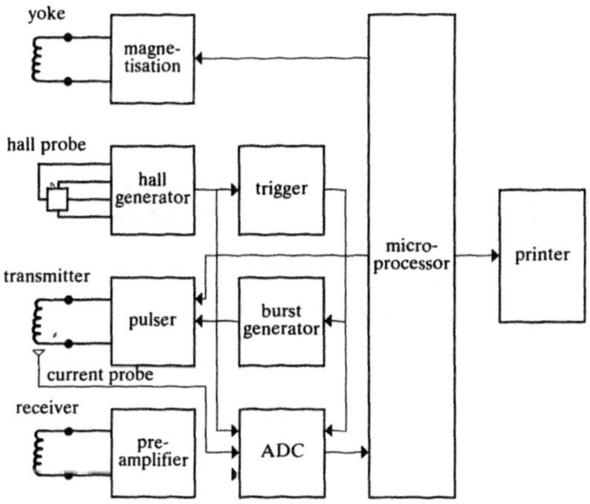

Fig. 2 blocdiagram of the 3MA dynamic magnetostriction modulus

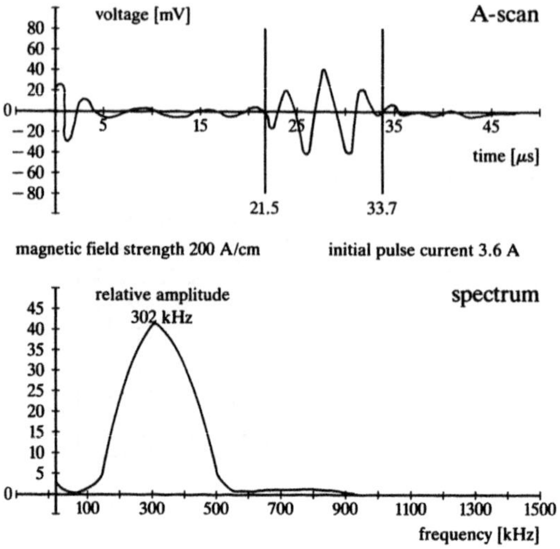

Fig. 3 A-scan and spectrum (rayleigh wave), specimen of a steel type: St 37, 100 mm thick

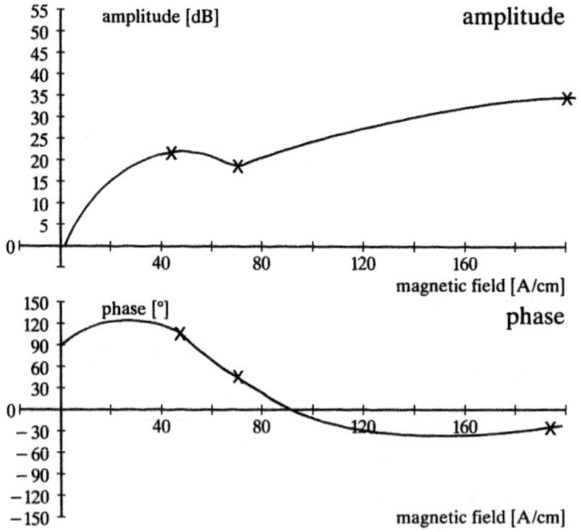

Fig. 4 amplitude and phase curves, specimen of a steel type: St 37, 100 mm thick

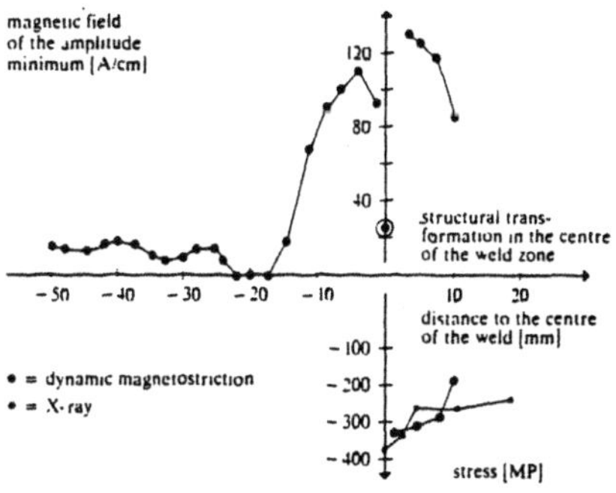

Fig. 5 measurements of residual stress perpendicular to the weld specimen of steel type: 22 NiMoCr3 7, 4 mm thick

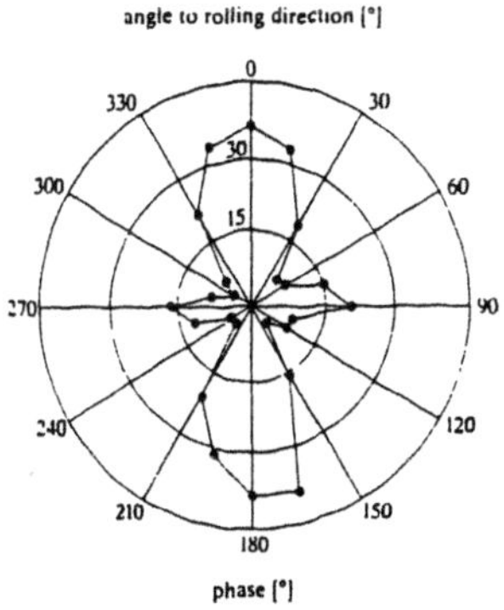

Fig. 6 "pole figure of the dynamic magnetostriction" measuring the phase for a magnetic field strength of 200 A/cm, specimen of cold rolled strip, steel type: St 12, 0.77 mm thick

Acknowledgement

This work was supported by the Arbeitsgemeinschaft Industrieller Forschungsvereinigungen (AIF).

References

1. Thompson, R.B.: Detection of strain by measurement of the efficiency of the magnetostrictive generation of ultrasonic surface waves, 1976 Ultrasonic Symposium Proceedings, p.585.
2. Theiner, W., Crossmann, J., Repplinger, W.: Nondestructive evaluation of stresses especially residual stresses by magnetostrictive excitation of ultrasonic waves, by magnetostriction measurements with strain gauges and especially by Barkhausen-noise, DGZfP, Int. Symp., 1979, Saarbrücken.

Nuclear Magnetic Resonance (NMR)

A Review of Nondestructive Characterization of Composites Using NMR

George A. Matzkanin

Texas Research International, Inc.
Austin, Texas, 78733 U.S.A.

Introduction

Nuclear Magnetic Resonance (NMR) is a nondestructive evaluation technique useful for characterizing organic matrix composites and other polymer based materials. NMR depends on the interaction between the nuclear magnetic moment and a magnetic field and thus it is sensitive to localized field variations caused by molecular motions, changes in molecular or crystal structure, and chemical composition differences. Application of NMR to composites and polymers involves measurement of the hydrogen nucleus (proton) NMR signal. Fortunately, the proton NMR signal is very strong and easily measured. Much of the physical and chemical information available through the use of NMR is associated with the relaxation characteristics of the nuclear magnetic moments, which can be measured using pulsed NMR techniques. The energy exchange between nuclear moments and the surrounding lattice is characterized by the spin-lattice relaxation time, T_1, while the energy exchange among nuclear magnetic moments is described by the spin-spin relaxation time, T_2. These relaxation times are very sensitive to molecular motions and structural changes and can be used to provide both qualitative and quantitative information on the dynamic environment in which the nuclei are located. For application to composites and polymers, hydrogen NMR has been used to characterize water absorption, molecular diffusion, environmental degradation, aging, degree of cure, and modulus variations. Composite materials which have been evaluated using NMR include graphite, glass and Kevlar reinforced composites in both cured and prepreg conditions, neat resins, and adhesives. Instrumentation approaches have been developed for practical application of NMR to nondestructive inspection problems involving composite components and structures. With this instrumentation, NMR measurements are made from a single surface and by controlling the magnetic field gradients, spatially varying properties such as moisture gradients and material inhomogeneities can be mapped out. In addition, the rapid, recent development of magnetic resonance imaging (MRI) for medical diagnostics has opened up new possibilities for the potential development of MRI for the NDE of advanced materials and composites. The ability of NMR to sense the local magnetic field at the molecular level provides a tool for nondestructively determining the chemical state of individual components within the material. Although the principal application of MRI thus far has been to image biological samples containing water, work is underway to develop NMR imaging for solids. NMR imaging can then potentially produce internal maps in solids of chemical state variations associated with internal discontinuities such as cracks, disbonds, fiber-matrix interfaces, and chemical variations associated with cure, strain, temperature, and density gradients. In the following sections reviews, including an extensive list of references, of recent and current work on nondestructive characterization of composites using NMR and developments of NMR imaging for solids will be presented.

Characterization of Composites and Polymers Using NMR

A brief computerized literature review was conducted in several data bases to identify journal articles and other documents dealing with the use of NMR and NMR imaging for the nondestructive evaluation or characterization of composites and polymers. The results of this search are summarized in Table 1, which lists the computer files searched, the time span involved and the number of documents found. After eliminating duplicates, a total of approximately 75 relevant documents resulted from this search. In some cases, the use of NMR techniques was peripheral to the main thrust of the reported work while in other cases, NMR was the primary focus. Recent, key articles in various

Table 1. NMR bibliographic literature search

Search Strategy:
1. Combine NMR or MRI terms with composite or polymer terms
2. Combine the results of step 1 with nondestructive (NDT) or material characterization terms

Results:

Files	Time Span	No. of Documents
Aerospace	1962-1988	32
Engineered Materials Abstracts	1986-1988	19
INSPEC	1977-1988	21
NTIS	1964-1988	18
	TOTAL	**90**

Estimated number of non-duplicates approximately 75

NMR areas are referenced in subsequent tables and a good recent review of NMR for chemical spectroscopy of composites and polymers can be found in Reference 1.

In reviewing the literature, one finds that the most investigated area from an NDE standpoint is the detection and characterization of moisture in composites and polymers. As noted on the list included in Table 2, most of the principal organic matrix composite systems have been investigated, as well as some related materials. It is known, of course, that moisture intrusion can have an adverse effect on the structural integrity of organic matrix composite materials. [14] Not only does moisture migration along the fiber-matrix interface weaken the interface bond, but moisture also can react with the organic matrix leading to degradation of the polymeric structure. Both interactions result in a deterioration in mechanical properties of the composite. In some cases, such as Kevlar reinforced composite, this potential problem is complicated by the fact that the Kevlar fiber itself is hygroscopic and absorbs moisture. In addition to the nondestructive measurement of the amount of absorbed moisture, NMR signals have been analyzed to determine the feasibility of associating certain signal characteristics with moisture absorbed in different physical states within the composite. [2]

Table 2. Application of NMR for moisture measurement in composites

1. Glass/Epoxy [2,3,4]
2. Kevlar/Epoxy; Kevlar Fiber [4,5,6]
3. Graphite/Epoxy [6,7,8,9]
4. Adhesives [10]
5. Miscellaneous - Circuit Boards; Wire Cable; Epoxy [2,11,12,13]

Typically, for a composite containing absorbed moisture, a multiplicity of NMR signal components is observed. [2,3] Shown in Figure 1 is a schematic drawing of a typical free-induction-decay signal following a transmitted RF pulse of approximately 15 μs duration. Following the end of the pulse is a period of time (about 10 μs) where the residual energy of the intense RF pulse is dissipated and the receiver circuits recover. Detection of signals produced by all the hydrogen nuclei within the NMR sensor coil occurs in the subsequent time interval, i.e., 25 to 100 μs. The signals generally consist of an initial large-amplitude, fast-decaying component associated with the chemically bound, structural hydrogen in rigid polymer molecules, and a lower-amplitude, slower-decaying component associated with hydrogen in more mobile absorbed moisture molecules. An example of an NMR free-induction-decay signal obtained at 30 MHz from the hydrogen in an organic matrix composite containing absorbed moisture is shown in Figure 2 for a Kevlar-reinforced epoxy composite. In this Figure, the fast-

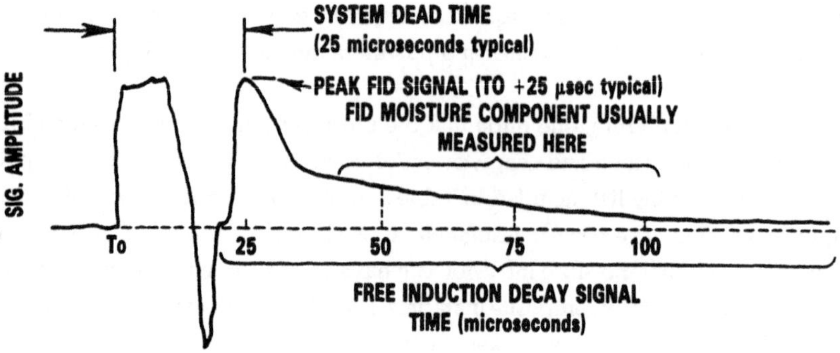

Fig. 1. Typical FID signal from composite with absorbed moisture

decaying solid component has a relaxation time of T_{21} = 6 μs. Association of the remaining slower-decaying component with absorbed moisture is easily verified by the absence of this component in oven-dried specimens. Close examination of the absorbed moisture component of the NMR signal shows that this signal consists of distinct multiple components indicating that the absorbed moisture exists in several states of molecular binding, such as, moisture at fiber-matrix interfaces, moisture in voids or capillaries, moisture absorbed into the matrix, and moisture absorbed into the Kevlar fibers. The spin-spin relaxation times, T_2, associated with these various components of the NMR signal are shown in Figure 2. For quantitative determination of the moisture content in a composite, the amplitude of the free-induction-decay signal can be measured at a selected time following the RF pulse. By properly choosing the measurement time, the NMR signal component associated with the absorbed moisture

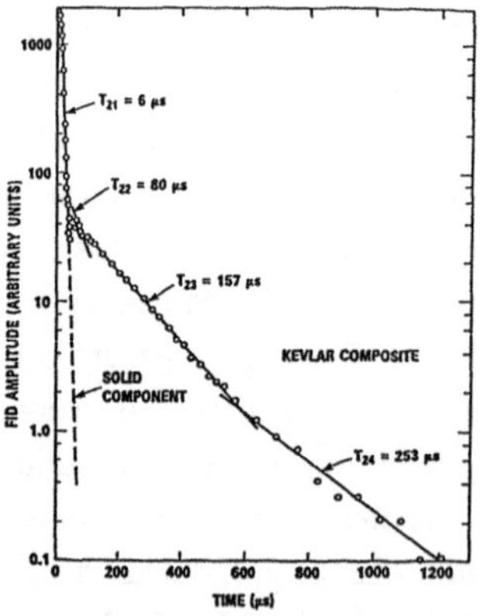

Fig. 2. Free-induction-decay (FID) for Kevlar composite (moisture=1.36%)

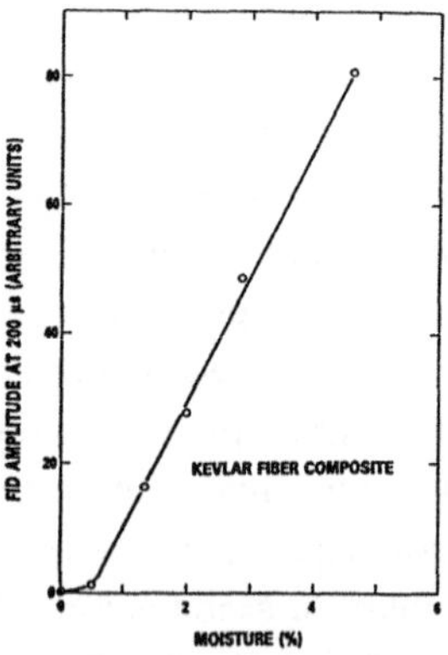

Fig. 3. Free-induction-decay vs. moisture for Kevlar composite

may be measured separately from the solid component. An example is shown in Figure 3 of the NMR free-induction-decay amplitude measured at 200 μs vs. moisture for a Kevlar composite. [2]

More extensive studies of NMR signal characteristics and relaxation times of Kevlar composites have been performed to obtain information on the characteristic sates of moisture in the constituent materials and to determine the capabilities of NMR for characterizing degraded resin. [5] In this work, NMR measurements were performed on Kevlar fiber, neat resin (both cured and uncured), Kevlar prepreg, and Kevlar composite. Results of studies on Kevlar fiber indicated that water molecules in the fiber exist in a phase distinct from other hydrogen atoms in the fiber molecular structure and that the water molecules may be formed into tightly bound clusters. [15] The distinctive NMR free-induction-decay signal observed from moisture in the Kevlar fiber provides the opportunity for characterizing and measuring the moisture in the fiber even in the presence of uncured resin as in prepreg, or in a cured composite. Analysis of relaxation time measurements on prepregs and cured composites in which varying amounts of water were added to the resin before cure indicated that moisture migrates between the resin and the fiber during cure with the direction of the migration depending on the relative amounts of moisture in the fiber and the resin. Such migration can potentially degrade the resin by altering the polymer cross-link structure.

In addition to Kevlar composite and its constituent materials, NMR has been used to characterize moisture absorption in other composites and polymers including glass, graphite and polyester reinforced materials, and adhesives (see Table 2). In all of these materials, the NMR signal amplitude was found to correlate linearly with the amount of absorbed moisture over the range studied.

In addition to moisture measurement and characterization, NMR has been used to study a variety of other aspects of composites and polymers. Some of these studies are listed in Table 3 along with appropriate references. In one recent report, results were presented on using NMR and electron spin resonance (ESR) to characterize certain aspects of solid structures of rigid rod polymers and their blends with flexible polymers. [33] It was found from the experimental results that these materials form primarily phase segregated structures, a condition which limits the reinforcing potential of rigid rod polymers in molecular composites. Most specimens exhibited NMR signals indicative of mobile species postulated to be either water or residual acid from which the materials are processed. Information on the location of the mobile species in the structure can be deduced from the NMR results. ESR measurements indicated that free radicals are generated when the rigid rod polymer fibers are mechanically stressed. It was found by ESR that the as-processed fibers contain appreciable levels of free radicals and that annealing at elevated temperatures caused a decrease in the free radical concentration by a factor of three.

Table 3. Characterization of composites (and polymers) using NMR

1. Degree of Cure; Prepregs [5,6,16,17,18]
2. Glass Transitions; Crosslinking [19,20]
3. Physical Aging; Degradation [16,21]
4. Environmental-Stress-Cracking in Polyethylene [22]
5. Nylon Rope [23]
6. Synthetic Fibers and Textiles [2,24]
7. Liquid Crystal Polymers [25]
8. Cyrstalline/Amorphous Determination [26,27,28,29]
9. Block Copolymers [30]
10. Polyurethanes [26,31]
11. Internal Strain by NQR [32]

Another example of the use of NMR techniques to characterize the properties of non-metallic materials is a study of chain dynamics and structure property relations in high impact strength polycarbonate plastic. [34] NMR lineshape and relaxation time analyses using both carbon-13 and deuterium as probes were used to quantitatively characterize both the structural geometry and the energetics of local motions in order to establish the structural origin of the bulk properties of these materials. Such information is needed to understand the dynamical mechanical properties which form the basis for characterization and evaluation of high impact materials.

Solid state carbon-13 NMR has been used to qualitatively characterize textiles of various compositions. [24] Qualitative identification was demonstrated for various fiber blends including mislabeled samples; in particular, rapid and accurate determination of cotton/polyester blends was accomplished with minimal sample preparation. These NMR techniques successfully applied to fabrics should be extendable to composite reinforcing materials and prepregs.

Instrumentation Developments for Practical Application of NMR

Although conventional NMR measurement configurations involve use of an encircling coil to apply RF fields to a specimen located in a homogeneous magnetic field, instrumentation developments in recent years have demonstrated the feasibility of NMR measurements on materials external to both the RF coil and the magnet structure.

[2,35] The basic configuration utilizes a U-shaped magnet and a flat spiral-coaxial RF coil as shown schematically in Figure 4. The U-shaped magnet provides a static magnetic field H_0 which extends from one pole to the other as shown by the dashed lines. The RF coil provides a field H_1 which is perpendicular to the plane of the coil. In that region where the frequency and static field intensity have the proper relationship an NMR response will be detected by the coil. By varying the magnetic field strength, the sensitive region can be moved closer to, or farther away from, the magnet, thus providing a means for obtaining NMR information as a function of depth. Based on this approach a fieldworthy NMR moisture measurement system for nondestructively measuring the amount of moisture in reinforced concrete bridge decks has been designed and fabricated. [36] This system utilizes a U-shaped magnet and a flat RF coil to obtain NMR signals at depths down to 9.5 cm from the top surface of a bridge deck. Profiles of the moisture gradient in bridge decks have been measured by using magnetic field gradients to obtain spatially localized NMR measurements with a resolution of approximately 3 mm. [37]

To achieve the measurement depth required for measuring moisture in concrete bridge decks, the NMR sensor for the inspection system described above needed to be rather large (about 272 kg). However, by using recently developed rare-earth permanent magnet technology, a considerably smaller, hand-portable one-sided NMR sensor is feasible for near-surface (e.g. depths of 3 mm) measurements on composite or plastic materials and structures. [38] A conceptual sketch of such a sensor is shown schematically in Figure 5.

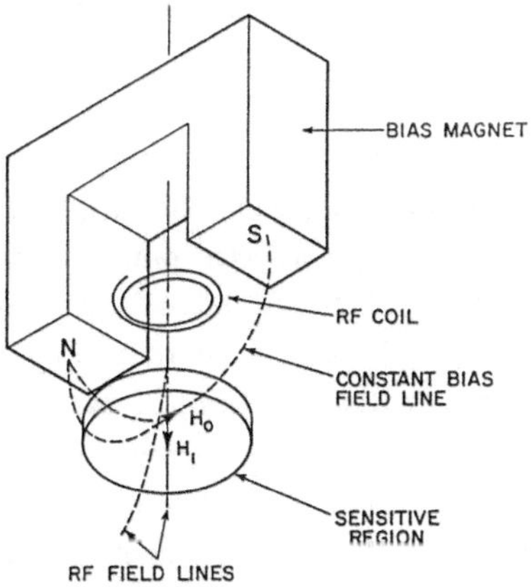

Fig. 4. Schematic illustration of one-sided NMR detection for a specimen external to the RF coil and the magnet

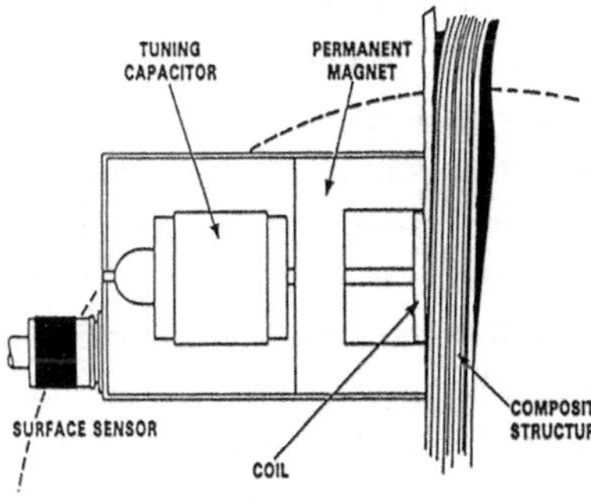

Fig. 5. Conceptual sketch of single-sided NMR sensor

Application of NMR Imaging to Composites and Polymers

There are many problem areas in nondestructive inspection where imaging methods can be valuable in providing a more complete evaluation. As pointed out in the Introduction, in recent years there have been spectacular advancements in the development of NMR methods and apparatus for medical imaging. These NMR methods provide data on the spatial distribution of a selected nuclear specie (usually hydrogen) and, in more advanced apparatus, the distribution of the selected specie in particular molecular binding states. These NMR capabilities have been found to be very valuable as an aid in medical diagnosis and could similarly be advantageous in non-medical inspections of non-metallic materials. Unfortunately, the present NMR imaging systems and techniques are not directly useable for this purpose since the range of NMR properties of solid materials that would be of interest in nondestructive evaluation are substantially different from body fluids and tissue. Some of the limitations are associated with apparatus engineering details but others are more basic and suggest the need for different approaches to implement NMR imaging for the inspection of solids. Systems presently used for medical imaging are susceptible to interference from metal items or external magnetic fields in the vicinity of the apparatus and must be used in a very carefully controlled laboratory if acceptable results are to be obtained. In addition, the field intensity in the sampling regions must be maintained constant within 1 part in 10^5. This limitation must also be eased if NMR imaging is to be generally useful for industrial applications of NDE.

In NMR imaging, different regions of the sample can be made to satisfy the resonance condition at any one time by varying the intensity of the externally applied, non-uniform (gradient) magnetic field. By suitably detecting the NMR signal and processing the spatial and temporal data, an NMR image can be created. Such images primarily show variations of the number density of the detected nuclear specie but the display can be further restricted to only image components having pre-selected ranges of spin lattice relaxation time, T_1, spin-spin relaxation time, T_2, chemical shift or other NMR parameters. The inherent spatial resolution of an NMR image is determined by the strength of the applied magnetic field gradient and the natural linewidth of the nuclear species being imaged. Currently, by far the greatest usage of NMR imaging, now more commonly called magnetic resonance imaging (MRI), is in the medical industry where magnetic fields of up to 1.5 T are used to image sections through the human body for diagnostic purposes. These NMR systems are almost always used to detect hydrogen (proton) NMR signals from body fluids, which, since the body is largely water, is what gives these instruments their utility. Current MRI scanners can collect and process the data for images with submillimeter resolution (typically 600 μm) in times of a few minutes. [39] Since in these applications the process is essentially imaging fluids, the NMR absorption lines are sharp (widths ~5mG) and T_2's are relatively long, typically 100-1000 ms. These medical MRI scanners apply magnetic field gradients ~100 mG/cm which gives a spatial resolution of ~0.5 mm for an NMR line of width 5 mG. These machines are easily able to "see" protons in tissue fluids or in soft materials such as fats, where the NMR absorption lines are fairly narrow (~20 mG).

In solids however, the proton resonance absorption lines are much broader and the T_2's correspondingly shorter (≤1ms). Such materials cannot be imaged with conventional medical MRI scanners since these machines have neither: (1) the short pulse capability required for detection of the rapidly decaying signal from protons in solids, nor (2) large enough magnetic field gradients to provide useful spatial resolution. Consequently the use of MRI in industrial applications has been limited to a few specialized cases such as fluid transport in solid materials. [40,41] A few small-bore NMR systems with shorter pulse durations, a compatible receiver bandwidth, and field gradients ~20 G/cm have been built to make measurements on soft solids having T_2 on the order of 1.0 milliseconds. These systems are intended for the industrial market but the high cost (~$500,000) and limited application (most solids, including some epoxy resins, have T_2 relaxation times much shorter than 1 ms) have discouraged their widespread application. Nevertheless these machines have been used in combination with multipulse line narrowing techniques to produce images of some soft solids. [42,43]

Successful development of NMR imaging for non-metallic solid materials would have very important consequences for NDE and materials characterization, in particular for composite materials. The capability of NMR to sense signals at the molecular level within localized regions of a material provides a potential tool for nondestructively determining the chemical state of individual components within the material. If the full potential of NMR imaging could be realized, internal maps of such chemical information within a composite could be generated. This information could possibly include chemical state gradients in the vicinity of internal discontinuities such as cracks, disbonds, fiber-matrix interfaces, and internal flaws. A listing of some applications of NMR imaging to composites and polymers is given in Table 4.

In work currently in progress under an Air Force contract, A.C. Lind and C. Fry of the McDonnell Douglas Research Laboratories are investigating applications of NMR imaging to graphite/epoxy composites. [55] One question addressed concerns the attenuation of NMR signals caused by the conductive carbon fibers. Since carbon fibers have good electrical conductivity along their length, but the resin between fibers in a composite has large resistance to current flow between fibers, the attenuation of NMR signals is strongly dependent upon the orientation of the NMR fields relative to fiber direction. From measured data and theoretical considerations, Lind and Fry have determined the decrease in NMR signal strength and the increase in RF transmitter power required to obtain NMR images at various depths for different NMR frequencies. Although spatial resolution and the time required to produce an image are factors, it appears that 1 mm is the greatest practical depth for imaging a carbon-fiber crossply composite when the NMR fields are oriented perpendicular to the fibers. For NMR field orientation parallel to the fiber direction, the practical limit for imaging is about 3 cm. These limitations do not apply in the case of non-conducting fibers, such as glass or Kevlar. Preliminary carbon-13 chemical shift images have been obtained by Lind and Fry of a solid sample consisting of 3-mm thick layers of PEEK (poly ether ether ketone), DER/DETA (diglyridyl either of bisphenol-A, cured with diethylene triamine). The carbons in the two resins were easily differentiated through their chemical shift spectra.

Table 4. Application of NMR imaging to composites and polymers

1. Porosity and Binder Distribution in Green-State Ceramics [44,45,46,47]
2. Distribution and Diffusion of Absorbed Fluids [48,49]
3. Blending of Low T_g and High T_g Polymers [41]
4. Cure and Strain Gradients [50]
5. Solid State Imaging Techniques [50,52,53]
6. One-Sided Measurement Techniques [37,54]

As pointed out earlier in this section, solid state imaging is difficult because of the very short T_2 relaxation times typically encountered in solids. Work is currently underway by Miller and Garroway to address this problem and several promising approaches have been reported. [51,53] The general approach taken by these investigators is to apply specialized sequences of RF pulses in order to effectively narrow the resonance lines by removing line-broadening mechanisms. Good success has been achieved using a refocused gradient imaging (RGI) technique to obtain images of an engineering polymer, i.e., polycarbonate. [52]

In work reported by Ellingson and his co-workers, [47,52] NMR imaging has been applied to imaging the distributions of organic binder/plasticizers and porosity in green state ceramics. Solution NMR imaging was used to image the open, surface-connected porosity in dewaxed parts. To accomplish this imaging, benzene was used as a filler fluid introduced into the sample via vacuum impregnation. To image the "soft-solid" binder/plasticizer, a specially designed NMR probe with field gradients of ~10 G/cm was used. With this approach, the distribution of binder/plasticizer was successfully mapped with a resolution of ~300 μm in cold pressed samples of Si_3N_4, SiC, and Al_2O_3.

Conclusions

As this review shows, there is currently a good deal of work in progress related to the application of NMR for the nondestructive evaluation and characterization of composites and polymers. Results to date show that NMR is capable of nondestructively characterizing composites and polymers in both cured and prepreg conditions. Certain features of NMR signals can provide information on moisture in different chemical and physical states within the material. In addition to moisture, NMR is capable of providing information on environmental degradation, the degree of cure and crosslinking, aging, molecular structure and mobility, and material validation. Surface sensor technology can be used to obtain spatially localized NMR information, and instrumentation using NMR surface sensor technology is under development for practical NDE measurements of composite materials and structures.

With respect to NMR imaging, equipment is commercially available for medical applications where the best reported spatial resolution to date is ~0.1 mm. Medical imagers can be used for certain applications involving somewhat mobile molecules in polymeric materials (e.g., elastimers and absorbed fluids in porous solids). However, promising approaches are currently under development for solid state imaging of short T_2 materials (e.g., composites) where it may be possible to circumvent the need for strong magnetic fields and/or gradients for imaging in solids.

References

1. Summerscales, J.; Short, D.: Chemical spectroscopy. Non-Destructive Testing of Fibre-Reinforced Plastics Composites. London: Elsevier Applied Science, (1987) 207-270.
2. Matzkanin, G.A.: Applications of nuclear magnetic resonance to the NDE of composites. Proceedings of the 14th Symposium on NDE, Nondestructive Testing Information Analysis Center, San Antonio, Texas (1983) 270-286.

3. Batra, N.K.; Graham, T.P.: Measurement of moisture in hygrothermally degraded fiber-reinforced epoxy composites by continuous wave (CW) nuclear magnetic resonance (NMR). Br. J. Non-destr. Test 25 (Jan. 1983) 21-23.
4. Shuford, R.J.; Murray, T.J.; Hinton, Y.L.; Brockelman, R.H.: Advanced NDE techniques for composites. 1986 SME Manufacturing Technology Review, May 28-30, 1986 (1987) 101-112.
5. Matzkanin, G.A.; De Los Santos, A.: Nondestructive characterization of Kevlar composites using pulsed NMR. Nondestructive Characterization of Materials II. New York, New York: Plenum Press (1986) 29-37.
6. Shuford, R.J.; Murray, T.J.; Bostic, M.T.; Brockelman, R.H.; Hinton, Y.L.: Evaluation of nondestructive evaluation techniques for detecting moisture in composite materials. Proc. of the 11th World Conference on NDT, Las Vegas, NV, 3-8 Nov. 1985, 1366-1473.
7. Matzkanin, G.A.; De Los Santos, A.: NDE of moisture absorption in composites using nuclear magnetic resonance. Composites Evaluation (1987) 23-27.
8. Fornes, R.E.; Memory, J.D.; Gilbert, R.D.: Studies of epoxy resin - grpahit fiber composites using NMR, ESR, and DSC. Third Annual Army Composites Research Review, Williamstown, MA, October 29-31, 1980.
9. Lawing, D.; Fornes, R.E.; Gilbert, R.D.; Memory, J.D.: Temperature dependence of broadline NMR spectra of water-soaked epoxy-graphite composites. J. Appl. Phys. 52 (Oct. 1981) 5906-5907.
10. Matzkanin, G.A.; De Los Santos, A.; King, J.D.: Nondestructive evaluation of adhesive bonds using nuclear magnetic resonance. Proceedings of the Fifth International Joint Military/Government-Industry Symposium on Structural Adhesive Bonding, Dover, New Jersey, November 3-5, 1987 (1987) 397-403.
11. Memory, J.D.; Lawing, D.: Broadline NMR of polymers. Mag. Resonace Rev. 5 (1979) 69-73.
12. Fuller, R.T.; Fornes, R.E.; Memory, J.D.: NMR study of water absorbed by epoxy resin. J. Appl. Polymer Sci. 23 (1979) 1871-1874.
13. Lind, A.C.: NMR study of inhomogeneities in amine cured epoxies. Polymer Preprints 103 (1981) 333-334.
14. Vinson, J.R. (ed.): Advanced composite materials-environmental effects. ASTM STP 658, American Society for Testing and Materials (1978).
15. Garza, R.G.; Pruneda, C.O.; Morgan, R.J.: Polymer Preprints 22 (1981).
16. Carlson, D.J.; Wiles, D.M.: Degradation. Encyclopedia of Polymer Science and Engineering Vol. 4. New York, New York: Wiley-Interscience (1986) 630-696.

17. Kowalska, M.; Wirsen, A.: Chemical analysis of epoxy prepregs—market survey and batch control. 25th National SAMPE Symposium, San Diego, May 1980, 389-402.

18. Happe, J.A.; Morgan, R.J.; Walkup, C.M.: NMR characterisation of BF_3-amine catalysts used in the cure of carbon fibre/epoxy prepregs. Composites Technology Review 6 (Summer 1984) 77-82.

19. Brown, I.M.; Lind, A.C.: Study of the glass transition in crosslinked polymers using nuclear and electron magnetic resonance. Final Report (1977), Contract No.: N00019-76-C-0565.

20. Banks, L.; Ellis, B.: The glass transition temperatures of highly crosslinked networks: cured epoxy resins. Polymer 23 (Sept. 2981) 1466-1472.

21. Bryden, W.A.; Poehler, T.O.: Nondestructive testing of nylon ropes using magnetic resonance techniques. Proc. of the 11th World Conference on NDT, Las Vegas, NV, 3-8 Nov. 1985, 1746-1750.

22. Brown, H.R.: A theory of the environmental stress cracking polyethylene. Polymer 19 (1978) 1186-1188.

24. Colletti, R.F.; Mathias, L.J.: Solid state C-13 NMR characterization of textiles: Qualitative and quantitative analysis. Technical Report No. 3, Aug. 1986 - Aug. 1987 (Report No. AD-A184349): Contract No. N00014-86-K-0659.

25. Muller, K.; Schleicher, A.; Kothe, G.: Two-dimensional NMR relaxation spectroscopy of liquid crystal polymers. Mol. Cryst. Liq. Cryst. 153 (Dec. 1987) 117-131.

26. Kintanar, A.; Jelinski, L.W.; Gancarz, I.; Koberstein, J.T.: Complex molecular motions in bulk hard segment polyurethanes: A deuterium NMR study. Technical Report, Sept. 1985 (Report No. AD-A159760; TR-2): Contract No. N00014-84-K-0534.

27. Runt, J.: Crystallinity determination. Encyclopedia of Polymer Science and Engineering Vol. 4. New York, New York: Wiley-Interscience (1986) 482-519.

28. Havens, J.R.; Vanderhart, D.L.: Multiple-pulse proton NMR of pressure-crystallized linear polyethylene. J. Magn. Resonance, 61 (Feb. 1985) 389-391.

29. Zupancic, I.; Lanhajnar, G.; Blinc, R; Reneker, D.H.; Peterlin, A.; NMR studiy of diffusion of butane in linear polyethylene. J. Poly. Sci., Poly. Phys. Ed. 16 (1978) 1399-1407.

30. Tanaka, H.; Nishi, T.: Spin diffusion in block copolymers as studies by pulsed NMR. Phys. Rev. B 33 (Jan. 1986) 32-42.

31. Brown, I.M.; Lind, A.C.; Sandrecski, T.C.: Magentic resonance studies of epoxy resins and polyurethanes. Final Report (1979) Contract No.: N00019-78-C-0031.

32. Hewitt, R.R.; Mazelsky, B.: Nuclear quadrupole resonance as a nondestructive probe in polymers. J. Appl. Phys. 43 (Aug. 1972) 3386-3392.

33. Vanderhart, D.L.; Wang, F.W.; Eby, R.K.; Fanconi, B.M.; Devries, K.L.: Exploration of advanced characterization techniques for molecular composites. Final Report, Feb. 1984 - Jan. 1985 (Report No. AFWAL-TR-85-4137): Contract No. MIPR-FY1457-84-N-5019.
34. Inglefield, P.R.; Jones, A.A.: Chain dynamics and structure property relation in high impact strength polycarbonate plastic. Final Report, Nov. 1981 - Nov. 1984 (Report No. ARO-18501.7-CH-H): Contract No.: DAAG29-82-G-001.
35. Paetzold, R.F.; Matzkanin, G.A.; Los Santos, A. De: Surface soil water content measurement using pulsed nuclear magnetic resonance techniques. Soil Sci. Soc. Am. J. 49 (May-June 1985) 537-540; also,
Paetzold, R.F.; Los Santos, A. De; Matzkanin, G.A.: Pulsed nuclear magnetic resonance instrument for soil-water content measurement: sensor configurations. Soil Sci. Soc. Am. J. 51 (March-April 1987) 287-290.
36. Matzkanin, G.A.; Los Santos, A. De; Whiting, D.A.; Determination of moisture levels in structural concrete using pulsed NMR. FHWA/RD-82/008 (April 1982).
37. Matzkanin, G.A.: Application of spatially localized NMR to nondestructive evaluation. Proceedings of the 11th World Conference on NDT, Las Vegas, Nevada, November 1985 (1985) 1607-1614.
38. Matzkanin, G.A.: Nondestructive evaluation of composites using pulsed NMR. Proceedings of the Society of Plastics Engineers 45th Annual Technical Conference, Los Angeles, California, May 4-7, 1987 (1987) 1060-1063.
39. Glover, G.H.: Magnetic resonance imaging in medicine: quantitative tissue characterization. Rev. of Prog. in Quantitative NDE 4A (1985) 1-10.
40. Vinegar, H.J.: Use of MRI and x-ray CT in core analysis. S.P.E.J., February 1986.
41. Rothwell, W.P.; Gentempo, P.P.: Non-radial applications of NMR imaging. Bruker Report (Jan. 1985) 46-51.
42. Garroway, A.N.; Baum, J.; Murowitz, M.G.; Pines, A.: NMR imaging in solids by multiple quantum resonance. J. Mag. Res. 60 (1984) 337-341.
43. Wind, R.A.; Yannoni, C.S.: Selective spin imaging in solids. J. Mag. Res. 36 (Oct. 1979) 269-272.
44. Ellingson, W.A.; Ackerman, J.L.; Gronemeyer, S.; Garrido, L.: Nuclear magnetic resonance imaging for detecting binder/plasticizers in green-state structural ceramics. Proceedings of 16th Symposium on Nondestructive Evaluation, San Antonio, Texas, April 21-23, 1987 (1987) 231-240.

45. Ackerman, J.L.; Ellingson, W.A.; Koutcher, J.A.; Rosen, B.R.: Development of nuclear magnetic resonance imaging techniques for characterizing green-state ceramic materials. Proceedings of the 2nd International Symposium on the Nondestrucitve Characterization of Materials, Montreal, Canada, July 21-23, 1986 (1986), 129-137.
46. Welch, L.B.; Gronczy, S.T.; Mitsche, M.T.; Bauer, L.J.; Dworkin, J.; Giambalvo, A.: Proton NMR imaging of green-state ceramics. Proceedings Quantitative NDE Conference, La Jolla, CA, Plenum Press, New York (1987) 441-456.
47. Ellingson, W.A.; Ackerman, J.L.; Garrido, L.; Wong, P.S.; Gronemeyer, S.: Development of nuclear magnetic resonance (NMR) imaging technology for advanced ceramics. November 1988 (Report No. ANL-87-53): Contract No.: ACK-85234.
48. Blackband, S.; Mansfield, P.: Diffusion in liquid-solid systems by NMR imaging. J. Phys. C 19 (Jan. 1986) L49-52.
49. Rothwell, W.P.; Holecek, D.R.; Kershaw, J.A.: NMR imaging: Study of fluid absorption by polymer composites. J. Polym. Sci. Polym. Lett. (22) (May 1984) 241-247.
50. Crowley, J.D.; Branman, S.K.: Nuclear magnetic resonance imaging. 4th NASA Nondestructive Evaluation Workshop (March 7, 1985).
51. Miller, J.B.; Garroway, A.N.: Multinuclear NMR imaging of solids. Review of Progress in Quantitative NDE, Vol. 8A, New York, New York: Plenum Press (1989) 567-574.
52. Miller, J.B.; Garroway, J.: ^{1}H refocused gradient imaging of solids. Submitted to J. Magn. Resonance (1988).
53. Miller, J.B.; Garroway, A.N.: NMR imaging of solids. Review of Progress in Quantitative Nondestructive Evaluation 7A (1988) 287-294.
54. Miller, J.B.; Garroway, A.N.: NMR imaging of solids with a surface coil. J. Mag. Resonance 77 (1988) 187-191.
55. Lind, A.; Fry, C.: Private communication. McDonnell Douglas Research Laboratories, work funded by the Air Force Materials Laboratory.

Instruments and Systems Process Control

Instrument for the Automated Ultrasonic Time-of-Flight Measurement - A Tool for Materials Characterization

R. Herzer and E. Schneider

Fraunhofer-Institut für zerstörungsfreie Prüfverfahren,
Universität, Geb. 37, D-6600 Saarbrücken 11

Summary

The propagation velocities of ultrasonic waves are functions of the density, the elastic constants and the strain-state of the material under test. Techniques are developed to use ultrasonic velocity or time-of-flight measurements in order to determine elastic properties of materials and hence to characterize materials behavior.
A prototype is built which enables the automated time-of-flight measurement. In order to minimize the influences of amplitude variations, the system determines the zero crossings of the RF-signals and generates the start and stop signals for the counter. The programmable high performance counter has a temperature stabilized 500 MHz time base and thus a single shot resolution of ± 2 ns. Usual repetition rates of ultrasonic pulsers enable a time-of-flight resolution of some hundred picoseconds by simple averaging procedures. The measured times-of-flight are read into the personal computer by an IEEE 488 bus with a transfer rate of 30 measured data per second. All operational functions are completely controlled by the computer system. Besides the time-of-flight, the coordinates of the measuring point and the temperature are also taken. The software, so far implemented in the prototype enables the evaluation of surface and bulk stress states. In order to evaluate the stress state, the particular evaluation equations are taken from the storage; the processor proves whether the appropriate data are measured and performs the evaluations. If necessary, the influence of temperature on the measuring quantities is corrected, using stored temperature coefficients. The measured data, and also the results can be displayed on the graphic-screen and on the printer, and they are stored on disks. The system is housed in two dustproof, portable boxes.

1. Introduction

The nondestructive characterization of materials with ultrasonic techniques takes advantage of the elastic and anelastic interactions between the ultrasonic wave and the material: Ultrasonic attenuation measurements are used to determine mechanical properties governed by the microstructure. Because of the complex interactions between the ultrasonic wave and the microstructure, calibration experiments are needed. Theoretical models and improved measuring techniques are available to exploit ultrasonic scattering and backscattering techniques in order to characterize microstructural properties and to evaluate hardening depths.

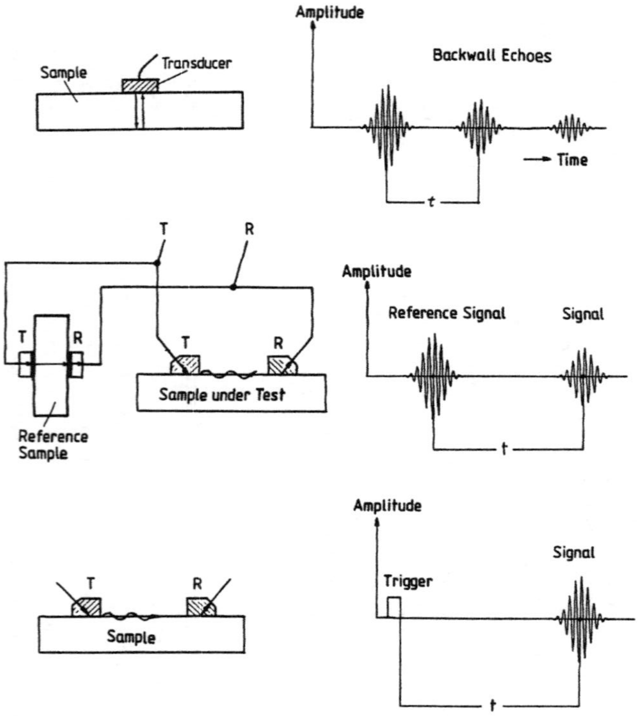

Fig. 1: Sketch of the pulse-echo-, the reference- and the transmitter-receiver-technique to measure ultrasonic time-of-flight.

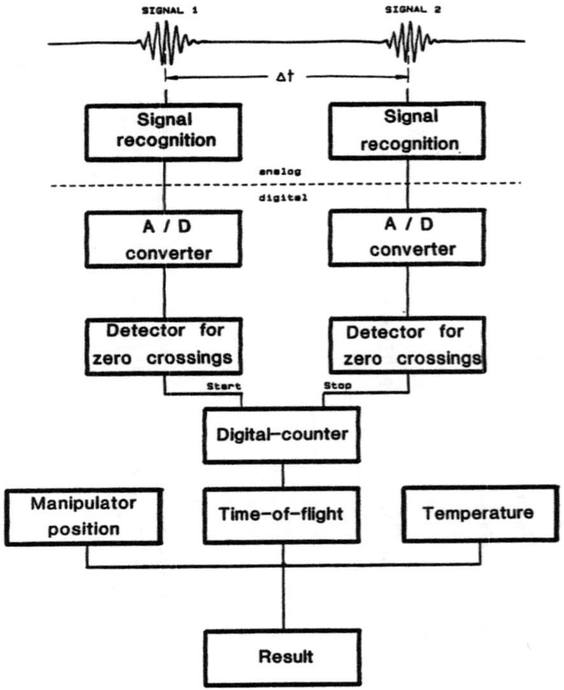

Fig. 2: Block-diagramm of the system.

Ultrasonic velocity measurements are used to characterize microstructural quantities and material states influencing the elastic behavior of materials. The velocities of ultrasonic waves are dependent on the density, the elastic constants and the strain state; hence the determination of density, porosity are possible as well as the evaluation of second and third order elastic constants and microstructural parameters influencing these constants. Furthermore, the elastic anisotropy, the strain or stress state can be evaluated.

In order to evaluate one of these quantities or to characterize elastic properties particular preassumptions have to be fulfilled. But in each case, a very precise time-of-flight measurement is needed. Increasing industrial interest supported the development of ultrasonic techniques to evaluate residual stress states or to characterize rolling texture and hence the deep drawability of rolled sheets. The accurate time-of-flight measurement and the automated data acquisition and evaluation are strong demands for the in-field application of the ultrasonic techniques.

The contribution describes a prototype, developed for the in-field measurement of ultrasonic times-of-flight and to evaluate residual stress states in metallic components.

2. Method

Almost all set-ups used for the applied ultrasonic testing measure the ultrasonic time-of-flight between the trigger pulse and that point, where the amplitude of the pulse echo reaches a certain level. Variations of coupling conditions e.g. during the manipulation of the ultrasonic probe cause significant changes of the echo amplitude and thus inaccuracies of the time-of-flight measurement. In order to minimize the influences of amplitude variations, the developed set-up measures the time between the corresponding zero crossings of two RF-signals. The two signals, as well as the corresponding zero crossings of these signals are selectable by the operator. This holds for the application of the pulse-echo technique and for the technique using an ultrasonic reference signal. The upper part of Fig. 1 shows sketches of these configurations, the lower one sketches the transmitting-receiving technique. In that case, the time-of-flight (TOF) is measured between a constant signal, artificially derived from the clock signal of the used ultrasonic set-up and a zero crossing of the ultrasonic RF-signal. As shown in Fig. 2, two signals are taken by suitable gates and separately amplified. Each signal can be amplified up to 64 dB. The signals are digitized and two zero crossing detectors create a rectangular pulse at

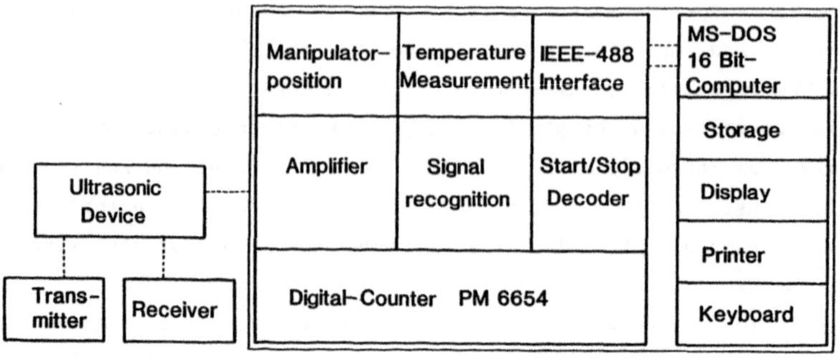

Fig. 3: Components of the system.

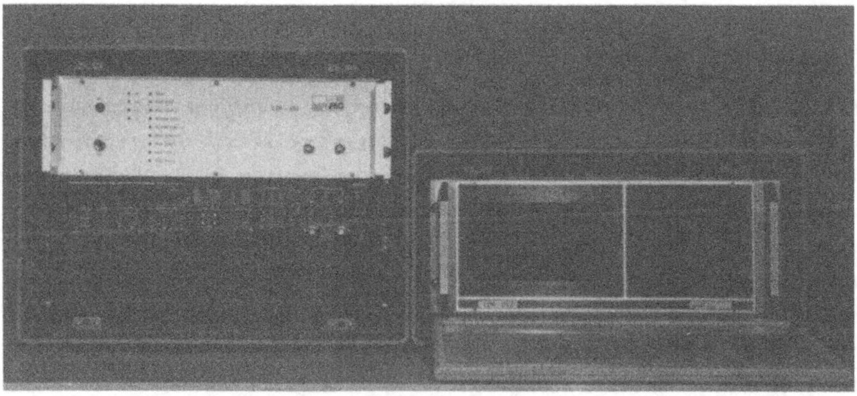

Fig. 4: Photograph of the system for the automated measurement of ultrasonic time-of-flight.

each zero crossing. START and STOP signals for the counter are deduced from two particular zero crossing pulses, choosen by the operator. The high performance counter has a temperature stabilized 500 MHz time reference and thus a single shot resolution of 2 ns. By averaging of n measurements, the accuracy can be improved by the factor of $1/\sqrt{n}$ and reaches a few hundred ps. The repetition rates of usual ultrasonic systems are some hundred Hz, so that the mentioned precision of the time-of-flight measurement can be reached on-line. The temperature of the sample and the ultrasonic probe and the coordinates of the measuring point are registered as shown in Fig. 2.
Fig. 3 displays a sketch of the prototype together with an ultrasonic device which delivers RF-signals with center frequencies between 1 and 10 MHz and a signal to noise ratio better than 15 dB. One component of the prototype contains the amplifiers, the analoge signal treatment, A/D converters, the generators for the START and STOP signals and the digital counter. The coordinates of the measuring points and the temperatures of the probe and of the sample under test are taken by two further moduls. An IEEE 488 interface connects the system with an MS-DOS-AT computer, housed in a second component. The personal computer is used to control the system and to evaluate the data. It has a 20 MByte Hard Disk and a 1.2 MByte Floppy Disk to store the data. The measured data as well as the evaluated results can be displayed as tables or graphs on the screen, or they can be plotted by an external printer or plotter.
The prototype is housed in two dust- and shockproof metallic boxes. The time-of-flight measuring unit is in one box with a width of 534 mm, a height of 394 mm, a depth of 690 mm and a weight of 30 kg. The second box with the dimensions 534 mm, 260 mm, 690 mm and a weight of 22 kg contains the PC, the keyboard and a screen. A photograph of the prototype is shown in Fig. 4.

3. Operational Functions

All operational functions of the prototype, the data acquistion and their evaluation are completely controlled by the computer. Guided by a menue, the following funtions are adjustable:

* choice of the signals
* preamplification of each of the two signals
* choice of the trigger edge
* choice of the zero crossings
* averaging time between 0.1 ms and 96 s

* choice of the application mode:
  - measurement between two ultrasonic pulse echoes
  - measurement between a reference signal and an ultrasonic signal
  - measurement between a given signal and an ultrasonic signal
* choice between three data acquisition modes

The operational functions, optimized for a particular case of application can be stored, so that repetition measurements can be performed with the same system conditions.

Three data acquisition modes are realized:

* Time continuous measurements, independent of the position of the probe. The measuring time for each single time-of-flight result (TOF) is given by the pulse repetition rate of the used ultrasonic set-up and the averaging time, which can be varied between 0.1 ms and 96 s.
  The maximum data transfer rate of the IEEE bus is 30 TOF-results per second, it can be increased up to 420 results per second.

* Continuous locus dependent measurements. The coordinates of the measuring point and the time-of-flight measured at that point are taken and stored. The position of the probe should be given by the manipulator in the voltage range between 0 and +10 V. 512 measured TOF-results can be taken for each trace of measurements. The averaging time and the data transfer rate are mentioned above.

* Single, locus dependent measurements. The single TOF-result and the coordinates of the measuring point are taken by pushing a button.

4. Data Evaluation

The measured TOF results, together with the corresponding coordinates can be displayed on the screen to enable a first overview. The influence of temperature on the TOF is corrected by the use of stored temperature coefficients of ultrasonic longitudinal and shear waves as well as of some special transmitter-receiver-configurations for the application of skimming longitudinal, SH- and Rayleigh waves. The temperature coefficients of the mentioned waves are taken for steel, cast iron and cast steel, for Al-alloys and ceramic samples.

The TOF results versus the locus or versus mechanical or heat treatments of the sample under test can be used for the evaluation of materials properties.

The evaluation software, already implemented into the system enables the evaluation of strain or stress states in the surface as well as in the bulk of metallic and ceramic components. The software to evaluate rolling texture in steels or Al-alloys and hence to characterize the deep drawability of rolled sheets is under work.

In order to evaluate surface strain or stress states, it can be choosen between the application of skimming longitudinal waves, SH- and Rayleigh waves. The birefringence effect of shear waves, as well as free ultrasonic longitudinal waves can be used for the characterization of strain or stress states in the bulk. Combinations of both, birefringence effect and longitudinal wave technique are possible as well.

The evaluation of the data is also guided by the menue. After the choice of the particular technique is made, the computer controls whether the appropriate data are taken. If necessary the influence of the temperature on the time-of-flight data is corrected. The strain or stress state is evaluated using the appropriate evaluation equations together with the elastic constants of the material. The results are given in tables or graphs on the computer screen or in a hard copy. The measuring data and the results can be stored on the Floppy Disk and/or transfered to another AT-compatible personal computer.

5. Experimental Results

The prototype was tested in different in-field applications in order to evaluate the stress states in rolls and rotors. Fig. 5 displays a comparison between time-of-flight results, measured manually and with the prototype. In a lab application the stress in an Al-plate with a shrink fitted Al-bolt was evaluated. Fig. 6 shows the change of stress in the specimen and along the center line through plate and bolt.

6. Acknowledgement

The development of the prototype was partly supported by the European Community for Carbon and Steel and by the Arbeitsgemeinschaft Industrieller Forschung e.V.

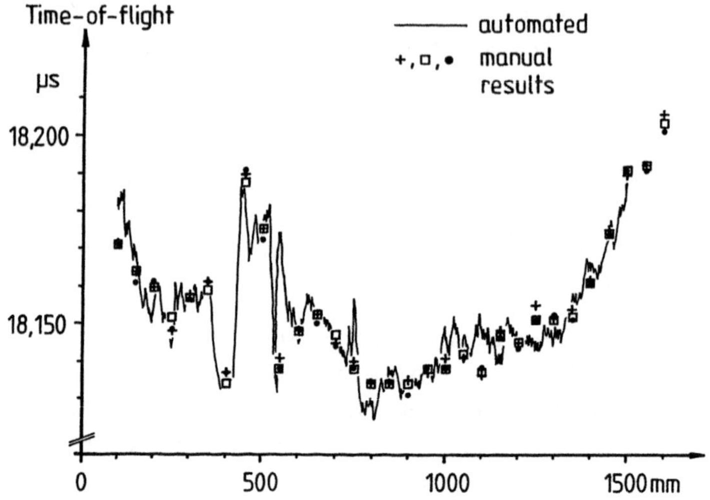

Fig. 5: Ultrasonic time-of-flight along the length of a turbine rotor.

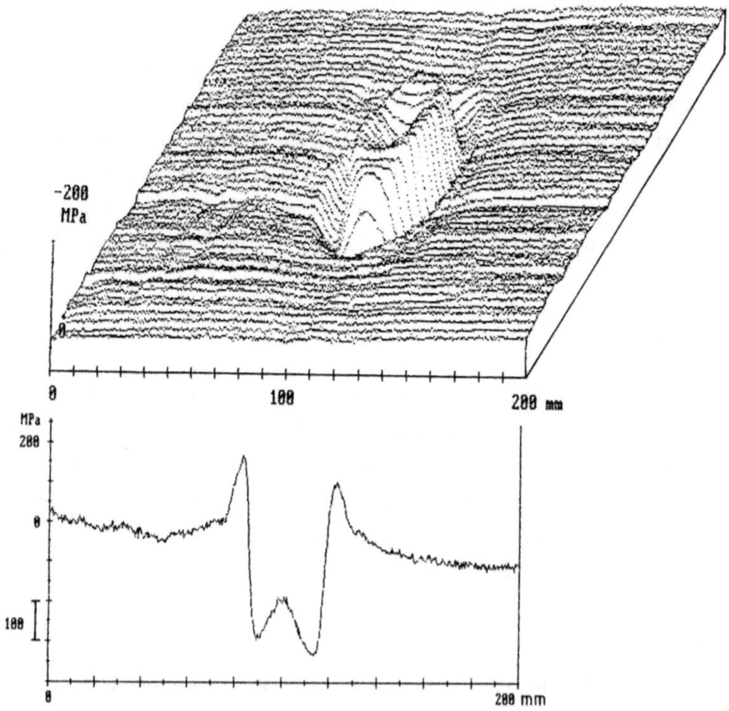

Fig. 6: Stress profile in an Al-plate with a shrink fitted Al-bolt and along a trace through the center of the arrangement.

EMUS-systems for Stress and Texture Evaluation by Ultrasound

A. Wilbrand, W. Repplinger, G. Hübschen, H.-J. Salzburger

Fraunhofer-Institut für zerstörungsfreie Prüfverfahren,
Universität, Geb. 37, Saarbrücken, West Germany

Zusammenfassung

Elektromagnetische Ultraschall-(EMUS-)Systeme eignen sich in besonderer Weise für die Ultraschall-Laufzeitmessung zur Spannungs- und Textur-Ermittlung. Sie stellen derzeit die einzige Möglichkeit dar, Messungen mit senkrecht einfallenden Transversalwellen und horizontal polarisierten Transversalwellen vor Ort rasch durchzuführen und bieten wegen der koppelmittelfreien Ultraschallwandlung gute Reproduzierbarkeit und Einsatz auch bei hohen Vorschubgeschwindigkeiten und erhöhten Temperaturen. Diverse EMUS-Prüfköpfe für den genannten Anwendungszweck werden vorgestellt. Es handelt sich um einen Permanentmagnet-Prüfkopf für linear polarisierte Transversalwellen, einen Prüfkopf mit wassergekühltem Elektromagneten für die Anregung von senkrecht zueinander polarisierten Transversalwellen in Turbinenwellen, einen Prüfkopf mit kombinierter piezoelektrischer bzw. elektromagnetischer Anregung von Longitudinal- bzw. Transversalwellen in Schrauben und Wandlersysteme für geführte Lamb- bzw. SH-Moden in Walzblech. Durch den Einsatz in Verbindung mit einem EMUS-Sende/Empfangs-Elektronik-Prototyp mit hohen Sendeleistungen und rauscharmen Empfangsverstärkereingängen werden die Anforderungen an die Meßgenauigkeit voll erfüllt.

Introduction

Applied and residual stresses as well as crystallographic texture produce elastic anisotropy in structural materials which results in a variation of ultrasonic wave velocities as a function of polarization and propagation direction. Evaluation of these variations is a nondestructive technique for determining the underlying stress and texture state. Various combinations of wave modes with specific polarization and propagation directions have been applied, among which horizontally polarized shear waves at normal or oblique incidence play an important role [1-8]. This wave type can be excited with piezoelectric probes only under rather inconvenient conditions, since couplants of high viscosity are required, and hence measurements using these devices are time-consuming and not suitable to mechanized scanning. These detractions are omitted when electromagnetic-ultrasonic (EMUS) probes are used. Other advantages of EMUS probes are a good reproducibility of the transduction conditions and their allowance for use at elevated temperatures. The dry contact allows operation at high scanning speeds. EMUS probes require some specific design considerations, among these is the connection of a small coil lift-off and a good wear resistance. A complete EMUS system consists of the probe and well-designed transmit/receive electronics which overcome the larger insertion loss, when compared to piezoelectric probes, by a high transmitter output power (typically 1.5 kW into an inductive load at series resonance) and a low preamplifier input noise (typically 10 μV_{p-p}).

This contribution presents EMUS probes developped for residual stress evaluation in turbine shafts, for applied stress evaluation in screws and bolts, a multi-purpose probe for stress evaluation in components with parallel front and back surfaces and a system for evaluation of stress and texture with guided SH and Lamb waves in rolled sheet. The probes are connected to a universal transmit/receive electronics unit which can be tuned to optimal performance with each individual probe.

Multi-purpose probe

For routine measurements of the acoustic birefringence effect on components or specimens with parallel front and rear surfaces and wall thicknesses typically between 5 and 100 mm there is need for a normal probe for linearly polarized shear waves which is light-weighted and easy to handle. Such a probe designed for application on ferromagnetic material is depicted in Fig. 1. Owing to its compact magnetic yoke with high-energy-product permanent magnets and flux concentrators it has dimensions of only 35×40×45 mm. The probe has two individual flat rectangular rf coils for transmission and reception, which are attached one upon the other. It operates at a center frequency of 2.5 MHz and is excited by a burst signal of 4 cycles duration. The transmitter current amplitude is typically 5 A_{p-p}. The oscilloscope traces in Fig. 1 show a backwall echo sequence obtained with this probe on a tensile specimen from 22 NiMoCr 37 of 30 mm thickness and the second backwall echo from this sequence on an expanded time scale. The signal-to-noise ratio which was achieved without signal averaging is about 25 dB. The performance of the probe is demonstrated by the evaluation of the acoustic birefringence effect in a weld specimen (Fig. 2). The diagram shows the variation in the difference of the principal stress components parallel and perpendicular to the weld axis along a line perpendicular to the weld seam (line 2 in the figure). To eliminate the influence of microstructure, an analogous measurement has been performed along a line close to the edge of the specimen (line 1), and the data have been subtracted from each other. The reproducibility of the time-of-flight data was typically ± 2 ns, i.e. the measurement accuracy is fairly good.

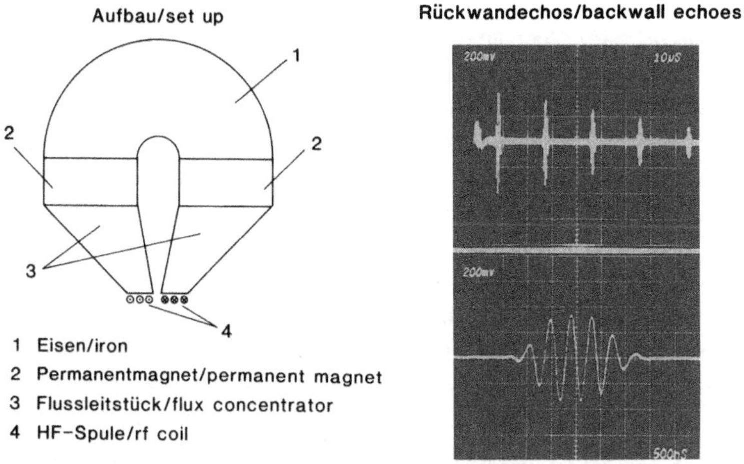

Fig.1. EMUS probe for linearly polarized shear waves

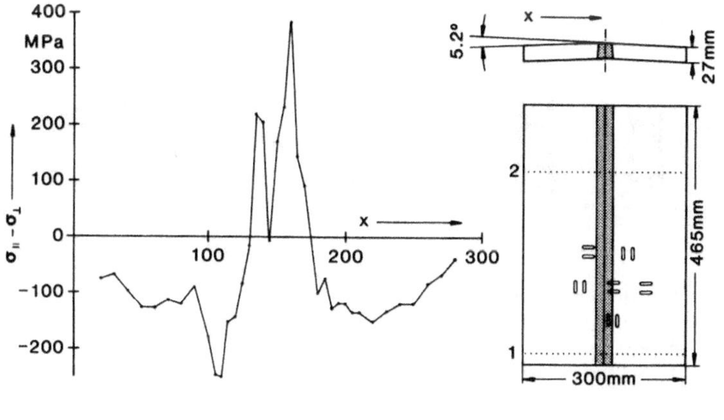

Fig.2. Measurement of residual stress in a weldment

Residual stress in turbine shafts

Inappropiate stress distributions in turbine shafts can result in bending moments and subsequent unbalanced mass or even cracking especially when thermal loading is superposed in service or during heat treatment in the manufacturing process. Therefore a technique is required to evaluate non-destructively the residual stress distribution at several states in the manufacturing process. Substantial information about the residual stress field can already be obtained from the acoustic birefringence effect of normally incident shear waves travelling through the center of the shaft, when this effect is investigated as a function of circumferential and axial position of the ultrasonic probe. Since a large amount of data has to be gathered even from a single turbine shaft only such a transduction technique can be employed that allows to scan the probe quickly along the surface, i. e. at the present time the EMUS technique is the only applicable method for this task. For the sake of rapid data acquisition and also considering the curvature of the surface it was decided to develop a transducer with two pairs of rf coils which superposes shear waves with two orthogonal polarizations and receives the echoes with the two polarizations in two separate coils (Fig. 3). In this way turning the transducer in order to measure the birefringence effect is omitted. Since turbine shafts often have diameters between 0.5 and 1 m the diffraction loss even of the first

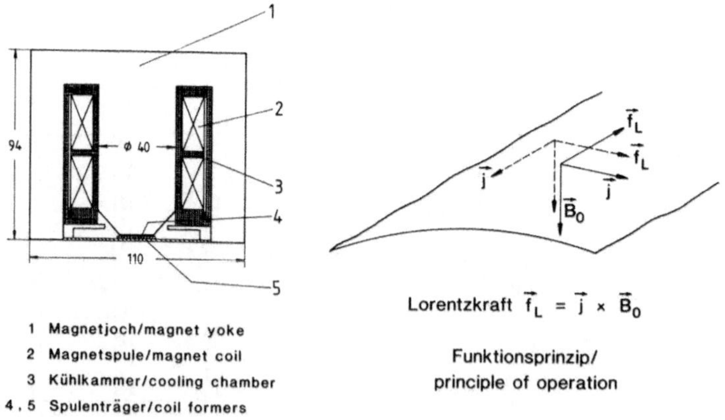

Fig.3. Combined EMUS probe for shear waves with two orthogonal polarizations

backwall echo is rather large and has to be compensated by a strong magnetic bias field and high transmitter currents. Therefore a transducer with a water-cooled magnet coil was constructed producing continuously a flux density of 1.0 Vs/m^2 in an air gap of about 3.5 mm, and currents with an amplitude of 30 A_{p-p} were sent through the transmitter coils, which have an ohmic resistance of about 15 Ω.

The birefringence effect is measured from the difference in time-of-flight of the first backwall echoes with axial and transverse polarization, respectively. Since the measured value includes also electronic differences between the two measuring channels and differences due to different lift-offs of the two rf coil systems, this electrical phase shift has to be determined in a calibration step and will be subtracted from all measured data as a correction. The calibration step involves measurements of the time-of-flight between the first and the second backwall echo for both polarizations and calculating the difference between both. This value is intrinsically independent from electrical phase shifts. Fig. 4 shows examples for these measurements of time-of-flight differences by means of oscilloscope traces obtained on a part from a turbine shaft with 0.5 m diameter. The difference of time-of-flight between transverse and axial polarization is about 75 ns whereas the measurement on the first backwall echo yields -260 ns. That means the value for phase shift correction is 335 ns, which

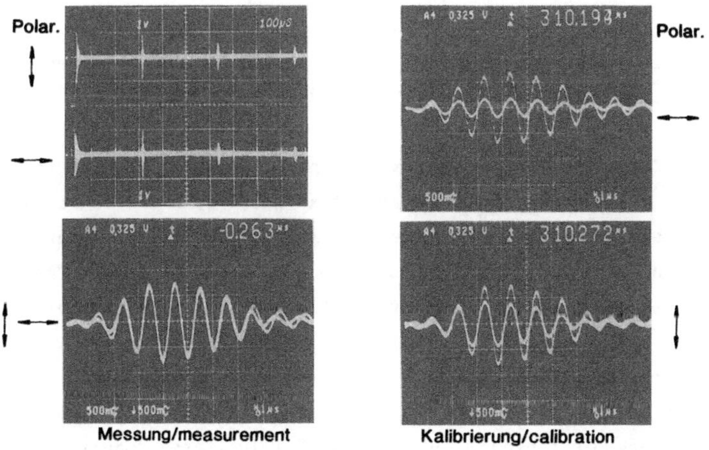

Fig.4. Time-of-flight measurement on a turbine shaft

consists of 250 ns deliberate delay between the trigger signals in the two channels and 85 ns from other phase shifts, mainly due to different lift-offs of the two coil systems.

Measurements with the laboratory system on a complete turbine shaft are planned to be performed until the end of 1988.

Axial bolt stress

Bolts and screws in machine assembly and steel constructions are often required to be installed with specified preloads. In some cases, e. g. in diesel engines, preloads during operation may differ from initial preloads because of an inevitable increase in temperature. The torque to be applied to the bolt at installation is used as a quantity to estimate the amount of preload, but in this correlation there are considerable uncertainties [9]. This circumstance has motivated the application of acoustoelasticity to determine axial bolt stress in-situ. With the method that uses both shear and longitudinal waves propagating along the bolt axis it is neither necessary to know the bolt's initial length nor to have access to more than one end of the bolt [10,11]. In order that the measurement can be performed conveniently and rapidly it is necessary to have a transducer that excites simultaneously longitudinal and shear waves at normal incidence. Transduction of the shear wave by the EMUS technique is most favourable all the more since use at elevated temperatures is a case of interest and the handling of suitable couplants is difficult, whereas for transduction of the longitudinal wave a piezoelectric element should be used in order to accomplish sufficient signal-to-noise ratio [12].

These considerations led to the development of a probe in which the constituents of a piezoelectric probe and of an EMUS probe are combined so that simultaneously longitudinal waves and radially polarized shear waves are excited and received. Fig. 5 shows the set-up of the probe. The rf coils for EMUS excitation and reception are flat spiral coils concentric to the ring-shaped piezoelectric ceramic. The bias magnetization is provided through a cylindrical permanent magnet from neodyme-iron-boron with a flux concentrator and a pot-shaped yoke. The signal-to-noise ratio of the first shear wave echo is typically 30 dB. Fig. 6 illustrates results from a feasibility trial in which a comparison of strain data measured on different screws at different temperatures up to 120°C by the ultrasonic method and with strain gauges, respectively, has been performed [11]. The calculation of the ultrasonic data includes corrections for the temperature dependence of the ultrasonic velocities and for the thermal expansion. Even at a temperature of 120°C the ultrasonic data and the strain gauge data differ by not more than 6%. These results demonstrate, that the implementation of an EMUS transducer and a piezoelectric element into a single probe for the described application has successfully been realized.

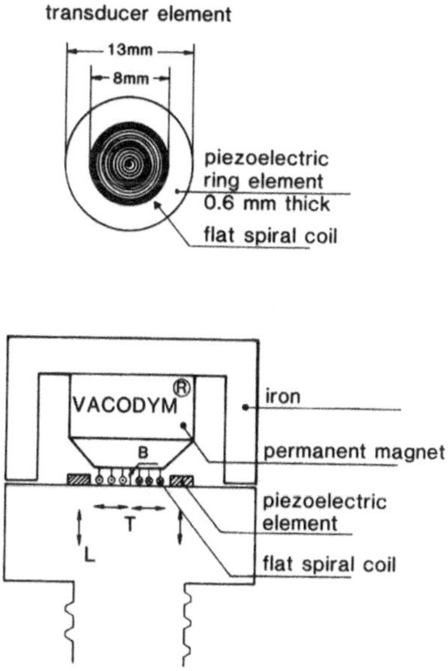

Fig.5. Combined piezoelectric/EMUS probe for compressional and shear waves

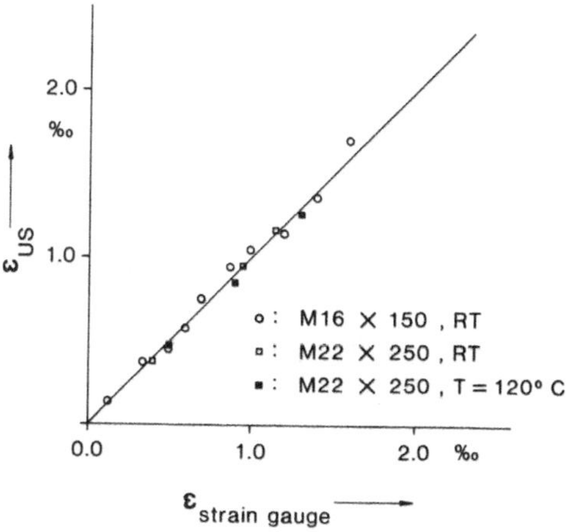

Fig.6. Measurement of applied strain in screws

Stress and texture in rolled sheet

Formability is a crucial property of thin metal sheet assigned to being shaped by deep drawing. It has been shown experimentally [13] as well as established theoretically [14,15] that the anisotropy of plastic strain on deep drawing is significantly correlated to elastic anisotropy produced by crystallographic texture. Therefore evaluation of texture can be used as a nondestructive method to characterize plastic anisotropy. Performing this task via ultrasonic measurements is a technique that is carried out rapidly, samples the bulk of the test material and avoids potentially hazardous radiation. To be more specific stress and texture in rolled steel sheet can be inferred from the angular dependence of ultrasonic plate mode velocities. In order to obtain the maximum information available from ultrasonic measurements both Lamb and SH modes have to be included [16,17].

Both wave types are conveniently excited and received by EMUS transducers. Fig. 7 shows designs of transducers for both wave types making use of periodic permanent magnets. A transmitter and one or two receivers are collinearly hold in position by rigid spacers and the time-of-flight is measured in through-transmission between transmitter and receiver or between the two receivers. The additional expenditure of a second receiver yields the facilitation that the time-of-flight is measured between two signals the phases of which are uniquely related to each other. The texture is characterized by coefficients of the crystallite orientation distribution function (CODF), which are determined from the times-of-flight for different orientations of the transducer array along certain directions of symmetry [16]. An example of experimental results is presented in Fig. 8, where for a set of cold-rolled steel sheet specimens the coefficient w_{400}

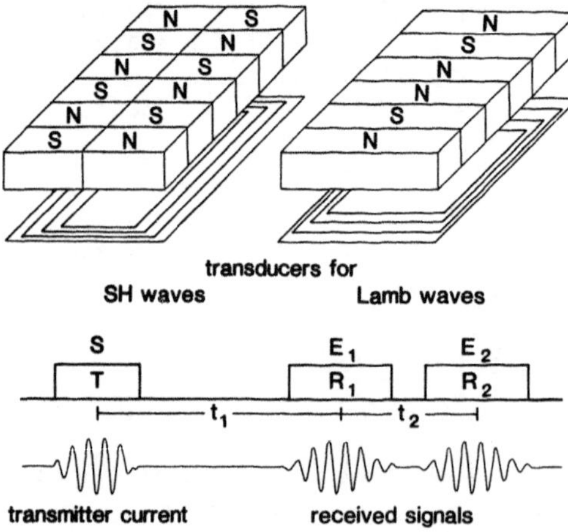

Fig.7. Time-of-flight measurement with guided waves in rolled sheet

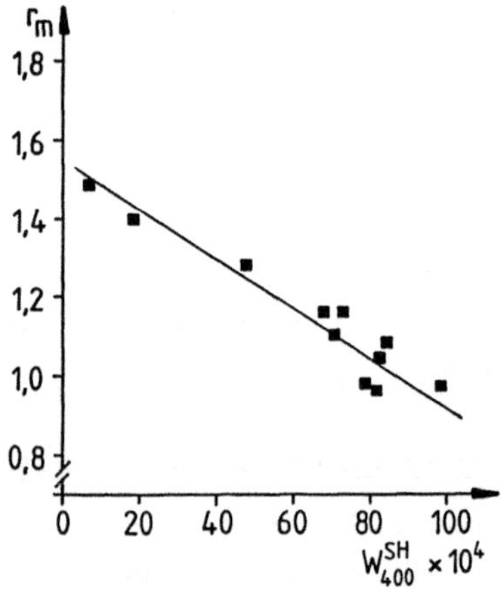

Fig.8. Correlation between r_m-value and w_{400} on cold rolled steel specimens

of the CODF is correlated to the destructively measured r_m-value, which is a measure of plastic anisotropy [17]. It is planned to further develop the technique to on-line use in the rolling mill.

Conclusions

The developments presented above show that EMUS systems are excellently suited for the measurement of stress and texture by ultrasound. By avoidance of viscous couplants measurements by hand can be performed in less time and mechanized scanning of shear wave probes is accomplished. The dry transduction guarantees good reproducibility of rf signal shape, phase and amplitude, and the demands which have to be put upon measurement accuracy are fulfilled by the systems presented. After laboratory systems for several different applications have now been realized, the development of prototypes for industrial applications is taken up.

Acknowledgement

This work was supported partly by the European Coal and Steel Community and partly by the Arbeitsgemeinschaft Industrieller Forschungsvereinigungen (AIF).

References

1. Schneider, E.; Pitsch, H.; Hirsekorn, S.; Goebbels, K.: Nondestructive detection and analysis of stress states with polarized ultrasonic waves. Review of Progress in Quantitative NDE, Vol. 4B, p. 1079-1088, Thompson, D.O. and Chimenti, D.E. (eds.) New York, London: Plenum Press 1985.

2. King, R.B.; Fortunko, C.M.: Determination of in-plane residual stress states in plates using horizontally polarized shear waves. J. Appl. Phys. 54 (1983) 3027-3035.

3. Clark, A.V.; Fukuoka, H.; Mitrakovic, D.V.; Moulder, J.C.: Characterization of residual stress and texture in cast steel railroad wheels. Ultrasonics 24 (1986) 281-288.

4. Chatellier, J.-Y.; Touratier, M.: A new method for determining acousto-elastic constants and plane stresses in textured thin plates. J. Acoust. Soc. Am. 83 (1988) 109-117.

5. MacLauchlan, D.T.; Burns, L.R.; Alers, G.A.: Measurement of stress in steel structures with SH wave emats. Review of Progress in Quantitative NDE, Vol. 7B, p. 1399-1404, Thompson, D.O. and Chimenti, D.E. (eds.) New York, London: Plenum Press 1988.

6. Clark, A.V.; Blessing, G.V.; Thompson, R.B.; Smith, J.F.: Ultrasonic methods of texture monitoring for characterization of formability of rolled aluminum sheet. Review of Progress in Quantitative NDE, Vol. 7B, p. 1365-1373, Thompson, D.O. and Chimenti, D.E. (eds.) New York, London: Plenum Press 1988.

7. Hirao, M.; Hara, N.; Fukuoka, H.; Fujisawa, K.: Ultrasonic monitoring of texture in cold-rolled steel sheets. J. Acoust. Soc. Am. 84 (1988) 667-672.

8. Cassier, O.; Donadille, Ch.; Bacroix, B.: Drawability assessment of steel sheets by an ultrasonic method. Review of Progress in Quantitative NDE, Vol. 8, Thompson, D.O. and Chimenti, D.E. (eds.) to be published.

9. Heyman, J.S.: A cw ultrasonic bolt-strain monitor. Experimental Mechanics 17 (1977) 183-187.

10. Cunningham, B.; Holt, A.C.; Johnson, G.C.: Sensitivity of an ultrasonic technique for axial stress determination. Review of Progress in Quantitative NDE, Vol. 7B, p. 1405-1412, Thompson, D.O. and Chimenti, D.E. (eds.) New York, London: Plenum Press 1988.

11. Schneider, E.; Repplinger, W.: Bestimmung der Lastspannungen in Schrauben mittels Ultraschallverfahren. Abschlußbericht zum AIF-Vorhaben Nr. 86, to be published.

12. Wilbrand, A.: Interaction mechanisms of electromagnetic-ultrasonic transduction. Materialprüf. 26 (1984) 7-11 (in german).

13. Stickels, C.A.; Mould, P.R.: The use of Young's modulus for predicting the plastic strain ratio of low-carbon steel sheets. Met. Trans. 1 (1970) 1303-1311.

14. Bunge, H.J.: Über die elastischen Konstanten kubischer Materialien mit beliebiger Textur. Kristall & Technik 3 (1968) 431-438.

15. Bunge, H.J.: Texture Analysis in Materials Science. London: Butterworth Publ. 1982.

16. Thompson, R.B.; Smith, J.F.; Lee, S.S.: Inference of stress and texture from the angular dependence of ultrasonic plate mode velocities. NDE of Microstructure for Process Control, p. 73-79, Wadley, H.N.G. (ed.) Metals Park, Ohio: The American Society for Metals 1985.

17. Schneider, E.; Spies M.: Nondestructive analysis of the deep-drawing behaviour of rolled sheets by ultrasonic techniques. These procedings.

Fiber-optic Based Heterodyne Interferometer for Noncontact Ultrasonic Determination of Acoustic Velocity and Attenuation in Materials

James B. Spicer
James W. Wagner

The Johns Hopkins University
Center for Nondestructive Evaluation
Maryland Hall 102
Baltimore, Maryland 21218
U.S.A.

Summary

Optical methods offer several potential advantages over contact transducer methods for accurate determination of acoustic velocity and attenuation in solids. Their noncontact nature eliminates inaccuracies introduced by variations in coupling efficiencies and delays. Broad bandwidth detection by optical means permits extreme fidelity of displacement measurement thus eliminating the need to deconvolve the transducer characteristics from the recorded signal. Finally, the effective "contact" area of optical transducers can be made to be smaller than 10 microns avoiding acoustic phase cancellation problems and permitting accurate measurements of surface wave velocity.

A heterodyne fiber optic interferometer system has been designed and constructed taking advantage of a unique input coupler and design which decreases system complexity and power losses. An absolute calibration scheme has been used to demonstrate system sensitivity to 5.6x10-5 Angstroms per square root Hertz of system bandwidth. A dual probe variation of the system has been implemented which permits direct measurement of surface wave velocities and attenuation. Laser generated surface waves in rolled aluminum alloy have been detected using the dual probe interferometer to measure rolling texture.

Introduction

Optical methods for generation and detection of ultrasound are being used with increasing frequency in situations where more conventional contact or near contact methods are inappropriate or impractical. Although optical methods are still in general less sensitive than contact techniques, there are certain attributes of these newer methods which offer the potential means for new forms of ultrasonic testing. For example, the small optical probe beam diameter (often less than 10 microns) permits detection of ultrasonic signals from a correspondingly small area of the test specimen. As a result, higher spatial scanning resolution may be obtained and phase cancellation issues are less of a concern. Since there is no effective mechanical interaction of optical probes with the sample surface, multiple probe beams may be used as spatial arrays without perturbing the propagating ultrasonic field.

Both of the features discussed above were used to advantage in the dual probe fiber optic heterodyne interferometer system to be described. Coupled with pulsed laser generation of surface acoustic waves, the dual probe interferometer has been used to measure velocity differences as a function of direction in rolled aluminum plate. In addition, a signal analysis method has been developed which may permit accurate measurements of surface acoustic wave attenuation independent of variations in detector "coupling" (ie. surface roughness and reflectivity).

Dual Beam Fiber Optic Heterodyne Interferometer

A heterodyne interferometer system was constructed in accordance with the design shown schematically in Figure 1. Unlike many other fiber based interferometric designs, this system requires only a single input coupler between the laser and the fiber thus improving alignment stability and reducing losses. After proper selection or tailoring of the input fiber diameters, a lens system is used to focus into corresponding optical fibers both the diffracted and undiffracted beams leaving the acousto-optic cell. The two input fibers are in fact opposite ends of a single fiber loop which passes through a 3dB directional fiber coupler. Light propagating in this loop in a clockwise direction is undiffracted by the acousto-optic cell and is, therefore, of the same frequency as the laser light. It passes through a delay loop and a polarization adjuster before coming into the coupler. Once in the coupler, the light is split with half of its power going into the optical detector. (The light which is not split out at the coupler need not be considered further.) Light which is diffracted by the acousto-optic cell is shifted in frequency by

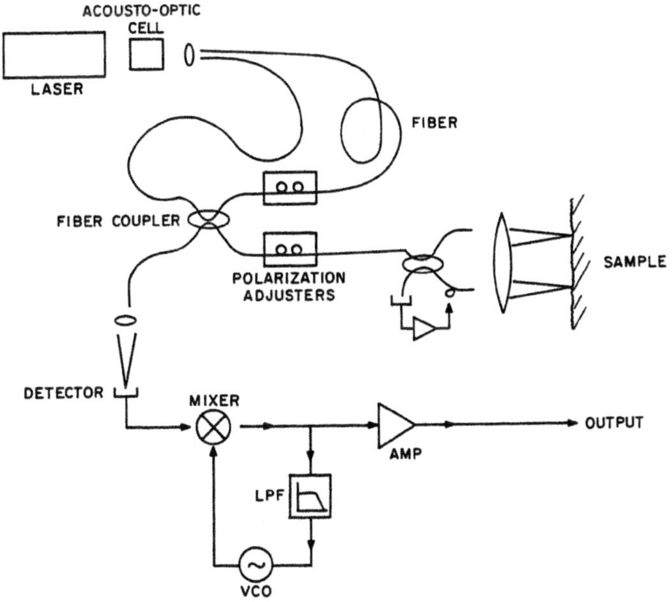

Figure 1. System Schematic

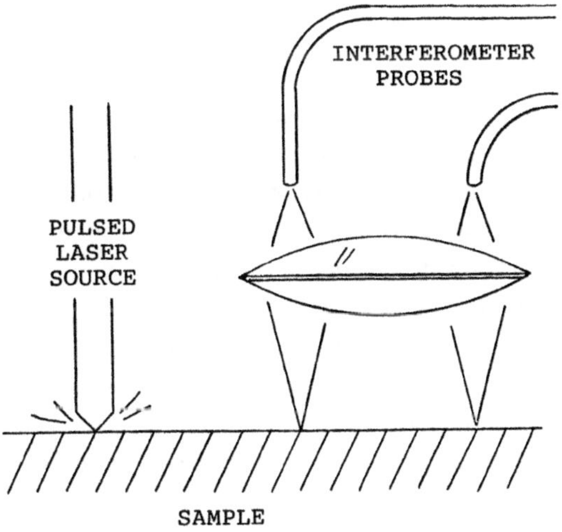

Figre 2. Test configuration.

40MHz relative to the undiffracted beam. It is coupled into the other path through the loop so that it propagates in a counter-clockwise direction. Light from this path is split at the coupler and subsequently reflected from the sample surface. Once coupled back into the fiber, this light recombines with that from the other path so that light from both paths interfere at the detector. In the absence of sample motion, the intensity of the light falling on the detector varies sinusoidally at the acousto-optic cell drive frequency, 40MHz. Vibration of the sample causes a phase (or frequency) shift in the detected signal. With the phase demodulation scheme shown, detection sensitivities to 5.6 X 10-5 Angstroms per square root Hertz of system bandwidth have been demonstrated for a prototype system [2].

Note from the schematic that the beam which leaves the first coupler passes through a second coupler en route to the sample. It is through the use of this second coupler that two surface probe beams are derived from a single output beam. In order to maximize the intensity of the light returned to the detector from the two probes, the optical path length in one probe must differ from the other by some multiple of one half of the optical wavelength. For this purpose, a feedback circuit incorporating a piezoelectric "stretcher" may be used as shown to null the intensity at the feedback detector.

Surface Wave Velocity Measurements

The test configuration shown in Figure 2 was used to measure surface wave velocity. The probe spot separation was about 6mm. An air cooled Nd:YAG laser rated at 19mJ per pulse with a 4ns pulse duration was focussed with a cylindrical lens to provide an acoustic source. The specimen used for these experiments was a 1/4" thick 2024-T351 aluminum alloy plate machined from the center of a 2" thick rolled plate. By simply rotating the sample, it was possible to measure surface wave velocities perpendicular and parallel to the rolling direction.

Figure 3 is the result of averaging the surface displacements as a function of time for 16 pulses of the laser

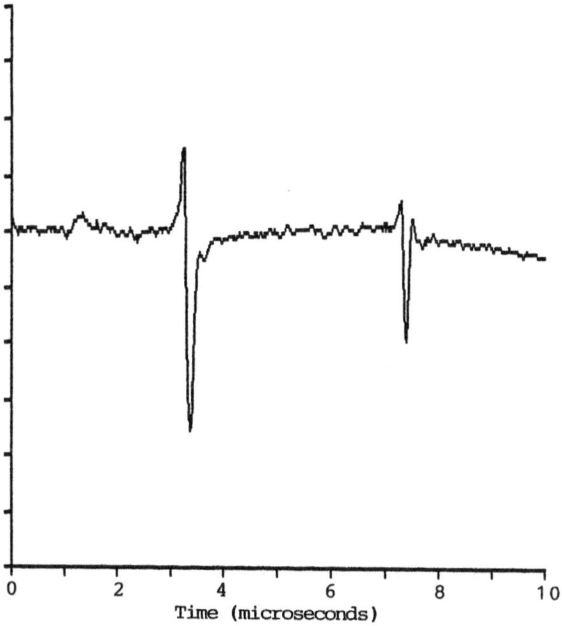

Figure 3. Laser generated Rayleigh spike detected with dual beam interferometer.

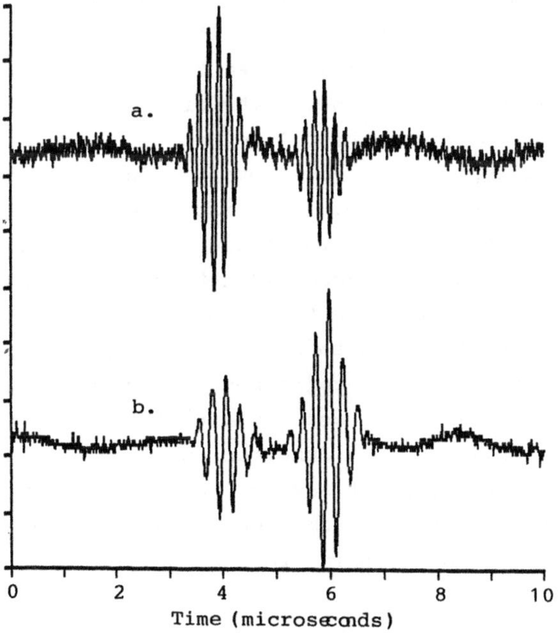

Figure 4. Dual probe "attenuation" data.

source. Two acoustic pulses are observed in the trace as the wave passes first beneath one probe and subsequently beneath the second. The time required for the wave to transit between the two probes was obtained after autocorrelation of the data. The autocorrelation procedure is especially helpful in cases where the signal has significant levels of noise which may lead to greater measurement uncertainty if one attempts to measure time delays directly from the raw (uncorrelated) data. The accuracy of the measurement is limited by the digitizing rate of the data acquisition system. Increases in time resolution have been obtained, however, by interpolating between sample points in the original data set [1]. For this particular test case, a 10ns sampling period was used and data interpolation was not performed. Test results showed a 1% variation in surface wave velocity between surface waves propagating in the rolling direction compared with those propagating in the direction perpendicular to the rolling direction.

Surface Wave Attenuation

Preliminary work has shown that a test configuration similar to that used for velocity measurements may be used to measure attenuation as well. In initial experiments, however, a piezoelectric transducer was used to generate surface waves using a mode conversion block. The transducer was driven with a 5MHz RF tone burst of about 1.5microseconds duration. Two signal traces typical of those obtained for these measurements are shown in Figure 4. Note that from the relative amplitudes of the two bursts recorded in Figure 4a, one could easily deduce that considerable acoustic attenuation had been suffered by the surface wave between the probe beams. From Figure 4b, however, one is at a loss to explain the apparent increase in ultrasonic wave amplitude corresponding to the measured increase in signal amplitude. Unfortunately, the measured signal peak amplitude is a function not only of the displacement amplitude but also of the surface reflectivity. Thus even optical methods for acoustic attenuation measurements suffer from apparent coupling variations.

Fortunately, one may take advantage of the nonlinear response of interferometric displacement detection in order to separate the effects of surface displacement from sample reflectivity. The effect of this nonlinearity may be seen by looking at the frequency spectrum of the signal at the optical detector as a result of a sinusoidal disturbance (Figure 5). As can be seen, the spectrum has a strong component at the carrier frequency with sidebands separated from the carrier at multiples of the acoustic displacement frequency. Prior work has shown that the first harmonic increases linearly with displacement amplitude while the second harmonic increases as the square of the displacement [2,3]. However, both the first and second harmonics scale linearly with surface reflectivity. Therefore by computing the ratio of the amplitudes of the first and second harmonics in the tone burst as recorded by the two probes, the degree of attenuation may be calculated directly. A similar calculation may be performed using the carrier amplitude relative to the first harmonic so long as heterodyne interferometry is used. In fact these measurements can be made with more certainty than the first/second harmonic measurements since the signal to noise ratio at the first harmonic is greater than that at the second harmonic. Unfortunately, a carrier component is not present in the signal from a homodyne system, and neither is it present once the heterodyne signal has been detected and downshifted. Owing to the relatively poor signal strength in the second harmonic, it is necessary that comparatively large acoustic waves be generated to measure accurately surface wave attenuation.

Conclusions

A dual beam fiber optic based heterodyne interferometer has been designed and constructed for the purpose of measuring surface acoustic wave velocity and attenuation. Surface velocity has been measured with a 10ns resolution over a 6mm gage length. Preliminary studies show that attenuation values may be obtained independent of variations in surface

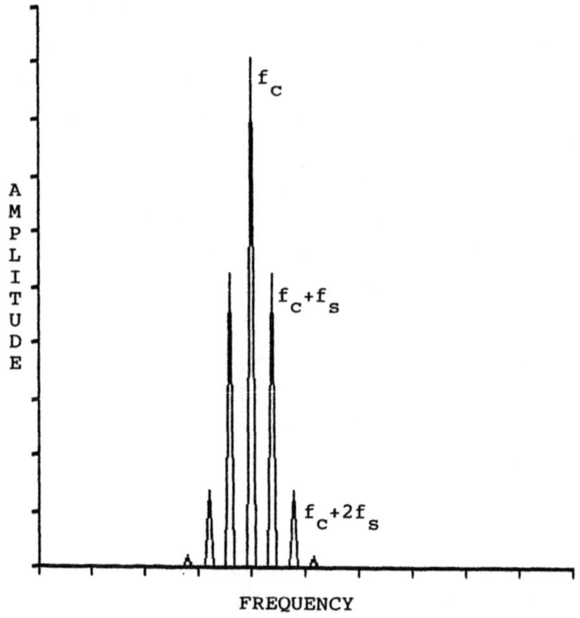

Figure 5. Frequency spectrum of heterodyne signal corresponding to sinusoid displacement at f_c.

reflectivity by proper calculation from parameters obtained from the frequency spectra of the detected signals.

References

1 Aussel J.-D., Monchalin J.-D., Precision acoustic velocity measurement and elastic constant determination by laser-ultrasonics, Ultrasonics (in press) 1988.

2 Spicer J.B., Wagner J.W., Absolute calibration of interferometric systems for detection and measurement of acoustic waves, Applied Optics 27(16), 3561-3566, (1988).

3 Wagner J.W., Spicer J.B., Theoretical noise-limited sensitivity of classical interferometry, J. Opt. Soc. Am B, vol 4, 1316-1326 (1987).

Samples were provided by the U.S. Naval Research Laboratory.

The 3MA-testing Equipment, Application Possibilities and Experiences

W.A.THEINER, B.REIMRINGER, H.KOPP, M.GESSNER

Fraunhofer-Institut für zerstörungsfreie Prüfverfahren (IzfP), Saarbrücken

Summary

The 3MA- (micromagnetic multiparameter microstructure and stress analysis) testing unit can be used for residual stress and microstructure analysis in the surface near zone. The equipment is built up in a modular technique. The modules are: the magnetic Barkhausen noise M, the acoustic Barkhausen noise A, the incremental permeability μ_Δ and the dynamic magnetostriction E_λ. The used sensors - transducers are matched to the used methods. The internal microcomputer controls the testing unit. This internal computer respectively the host computer are responsable for the evaluation of measuring datas. Up to know the equipment is used for hardness tests, anisotropy measurements, residual stress measurements, for the characterisation of microstructure states and for the determination of microstructure gradients as they are produced by surface hardening processes.

Zusammenfassung

Das 3MA- (Mikromagnetische Multiparameter Mikrostruktur und Spannungsanalyse) Gerät kann zur Eigenspannungs- und Gefügeanalyse im oberflächennahen Bereich eingesetzt werden. Das Gerät ist modular aufgebaut und besteht aus den Modulen: magnetisches Barkhausenrauschen M, akustisches Barkhausenrauschen A, der Überlagerungspermeabilität μ_Δ und der dynamischen Magnetostriktion E_λ. Beim Einsatz der einzelnen Verfahren kommen entsprechend angepaßte Sensoren bzw. Aufnehmer zum Einsatz. Der im Gerät eingebaute Mikrocomputer überwacht die einzelnen Gerätefunktionen. Dieser Mikrocomputer bzw. ein externer Rechner übernimmt auch die Auswertung der jeweiligen Meßkurven. Das Prüfsystem wurde bisher für die Härtemessung, für die Anisotropiebestimmung in Blechen, bei der Eigenspannungsmessung und Gefügecharakterisierung sowie für die quantitative Bestimmung von Gefügegradienten, wie sie bei den einzelnen Härteprozessen auftreten, eingesetzt.

3MA-Testing-Unit

The 3MA testing unit consists of the basic equipment, different transducers and the yoke. Transducer and yoke can be integrated in a hand operated probe. The sensor can also be mounted on a manipulator-unit.

The different non-destructive methods which can be used by the 3MA set-up are: the magnetic Barkhausen noise (M), the acoustic Barkhausen noise (A), the incremental permeability (μ_Δ), and the electrical conductivity (σ) /1/, the dynamic magnetostriction (E_λ), and the coercivity H_c, which can be derived from the methods mentioned above or from the time signal of the tangential field strength H_t /2/. Additionally the harmonic analysis of all these methods can be used for further correlations to microstructure states. All quantities with the exeption of some eddy current quantities and the electrical conductivity, are measured during magnetic excitation. The magnetic excitation of the test specimen component is carried out by a yoke. This yoke must be adapted to the geometry of the specimen respectively to the component to be tested. The tangential magnetic field strength H_t is measured by a Hall probe which is normally integrated into the sensor. The maximum of the exciting magnetic field strength as well as the frequency are controlled by the 3MA microcomputer.

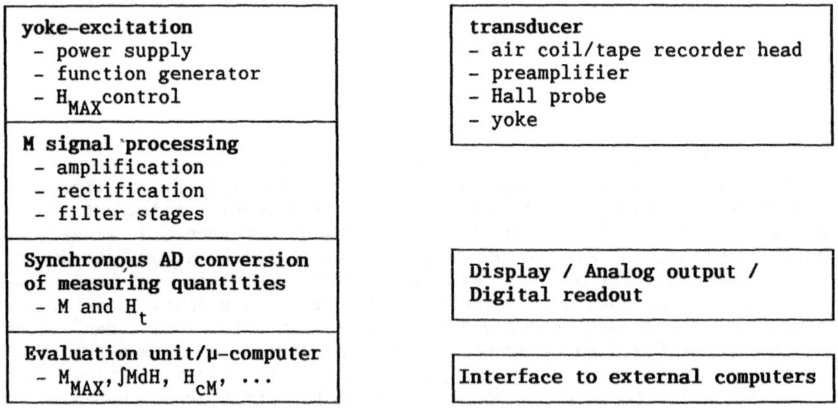

Fig.1: 3MA-Modul "magnetic Barkhausen noise"

In Fig.1 the 3MA-modul "magnetic Barkhausen noise" is sketched. The Barkhausen events are picked up with an air coil or tape recorder head and amplified by different stages between 60 dB and 100 dB.

These signals are then rectified and after a synchronous analog digital conversion of the Barkhausen signal M and the tangential field strength H_t different nd quantities can be evaluated from this envelope. The most common quantities are the coercivity value H_{CM}, the maximum of the Barkhausen noise signal M_{max} and the total irreversible content of Barkhausen events $\int MdH$. The measuring range lies between 100 Hz and 10 MHz. Below 160 kHz variable filter stages can be used for analyzing the Barkhausen noise whereas between 160 kHz and 10 MHz fixed band bass filters are applied. The yoke exciting frequency f_E can be changed between 0.1 Hz and 1.6 kHz.

In Fig.2 the 3MA incremental permeability modul and in Fig.3 the dynamic magnetostriction E_λ modul are sketched.

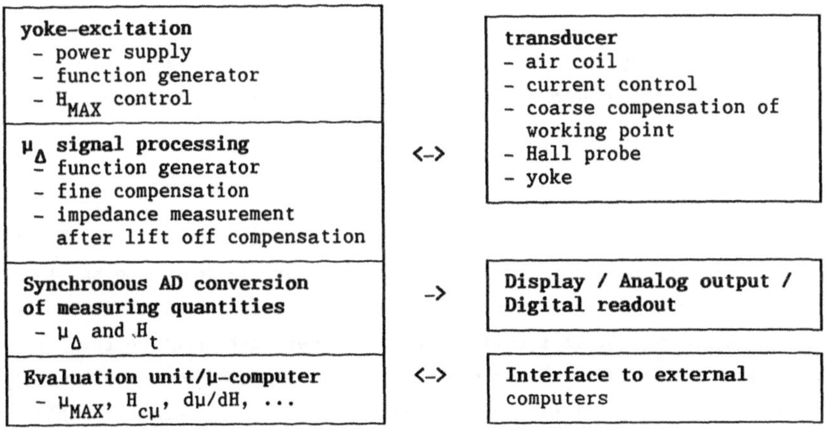

Fig.2: 3MA-Modul "incremental permeability"

The exciting frequency f_Δ of the incremental permeability probe can be chosen between 1 kHz and 30 MHz. The frequency in the case of the dynamic magnetostriction is fixed and can be matched to the testing problem in the range between 0.5 and 2 MHz.

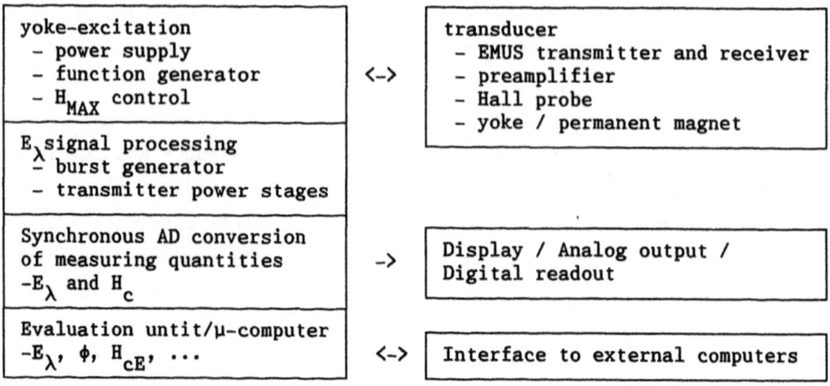

Fig.3: 3MA-Modul "dynamic magnetostriction"

By the exciting frequency f_Δ, the analyzing frequency f_A and yoke exciting frequency f_E the penetration depth respectively the analyzing depth can be chosen in a wide range corresponding to about 10 µm up to several mm.

The measuring curve as well as the evaluated values are displayed on the 3MA screen or by analog/digital outputs. Furtheron the whole equipment, the testing procedures as well as the evaluation of testing results can be controlled or performed by an external computer.

Results

Nondestructive hardness tests: For direct correlations, one has to use the coercivity H_c. The coercivity value can be derived from all methods mentioned above. After a calibration step where different coercivity values are correlated with different hardness values of one steel quality, quantitative measurements can be performed. Fig. 4 gives an example for different microstructure states respectively hardness values of the steel 22 NiMoCr 3 7; in Fig. 5 results of a conventional (HV10) and of a nondestructive hardness test (HV_{nd}) over a weld is given. One can see that the heat affected zones with the great hardness values are recognized in a very good way by the used sensor whereas the hardness value of the base material did not assure a good correlation to the coercivity values. If one applies a linear regression, one obtains for the examined region $HV_{nd} = 48.8+28H_{cM}\pm25$. Further results and evaluation algorithms are discussed in /3/. Investigations up to now have shown that in the case of direct correlations between hardness and coercivity testing results compare well with conventional hardness tests whenever no microstructure gradients, no residual stresses respectively stress gradients and no anisotropic material behaviour influence the coercivity measurement.

	nd quantities
	H_{cM}, $H_{c\mu}$, H_{cE_λ}, H_{co}
	$\mathbf{HV = f(H_{c,n})}$ $\quad n = M/\mu/E_\lambda/0$

Tab.1: nd hardness test

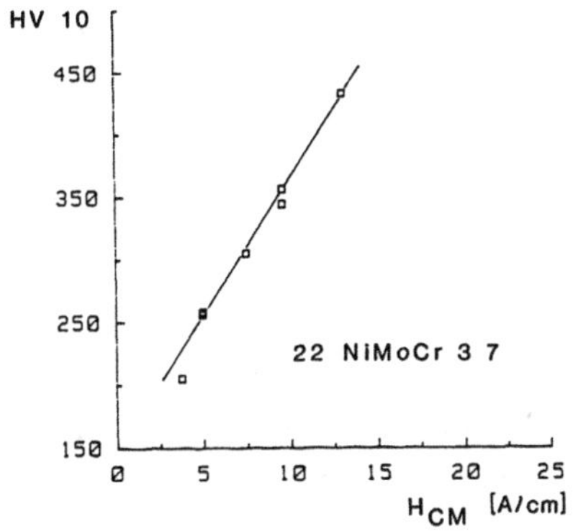

Fig.4: Calibration curve coercivity H_{cM} - hardness HV10

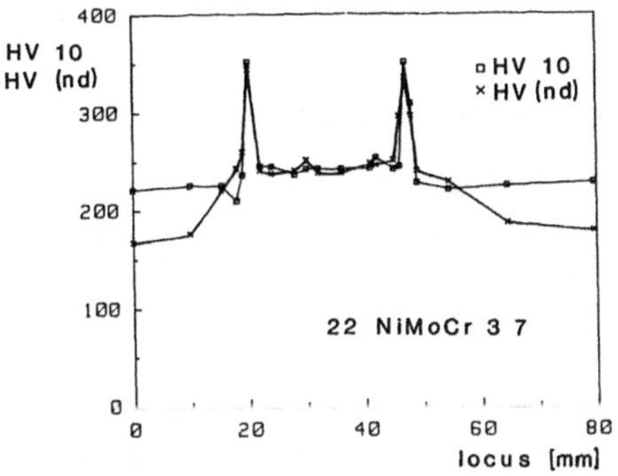

Fig.5: Hardness test perpendicular to a weld seam
HV10-conventional hardness test, HV_{nd} nondestructive value

Hardness and residual stress measurement: In this case quantitative measurements can only be performed by using one nondestructive quantity, if either the microstructure state or the residual stress state becomes dominant. In all other cases one has to use at least two independent nondestructive quantities as described e.g. in /4/ for the testing procedure of low pressure turbine blades. To get quantitative results, the used nondestructive quantities have to be calibrated. These quantities must be chosen in such a way that one can combine microstructure sensitive and stress sensitive quantities (tab.2). Microstructure sensitive quantities are e.g. the coercivities, the electrical conductivity σ, the distortion factor K, whereas stress sensitive quantities are e.g. the M_{MAX}-value, the broadening of the incremental permeability $\Delta\mu_{\Delta}$ or the quantity $\int dMH$. In the case of the dynamic magnetostriction one has to use the derived coercivity H_{cE} and the amplitude of the excited ultrasonic wave E_{λ} at defined working point at the magnetostriction loop.

To perform quantitative measurements one can use after the calibration step numerical iterative methods. In this case one starts with the microstructure sensitive quantity, e.g. the coercivity, and after this testing step one can evaluate in a first approximation the residual stress level. In a second approximation, one can compare these stress and hardness values with the real values of the calibration curves and so on.

nd quantities

microstructure sensitive quantities

H_c, μ_{Δ}, σ_{el}, K, ...

stress sensitive quantities

M_{MAX}, $\int MdH$, ΔM, A, E_{λ}, $\Delta\mu_{\Delta}$

$HV = f(H_{cM}, M_{MAX})$ or $HV = f(H_{c\mu}, \Delta\mu_{\Delta})$

$\sigma = f(H_{cM}, M_{MAX})$ or $\sigma = f(H_{cM}, E_{\lambda}(Hi))$

Tab. 2: nd hardness and residual stress test

Surface hardness and surface residual stress and microstructure gradient test: For the surface hardness and the surface residual stress determination the procedure described above can be used. To determine microstructure gradients normal to the surface one has to change the penetration depth of the used method. This can be done e.g. by the analyzing frequency f_A of the magnetic Barkhausen noise, the exciting frequency f_Δ in the case of the incremental permeability or the eddy current method, and by the macroscopic exciting frequency f_E of the yoke. Whenever the magnetic properties of the bulk and surface states are different, very sensitive quantities can be derived e.g. from the magnetic Barkhausen noise /5/ and other techniques mentioned above. In the sensor encycling method also eddy current quantities can be used. In the set-up technique the analyzing depth limit of micromagnetic quantities is about 3 mm. If the hardening depth becomes greater than 2 mm, in many cases also the ultrasonic backscattering method is a very useful tool. Quantitative results have been obtained of case-, inductive- and laser hardened materials, for grinding states /6/ and shot peened surface states. The results compare in many cases very well with the metallographically determined hardening depths.

nd quantities

M, dM/dH, A, μ_Δ, $d\mu_\Delta/dH$, σ_{el}, ...

$\sigma = f(H_c, M_{MAX})_{f_A \hat{=} S}$; ...

$HV = f(H_c, M_{MAX})_{f_A \hat{=} S}$; ...

$d = f(H_{cb}, H_{cS}, M_b, M_S)_{H, f^*}$

d = case hardening depth, inductive hardening depth

b = bulk material, S = surface

$f^* = f_A$ or f_Δ or f_E

Tab. 3: nd surface hardness and nd surface residual stress and nd microstructure gradient tests

Anisotropy/texture: For these investigations nondestructive measuring quantities must be used which are sensitive to crystal, grain size or precipitation textures. Grain size and precipitation textures can be indicated by direction dependent dynamic micromagnetic quantities (e.g. M, A, μ_Δ) and coercivity measurements. In the case of crystal anisotropy very sensitive quantities can be derived from the dynamic magnetostriction as well as from ultrasonic time of flight measurements /7/. In this area especially micromagnetic techniques must be developped furtheron for getting quantitative results.

nd quantities

E_λ, M, μ_Δ, H_c, ... and ultrasonic measuring quantities

+

H-exciting field direction

+

probe characteristics

Tab. 4: nd determination of material anisotropy

References:

/1/ R.Becker, Ch.Rodner, K.Betzold
Werkstoffsortierung und Bestimmung der Rand- und Einhärtetiefe; σ,μ und Impedanzspektren als Prüfgrößen
DGzfP Berichtsband 10 (1987), 683-693

/2/ G.Dobmann, H.Pitsch
Magnetic tangential field-strength-inspection, a further ndt-tool for 3MA, see these proceedings.

/3/ W.A.Theiner, P.Deimel
Nondestructive Testing of Welds with the 3MA-Analyzer
Nuclear Engineering and Design 102 (1987) 257-264

/4/ E.Stücker, G.Gartner
Quality Control of Last Stage Blades with Flame Hardened Leading Edges. ASME/IEEE Power Generation Conference. Miami Beach, Florida, October 4-8, 1987 JPGC-Pwr 56

/5/ W.A.Theiner, R.Kern, R.Conrad
Determination of Surface Integrities by Ferromagnetic Quantities. First International Conference on Surface Engineering, Brighton, England, 26-28 June 1985, P25

/6/ R.Conrad, R.Jonck, W.A.Theiner
Zerstörungsfreie Ermittlung von wärmebeeinflußten Randschichten und deren Dicke.
HTM 41, 1986, 4, 213-218

/7/ E.Schneider, M.Spies
Nondestructive analysis of the deep drawing behaviour of rolled sheets by ultrasonic techniques (see these proceedings)

Small Neutron Radiopgraphy Systems and Their Applications*

H.-U. MAST, T. BRANDLER, E. KNORR, P. STEIN

Industrieanlagen-Betriebsgesellschaft mbH (IABG)
Ottobrunn, Fed. Rep. Germany

Summary

Radiography with neutrons is a nondestructive testing method of growing importance for the inspection of aerospace and electronics components as well as ceramics. Potential applications of neutron radiography that have been investigated as part of a current research project include the detection of corrosion on metals, entrapped moisture and adhesive defects in metal, composite and metal-composite structures, defects in composite materials, flaws in the potting of electronics components and porosity/density fluctuations in ceramics. However, the large-scale use of this NDT method requires the availability of small mobile or stationary in-house neutron radiography systems. A comparative review of existing and possible future small neutron sources shows that sealed-tube neutron generators in particular represent a good compromise with respect to neutron yield, portability, reliability, safety and costs.

1. Introduction

Neutron radiography is a nondestructive testing (NDT) method similar and complimentary to radiography with X-rays. Thermal neutrons are predominantly scattered or absorbed by nuclei of some elements with low atomic numbers, in particular hydrogen, carbon and boron, whereas most metals and materials opaque to X-rays are comparatively transparent to neutrons. Hence neutron radiography can be used to detect details in hydrogeneous or other light materials, even when these are surrounded by metals. The first major applications of neutron radiography outside the nuclear industry were the detection of defective spacecraft ordnance [1] and of residual ceramic core in cast turbine airfoils [2]. Yet only test or research reactors could provide neutron beams of sufficient intensity to produce radiographs with a high ratio of collimation, i.e. a geometrical unsharpness as low as that of X-radiographs, within an exposure time of a few minutes.

For years the large-scale use of neutron radiography was delayed by the lack of small and inexpensive but powerful neutron sources. However, recent progress in this area led to novel applications [3,4] particulary in the maintenance of aircraft structures, which, in turn, accelerated the present development of the second generation of small neutron sources and radiography systems.

* Work supported by the Bundesministerium für Forschung und Technologie

2. Applications of neutron radiography

In view of a growing need for inspection methods for modern materials a joint research project between four German institutions was set up in order to identify and investigate promising applications of neutron radiography [5]. Neutron radiographs of appropriate test objects were produced at a test reactor and by means of a new neutron generator-based mobile radiography system now in operation at IABG or similar small systems and compared to results obtained by conventional NDT methods. In the course of this project neutron radiography was successfully applied to the detection of

- corrosion products on metals
- entrapped moisture, water or liquids in honeycomb structures
- adhesive defects (voids in the adhesive, too much or too little adhesive)
- cracks and delaminations in the potting of integrated circuits and similar electronic components, made visible by a liquid penetrant opaque to neutrons
- porosity and density fluctuations in ceramics, particulary silicon carbide
- various flaws and defects in composite materials
- missing or defective O-ring seals in metallic joints

All of these applications can be realised with small neutron radiography systems.

Three examples are presented below. Figure 1 shows the neutron (left) and X-radiograph (right) of the section of a helicopter rotor blade with an aluminium honeycomb core. A part of the honeycombs was removed and replaced by corroded material (1), visible in the neutron radiograph due to the hydrogen content of the corrosion products. The honeycomb image seen in the neutron radiograph does not, as in the X-radiograph, originate from absorption in the aluminium (which is transparent to neutrons), but from compressed layers of adhesive between the honeycomb core and the outer aluminium profile (2). Entrapped water in the honeycombs produces high-contrast images (3) in the neutron radiograph. Also visible in the neutron radiograph are voids in the adhesive (4). Aluminium wedges containing steel (5) and aluminium (6) corrosion products are distinguishable only by X-radiography, since the hydrogen content is similar. Wedges of polythene (7) and sealant (8) were added for calibration purposes. This example indicates that the comparison of the X- and neutron radiographs of a given object is particulary useful for the characterization of the materials included. Figure 2 shows the neutron radiograph of a carbon-fiber-composite surrounded by steel. The ply orientation is clearly recognizable. Figure 3 shows the neutron radiograph of a ring of silicon carbide ceramics with an inhomogeneous distribution of silicon and carbon, visible due to the large difference between the attenutation coefficients of both elements for neutrons. The neutron radiographs described above were produced with small, partially mobile neutron radiography systems.

3. Present and future small neutron sources

Small neutron sources may be classified in five groups in descending order of neutron yield:

- Cyclotrons

For a few years small compact cyclotrons [6] and even smaller cyclotrons with superconducting magnets [7] have been available. Using the the Be(p,n) nuclear reaction or the equivalent reaction in lithium targets at bombarding ion energies of up to 18 MeV and beam currents of 50-200 μA a neutron yield of several 10^{12} up to 10^{13} n/s is achievable. Hence high-quality radiographs can be produced within an exposure time of several minutes.

Even superconducting cyclotrons cannot be considered as mobile, e.g. as required for aircraft maintenance, but they are transportable.

Cyclotrons will for some time be the most powerful non-reactor neutron sources available, but their cost of several million US-$ and operating complexity restrict their employment to service centres with a high throughput of test objects.

- Van de Graaff accelerators

Stationary Van de Graaff acceletators were employed very early, by using the Be(d,n) nuclear reaction, as in-house neutron radiography systems, for they are simple accelerators capable of providing accelerating potentials of more than 2.5 MV, above which the Be(d,n) neutron yield is higher than that of the T(d,n) reaction used in neutron generators. However, Van de Graaff accelerators may now be considered as less attractive, since small cyclotrons with a higher neutron yield are available. A project to develop a mobile Tandem - Van de Graaff accelerator in the USA was terminated in 1987 for apparent lack of success.

- Neutron generators

The nuclear reaction T(d,n) shows a strong resonance in the cross-section for deuteron energies of 120 keV, hence accelerating potentials of only a few hundred kV are required to obtain powerful neutron sources. Whereas early neutron generators using this reaction needed continuous pumping to maintain a vacuum and had very short-lived targets, modern systems are built as sealed tubes with a weight of a few kilograms, containing a deuterium and tritium (for replenishment of the target) reservoir, an ion source, a small accelerator and a target. The high voltage and cooling supply units are separate and readily transportable. The small size and weight of these sealed-tube neutron generators make them ideal for mobile radiography systems.

Major drawbacks of sealed-tube generators are as yet the low neutron yield of approximately 10^{11} n/s, further enhanced by the high energy of 14 MeV of the neutrons emitted and therefore low fraction of neutrons thermalized in the moderator, as well as a short target lifetime or a few hundred hours with a continuous drop in the neutron yield. In order to overcome these remaining drawbacks a Eureka-project with the aim to develop a powerful mobile neutron radiography system, "DIANE" (Dispositif Integre et Automatique de NEutrographique) [8], was initiated by SODERN (France), SENER (Spain) and IABG in 1987. The essential part of the system is a new sealed-tube neutron generator, "GENIE 46", to be developed by SODERN. The generator will have a neutron yield of 10^{12} n/s and a target life of more than 1000 hours at full output. The first "GENIE 46"-prototype will be incorporated into the IABG-system in late 1989.

Primarily designed for mobile use, "DIANE" will also be attractive as a stationary system because of the high neutron yield, the comparatively low cost of a few hundred thousand US-$ for the neutron source "GENIE 46 " and because of the simple, essentially maintenance-free operation.

- Electron linear accelerators

Several electron linear accelerators (linacs) that were originally built to produce γ-radiation have been used as neutron sources by exploiting reactions such as Be(γ,n). Sufficient experience has been gained to conclude now that electron linacs are quite unsuitable for routine neutron radiography. Due to the simultaneous emission of γ-rays in the accelerator target neutron images must be recorded by indirect techniques: neutrons first activate a foil, which is subsequently used to expose a conventional film with radiation emitted by radioactive decay. This procedure, requiring several hours up to days, is too time-consuming for all work except research.

- Radioisotope sources

Radioisotope facilities have so far been the most widely used neutron sources for radiography, because they are simple to operate and require no maintenance.

Of all radioisotope neutron sources available, californium (Cf252) has by far attracted the most attention due to the high yield per unit cost, the small size and the low γ-contamination of the neutron beam. Typical mobile radiography systems contain several milligrams of californium, surrounded by a moderator and a biological shield. Portable facilities with fifty milligrams of californium, as planned for aircraft maintenance in the USA [9], have a biological shield with a weight of several tons.

Apart from safety aspects, which would make licensing of a californium facility in many countries difficult if not impossible, a disadvantage of californium is the short half-life of 2.65 years. One milligram of californium produces 2.3 $\times 10^{9}$ n/s at a cost of approximately 10000 US-$. In order to replace for example a "GENIE 46" - neutron generator, an equivalent source of about 100 mg of californium may be chosen, which requires an annual replenishment of 26 mg to maintain the initial neutron yield.

Furthermore, the future use of californium is questionable, for the production of this radioisotope was discontinued in the USA in 1987 and its resumption is uncertain.

The above comparison between neutron sources shows that small cyclotrons and neutron generators will be dominant for neutron radiographic purposes in the foreseeable future. The former may be installed as stationary systems where the throughput justifies the high expenses, while the latter represent simple to operate neutron sources for mobile and stationary use that will be sufficiently powerful for most applications.

4. The future

Taking into account preliminary results of current research projects, there can be little doubt that the demand for neutron radiography will increase in the future. With regard to the radiography systems employed, it appears unlikely that more powerful neutron sources than those now in operation or development will be useful in view of stringent licensing conditions and high cost of adequate shielding. However, the present approach of reducing the geometrical unsharpness in neutron radiographs by increasing the ratio of collimation at the expense of neutron flux, which varies as an inverse square function of the ratio of collimation, may become obsolete due to the possibility of reducing the geometrical unsharpness by digital image processing.

It is feasible to develop a computer-based image processing system to reduce the geometrical unsharpness of film or electronically recorded radiographs by digitally eliminating in the Fourier spectrum of the radiographs the Fourier-transformed point spread function (PSF) originating from the finite diameter of the neutron collimator. Hence neutron radiographs with a given quality can be produced in shorter exposure times, reducing the need for the development of more powerful neutron sources that also require a more extensive shielding.

References

[1] Golliher, K.G., Neutron radiography of Apollo ordnance, Proc. of the First World Conference on Neutron Radiography, San Diego (1981) pp 325-332

[2] Edenborough, N.B., "Neutron radiography to detect residual core in investment cast turbine airfoils", Neutron Radiography and Gaging (Berger, H., Ed.), STP 586, ASTM, Philadelphia (1976) pp 152-157

[3] John, J., "The role of neutron radiography in a maintenance environment", Innovation for maintenance technology improvements, 33rd Meeting of MFPG-Mechanical Failures Prevention Group (Shiver, T.R., Ed.), Washington (1982) pp 417-453

[4] Dance, W.E., "Neutron radiographic evaluation of aerospace structures", Neutron Radiography and Gaging (Berger, H. Ed.), STP 586, ASTM, Philadelphia (1976) pp 137-151

[5] Mast, H.-U., Untersuchungen zur Leistungsfähigkeit der Neutronenradiographie in der zerstörungsfreien Werkstoffprüfung, Berichtsband zum Symposium Materialforschung 1988 des Bundesministers für Forschung und Technologie, Hamm (1988) pp 1234-1249

[6] Fukushima, Y., Nakamura, T., Hiroaka, E., Sekita, J., Yokochi, H., Yamada, T., Yamaki, S., Neutron radiography using ultra-compact cyclotron, Proc. of the Second World Conference on Neutron Radiography, Paris (1986) pp 215-221

[7] Wilson, M.N., Finlan, M.F., The superconducting cyclotron as a tranportable neutron source, Proc. of the Second World Conference on Neutron Radiography, Paris (1986) pp 199-206

[8] Dubouchet, M., Bach, P., Cluzeau, S., Les tubes neutronique scelles et la neutrographie mobile, Proc. of the Second World Conference on Neutron Radiography, Paris (1986) pp 175-182

[9] Froom, D.A., Barton, J.P., Plans for aircraft maintenance neutron radiography systems, Proc. of the Second World Conference on Neutron Radiography, Paris (1986) pp 431-438

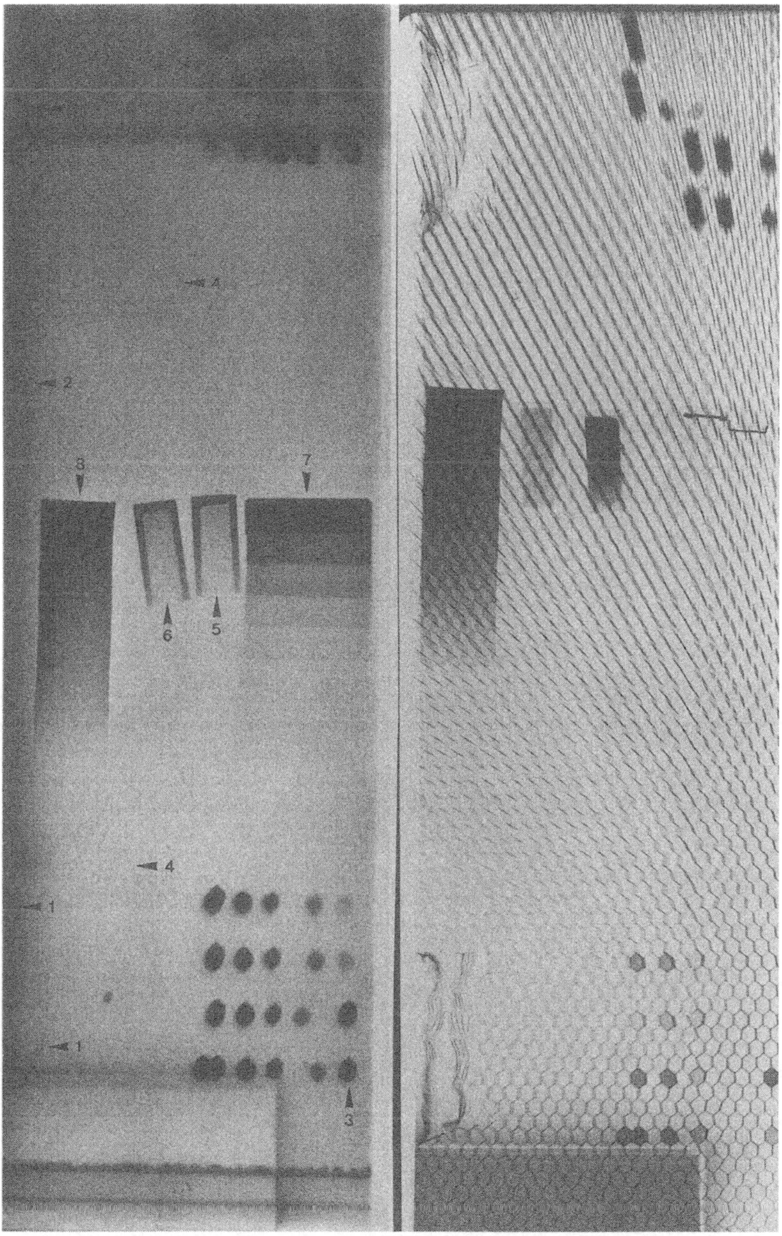

Figure 1   Neutron (left) and X-radiograph (right) of a section of a helicopter rotor blade with various simulated defects

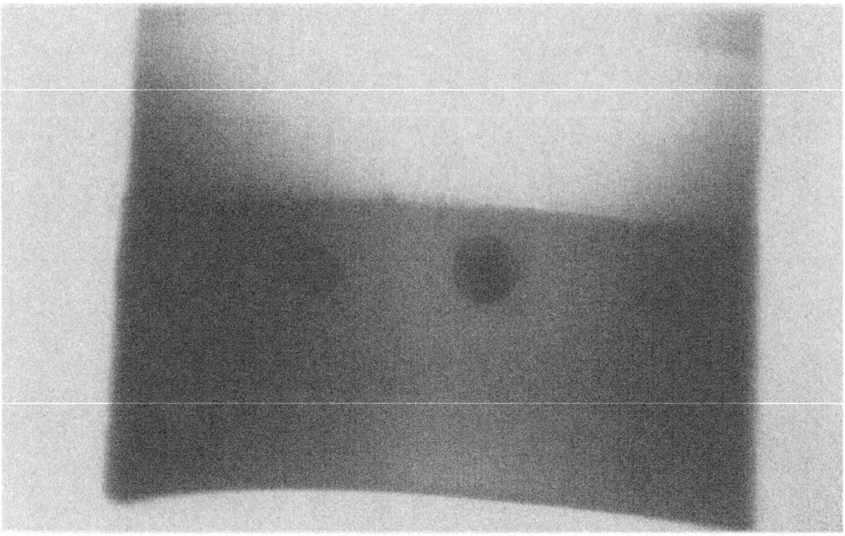

Figure 2 Neutron radiograph of a carbon-fiber-composite surrounded by steel

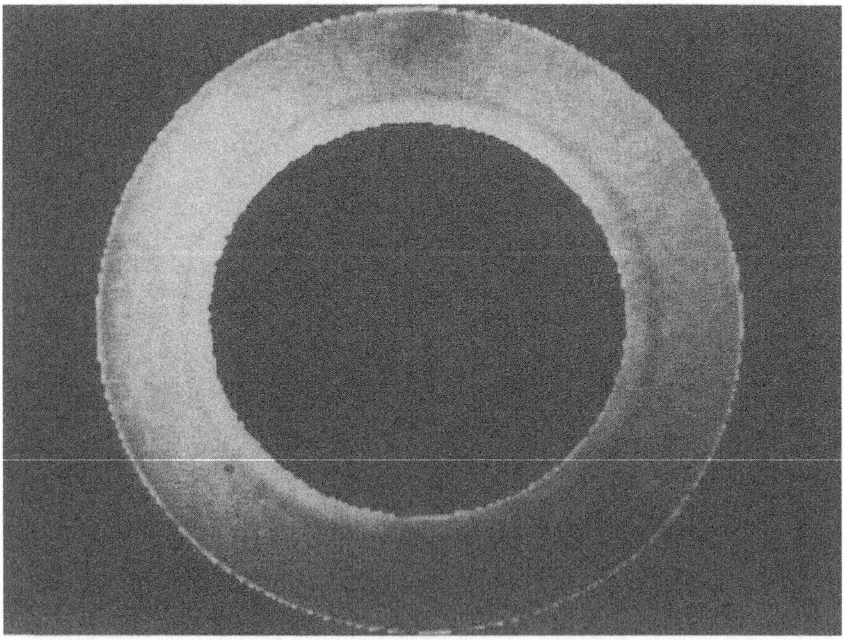

Figure 3 Neutron radiograph of a ring of silicon carbide ceramics with an inhomogeneous distribution of silicon and carbon

Multiparameter Magnetic Inspection System for NDE of Ferromagnetic Materials

D. C. JILES

Center for NDE and Department of Materials Science and Engineering
Iowa State University
Ames, IA 50011

Summary

This paper describes the development and application of a multiparameter magnetic inspection system for nondestructive characterization of the properties of ferromagnetic steels. The system makes simultaneous measurements of magnetic field H, magnetic induction B, magnetostriction λ and either magnetic Barkhausen effect emissions (MBE) or magnetoacoustic emissions (MAE). All measurements are made under computer control and results are automatically recorded and stored on magnetic disc. Software packages available for the system include both control and data analysis functions. The control functions change the magnetic field and record results. The data analysis functions convert the raw data into important magnetic properties such as coercivity, permeability and hysteresis loss which can be used to evaluate the mechanical condition of the specimens.

Introduction

Various NDE techniques can be used on steels, however magnetic methods, which rely on the inherent ferromagnetic properties for evaluation of a wide range of materials properties [1], from flaws to residual strain, are unique to steels. In general, the changes in magnetic properties that are observed are large and therefore easily measurable. Despite these obvious advantages, magnetic methods have yet to be fully exploited compared with other NDE techniques.

The various magnetic measurements made with this system generate a range of independent parameters which can be used to quantify the condition of the material. The magnetostriction is measured as a function of both field and induction. Hysteresis curves allow determination of coercivity remanent induction, hysteresis loss, initial differential permeability, and anhysteretic differential permeability. The Barkhausen effect and MAE each give count rates, total number of pulses and the pulse height spectra.

We consider that it is important to measure these various properties simultaneously, not only to expedite the measurements, but to ensure that the various independent parameters are measured under identical conditions.

Background

Microcomputer controlled systems for measuring the magnetic properties of materials have been described previously [2]. Earlier systems have concentrated mainly on magnetic hysteresis measurements, although a system for simultaneous application of magnetic field H and stress σ [3] has also been reported as part of an investigation into NDE of stress.

Theiner and coworkers [4,5] have shown that, in view of the number of factors influencing the magnetic properties of steels, it is necessary to take a number of independent measurements. Their "3MA" system measures the Barkhausen effect, dynamic magnetostriction and incremental permeability. From these three measurements, it is possible to unravel the competing influences of several factors in order to determine levels of stress [6]. The early development of the present system has been described previously by Habermehl and Jiles [7].

Progress to Date

The system contains completely integrated instrumentation for measurement of magnetostriction, magnetic hysteresis and Barkhausen effect. We are planning the integration of the magnetoacoustic emission hardware in the near future. These four magnetic measurements are the most widely used magnetic NDE techniques for evaluation of intrinsic magnetic properties of materials [1]. A schematic diagram for the system is shown in Fig. 1.

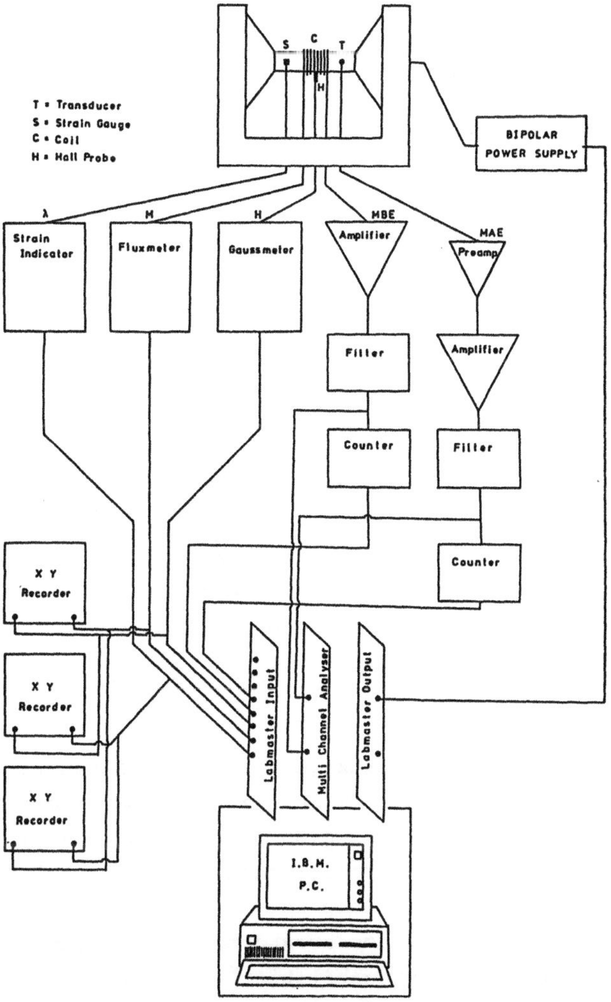

Fig. 1. Diagram of the magnetic inspection system.

i) Control functions

The field control functions are operated using one of the two digital to analog output lines on the Labmaster board. The programs are called from a master control program, which assumes control again once the operations are complete. The master control program provides a menu of available operations. These include demagnetization, initial magnetization, anhysteretic magnetization and hysteresis loop algorithms.

ii) Data acquisition

The data acquisition system used is part of the Labmaster board which has a 12 bit analog to digital converter with eight separate input lines. Typical operation of the instrument gives a field resolution of 0.0048 Oe with an accuracy of 2% for H, and a resolution of 0.48 Gauss for B with a similar accuracy to that of H. Magnetostriction measurements are taken using a strain gauge bridge. The strain sensitivity is 1 mv per 10^{-6} strain and with 0.48 mV resolution this gives a strain resolution of 0.48×10^{-6}. The Barkhausen emissions are detected by a second flux coil wound on the specimen. The signal is amplified using a preamplifier and then passed through a band pass filter. The count rate is measured by a universal pulse counter. The output from this counter is connected to the computer via an IEEE-488 interface.

iii) Analysis

A data analysis software package has been written for use with the system which calculates magnetization, permeability and differential permeability at each point. Coercivity and remanence are calculated by interpolation of results close to B=0 and H=0 on the hysteresis loop to saturation. Hysteresis loss is calculated by integration around the hysteresis loop. Initial and anhysteretic permeabilities are obtained using linear least squares analysis of data at the origin H=0, B=0. The maximum differential permeability is calculated from the slope of the hysteresis loop at H = Hc, B = 0.

Results

i) Applied Stress

The effect of an applied stress on the magnetic properties of ferromagnetic materials such as steels has been of continued interest because of the possibility of using magnetic measurements for nondestructive evaluation of stress. It is known, for example, that the application of stress causes a change in orientation of the hysteresis loop [8].

Measurements taken with the multiparameter magnetic inspection system have been used to test the theory of Sablik et al. [9] which is a generalization of the theory of ferromagnetic hysteresis. This model predicts the variation anhysteretic susceptibility with stress as discussed in the companion paper [10].

ii) Plastic Deformation

Results taken on unstressed specimens of both AISI 4130 and AISI 4140 steels before and after plastic deformation have revealed very different effects from those observed under stress. Whereas stress causes a perturbation of the anisotropy leading to changes in the differential susceptibility, plastic deformation leads to an increase in dislocation density and hence to increased coercivity and hysteresis loss.
The increase in coercivity with plastic deformation for a single specimen of AISI 4140 has been shown previously, while the increase in hysteresis loss of several specimens of AISI 4130 is shown in Fig. 2. Similarly corresponding decreases in the initial susceptibility of these specimens with plastic deformation have been observed.

iii) Fatigue Damage

Materials subjected to cyclic stresses over a prolonged period will fail spontaneously at stress levels below the yield strength. Such modes of failure are very common in industry and represent a major challenge for NDE.

Results on a range of specimens have shown that the progress of fatigue damage can be monitored by magnetic methods similar to those used for plastic deformation; that is by measuring changes in coercivity or hysteresis loss. The changes in these

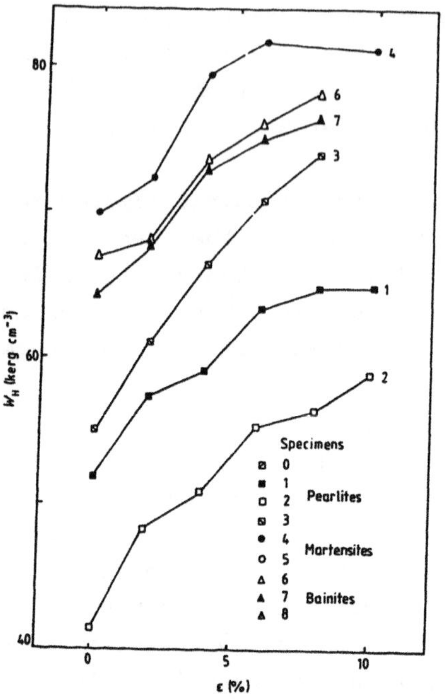

Fig. 2. Dependence of hysteresis loss on plastic deformation for various heat treatments of AISI 4230 steel.

parameters are very large at low levels of expended fatigue life, but as shown in Fig. 3, they seem to saturate as the material approaches the end of its fatigue life.

iv) Magnetostriction

The system has been used to characterize highly magnetostrictive alloys for transducer applications as well as to determine the magnetostriction for stress evaluation as described above. A single example of the most highly magnetostrictive alloy of terbium-dysprosium-iron is shown in Fig. 4. This has a strain amplitude of 2250×10^{-6}, which is the largest ever observed at room temperature. This may be compared with the saturation magnetostriction of iron which is only -7×10^{-6}.

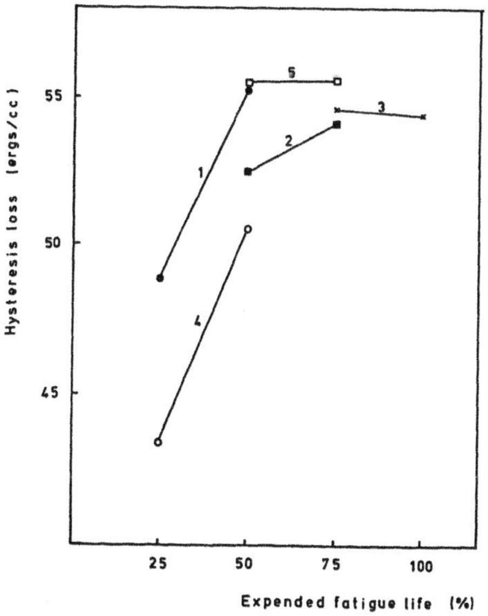

Fig. 3. Variation of hysteresis loss with expended fatigue life for five samples of AISI 4140 steel.

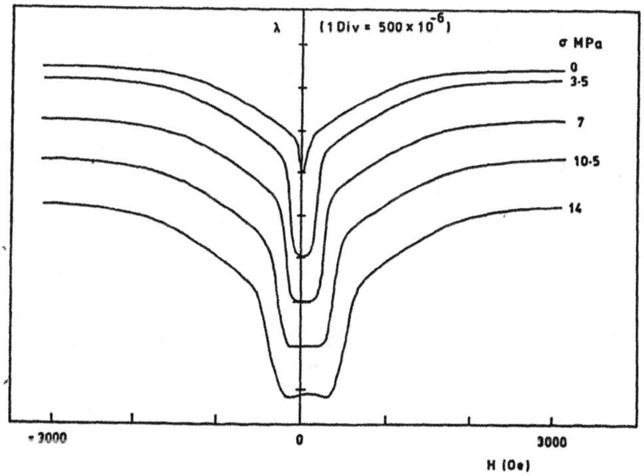

Fig. 4. Dependence of the bulk magnetostriction λ on the magnetic field H for a specimen of Tb-Dy-Fe under different amounts of compressive stress. The magnetic field is being swept from -3000 Oe to +300 Oe. In order to make the results clearer, the successive curves have been displaced relative to one another along the λ axis. Each curve has a value of λ = 0 at H = 0.

Acknowledgement

This work was sponsored by the Center for NDE at Iowa State University and was performed at the Ames Laboratory. Ames Laboratory is operated for the U.S. Department of Energy by Iowa State University under Contract No. W-7405-ENG-82.

References

1. D. C. Jiles, "Review of Magnetic Methods for NDE", to be published in NDT International (1988).

2. E. Martin, K. M. Zamarro and J. Rivas, J. Phys. E (Sci. Instrum.) 15, 539 (1982).

3. D. C. Jiles, D. L. Atherton, H. E. Lassen, D. Noble, J. deVette and T. Astle, Rev. Sci. Instrum. 55, 1843 (1984).

4. W. A. Theiner and P. Deimel, "Nondestructive Testing of Welds with 3MA Analyzer", Twelfth MPA Seminar, October 1986.

5. W. A. Theiner, E. Brinksmeier and E. Strucker, "Stress Measurements on Components with Nondestructive Ferromagnetic Methods", International Conference on Residual Stresses, Garmisch-Partenkirchen, Germany, October 1986.

6. W. A. Theiner and I. Altpeter, "Determination of Residual Stresses Using Micromagnetic Parameters", in "New Procedures in Nondestructive Testing". Edited by P. Holler, Springer-Verlag, Berlin (1983).

7. S. Habermehl and D. C. Jiles, "An Automated Control and Data Logging System for the Determination of Magnetic Properties of Materials for NDE", Published in "Review of Progress in Quantitative NDE, 5A, D. O. Thompson and D. E. Chimenti, Eds., (Plenum Press, New York, 1986), p. 843.

8. R. M. Bozorth, "Ferromagnetism", Van Nostrand, New York, 1951, Chapter 13.

9. M. J. Sablik, H. Kwun, G. L. Burkhardt, and D. C. Jiles, J. Appl. Phys. 63, 3930, (1988).

10. D. C. Jiles and P. Garikepati. See these conference proceedings.

Development of New Quantitative SLAM Techniques for Material Evaluation

A. C. Wey, L. W. Kessler
Sonoscan, Inc., 530 E. Green St., Bensenville, IL 60106, USA

R. Y. Chiao
Department of Electrical and Computer Engineering
University of Illinois at Urbana-Champaign, IL 61801, USA

Abstract
Scanning laser acoustic microscopy (SLAM) has demonstrated capability of real-time detection of flaws in a variety of materials. The primary activities in the field today involve developing methods for image generation, image improvement, and applications of the technology as an analytical method. The near revolutionary, recent advances in video acquisition, image processing, and high speed computational hardware permit SLAM to be a very promising and powerful NDT tool. This paper presents the newly developed computer routines for specific SLAM applications in the areas of automatic defect detection, evaluation of bonding and joining, quantitative material characterization, holographic image restoration, and tomographic image reconstruction.

Introduction

Scanning laser acoustic microscopy (SLAM) has received considerable attention as a promising NDT technique [1]. SLAM has demonstrated capability of real-time detection (30 images per second) of flaws in a variety of materials [2,3]. The primary activities in the field today involve developing methods for image generation, enhancement, and applications of the technology as an analytical method. This paper describes the development of SLAM signal analysis techniques for the specific SLAM applications in defect detection, evaluation of bonding and joining, and material characterization.

In addition, we present an image restoration technique by which the acquired SLAM data can be digitally backpropagated to a chosen plane in the sample to eliminate the diffraction effect in SLAM imaging. An image reconstruction algorithm for the development of scanning tomographic acoustic microscopy (STAM) [4] is also presented. STAM overcomes the shadowgraphic effect in SLAM images and produces tomograms of the internal structure of microscopic specimens.

The SLAM Imaging Technique

The SLAM uses high frequency ultrasound, ranging from 10 - 500 MHz. The ultrasound is transmitted through the material and is reflected, refracted or absorbed by internal discontinuities such as cracks, voids, inclusions or delaminations.

The SLAM imaging technique was developed on a digital image analysis system which combines the image digitizer/display hardware and software. Our host computer was an IBM Personal Computer AT compatible with a 10 MHz, 32 bit coprocessor board. The control program was written in the C programming language and ran on the 32-bit coprocessor which acts as the main computer using the MS-DOS computer as an I/O controller. Included also was a parallel digital I/O interface for controlling the real-time integrator.

The input of the digital image analysis system can be any standard video source (1 volt p-p into 75 ohms). The output of the digital image system was directly connected to a high resolution, analog RGB video monitor. This image display monitor was used to display images in monochrome and pseudocolor. Highlighted, intensified and colored graphic overlays were also displayed with the image when using a data window or when enhancing or analyzing the image.

We have used the digital image analysis system and SLAM (see Figure 1) to develop the routines for specific applications:

Defect Detection

Extracting a defect from random background texture is very difficult because of many orientations and a significant randomness associated with the texture. To simplify the detection analysis, we developed the variance filter method. It is known that the variance in a defined domain is invariant in orientation and translation [5]. Our variance filter method of recognizing defects was therefore based on calculating the typical standard deviation of the specimen background texture and comparing this to the standard deviation in the area containing defects.

The variance filter method requires a calculation window of a specific size, or n x n pixels. The window must be roughly equal to at least a single cycle of the lowest, significant, spatial frequency in the background texture. We calculated the standard deviation of the window and then mapped the value to the window's center pixel. The window was then hopped to the next pixel and repeated the procedure.

Due to the degree of randomness in SLAM image background texture of a material, the standard deviation of one window may differ from that of the other. The collection of window standard deviation values defines a statistical distribution for an image. The Defect Detection algorithm compares the absolute difference between the mean value of all of the standard deviations in an image and the standard deviation in a specific window. If the difference is larger than a chosen threshold value, then a defect is said to be detected.

The Defect Detection algorithm has been applied to the automated NDT of a large niobium plate. Defects larger than 50 um could be easily identified throughout the volume of 23.4 cm x 23.4 cm x 0.3 cm plate using 60 MHz SLAM images [6]. The inspection routine is summarized in Figure 2. A video image from SLAM passed through a temporal averaging filter (30 frames in our experiment) to increase the signal-to-noise ratio. A spatial averaging filter combined four pixels to one to reduce the data set size. Then the above mentioned variance filter was used to obtain the probability density function of the window standard deviation, S. The standard deviation of the distribution could be obtained by Gaussian Fitting method and a threshold value was determined. Defects were identified by the threshold detection, and their location and sizes were indicated in the display.

Bonding and Joining

In SLAM images, the good bonds appear bright because the

ultrasound propagates across the interface with minimum attenuation. Conversely, the dark areas represent the poor bond or disbond conditions due to high attenuation caused by air gaps or discontinuities.

To analyze acoustic images, a data window was established whose size represented the 100 % bond condition. The grey level threshold was established to distinguish the bond and poor-bond or no-bond conditions. Each acoustic image could then be converted to a binary format where the image percentage of area bonded could be easily calculated. The resulting percentage of the total bond area could be used as an accept/reject parameter. In addition, the digitized image data could be displayed in histogram format to represent the degree of bond. This data could be used to correlate with other data such as pull tests.

The applications of the Bonding and Joining routine are related to electronic device packaging such as die attach and TAB bond evaluation. Preliminary results [7,8] indicate that SLAM imaging technique appears to be useful for nondestructively measuring the bond sizes.

Material Characterization

The spatial variations in SLAM image can be related to the microstructure of a material. We have developed a system to measure spatial variations in ultrasonic transmission. The dynamic range of measurements was over 40 dB.

A computer program was developed to calculate attenuation coefficients (dB/mm). Corrections were made for insonification angle, reflection and mode conversion losses at sample interfaces. The computer program provides the spatial frequency spectrum of a given sample by use of one and two dimensional FFT's. Variations in image texture can then be correlated to material microstructural variations through the use of spatial frequency analysis.

Experiments were performed to characterize the microstructure of titanium [9]. Results showed large variations in average attenuation and image texture for different alloys. For example, the differentiation of a gas-contaminated weld and a noncontaminated weld was achieved.

Image Restoration

When SLAM operates in the transmission mode, the images produced are 2-D shadowgraphic views of 3-D objects. And when applications involve imaging structures comparable in size to the acoustic wavelength used, overlapping and diffraction cause difficulty in interpreting the image. This difficulty is removed by using holographic image restoration.

To perform holographic reconstruction, phase information of the wavefield must be preserved so that a quadrature detector is needed. Quadrature detection involves multiplying the signal with two coherent electronic references which are 90 degree out of phase to each other. As signals detected in SLAM always contain both positive and negative spatial frequencies, there are four output signals from the modified SLAM which constitute the holographic data.

A linear system model representing SLAM imaging process is shown in Figure 3(a), while 3 (b) shows a model corresponding to image reconstruction by inverse filtering. In reconstruction process we first compensate for the imperfect frequency response of the knife-edge detection and then compute the corresponding backward propagation to correct the diffraction experienced by wavefield propagation. In order to decrease the computation time and the required memory space, a thin lens approximation is used to represent the wedge shaped water path between the coverslip and sample.

The holographic image restoration technique can be used to 3-D localization of subsurface flaws in material such as ceramics, metals and composites. In our experiments, defects in the range of 50 to 500 um in a 5 mm thick silicon nitride disk were successfully identified. The size, shape, and depth of subsurface defects were actually determined. These data could be correlated with the performance-related properties of the material such as service strength, toughness, etc..

Image Reconstruction

As described in the previous section, holographic data can be acquired by a modified SLAM system. In general, many of these holographic data, corresponding to different illumination directions of the object, are needed to exactly reconstruct the 3-D structure. To obtain a number of projections for a tomogram, we can rotate either the transducer or the specimen. Many sets of holographic data, each corresponding to a projection, can be obtained in sequence by the rotation. Coherently summing the propagation over the sets of data can be solved to form the object profile. However, there are different problems associated with these two schemes. Major difficulties faced by rotating the transducer are the nonuniformity of knife-edge detection sensitivity of SLAM and phase errors introduced when the distance between the object and tne rotation axis varies. The problem to be overcome in the rotation of the specimen is to keep the specimen flat and at the same horizontal level during the rotating process to avoid phase errors. Although coordinate transformation is required, the latter is more practical than the former. An alternative and preferable scheme is to keep both transducer and specimen stationary but simply sum the complex wavefields of a given plane over a number of acoustic frequencies.

To demonstrate the feasibility of multifrequency tomographic reconstruction, a limited frequency range of 93 - 108 MHz was used to obtain nolographic image of a honeycomb finder grid. Preliminary results showed significant improvement in minimizing overlapping effects. The scanning tomographic acoustic microscope (STAM) could produce tne unambiguous images for quantitative NDT applications such as detailed material characterization, statistical process control, inspection of microelectronic components, non-invasive imaging of biological samples, and so on.

Summary

In this paper we have presented several application-specific analysis techniques in scanning laser acoustic microscopy (SLAM). The Defect Detection algorithm combining the variance filter method and thresholding can automatically distinguish and locate flaws from random background texture. The Bonding and Joining routine uses a data window and thresholding to differentiate good bonds, poor bonds and disbonds. The Material Characterization routine utilizes spatial attenuation and spatial frequency analysis to characterize material microstructural variations. The Holographic Image Restoration algorithm involes modification of SLAM and the use of backpropagation of the complex wavefield to eliminate diffraction effect in SLAM images. The multi-frequency Tomographic Image Reconstruction software overcomes the diffraction and overlapping effects in SLAM imaging.

References

1. L.W. Kessler, J.E. Semmens, F. Agramonte, "Scanning Laser Acoustic Microscopy: A New Tool for NDT", 11th World Conference of NDT, pp. 995-1002, 1985.
2. L.W. Kessler, M.G. Oravecz, H. Padamsee, "Nondestructive Inspection of Niobium to Improve Superconductivity", Proc. of IEEE Ultrasonic Symp., pp. 547-552, 1985.
3. D.J. Roth, E.R. Generazio, G.Y. Baaklini, "Quantitative Void Characterization in Struatural Ceramics by Use of SLAM", Material Evaluation, vol. 45, pp. 958-966, 1987.
4. Z.C. Lin, H.Lee, G. Wade, "Scanning Tomographic Acoustic Microscope: A Review", IEEE Trans., vol. SU-32, no. 2, pp. 168-180, 1985.
5. S.Haykin, Communication Systems, pp. 236-238, Wiley, 1983.
6. L.W. Kessler, Final Report to US DOE, SBIR Contract No. DE-AC02-84ER80180.A003, 1987.
7. L.W. Kessler, "Acoustic Microscopy: A Nondestructive Tool for Bond Evaluation on Tab Interconnections", Proc. of ISHM Symp., pp. 79-84, 1984.
8. L.W. Kessler, J.E. Semmens, F. Agramonte, "Nondestructive Die Attach Bond Evaluation Comparing SLAM and X-Radiography" IEEE Proc. of ECC, pp. 250-258, 1985.
9. D.E. Yuhas, M.G. Oravecz, "Microstructural Characterization of Titanium by Acoustic Microscopy", Materials Evaluation, vol. 41, no. 11, pp. 1304-1309, 1983.

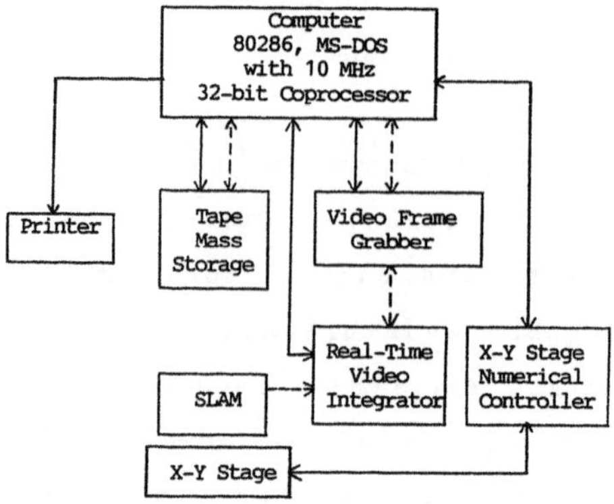

Figure 1. Hardware block diagram for digital image analysis system in SLAM.

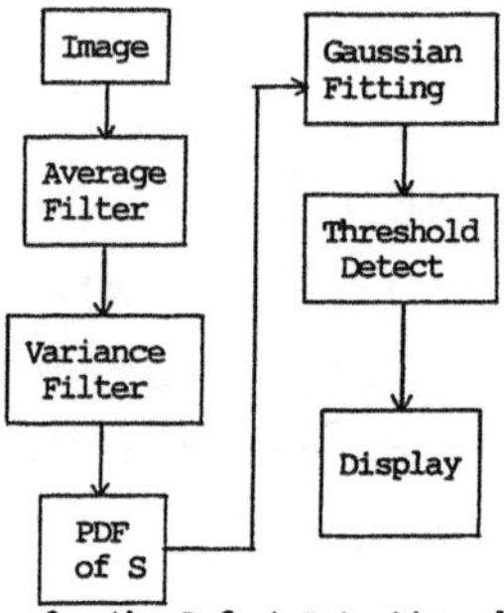

Figure 2. Diagram for the Defect Detection algorithm.

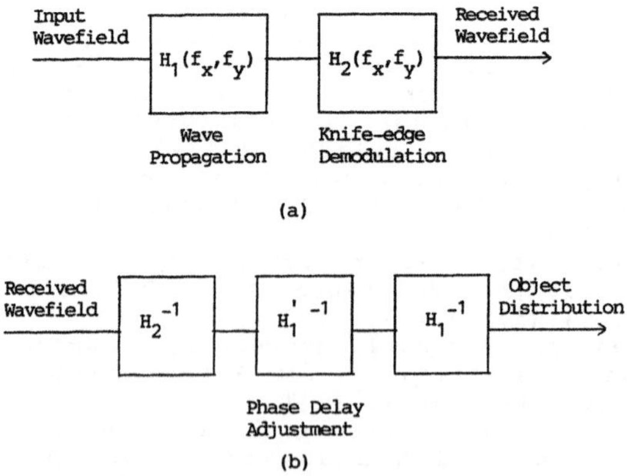

Figure 3. Models corresponding to (a) SLAM imaging system, and (b) image reconstruction by inverse filtering.

Nondestructive Characterization of Material Properties by an Automated Ultrasonic Technique

J.P. Panakkal*, H. Peukert**, and H. Willems

Fraunhofer-Institut für zerstörungsfreie Prüfverfahren
Universität, Gebäude 37, D-6600 Saarbrücken, FRG

* Permanent Address: Bhabha Atomic Research Centre, Bombay, 400085, India
** Permanent Address: Rheinisch-Westfälischer TÜV, D-4300 Essen, FRG

SUMMARY

Ultrasonic velocity and/or attenuation measurement is a versatile tool for nondestructive characterization of material properties such as residual stress, texture, porosity, elastic constants, microstructure etc. In this contribution, an automated, computer-controlled system is described which enables fast and accurate measurement of the time-of-flight of ultrasonic signals as well as the frequency dependent attenuation $\alpha(f)$. A high speed digitiser is used for signal recording of the ultrasonic backwall-echoes. The evaluation of the recorded signals is performed involving Fast Fourier Transforms (FFT) for attenuation measurement and cross-correlation technique for time-of-flight measurement. Time resolution better than 1 ns is achieved by applying additional data interpolation. As examples of practical application, the determination of porosity in sintered material and the measurement of grain size in thin ferritic sheets are demonstrated.

ZUSAMMENFASSUNG

Messungen von Ultraschallgeschwindigkeit und/oder Ultraschallschwächung bieten vielfältige Möglichkeiten zur zerstörungsfreien Charakterisierung von Materialeigenschaften wie z.B. innere Spannungen, Textur, Porosität, elastische Moduln, Korngröße u.a. In diesem Beitrag ist ein automatisiertes, rechnergesteuertes Meßsystem beschrieben, welches schnelle, hochgenaue Messungen der Laufzeit von Ultraschallsignalen sowie die Messung der Frequenzabhängigkeit des Schwächungskoeffizienten α ermöglicht. Zur Aufnahme der US-Signale wird ein schneller Transientenrecorder mit einer maximalen Abtastrate von 200 MHz benutzt. Anschließend werden die digitalisierten Signale mittels Kreuzkorrelationsanalyse (Geschwindigkeitsmessung) bzw. schneller Fourier-Transformation (Schwächungsmessung) weiterverarbeitet. Durch zusätzliche Anwendung von Interpolationsverfahren wird eine Zeitauflösung von ~ 1 ns bei der Laufzeitmessung erreicht. Als Beispiel praktischer Anwendungsmöglichkeiten werden die Bestimmung der Porosität in Sinterwerkstoffen sowie die Messung der Korngröße in dünnen, ferritischen Blechen demonstriert.

INTRODUCTION

Ultrasonic propagation in solids is closely related to both elastic and anelastic properties of the material leading to a number of applications in characterizing material states nondestructively. Ultrasonic velocity meas-

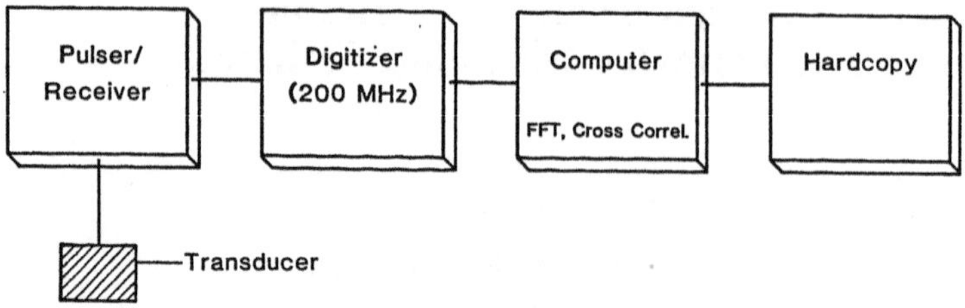

Fig.1. Schematic diagram of the measuring system

urement is used in order to determine elastic moduli, residual stresses, texture, porosity, etc. whereas ultrasonic attenuation measurement is exploited for the determination of grain size in polycrystalline materials [1,2,3]. Velocity measurement especially requires high-resolution techniques, because some of the effects to be measured are of the order of 10^{-3}. Additionally, measuring systems with a large degree of automation are needed in case of practical application in modern industry. We have built up such a system using cross-correlation technique for high-precision velocity measurement and FFT-technique for attenuation measurement. As examples of application, the characterization of sintered iron compacts and the determination of grain size in ferritic sheets are demonstrated.

DESCRIPTION OF MEASURING SYSTEM

The block diagram of the measuring system is shown in Fig. 1. For ultrasonic excitation, commercially available transducers in the frequency range from 2 to 50 MHz are used. Measurement is carried out in the pulse-echo mode. The received ultrasonic signal is digitised with a maximum sampling rate of 200 MHz by means of a programmable digitiser (TEK 7612D). Data acquisition and evaluation are performed by means of a mini-computer (TEK 4054) allowing also graphic display and hardcopy output. Data processing techniques such as cross-correlation (CC) and fast Fourier transform (FFT) are performed by an implemented software package. Diffraction corrections to velocity and attenuation are also included in the software.

MEASUREMENT TECHNIQUE

Ultrasonic velocity

Several methods are in use for accurate time-of-flight measurement of ultrasonic pulses. Usually, the overlap method [4] is applied where two back-

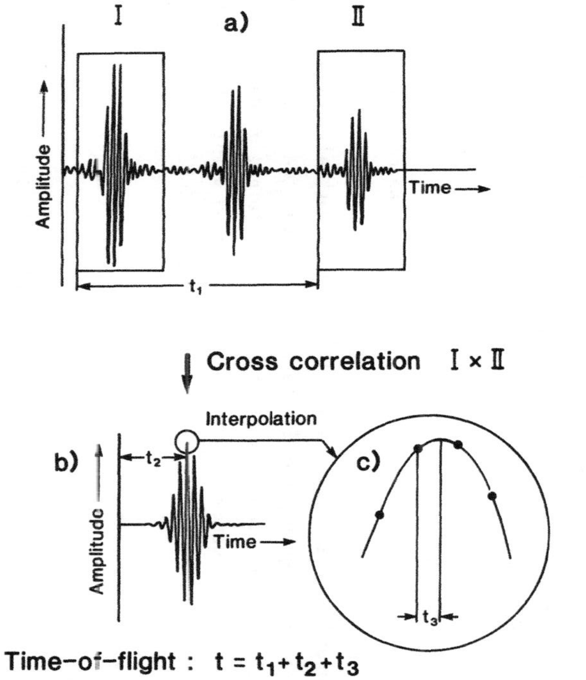

Fig.2. Time-of-flight measurement by means of cross correlation method - schematic

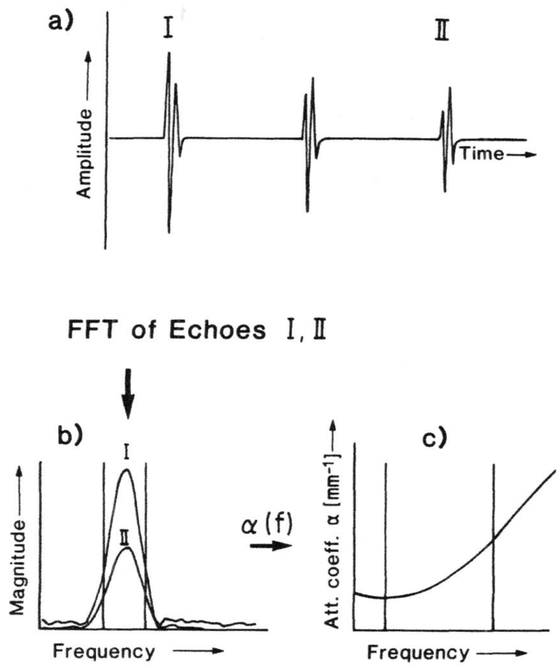

Fig.3. Attenuation measurement by means of FFT-technique - schematic

wall echoes are brought into coincidence and the time between them is measured to a resolution of ~ 1ns. This method can be automated, for example, by tracking electronically the position of a selected peak or zero crossing [5]. In the case of strongly dispersive materials or noisy signals, the application of the overlap method can become ambiguous. Here, cross-correlation techniques offer some advantages because the whole ultrasonic pulse is taken into account instead of only one point (peak amplitude, zero crossing). A comparison of different techniques was reported in [6] showing the advantage of cross-correlation technique in case of distorted echoes.

In order to perform time-of-flight measurement by cross-correlation technique, a computer program was developed. The main steps are explained in Fig. 2. Fig. 2a shows the digitised high-frequency signal with several backwall-echoes. Two backwall echoes are gated via the graphic display and a rough value (t_1) for the ultrasonic time-of-flight is obtained. Now, the cross-correlation function of the two gated signals is calculated, yielding the time shift t_2 (Fig. 2b). t_2 corresponds to the maximum amplitude of the cross-correlation function. At this stage the time resolution is 5 ns at a sampling rate of 200 MHz. Electronic jitter is below 1 ns. By applying cubic spline fitting to the peak part of the cross-correlation function, its peak position is determined with a resolution of about 1 ns (Fig. 2c). The total time-of-flight corresponding to the maximum cross-correlation of the selected backwall-echoes is now given by $t=t_1+t_2+t_3$.

Ultrasonic attenuation

For measurement of the frequency dependence of ultrasonic attenuation, FFT-technique is applied. To cover a large frequency range, it is necessary to use broadband pulses (Fig. 3a). Two selected backwall-echoes are transformed into the frequency domain via FFT (Fig. 3b). The ultrasonic attenuation coefficient as a function of frequency is obtained as described later. The frequency dependence of ultrasonic attenuation within the bandwidth of the probe is used for microstructural characterization.

APPLICATIONS

Sintered materials

This system was used to measure ultrasonic velocity and to evaluate the elastic parameters of sintered iron powder compacts supplied by Dept. of

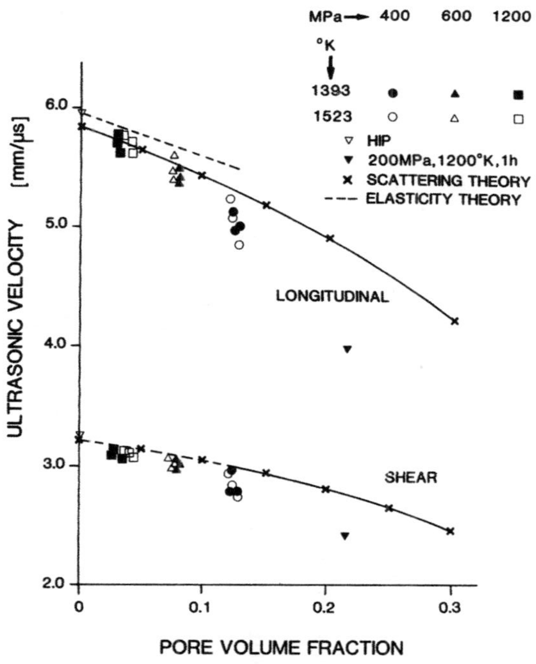

Fig.4. Variation of ultrasonic velocity of sintered iron compacts with porosity

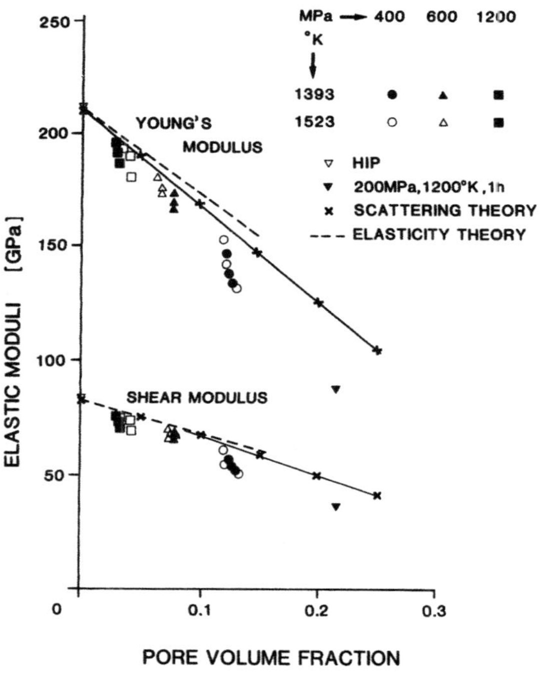

Fig.5. Elastic moduli of sintered iron compacts determined by ultrasonics as a function of pore volume fraction: comparison with theory [7,8]

Physics, Vienna University. Cut specimens of ~ 9 mm and thickness ~ 3 mm having a plane parallelism better than 2 μm were used for the evaluation.

The specimens were fabricated by compacting at different pressures (400, 600 and 1200 MPa) and sintering at two different temperatures (1393 K and 1593 K) for different time (0,5 h, 2 h and 8 h). Longitudinal ultrasonic velocity and the shear velocity were measured using transducers with a centre frequency of 5 MHz. Figure 4 presents the plot of ultrasonic velocity vs pore volume fraction p and the data were fitted into a second degree polynomial given by

$$v_L = 5{,}908\cdot(1 - 0{,}80\cdot p - 0{,}032\cdot p^2)$$
$$v_S = 3{,}225\cdot(1 - 0{,}62\cdot p - 0{,}0024\cdot p^2) \qquad (1)$$

with a coefficient by correlation better than 0,99. The theoretical values calculated from elastic and scattering theories are also presented in Fig. 4, showing good agreement at lower porosities [7,8].

Using the ultrasonic velocities measured, elastic moduli viz. Young's modulus E and shear modulus G were calculated using the well-known formulae:

$$E = \rho\cdot v_L^2\,(3\cdot v_L^2 - 4\cdot v_S^2)/(v_L^2 - v_S^2)$$
$$G = \rho\cdot v_S^2\ . \qquad (2)$$

Figure 5 presents the plot of elastic moduli against pore volume fraction along with the theoretical curves. Both the moduli can be fitted into the empirical relation reported for ceramic materials given by $E = E_0\,\exp(-bp)$ where E_0 is the nonporous value and b a constant. More details of this study are being reported elsewhere [9].

Grain size determination

The grain size of thin sheets is determined from the frequency dependence of attenuation α given by [10] assuming Rayleigh-scattering:

$$\alpha(f) = a_0 + a_1\cdot f + S\cdot d^3 f^4\ . \qquad (3)$$

a_0 represents coupling loss, a_1 is the absorption coefficient, S is the scattering parameter, d is the grain size and f is the frequency. The dependence of scattering contribution $\alpha_S = Sd^3f^4$ on d is only valid in the Rayleigh region defined by $d/\lambda \ll 1$ (λ - ultrasonic wavelength). Based on this equation, backscattering measurements with the two-frequency method [11,12] have been used successfully in determining the grain size.

Neglecting coupling losses, the frequency spectrum $U_i(f)$ of the i-th backwall-echo is given by

$$|U_i(f)| = |G(f)| \cdot |I(f)| \cdot \exp\{-\alpha(f)\cdot x_i\} \tag{4}$$

with $|I(f)|$: = transfer function of the transducer

$|G(f)|$: = excitation spectrum

x_i : = distance travelled by the ith echo.

Comparing $|U_i(f)|$ with $|U_j(f)|$, the frequency spectrum of the j-th backwall echo, $\alpha(f)$ is determined by using the relation:

$$\alpha(f) = (\ln |U_i(f)| - \ln |U_j(f)|) \;/\; (x_j - x_i) \tag{5}$$

For evaluation two selected backwall-echoes in the time domain ($u_i(t)$, $u_j(t)$) were Fourier transformed employing FFT-Routines resulting in $U_i(f)$ and $U_j(f)$ in the frequency domain. The parameters a_o, a_1 and d are determined from the total spectral attenuation using equation 3. The fitting is carried out within the range of frequencies which lie in the bandwidth of the transducer and in the Rayleigh-region ($d/\lambda \lesssim 0.2$).

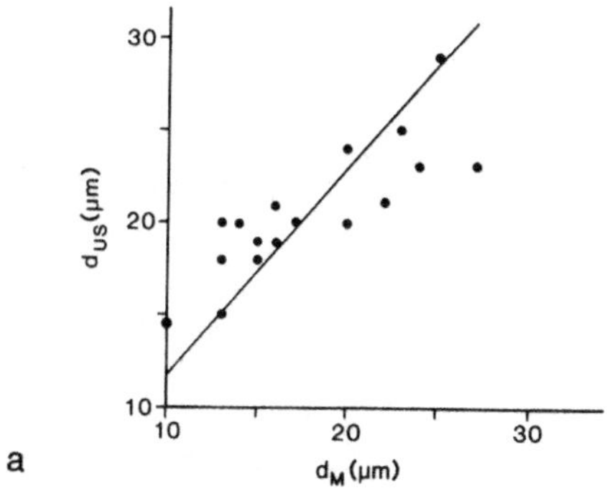

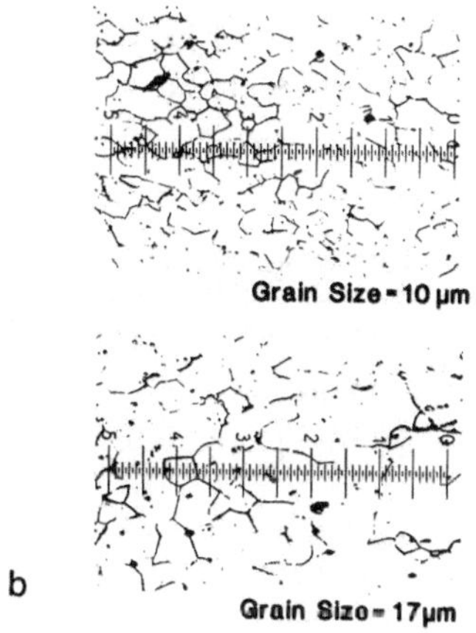

Fig.6. a) Comparison between grain size measured by ultrasonics (d_{us}) and by metallography (d_M)

b) Examples of microstructure of investigated sheets

19 ferritic sheets with thickness from 0,69 to 1,48 mm and metallographically determined grain-size from 10 to 27 µm were investigated. Fig. 6a shows a comparison of grain sizes determined by ultrasonics and metallography. The agreement is quite good. Fig. 6b presents micrographs of two typical specimens. Based on our laboratory results, it is planned to develop a noncontact technique for determining the grain size in thin metal sheets by using laser generated ultrasound and its interferometric detection [13,14].

CONCLUSION

An automated system for measuring ultrasonic velocity by cross-correlation and attenuation from the FFT's is described. Applications in characterizing sintered iron compacts and measuring grain size in thin ferritic sheets are demonstrated. The system can be used for a fast and automated nondestructive determination of porosity in sintered materials using ultrasonic velocity measurement. As practial application of ultrasonic attenuation measurement, the determination of grain size in thin ferritic sheets seems to be feasible.

ACKNOWLEDGEMENT

The sintered iron compacts were supplied by Dr. R. Hessler from Institut für Physikalische Chemie of University of Vienna, Austria. The financial support received by the first author from Kernforschungsanlage (KFA) Jülich is gratefully acknowledged.

REFERENCES

1. Bussière, J.F. et al. (eds.): Proceedings of the Second Int. Symp. on Nondestructive Characterization of Materials. July 21-23, 1986, Montreal, Plenum Press, New York 1987.

2. Goebbels, K.: Structure analysis by scattered ultrasonic radiation. In: Research Techniques in Nondestructive Testing, (ed. R.S. Sharpe), Vol. IV, Academic Press 1980, 87-157.

3. Panakkal, J.P.; Ghosh, J.K.; Roy, P.R.: Use of ultrasonic velocity for measurement of density of sintered fuel pellets. J. Physics D. Appl. Phys. 17 (1984) 1791-1795.

4. Papadakis, E.P.: Ultrasonic velocity and attenuation measurement methods with scientific and industrial applications. In: Physical Acoustics (eds. W.P. Mason, R.N. Thurston). Academic Press 1976, 277-374.

5. Schneider, E.; Herzer, R.: Automated analysis of stress states using ultrasonic techniques. 13. MPA-Seminar, Stuttgart, 8.-9.Okt. 1987.

6. Hull, D.R.; Kautz, H.E.; Vary, A.: Measurement of ultrasonic velocity using phase-slope and cross-correlation methods. Mat. Evaluation 43 (1985) 1455-1460.

7. Ondracek, G.: On the relationship between the properties and the microstructure of multiphase materials. Z. Werkstofftechnik 9 (1978) 831-836.

8. Sayers, C.M.; Smith, R.L: The propagation of ultrasound in porous media. Ultrasonics 19 (1982) 201-204.

9. Panakkal, J.P.; Willems, H.: Nondestructive evaluation of elastic parameters of sintered iron powder compacts (to be published).

10. Bhatia, A.B.: Ultrasonic Absorption. Oxford: Clarendon Press 1967.

11. Goebbels, K.: Gefügebeurteilung mittels Ultraschall-Streuung. Materialprüfung 17 (1975) 231-233.

12. Arnold, W.; Willems, H.: Korngrößenbestimmung in polykristallinen Werkstoffen mit Ultraschallverfahren. In: Walzen von Flachprodukten (eds. H. Calla and H. Jung) Verlag Informationsgesellschaft (1988) S. 225.

13. Faßbender, S; Kulakov, M.; Hoffmann, B.; Paul M.; Arnold W.: this Proceedings.

14. Paul, M.; Peukert, H.; Arnold, W.: to be presented at the 12th WCNDT, Amsterdam, April 1989.

Industrial Application of On-line Texture Measurement

H.-J. KOPINECK
Hoesch Stahl AG, Dortmund, Federal Republic of Germany

Summary

Various technological data of cold rolled and annealed steel strips are influenced by the texture of the material. For the on-line measuring some of their technological data it is necessary to determine the texture. Hoesch Stahl AG developed a new measuring procedure, which allows to measure on-line the desired texture data of steel strips. That means, this new measuring procedure allows the industrial application of texture investigation. It takes advantages of an X-ray transmission technique using the high energy bremsspectrum of an tungsten tube and of energy-dispersive detectors. It permits simultaneous detection of several X-ray reflexions.
Since pole intensities of the texture field are correlated to the r_m-values of cold-rolled and annealed steel strips it can be used for on-line estimation of r_m-values of these steel strips.

Two instruments are installed for measuring the anisotropy coefficient r_m of cold rolled and annealed steel strips. One of them is operating since Dec. 85 in the Hoesch-Continuous-Annealing-Line. This line produces steel qualities with different texture types. Different correlation functions and an additional selection criterion therefore proved necessary in the design of this on-line r_m-value estimating device. The system measures the r_m-values over the entire length of steel strips, passing the CAL. It works very accurately and reliably.

The measured results of the on-line instruments are helpful and important for quality control and quality assurance. Much information for the improvement of the cold-rolled and annealed material has been and can furthermore be obtained. They demonstrate, that these new on-line texture measuring devices are reliable industrial instruments.

Introduction

The importance of non-destructive evaluation of characteristic technological values of materials has significantly increased in recent years. In rolled strips, e. g. in cold rolled steel- or Al-strips, some of these technological properties are correlated with the crystallographic microstructure, i. e. with the texture.

Therefore it was of fundamental importance, to search for theories and ways to determine these technological values by texture measurements. Laboratory investigations, especially by ODF-analysis, have demonstrated, that it will be possible. That means it will be of great importance for the industry, if we succeed in determining these values by an on-line measuring procedure at steel- or Al-strips during production processes.

Therefore the Hoesch Stahl AG, Dortmund, conducted research and developed an X-ray texture analysing method for non-destructive measuring of the texture of steel strips 10 years ago [1] [2]. This new method is especially suitable for on-line inspection of cold rolled steel strips over the entire length. Of course it is possible to apply the same to Al-strips.

These new systems for on-line measuring the r_m-value in the production line have been constructed and installed. The first complete system was built seven years ago as a pilot system into the inspection line of a cold rolling mill of the Hoesch Stahl AG [3] [4]. It is working successfully up to now. A second system, improved by the experience with the first instrument, was installed at the end of 1985 in the run-out part of the new Hoesch-continuous annealing line.

These two instruments have been proved as reliable and important measuring systems, within the production quality assurance system [5] [6] [7].

As these instruments are the only industrially applied on-line texture devices, I can only report in this paper about their specific application. The first part gives a short description of the measuring technique and the equipment. The second part will inform of some results and of the long time experience in industrial application.

Measuring principle

The new measuring principle has two main points:

1. The steel strips for testing are irradiated by the X-ray-bremsspectrum in transmission technique.

2. The characteristic pole-intensities of the texture field are detected by energy dispersive detectors. The values are transferred to a data processing system.

What does that mean?

To the first point: The upper part of fig. 1 shows the X-ray reflexion technique normally used in laboratories. The lower part of the figure illustrates the important transmission technique. This technique has the great advantage to use the complete thickness of the material for taking measurements and not only the small areas near the surface. This method can therefore be applied without special surface treatment. Using the high energy radiation, for instance the 60 keV bremsstrahlung, the transmission method can be applied to steel strips up to 2 mm thickness. This new method allows to measure all net planes vertical to the rolling plane of the strips. Moreover it makes it possible, to consider all parts of the net planes across the total sheet thickness at the same time. Investigation have been done in the last years to extend the measuring range to more than 4 mm thickness.

The second feature of this newly developed procedure is the use of a special measuring technique, the energy-dispersive technique. It allows, to get a relation of the different lattice data to the discrete energy values of the X-ray radiation.

Using the white X-ray light, as it is emitted in the bremsspectrum of an X-ray tube - this is an important feature of this technique - the individual diffraction lines can be associated with corresponding energy levels. This means, is is possible, to measure the intensities of the different energy levels, selected according to the Bragg-law. Semiconductor detectors allow to measure simultaneously numerous different energy levels. The energy resolution of semiconductor detectors is high enough for the required separation of the individual lattice reflexes. Hence it follows that a method using these main points allows to measure several lattice reflexes simultaneously and thus the pole figures of several lattice planes and that this measuring principle may be the basis for texture-measuring-installations in a plant.

The conclusion is: The energy dispersive measuring device together with the transmission technique and the radiation of a white X-ray spectrum is well suited for industrial application in a texture measuring system. It is especially suited to determine the anisotropy coefficient r_m. The main condition for the equipment is, that it has to measure in the very rough environment of a rolling production line for steel strips.

Fig. 2 shows the structural principle of the equipment. The steel strip under test, running from the left to the right, is irradiated by the bremsstrahlung of an X-ray tube under a fixed angle. The X-rays scattered into the texture cone are measured here in two positions. These two positions are at characteristic poles of the (220)- and (211)-texture field. It means, that 4 measuring values can be detected simultaneously. The following paper [8] informes on the next step of our investigation, measuring more than 4 - up to 25 pole values.

In the start-up period during laboratory research an intrinsic germanium detector with an usual liquid nitrogen cooling system was used. For the on-line application in a production line have needed germanium detectors too; but with a cooling system operating in a steel production plant round the clock. In cooperation with a detector producer it was possible to develop and to get a small closed loop-Helium-cooling system for Ge-detectors and to apply it in the new r-measuring system in the continuous annealing line.

These new closed loop cooled Ge-detectors are important parts for the reliability and accuracy of the industrial measuring system.

I should like to make a supplementary remark:
The first pilot device for on-line estimation of the r_m-value had only one correlation function, because it was only used for batch annealed aluminium killed steel. In general, this sheet quality has always the same type of texture over a wide range.

Measuring in the new continuous annealing line required the design of a r_m-value estimating device for different types of texture. For this device it was necessary to work out two different correlation functions for different qualities. The automatical measuring device has at first to estimate the texture type of the strips and then to find out which correlation function must be used. Therefore a special selection criterion was developed. It is a mathematical equation using the intensities of the 4 different measured pole values [6] [7].

The on-line measuring device

For the design of the r_m-value estimation device for the CAL, it was possible to refer to the long time experience gained from the prototype measuring system.

The original construction principle was retained as shown in Fig. 2. The most important components here are the X-ray tube above the steel strip, the detection unit below the steel strip and the electrical analysis and control systems integrating the measuring device in the production process.

The above mentioned new features are integrated in this measuring equipment. The new detector system can be seen installed in the measuring device in Fig. 3. The two secondary collimators, through which the X-rays reach both detectors are in the upper middle of the figure. The refrigeration unit is supplied with precooled helium by flexible copper tubes.

Fig. 4 shows the new device during on-line measurement in the inspection area at the end of the continuous annealing line. The steel strip runs from left to right between the X-ray tube (above the strip) of the measuring device. The strips are stabilized through the inspection area of the CAL by the roll arrangement.

It has already been mentioned above that the new device estimates r_m-values of different texture types and identifies these texture types by the measurement itself, based on a selection criterion. This is necessary, since steel strips with given quality properties can be assigned to another quality group. In such cases, the quality group in process control would not agree with the "texture quality", and the measurement would lead to false results. The selection criterion, however, results in a correct r_m-value estimation.

The whole measurement is controlled by computer. With permanent logical controls in the computer program interruptions are registered immediately. Data exchange with the process-control-computer of the CAL allows all measured and evaluated r_m-values to be transfered to the quality assurance system.
The estimated r_m-values are recorded over the whole strip length. The lowest and highest estimated r_m-values and the strip lengths belonging to these values are also displayed on a monitor on the control platform. A direct response to the measured results is therefore possible.

The design of the new in-plant-device permits an easy maintenance. The C-frame with X-ray tube, detectors, amplifyers, cooling device and control system can be used also for other applications of this measuring technique, e. g.: for application during

the production and the processing of hot and cold rolled steel- or non-iron metal strips. It is possible to determine the different $r(\beta)$-values and also further technological values, e. g. the E-modul, the earings of very thin rolled material, the degree of recristallisation and the magnetic technical values of special steel qualities [8], [9].
Hoesch Stahl AG developed the procedure determining the earings of thin steel strips by on-line texture analysis.

The basic part of the computer software is identical for all applications. Only supplementary software parts for the individual application form are necessary and can be exchanged easily. It means, this new type of on-line texture measuring device is a well tested industrial instrument.

Long time experience

The on-line measurements in two production lines have shown that the complete systems work accurately and reliably. No component of the equipments has had any serious failure during the whole time.

The measured results are of high interest. They are plotted versus strip length and transmitted to the quality assurance department. These plots provide condensed information in a clear form. The accuracy of the r_m-data estimated by X-ray diffraction is better than $\pm$ 0,1.

Figure 5 gives an example of an r_m-value recording. The value is plotted versus the strip length. The mean estimated value of r_m over 3 km strip length is 2,093. In this example, the variation σ is $\pm$ 0,033, whereby the single measuring data are integrated values over several meters strip length.
The observation of cold-rolled steel strips shows in general that the standard deviation, σ, has to be between 0,03 and 0,05. These "values" characterize steel strips with excellent homogenity and consistent deep-drawing properties. Deviations of this given uniformity standards were investigated and gave interesting information.

The results during the first weeks when the new measuring system was used were very exciting. In that time it could be learned that much detailed information is available.
Fig. 6 shows a plot from 1981. The upper example gives the r_m-values of a cold rolled strip, welded together from two hot strips, rolled under different conditions. The step in the r_m-value record at the welding zone is an indication how useful this measurement can be for the quality assurance system. The second example below indicated at that time the importance of homogeneous r_m-values over the strip length.

In the last years it was possible to learn about the influence of variations in the given rolling parameters of the hot and cold strip mill, e. g. the influence of the final rolling temperature, of the degree of cold rolling.

The measuring results gave also informations on how to control the production process to get the desired uniformity of the deep drawing properties over the whole strip length.

Fig. 7 shows an example with deviation of the uniformity on one end of the strip. The reason is a change in the coiling temperature at this end of the hot rolled strip.

Fig. 8 gives some examples of cold rolled and annealed strips with homogeneous, good r_m-values over the whole strip length.

Today it is possible to state:
The on-line r_m-value measuring devices used in the cold rolling plant, especially in the continuous annealing line of Hoesch Stahl AG are able to evaluate reliably the deep drawing quality parameter of steel strips. They are important devices for non-destructive measuring of this technological value under production conditions.

Outlook

An extension of the measuring details of the texture analyzer is given in the paper of H.-J. Bunge, H.-J. Kopineck and H. Otten [8]. With these new results it will be possible to measure the texture of hot rolled strips over the whole lengths, according to the crystallographic structure in a very early state of production. This seams to be very important, because the texture of cold rolled strips are influenced by the texture of the hot rolled strips in a decissive grade.

Fig. 9 will demonstrate it. It shows the correlation between the r_m-value of cold rolled strips and the intensities of one Bragg peak of the texture of the adequate hot rolled strips for two qualities. It shall demonstrate, that texture values of the hot rolled strips are correlated to the deep drawing values of the cold rolled and annealed steels strips. Further it will be possible to determine other technological values too.

References

1. Maurer, A.; Böttcher, W. and Kopineck, H.-J.: Vorzugsorientierungen kaltgewalzter Bänder im Verlauf des Walzprozesses, Abschlußbericht zum Forschungsvertrag Nr. 6210-GA04 (August 1978). Herausgegeben von der Kommission der Europäischen Gemeinschaften unter Nr. EUR 6291 DE, 1980.

2. Maurer, A.; Böttcher, W. and Kopineck, H.-J.: Bestimmung von Werkstoffkennwerten an kaltgewalzten Blechen mittels Röntgendurchstrahlung und energiedispersiver Meßmethode. In: Vorträge der 1. Europäischen Tagung für Zerstörungsfreie Materialprüfung, Bd. 2, S. 421 - 26. Herausg.: Deutsche Gesellschaft für Zerstörungsfreie Prüfung e.V., Berlin, 1978.

3. Maurer, A.: Ein neuartiges Untersuchungssystem zur zerstörungsfreien Bestimmung von Werkstoffkennwerten an Feinblechen mittels Röntgenfeinstrukturverfahren. Münster 1982, Dissertation zum Dr. rer. nat. der Mathematisch-Naturwissenschaftlichen Fakultät.

4. Böttcher, W.; Kopineck, H.-J.: Über ein Röntgentexturmeßverfahren zur zerstörungsfreien On-line-Bestimmung technologischer Kennwerte von kaltgewalzten Stahlbändern, Stahl und Eisen, 105 (1985), 509 - 516.

5. Kopineck, H.-J.; Otten, H.: Texture Analyzer for on-line r_m-Value Estimation. Textures and Microstructures, 1987, Vol. 7, pp. 97 - 113.

6. Kopineck, H.-J.; Böttcher, W.; Otten, H.; Trültzsch, K. L.: The on-line r_m-value measuring of cold rolled steel strips in the exit of an continuous annealing line. Preprints "4th International Steel Rolling Conference, Deauville, June 1 - 3, 1987.

7. Kopineck, H.-J.; Böttcher, W.; Otten, H.: On-line Meßanlage zur r_m-Wert-Bestimmung mittels Röntgenstrahlbeugung an kaltgewalzten Stahlbändern. Berichte aus Forschung und Entwicklung unserer Gesellschaften der Hoesch AG, Heft 1/87, S. 15 - 18.

8. Kopineck, H.-J.; Otten, H.; Bunge, H.-J.: On-line determination of technological characteristics of cold and hot rolled steel strips by a fixed angle texture analyzer. 3rd International Symposium on Nondestructive Characterization of Materials, October 2 - 6, 1988, Saarbrücken.

9. Wagner, F.; Otten, H.; Kopineck, H.-J.; Bunge, H.-J.: Computer aided optimization of an on-line texture analyzer. 3rd International Symposium on Nondestructive Characterization of Materials, October 2 - 6, 1988, Saarbrücken.

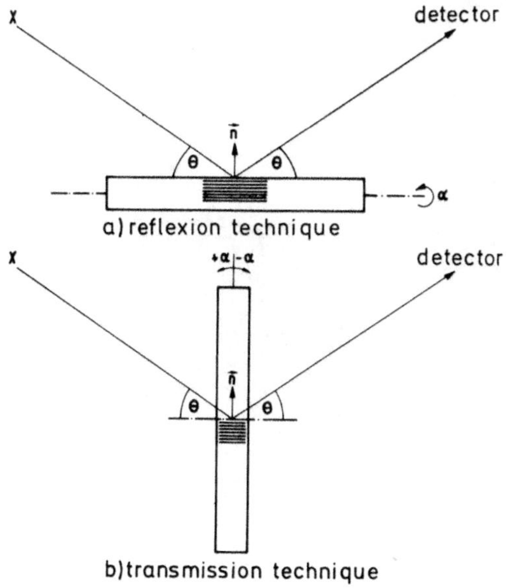

Figure 1:

Methodes of X-ray fine structure measuring (schema)

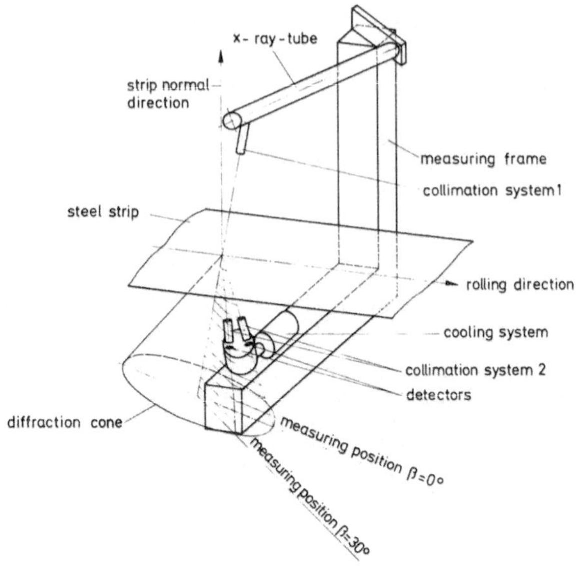

Figure 2: System for r_m-value estimation

Figure 3: Closed-loop cooled germanium detector system of the on-line r_m-value measuring device

Figure 4:

On-line r_m-value measuring device in the steel strip production line (CAL)

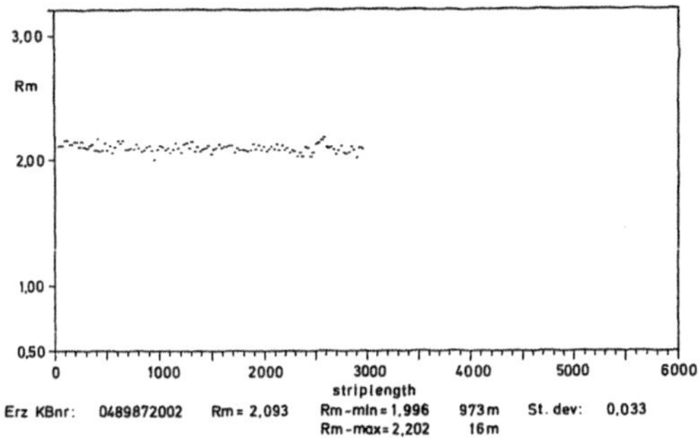

Figure 5:

Example of a measured result demonstrating the homogenity of the cold rolled and annealed steel strip

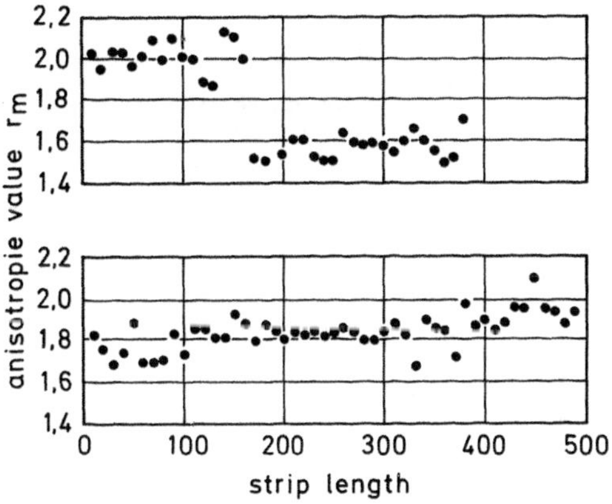

Figure 6:

Examples of measured results from the first weeks of measuring (1981)

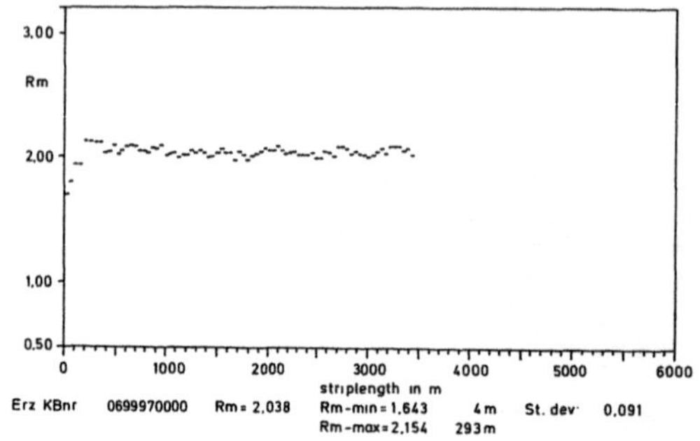

Figure 7:

Example of a measured result influence of non uniform coiling temperature

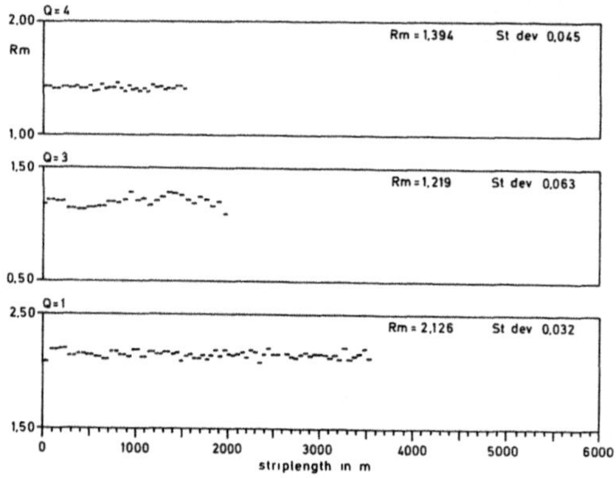

Figure 8:

Examples of measured results from homogeneous steel strip

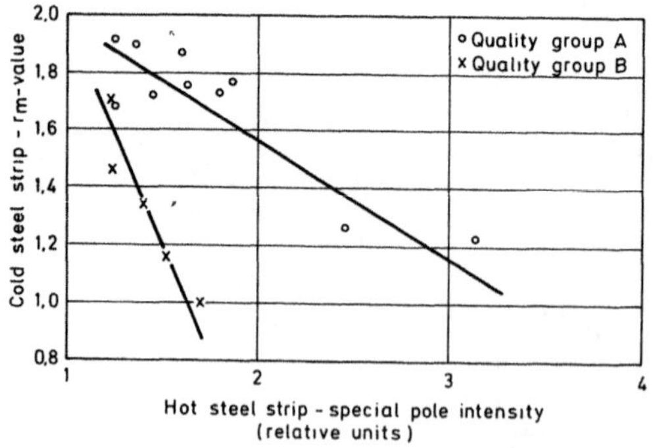

Figure 9:

Relation between texture measurements of hot and cold rolled steel strips

On-line Measuring of Technological Data of Cold and Hot Rolled Steel Strips by a Fixed Angle Texture-analyzer

H.-J. KOPINECK, H. OTTEN
Hoesch Stahl AG, Dortmund

H. J. BUNGE
Technische Universität, Clausthal-Zellerfeld

Summary

The r_m-value has been estimated on-line from the intensities of diffraction peaks of different texture types of sheets in a production line at the Hoesch Stahl AG for several years. For this measuring process it is necessary to find out correlations between intensities of Bragg peaks and the r_m-values by laboratory research.

This special on-line r_m-value measuring device has been developed up to now into an on-line texture measuring device with which the low order texture coefficient can be measured unconditionally at steel strips up to a thickness of 4 mm. Based on this texture coefficients material properties which depend on the texture can be calculated. These are, for example, mangetic properties or the $r(\beta)$ value. By means of this research an unconditional on-line texture measuring has been developed from the special on-line r_m-value measuring device for different texture types.

Introduction

The previous paper titled 'Industrial application of on-line texture measurement' described the results of on-line texture measuring of steel strips in order to estimate the r_m-value at Hoesch Stahl AG. The measuring principle is characterized by the use of the energy dispersive measuring technique in connection with white X-rays and by the transmission technique. With this device the r_m-value is estimated at two texture types of sheets. Therefore it is necessary that for each texture type a correlation is found between the intensities of Bragg peaks and the r_m-value [1],[2]. For this research a lot of samples of each texture type are needed. A changing in the texture type, because e. g. of variations in the products, means that new correlations between the intensities of Bragg peaks and the r_m-values have to be found. In Fig. 1 the correlation between the intensity of one texture measuring point, in this case it is the (220) peak at a special point in the pole figure, and the r_m-value is shown. If there is a change in the texture type the intensity of the (220) peak does not correlate any longer with the r_m-value. Fig 2 shows this very clearly. For this texture type it is necessary and also possible to find a new correlation [3]. This has been done for the device which is working at the exit of the 'Continuous Annealing Line' (CAL). With this measuring device the r_m-value is measured under production conditions. In most cases this kind of r_m-value calculating is satisfactory.

An estimate of the r-value in relation to the angle β to the rolling direction, on the basis of which the r_m-value is calculated, has up to now not been possible with the method used.

To measure the $r(\beta)$-value independent of the texture type, and by doing so independent of product, was the aim of further research. In addition to that the measuring technique should be extended up to strip thickness of 4 mm by using an X-ray equipment with higher energy. Normally X-ray energies up to 60 keV are used for X-ray diffraction.

In the paper with the title 'Computer aided optimization of an on-line texture analyzer' Wagner described a mathematical method to calculate the low order texture coefficients approximately on the basis of only a few measuring points [4]. From these coefficients a lot of material properties can be calculated by using different theories [5], [6], [7], [8] . The accuracy of the mathematical approximation method depends on the number of measuring points, on their position in the pole figures and on the number of the Bragg peaks which are used.

Research into the available total number of measuring points in the texture

To verify if the approximation method is suitable for the on-line texture measuring it has to be investigated how many diffracted peaks can be measured by the energy dispersive method with regard to the thickness of the strip. Also it was necessary to verify how many single germanium-detectors can be installed in one detector system with one closed-loop cooling system. The multiplication of the number of detectors and the number of Bragg peaks for each single detector is the total number of measuring points which can be used by the mathematical approximation method.

The number of the diffracted peaks which can be measured simultaneously depends on the width of the energy window in which the peaks are measured with enough intensity. The low energy side of the useable energy window is given by the absorption of the X-rays in the steel strip. On the high energy side the barrier edge of the energy window is given by the efficiency of the semiconductor detectors as well as by the Compton-scattering which increases with higher energy. Because of this an increase in the X-ray energy from 60 keV to over 100 keV is not useful.
Fig. 3 shows the relation between the intensity of a diffracted peak and the used energy. The curve in Fig. 3 is obtained by laboratory research. In this particular case the thickness of the steel sample was 3 mm and the measuring time was 30 seconds. The curve shows that the diffracted peaks of a 3 mm thick sample can be measured with a sufficient intensity between 55 keV and 95 keV. Within an on-line texture analyzer the measuring time of 30 seconds can be substantially considerably reduced by using several parallel fixed collimators.

Below the curve the positions of the five diffraction peaks within the available energy range are marked. These five Bragg peaks can be measured with semiconductor detectors simultaneously up to a sheet thickness of 4 mm with a sufficient statistical accuracy. The vertical broken lines show that the emission energies of the K series of the tungsten tubes are within the energy resolution of the first two diffraction peaks. Because of this the intensity and as a result of that the accuracy of these two peaks increase. Fig. 4 shows the five diffraction peaks registered with a multichannel analyzer. The intensity of the peaks in relation to their individual position in the available energy range is to be seen. The diagram of the measuring result shows, that the Bragg peaks can be separated clearly enough also by using higher X-ray energy and as a result of that a smaller Bragg-angle.

The detector system which is used now for on-line measuring can be extended up to five single germanium-detectors without changing the closed-loop cooling system. If more measuring points are needed, a second detector system has to be used within one on-line texture analyzer. This would complicate on-line measuring and would increase costs.
The multiplication of the number of diffraction peaks and the number of the single germanium-detectors results in a total number of 25 measuring points with fixed diffraction peaks.

Determining of the measuring positions in the pole figures

The next task was to find out the best positions of the fixed germanium-detectors in the pole figures. Taking the principal texture orientations of the pole figures into consideration seven combinations of measuring points were chosen. Then the first three texture coefficients were determined by using the seven combinations of measuring points with a large number of samples which consisted of hot rolled, cold rolled and annealed sheet samples. With this coefficient and by using the Taylor theorie the angle-dependent $r(\beta)$-values were calculated. In addition to that the $r(\beta)$-values were measured mechanically.

Fig. 5 and Fig. 6 show two examples of the results of this research. In the figures the r-value calculated on the basis of the X-ray measuring at 0°, 45°, 90° according to the rolling direction and the r_m-value are plotted in relation to the mechanically measured r-values. Fig. 5 shows the results of a suitable combination of measuring points and Fig. 6 shows the result of an unsuitable combination. By comparing these two results, the importance of the positions of the measuring points in the pole figures becomes clear.

Because of the increase in the technical devices used in the on-line measuring process brought about by the numbers of detectors, combinations of only 4 measuring points were composed on the basis of the suitable combinations with five measuring points.

The results in Fig. 7 show, that it is possible to calculate the r-values without any pre-conditions from 4 measuring points and five peaks at each detector with a sufficient accuracy. By comparing these results with those shown in Fig. 5 where five measuring points were used it is to be seen that the fifth measuring point doesn't improve the results in this case.

In Fig. 8 the mechanically measured r-values and the r-values calculated on the basis of four measuring points with five Bragg peaks, each are shown in relation to the angle to the rolling direction. Both measuring methods lead to the same results.

On-line texture-analyzer

The transfer of these laboratory results into an on-line measuring device is shown in Fig. 9. This figure shows that the principle has been adopted from the on-line measuring device which is installed in the continuous annealing line. Above the steel strip a 100 keV X-ray tube is installed. From the X-rays emitted by the X-ray tube our single X-ray beams are separated. Below the steel strip there is the detector system consisting of 4 germanium crystals with the closed-loop cooling system. Other than in the case of the already built on-line r_m-value device several primary X-ray beams are used by this new on-line texture analyzer.

With this developed on-line texture analyzer technological data, which depend on the texture of the material can be measured without any pre-conditions. The applications of this on-line texture analyzer lie in the measuring of elastic plastic, magnetic material properties but also in the measuring of the amount of recrystalisation. At present one of these new developed on-line devices is being built at the Hoesch Stahl AG for measuring the texture of hot rolled steel strips. On the basis of the texture measuring results of the hot rolled strip conclusions can be drawn regarding to the hot rolling process and useful information can be derived for the following production processes.

References

1. Böttcher, W.; Kopineck, H.-J.: Über ein Röntgentexturmeß-verfahren zur zerstörungsfreien on-line Bestimmung technologischer Kennwerte von kaltgewalzten Stahlbändern Stahl und Eisen 105 (1985) 509 - 516.

2. Maurer, A.: Ein neuartiges Untersuchungssystem zur zerstörungsfreien Bestimmung von Werkstoffkennwerten an Feinblechen mittels Röntgenfeinstrukturverfahren" Münster 1982 (Thesis University Münster).

3. Kopineck, H.J.; Otten, H.: Texture Analyzer for On-Line r_m-Value Estimation. Textures and Microstructures, 1987, Vol. 7, pp. 97 - 113.

4. Bunge, H.J.; Wang, F.: Computational Problems in Low Resolution Texture Analysis. Theoretical Methods of Texture Analysis (H. J. Bunge) DGM Informationsgesellschaft 1987, pp. 163 - 173.

5. Bunge, H.J.: Texture Analysis in Materials Science. Butterworths Publ. London 1982

6. Bunge, H.J.: Technological Application of Texture Analysis. Z. Metallkunde 76 (1985) 457 - 470.

7. Bunge, H.J.: Textur und Anisotropie. Z. Metallkunde 70 (1979) 411 - 418.

8. Bunge, H.J.; Schulze, M.; Grzesik, D.: Calculation of the Yield Locus of Polycrystalline Materials According to the Taylor Theory. Peine + Salzgitter Berichte, Sonderheft 1981.

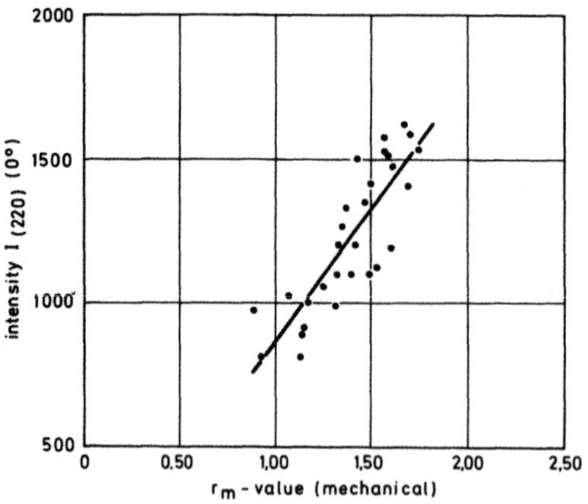

Figure 1:

Correlation between measured texture intensities and mechanical r_m-values

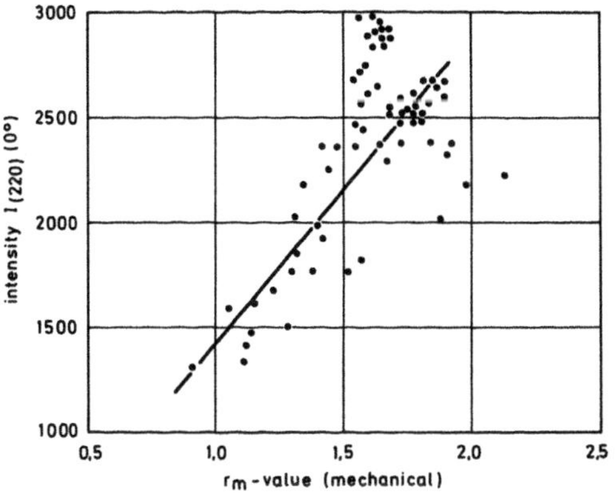

Figure 2: $I_{(220)}$-intensity for $\alpha = \beta = 0°$ in relation to the mechanically measured r_m-value

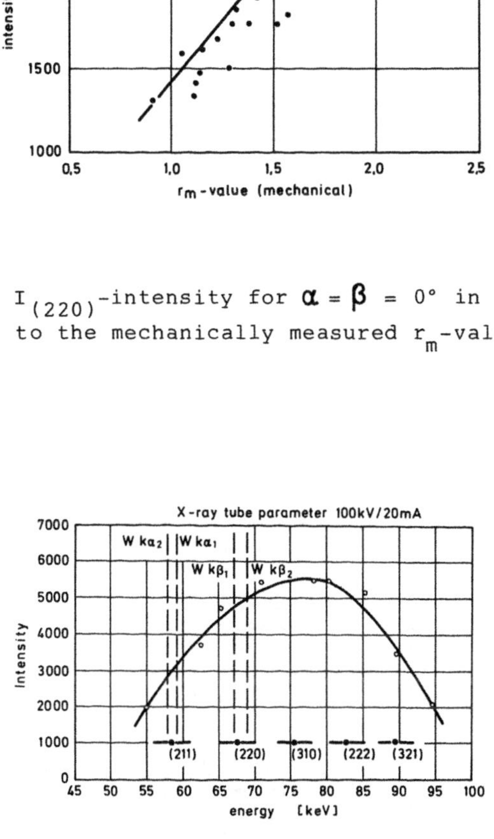

Figure 3: Dependence of the intensity of the diffraction peaks on the used energy

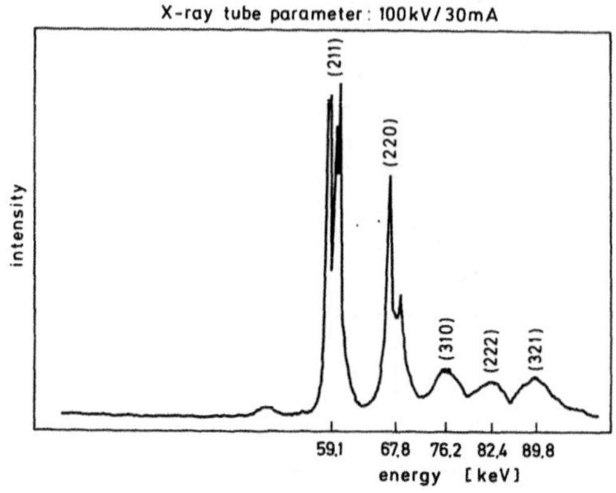

Figure 4: Resolution of diffraction peaks by using a 100 keV X-ray bremsspectrum

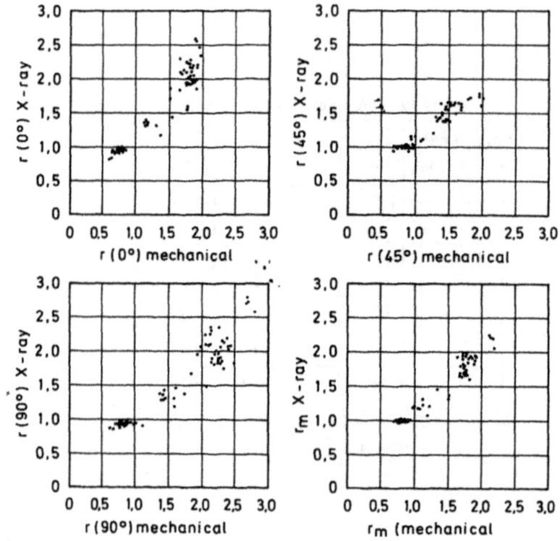

Figure 5: Correlation between X-ray measured $r(\beta)$-values and mechanically measured values (25 texture measuring points)

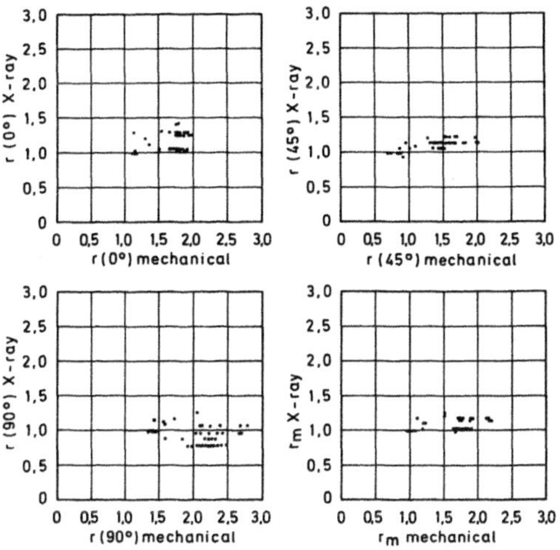

Figure 6: Correlation between X-ray measured r(β)-values and mechanically measured values (25 texture measuring points)

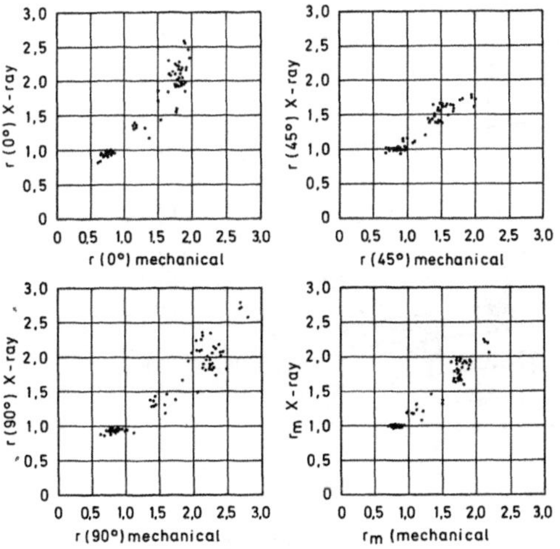

Figure 7: Correlation between X-ray measured r(β)-values and mechanically measured values (20 texture measuring points)

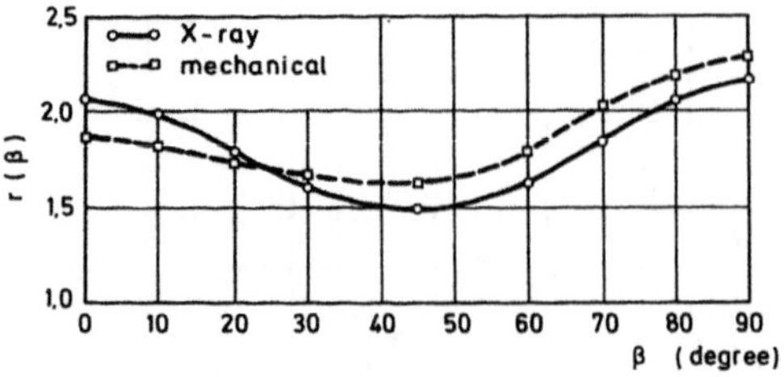

Figure 8: Comparison between the r(β)-value measured with X-rays and mechanically

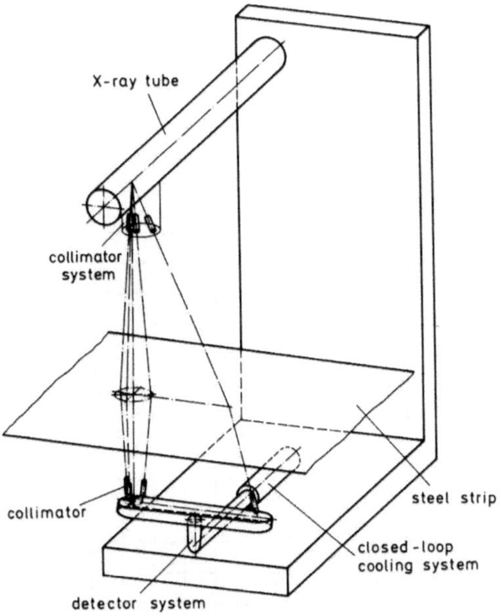

Figure 9: Schema of the on-line texture analyzer

Ultrasonic Measurements of Surface Roughness

G.V. Blessing and D.G. Eitzen

National Institute of Standards and Technology
Gaithersburg, MD 20899

Summary

Ultrasonic reflectance/scattering measurements have been made on metal samples possessing a large range of surface roughness values. The root-mean-square roughnesses R_q ranged from 0.3 to nearly 40 μm on the mostly periodic surfaces. The echo amplitude from short incident pulses of ultrasound in the frequency range of 1 to 30 MHz was used, in the manner of a comparator, to measure relative roughnesses with an area-averaging approach defined by the ultrasonic beam spot size. Ultrasonic wavelengths ranged from about 50 to 300 μm at these frequencies, and the beam spot sized varied from 0.2 to 5 mm in diameter. Both air and fluid coupling techniques were used between the sensor (transducer) and surface, on both static and rapidly (in excess of 5 m/sec surface speed) moving parts. On static surfaces, a resolution of better than 1.0 μm R_q was achieved at the higher ultrasonic frequencies. By focusing the ultrasonic beam at 30 MHz, a profilometry capability was demonstrated on a 1 μm R_q sinusoidal specimen of 800 μm wavelength.

1. INTRODUCTION

Frequently in material and parts manufacture, the surface roughness must be determined[1]. In addition to quality control of the material or the part itself, this knowledge may be useful for process control, such as monitoring cutting-tool condition. For this purpose, the capability of ultrasonic sensors to monitor surface roughness by measuring the reflected/scattered ultrasound have been studied. Emphasis has been placed on those approaches suited to in-situ real-time applications, with a feedback capability for machine control. We conceptualize an approach to use the existent coolant/lubricant stream to couple ultrasound from an electromechanical transducer positioned up-stream[2]. (See Ref. 3 for an example of the detection of surface effects due to a defective tool in a milling center.)

2. EXPERIMENTAL

Ultrasonic pulse-echo measurements were made as a function of many of the variables affecting the echo amplitude. Flat and curved surfaces were investigated, both while stationary and while rapidly moving. Commonly employed coolant/lubricant tool-cutting fluids were successfully tested as coupling media for on-line applications. Ultrasonic frequencies from 5 to 30 MHz were used in liquid media, and from 1 to nearly 5 MHz in air. In both coupling media, these frequency ranges corresponded to a wavelength range of about 300 to 50 μm.

If it is desired to guide the coupling stream and ultrasound through a curved path to the part surface for convenience of remote sensor location, a tubular extension may be attached to a squirter system as illustrated in Fig. 1.[3] Tube curvatures as large as 180 degrees were successfully demonstrated, with the tube diameter and its acoustic impedance significantly affecting the transmission efficiency, in agreement with theory[4].

3. THEORY

Much theoretical work has addressed acoustic wave scattering from both regular (periodic) and irregular (random) roughness surfaces (see e.g., refs. 5-7 and references therein). In addition, analyses taken from the electromagnetic scattering literature (see, e.g., refs. 8 and 9) have shed significant light on elastic wave scattering mechanisms. The Kirchhoff model [6] has been successfully applied to experimental ultrasonic results obtained on both random[10] and periodic[11] liquid-solid interfaces. This model yields a simple but useful expression for echo amplitude as a function of the ultrasonic wavelength and the surface roughness, assuming plane wave incidence and coherent scattering in the far field[6]:

$$A = A_o e^{-2k^2 R_q^2} \qquad (1)$$

Here A_o is the smooth-surface amplitude, $k=2\pi/\lambda$ is the wave number, and λ is the wavelength.

4. RESULTS AND DISCUSSION

4.1. Flat Bar

Echo-amplitude scans were made as a function of sample position on a flat aluminum bar sample with one ground and four fly-cut-generated roughness sections having a wide range of R_q values from less than 1 to nearly 40 μm, and surface periodicities Λ ranging from 25 (ground) to 800 μm.

The sample was scanned perpendicular to the surface lay with different coupling techniques using both air and liquids, and a range of ultrasonic frequencies. Here we present data taken at 6 MHz with 5 mm diameter liquid stream coupling, and a beam dimension much greater than surface feature dimensions for an area-averaging approach.

Using eq. 1 and the mean echo amplitude levels of the scanned roughness sections for the discrete data points, Fig. 2 illustrates good agreement between theory and experiment for a large range of roughness values in the nominally coherent scattering regime. Similar results were obtained at 1 MHz with ambient air coupling[12], as might be expected based on the comparable wavelengths of ultrasound in the respective media.

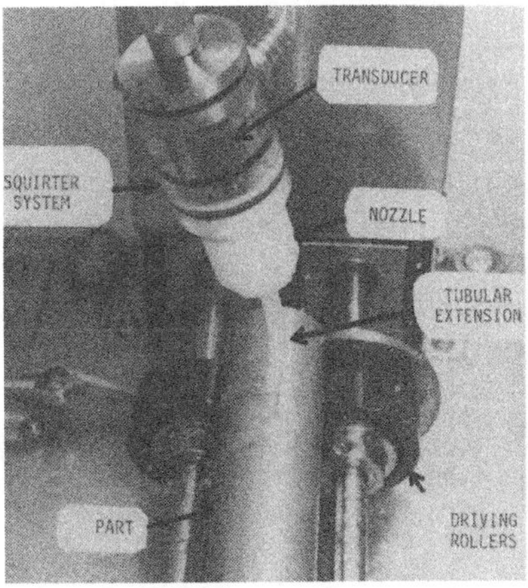

Fig. 1. A squirter system with an arcuate tubular extension for stream coupling ultrasound to a rotating part.

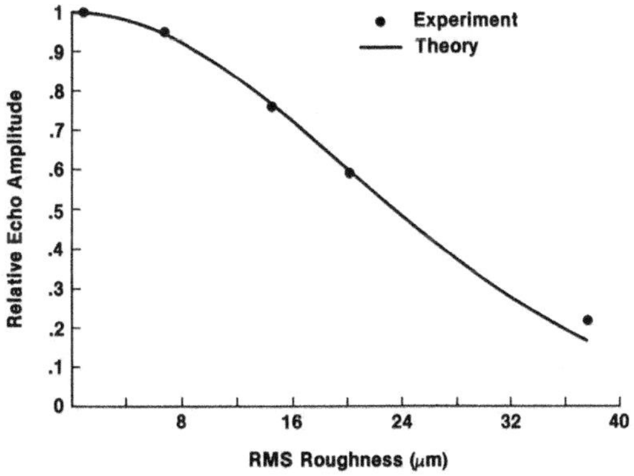

Fig. 2. Relative echo-amplitude (experiment and theory) versus surface roughness at 6 MHz with stream coupling.

4.2. Cylinder

The feasibility for in-process surface monitoring of parts on a turning center was evaluated using the laboratory rotating device illustrated in Fig. 1. The tubular extension was polyethylene tubing 8 cm in length and 0.5 cm inner diameter, with about a 25 degree net curvature from the nozzle center axis to the surface normal. Tests for non-normal stream incidence on the two-section cylinder showed that the ratio of rough to smooth back-scattered amplitudes at 6 MHz was encouragingly insensitive to the angle of incidence up to 15 degrees from the normal.

Figure 3 shows representative amplitude data taken as a function of time at 6 MHz on the two-section roughness aluminum cylinder 7.5 cm in diameter. The data were taken with the cylinder (a) stationary, and (b) rotating with a surface speed of 2 m/sec. The R_q values were nominally 1 and 13 μm, and the Λ about 30 and 650 μm, respectively. The flow rate was about 3 L/min, corresponding to an exit stream velocity of 0.6 m/sec. The sensor was angulated for normal stream incidence (maximum echo amplitude on the smooth section), with an exit stream length of 1 cm. The peak-detected and digitized amplitudes of individual echoes are shown in the figure. The sensor was positioned for a number of seconds duration over the smooth section to produce the left-hand-side data, and then over the rough section for the (lower amplitude) right-hand-side data.

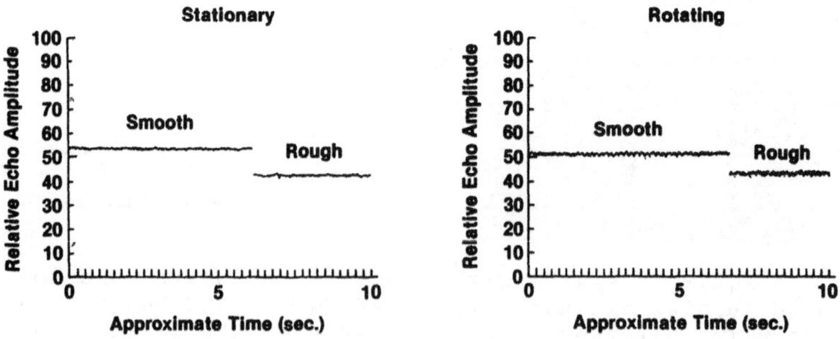

Fig. 3. Relative echo amplitude measurements at 6 MHz with stream coupling on a two-section-roughness aluminum cylinder (a) stationary, and (b) rotating.

The data in Fig. 3(a) are a measure of the scattering at a given (stationary) surface spot, determined in this case by the stream contact area. By contrast, the data in Fig. 3(b) are a measure of scattering as a function of position also, with the insonified spot sweeping out a circumferential band on the rotating cylinder. As a result, the scan of seconds at 500 rpm includes multiple cylinder rotations in the data of Fig. 3(b), which acted to superpose a periodic structure of real surface variations as a function of rotation angle on the static data.

Data were also taken at 15 MHz with the same rotating system and sample, and at 1 MHz by means of ambient air coupling (without nozzle, tube, etc.). The 15 MHz transducer was 3 mm in diameter and the 1 MHz unit was 12 mm in diameter. The transducer-surface separation was 1 cm for the air-coupled measurements. Table 1 summarizes averaged data (mean level for a given surface and coupling condition), reduced from digitized raw data like that illustrated in Fig. 3. Air turbulence generated at the surface during rotation caused such severe signal distortion as to preclude a meaningful interpretation.

Table 1. Reduced amplitude data taken on the rough (A_R) and smooth (A_S) cylinder sections as a function of frequency and coupling technique, while stationary and while rotating. The two standard deviation values 2σ are for the amplitude ratios of multiple-second data records like those in Fig. 3.

f(MHz)	Sample	A_R/A_S	2σ	Coupling
6	Stationary	.80	.03	water stream
6	Rotating	.80	.04	water stream
15	Stationary	.66	.02	water stream
15	Rotating	.60	.05	water stream
1	Stationary	.85	.02	ambient air

4.3. Sinusoids

Precision sinusoidal surfaces, faced by single-point diamond turning of nickel-plated surfaces, were used to evaluate the ultrasonic technique as both a scatterometer and a profilometer in the pulse-echo mode at normal incidence. Area-averaging scatterometer measurements were made at 30 MHz in water (λ = 50 μm) and at 4.4 MHz in air (λ = 75 μm) on the three-section

100 μm Λ sinusoidal specimen. The three R_q values were 0.33, 1.1, and 3.3 μm.[13] The beam width of the incident 30 MHz ultrasound was 220 μm, and that of the 4.4 MHz was approximately 1 cm. With a target distance of 20 mm in water, the relative echo amplitude values at 30 MHz were respectively 1.00, 0.95, and 0.83 $\pm$ .004, normalized to the smoothest surface. Similarly, with a target distance of 3 mm in air, the relative values at 4.4 MHz were 1.00, 0.95, and 0.78 $\pm$ .01.

Echo-amplitude profilometer measurements were made using 30 MHz focused ultrasound on the 800 μm Λ surface with R_q=1.1 μm, scanning perpendicular to the lay with the sample immersed in water. The transducer was a broad-band 6 mm diameter spherically focused unit with a focal length F of 20 mm. Pulse-echo measurements on a 0.76 mm diameter glass target sphere determined the half-amplitude beam width at the transducer focus to be 220 μm. Figure 4 compares the results of (a) a contacting stylus scan[14], with (b) the ultrasonic echo-amplitude scan. While the comparison is most encouraging for the ultrasonic technique, we note that this signature is a convolution of the incident beam size and the surface topography. For example, Fig. 4(b) may provide a measure of the surface periodicity, but does not provide surface height information without a reference standard. (See ref. 12 for transit-time data between the sensor and surface, which is proportional to surface height.)

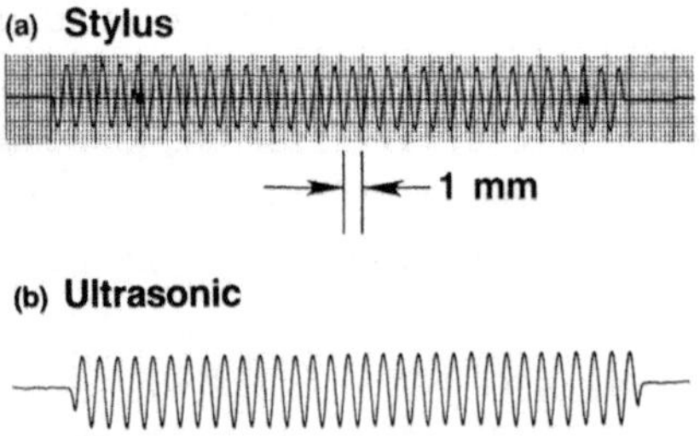

Fig. 4. Profilometer scans of a 1.1 μm R_q sinusoidal surface by (a) stylus, and (b) a noncontacting ultrasonic echo-amplitude technique.

5. CONCLUSIONS

Ultrasonic sensor techniques were studied in both the scatterometer and profilometer modes to evaluate the capability for measuring surface roughness features. As a scatterometer, area-averaging echo-amplitude measurements were made on flat and cylindrical metal samples possessing a wide range of surface roughness values ($40 > R_q > 0$ μm) representative of those to be found in various manufacturing applications. All samples possessed a unidirectional periodic structure formed by shaping or turning. Ultrasonic frequencies from 1 to 30 MHz provided the incident wave pulses with a spectrum of wavelengths from about 50 to 300 μm in water and air. Stationary samples were scanned with several means of coupling: immersed in water, by a liquid stream, and in air. The relative amplitude data generally showed good agreement with a Kirchhoff model for coherent wave scattering at low frequencies, where the ultrasonic wavelengths were much greater than R_q. Stream coupling was shown to work well using water or the coolant/lubricant fluids employed with cutting tools on turning and milling centers. This coupling was extended to rapidly rotating cylindrical samples to evaluate the suitability of the ultrasonic technique for surfaces moving at speeds (up to 5 meters/sec) that might be encountered in manufacturing.

Scatterometer and profilometer techniques were applied to precise sinusoidal surfaces with R_q values of 0.33, 1.1, and 3.3 μm, and Λ values of 100 and 800 μm. Ultrasound at 30 MHz clearly resolved these roughness values. In addition, with the ultrasonic beam focused to a spot size of 220 μm diameter, a profilometry capability was demonstrated on the 1.1 μm R_q sinusoid with Λ equal to 800 μm.

6. ACKNOWLEDGEMENTS

The authors thank T. Vorburger and colleagues for many helpful discussions on surface roughness parameters and their measurement. They also thank students R. Tregoning (Johns Hopkins University) and H. Ryan (University of Maryland, Baltimore Campus) for laboratory measurements and analysis, and

D. Neal for hardware design and fabrication. This work was supported by the Office of Nondestructive Evaluation, and by the Automated Manufacturing Research Facility Project (AMRF) at the National Institute of Standards and Technology.

7. REFERENCES

1. E.G. Thwaite, Prec. Eng. 6 (1984) 207-217.

2. G.V. Blessing and D.G. Eitzen. U.S. Patent 4,738,139 (19 Apr 88).

3. D.G. Eitzen and G.V. Blessing, MRS Bulletin XIII Vol. 4 (1988) 49-52.

3. G.V. Blessing and D.G. Eitzen. U.S. Patent 4,738,139 (19 Apr 88).

4. A. Sataai-Jazi, C.K. Jen, G.W. Farnell, and J.D.N. Cheeke, J. Acoust. Soc. Am 81 (1987) 1273-1278.

5. W.G. Neubauer, J. Acoust. Soc. Am. 35 (1963) 279-285.

6. N.F. Haines and D.B. Langston, J. Acoust. Soc. Am. 67 (1980) 1443-1454.

7. J.A. Ogilvy, J. Phys. D, Vol. 21 No. 2 (1988) 260-277.

8. P. Beckmann and A. Spizzichino, The Scattering of Electromagnetic Waves from Rough Surfaces (Pergamon, New York, 1963).

9. H.E. Bennett and J.O. Porteus, J. Opt. Soc. Am. 51 (1961) 123-129.

10. M. deBilly, F. Cohen-Tenoudji, G. Quentin, K. Lewis, and L. Adler, Review of Progress in Quantitative NDE (July 1980) 320-329.

11. G.V. Blessing, P.P. Bagley, and J.E. James, Matls. Eval. 42 (1984) 1389-1400.

12. G.V. Blessing and D.G. Eitzen, Fourth International Conference on the Metrology and Properties of Engineering Surfaces, Proceedings, 13-15 APR 88, Washington, D.C. (to be published: Kogan Page, London; and in Surface Topography, same publisher, 1989).

13. E.C. Teague, F.E. Scire, and T.V. Vorburger, Wear, 83 (1982) 61-73.

14. T.V. Vorburger, Nat. Bur. Stand. (U.S.) (private communication of sample and data).

Characterization of Industrially Important Materials Using X-Ray Diffraction Imaging Methods

J. M. WINTER, JR, and R. E. GREEN, JR.

Center for Nondestructive Evaluation
The Johns Hopkins University
Baltimore, MD. 21218, U.S.A.

Summary

The ability to obtain x-ray diffraction images quickly can enable improved process control through on-line materials characterization. The direct measurement of grain boundary velocities **in situ** at elevated temperatures, diffraction from melt spun alloy ribbons solidified only milliseconds earlier, or diffraction from shaped charge liner jets while in flight are shown as examples of real time dynamic characterization. Imaging of single crystal nickel alloy jet engine turbine blades using monochromatic asymmetric crystal topography in back reflection is examined. White beam synchrotron transmission topography of quartz precision oscillator crystals, ZnCdTe crystals used as substrates for HgCgTe, and GaAs substrates used for microwave integrated circuits is discussed.

General

Characterization of materials by x-ray diffraction can readily be performed rapidly or even in real-time using a variety of hardware and techniques. Generally such measurements are arranged to generate data in the form of images, which in turn may be digitally processed to extract details of interest not apparent in the raw image. The ability to obtain such images quickly opens up new opportunities for nondestructive testing, particularly in providing timely information for materials process control.

Examples using conventional and flash x-ray sources

A clever arrangement for measuring grain boundary migration rates in deformed metal single crystals has been shown to work effectively with a conventional x-ray source [1]. A vertical line x-ray source is collimated and then interrupted by a wire grid just before impingment on the specimen. The result is illumination of a vertically suspended strained wire specimen by a vertical dashed line of x-rays. A typical diffraction pattern is shown in Fig.1. The specimen is in a furnace maintained at a preselected temperature, and the various diffracted x-ray beams pass out of the furnace through a thin aluminum heat reflecting foil. They are imaged by an x-ray image intensifier system coupled to a video camera and a video cassette recorder. The image has two sets of Laue patterns, one from each side of the grain boundary. Measurement of the time required for any one Laue line segment to increase of decrease in length by a dash length can be used to determine grain boundary velocity if the spacing between wires in the grid in front of the beam is used as calibration.

A system which dynamically characterizes material moving at 50 meters/second is shown in Fig.2. A conventional x-ray source is combined with a rapid imaging system to quickly record diffraction patterns from melt spun ribbons within a few milliseconds of when they first solidify. These alloys undergo quench rates in excess of 10^6 ^oC per second and are often amorphous. Later static measurements on extent of subsequent crystallization as a function of storage time and temperature have led to the concept of a "half-life for amorphism".

Flash (or pulse) x-ray systems are triggerable and the units suitable for diffraction can generate a single output pulse which typically might span 40ns to 80ns, but with anode currents of the order of 4kA to 5kA, (compared to 50mA in a conventional tube). Figure 3(a) shows the schematic arrangement for using a flash x-ray and image intensifier system to look for a diffraction pattern from the metallic jet formed upon detonation of a shaped charge anti-tank round [2]. The jet moves at hypervelocity with extremely high kinetic energy. The original question was whether the metal is still partially crystalline or entirely liquid while in flight. Figure 3(b) shows the result. The diffraction pattern shows crystallinity.

The next group of techniques are devoted to evaluation of single crystals. Perhaps they are best considered off-line techniques for rapidly imaging defect structures as part of an evaluation or sorting operation between process steps.

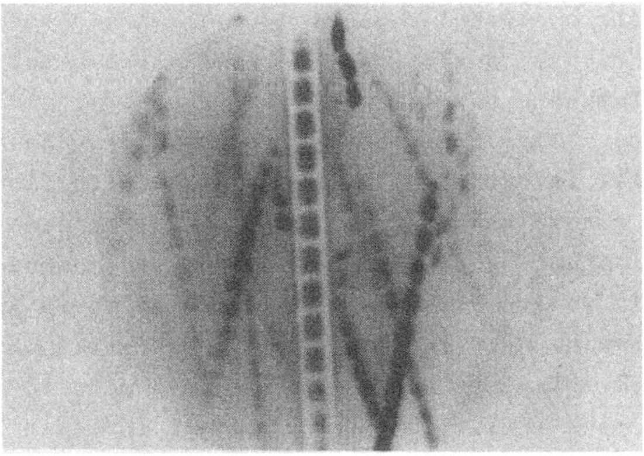

Fig.1. Diffraction from a strained single crystal wire suspended vertically in a furnace and illuminated by an x-ray beam arranged as a dashed vertical line. As a grain boundary propagates during annealing, the new grain diffracts a new pattern of dashed lines. Rate of addition of new dashes gives growth velocity.

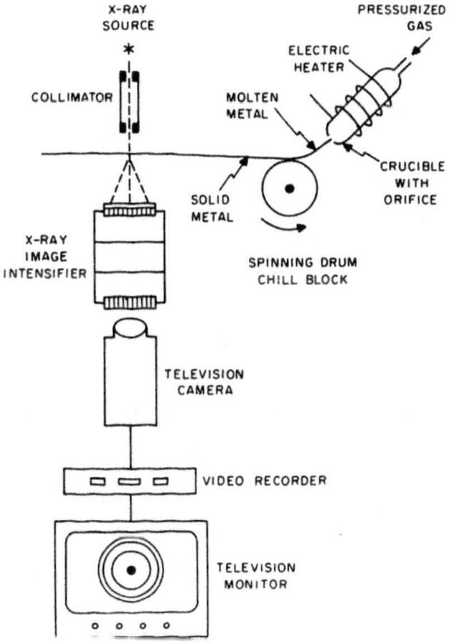

Fig.2. Schematic of real time imaging system for recording x-ray diffraction patterns from chill block melt spun ribbon alloys within milliseconds after solidification. Patterns taken later document rate of crystallization of alloys which are amorphous.

Asymmetric crystal topography (ACT) has been described elsewhere [3]. A conventional vertical line x-ray source is used to illuminate a silicon single crystal at a very low angle of incidence. The "footprint" of the incident beam on this first crystal is expanded in the downbeam direction. The crystal is cut so that diffraction from a low order set of planes will occur at a high angle, creating a monochromatic beam with roughly square cross-section. The expanded monochromatic beam may then be used to illuminate the crystal under evaluation. This crystal is brought into an orientation which projects a back reflection Laue image of interest into the input of an image intensifier system. A back reflection topograph will be generated by this twice diffracted beam, and the image can then be recorded. Note that this technique permits observation of only one Laue image at a time and requires fine control of specimen orientation. Figure 4 is an example of such a topograph, showing part of a cross-section of a single crystal turbine blade for an aviation jet engine. Several sections don't share the same crystallographic orientation.

Examples using a synchrotron x-ray source

Figure 5(a) shows a white beam transmission topograph recorded on film using a synchrotron x-ray source. Note the large number of Laue images recorded all at one time. In this particular case, the 15mm diameter crystal was tilted about a horizontal axis orthagonal to the axis of the x-rays, so the 6mm high incident beam can illuminate the entire surface of the quartz disk. This technique effectively expands the incident beam vertically. Figure 5(b) shows an enlarged view of one of the Laue images from Fig.5(a). An entire section of the crystal is oriented differently, and its Laue image is somewhere else. Figure 6 shows similar enlargements, only in this case for alloys which serve as substrates for HgCdTe epitaxial films. Both specimens were grown from the melt and these topographs were recorded before homogenization. Note that the band in the image in Fig.6(b) apparently represents a growth twin. Figure 7(a) is a topograph similar to the one shown in Fig.5(a), except the specimen is GaAs, not quartz. In this case the substrate is a 76mm diameter cut-and-polished wafer. It is orthogonal to the

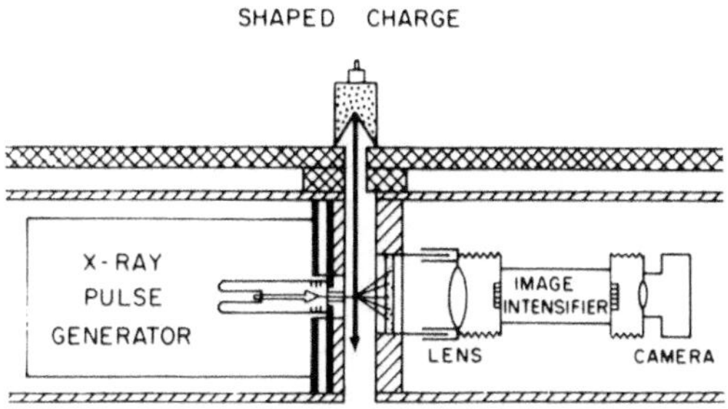

(a)

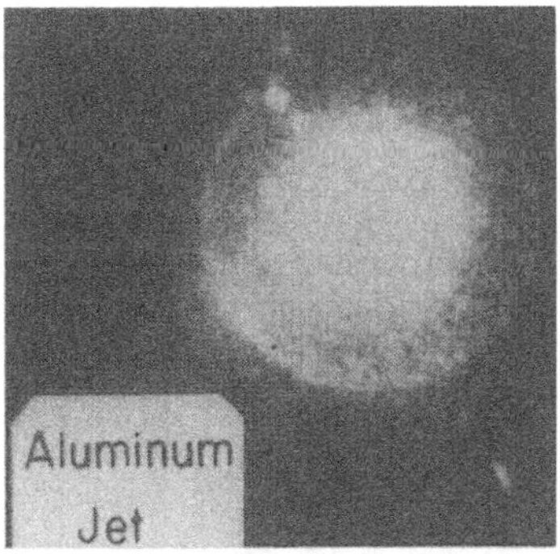

(b)

Fig.3. (a) Schematic of experiment using pulsed x-ray to record diffraction pattern of hypervelocity jet formed by shaped charge liner, (b) Resulting diffraction pattern showing crystallinity

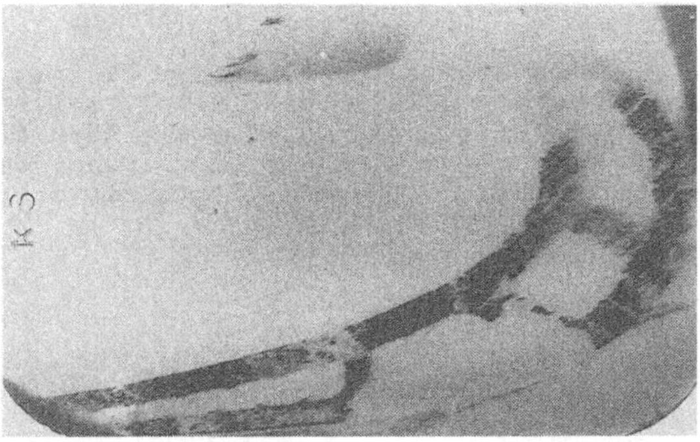

Fig.4. Reflection topograph of a single crystal turbine blade shows some sections are not the same orientation. Imaged using asymmetric crystal topography and conventional x-ray source.

incident beam, so the size and shape of the Laue images shown represent the cross-section of the incident beam quite accurately. Each Laue image in the topograph carries with it information on the defect structure of the wafer (as projected into that particular Laue image's lattice planes), but only from the limited volume of specimen illuminated by the incident beam. To cover more of the specimen, a master series of topographs is recorded, each with an incident beam translated one beam height in a vertical direction [4]. If a particular Laue image is

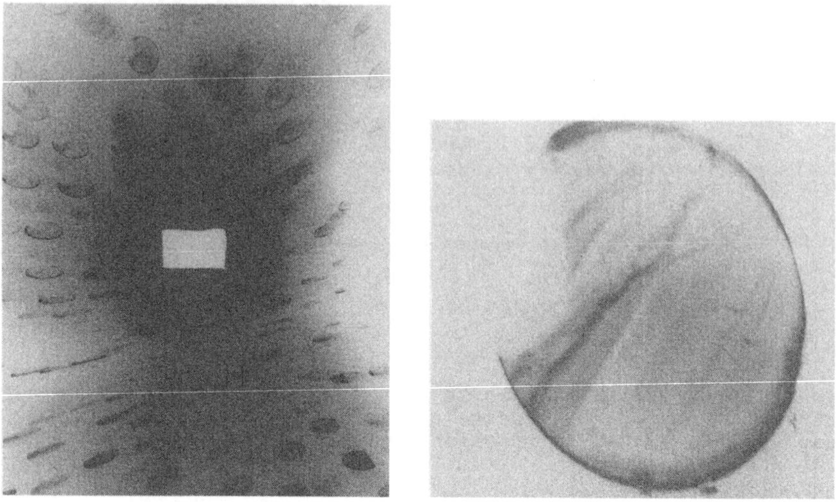

(a) (b)

Fig.5. (a) White beam transmission topograph of 15mm diameter quartz crystal used in a precision oscillator, showing the multitude of Laue images recorded on a 20cm x 25cm film placed near specimen. (b) Enlargement of one Laue image, showing crystal has sector of different crystallographic orientation.

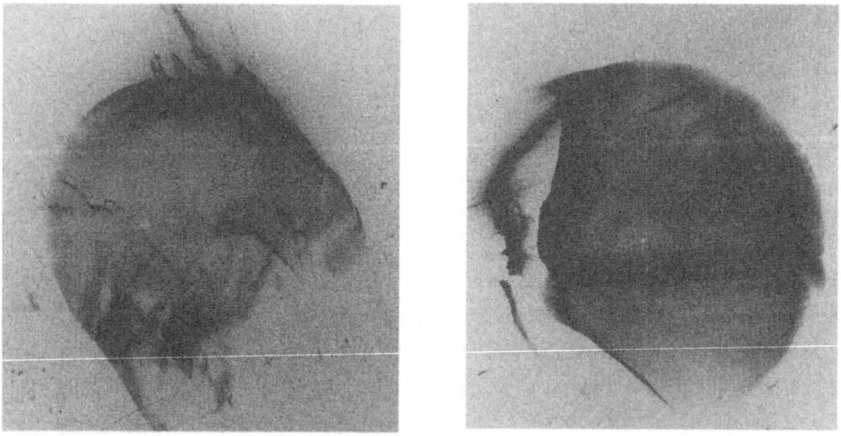

(a) (b)

Fig.6. (a) Enlargement of single Laue image from white beam transmission topograph of single crystal of $Zn_{0.2}Cd_{0.8}Te$ used as substrate for HgCdTe epitaxial films.
(b) Enlargement of single Laue image from white beam transmission topograph of single crystal of $Zn_{0.4}Cd_{0.6}Te$.

chosen to be enlarged from each of these topographs, the resulting images can be assembled in a montage such as that shown in Fig.7(b). The montage covers the 76mm diameter specimen from center to edge.

An improvment on the method just described is indicated in Fig.8. In this example, the Laue image for a partcular set of planes is isolated from its neighbors by moving the film far from the specimen. This provides enough room to scan the specimen without generating overlapping images. Figure 9 shows scanning white beam transmission topographs of three different GaAs substrates, two from the same boule and one from an entirely different supplier. All three substrates are typical of material used in regular production of microwave integrated circuits.

Acknowledgments

The synchrotron white beam transmission topography was performed at Beamline 19C of the National Synchrotron Light Source, Brookhaven National Laboratory, which is supported by the U.S. Dept. of Energy under Grant No. DE-FG-02-84ER45098. The work on gallium arsenide was supported by Westinghouse Electric Corporation, Advanced Technology Division, Balto. Md.

References

1. Robert E. Green, Jr., An electro-optical x-ray diffraction system for grain boundary migration measurements at temperature. Adv. in X-Ray Analysis 15 (1972) 435-445.

2. Robert E. Green, Jr., First x-ray diffraction photograph of a shaped charge jet. Rev. Sci. Instr. 46 (1975) 1257-1261.

3. W. J. Boettinger, H. E. Burdette, M. Kuriyama, and R. E. Green, Jr., Asymmetric crystal topographic camera. Rev. Sci. Instr. 47(8) (1976) 906-911.

4. J. M. Winter, Jr., R. E. Green, Jr., and W. S. Corak, White beam synchrotron x-ray toporaphy of gallium arsenide. Rev. of Progress in Quantitative NDE 7B (1988) 1153-1160.

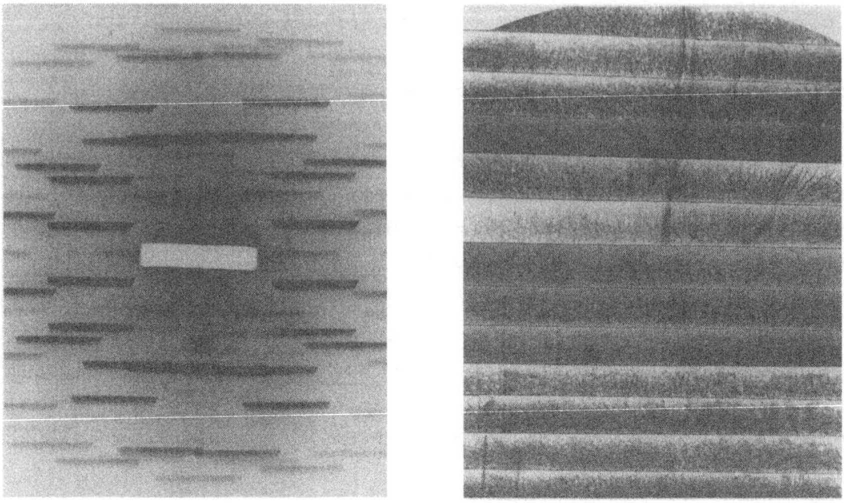

(a) (b)

Fig.7. (a) White beam transmission topograph of 76mm diameter GaAs substrate, showing the numerous Laue images recorded on a 20cm x 25cm film placed near specimen. Both film and specimen are orthogonal to axis of incident x-ray beam. The incident x-rays cover only a small section of the wafer. (b) A montage of Laue images (all of the same Miller indices) selected and enlarged from a master series of topographs like (a). The master series is arranged in steps to cover the 76mm diameter substrate from center to edge.

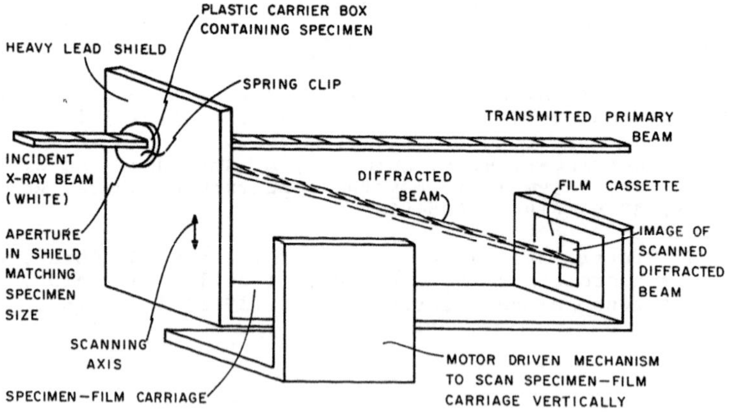

Fig.8. Schematic of arrangement for scanning white beam transmission topography

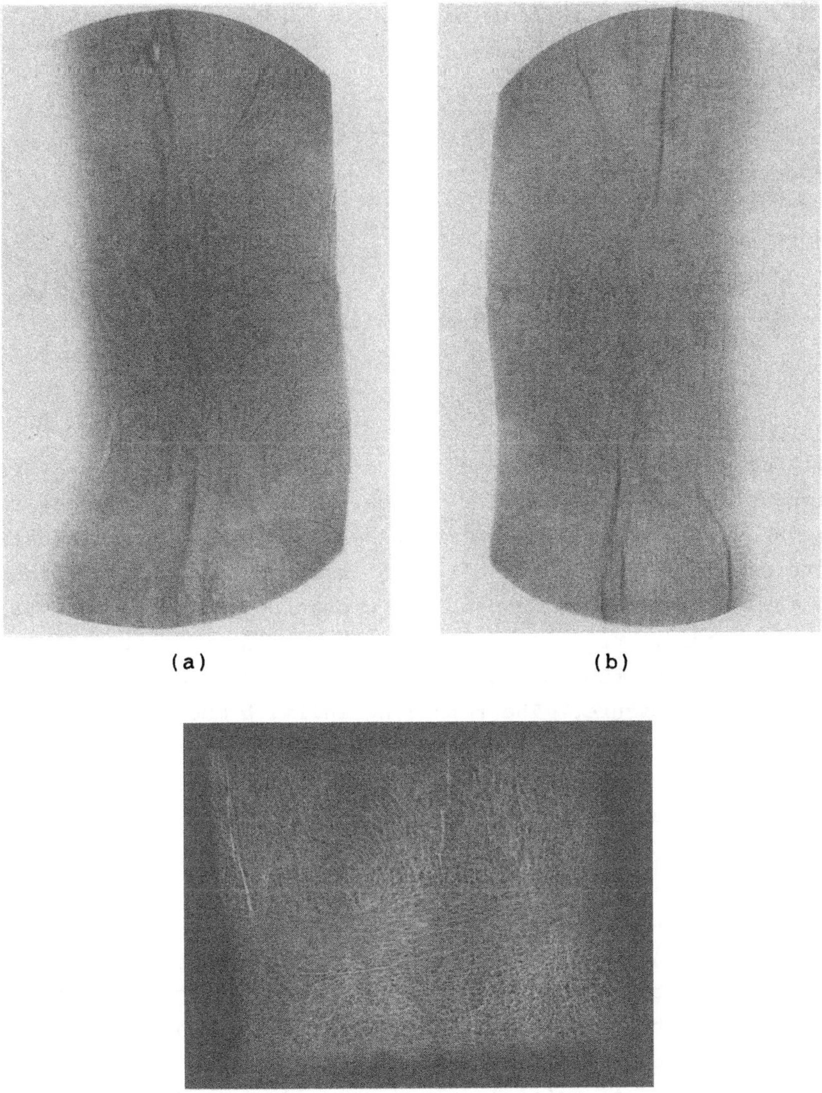

(a) (b)

(c)

Fig.9. (a) Continuous scan white beam tranmission topograph of one Laue image from 76mm diameter wafer of GaAs. (b) Same for a different wafer from the same boule, 13 slices closer to the seed end, (c) Continuous scan white beam tranmission topograph from 50mm diameter GaAs wafer from different supplier.

Ultrasonic Characterization of Cold Welds

PETER B. NAGY, ARNON WEXLER, AND LASZLO ADLER

Department of Welding Engineering
The Ohio State University, Columbus

MARYLINE TALMANT

Groupe de Physique des Solides
Universite Paris VII, Paris

Introduction

Cold weld is a fairly common type of defect in inertia and friction welds and, to a smaller degree, in diffusion bonds and resistance welds as well. It occurs as a combined result of strong compressive stress and insufficient heating when the elevated interface temperature is not high enough to produce good metallurgical bond, but at least one of the contacting parts is softened enough by the heat to reach plastic deformation at that particular pressure. The resulting intimate mechanical contact between the compressed surfaces causes small reflection and high transmission, i.e. the bond appears to be flawless for ultrasonic inspection, but the joint strength is zero in this cold welded region [1].

Fig. 1 shows the schematic diagram of the axial cross-section of a defective inertia weld. The friction heat is generated mainly at the perimeter of the sample where the relative velocity between the parts is maximum. The center part is heated via conduction only and the interface temperature lags behind that of the periphery. At sufficiently high welding pressures, the bond is very good at the circumference of the joint, but usually contains a large cold weld spot at the center. What happens is that the best part of the heated material is extruded from the welded region and the motion stops in a very short time, therefore there is not enough heat conducted to the center part to raise the interface temperature to the necessary level.

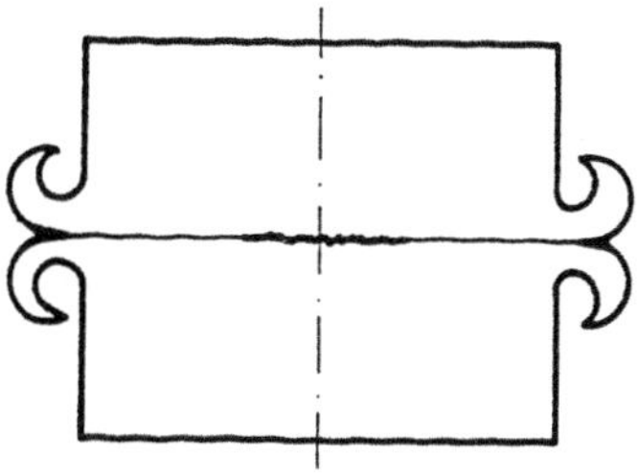

Fig. 1. Schematic diagram of an inertia friction weld with a cold welded region at the center.

In order to demonstrate the difficulty encountered in the inspection of such cold welded interfaces, Fig. 2 shows the typical ultrasonic reflection and transmission profiles of a 1" diameter stainless steel-copper inertia weld made at 2000 psi axial pressure. The shaded area indicates the cold weld location. The fracture surface of these welds revealed that there was no

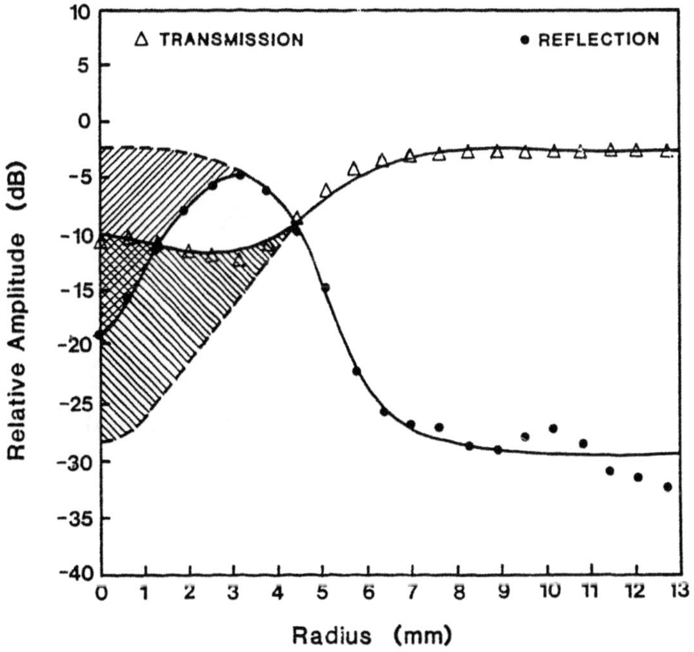

Fig. 2. Reflection and transmission profiles for a 1" diameter, 2000 psi stainless steel-copper inertia weld at 10 MHz.

metallurgical bond whatsoever at the center within a 5 mm radius, while the bond was apparently flawless in an approximately 5 mm wide ring at the circumference of the joints. Especially at higher pressure, the ultrasonic transmission is greatly enhanced while the reflection drops to a very low level close to that of a perfect bond at the center.

Plastic Contact Between Rough Surfaces

From numerous imperfect boundary models, Haines' model for plastic deformation of contacting rough surfaces seems to be the best suited for cold welds [2]. The schematic diagram of plastic contact between flat, rough surfaces is shown in Fig. 3. For any two real surfaces, the initial contact area comprises a number of isolated islands at the highest points above the mean

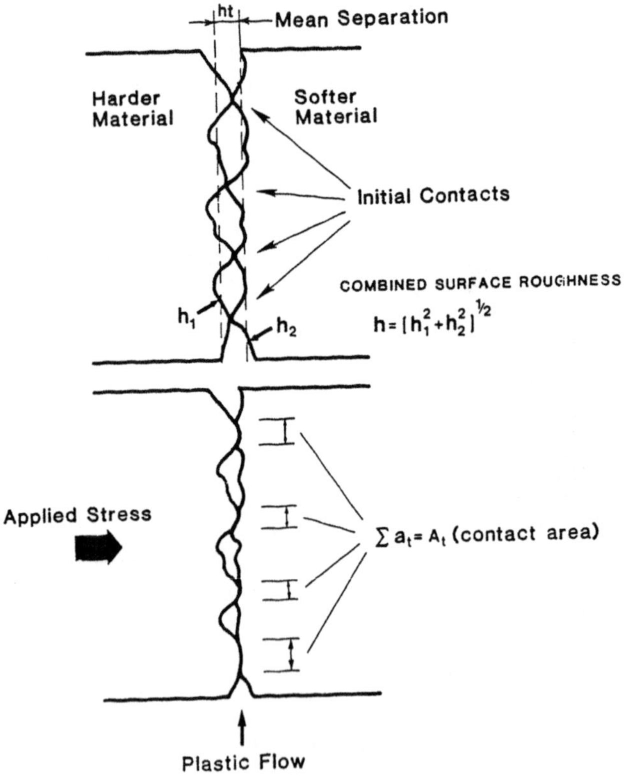

Fig. 3. Plastic contact between flat, rough surfaces.

planes of the rough surfaces. Upon applying a compressive stress s, the apparent contact area A_t increases via plastic flow in one or both materials at the contacting peaks. Haines obtained very simple formulae for the ultrasonic reflection R and transmission T coefficients for interfaces between compressed rough surfaces. Although Haines derived his final formulae for similar interfaces only, his approach can be readily applied to the more general dissimilar case as well.

$$T = \frac{T_o}{1 + \frac{i\omega}{\Omega}}, \tag{1}$$

and

$$R = \frac{R_o - \frac{i\omega}{\Omega}}{1 + \frac{i\omega}{\Omega}}, \tag{2}$$

where R_o and T_o denote the reflection and transmission coefficients of the perfect interface. The frequency dependence of both coefficients is given by a single characteristic angular frequency Ω which increases with pressure s:

$$\Omega = \frac{s}{p_m} \frac{1}{\bar{r}} \frac{4c}{\Pi}, \tag{3}$$

where p_m is the flow pressure, $\bar{r}$ is the average radius of the contact areas, and c is the sound velocity. $\bar{r}$ is a function of the normalized applied pressure s/p_m and the combined rms roughness $h = (h_1{}^2 + h_2{}^2)^{\frac{1}{2}}$ of the contacting surfaces. Haines derived the following approximate formula for the mean contact radius $\bar{r}$

$$\bar{r} \simeq \frac{2}{\Pi t} (3.48h + 4.69 \text{ microns}), \tag{4}$$

where t is the mean separation of the rough surfaces normalized to h. For normally distributed surfaces t can be determined from the probability function:

$$\frac{s}{p_m} = \frac{1}{\sqrt{2\Pi}} \int_t^\infty \exp[- \frac{t^2}{2}] \, dt. \tag{5}$$

Fig. 4 shows the frequency dependent reflection and transmission coefficients of contacting stainless steel surfaces for different compressive pressure to flow pressure ratios. As can be expected from simple physical consideration, the reflection increases while the transmission decreases with increasing frequency, and the turning point of the spectra moves upward with increasing compressive pressure.

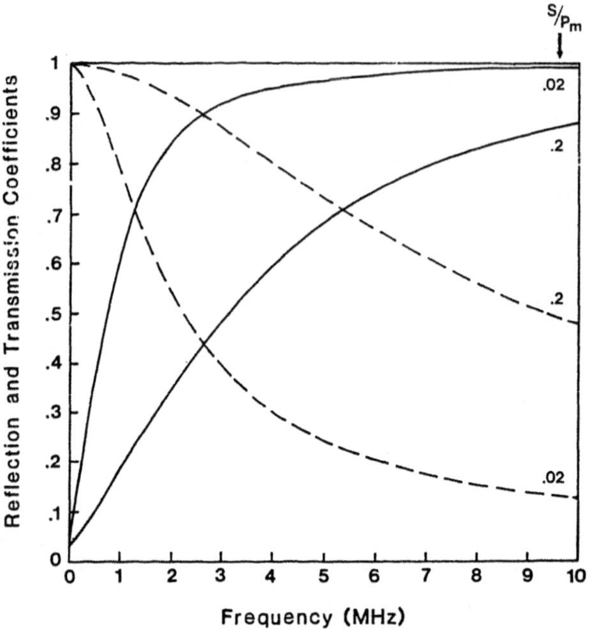

Fig. 4. Reflection (solid lines) and transmission (dashed lines) coefficients versus frequency for a stainless steel-copper interface at 2% and 20% compressive pressure to flow pressure ratios ($h_1 = h_2 = 10\ \mu m$).

It is well known that extremely high compressive pressures are needed to achieve good ultrasonic contact between dry solid surfaces [3]. This is in good accordance with predictions based on Fig. 4 indicating that at as high as 10% of the flow pressure the interface is still rather poorly transmitting and strongly reflecting. So, how is it possible that welding pressures around 2 ksi can result in almost perfect ultrasonic contact between materials with flow pressures of 30 ksi or even higher? The answer is, of course, that the temperature of the interface increases to a maximum just below the melting point where the flow pressure drops to the level of the applied compressive stress.

Experimental Results

Cold weld is a major problem in dissimilar inertia welds such as the commonly used copper-stainless steel combination. Copper is a very good heat conducting and ductile material with a relatively low melting point around 1080°C, while stainless steel is a much poorer heat conductor with less ductility and a higher melting point around 1500°C. At higher welding pressures, the hot, softened copper is almost instantaneously extruded from the interface region leaving the center part of the stainless steel too cold to form good metallurgical bond. At the same time, the extensive plastic deformation in the copper results in a very "strong" cold weld of low ultrasonic reflectivity. Fig. 5 shows the experimentally determined reflection spectra from cold welded center parts of stainless steel-copper inertia welds made at different compressive pressures. Due to certain approximations used in Haines' method, the above simple formulae for reflection and transmission coefficients of contacting rough surfaces are valid

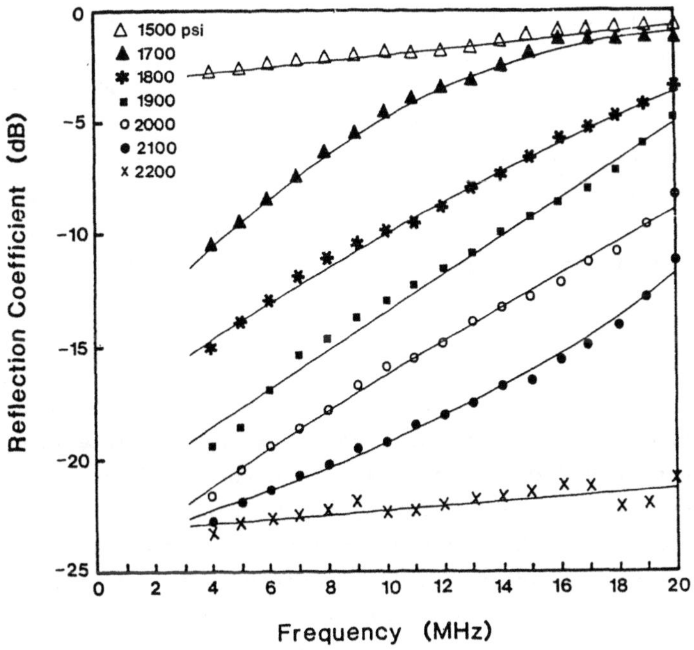

Fig. 5. Reflection spectra from the cold welded center part of stainless steel-copper inertia welds made at different pressures between 1500 and 2200 psi and constant rotational speed of 2000 rpm.

for relatively weak plastic deformations only when the compressive perssure does not exceed 10 - 20% of the flow pressure. In spite of the considerable plastic flow during the welding process, the measured spectra still exhibit the main features predicted by Haines. In particular, from low to high frequencies the curves rise from the low reflection coefficient of a perfect interface to the full reflection of total misbond, and the transient frequency moves upward with increasing pressure. At 1500 psi this turning point is below the measuring frequency range and the cold weld looks like a strong misbond. At the other end of the scale, at 2200 psi, the turning point is above 20 MHz, therefore this very strong cold weld cannot be detected in the applied frequency range.

The other principal welding parameter, the rotation speed of the sample has a similar effect on the reflection spectrum too. At higher speeds, more kinetic energy is dissipated at the interface and it takes a little longer to stop the rotation, which leaves more time to conduct the friction heat to the center part. Fig. 6 shows the measured reflection spectra from the cold welded center part of stainless steel-copper inertia welds made at different rotational speeds between 1800 and 2300 rpm and constant welding pressure of 1800 psi.

Conclusions

The low ultrasonic contrast of cold welded interfaces was shown to be caused by plastic deformation at the contacting rough surfaces. Strong correlation was found between the reflection spectra of such defective interfaces and the axial pressure to flow pressure ratio reached during the welding process. This ratio was changed both directly and indirectly through the welding pressure and rotational speed, respectively. Quantitative evaluation of the reflected spectrum necessitates the further development of Haines' model to include very strong plastic deformations as well.

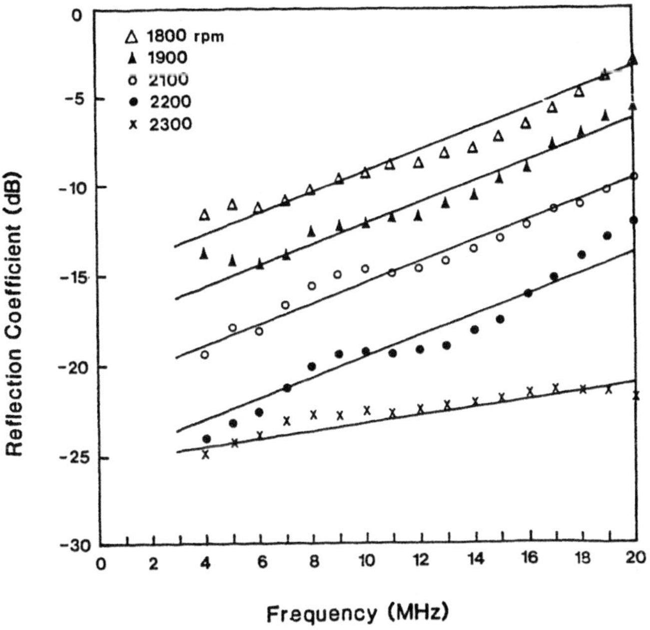

Fig. 6. Reflection spectra from the cold welded center part of stainless steel-copper inertia welds made at different rotational speeds between 1800 and 2300 rpm and constant welding pressure of 1800 psi.

Acknowledgement

This work was supported by the NATO Scientific Affairs Division through Grant No. 0131/87.

References

1. Bell, R.A.; Lippold, J.C.; Adolpson, D.R.: An evaluation of copper-stainless-steel inertia friction welds. Weld. Journ. 63 (1984) 325s-332s.

2. Haines, N.F.: The theory of sound transmission and reflection at contacting surfaces. Report RD/B/N4744, Central Electricity Generating Board, Berkeley Nuclear Laboratories, (1980).

3. Arakawa, T.: A study on the transmission and reflection of an ultrasonic beam at machined surfaces pressed against each other. Mater. Eval. 41 (1983) 714-719.

Ultrasonic Monitoring of the Molten Zone During Float Zone Refining of Single Crystal Germanium

C.K. JEN, J.F. BUSSIÈRE, PH. DE HEERING* AND P. SUTCLIFFE*

Industrial Materials Research Institute, NRCC
Boucherville, Québec, CANADA J4B 6Y4

* Canadian Astronautics Ltd.
Ottawa, Ontario, CANADA K2H 8K7

Summary

An ultrasonic pulse-echo method was investigated to monitor the float zone in a Ge rod during float zone refining. The feasibility of this method has been experimentally demonstrated for small diameter Ge rods, by using shear wave transducers cooled at about 30°C. It was shown that arrival times of the reflected signals from the solid-liquid Ge interface can be used for monitoring purposes and good resolution can be obtained for both zone position and thickness. Acoustic energy transfer at the solid-liquid interface, observed in the case of shear waves, is discussed in terms of multiple mode conversion. Effects of diffraction on received echoes of longitudinal and shear waves are described and analysed in terms of acoustic wave propagation in anisotropic Ge rods.

Introduction

Large, high-purity single crystals of Ge are used for high sensitivity, high resolution gamma ray detectors. Single crystals of the desired size can be obtained by the standard Czochralski crystal growth method. However, maximum size limitations are introduced in the purification process - float zone refining - because of instabilities of the molten zone under normal gravity conditions. It has, therefore, been proposed to perform the float zone refining under microgravity conditions, in space.

Since the thickness of the molten zone and the shape of the liquid-solid Ge interface affect the quality of the Ge rod and also minimum energy consumption is desirable for any experiment performed in space, a monitor for this float zone may be desired to improve process control. The purpose of the monitor is (i) to detect whether the melt is complete throughout the diameter, (ii) to determine the location of the melt with high ranging resolution and (iii) to obtain more information on the melt such as its thickness and the shape of the solid-liquid interface. An ultrasonic method was selected because the ultrasonic attenuation is low for Ge, the technique is simple and real time monitoring may be achieved. In addition, it will not deteriorate the crystal quality.

Experiments

Float zone refining consists of slowly moving a narrow section of molten (liquid) Ge along a given direction of the single crystal to be purified. The heat is supplied by a high power radio frequency (RF) source through conducting coils which encircle but do not contact the Ge rod, as shown in Fig. 1. Because of the high melting point (936°C) of Ge, the temperature at the end of the Ge rod is found to be approximately 500°C under the present experimental conditions, i.e. a rod of 10 mm diameter and 300 mm length. At such high temperature it was necessary to water cool the transducers, which were bonded at each end of the Ge rod. The ultrasonic system was wideband pulse-echo using commercial electronics and transducers and included a DATA 6000 from Analogic Corp. for data acquisition.

Figures 2 and 3 show the reflected signals from the end of a 10 mm diameter, 350 mm long Ge rod oriented in the $\langle 100 \rangle$ direction using a 5 MHz shear and a 2.25 MHz longitudinal transducer respectively. The measurements were recorded every one, two, or four minutes and time is indicated at the righthand side of each trace. The rod, initially entirely at room temperature (lowest trace) was heated near one half of its length causing a localized molten region which was then continuously moved at a speed of 50 mm/hr.

In Fig. 2 the early arrived dominant echo is from the solid-liquid Ge interface and the last large echo is from the end of the rod. The solid-liquid interface echo is clearly observable as soon as melting begins and during the entire presence of a molten zone. The time delay of the received signal from the end of the Ge rod is seen to increase during heating and was also found to decrease during cooling (not shown here). Furthermore, during the formation of the float zone, the time delay is further decreased (compare the bottom trace and one next to it) and vice versa. This observation can be explained by the data listed in TABLE I, where the shear wave velocity decreases as the temperature increases and the fact that the longitudinal velocity in liquid Ge is lower than that of shear waves in solid Ge. The data shown in TABLE I were obtained mainly from Refs. [1,2] and a few of them from interpolation.

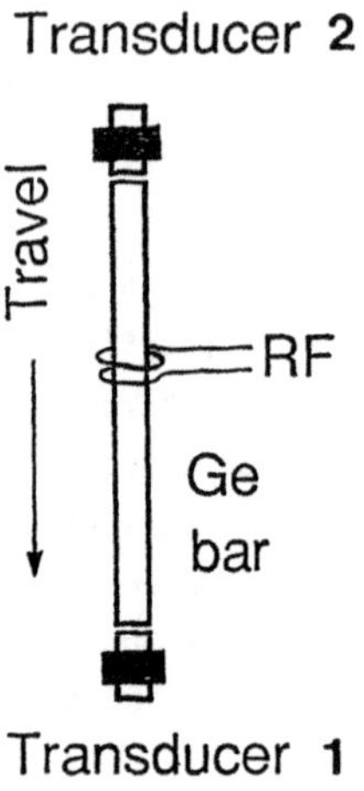

Fig. 1: The experimental setup.

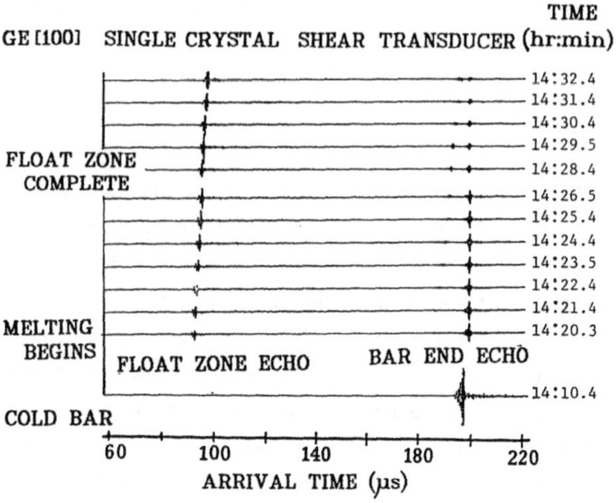

Fig. 2: The reflected signals from the solid-liquid interface and the end of the rod with a 5 MHz, 12.7 mm diameter shear wave transducer performed on a 10 mm diameter 350 mm long Ge rod oriented in the <100> direction.

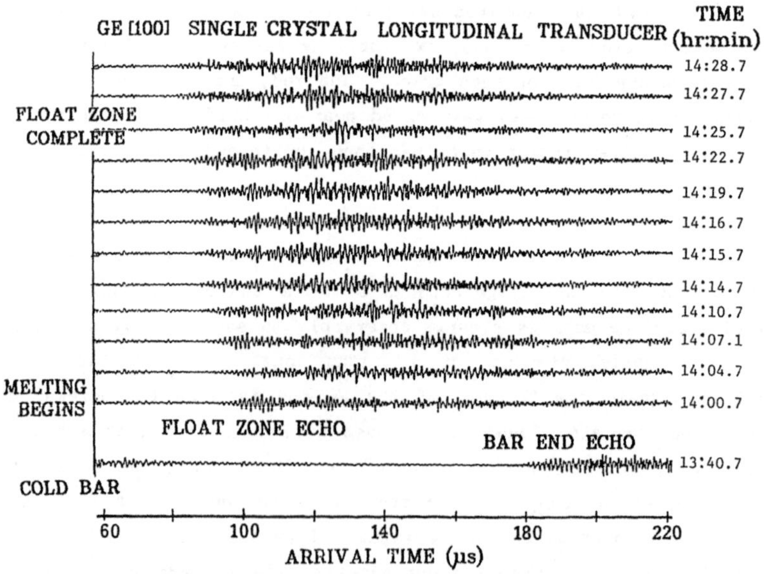

Fig. 3: The reflected signals from the solid-liquid interface and the end of the rod with a 2.25 MHz, 12.7 mm diameter longitudinal wave transducer performed on a 10 mm diameter 350 mm long Ge rod oriented in the <100> direction.

TABLE I - ACOUSTIC PROPERTIES OF Ge AT ELEVATED TEMPERATURES

TEMP (K°)	DENSITY (Kg/m^3)	V_L (m/s)	V_S (m/s)
298	5323	4826	3582
573	5297	4761	3527
973	5257	4617	3436
1155	5240	4552	3372
1209	5235	4529	3351
LIQUID	5510	2710	

We also note that reflected shear wave signals from the end of the Ge rod are observed even in the presence of a complete molten Ge zone (verified by twisting one end during melting). In general, if an ideal shear wave (only having the transverse particle displacements) strikes the solid-liquid boundary with normal incidence, all its energy will be reflected. Here, however, the shape of the liquid-solid interface is not smooth and the shear waves propagating in the Ge rod used were not ideal plane waves, and nearly all of them have a small longitudinal component along the rod axis [3].

This means that mode conversion, as shown in Fig. 4, will take place. The incident shear waves from the solid Ge will be partially reflected into one longitudinal and two shear waves, and partially transmitted into the longitudinal wave in the liquid Ge. The solid lines in Fig. 4 indicate this process. When the longitudinal waves in the liquid Ge strike the liquid-solid boundary, mode conversion will also happen as indicated by the dashed lines. Although the longitudinal waves are generated via mode conversion and reflected back to the shear wave transducer, no longitudinal arrival has been observed, since the shear wave transducer is insensitive to longitudinal waves. Also, because the mode conversion, longitudinal waves propagate in the liquid Ge region, causing an increase in time delay because of the slower velocity compared to that of shear waves in solid Ge. This agrees well with the above mentioned time delay observation. As indicated in TABLE I, the acoustic impedance of the shear waves in the solid Ge is only about 1.2 times that of the longitudinal waves in the liquid. It is expected that large amount of energy will be converted from shear waves in the solid Ge into longitudinal waves in the liquid Ge and vice versa. This is the reason why the echo reflected from the end of the rod can be observed even if it travels through the melt twice.

Figures 5a and 5b show the measurement results, obtained by the transducer 1 and 2 indicated in Fig. 1 respectively, of the melt position. Since the RF heating source was not regulated, the curves are not smooth, and the variation was consistently observed by both transducers. From Figs. 5a and 5b it can be concluded that, with the commercially available shear wave ultrasonic transducers water-cooled below 30°C, the relative float zone position can be determined in a 10 mm diameter Ge rod oriented in the ⟨100⟩ direction and with a resolution of about 0.3 μs.

The thickness of the float zone is also an important factor to determine the quality of the crystal growth. It may be determined using the two transducer geometries shown in Fig. 1. Using transducers 1 and 2 one can obtain t_1 (refl.), t_2 (refl.) and t_T (pitch-catch). Since t_T is equal to $t_1 + t_2 + t_\ell$, t_ℓ, which is the propagation time of the longitudinal waves through the liquid Ge (float zone) of thickness h, can be obtained. Figure 6 shows the measured result of melt thickness for the measurements presented in Fig. 5, using $t_T - t_1 - t_2$. If the velocity profile of liquid Ge is known, then the absolute thickness, h, may be obtainable. Again, the unregulated heating source is responsible for the thickness fluctuations in Fig. 5. Information concerning the shape and the completeness of the melt could be obtained from the amplitudes of echoes reflected from the liquid-solid interface and the end of the rod. These are still under our study.

Acoustic wave propagation in Ge rods

Figure 3 obtained from a longitudinal wave transducer indicates that the reflected echoes from liquid-solid Ge interface are more dispersed compared to those in Fig. 2. This is partly due to the larger diffraction loss associated with elastic anisotropy, as explained below.

Figure 7 shows the slowness curves [4] of Ge single crystal in the ⟨100⟩ [010] plane. The slowness curves are identical to wavenumber k curves for a given operating frequency. The direction of the Poynting vector p (energy flow) is always perpendicular to the tangential plane (here, only lines are shown). Curve B represents the shear wave, S2, which is isotropic. For the isotropic case, the direction of p is the same as that of k. In anisotropic crystals, indicated as curve A representing the quasi longitudinal wave, L, and C representing the quasi-shear wave, S1,

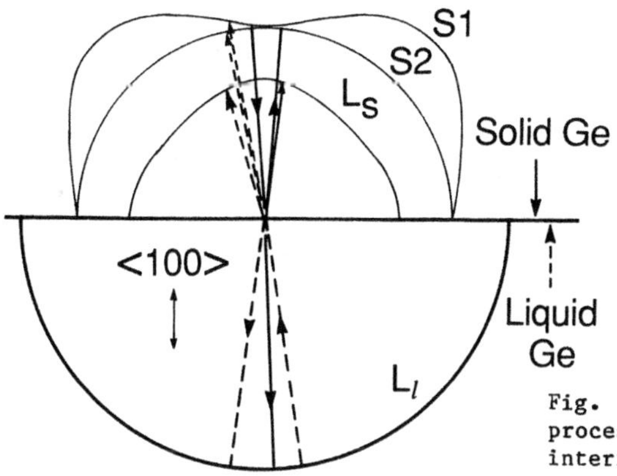

Fig. 4: Multiple mode conversion processes at the solid-liquid interface.

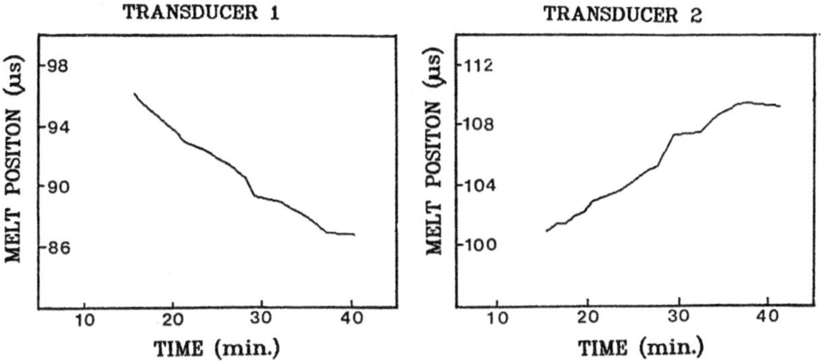

Fig. 5: Melt position measurement results obtained with transducer (a) 1 and (b) 2 shown in Fig. 1.

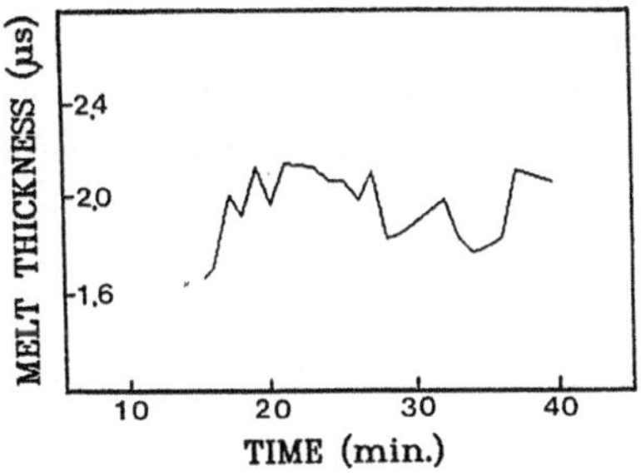

Fig. 6: Melt thickness measurement results for experiments performed in Fig. 5.

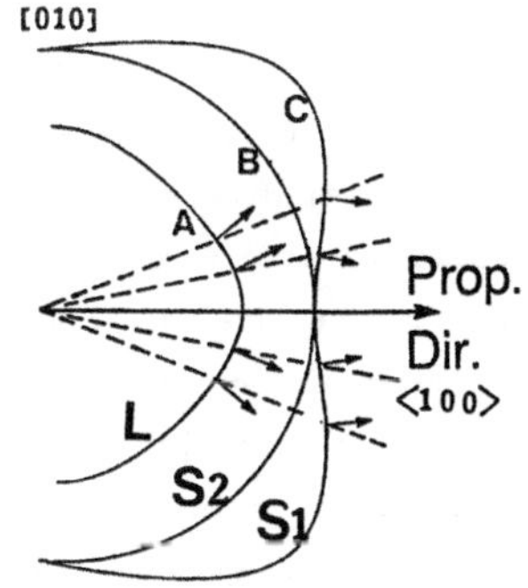

Fig. 7: The slowness curves of Ge crystal in the <100> [010] plane.

the direction of p is very often different with that of k. In analogy to ray optics the plane acoustic waves expressed by acoustic rays can exhibit diverging (Curve A) and focusing (Curve C) phenomena, thus their diffraction loss may be larger or smaller than that of the isotropic case (Curve B) [5]. After calculating slowness curves in the propagating <100> direction and many other directions perpendicular to <100> for Ge, it is concluded that the diffraction loss for shear waves is substantially less than for the longitudinal wave. It is noted that diffraction phenomena of the acoustic wave propagating in the Ge rod is very complicated [5] especially for shear waves. The analysis given here using slowness curve is only a rough approximation.

Due to large diffraction, most longitudinal waves are mode converted into many flexural acoustic waves [3], which have velocities closer to that of shear waves, at rod peripheres along the Ge rod, therefore more dispersed waveforms were observed. Dispersive waveform implies less ranging resolution. Therefore it is concluded that shear wave transducers should be used for the ultrasonic monitoring. It is interesting to note that with shear wave excitation, the float zone in polycrystal Ge rod with similar dimensions as those in Figs. 2 and 3, can also be monitored ultrasonically.

Considerations for high temperature operations

As mentioned earlier, the float zone detection is also intended to be performed in space. Because of the potential limitation on the weight and power consumption, the ultrasonic transducers may have to operate at around 500°C. Because no such transducers are currently available on the market, a preliminary study was performed at IMRI. For piezoelctric transducer materials with high coupling constant lithium niobate crystal was our choice due to its high Curie temperature (1210°C). After a lengthy test of many coupling materials between the transducer and the Ge rod, only the ceramic cement Cotronics 989 (From Cotronics, Brooklyn, N.Y.) was found promising at the present time. This ceramic cement was cured at room temperature for about 48 hours. Figure 8(a) shows the transmitted longitudinal signals through a 10 mm diameter 27 mm long Ge rod using a pair of 5 MHz longitudinal disc $LiNbO_3$ transducer with a 6.35 mm diameter at 350°C. The reflected shear wave signals through a 10 mm diameter 20 mm long Ge rod using a 5 MHz shear wave transducer with a 6.35

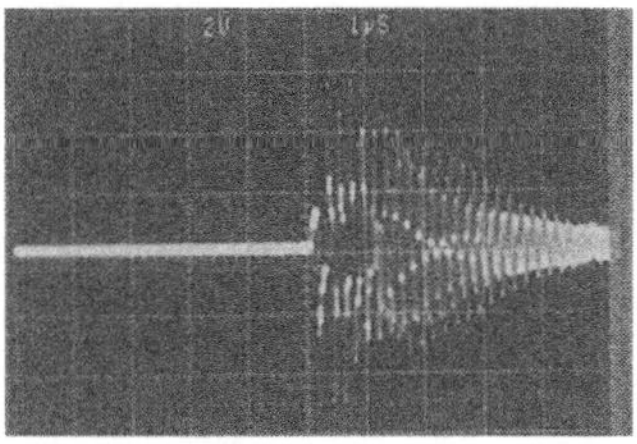

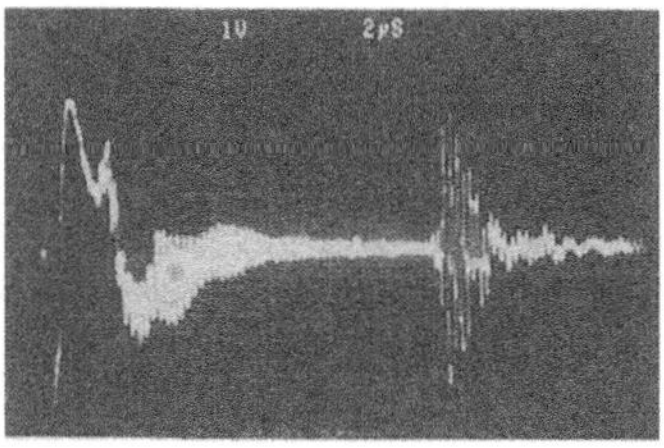

Fig. 8: The received signal through a short section of Ge rod at 350°C using (a) longitudinal and (b) shear wave transducer.

mm diameter at 350°C is given in Fig. 8(b). From Figures 8(a) and (b) it can be seen that the damping of the acoustic signal is not very high, implying that the range resolution will be reduced.

Conclusions

An ultrasonic pulse echo method has been presented for monitoring the float zone of Ge rods. For Ge rods oriented in ⟨100⟩ direction shear waves are preferred over longitudinal waves. Real time monitoring of the position and thickness of the melt was proved feasible. Considerations and preliminary results related to high temperature measurements were presented.

References

1. Burenkov, Y.A.; Nikanorow, S.P. and Stepanov, A.V., "Elastic Properties of Germanium", Sov. Phys. Solid State, 12, (1971) 1940-1942 and Glazov, V.M., Aivazov, A.A. and Timoshenko, V.I., "Temperature Dependence of the Velocity of Sound in Molten Germanium", Sov. Phys. Solid State, 18, (1976) 684-685.

2. Parker, R.L. and Manning, J.R., "Application of Pulse-echo Ultrasonics to Locate the Solid/Liquid Interface During Solidification and Melting", J. Crystal Growth, 79, (1986) 341-353.

3. Meeker, T.R. and Meitzler, A.H., "Guided Wave Propagation in Elongated Cylinders and Plates", Physical Acoustics, Vol. I, Part A, Academic Press, Chapter 2, (1964) 111-167.

4. Auld, B.A., "Acoustic Fields and Waves in Solid", John Wiley & Sons, Vol. I, Chap. 7, (1973) 265-348.

5. Papadakis, E.P., "Ultrasonic Diffraction Loss and Phase Change in Anisotropic Materials", J. Acoust. Soc. Am., 40, (1966) 863-876.

Optical and Thermal Properties, Special Techniques

Optical Parameter Effects on Laser Generated Ultrasound for Microstructure Characterization

K. L. Telschow
Idaho National Engineering Laboratory
EG&G Idaho, Inc. Idaho Falls, ID 83415
and
R. J. Conant
Department of Mechanical Engineering
Montana State University, Bozeman, MT 59717

Summary

Under appropriate conditions a significant feature of the laser generated elastic waveform in the thermoelastic regime is a precursor (sharp spike) signaling the arrival of the longitudinal wave. This paper shows that the observed precursor can be understood through the use of one-dimensional models that account for optical penetration and thermal diffusion into the material. Experimental results are compared with a two-dimensional analytic calculation including optical penetration.

Introduction

In order to optimize the use of ultrasonics for characterization of microstructural features in materials, particularly ceramics, high frequencies (small wavelengths) must be employed to achieve sufficient scattering from the often small grain size and porosity. Previous work using a laser source for microstructure characterization has concentrated on exploiting the normal force behavior of ablation from the material [1,2]. The ablation source produces a sharp longitudinal waveform with a large bandwidth. This paper investigates the effects of subsurface ultrasonic wave sources produced by the laser in the thermoelastic regime. It is found that when the laser beam penetrates sufficiently into the material, even the thermoelastic source mechanism produces a sharp longitudinal waveform similar to that obtained from an ablation source [3-5]. This precursor has substantial high frequency content, which can be exploited for material microstructure characterization. Analysis shows that this precursor is present whenever subsurface sources exist caused by optical penetration or thermal diffusion.

One-Dimensional Models

The origin of the precursor signal from a nonablative source can be understood by using one-dimensional models describing the laser generated thermoelastic signal. First, a single point source buried at the depth h below the surface is considered, and a positive precursor signal is produced. This model serves to illustrate that any source below the surface will generate the precursor signal. Next, optical penetration is taken into account by distributing point sources with an exponentially decaying magnitude with depth into the material. This model produces a precursor signal whose shape reflects the temperature with depth profile. Finally, a third model treats only thermal diffusion into the material. It is shown that again a precursor signal results when the diffusivity is large enough to distribute the heat into the material during the experimental measurement time.

Single Buried Point Source

Consider a half-space with surface at $z = 0$, as depicted in figure 1. The exponentially absorbed laser beam can be modeled as a series of point temperature sources, whose magnitudes decrease with depth. In the absence of diffusion, a single point source at the depth h produces the distribution $T(z,t) = T_h h\delta(z-h) H(t)$ where δ is the Dirac delta function and $H(t) = 1$, for $t > 0$ and zero otherwise. The differential equation describing the displacement W in the z direction is $W,_{zz} - (1/c^2)W,_{tt} = \xi T,_z$ where $\xi = [(1 + \nu)/(1 - \nu)]\alpha$, c is the longitudinal wave speed, α is the material's linear coefficient of thermal expansion and ν is Poisson's ratio (see Ref. 6). The notation $W,_{zz}$ stands for the second partial derivative of W with respect to z. The surface is traction free so that $W,_z - \xi T = 0$ @ $z = 0$; the displacement also must be bounded at infinity. The solution, obtained by Laplace transform techniques, is

$$W(z,t,h) = (\xi T_h h/2)\ \{H(z-h)H[t-(z-h)/c] - H(h-z)H[t-(h-z)/c] - H[t-(h+z)/c]\}$$

From this it is seen that the solution consists of three waves, each propagating with the longitudinal velocity c. The first term represents a wave starting from $z = h$ at $t = 0$ and propagating into the half-space ($z > 0$); the second term represents a wave starting from $z = h$ at $t = 0$ and propagating toward the surface; the last term represents a wave starting from the surface at $t = h/c$ and propagating into the half space. The wave represented by the last term is thus the reflection of the wave represented by the second term after it strikes the surface. These waveforms are shown in figure 1. It is interesting to note from the above solution that

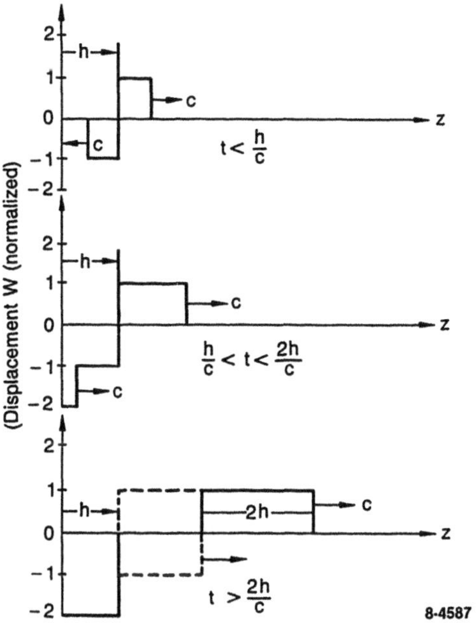

Fig.1. Visualization of the optically penetrating laser source as a point temperature source in an infinite half-space, along with the three waveforms produced.

$W(z,t,h = 0)$ is zero; a source point on the surface produces a reflected wave propagating immediately from the surface that cancels the incident wave propagating into the half-space.

Optical Penetration Model

The generalization to illumination by a laser is given by considering a continuous exponential decrease in temperature with depth. Assume that the laser pulse is of sufficiently short duration to be represented by a delta function in time and that diffusion of heat into the half-space can be neglected. The temperature distribution is $T(z,t) = T_0 \exp(-bz) H(t)$ where T_0 is the surface temperature and b is the reciprocal of the optical penetration depth of the material. The resulting displacement is

$$W(z,t) = (\xi T_0/b) \{-H(t-z/c)\sinh[b(ct-z)]+[\cosh(bct)-1]\exp(-bz)\}$$

Figure 2 shows this solution, which consists of a positive pulse, symmetrical about the arrival position in time. Both the rise and fall

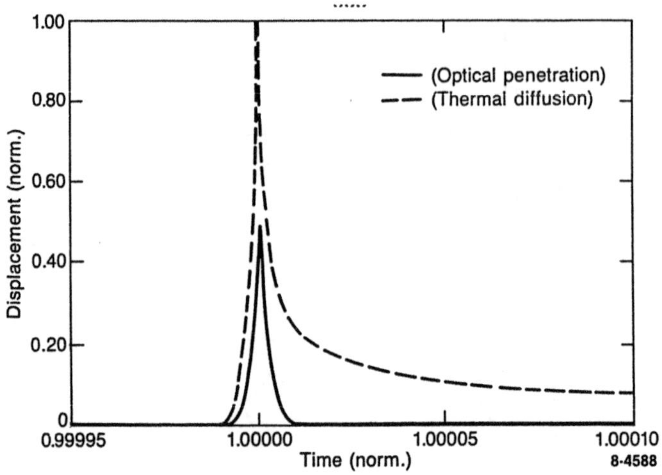

Fig.2. Precursor signals produced in the one-dimensional optical penetration and thermal diffusion models. For the purpose of comparison, the curves are normalized (in terms of parameters defined in the text) by letting $T_0/b = E_0D/K$ and $b = c/D$, where $c = 5$ mm/μs and $D = 1$ cm^2/s.

shapes of this pulse reflect the exponential decay with depth of the temperature distribution. Inspection of this solution shows that the magnitude of the precursor depends on the optical penetration depth in that as the penetration depth becomes very small, the magnitude of the signal becomes very small also. Thus, again it is seen that for a surface source there is no waveform generated with the one-dimensional model. The two-dimensional solution, described below, also predicts that the precursor magnitude decreases to zero as the optical penetration depth decreases to zero. This result is consistent with the two-dimensional solutions of References 3 and 7, both of which employ point surface sources and neglect thermal diffusion and produce no precursor signal.

Thermal Diffusion Model

To investigate the effects of thermal diffusion, assume that all of the laser energy is absorbed at the surface and that the time dependence for the heat source is a delta function as before. Solution of the heat conduction equation with the heat pulse boundary condition yields a diffusive temperature distribution in the half-space $T(z,t) = (2E_0D/K)/\sqrt{(4\pi Dt)}\,\exp(-z^2/4Dt)$ where D is the material diffusivity, K the thermal conductivity, and E_0 the absorbed laser pulse energy. The displacement waveform is found by solving the wave equation with this temperature source by Laplace transforms to yield $W(z,t) = [w_1(z,t) + w_2(z,t) + w_3(z,t)]$, where

$$w_1(z,t) = (E_0 D\xi/2K)\{\exp[(c^2/D)(t-z/c)]\ \mathrm{erfc}[-(t-z/2c)\sqrt{c^2/Dt}] + \exp[(c^2/D)(t+z/c)]\ \mathrm{erfc}[+(t+z/2c)\sqrt{c^2/Dt}]\}$$

$$w_2(z,t) = -(E_0 D\xi/K)\ H(t-z/c)\ \exp\{[c^2(t-z/c)/D]\}\ \mathrm{erf}\ \sqrt{c^2(t-z/c)/D}$$

$$w_3(z,t) = -(E_0 D\xi/K)\ \mathrm{erfc}[z/\sqrt{4Dt}]$$

Figure 2 shows the waveform calculated from this model. A positive precursor signal is observed, even though the heat source is confined to the surface [4].

It is apparent that the precursor signal arises from the penetration of heat into the material and is dependent on material properties. Only a positive displacement spike is produced as a result of the heating with depth. Ceramic materials will most likely be dominated by the effects of optical penetration, as these materials tend to be nonmetallic and have small thermal diffusivity. On the other hand, opaque materials or metals that have very small values for the optical penetration depth are probably dominated by thermal diffusion to produce the subsurface sources. For common values of material properties, thermal diffusion produces extremely narrow precursor signals in time. These signals are difficult to record experimentally as the detector bandwidths available are limited and this signal is highly attenuated by microstructural scattering.

Experimental Measurements

The sample materials were stainless steel, which served as a material with essentially zero optical penetration depth, and neutral density filters (NDF) with a variety of optical penetration depths. All the samples were flat plates approximately 2.5 to 3.5 mm thickness. The detector used was a capacitive transducer, which directly recorded the surface displacement on the opposite side from where the laser pulse was absorbed. All the NDF glass plates were coated with platinum on the capacitive detector side to serve as the grounding electrode. The results reported here are for the detector on epicenter from the laser beam.

Figure 3 shows the measured waveforms for the stainless steel sample [SS] and the NDF samples [A-E]. The primary effect of the laser source on sample [SS] is to produce a negative displacement starting at the longitudinal wave arrival time. This motion is brought back toward the equilibrium line by the arrival of the transverse wave motion of the initial expansion. For relatively thin plates, as used here, the subsequent motion represents a significant excitation of low frequency plate mode vibrations, which are only lightly damped. Sample [A] exhibits an optical penetration depth of about 11.4% of its thickness. Its waveform is similar to that for [SS] except that now a positive spike appears in the waveform somewhat before the longitudinal arrival time. This is the precursor signal.

No observable precursor was recorded for the stainless steel sample. This material has negligible optical penetration depth and significant scattering due to grain structure; therefore, the limitation was most likely the sensitivity and bandwidth of the capacitor detector. For the materials with larger penetration depths [B-E], the shape of the precursor is altered markedly (figure 3) and for significant optical penetration through the plate a positive displacement is seen immediately when the laser pulse is absorbed [D-E]. The shape of this precursor signal is strongly determined by the depth dependence of the optical absorption in the material.

<u>Two-Dimensional Model</u>

The more general problem of thermoelastic generation with optical penetration in a plate (without thermal diffusion) can be treated as two-dimensional by taking the radial heating function as Gaussian and the temperature as exponentially decaying with depth. The solution to the thermoelastic equations follows the treatment of reference [8], except that now concentration is on the waveform displacements. The results are obtained as an infinite series of plate propagation modes, both symmetric and antisymmetric and are described in more detail elsewhere [5].

A direct comparison between the experimental measurement for the NDF sample [A] and the appropriate two-dimensional calculation is shown in figure 4. Here, the two graphs are normalized by overlapping the precursor peaks in time and adjusting the vertical scale so the maximum negative displacements agree. Most of the essential features of the two graphs agree. The

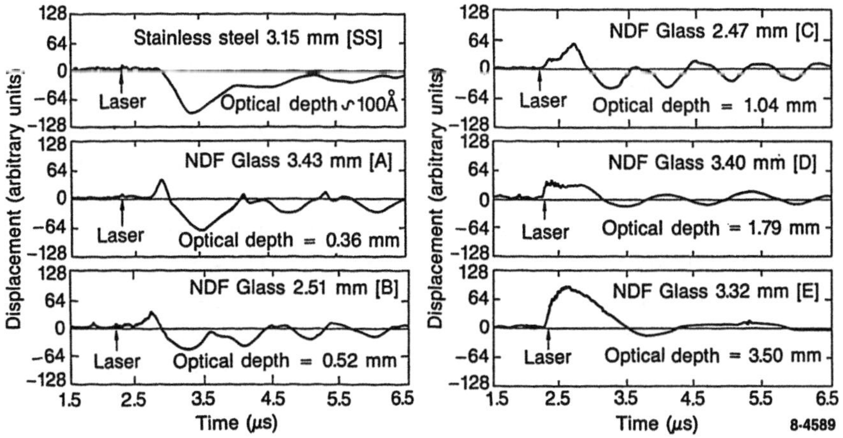

Fig.3. Normal surface displacement from absorption of the laser pulse (arrow). Note the sharp positive displacement (precursor) for the neutral density filters (NDF) plates [A-E] as compared with the relatively nonabsorbing stainless steel plate [SS].

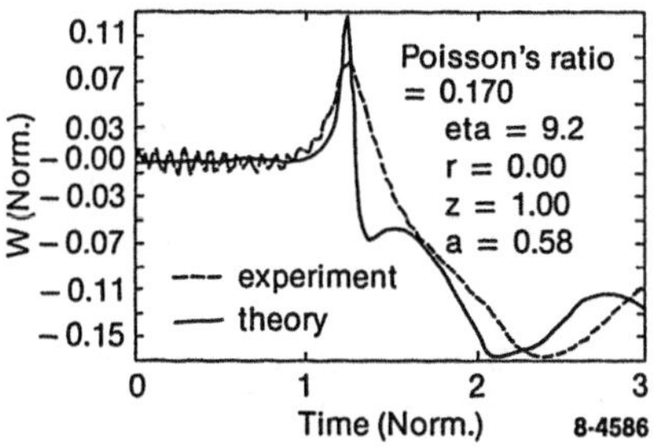

Fig.4. Comparison between the experimentally measured surface displacement and that calculated for a NDF with optical penetration depth of 11.4% of the plate thickness. The signal amplitudes are normalized so that the maximum negative displacements are equal.

precursors rise approximately exponentially and fall abruptly to be followed by the negative thermoelastic expansion. Both curves show a shift to later times for the arrival of the transverse wave than would be expected from a point source thermoelastic expansion model, which predicts an arrival time of t = 2 in normalized units. The origin of the broader precursor peak and the later transverse wave arrival time is thought to be from the finite optical beam width and the limited bandwidth (~70 MHz) of the capacitive detector.

Conclusions

Ultrasonic waveforms generated from a thermoelastic laser source have been investigated with one- and two-dimensional analytical models. A positive precursor signal was found that results from subsurface sources, which arise from optical penetration and/or thermal diffusion into the material. The net waveform produced by this thermoelastic source was measured experimentally by utilizing glass plates with varying optical penetration depths and stainless steel, as an example of a material with an essentially zero penetration depth. Experimental and calculational results agree for a sample with an optical penetration depth of 11.4% of its thickness. The fact that the positive displacement precursor signal can be produced even in the thermoelastic laser generation regime yields an additional method to the ablation technique for producing high frequency ultrasonic waves in materials.

Acknowledgment

The authors would like to thank J. Jolley for assistance in preparing the platinum coated samples, N. Boyce for constructing the capacitive detector and T. O'Brien for recording the data. This work was supported by the Department of Interior's Bureau of Mines under Contract No. J0134035 through Department of Energy contract No. DE-AC07-76ID01570.

References

1. C. B. Scruby, R. L. Smith and B. C. Moss, "Microstructural Monitoring by Laser-Ultrasonic Attenuation and Forward Scattering," NDT International, 19, 307-313, 1986.
2. K. L. Telschow, "Microstructure Characterization with a Pulsed Laser Ultrasonic Source," Reviews of Progress in QNDE, 7B, D. O. Thompson and D. E. Chimenti (eds.), 1211-1218, 1988.
3. C. B. Scruby, R. J. Dewhurst, D. A. Hutchins, and S. B. Palmer, "Quantitative Studies of Thermally Generated Elastic Waves in Laser-Irradiated Metals," J. Appl. Phys., 51, 6210-6216, 1980.
4. P. A. Doyle, "On Epicentral Waveforms for Laser-Generated Ultrasound," J. Phys. D: Appl. Phys., 19, 1613-1623, 1986.
5. R. J. Conant and K. L. Telschow, "Longitudinal Wave Precursor Signal from an Optically Penetrating Thermoelastic Laser Source," Reviews of Progress in QNDE, D. O. Thompson and D. E. Chimenti (eds.), 1989, to be published.
6. W. Nowacki, Thermoelasticity, 2nd edition, Pergamon Press, New York, 1986.
7. L. R. F. Rose, "Point-Source Representation for Laser Generated Ultrasound," J. Acoust. Soc. Am., 73, 723-732, 1984.
8. C. Sve and J. Miklowitz, "Thermally Induced Stress Waves in an Elastic Layer," J. Appl. Mech., 40, 161-167, 1973.

Determination of the Thermal Properties of Thin Layers by a Photothermal Technique

M. BEYFUSS AND J. BAUMANN

Siemens AG, ZT ZFA PTE 22, Otto-Hahn-Ring 6, D-8000 München 83

Summary/Zusammenfassung

The determination of the thermal properties of thin layers is mainly due to the miniaturisation of electronic devices of growing interest. Nondestructive and noncontacting measurement-techniques are required for this purpose. We show by examples of practical interest, that applying a photothermal technique the thermal properties of thin coatings can be determined very accurately, supporting the evaluation by model calculations. A measurement-system based on this technique can be integrated into the process line.

Die Bestimmung der thermischen Eigenschaften von dünnen Schichten gewinnt vor allem wegen der Miniaturisierung elektronischer Bauteile zunehmend an Wichtigkeit. Zerstörungsfreie und berührungslose Meßverfahren sind zu diesem Zweck gefordert. Wir zeigen anhand von Beispielen mit praktischer Bedeutung, daß mit einem photothermischen Verfahren die thermischen Eigenschaften von dünnen Beschichtungen -unterstützt durch Modellrechnungen- sehr genau bestimmt werden können. Ein entsprechendes Meßsystem läßt sich in die Prozeßlinie einbinden.

1. Introduction

In industrial development as well as in manufacturing it is of great importance to know accurately the specific properties of materials. Materials with required properties are to be selected in order to get best performance. The thermal properties of materials (the thermal conductivity λ, and the product density $\varrho \cdot$ specific heat c, i.e. the heat capacity per unit volume) are of interest, due to the high degree of integration in microelectronics and microsystem-technique. The stationary and dynamic heat transfer has to be sufficiently large to limit the device temperatures to a certain range.

The photothermal technique described in this contribution determines the thermal properties of thin layers in a noncontacting and nondestructive way, ensuring high accuracy and high lateral resolution as well. These factors are of considerable advantage when compared to the standard method of stationary heat flux measurements.

2. Experimental setup

In the photothermal experiments we present, a 3W Ar^+-laser is used to illuminate the sample's surface periodically (Fig. 1). A mechanical chopper-wheel modulates the laser beam. The thermal waves produced in this way are reflected at each interface of a layered sample. The resulting temperature-oscillation at the surface is measured by focussing the infrared-radiation with a concave mirror into a HgCdTe-infrared-detector. Therefore, only the front side of the sample has to be accessible. A lock-in-amplifier enables the measurement of the amplitude of the temperature signal and its phase shift relative to the periodical laser illumination. A computer controls the experiment, sweeps the frequency, acquires the digitized data and plots the results.

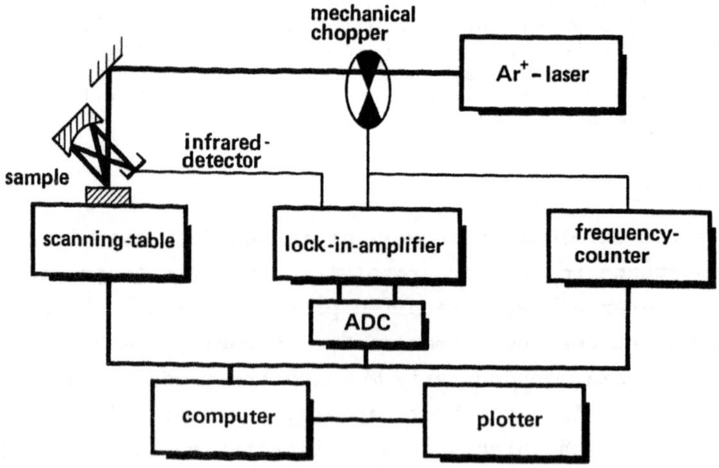

Fig. 1. Experimental setup

3. Samples under investigation

The first group of samples were thin glazes (10-15 μm) basically consisting of SiO_2. They were put on ceramic substrates by a screen-printing-technique. The glazes may be used in thermotransfer-printheads, where an optimized dynamic temperature behaviour is required [1]. Therefore the thermal properties of suitable materials are allowed only to vary within a given small range.

Secondly, we investigated several high-temperature-resistant organic dielectrics, which are used for microelectronic device processing. The thin dielectrics (2-15 μm), basically consisting of polymers, were put on silicon-wafers by a spin-on-technique.

Because the glazes, as well as the dielectrics were optically transparent, they were coated with a thin (0.5 μm), black, optical opaque TiN_x-layer (cf. chapter 5).

4. Experiment and evaluation

As shown by several authors, the photothermal technique described here is sensitive to a change in the thermal properties [2-6] resp. the thicknesses [7-11] of layers. Therefore, applying mathematical models of thermal-wave-propagation [12,13], the thermal properties of a thin layer can be determined. For this purpose we extended the one-dimensional one-layer-model of ROSENCWAIG and GERSHO [12] to three layers [10]. Enlarging the diameter of the laser beam sufficiently, a one-dimensional heat flow within the sample was ensured [6]. As the phase of the surface-temperature-oscillation is more sensitive than the amplitude, the evaluation of the thermal data was based on phase measurements.

The resolution of the method is shown in Fig. 2 and Fig. 3. Because the typical uncertainty in phase-measurement is only ± 0.2°, variations of λ can be measured very accurately for both glazes and dielectrics. For the glazes, also $\varrho \cdot c$ can be determined with good accuracy at the higher end of the modulation frequency range. For the dielectrics, ϱ and c were known from separate gravimetric and DSC-measurements, so that the thermal conductivity was the only interesting quantity.

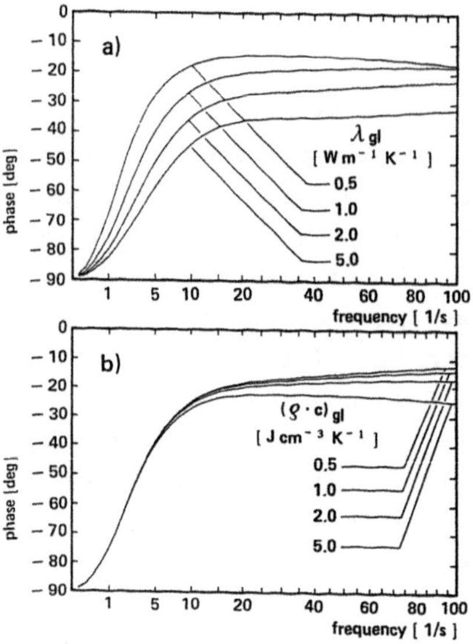

a) variation of λ_{gl}
$(\varrho \cdot c)_{gl} = 2.0\ J\ cm^{-3} K^{-1}$

b) variation of $(\varrho \cdot c)_{gl}$
$\lambda_{gl} = 1.0\ W\ m^{-1} K^{-1}$

Fig. 2. Calculated phase versus modulation frequency for a typical glaze (14 µm) on a Al_2O_3 ceramic substrate (620 µm).

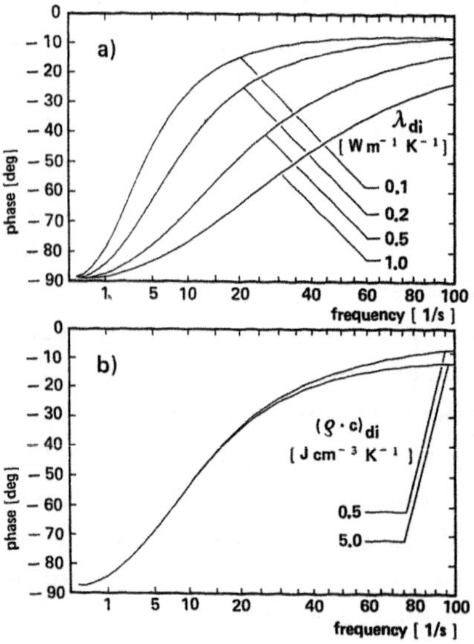

a) variation of λ_{di}
$(\varrho \cdot c)_{di} = 1.7\ J\ cm^{-3} K^{-1}$

b) variation of $(\varrho \cdot c)_{di}$
$\lambda_{di} = 0.3\ W\ m^{-1} K^{-1}$

Fig. 3. Calculated phase versus modulation frequency for a typical dielectric (5 µm) on a silicon-wafer (380 µm).

To obtain the thermal data the calculated curves have to be fitted to the experimental results. Minimizing the deviation of the calculated curves from the experimental dots (least-squares-fit) the thermal data λ and $\varrho \cdot c$ were found as the parameters of the best fit. Experimental results and best fits are shown in Fig. 4.

The accuracy of the thermal data, only determined by the uncertainty in phase measurement, would be about 2% for our samples as long as the layer-thickness and the thermal data of the substrate and the TiN_x-absorber would be known exactly. As, of course, these data themselves have some uncertainty, the real accuracies are 7% for λ_{gl}, 10% for $(\varrho \cdot c)_{gl}$, and 8% for λ_{di}. This can be seen from calculations using the mentioned uncertainty of the input-data. As far as we know, such high accuracies in the determination of thin layer's thermal properties can be obtained only by photothermal methods.

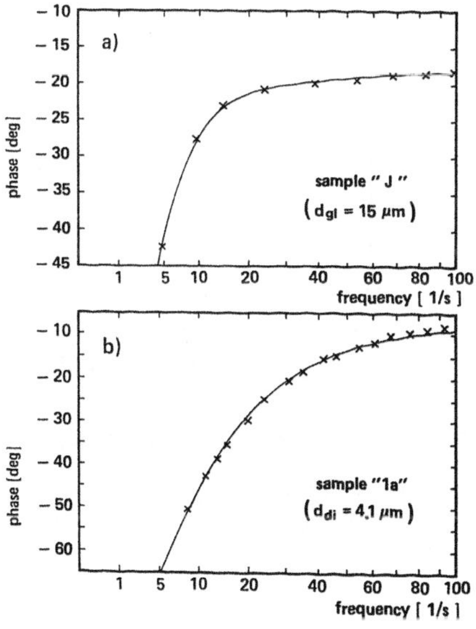

Data obtained from fit:

a) $\lambda_{gl} = 1.16 \ \mathrm{W\ m^{-1} K^{-1}}$
$(\varrho \cdot c)_{gl} = 2.07 \ \mathrm{J\ cm^{-3}\ K^{-1}}$

b) $\lambda_{di} = 0.205 \ \mathrm{W\ m^{-1} K^{-1}}$

Fig. 4. Phase versus frequency for a typical glaze on the ceramic (a,) resp. a typical dielectric on the silicon-wafer (b,). The crosses are experimental results, the solid lines are the best fitted curves.

A summary of the resulting fit parameters of all the samples is shown in Fig. 5. The thermal data of the glazes are in the order of the values for glasses [14]. Looking at the dielectrics we see different types: Due to its large content of SiO_2 the type 3 is the best thermal conductor of all the dielectrics investigated. The values obtained for the other types are typical for polymers [14]. The systematic difference in the thermal conductivity of the types 1 and 2 is due to the different chemical structures of the polymers.

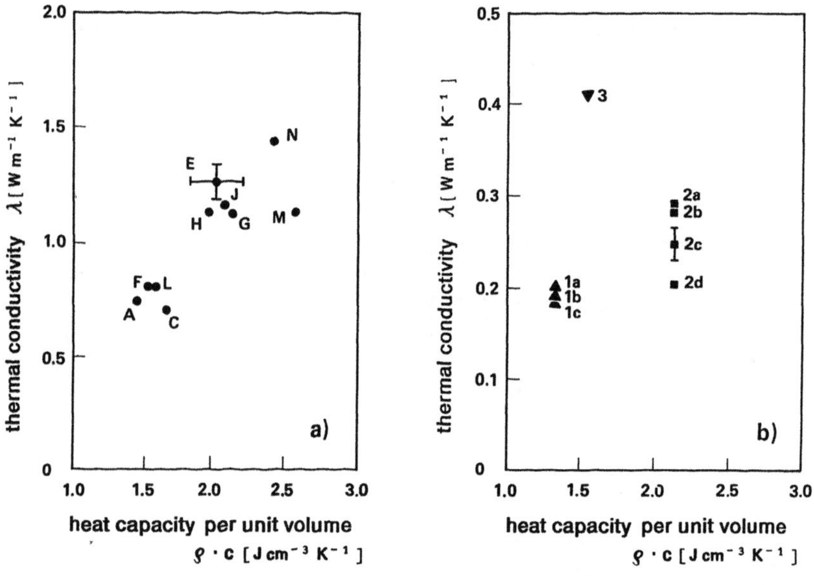

Fig. 5. Summary of the thermal data of all glazes (figure a,) and dielectrics (figure b,) under investigation. The heat capacities per unit volume of the dielectrics were measured by the producer applying gravimetric and Differential-Scanning-Calorimetry methods.

5. Optimizing the experimental parameters

We demonstrated, by discussing two examples, the performance of the photothermal method when applied to the determination of thermal properties. The accuracy in the thermal data obtained by this technique depends on the experimental parameters. In order to state the conditions for optimizing these parameters, we calculate a thermal-wave propagation-model corresponding to a thin coating (layer 2) on a substrate (layer 1) [5].

At first, the sensitivity of the method is based on the interference of the thermal waves produced at the sample's surface and reflected at the interface between coating and substrate. A high accuracy is obtained at a high value of the thermal-wave reflection-coefficient $r_{12} = (e_1/e_2 - 1)/(e_1/e_2 + 1)$ and, therefore, at a strong contrast of the thermal effusivities of coating and substrate, given by $e_{2,1} = (\lambda_{2,1} (\rho \cdot c)_{2,1})^{\frac{1}{2}}$. The same behaviour occurs in layer-thickness-measurement [8].

Secondly, the accuracy of the method can be optimized matching the modulation frequency f. The length μ_2, where the amplitude of a thermal wave is damped to its 1/e part (thermal diffusion length) is proportional to the inverse squareroot of the frequency. This dependence leads to a corresponding behaviour of the sensitivity in the determination of the thermal data.

The sensitivity can be demonstrated by calculating the phase for slightly different values (for example for a variation of ± 10% around a mean value) of one thermal parameter while the other one is fixed (Fig. 6). A high sensitivity corresponds to a large splitting of the phase curves for the different parameters.

At lower frequencies, where the coating-thickness d_2 is small compared to the thermal diffusion length μ_2, only the thermal conductivity λ_2 can be determined with high resolution (see range 1). Here, variations of $(\rho \cdot c)_2$ do not affect the phase behaviour. In this range the dielectrics were investigated. At a slightly higher frequency and, therefore, higher ratio d_2/μ_2, $(\rho \cdot c)_2$ can also be determined (range 2). The glazes were investigated in this range.

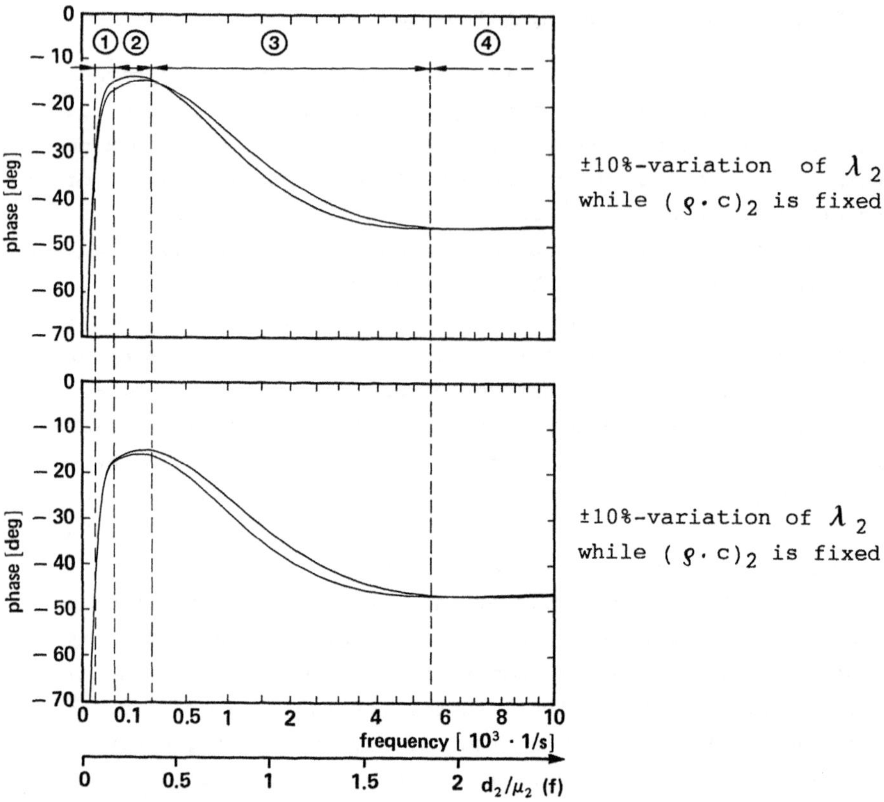

Fig. 6. Matching the layer-thickness to the thermal diffusion length for a thin (10 µm), optical opaque coating on a substrate (1 mm). Data assumed for calculations: λ_1 =50 W $m^{-1}K^{-1}$, $(\varrho \cdot c)_1$ = 1.5 J cm^{-3} K^{-1}, λ_2 =0.75 W m^{-1} K^{-1}, $(\varrho \cdot c)_2$ =1.5 J cm^{-3} K^{-1}.

If the thermal diffusion length μ_2 comes in the order of the coating thickness d_2, both λ_2 and $(\varrho \cdot c)_2$ can be measured with best resolution (range 3). This behaviour is similar to layer-thickness-measurement [8]. At higher frequencies, where the coating-thickness d_2 becomes more than about double the diffusion length, an accurate determination of the thermal data is impossible (range 4).

Sometimes the light absorption and/or the infrared emission of the coating is weak. Here, the infrared signal is poor and the simple mathematical model of exponential light absorption and infrared emission only at the coatings surface is no longer correct. In this case one has to coat the sample's front side with thin, highly absorbing layers. The influence of additional absorber layers is well-known [5].

If the thermal data of thin layers are to be determined with high lateral resolution, the diameter of the light-beam has to be very small, which causes an additional lateral heatflow within the sample. For a correct evaluation one has to calculate the three-dimensional thermal-wave propagation-model, if the light-beam diameter 2R is less or in the order of one of the thermal diffusion lengths $\mu_{1,2}$ of coating and substrate. The simple one-dimensional treatment is valid as long as $2R/\mu_{1,2} \gtrsim 20$ [6].

6. Conclusion and outlook

The determination of thermal properties is of growing importance in industrial development and manufacturing of microelectronic and micro-system devices. Noncontacting and nondestructive testing methods guaranteeing a high measurement-accuracy are required. Wellknown nondestructive testing methods such as X-ray-absorption and ultrasonic-testing are not direct methods for thermal data measurements. Standard stationary heat-flow-methods are contacting and not sensitive enough when applied to thin layers.

We determined the thermal data of thin layers with high accuracy, using a photothermal method. A testing-system based on this technique, which is also suitable for wet layers, can be integrated in the process line.

To obtain results with high accuracy the evaluation of the experimental data has to be based on correct mathematical models of thermal-wave-propagation in layered samples. Optimizing the method with respect to the thermal-wave-reflexion-coefficient, the coating thickness in relation to the thermal diffusion length, the absorber-layer and the light-beam-diameter, there is no other method of compatible performance in the determination of thermal properties of thin layers.

References

1. Dress, F.W; Nisius, B.M; Pekruhn, W. in: Proceedings of the 4th international conference on advances in nonimpact printing technologies, New Orleans/LA, 1988 (accepted for publication).

2. Vidberg, H.J.; Jaarinen, J.; Riska, D.O.: Can. J. Phys. 64 (1986) 1178-1183.

3. Morris, J.D.; Almond, P.M.; Patel, P.M.; Reiter, H. in: Hess, P.; Pelzl, J. (eds.): Springer series in optical sciences Vol. 58. Berlin, Heidelberg: Springer-Verlag 1988, 424-426.

4. Korpiun, P.; Merté, B.; Fritsch, G.; Tilgner, R.; Lüscher, E.: Coll. Polym. Sci. 261 (1983), 312-318.

5. Tilgner, R.; Baumann, J.; Beyfuß, M.: Can. J. Phys. 64 (1986) 1287-1290.

6. Beyfuß, M.; Tilgner, R.; Baumann, J. in: Hess, P.; Pelzl. J. (eds.): Springer series in optical sciences Vol. 58. Berlin, Heidelberg: Springer-Verlag 1988, 392-395.

7. Busse, G.; Eyerer, P.: Appl. Phys. Lett. 43 (1983) 355-358.

8. Beyfuß, M.; Tilgner, R.; Baumann, J.; Pape, H. in: Vorträge und Plakatberichte, Jahrestagung 1987, Zerstörungsfreie Materialprüfung, Deutsche Gesellschaft für Zerstörungsfreie Materialprüfung e.V., Berlin 1987, 764-772.

9. Tam, A.C.; Sontag, H.: Appl. Phys. Lett. 49 (1986) 1761-1763.

10. Baumann, J.; Tilgner, R.: Can. J. Phys. 64 (1986) 1291-1292.

11. Favro, L.D.; Kuo, P.K.; Thomas, R.L. in: Mandelis, A. (ed.): Photoacoustic and thermal wave phenomena in semiconductors. New York, Amsterdam, London: North-Holland 1987, 69-96.

12. Rosencwaig, A.; Gersho, A.: J. Appl. Phys. 47 (1976) 64-69.

13. McDonald, F.A.: J. Appl. Phys. 52 (1981) 381-385.

14. Touloukian, Y. (ed.): Thermophysical properties of matter. New York: IFI/Plenum Press 1970.

Laser Speckle Photography for the Measurement of Changes in Refractive Index in Phase Media

B.S. Ramprasad, T.S. Radha and E.S.R. Gopal

Instrumentation and Services Unit,
Indian Institute of Science,
Bangalore 560 012, India.

Summary

A technique of measurement of changes in refractive index in liquids and gases using laser speckle photography is presented. By analysing Young's fringes from the double exposed specklegrams changes in refractive index of liquid mixtures as small as 0.0005 can be measured. The application of this technique for the nondestructive testing of incandescent lamps is described.

Übersicht

Es wird ein Verfahren für die Messung von Änderungen des Beugungsgrades in Flüssigkeiten und Gasen mit Hilfe der Laser-Speckle-Fotografie vorgestellt. Durch die Auswertung der Youngschen Interferenzstreifen auf den doppelbelichteten Specklegrammen lassen sich die kleinsten Änderungen des Beugungsgrades in flüssigen Gemischen bis zu 0,0005 messen. Die Anwendung dieses Verfahrens auf die zerstörungsfreie Prüfung von Glühlampen wird beschrieben.

1. Introduction

The knowledge of refractive index of phase materials is important for characterization of materials. Many instruments and techniques exist for the measurement of refractive index like refractometers and interferometers. Refractive index distribution is also important in phase media like liquids and gases for characterising flow [1]. Small changes in refractive index is usually measured by conventional interferometry. Some recent techniques include holography and moiré deflectiometry. Speckle techniques which are complementary to holography are well known for metrology applications [2]. However the use of speckle for the study of transparent or phase objects has not been so common. Köpf [2] suggested the application of speckle for the deflection of laser light by phase objects. In this paper a technique developed for the measurement of changes in refractive index in liquids and gases using speckle photography is described. It is shown that, in spite of some limitations, accurate measurements are possible.

Speckle photography

When laser light falls on a diffusely reflecting or transmitting object a shimmering pattern of dark and bright spots are seen. These spots are called speckles. When one looks at this pattern through an aperture the dark spots become fewer and the bright spots become larger as the aperture decreases. This demonstrates that the laser speckles carry information.

Speckle photography is a method of recording a speckle pattern which is characteristic for an object. If two speckle patterns of an object displaced between two exposures, is recorded on a high resolution photographic plate, the displacement information will be contained in the developed photographic plate, which is now called as a specklegram. This information can now be extracted by shining an unexpanded laser beam on the specklegram. On a screen away from the specklegram a diffraction halo is seen modulated by dark and bright straight interference fringes. These fringes are called Young's fringes. The spacing between the fringes is a measure of the displacement of an object.

2. Experimental details

Laser speckle can be generated by shining a laser beam on a ground glass plate. It can also be generated by passing a laser beam through a multimode optical fiber. The application of speckle photography for measuring changes in phase media is experimentally carried out as follows.

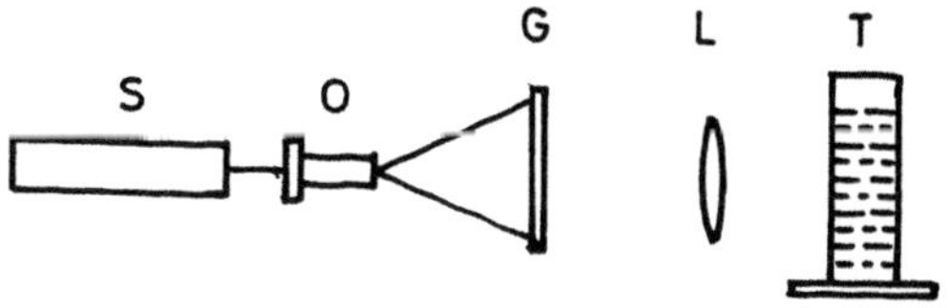

Fig.1 Speckle photography set-up for measuring changes in refractive index of liquid mixtures. S-Laser Source, O-microscope objective, G-ground glass, L-collimating lens, T-liquid cell, P-photographic plate.

Referring to Fig.1, a laser beam is expanded using a microscope objective. The expanded laser beam falls on a ground glass plate. The speckle pattern generated is collimated using a lens. The collimated speckle pattern passes through a transparent liquid cell containing a liquid, in this case, water. A high resolution photographic plate (Agfa 10E75 plates used in the experiment) is placed in a position where a sharp image of the speckle pattern is seen. First exposure of the photographic plate is made with water in the cell. A known quantity of a liquid with a refractive index other than that of water is added to the cell. In this experiment ethanol was used. After a short time interval during which the liquid mixture settles, the second exposure is made. The double exposed photographic plate is developed in the normal way. This double-exposure specklegram contains the information about speckle displacement.

When the developed specklegram is placed in an unexpanded laser beam Young's interference fringes are seen. The spacing between the fringes is given by

$$d_s = \lambda D/mp$$

where d_s = in-plane displacement of the speckle at the cell, D is the distance between the specklegram and the screen, p is the distance between the fringes, λ the wavelength of the laser light and m the magnification of the recording system. For different concentrations of ethanol different double exposure specklegrams are made. In each case the in-plane speckle displacement is calculated. The refractive index of the mixture is measured using the Abbe refractometer for each experiment. The speckle displacement can be compared to the lateral shift of a ray passing through the liquid cell. The shift due to a change in the refractive index of the liquid from n_1 to n_2 is given by the relationship (Fig.2)

$$d_s = t \sin i \left\{ \left(\frac{1}{\sqrt{n_1^2 - \sin^2 i}} - \frac{1}{\sqrt{n_2^2 - \sin^2 i}} \right) \right\}$$

where i is the angle of incidence of the light beam on the face of the cell and t is the width of the cell through which the light beam passses.

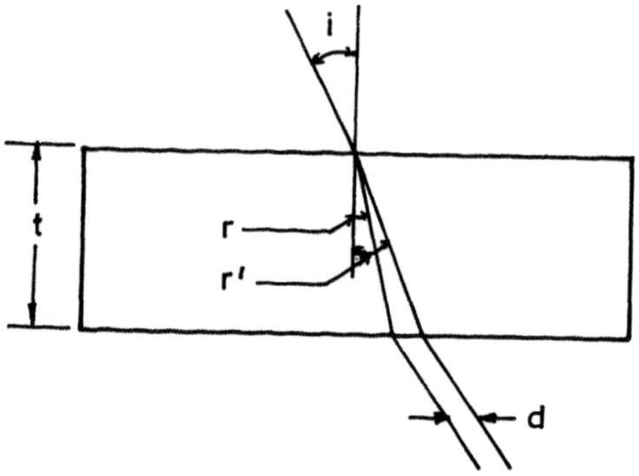

Fig.2 Lateral shift of a ray of light through a medium. t - thickness of cell, i - angle of incidence of collimated laser beam, r - refracted ray for liquid with refractive index n_1, r' - refracted ray for liquid mixture with refractive index n_2.

It is observed that as the refractive index of the mixture increases the fringe spacing becomes closer. Table 1 gives the results of the experiments. The experiments demonstrate that changes of refractive index as small as 0.0005 can be detected by the technique of speckle photography.

TABLE 1

Calculated and measured speckle displacement ethanol - water mixture

Concentration percent	Measured n_1	n_2	n n_2-n_1	Displacement in microns calculated	measured
10	1.3320	1.3382	.006	7.435	6.5
20	1.332	1.3443	.0123	14.636	13.447
30	1.332	1.3502	.0182	21.629	19.210
40	1.332	1.3552	.0232	27.432	26.500

As seen from equation 2 the shift depends on the angle of incidence and the width of the cell. It is seen that the sensitivity increases with increasing angle of incidence. Experiments have been conducted for various angles of incidence from 0° to 30°. In this experiment a quartz liquid cell of 30 x 10 x 10mm is used at an angle of incidence of 10°. Agfa 10E75 Holotest plates have been used for recording the specklegrams. The discrepancy between the calculated and experimental shifts is to a large extent due to the error in the positioning of the cell at the correct angle of incidence. It can be calculated that an error in the incident angle of 6 minutes gives rise to an error in the shift of approximately 1 micron.

3. Testing incandescent lamps

Attempts have been made to use speckle photography for studying changes in gases also. For example, Farrel and Hofeldt [4] measured temperature distribution in gases using speckle photography. In this paper a successful attempt to test incandescent lamps using speckle photography is reported. The experimental set up is shown in the schematic in Fig.3. An expanded laser beam illuminates a ground glass plate. The diffuse laser beam now passes through an incandescent lamp. A camera is focussed on the lamp. The lamp

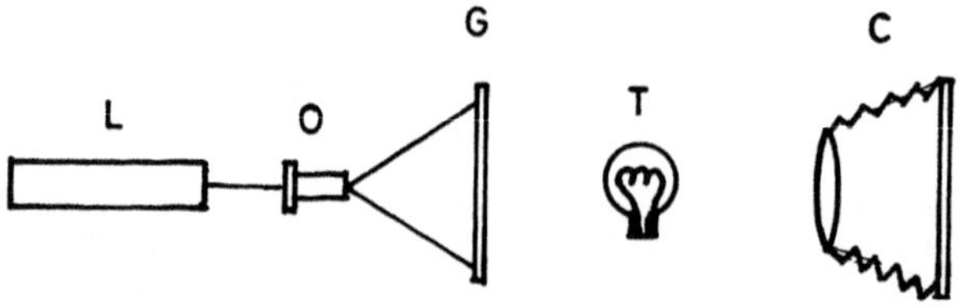

Fig.3 Schematic for testing incandescent lamps. L-laser source, O-microscope objective, G-ground glass, T-test lamp, C-camera.

is switched on for a known amount of time and switched off. The first exposure is made immediately. The second exposure is made after the lamp has cooled. When the specklegram is placed in an unexpanded laser beam Young's fringes are seen. This is due to the convection in the lamp. When a defective lamp of the same wattage is used and a specklegram is made under identical conditions it is found that the fringe spacing is different. Several good and defective lamps, given by a local lamp inducstry, with different defects, like defective filaments, wrong pressure or micro-leaks, were tested. While lamps of a particular wattage without defects showed consistently the same fringe spacing, defective lamps showed distinctly different fringe spacing. Some results are illustrated in Fig.4.

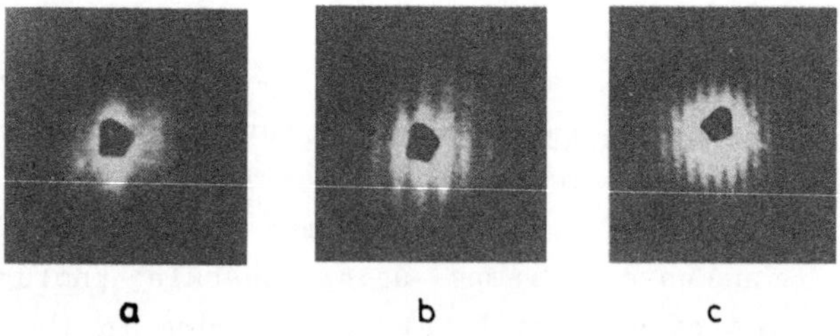

Fig.4 Testing of incandescent lamps. Young's fringes from a double exposed specklegram of a 100 watt lamp. a) good lamp b) defective lamp with an improper resistance c) lamp with a microleak.

The normal filament of the 100 watt bulb has a resistance of 40.86 ohms and the defective lamp whose fringes are shown in Fig.4b has a resistance of 42.98 ohms. As expected the fringes are closer. Fig.4c shows the fringes of a bulb which was found to have a microleak. In fact the filament fused after switching on and off a few times.

An attempt was made to see if speckle photography could give information about the refractive index distribution in the bulb. Controlled experiments using defocussed speckle photography were made. The camera is focussed 2 cms away from the filament of the lamp. The specklegram can be analysed using whole field filtering as shown in Fig.5. The refractive index distribution due to convection in the bulb is

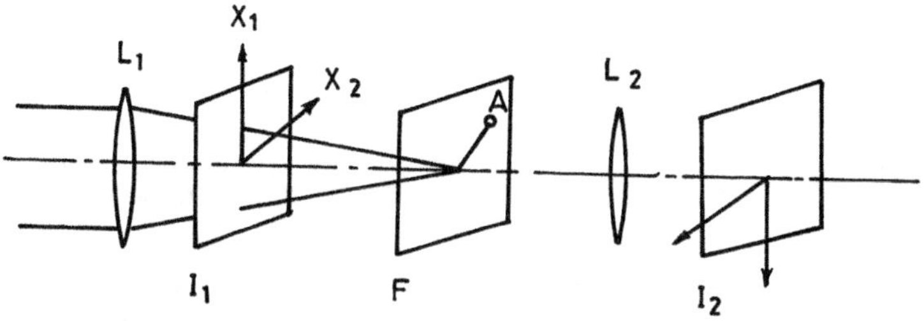

Fig.5 Whole field filtering for double exposure specklegrams. L_1-transforming lens, L_2-imaging lens, I_1-input plane where specklegram is placed, I_2-imaging plane, F-Fourier filtering plane, A-filtering aperture. The combination L_2I_2 can be a Camera.

represented by fringe contours usually called isothetic fringes. Fig.6 shows these fringes at two different filtering positions.

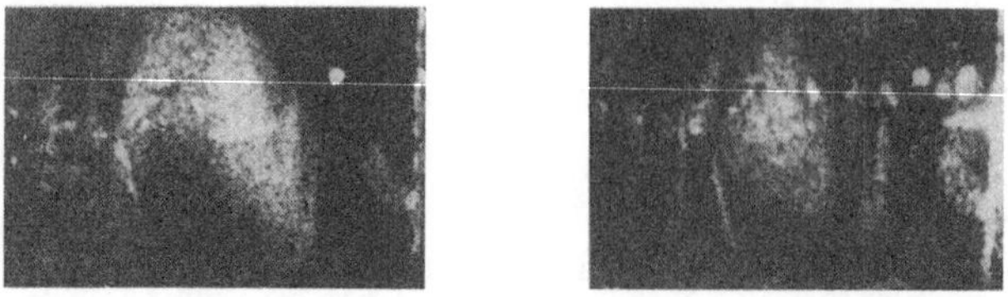

a b

Fig.6 Isothetic fringes of a good 100 watt lamp at two filtering positions. a) 1.5 cms b) 3 cms.

Though the application of the technique has been demonstrated successfully, it suffers from the disadvantage that it is a two-step photographic process.

References

1. Merzkirch, W.; Flow Visualization. New York, London: Academic Press 1974.
2. Jones, R. and Catherine Wykes.; Holographic and Speckle interferometry. Cambridge: Cambridge University Press 1983.
3. Köpf, V.; Application of speckling for measuring the deflection of laser light by phase objects. Opt.Comm.5 (1972) 347.
4. Farrel, P.V. and Hofeldt, D.L.; Temperature measurement in gases using speckle photography. Appl. Opt. 23 (1984) 1055-1059.
5. Ramprasad, B.S.; Ph.D. Thesis. Indian Institute of Science, Bangalore 1987.

Testing of Coatings by Means of Acoustic Emission

Testing of Coatings by means of Acoustic Emission

H.-A. Crostack, V. Beckmann, H.-J. Storp

Summary

Investigations are reported on the detection of microdamage in different coatings under load. Acoustic emission analysis allows not only microcracks but also crack propagation to be detected. The relationship between acoustic emission analysis and the type of damage is established metallographically.

Zusammenfassung

Es wird über den Einsatz zerstörungsfrei arbeitender Prüftechniken beim Nachweis von Zeitstandschäden berichtet. Hierzu werden die Wechselwirkungen zwischen dem Gefügeaufbau und den Prüfanzeigen eingehend diskutiert. Darüber hinaus wird auf die Randbedingungen bei der Prüfung hinsichtlich der Übertragbarkeit der Ergebnisse auf Bauteile detailliert eingegangen. Als Prüfverfahren wurden die magnetische Prüfung, die Ultraschallprüftechnik und ein thermoelektrisches Meßverfahren eingesetzt.

1. Problem

Components are given problem-oriented coating particularly for reasons of corrosion protection and thermal insulation. The use of this kind of composite materials requires detailed knowledge of the mechanical properties of the coatings, in particular those of adhesion, crack formation and crack propagation. This information is necessary to optimize the forming parameters of electroplated sheets on the one hand and to avoid critical load courses in multiple-layer coatings manufactured by thermal spraying. Moreover, the information is required to judge the quality of the coatings produced.

The relationship between the kind of failure within the coating and the load is established by acoustic emission analysis. Analysis of the intensity and amplitude of the acoustic emission signals allows differentiation of the types of damage.

Two completely different coating kinds are investigated, namely:

- electroplated sheets for automobile construction and
- heat-insulating coatings of $ZrO_2 7Y_2O_3$ applied by thermal spraying (multiple-layer coatings).

No detailed information was available on the mechanical properties, and on the strength properties in particular. To obtain more detailed information, the composites are mechanically loaded in a three-point bending test in order to investigate the time points for beginning crack initiation, crack propagation und possible local spalling of the coating.

2. Test technique

Fig. 1 shows the experimental apparatus. To suppress frictional noise during the bending test, the specimen is shielded with teflon against the loading device. The acoustic emission transducer (140 kHz, resonant) is applied to the substrate opposite to the coated side, thus ensuring that acoustic emission signals can be received even when cracks are already present in the coating. A two-channel measuring system is used for AEA. The acoustic emission signals registered by the transducer are electronically preampliefied and filtered. The band pass filter suppresses background noise from the environment and from electrical interferences. After further amplification, the acoustic emission signals are analysed for number of counts and intensity.

Results

Setting data for AEA are first determined in preliminary tests. Uncoated specimens are investigated to ensure that machine-specific background noise und frictional noise are not registered or that their signal characteristics are known.

3.1 Electroplated sheets

For the testing of plated sheets, a band pass between 50 and 2000 kHz and a discriminator threshold of 44 mV_{pp} were selected for the number of counts on the basis of the results from the preliminary experiments. The discriminator threshold for the peak amplitude was 680 mV_{pp}, and the high-pass filter was set to 50 kHz. The AEA results for bending a ZnFe coating (sheet thickness 0.8 mm, coating thickness 15 µm) are plotted in Fig. 2. The registered acoustic activity intensifies abruptly increasing from about 1 mm bending onwards. This applies equally to the pulse sum and peak amplitude, i.e. the intensity of the acoustic emission signals. From about 5 mm bending onwards the initiation of microcracks within the coating in the region of maximum load is complete. Upon further load the microcracks run into each other, whereby on the one hand the intensity and the event frequency of the emitted acoustic signals decrease. In the case of sheets coated on both sides, as shown in Fig. 2, it must however be taken into account that the acoustic transmission is increasingly hindered by crack formation and that not all signals are registered by the transducer after the crack initiating phase. The test of Zn and ZnNi-coated sheets result also in the relationships shown in Fig. 2. A significantly lower acoustic activity is observed with the ZnAl coatings. Fig. 3 shows a comparision between the results obtained from the different composites. At equal degree of formation, the value of the pulse sum depends strongly on the coating type. A lower acoustic acitivity does not mean a reduction of microcrack formation. Rahter, the acoustic activity depends strongly upon the coating material. This is confirmed by the metallographic findings, Fig. 4. Despite less acoustic activity in the Zn coating, microcracks are formed to the same extent as in the ZnFe coating. This applies to the bending radii R of 1 and 10 mm alike.

3.2 Thermal barrier coatings

In analogy to the procedure of section 3.1, the time point of initial crack formation in thermally-sprayed coatings is determined as a function of load within the coating by means of acoustic emission analysis. To characterize the coating, the bending stresses for the composite are determined for crack initiation (AEA), for the elasto-plastic transition, for the occurrence of macroscopic cracks at the coating surface and for the beginning of local spalling of the coating. Results for two substrates and different coatings as well as the substrate material are illustrated in Fig. 5. For Cr-steel (17 % Cr) as a substrate, the differences between the bending-stress values for initial crack formation on the surface, for partial spalling of the coating and for the elasto-plastic transition are relatively low. However, AEA indicates a significantly earlier begin of damage within the triplex coating. Accompanying metallographic tests showed that initial cracks in the triplex coating occur in the intermediate layer and in the bond layer resulting from the bending load. For the specimens with AlSi12 as substrate, the values - with the exception of the elasto-plastic transition - for each bending stress of the coated specimens shifted to higher values. The difference for the triplex coating is the most conspicuous.
The application of thermally sprayed coatings onto AlSi12 ameliorates the rigidity of the entire composite, with reduction of the deformability,
This tendency increases at the transition from the duplex to the triplex coating. For the AlSi12 specimens, the AE indications have their origin in crack formation in both substrate and multi-layer coating. For the coated steels tested, initial AE indications arise from crack formation within the coating.
The influence of the bonding layer on the static rigidity is shown in Fig. 6. If the bonding layer NiCoCrAlY is used instead of NiCrAlY, smaller stress values result in view of the abovementioned criteria. This is particulary true for the bending-stress values at the occurrence of initial damage detected by AEA.

4. Conclusions

The use of AEA enables the early detection of the beginning of damage to composite materials within the individual components of the composite. This is of particular significance when damage already occurs at loads where the mechanical parameters force and bending still indicate elasticity of the material. The use of AEA hence allows the determination of critical loads and deformations below which no damage occurs within the coating. The composites can therefore be used optimally without premature loss of their protective function.

Thanks

The investigations were funded by the Deutsche Forschungsgemeinschaft within the scope of SFB 316. We wish like to express our thanks to the DFG here.

Literature

/1/ TRD 508 (Technische Regeln für Dampfkessel): Zusätzliche Prüfungen an Bauteilen, berechnet mit zeitabhängigen Festigkeitskennwerten, Okt. 1978, Heymanns Verlag, Köln, Beuth-Verlag, Berlin

/2/ Crostack, H.-A., J. Nehring, W. Oppermann: Einsatz von Korrelationsimpulsen (CS-Technik) zur Wirbelstromprüfung, Materialprüfung 26 (1984) 1/2, S. 16-20

/3/ Crostack, H.-A., W. Roye: Profile and deformation measurements with high axial resolution using the acoustic holography, Nuclear Engineering and Design 102 (1987), p. 313-317, North Holland, Amsterdam

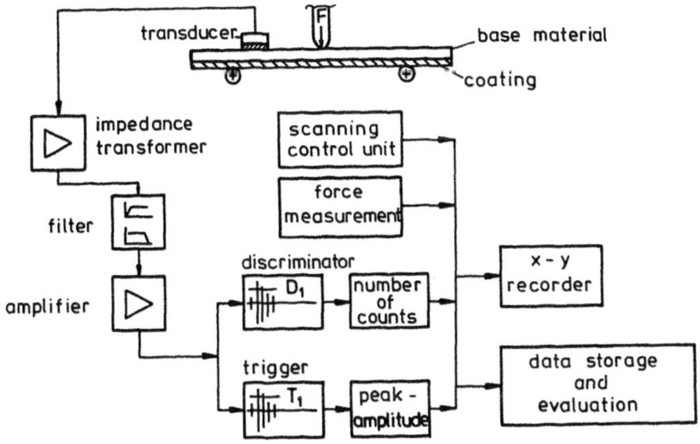

Fig.1:Principle of acoustic emission measurement during bending test of coatings

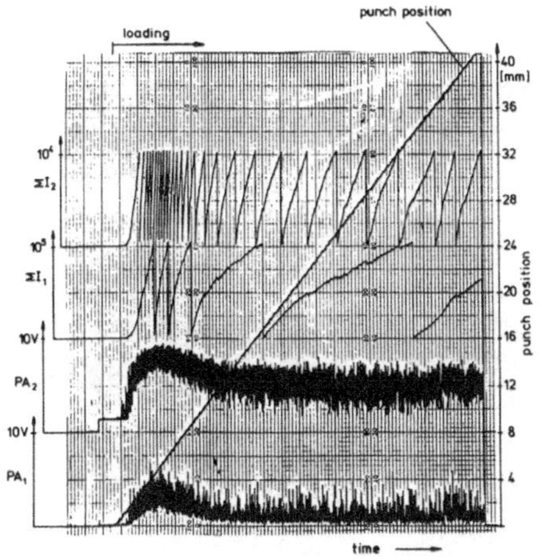

Fig.2:Acoustic emission analysis during bending test of a ZnFe-coating, base material St 1403

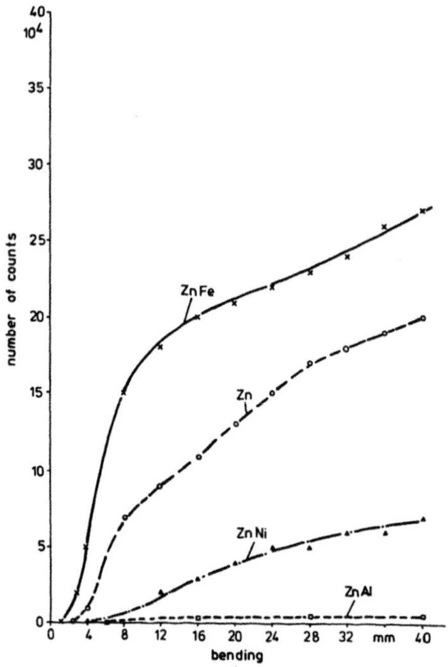

Fig.3:Number of counts (AE) as a function of deflection for differnt Zn-coatings, base material St 1403

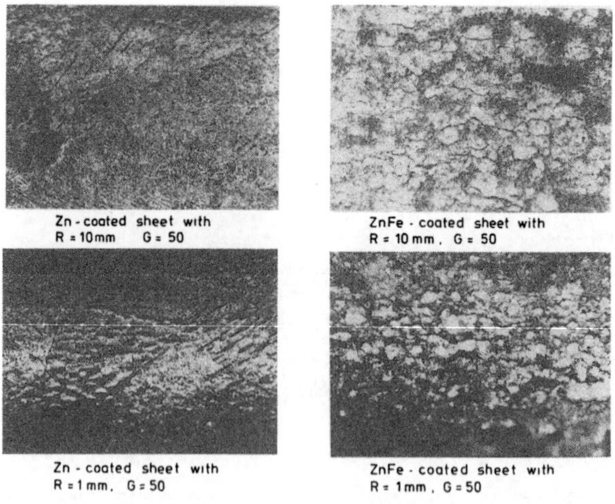

Fig.4:Influence of bending angle and surface finish on failure mechanism

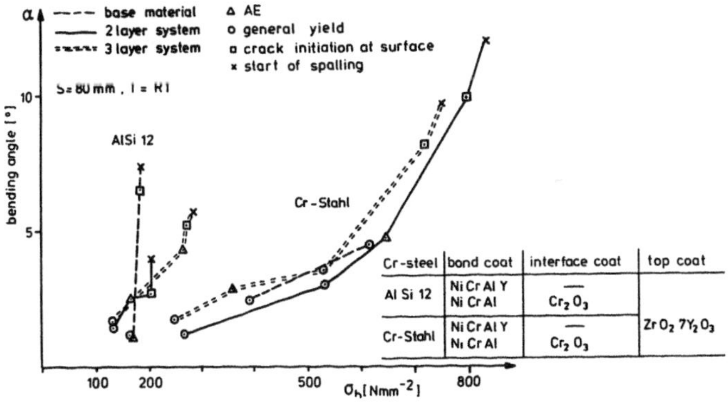

Fig.5: Critical bending stresses of different multilayers (thermal sprayed coatings)

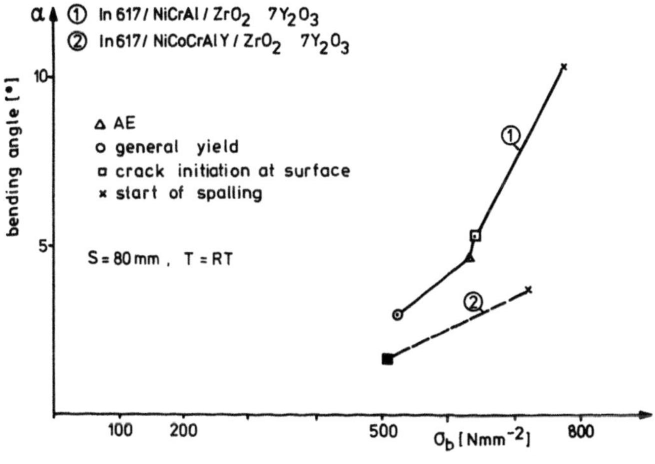

Fig.6: Influence of adhesion coat on critical bending stresses and start of microdamage (AE)

Nondestructive Characterization of Mixed Oxide Pellets in Welded Nuclear Fuel Pins by Neutron Radiography and Gamma-autoradiography

J.P. PANAKKAL, J.K. GHOSH, P.R. ROY

Bhabha Atomic Research Centre
Bombay India 400085

Abstract

Nondestructive evaluation of nuclear fuel pellets after the welding of fuel pins plays a vital role in assuring a safe and reliable operation of reactors. Some of the important characteristics to be monitored in low plutonium enriched mixed oxide fuel pellets are plutonium enrichment, size of plutonium dioxide agglomerates, incorrect loading and geometric shape. Experiments were carried out at Bhabha Atomic Research Centre, Bombay on experimental fuel pins containing mixed oxide pellets of different geometry (solid and annular), of different plutonium enrichment (0-6 w% of plutonium dioxide) and containing PuO_2 agglomerates of size 125-2000 microns to evaluate these characteristics nondestructively. Neutron radiography of these fuel pins was carried out using a swimming pool type reactor "APSARA". Results of quantitative evaluation of the neutron radiographs and a simple model correlating neutron interaction probability and the optical density are presented. Gamma autoradiography of these fuel pins showed that these parameters could be evaluated with a few limitations. This paper presents the experimental details, quantitative analysis of the radiographs by microdensitometry and merits and demerits of neutron radiography and gamma autoradiography for nondestructive characterisation of nuclear fuel pellets.

Introduction

Quality control personnel in nuclear industry have always felt the need for a more complete characterisation of nuclear fuel pellets in as-fabricated fuel pins containing plutonium. It is necessary to ascertain geometric shape, physical integrity and correct loading of the pellets, maximum size of plutonium agglomerates present and plutonium enrichment in the welded fuel pins. Experiments were carried out at Bhabha Atomic Research Centre, Bombay on specially fabricated fuel pins containing mixed oxide pellets of different plutonium enrichment (0-6 w%) and geometry and with PuO_2 agglomerates (125-2000 μm). This paper presents the results of neutron radiography and gamma autoradiography of these fuel pins including quantitative analysis of radiographs by microdensitometry and compares the two techniques with respect to the characteristics of the pellets investigated.

Experimental

Fuel Pellets

Uranium plutonium mixed oxide fuel pellets (nominal diameter 12.37mm) with plutonium enrichment 0, 2.5, 4.0 and 6.0% and PuO_2 agglomerates

(a) photomicrograph

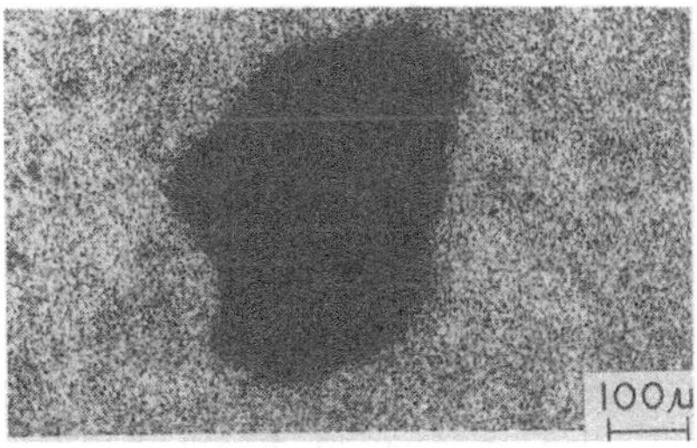

(b) alpha autoradiograph

Fig.1 PuO_2 agglomerates in a mixed oxide pellet.

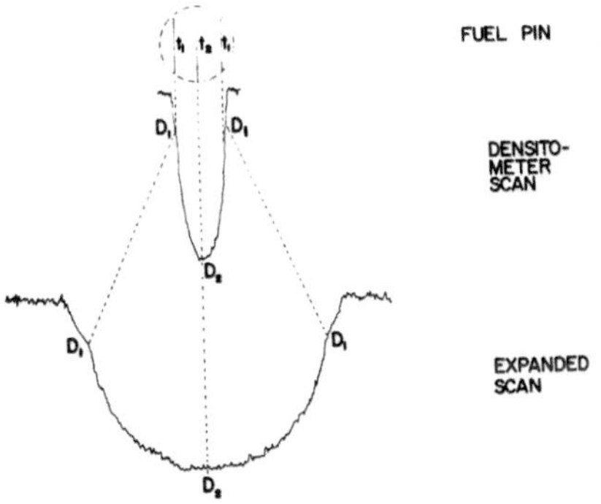

Fig.2 Expansion of microdensitometer scan over fuel pin.

(125-2000 μm) and different geometries (solid and annular) were encased in zircaloy-2 clad tubes of outside diameter 14 mm and nominal thickness 0.8 mm as reported elsewhere (1-3). The size of plutonium dioxide agglomerates in the pellets was confirmed by alpha autoradiography and ceramography of a few sample pellets (fig.1). The composition of the pellets was checked by chemical analysis.

Neutron radiography

Neutron radiography of the fuel pins was carried out in APSARA a swimming pool type reactor at Bhabha Atomic Research Centre using direct technique (25 μ thick gadolinium screens and Agfa Structurix D2 film) as reported earlier (1,2). The ratio of length L to diameter D of the Collimator was 90:1. The experimental parameters were adjusted so that the optical denisty of interest was in the range 0.8 to 2.5.

Gamma Autoradiography

A PVC cassette containing the X-ray film (Structurix D 10) was held in contact with the fuel pin containing pellets of different composition. An exposure time of 17 hours gave an optical density of 1.7 over pellet of 6% by weight PuO_2 . The PVC cassette was wrapped around a fuel pin containing pellets with agglomerates. The exposure time required for imaging the agglomerates was 17 hours for D10 film and 65 hours for D7 film.

Microdensitometry

The gamma autoradiograph was scanned using a microdensitometer across the image of the pellets with an aperture size of 15 x 1000 μ. The effective scanning aperture used for neutron radiographs was 10 x 500 μ. An expanded scan across the diameter of the pellets with a ratio arm of 1:100 was also made to measure optical density corresponding to different thickness segment of the pellet on the neutron radiograph (4,5) (Fig.2).

Results and discussions

Geometry and integrity of the pellets

The neutron radiographs revealed clearly the presence of annular pellets in the pin as shown in fig.3. The gamma autoradiographs did not distinguish between solid and annular pellets as expected because of self absorption of gamma rays in the pellets.

Plutonium dioxide agglomerates

Neutron radiographs revealed the presence of PuO_2 agglomerates down to 250 μ, distributed through out the pellets. Autoradiography could also reveal agglomerates of the order of 250 μ, lying in the outer annulus of the pellets. Figure 4 presents neutron radiograph and autoradiograph of the same fuel pin. In the gamma autoradiograph the image of the agglomerates lying closer to the surface is sharper than those lying deep in the pellet. Alpha autoradiography which is the standard quality control check for PuO_2 agglo-

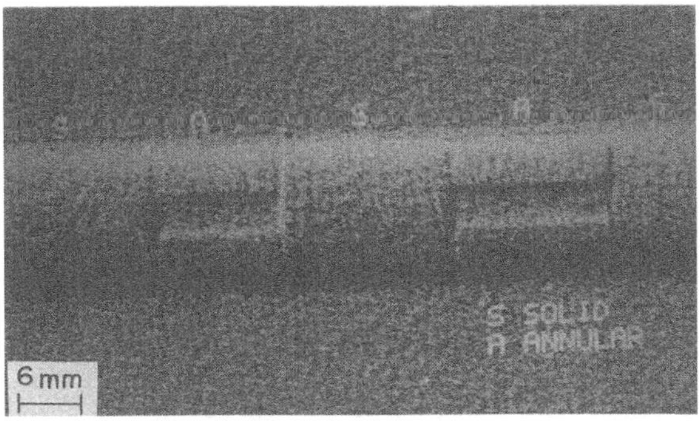

Fig.3 Solid and annular pellets in the neutron radiograph. (enhanced image)

(a)

(b)

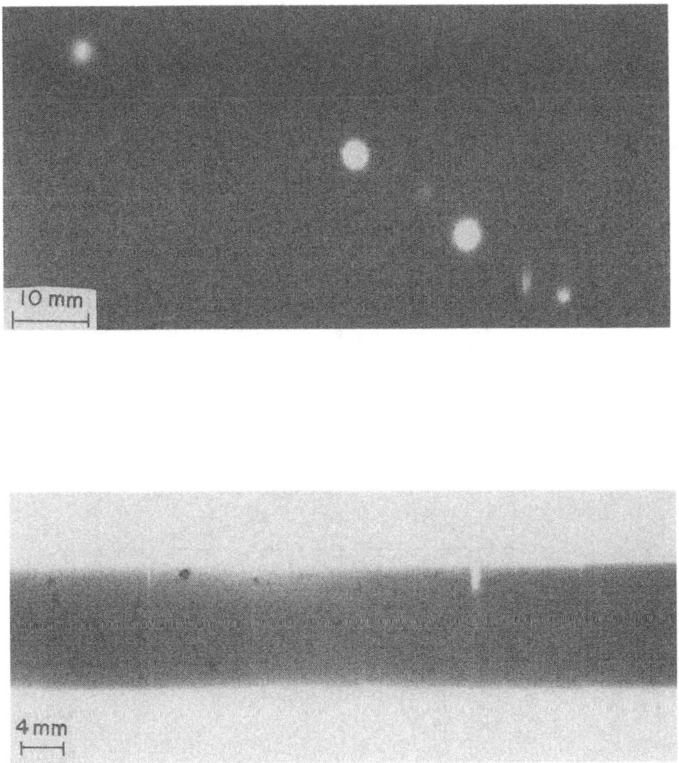

Fig.4 Fuel pin containing plutonium agglomerates
(a) autoradiograph (b) neutron radiograph

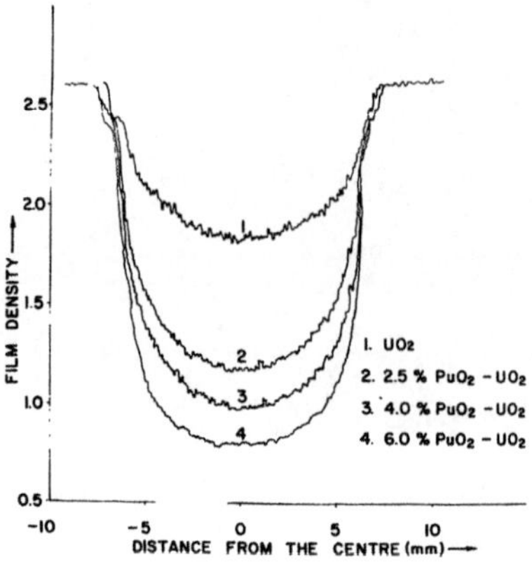

(a) on neutron radiograph

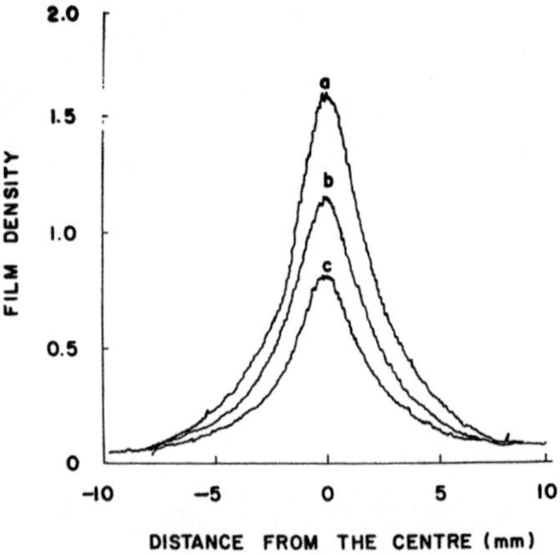

(b) on autoradiograph

Fig.5 Microdensitometric scan over fuel pin

merates is carried out on a sample basis. The occurence of agglomerates is statistical in nature and hence it is probable that some of them lie near the surface and will be detected by autoradiography. Autoradiography is simple and inexpensive to carry out on all the fuel pins compared to neutron radiogrphy which is not practical to carry out on a 100% basis. The exposure time of 17 hours is not very inconvenient since this test is to be carried out at the final stage of fabrication.

Pu enrichment

Microdensitometer scan of both the neutron radiograph and autoradiograph are presented in Fig.5. Both techniques have more or less the same sensitivity and may be used to monitor plutonium enrichment of the pellets in welded fueled pins as reported elsewhere (1,2).

From the expanded scans of the neutron radiographs, 56 data points were generated corresponding to different thickness segment of the pellets. The value of thermal interaction probability $p_j = \sum_{i=1}^{k} n_i \sigma_i t_j$ where n is the number of nuclei/cm of type of nuclei i, σ_i is the total microscopic cross-section of nuclei 'i' k is the number of species of nuclei and t_j is the distance traversed by neutrons in the pellet, was calculated (4,5). The data points were fitted into linear, exponential, logarithmic, power and polynomial. It was observed that the polynomial

$$\Delta D_j = a + bp_j + cp_j^2$$

was the best fit as evident from fig.6. This simple model would be useful for planning future neutron radiography experiments with a given set up.

Conclusion

A comparative study of neutron and autoradiographic techniques shows that the former gives complete information of the pellets in the fuel pins, Gamma autoradiogrphy provides information about the outer annulus of the pellets. Autoradiography, being simple and inexpensive still provides a lot of information about the characteristics of all the pellets which may go unnoticed in the absence of neutron radiography.

Acknowledgement

The authors wish to thank their colleagues in Radiometallurgy Division and Nuclear Physics Division for their active help and cooperation during this work.

References

1. Ghosh, J.K., Panakkal, J.P., Roy, P.R : Use of autoradiography for checking plutonium enrichment and agglomerates in mixed oxide fuel pellets inside welded fuel pins. NDT Int. 17(1984)269-271.
2. Panakkal, J.P., Ghosh, J.K., Roy, P.R. : Characterisation of

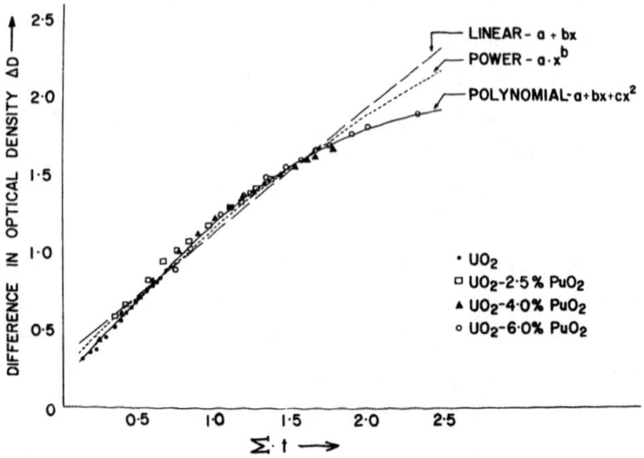

Fig.6 Film density ΔD verses thermal neutron interaction probability.

uranium-plutonium oxide nuclear fuel pins using neutron radiography Br. J. NDT. 27 (1985) 232-233.

3. Ghosh, J.K., Panakkal, J.P., Roy, P.R. : Monitoring plutonium enrichment in mixed oxide fuel pellets in sealed nuclear fuel pins by neutron radiography., NDT Int. 16 (1983) 275-276.
4. Panakkal, J.P., Ghosh, J.K., Roy, P.R. : Analysis of optical density data generated from neutron radiographs of uranium-plutonium mixed oxide fuel pellets inside nuclear fuel pins. Nucl. Inst. Meth. Phy. Res. B 14(1986) 310-313.
5. Panakkal, J.P., Ghosh, J.K., - A simple model for neutron radiography of uranium-plutonium mixed fuel pins, J.Nucl.Mat.153 (1988) 82-83.

Nondestructive Characterization of a Deformed Steel Using Positron Annihilation

Yong Ki PARK, Jae Ok LEE and Sekyung Lee

Nondestructive Evaluation Laboratory
Korea Standards Research Institute
Taedok Science Town, Republic of Korea

Summary

Positron annihilation spectroscopy is a relatively new NDE technique for characterizing the atomic scale lattice defects like vacancies and dislocations. Positron annihilation lifetime and Doppler broadening line-shape were measured and analyzed in cold rolled low carbon steel. Both positron lifetime and Doppler broadening line shape reflect the change in the microstructure and the substructure of the specimens sensitively.

Introduction

Positron annihilation spectroscopy is a relatively new technique for nondestructive characterization of materials. This technique is unique in that it provides us with the direct information about atomic scale defects. Especially positron annihilation is very sensitive to the vacancy type defect while it is insensitive to the interstitial [1] . In this paper the microstructures and substructures of cold rolled low carbon steel were characterized by the relatively new NDE technique, positron annihilation spectroscopy.

Experiment

The specimens were prepared by cold rolling low carbon steel plates for deep drawing. The chemical analysis of the specimen is in Table 1.

Table 1. Chemical analysis of the specimen. (wt.%)

C	Si	Mn	P	S	Sol. Al	Fe
0.05	0.01	0.23	0.015	0.015	0.045	Bal.

The starting plate was 2.4 mm thick and from this 10 sets of specimens with different amount of deformation were prepared. For the measurement of positron annihilation one surface of each specimen was polished.

Positron lifetime and Doppler broadening line shape were measured. Fig.1 shows the block diagram of the positron lifetime measurement system. The time resolution of the system measured with ^{60}Co was 296 ps. Doppler broadening line shape was measured with a high purity Ge detector of which energy resolution was 1.64 keV (FWHM) for the fiducial γ-ray of 1.274 MeV of ^{22}Na. The positron source was 9.3 μCi ^{22}Na supported by Kapton foil. The details of the measurement and data analysis are reported elsewhere [2,3].

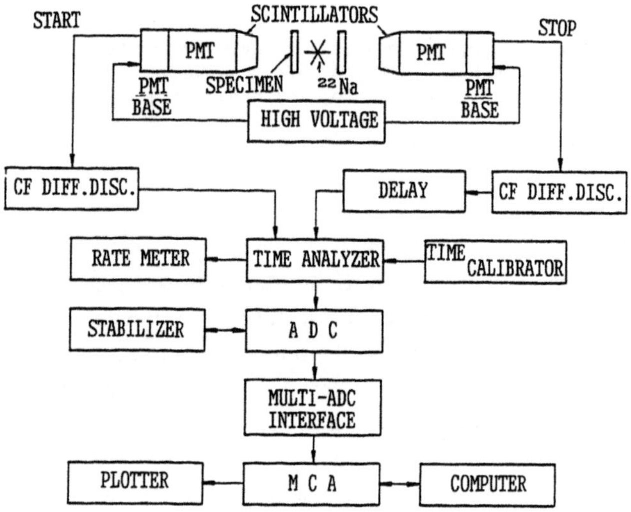

Fig.1. Block diagram of the positron annihilation lifetime measurement system.

Results and Discussion

The Doppler-broadening line shape was analyzed using the line shape parameters after subtracting the background expressed as an error function [4]. Fig.2 shows the typical line shapes for the undeformed and the deformed specimens (22.1 % cold rolled) after subtracting backgrounds. The absolute value of the difference in counts for each channel between two curves is plotted together. The peak parameter (P) and the wing parameter (W) are also defined in Fig.2.

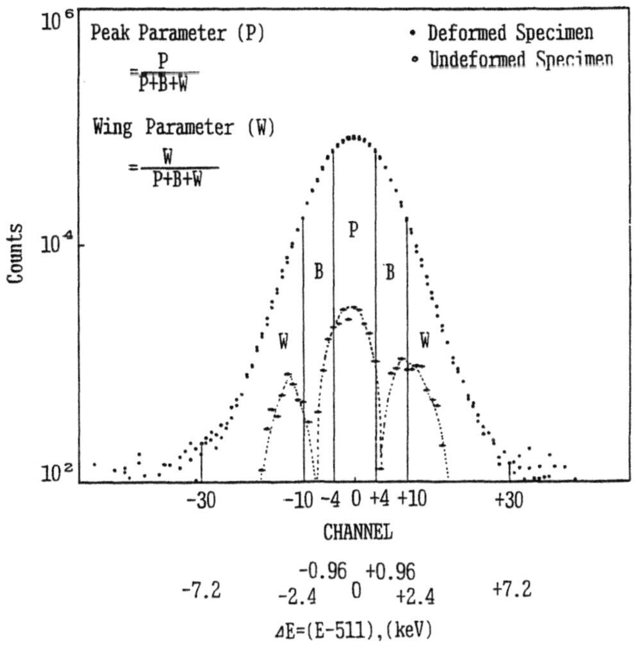

Fig.2. Doppler broadening line shapes of the undeformed and the cold rolled (22.1 %) specimens. The absolute value of the difference in counts for each channel between two curves is plotted inside of two curves. The line shape parameters are also defined.

Fig.3 shows the variation of the line shape P and W parameters as a function of the thickness reduction. At small amount of deformation, P parameter increases rapidly as the thickness reduction increases. The increase rate of P parameter, however, becomes smaller when the thickness reduction is bigger than 30 %. The wing parameter decreases rapidly up to about 20 % deformation and seems to be saturated when the deformation is bigger than 30 %.

Fig.3 shows that there are some indications of change in the major trapping defects at about 6 % and about 25 % of deformation. Mantl and Triftshäuser [5] suggested that a defect-specific parameter R can be used to resolve the information about the type of the defect. The R parameter is defined using P and W parameters as

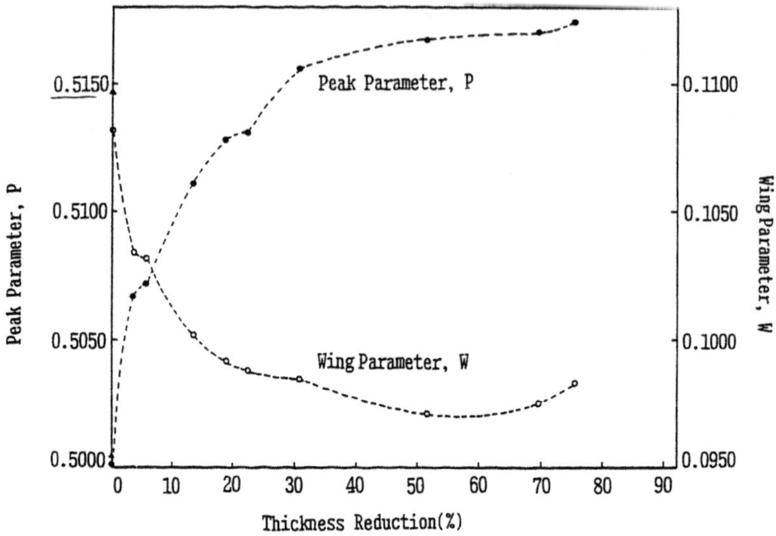

Fig.3. Variation of the line shape P and W parameters measured in the cold rolled low carbon steel specimens. The triangle is for the well annealed specimen.

$$R = \left| \frac{P - P_f}{W - W_f} \right| = \left| \frac{P_T - P_f}{W_T - W_f} \right| \tag{1}$$

R parameter is characteristic of the type of the defect involved and is independent of the trap concentration.

Fig.4 shows the change in R parameter. R parameter is constant up to about 20 % of deformation but it increases and also scatters much at bigger deformation. This implies that the major defect which traps positrons is the same up to 20 % deformation but two or more trap sites compete when the deformation is bigger than 20 %. These results of the line shape analysis are very coincident with the results of positron lifetime measurement.

Fig.5 shows the variation of the positron lifetime as a function of the thickness reduction. The lifetime spectrum was analyzed with single lifetime component. Although it was difficult to resolve into two or more lifetime components with good statistics, the single lifetime shown in Fig.5 reflects the change in the microstructures and substructures of the specimens very well.

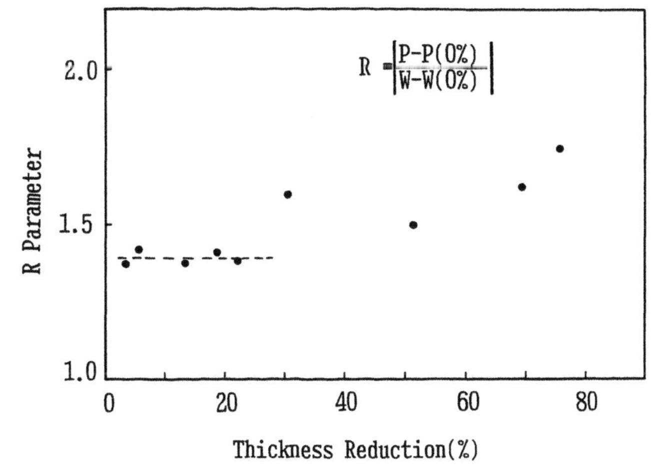

Fig.4. Changes of the defect-specific R parameters measured in the cold rolled low carbon steel specimens.

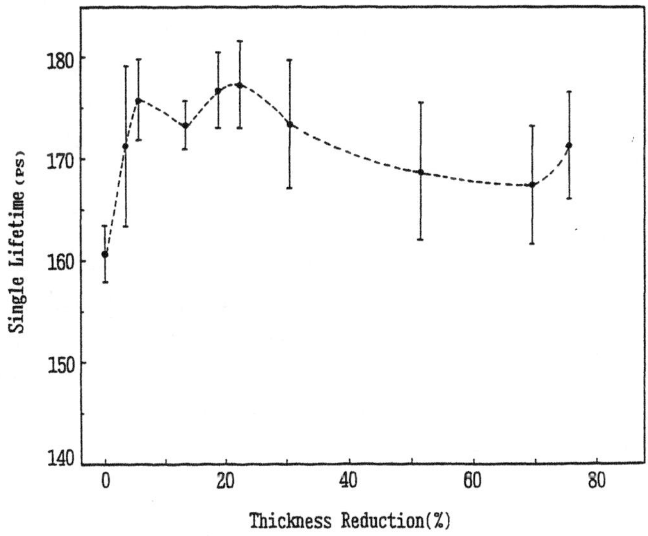

Fig.5. Variation of the positron lifetime (fit with single lifetime component) in cold rolled low carbon steel as a function of strain.

The lifetime increases rapidly with increasing strain up to about 6 %, because at this stage positrons are mainly trapped in the dislocations and the dislocation density increases rapidly at the beginning of deformation. When the strain is bigger than 6 %, partial cell structure starts to be formed and the regions of low dislocation density are developed among the regions of high dislocation density. Therefore even if the total dislocation density increases (as seen in Fig.3) the positron lifetime decreases slightly at around 10 % strain because the probability of positron annihilation at the defect free area increases. As the amount of deformation increases more dislocations are generated and the average distance between the regions of high dislocation density becomes smaller, and hence the positron lifetime increases again. Keh [5] reported in his TEM study of the deformed iron that the ferrite grains contained the complete cell structure after 30 % deformaiton and an average cell dimension became as small as 2 μm. Therefore at higher deformation above 30 %, positrons may be trapped in both free dislocations and cell walls, and the positron lifetime is the combination of the two lifetimes in these two trap sites.

The cell wall is very similar to the grain boundary and the positron lifetime there is shorter than that in the free dislocation. The decrease in positron lifetime above 30 % deformation can be explained thus. The error bar of the lifetime data represents the standard deviation of the lifetime data obtained from the different places of the same specimen. The uncertainty of each lifetime data was normally 2 or 3 ps. Therfore the big error bar reflects the inhomogeneity of the microstructures and substructures of the specimen. Alex et al. [6] reported the similar result in the deformed 4340 steel, but they briefly mentioned that the decrease in positron lifetime suggested a work softening effect.

It may be interesting to compare the results of positron annihilation spectroscopy which is a relatively new nondestructive characterizing technique with that of well established NDE techniques such as ultrasonic methods. Ultrasonic attenuation measurement in the same specimen is now being carried out and the comparison of those results will be reported in near future.

Acknowledgement

The authors are indebted to Dr. W.Y. Choo at Research Institute of Industrial Science & Technology for offering specimens. The assistance of T.S. Park at Korea Standards Research Institute in calibration of the measurement system is gratefully acknowledged.

References

1. J.T. Waber : Positron annihilation-A nondestructive probe for materials science, Conf. Proc. Novel NDE Methods for Materials, B.B. Rath (ed.), Met Soc. of AIME 1982.

2. Y.K. Park : Determination by positron annihilation and electrolytic hydrogen permeation of the dislocation and hydrogen trap densities in annealed and deformed pure iron single crystals, Ph. D, thesis, 1985, Northwestern University, Ill., U.S.A.

3. Yong-Ki Park, James T. Waber, Michael Meshii, C.L. Snead, Jr., and C.G. Park : Dislocation studies on deformed single crystals of high-purity iron using positron annihilation : Determination of dislocation densities, Phys. 34 B (1986) 823-836.

4. H.H. Jorch and J.L. Campbell : On the analytic fitting of full energy peaks from Ge (Li) and Si (Li) photon detectors, Nucl. Instr. Met. 143 (1977) 551-559.

5. S. Mantl and W. Triftshäuser : Defect annealing studies on metals by positron annihilation and electrical resistivity measurements, Phys. Rev. 17 (1978) 1645-1652.

6. A.S. Keh : Dislocation arrangement in alpha iron during deformation and recovery, Direct Observation of Imperfections in Crystals, J.B. Newkirk and J.H. Wernick (eds.), Interscience Publishers, New York (1962).

7. F. Alex, T.D. Hadnagy, K.G. Lynn and J.G. Byrne : Positron annihilation studies of hydrogen embrittlement, Int. Conf. on Effect of Hydrogen of Behavior of Materials, AIME (1975) 642-650.

Ambulante elektrochemische Charakterisierung metallischer und metalloider Festkörperoberflächen

K. Ibendorf[*], A. Hinz[*] und W. Schröter[+]

[*]Sektion Kriminalistik
der Humboldt-Universität zu Berlin, DDR.
[+]Sektion Chemie und Werkstofftechnik
der Technischen Universität Karl-Marx-Stadt, DDR.

Summary

Electrochemical reactions occurring on electron-conducting solid surfaces give information about the corrosion behaviour, the composition and the structure.
A stick-shaped measurement cell touching the test surface on a decided area of contact contains an aqueous solution as electrolyte, the reference electrode and the counter electrode of a 3-electrodes system.
The cell is connected with a portable device for polarization and current measurement.
Current-potential curves of nitrided steel surfaces, diverse gold and German silver alloys are presented.

1. Einführung

Zerstörungsfreie Prüfmethoden zur Beurteilung des Werkstoffzustandes mit den Möglichkeiten einer unkomplizierten Vor-Ort-Untersuchung stehen nur in begrenzter Zahl zur Verfügung /1/. Mit Hilfe der Tüpfelanalyse, Funkenprobe oder einfachen Metallspektroskopie sind zumeist nur beschränkte qualitative Aussagen möglich. Andererseits ist bekannt, daß sich alle maßgeblichen Kenngrößen von Metallen und Legierungen, die letztlich von ihrem chemischen, strukturellen und gefügemäßigen Aufbau abhängen, in bestimmten elektrochemischen Reaktionen widerspiegeln und somit quantitativ erfaßbar sind. Problematisch waren bisher die aufwendigen elektrochemischen Meßsysteme und die Anforderungen an den Zustand des Untersuchungsmaterials, die die Anwendung auf den stationären Betrieb beschränkten. Mit der Entwicklung des vorgestellten Meßgerätes (Bild 1) wurde das Ziel verfolgt, die Bestimmung des Strom-Potential-Verhaltens metallischer und metalloider Festkörper zu einer einfach durchzuführenden Methode der ambulanten Werkstoffuntersuchung zu machen.

2. Geräteaufbau und -funktion

Entgegen bisher üblichen elektrochemischen Meßanordnungen ist das Untersuchungsmaterial nicht als Probe mit vorgegebener Gestalt in eine Zelle einzubringen, sondern eine Stabmeßzelle wird auf die Oberfläche des zu untersuchenden Festkörpers wie ein Prüfstift aufgesetzt, wobei ein der Proben- bzw. Bauteilgestalt angepaßter Meßfleck (Reaktionsfläche) von max. 1 cm^2 erfaßt wird /2//3/. Nach dem Aufsetzen der Stabmeßzelle wird eine Spannung (das sogenannte Bezugspotential) von -1 bis +2 V in einer wählbaren Meßzeit durchgestimmt und ein dazugehöriger Zellenstrom gemessen. Die Stabmeßzelle enthält die Bezugs- und die Gegenelektrode eines 3-Elektrodensystems; das Untersuchungsmaterial ist als Arbeitselektrode geschaltet. An der unteren Stirnseite der rohrförmigen Zelle befindet sich eine bedingt permeable sensitive Wandung, die den elektrolytgefüllten Zellenkörper auslaufsicher verschließt und nur bei Berührung mit der Oberfläche des Prüflings einen vorgegebenen Meßfleck mit Elektrolytflüssigkeit benetzt.

Die Meßtechnik beruht auf folgendem Prinzip (Bild 2):
Zwischen der Bezugselektrode BE der Meßzelle und dem invertierenden Eingang eines Operationsverstärkers befindet sich eine durchstimmbare Spannungsquelle U_X. Innerhalb seines Aussteuerbereiches sorgt der Operationsverstärker für einen Zellenstrom I, der zwischen Bezugselektrode BE und Arbeitselektrode AE einen

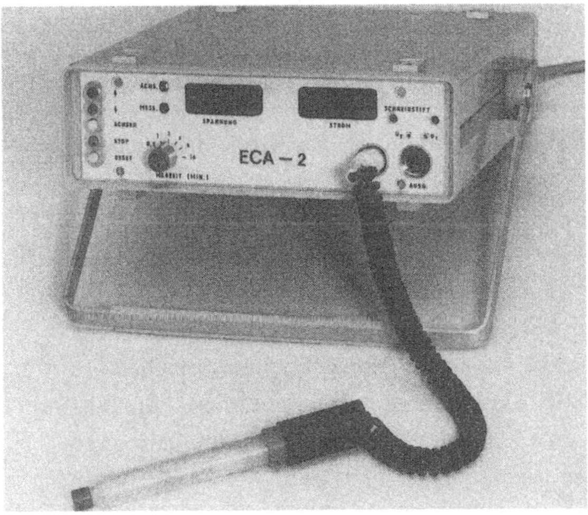

Bild 1:

Grundgerät mit Stabmeßzelle.

Basis device with stick measurement cell.

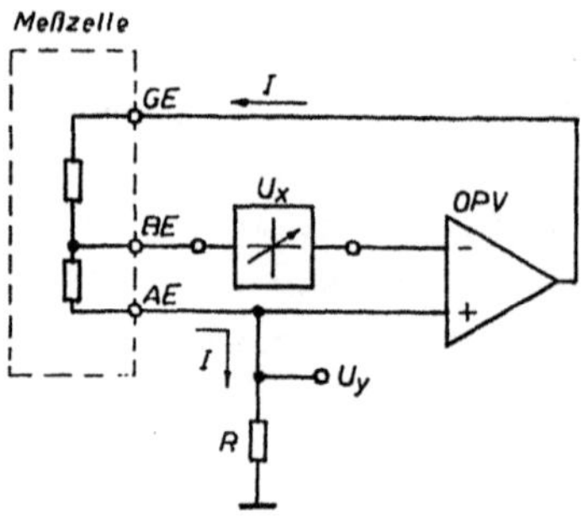

Prinzipschaltbild des EC-Analysators

mit Ersatzschaltung für die Meßzelle

Bild 2:

Prinzipschaltbild.

Wiring diagram on principle.

solchen Spannungsabfall hervorruft, der U_X gerade kompensiert und damit die Differenzeingangsspannung des Operationsverstärkers Null setzt.
Läßt man den Zellenstrom I über einen Widerstand R abfließen, ergibt sich eine stromproportionale Spannung U_Y.
U_X und U_Y können dann als unabhängige und abhängige Variable registriert werden und geben so das Strom-Potential-Verhalten der Arbeitselektrode (= Untersuchungsmaterial) wieder.
Die Meßwerte werden digital angezeigt, mittels XY-Recorder graphisch dargestellt oder im on-line-Betrieb mit einem Mikrorechner verarbeitet.

3. Anwendungsbeispiele

Beispiel 1: Charakterisierung thermo-chemisch erzeugter Randschichten

Problemstellung: Nitridhaltige Schichten auf un- und niedriglegierten Stählen werden industriell zunehmend zur Verbesserung des Bauteilverhaltens in bestimmten korrosiven Beanspruchungssystemen eingesetzt. Es ist üblich, das Korrosionsverhalten der auf unterschiedlichste Weise erzeugten Randschichten durch Korrosionsversuche (z.B. Wechseltauchversuch) festzustellen.
Um durch Anwendung elektrochemischer Untersuchungsmethoden schneller zu Aussagen über die Eigenschaften dieser Schichten zu gelangen, war bisher die Herstellung von Proben und deren zumeist aufwendige Präparation zur Einbringung in spezielle Meßzellen notwendig.

Anwendungsziel: Zerstörungsfreie Bestimmung des elektrochemischen Auflösungsverhaltens unterschiedlich erzeugter (gas-, salzbad-carbonitriert, oxinitriert) und nachbehandelter Nitridschichten. Eigenschaftsnachweis und Qualitätskontrolle an Bauteilen.

Untersuchungsgegenstand: Hydraulikkolbenstangen ∅ 12 mm / zylindrische Proben ∅ 8 mm (St 50 / C 45).

Meßbedingungen: Elektrolyt 0,025 M H_2SO_4/0,05 M Na_2SO_4-Lösung; Potentialdurchlauf mit 12 mV/s; Reaktionsfläche 1 cm^2 (gekrümmt).

Untersuchungsergebnisse:

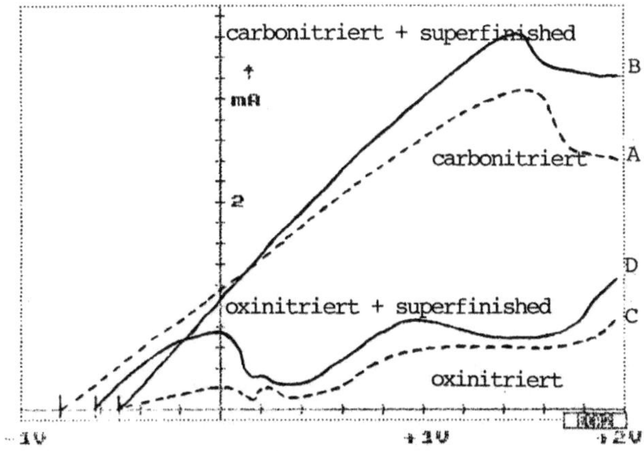

Bild 3:

a.

Strom-Potential-Kurven verschiedener Oberflächenzustände. (Proben wie b.)

Current-potential curves of various surface states. (testpieces as b.)

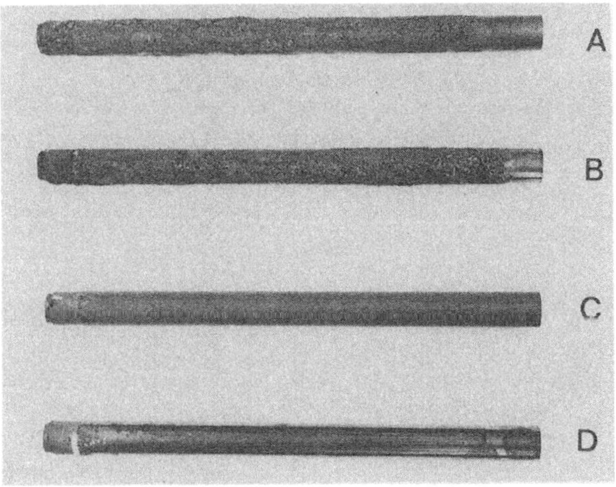

b.

Nitrierte Hydraulikkolbenstangen, korrosiv beansprucht. (Wechseltauchversuch)

Nitrided hydraulic rods, exposed to corrosion. (intermittent immersion test)

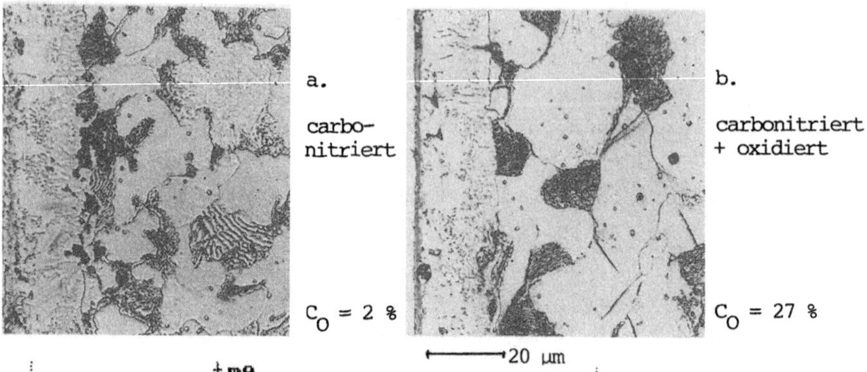

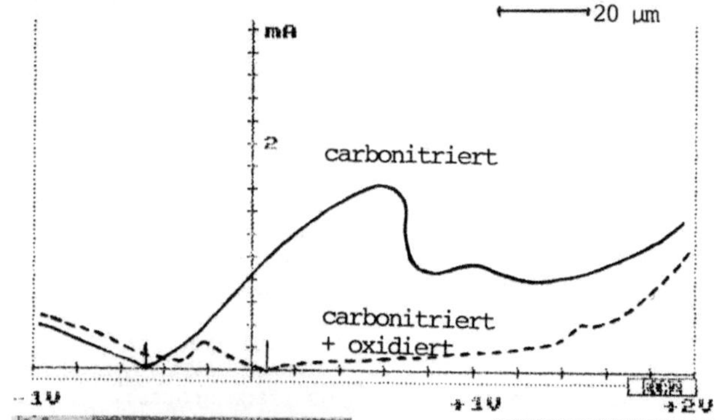

c.

Bild 4:

Mikrogefüge (Querschliff) und Strom-Potential-Kurven nitrierter Oberflächen.

Microstructure (transverse section) and current-potential curves of nitrided surfaces.

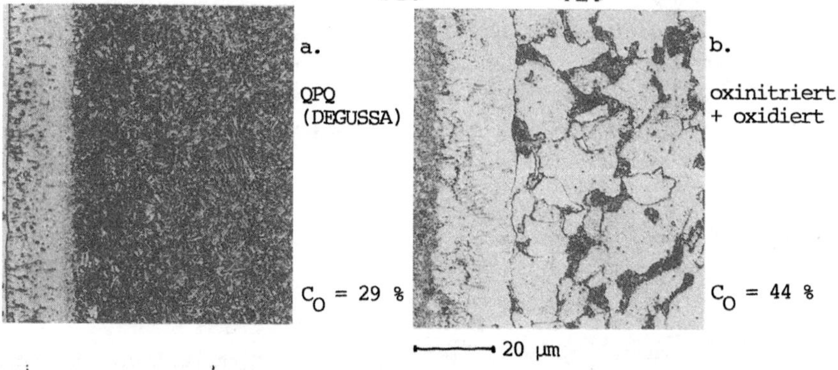

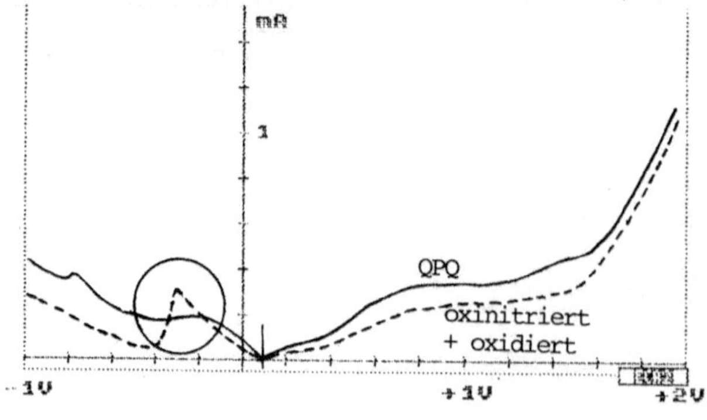

c.

Bild 5:

Mikrogefüge (Querschliff) und Strom-Potential-Kurven nitrierter Oberflächen.

Microstructure (transverse section) and current-potential curves of nitrided surfaces.

Mit der an die Testoberfläche angepaßten Reaktionsfläche der Stabmeßzelle sind spezifische Strom-Potential-Verläufe der unterschiedlichen (chemisch; strukturell; geometrisch) Oberflächenzustände nachweisbar.
Der Vergleich des im 1000-Stunden-Wechseltauchversuch erhaltenen Resultates mit dem in wenigen Minuten am Bauteil mittels Anwendung der vorgestellten Untersuchungsmethode bestimmten Strom-Potential-Verhalten (Bilder 3 a. und b.) veranschaulicht eine weitgehende Übereinstimmung in den jeweiligen Aussagen über das Korrosionsverhalten der untersuchten Schichten.
Mit zunehmendem Sauerstoffgehalt (C_O = 2...44 %) der Randschicht ergibt sich eine spürbare Veränderung: flacher werdender Kurvenverlauf, passives Verhalten. Oberhalb von etwa 27 % Sauerstoff (Phasengrenze Fe_2O_3 / Fe_3O_4 im Zustandssystem Fe - O) tritt eine charakteristische Unstetigkeit im katodischen Teilast der Strom-Potential-Kurve auf, die auf das Vorhandensein der sauerstoffreichen Oxidphase zurückzuführen ist (Bilder 4 und 5).

Beispiel 2:
Charakterisierung von Goldlegierungen

Problemstellung: Gegenstände aus goldhaltigen Legierungen müssen bekanntlich einen festgelegten Mindestgehalt an Gold aufweisen, mit dem sie zu kennzeichnen sind. Ein schneller Nachweis darüber, ob dieser Mindestgehalt tatsächlich vorhanden ist, gelingt mit der Tüpfelanalyse. Über die prozentuale Beteiligung anderer Legierungskomponenten gibt der Schnelltest zunächst keine Auskunft. Auch vom Hersteller sind dazu im allgemeinen keine Angaben erhältlich. Für technische Anwendungen (Kontaktwerkstoffe, Dentallegierungen) ist jedoch der Einfluß aller Elemente (insbesondere unedler Komponenten) von Interesse.

Anwendungsziel: Zerstörungsfreie Differenzierung von Legierungen mit gleichem Goldgehalt. Sortentrennung; Herkunftsbestimmung; Bestimmung von Gebrauchseigenschaften.

Untersuchungsgegenstand: 7 verschiedene Goldlegierungen, Bleche 0,5 x 0,5 cm^2; Au 33,3 %, Ag 0...54 %, Cu 13...43 %, Zn 0...19 %.

Meßbedingungen: Elektrolyt 0,05 M H_2SO_4/0,05 M Na_2SO_4-Lösung;

Potentialdurchlauf mit 24 mV/s; Reaktionsfläche 0,25 cm^2.

Untersuchungsergebnisse: Die untersuchten 333-Goldlegierungen besitzen ein unterschiedliches anodisches Auflösungsverhalten. Legierung 1 enthält kein Silber, dafür aber 12 % Zink. Der Steilanstieg setzt in einem vergleichsweise niedrigen Potentialbereich ein. Legierung 4 enthält dagegen 54 % Silber. Der Verlauf der Strom-Potential-Kurve ist durch eine ausgeprägte Schulter gekennzeichnet. Eine Unstetigkeit im gleichen Potentialbereich (+ 0,8 V) ist auch bei Legierung 3 (33 % Ag) zu beobachten. Legierung 2 ist hoch-Cu-haltig. Die dazugehörige Kurve zeigt ein Maximum bei + 0,3 V, das sich auch bei Kurve 3 andeutet. Legierung 5 besitzt den höchsten Zinkgehalt und ein dementsprechendes Auflösungsverhalten. Die Legierungen 6 und 7 haben annähernd die gleiche Zusammensetzung und somit auch einen sehr ähnlichen Kurvenverlauf.

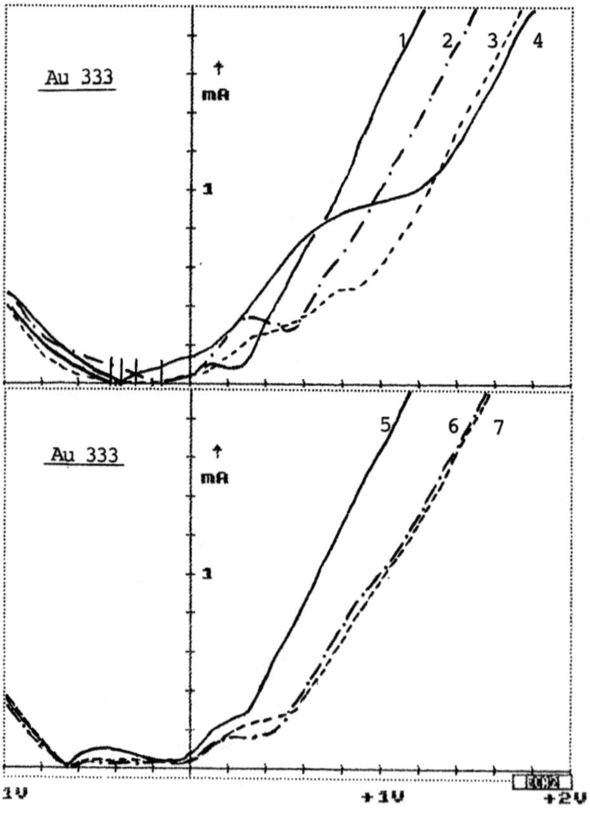

Bilder 6 und 7:

Strom-Potential-Kurven einiger Goldlegierungen.

Current-potential curves of several gold alloys.

Beispiel 3:
Charakterisierung von Neusilber (German silver)

Problemstellung: Neusilber, eine Mehrstofflegierung mit den 3 Hauptkomponenten Kupfer, Nickel und Zink, wird u.a. für die Herstellung gültiger Zahlungsmittel verwendet. Eine Münzlegierung muß besonderen Einsatz- und Beanspruchungsbedingungen entsprechen.

Anwendungsziel: Zerstörungsfreie Differenzierung verschiedener Cu-Ni-Zn-Legierungen; Feststellung des Verhaltens in bestimmten Medien; Echtheitsnachweis.

Untersuchungsgegenstand: Ronden ø 31,2 mm (Münzrohlinge), Cu-Gehalt 64,4...68,2 %.

Meßbedingungen: Elektrolyt 5 M NaOH; Potentialdurchlauf mit 12 mV/s; Reaktionsfläche 0,8 cm^2.

Untersuchungsergebnisse: Die untersuchten Legierungen verhalten sich im verwendeten Prüfmedium weitgehend passiv. Trotz der relativ geringen Unterschiede in der Legierungszusammensetzung zeichnet sich der zunehmende Cu-Gehalt im Verlauf der Strom-Potential-Kurven (Ausbildung eines Peaks bei +1,1 V) ab.

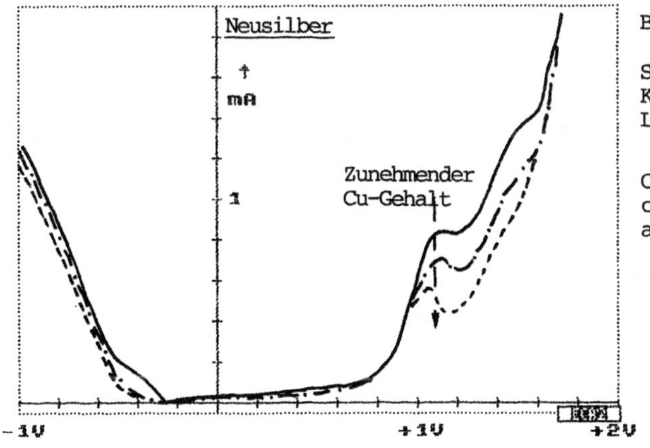

Bild 8:

Strom-Potential-Kurven von Cu-Ni-Zn-Legierungen.

Current-potential curves of Cu-Ni-Zn-alloys.

4. Literatur

/1/ Christian, H. u.a.: Möglichkeiten der ambulanten Werkstoffprüfung. Der Maschinenschaden 59(1986)5 S. 185-193.

/2/ DDR-Wirtschaftspatent 263 829 A1.

/3/ Franzke, B.; Ibendorf, K.; Hinz, A.: Tragbares Meßgerät zur Charakterisierung von Metallen und Legierungen. Forum der Kriminalistik 24(1986)1 S. 58-60.

Noncontact Ultrasonic Sensing of Weld Pools for Automated Welding

John A. Johnson and Nancy M. Carlson

Idaho National Engineering Laboratory, EG&G Idaho, Inc.
P.O. Box 1625, Idaho Falls, ID 83415-2209, USA

Abstract
Contacting ultrasonic techniques can determine the geometry of the molten/solid interface [1,2] and detect conditions which could lead to defect formation [3]. However, the couplant required is a potential source of contamination in the weld and may be difficult to use under industrial conditions. Stress waves can also be generated and detected using noncontacting techniques [4]. Ultrasonic stress waves are generated by focusing a beam from a pulsed laser on a stationary weld pool formed on the surface of a steel plate or on a weld pool in a fillet weld. The transmitted ultrasound is detected by an electromagnetic-acoustic transducer (EMAT). The signal received by the EMAT provides information about the properties of the pool, including the geometry and defect generating conditions. This information can be input to an intelligent controller. The ultrasonic system described in this paper is being developed as a portion of the sensing and control system for a completely automated welder [5].

Background

This paper discusses some recent work in the welding program being conducted at the Idaho National Engineering Laboratory (INEL) which involves modeling the gas metal arc welding (GMAW) process and sensing physical properties of the weld using optical and ultrasonic sensors [5]. The work investigates the potential of using a noncontacting system for ultrasonic sensing of GMAW.

In the previous ultrasonic work a gel-coupled piezoelectric transducer generating transverse waves [1,2] or longitudinal waves [3] and operating in the pulse-echo mode is used to interrogate the molten/solid interface of a root weld in a single bevel V-groove. Using the timing and amplitude of these reflected signals, the geometry of the interface and quality of the welding process can be determined. In the current work the timing and amplitude of the ultrasound generated at the surface of the weld pool by a laser and transmitted to the EMAT contain similar information. The noncontacting system has two advantages over the piezoelectric method. First, the laser can excite sound easily in a sample at any location.

Second, neither the transmitter (laser light) nor receiver (EMAT) requires any mechanical coupling to the part.

Laser Generation of Ultrasound

Many investigators [e.g., 6,7] have demonstrated that pulsed laser light is capable of generating ultrasound in liquids and solids. For this research, a pulsed Nd-YAG laser with a pulse length of about 10 ns and pulse energy of 10 mJ is used to generate sound in a molten weld pool during the welding process. The laser beam is focused on a predetermined position of the weld pool. At this energy level a transient stress is generated by two mechanisms, thermoelastic expansion and ablation. The dominant cause of the transient stress that leads to stress wave propagation in the part is the ablation of small amounts of the surface.

An early study of the generation of ultrasound in a fixed weld pool is described in Ref. 8. A piezoelectric transducer is used to receive the generated sound. Prior to welding, two distinct signals are observed at the times calculated using the longitudinal and shear sound speeds for the steel at room temperature. As the weld pool forms, each of the original signals separates into two signals. The first signals of each pair are the original longitudinal and mode-converted shear waves. Each arrives later due to the reduced sound speed in the weld pool. Modeling efforts using ray-tracing techniques and finite element techniques are under way in an attempt to understand the observed signals in detail. The effects of interface waves at the molten/solid boundary are also being considered. However, the increased time required for the wave to pass through the molten pool, indicated by both of the initial signals, provides information about the depth of penetration of the pool.

EMAT Receivers

The electromagnetic acoustic transducer (EMAT) can function as both a transmitter and receiver of ultrasound, but for this application the EMAT operates only as a receiver. The theory of the operation of EMAT is discussed in Reference 9. Two different permanent-magnet EMAT designs are being tested. The design, sensitivity, and directivity of the two EMATs to longitudinal and transverse waves vary. The first is a meanderline coil with dimensions appropriate to receiving 2.25 MHz, 45° longitudinal waves or 22° shear waves in steel. The meanderline coil EMAT is insensitive to transverse waves at 45°. The second EMAT is designed to receive shear waves at normal incidence but can also receive angle-beam shear waves with a cosine amplitude factor. Both EMATs are insensitive to the electromagnetic noise of the welding process, which is in a frequency range well below that of the EMAT tank circuit.

Welding Experiments

To evaluate the proposed laser/EMAT noncontact system as a sensor in a welding process, the EMAT is used in the position of the piezoelectric transducer and wedge of the earlier work [8]. The initial experiments show the capability of both EMATs to operate next to the arc without any noticeable increase in noise. In Figure 1, the A scans received before and during welding from one of the EMATs show no increase in noise during welding. (The amplifier gain is 14 dB higher for the second A scan during welding.)

The first signal before welding is mainly due to an initial longitudinal wave which mode-converts to a transverse wave (LT, where L is the longitudinal leg and T is the transverse leg) upon reflection off the bottom of the part. However, this transducer is also sensitive to a direct transverse wave from the source (T) and the Rayleigh wave (R) which also arrive about the same time in this geometry. The second signal (starting about 21.0 μs) is a combination of a head wave, a transverse wave reflecting off the bottom at 45° (TT), and a combination of three longitudinal legs with a final transverse leg (LLLT) to the transducer. The expected arrival times for each ray are marked on Figure 2.

During welding, any ray that begins with a transverse leg is not expected to be observed, so the first signal is probably only the mode-converted transverse ray. The low frequency signal at about 17.6 μs may correspond to the second signal following the mode-converted wave seen in the piezoelectric data. The signal at 22.56 μs is the LLLT wave.

The A scans acquired by the EMATs from the stationary spot welds indicate little variation in the depth of the spot weld. Destructive evaluation of these spot welds reveals that the pool depths vary little, only from 1.0 to 1.2 mm. To confirm that the laser/EMAT system is sensitive to depth changes in the pool, a steel sample is prepared with a series of five holes varying

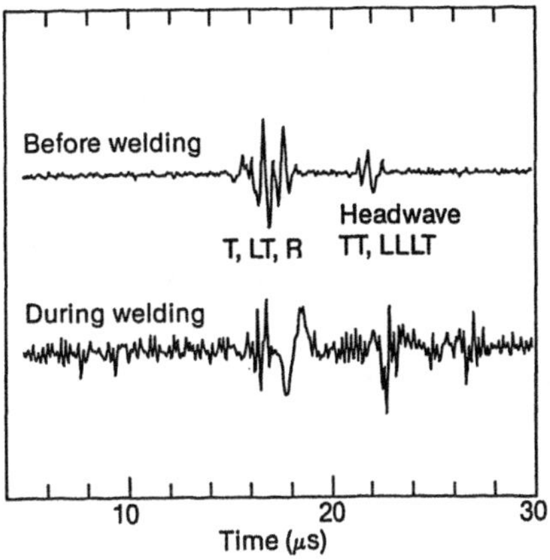

Fig.1. EMAT signals received during welding.

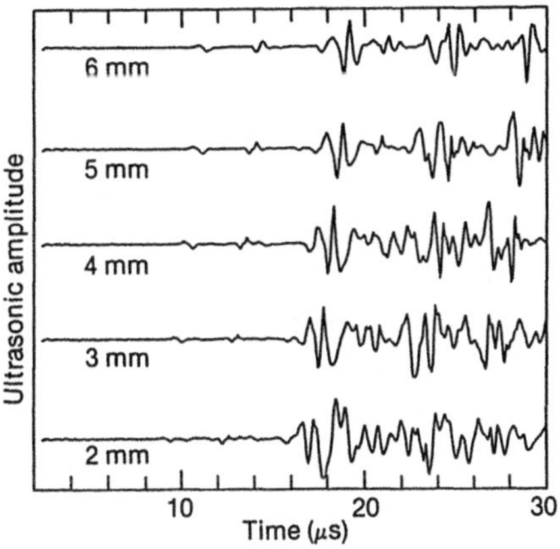

Fig.2. A scans of varying pool depths of glycerine filled holes.

in depth from 2 to 6 mm milled out on the surface using a 6.35 mm ball-end mill bit. The holes are filled with glycerine to simulate the molten/solid interface in a spot weld. The glycerin is dyed black so that the laser energy is absorbed at the surface of the liquid as it is in the molten steel pool. A scans acquired from the glycerine-filled holes have similar patterns to those obtained during welding, indicating that reflections are due to the presence of the liquid/solid interface rather than other variables such as temperature gradients or convection in the pool due to the welding process. Figure 2 shows the shift in signal arrival times as the depth of the hole and the amount of liquid travel decrease. These data correlate with data acquired using a piezoelectric receiver described in Reference 8. The first large signal is the LT ray which arrives earlier in time as the length of the ray path in liquid decreases.

A fillet-weld configuration is selected for the next phase of the work. The EMAT is positioned 25.4 mm from the bottom edge on a 25.4 mm thick carbon steel sample. Each EMAT is equipped with metal wheels that allow it to move down the weld sample aligned with the electrode. The focused laser beam is delivered to the molten fillet weld using mirrors and a 500 mm lens mounted on the side beam frame and translation fixture. Because of the method of laser beam delivery, the area of the weld pool that could be excited is restricted.

Analysis Methods

Computer codes are used to confirm the sound path of the data acquired from the stationary weld spots and the molten fillet. The ultrasound field produced by the laser can be calculated using finite element methods [10] and ray tracing methods. Both these methods are capable of accounting for the transmission through the complex geometry of the molten/solid interface. The finite element method provides complete information about the stress wave within the limits of the approximation used to model the laser source and of the size of the region that can economically be analyzed. However, this large amount of information is sometimes difficult to interpret and so a ray tracing technique is used to identify various wave modes that are produced.

Figure 3 shows the waves calculated by the finite element program DYNA2D [11] 2.3 μs after the initiation of the sound at the center of the pool. The left vertical axis is the centerline of the axially symmetric problem. The lines overlaying the finite element displacement contours are the positions

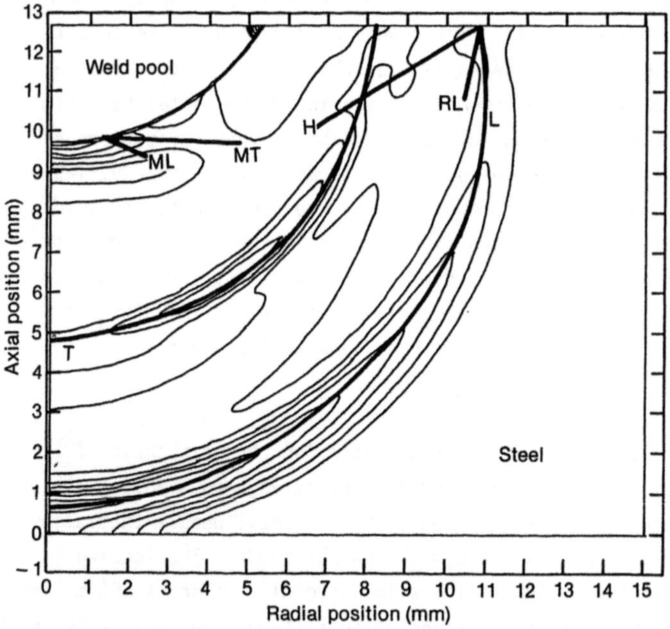

Fig.3. Contours of displacement calculated using the finite element method. Wave fronts calculated by ray tracing include the longitudinal (L), transverse (T), head wave (H), reflected longitudinal (RL), and longitudinal (ML) and transverse (MT) waves after multiple reflections in the pool.

of the wavefronts of the various modes as calculated by the ray tracing program. The longitudinal (L) and transverse (T) waves can be clearly seen in the figure along with waves due to multiple reflections (ML and MT) in the interior of the weld pool, the head wave (H) and reflected longitudinal (RL).

The finite element program has not been applied to the fillet weld but the ray tracing program has. Figure 4 shows the fillet geometry with the major ray traced from the laser-generation source to the position of the EMAT receiver. The initial ray in the pool is longitudinal. This ray passes through the interface and mode converts when it reflects off the bottom surface. The resulting transverse ray is then received by the EMAT, which is positioned on the top surface of the part 25.4 mm from the edge of the weld preparation.

Results

The noncontacting system is able to detect changes in the fill level and the slope of the fillet weld. When the fillet weld geometry is acceptable, sound generated by the laser is received by the EMAT. Figure 5 shows a sequence of A scans taken at 8 mm intervals along the weld. The signal at about 12 μs is present in the first three A scans. Starting with the fourth A scan, a severe drop of fill level results in the disappearance of this signal. This occurs because the change in the fillet weld geometry alters the sound path to the EMAT to such an extent that no sound is received. As an acceptable fill level is again achieved, the EMAT receives a signal from the weld as seen in the last three A scans of Figure 5.

Conclusion

A noncontact ultrasonic sensing system using laser sound generation and EMAT reception can be used to sense the welding process. The results of the feasibility work show that the laser/EMAT system is able to detect pool depth, fill level, and fill slope. The sensing system can acquire data from the molten weld when the EMAT is in a stationary position and when the EMAT is moving aligned with the electrode. The signals acquired by the laser/EMAT system contain only information about the weld geometry and weld condition. An expert system could use these signals to determine the quality of the welding process and provide an input signal to alter welding parameters if an unacceptable welding condition is detected.

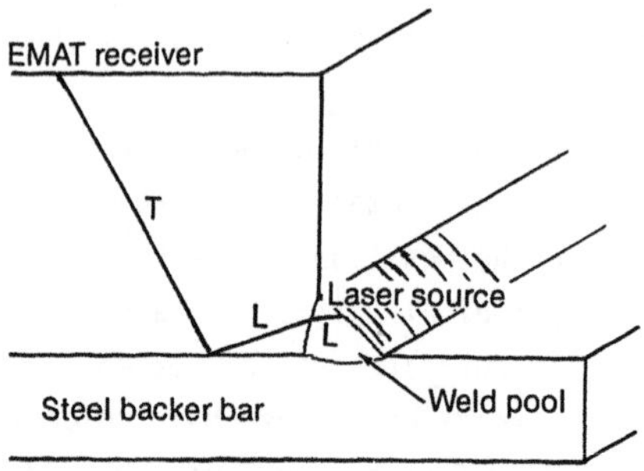

Fig.4. Ray trace of laser generated waves in a fillet weld pool.

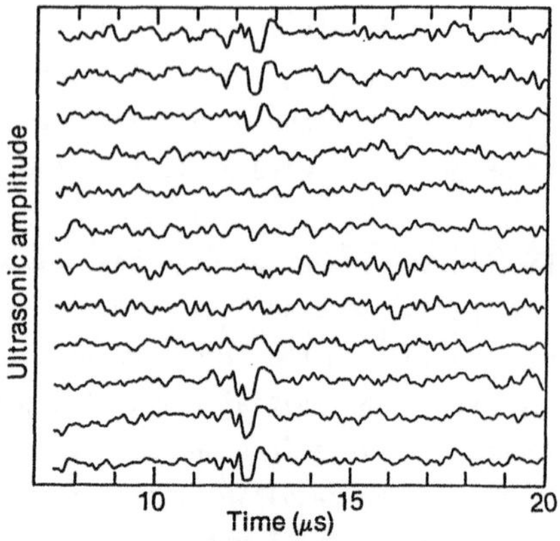

Fig.5. A scans acquired from molten fillet weld at 8 mm intervals. The signal due to the ray traced in Figure 4 disappears when the fill level is too low.

For the system to be fieldable, several modifications to the experimental system must be considered. The laser beam must be delivered to the weld via a fiber optic cable to reduce personnel hazards in the work area and to increase the area of excitation on the weld for inspection of the total weld. The directivity and sensitivity of the EMAT must be evaluated to determine the optimum parameters for each welding application. The use of computer field codes to simulate the welding configuration to be inspected allows the user to design the procedure to assure that the area of critical interest is being inspected.

Acknowledgments

This work is supported by the U.S. Department of Energy, Office of Energy Research, Office of Basic Energy Sciences under DOE Contract No. DE-AC07-76ID01570. Appreciation is expressed to U. S. Wallace, C. L. Shull, and G. L. Fletcher for technical assistance.

References

1. Carlson, N. M. and Johnson, J. A.: Ultrasonic detection of weld pool geometry. Review of Progress in Quantitative Nondestructive Evaluation, 6B. Thompson, D. O. and Chimenti, D. E. (eds.). New York: Plenum Publishing Corp. 1987, 1723-1730.

2. Carlson, N. M. and Johnson, J. A.: Ultrasonic sensing of weld pool penetration, to be published in Welding Journal.

3. Carlson, N. M.; Kunerth, D. C.; Johnson J. A.: Process Control of GMAW by detection of discontinuities in the molten weld pool. To be published in Review of Progress in Quantitative Nondestructive Evaluation 8, Thompson, D. O. and Chimenti, D. E. (eds.). New York: Plenum Publishing Corp. 1989.

4. Hutchins, D. A. and Wilkins, D. E.: Elastic waveforms using laser generation and electromagnetic acoustic transducer detection. J. Appl. Phys. 58 (1985), 2469-2477.

5. Johnson, J. A. et al.: Automated welding process sensing and control. Nondestructive Characterization of Materials II, Bussiere J. F.; Monchalin J. P.; Rudd C. O.; Green R. E., (eds.). New York: Plenum Press, 409.

6. Muir, T. G.; Culbertson, C. R.; Clynch, J. R.: Experiments on thermoacoustic arrays with laser excitation. J. Acoustical Society of America 59 (1976) 735-743.

7. Dewhurst, R. J.; Hutchins, D. A.; Palmer, S. B.: Quantitative measurements of laser-generated acoustic waveforms. J. Appl. Phys. 53 (1982) 4064-4071.

8. Carlson, N. M. and Johnson, J. A.: Laser sound generation in a weld pool. Review of Progress in Quantitative Nondestructive Evaluation 7. Thompson, D. O. and Chimenti, D. E. (eds.). New York: Plenum Publishing Corp. 1987.

9. Maxfield, B. W. and Fortunko, C. M.: The design and use of electromagnetic acoustic wave transducers (EMATs). Materials Evaluation 41, (1983) 1399-1408.

10. Johnson, J. A.: Modeling of laser sound generation in a weld pool. Review of Progress in Quantitative Nondestructive Evaluation 7. Thompson, D. O. and Chimenti, D. E. (eds.). New York: Plenum Publishing Corp. 1987.

11. Halquist, J. O.: User's manual for DYNA2D - An explicit two-dimensional hydrodynamic finite element code with interactive rezoning. Lawrence Livermore Laboratory Report UCID-18756 (Rev. 2) (1984).

Accurate Determination of the Focal Spot Size of a Microfocus X-Ray Tube

J. BAUMANN, P. KLOFAC and G. FRITSCH*

ZT ZFA PTE 22, Siemens AG, D-8000 München 83

*BAUV/I1 Physik, Universität der Bundeswehr München, D-8014 Neubiberg

Abstract:

The subject of this report is the accurate determination of the focal size of a microfocus X-ray tube. Some information on the angular intensity distribution and on the energy spectrum is also given. The potential of an application of energy-dispersive radiography in nondestructive testing with high spatial resolution is shown in two examples of electronic devices.

Zusammenfassung:

Es wird über die genaue Bestimmung der Brennfleckgröße einer Mikrofokus-Röntgenröhre berichtet. Zusätzliche Informationen betreffen die Intensitätsverteilung als Funktion des Abstrahl-winkels und das Energiespektrum. Schließlich zeigen zwei Beispiele die Möglichkeiten auf, die durch energieauflösende Radiographie für die zerstörungsfreie Werkstoffprüfung hoher Ortsauflösung gegeben sind.

I. Introduction

Usually the X-ray shadow imaging method is used to get information about the interior of optically nontransparent samples. Since the focal spot size of conventional X-ray tubes is in the order of some tenth of millimeter, the recording unit has to be placed immediately behind the object, otherwise the spatial resolution will be reduced evidently by the penumbral effect.

This situation is changed dramatically using a microfocus tube. This device offers a focal spot size in the range 1-20 μm effective diameter. Placing the specimen near the pointlike X-ray source, an enlarged image will be taken in a plane at some distance behind the specimen. With the aid of an X-ray image amplifier instead of a film, real-time micro-radioscopy is realized with high lateral resolution. The resolving power is mainly limited by the focal spot size; diffraction effects must be taken into account only for resolutions beyond about 1 μm [1].

Therefore, one of the most discussed question regarding

microfocus X-ray technique concerns the focal spot size and methods to determine this parameter [2-5]. The results of measured focal dimension published so far are obviously not consistent. They seem to depend on the method applied. The sensivity of the various methods for different energy ranges of the X-rays is an open question, too. So, for complete information the focal spot diameter must be determined as a function of energy.

Furthermore, an energy dispersive detector system allows a chromatic analysis and offers an improved image contrast exploiting the absorption edges of each element contained in the specimen. Depending on the thickness of the object, the heavier elements are especially well suited for detection.

II. Experimental Details

The arrangement described above has been constructed in our lab. A microfocus X-ray tube (Feinfocus FMF-160) is used for this investigation operating in most cases at 80 kV and 0.1 mA. The object, held by a manipulator, is positioned about 10 mm from the focal spot. The Germanium-detector (Canberra 71255P) is mounted on a scanning unit (Huber 5102/9002). The acceptance area for radiation is defined by an adjustable quadratic slit made out of tungsten, which is fixed directly in front of the detector.

The counts from the Ge-detector were stored in a multi-channel-analyser (Canberra S35 Plus/CI 3502) and normalized by the counts from a monitor counter (Reuter Stokes) . The spectra could be recorded on a plotter and on a floppy-disc for further evaluation. The full arrangement was controlled by a personal computer.

III. Determination of the Focal Spot Size

Since the target used in the X-ray tube is a tungsten cylinder with a horizontal axis, the vertical spot size is always smaller than the horizontal one. We use the test mask, shown in Fig. 1, for this purpose [2]. It contains gold-patterns (thickness s_{Au} about 0.8 µm) evaporated onto a polymer foil (polyimid, thickness about 3 µm). The smallest structures (both bars and spacings) are around 2 µm in width. We have performed scans across edges and across periodic patterns of four or five bars.

Let us discuss the edge-scans first. Some results are reproduced in Fig. 2. For simplicity, we assumed a rectangular source profile of width d, when analysing the data. Purely geometrical considerations yield the following expression for the quantity d:

$$d = (m\ V)^{-1}\ [1 - \exp\ (-\mu_{Au}\ s_{Au})], \qquad (1)$$

where m is defined by:

$$m = I_o^{-1} \cdot dI(x)/dx\,\big|_{x=x_o}\ .$$

Here, V denotes the magnification of the object and μ_{Au} the average absorption coefficient of gold in the energy range considered. If the intensity coming through the polymer foil is called I_o, then the one penetrating the foil plus the gold layer is $I_o\ \exp\ (-\ \mu_{Au}\ s_{Au})$. Thus, the exponential can be derived directly from the data. Finally, dI(x)/dx may be calculated from the slope of the experimental intensity versus distance I(x) in the center x_o of the slope (see Fig. 2). The results obtained for various energies are summarized in Table I.

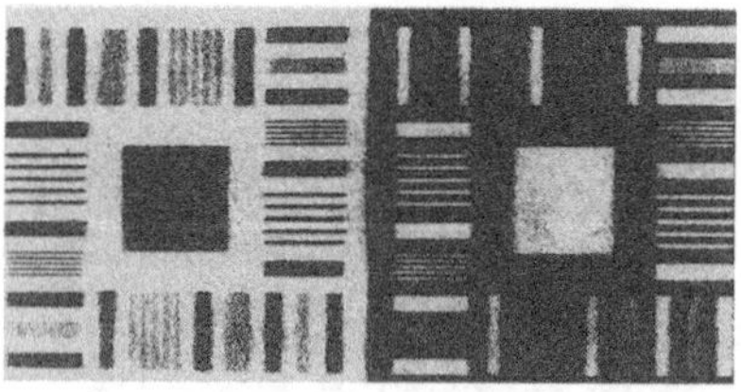

Fig.1. Test mask, used for evaluation of the focal size as seen by an X-ray picture [2].

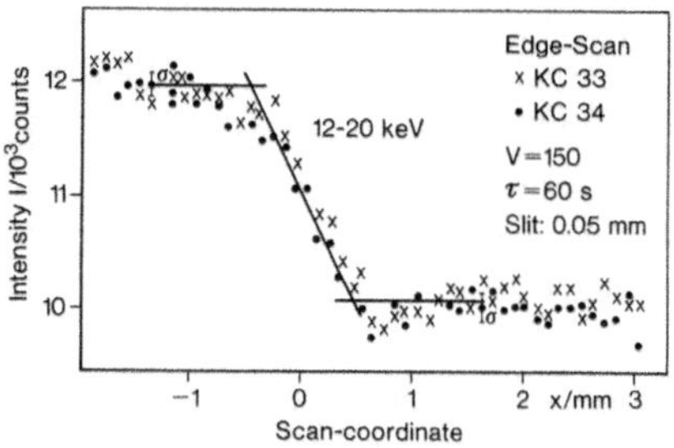

Fig.2. I(x) from scans across a gold edge. τ : counting time, σ : standard deviation.

We simulated the intensity distribution expected from a ruling- scan by assuming rectangular bars and a Gaussian source profile of width D at half height. The modulation M of the intensity distribution I(x) may then be expressed as:

$$M = M_1\ [1 - \exp\ (-\mu_{Au}\ s_{Au})]/[1 + (M_1/M_2)\exp\ (-\mu_{Au}\ s_{Au})]. \qquad (2)$$

Here, an absorber bar is assumed to be in the symmetric center of the pattern. The quantities M_1 and M_2 are the modulations for the intensity produced by a pattern with an absorber bar of width $(W - \varepsilon)$ and a spacing of the same width in the center, respectively, calculated for totally absorbing structures as a function of D. 2W describes the period of the ruling and ε its asymmetry. If $\varepsilon = 0$, then $M_1 = M_2$. A modulation M is defined as follows:

$$M = (I_{max} - I_{min})/(I_{max} + I_{min}). \qquad (3)$$

where the suffix "max" or "min" refers to maximal or minimal values of I(x).

Since M as well as $\exp(-\mu_{Au}\, s_{Au})$ can be derived directly from the data, the quantities M_1 and M_2 may be determined iteratively from Eq.(2) with the help of the diagrams $M_1(D)$ and $M_2(D)$ (Fig. 3). From this the width at half height of the Gaussian source distribution is deduced after two iterative steps starting with $M_1/M_2 = 1$ in the denominator of Eq.(2) (Table I, Fig.4).

Table I. Summary of Results for the Focal Spot Size

Energy Range keV	Edge-Scan with Rectangular Source Profile vertical d/µm	Ruling-Scan with Gaussian Source Profile vertical D/µm	Absorption factor $\exp(-\mu_{Au}\, s_{Au})$
7.5 - 12	5.9; 5.3; 5.2±0.2 4.8±0.2; 3.7±0.5	3.6; 3.6±0.2 3.8	0.837; 0.812 0.814; 0.827 0.822;
12 - 20	6.6±0.4; 5.6±0.4	3.9; 4.2; 4.0±0.3	0.839; 0.839
20 - 30	7.0±0.4; 7.2±0.9 6.8±0.9	3.5±0.6 3.4±0.6	0.937; 0.965 0.939
30 - 40	5.2±1.5	-	0.975
integrated photographically	-	2.5 ± 0.5	-
	horizontal d/µm	horizontal D/µm	
7.5 - 12	12.4; 12.0, 12.8±0.4; 14.7±0.4 12.5±0.3	10.2; 10.3	0.867; 0.842 0.852; 0.855 0.843
integrated photographically	-	10 ± 2	-

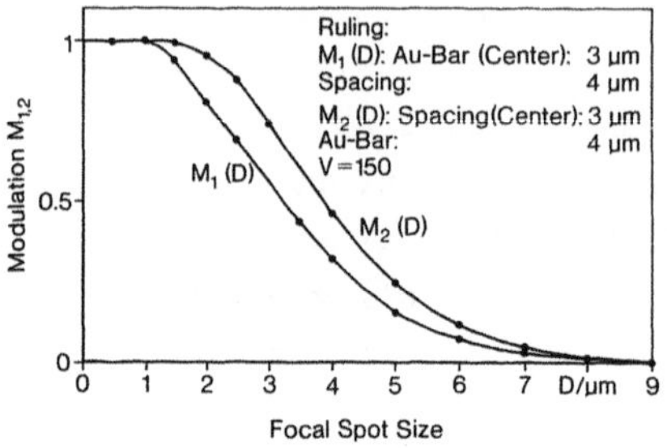

Fig.3. Example for the calculated modulations $M_1(D)$ and $M_2(D)$.

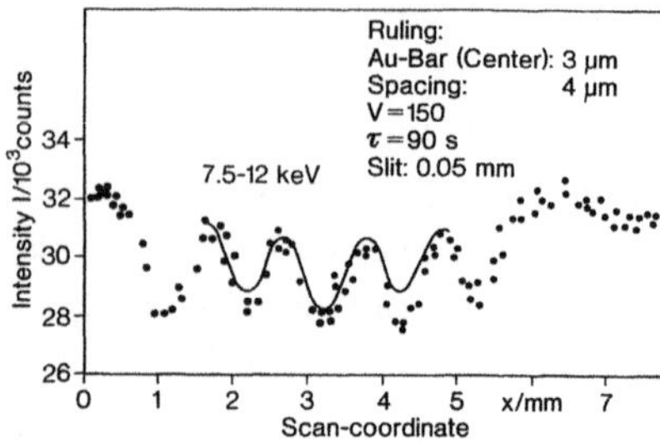

Fig.4. I(x) from a scan across a ruling of five gold bars. ●: data points, —— : simulation, τ : counting time.

It is worth mentioning, that many techniques have been described in the literature for the analog problem of measuring laser beam spot sizes. Thus, the knife-edge and a special ruling technique have been discussed in some publications [6].

We may state by inspecting Table I, that the vertical and the horizontal focus spot sizes are between 3 to 6 µm as well as between 10-13 µm, respectively. There is no significant energy dependence. These data are in agreement with results reported earlier from photographs of the test mask [2]. However, whereas the time scale of those data is in the range of one minute, the measurements reported now are averages over a time scale of about 100 minutes. Hence, the variation of the spot size over this period has to be in the order of the short-time spot size itself or smaller.

IV. Spectra and Intensity Distribution

Fig. 5 shows some spectra obtained with the microfocus tube. The spectrum shown in Fig. 5a displays the characteristic tungsten lines at about 8, 10 and 11 keV (L-lines) as well as at about 60 and 70 keV (K-lines). The spectrum decreases roughly linearly for high energies. The effect of an additional gold foil is shown in Fig. 5b. The Au K-lines and the K-absorption edge appear. Finally, the angular intensity distribution is reproduced in Fig. 6. At an angle of about 0.75° from the center a decrease in intensity of 5.5% is to be measured.

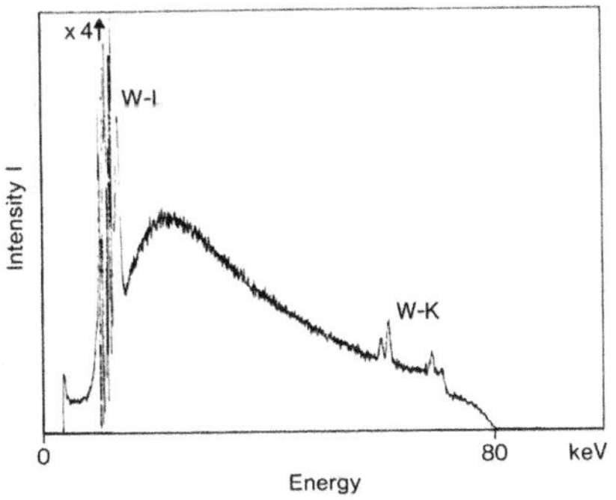

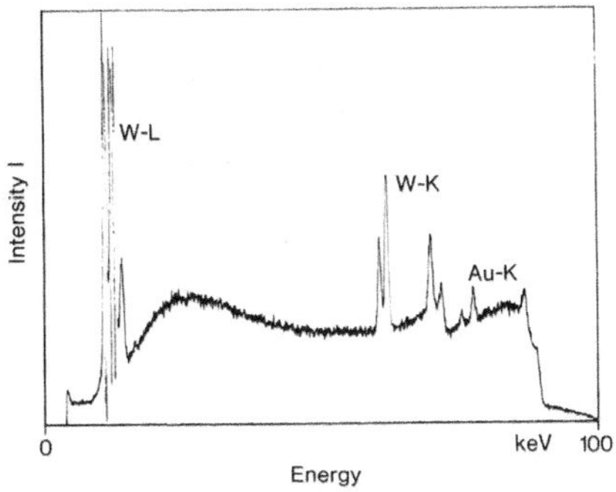

Fig.5. Energy spectrum emitted by a microfocus X-ray tube, a) excited with 80 kV and air absorption, distance 1.5 m), b) excited with 100 kV, in this case only, and gold foil in addition. (I(x) not to scale).

V. Some Applications

In order to give some indication of the potential of this method we present the results of line scans across electronic devices. Data obtained from a car ignition chip which is soldered to a ceramic and Al-backing material are shown in Fig. 7. The solder layer possesses some defects one of which is clearly visible in the figure. The contrast varies with the energy range and is largest just above the tin absorption edge (29.2 keV).

Figs. 8 and 9 give the results from a highly integrated computer chip. In the first one a scan across one of the legs is shown. An enormous contrast is achieved in the energy range from 7.7 to 12.7 keV. The intensity variation across the leg indicates some additional matter at one side. The second scan is across a solder joint. Here, the highest contrast occurs in the range 17.6 to 20 keV. More details, however, can be seen in the range 29.6 to 31.5 keV, which is just above the absorption edge of tin. Fig. 10 shows a spectrum taken directly at the solder joint.

VI. Conclusion

From the presented data, it is evident that both the edge-scan and ruling-scan technique yield consistent results for the focal spot size. With the edge-scan technique, it is necessary to know the accurate distance of the edge to the focal spot in order to determine the value of magnification. The ruling, on the other

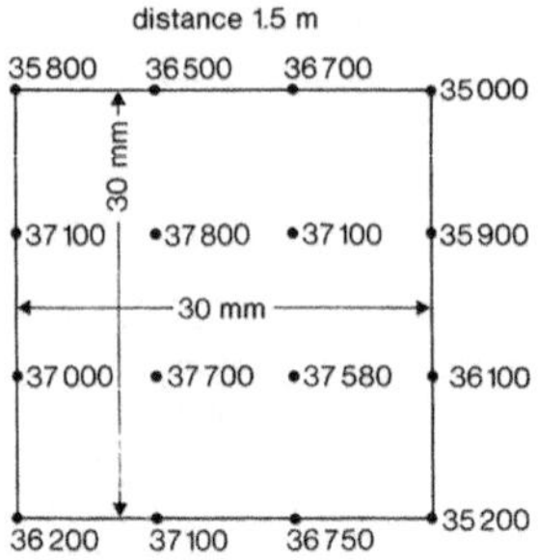

Fig. 6.

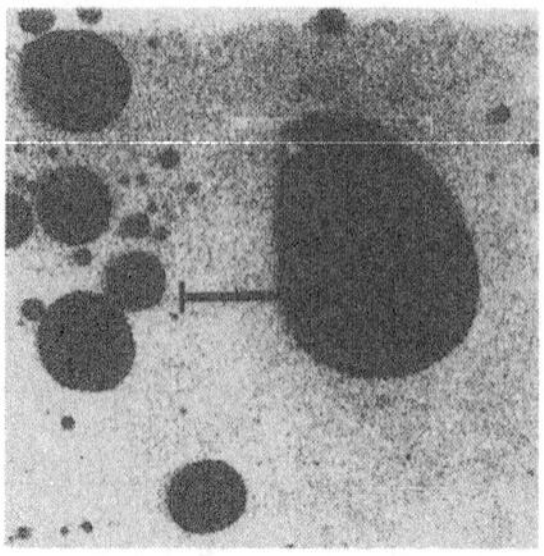

Fig. 7. a)

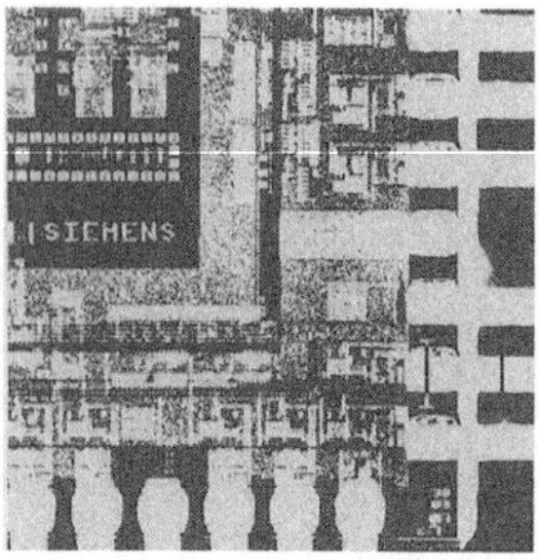

Fig. 8. a)

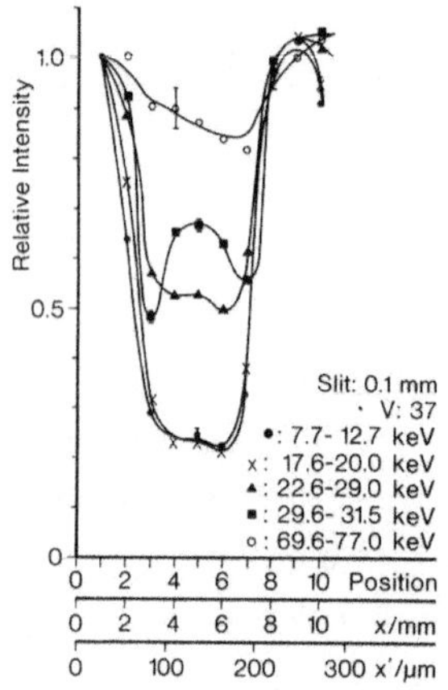

Fig. 9.

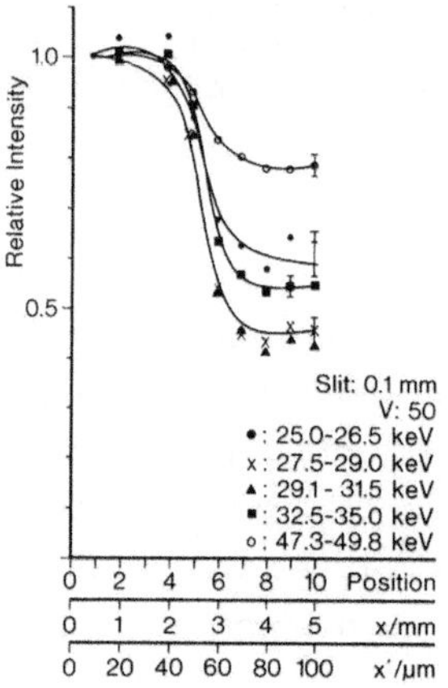

Fig. 7. b)

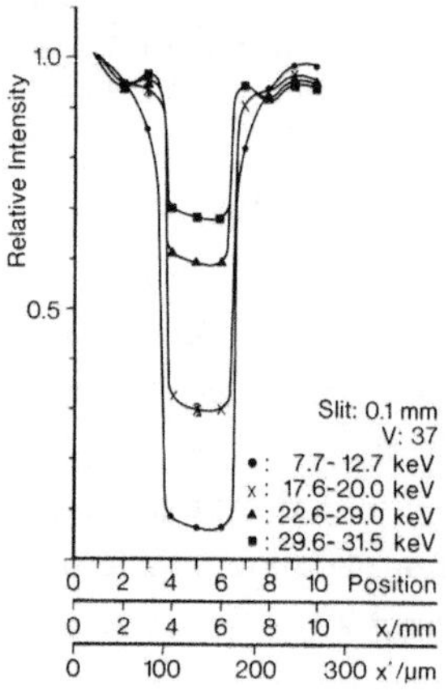

Fig. 8. b)

Fig. 6. Angular intensity distribution at 1.5 m distance, in the energy range 7.5 to 12 keV.

Fig. 7. a) Device with scan path. b) Line scan across a car ignition chip soldered to a ceramic and Al-backing. x': Scale on the device. Error bars: 3σ .

Fig. 8. a) Scan paths at the device. b) Line scan across the legs of a highly integrated computer chip at various energies. Error bars: 3σ .

Fig. 9. Same as Fig. 8b, across a solder joint (see Fig. 8a).

hand, has the required scale automatically. The period of the grid, however, has to be adapted to the expected focal size. Generally, the determination of the focal spot size of the microfocus tube by means of suitable rulings seems to be more accurate, and a simple visual estimation can be performed very

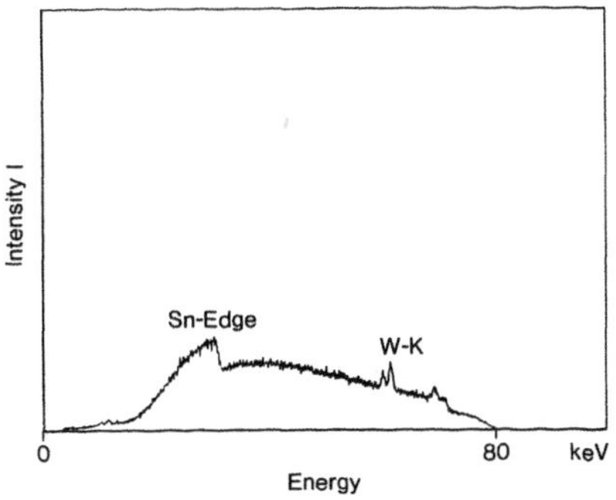

Fig. 10. Spectrum taken directly at the position of the solder joint, showing the absorption edge of tin at 29.2 keV.

quickly. Furthermore we could not find any significant energy dependence of the focal size. Finally, we could demonstrate the potential of energy resolved radiography for application in nondestructive testing.

References

1. Kirkpatrick, P.; Patty jr., H.H. in: Flügge, S. (ed.): Handbuch der Physik, Vol. XXX, Berlin, Göttingen, Heidelberg: Springer Verlag 1957, p. 305-336.

2. Baumann, J.: Kurzfassung der Vorträge bei der Jahrestagung "Zerstörungsfreie Materialprüfung" Essen 28. - 30. Mai 1984, Deutsche Gesellschaft für Zerstörungsfreie Prüfung e.V., Berlin 1984.

   Baumann, J.: Abstracts "Real-Time Imagery Topical", August 14-16, 1984, Atlanta/GA. American Society for Nondestructive Testing, Inc., Columbus/OH 1984.

3. Halmshaw, R.: Brit. J. of NDT 27 (1985) 220, ibid.: 27 (1985) 341-343.

4. Heidt, H.; Nabel, E.: Materialprüfung 28 (1986) 320-325. Nabel, E.; Heidt, H.: Vorträge und Plakatberichte, Jahrestagung 1987 Zerstörungsfreie Materialprüfung, Deutsche Gesellschaft f. Zerstörungsfreie Prüfung e.V. Berlin 1987, p. 675-682.

5. Grider, D.E.; Ausburn,P. K.: Brit. J. of NDT 29 (1987) 15-17.

6. Cohen, D.K.; Little, B.; Luecke F.S.: Appl. Opt. 23 (1984) 637-640.

Characterization of Ultrasonic Probes with Physical and Parametric Methods

U. Kiefer, K.-D. Becker
Lehrstuhl für Theoretische Elektrotechnik , Universität des Saarlandes, Bau 22, D-6600 Saarbrücken

W. Gebhardt, F. Walte
Fraunhofer-Institut für zerstörungsfreie Prüfverfahren, Saarbrücken

Abstract

System identification methods are applied to develop a parametric model of an ultrasonic inspection situation consisting of a normal probe and a specimen. To simulate the probe behaviour at varying physical properties, a theoretical model is calculated to generate synthetic data. The comparison of the properties of the parametric model with that of the real or simulated probe shows a good agreement.

1. Introduction

In the field of nondestructive ultrasonic testing of materials there is an increasing demand for an exact specification of the ultrasonic probes. Apart from the probe geometry the knowledge of the probe parameters, e.g. frequency response, center frequency and bandwidth, is necessary for a reproducible defect evaluation.

2. Procedure

The conventional determination of the probe parameters demands for a cumbersome electronics and is very time consuming. In the following, a probe characteriziation method is described, which necessitates only one measurement with a relative small amount of test equipment.

The ultrasonic probe is deposited on a specimen with plane parallel surfaces and is excited with an appropriate electrical signal. The input signal and the time shifted back wall echo are digitized and transfered to a computer. These input and output signals are used to generate an experimental model according to the methods of system and process identification. From the model parameters the following properties of the probe can be calculated :

- frequency response
- bandwidth
- sensitivity

- further model parameters, which allow to draw conclusions on the state of the probe

Apart from the measured input and output signals of real probes, simulated ones by theoretical modelling can be investigated too. The theoretical analysis, i.e. the generation of a theoretical-physical model of an ultrasonic probe to simulate its dynamic behaviour, is used to evaluate the influence of varying physical properties of the probe on the experimental analysis. In the following, ultrasonic transducers with piezoelectric discs vibrating in the thickness mode, which are in current use in nondestructive testing, are investigated and simulated.

3. Theoretical analysis

The theoretical modelling is based on the description of a multi-layer-probe by a transmission matrix with complex frequency dependent elements, as done by Sittig [1]. To this end, each layer, assumed to be homogeneous and laterally infintely extented, is described according to fig.1 by an impedance matrix (3.1)(3.2).

$$\begin{bmatrix} F_1 \\ F_2 \\ V \end{bmatrix} = \begin{bmatrix} \dfrac{Zo}{j\tan(\frac{\omega}{v_t^D}d)} & \dfrac{Zo}{j\sin(\frac{\omega}{v_t^D}d)} & \dfrac{h_{33}}{j\omega} \\ \dfrac{Zo}{j\sin(\frac{\omega}{v_t^D}d)} & \dfrac{Zo}{j\tan(\frac{\omega}{v_t^D}d)} & \dfrac{h_{33}}{j\omega} \\ \dfrac{h_{33}}{j\omega} & \dfrac{h_{33}}{j\omega} & \dfrac{1}{j\omega C_o} \end{bmatrix} \begin{bmatrix} U_1 \\ U_2 \\ I \end{bmatrix} \tag{3.1}$$

Impedance matrix of the piezoelectric thickness vibrator

$$\begin{bmatrix} F_1 \\ F_2 \end{bmatrix} = \begin{bmatrix} \dfrac{Zo}{j\tan(\frac{\omega}{v_t^D}d)} & \dfrac{Zo}{j\sin(\frac{\omega}{v_t^D}d)} \\ \dfrac{Zo}{j\sin(\frac{\omega}{v_t^D}d)} & \dfrac{Zo}{j\tan(\frac{\omega}{v_t^D}d)} \end{bmatrix} \begin{bmatrix} U_1 \\ U_2 \end{bmatrix} \tag{3.2}$$

Impedance matrix of an inactive layer

Z_o	acoustic impedance	ω	angular frequency
v_t	propagation velocity	F	force
d	thickness of the layer	U	displacement velocity
h_{33}	piezoelectric constant	V	electric voltage
C_o	capacitance	I	electric current

According to methods of the four-pole-theory, the matrices of the different layers are combined to a total transfer matrix (3.3), which connects the electrical quantities with the acoustical ones.

$$\begin{bmatrix} V \\ I \end{bmatrix} = \begin{bmatrix} q_1 & q_2 \\ q_3 & q_4 \end{bmatrix} \begin{bmatrix} F_2 \\ u_2 \end{bmatrix} \tag{3.3}$$

Transfer matrix of the multi-layer-probe

The circuit of Kojima [2] to measure the voltage transfer gain (VTG) is modelled in fig.2. The probe is assumed to work in the T/R-mode. According to fig.3, the transmitting and the receiving case can be seperated because there exists no interaction between transmitter pulse and backwall echo. The voltage transfer function (VTF) $T_o(\omega)$ calculated from fig.3 with the elements q_i of the transfer matrix and the acoustic impedance Z_M of the specimen according to equation (3.1) can be presented as a spectral distribution. This transfer function describes the ratio of the amplitudes of the open circuit receiver voltage to the transmitter voltage.

$$T_o(\omega) = \frac{V_o(\omega)}{V_i(\omega)} = \frac{2 \cdot Z_M}{(q_1 \cdot Z_M + q_2)(q_3 \cdot Z_M + q_4)} \tag{3.4}$$

As an example, the voltage transfer function in dependence on the thickness of the matching layer is shown in fig.4. The output signals for arbitrary input signals can be calculated with the aid of the inverse Fourier-transformation [3]. These output signals can be used for the experimental analysis as a synthetic data set.

4. Experimental analysis

The ultrasonic probe in the test arrangement - probe and specimen - is excited with a suited input signal. The measured input and output signals are digitized and recorded. From these sample values a digital computer generates an experimental model according to an identification algorithm. The algorithm is performed with the requirement, that the difference between the output signals of the system and the experimental model must be minimized [4]. The procedure of the experimental analysis is shown graphically in fig.5.

It is assumed, that the test agreement is a linear time-invariant system, so that the relation between the sampled input and output signal can be described in form of a linear difference equation according to equation (4.1). This equation represents the experimental model.

$$y(k) = \underline{z}(k) \cdot \underline{\Theta} \tag{4.1}$$

with

$$\underline{z}(k) = \Big(u(k-m), \ldots, u(k-1), u(k), y(k-n), \ldots, y(k-1) \Big) \quad \text{data vector} \tag{4.2}$$

$$\underline{\Theta} = \left(a_m, a_{m-1}, \ldots, a_o, b_n, b_{n-1}, \ldots, b_1\right) \qquad \text{parameter vector} \quad (4.3)$$

The data vector $\underline{z}$ contains the input and output signals u(i) and y(i), sampled at times i·T, where T denotes the sampling time. The vector $\underline{\Theta}$ contains the parameters which characterize the system. Performing the z-transfomation to equation (4.1) the transfer function

$$H(z) = \frac{\sum_{i=0}^{m} b_i \cdot z^{-i}}{1 + \sum_{j=1}^{n} a_j \cdot z^{-j}} \quad . \qquad (4.4)$$

is obtained.

To determine the parameter vector $\underline{\Theta}$, the linear equation system (4.5) is formed from the sample values at times k=1, 2,.... , N with $N \geq m+n+1$.

$$\underline{y} = \underline{Z} \cdot \underline{\Theta} \qquad (4.5)$$

with

$$\underline{y} = \left(y(1), y(2), \ldots\ldots, y(N)\right)^T \qquad (4.6)$$

$$\underline{Z} = \left(\underline{z}(1), \underline{z}(2), \ldots, \underline{z}(N)\right)^T \qquad (4.7)$$

The error of measurement e(k) of the kth measurement is given by

$$e(k) = y(k) - \underline{z}(k) \cdot \underline{\Theta} \qquad (4.8)$$

and

$$\underline{e} = \underline{y} - \underline{Z} \cdot \underline{\Theta} \text{ , resp.} \qquad (4.9)$$

This value is called equation error or residuum. The estimate value $\hat{\underline{\Theta}}$ of the parameter vector $\underline{\Theta}$ is given by the requirement of a minimization of the error, where the euclidian norm in equation (4.10) is used to quantify the error vector $\underline{e}$.

$$\|\underline{e}\|^2 = \left(\underline{y} - \underline{Z} \cdot \underline{\Theta}\right)^T \left(\underline{y} - \underline{Z} \cdot \underline{\Theta}\right) \qquad (4.10)$$

The necessary and sufficient condition, that the derivation of the norm with respect to the parameter vector $\underline{\Theta}$ vanishes, gives the non-recursive estimate equation

$$\hat{\underline{\Theta}} - (\underline{Z}^T \cdot \underline{Z})^{-1} \underline{Z}^T \cdot \underline{y} \, . \qquad (4.11)$$

From the parameters of the experimental model estimated in this way certain properties of the system can be obtained. The voltage transfer function calculated in section 2 corresponds to the frequency response of the experimental model. It is calculated from the transfer function (4.4). From the frequency response, center frequency, bandwidth and sensitivity of the probe can be determined. As a further characteristic feature of the model the distribution of the poles and zeros of the transfer function is calculated in the image space of the z-transform or in the image space of the Laplace-transform. As well as the coefficients of the difference equation, this distribution of poles and zeros characterizes each inspection arrangement can be used to check the specifications of the probe.

5. Results

The following investigations were performed with a normal probe consisting of a piezoelectric plate, a matching layer and a backing. The probe was deposited on a steel specimen with plane-parallel surfaces.
In a first step the voltage transfer function was determined evaluating the amplitudes of the backwall echo exciting the probe with signals of variable frequency (fig.6.a). The exciting signal was a sine burst of two periods. Fig.6.b shows a measured one.
The voltage transfer function calculated from the theoretical-physical model can be seen in fig.7.a. The calculated backwall echo of the theoretical model in response to the sine burst in fig.6.b was shown in fig.7.b.
From the excitation signal and the backwall echo, depicted in fig.6.b., an experimental model was generated, whose frequency response and backwall echo was shown in fig.8.a and 8.b, resp.. The corresponding distribution of the poles and zeros of the transfer function in the z-plane can be seen in fig.9.

6. Conclusions

As was shown in the preceeding section, the parameters of the experimental model are estimated in such way, that there is a good comparison between frequency response of the experimental model and the measured voltage transfer function of the real probe and the calculated one of the theoretical model, resp.. In contrast to the acquisition of the voltage transfer function the experimental analysis necessitates only one measurement.
The advantage of the parametric model of the test arrangement is the compression of the spectral distribution to a finite number of model parameters. To control the specifications of a probe during the manufacturing process, it suffices to compare its model parameters with the parameters of a given correct probe.

[1] Sittig, E.K. : Transmission Parameters of Thickness-Driven Piezoelectric Transducers Arranged in Multilayer Configurations. IEEE Trans. Sonics Ultrason. su-14 1967 S.167-174

[2]Kojma, T. : A New Method for Estimating the System-Independant Characteristics of an Ultrasonic Piezoelectric Transducer. IEEE Ultrasonics Symposium 1981 S.649-654

[3]Kiefer, U. : Modellierung und Simulation von Ultraschallprüfköpfen im Impuls-Echo-Betrieb. Diplomarbeit am Lehrstuhl für Theoretische Elektrotechnik , Universität des Saarlandes, 1987

[4]Isermann, R. : Prozeßidentifikation. Berlin: Springer-Verlag 1974

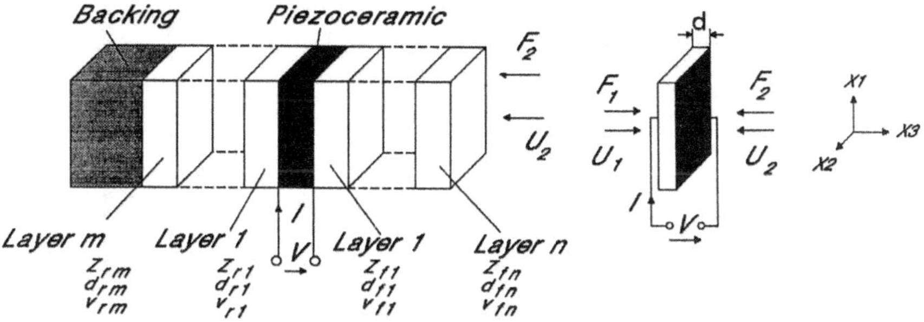

Fig.1 Multi-layer-probe with input and output signals / Mechanical loaded piezo-electric plate

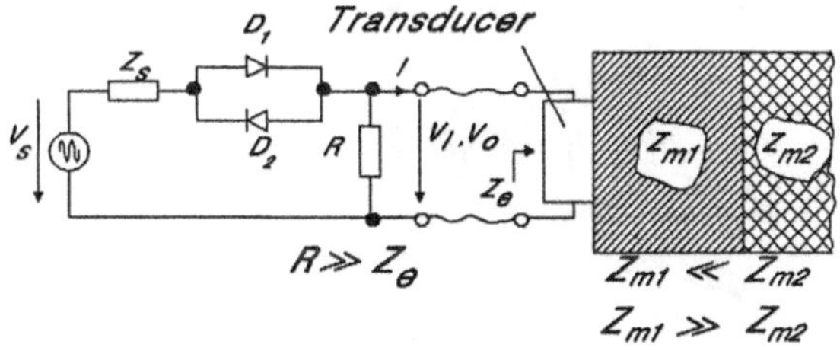

Fig.2 Measurement circuit of the voltage transfer function

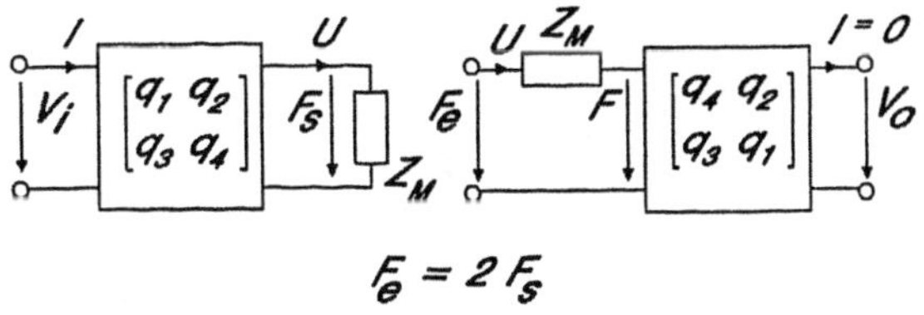

Fig.3 Equivalent circuit of the probe in the T/R-mode

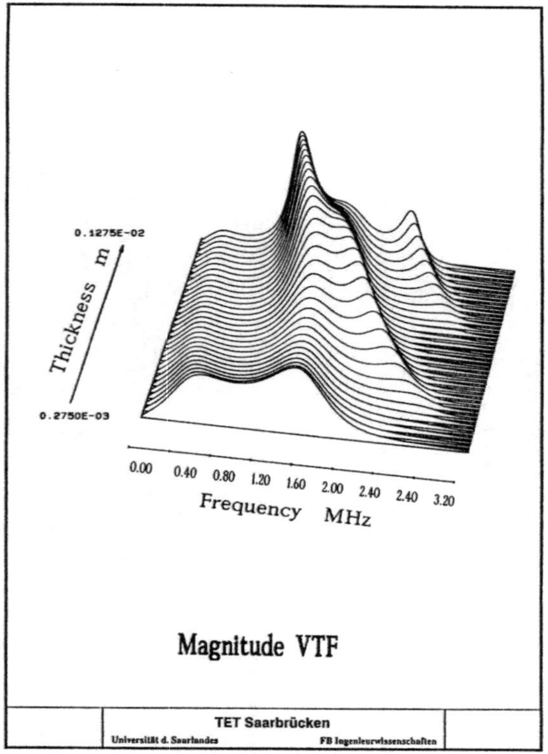

Fig.4 Voltage transfer function in dependence on the thickness of the matching layer

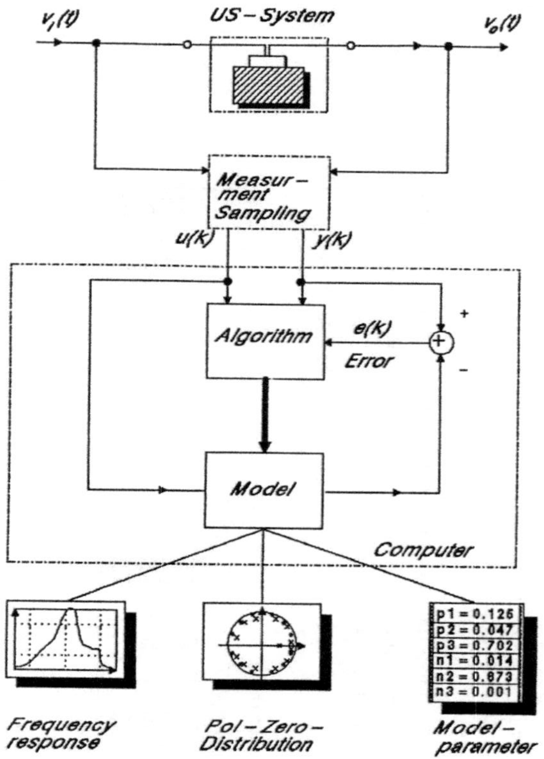

Fig.5 Block diagram of the experimental analysis

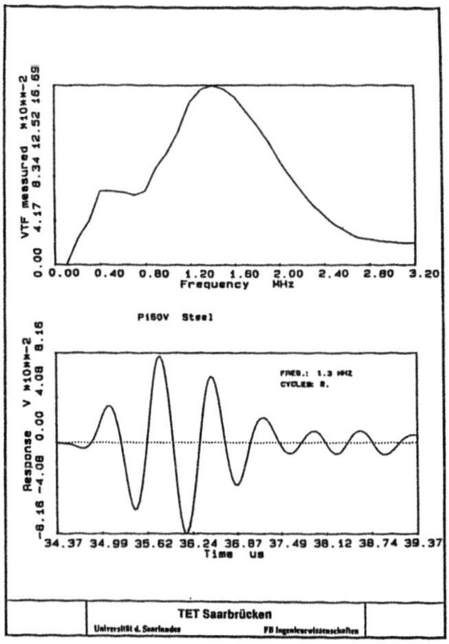

Fig.6 a/b

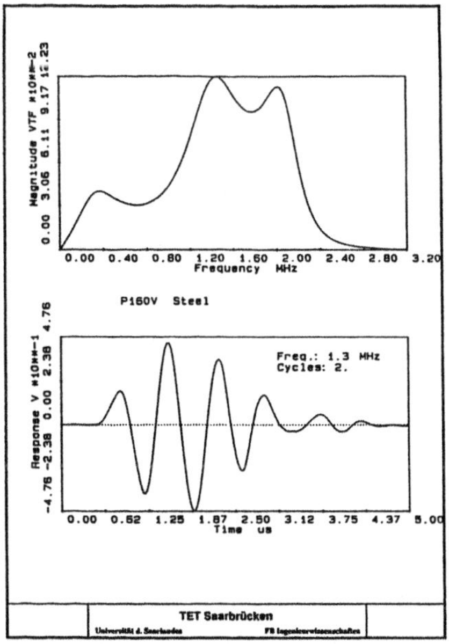

Fig.7 a/b

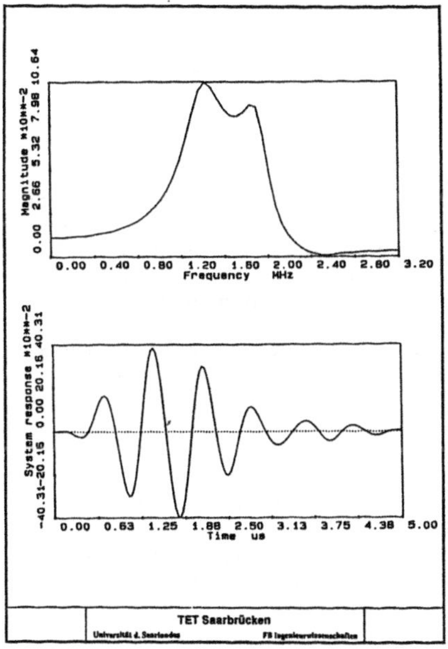

Fig.8 a/b

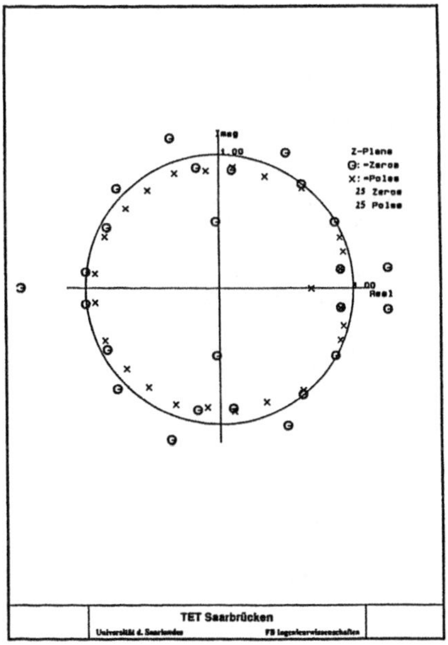

Fig.9

Mirage Effect and Optical Reflectance: New Improvements in Nondestructive Evaluation

A.C. Boccara, F. Charbonnier, D. Fournier,
E. Legal Lasalle, F. Lepoutre, J.P. Roger and J.P. Sachet
Laboratoire d'Optique - ER 5 CNRS
ESPCI, 10, rue Vauquelin
75005 PARIS - France.

A. Lemoine
DELAS (Groupe CGE Alsthom)
12-14, rue d'Alsace
92302 Levallois - France

P. Robert
SFENA
B.P. 128
86101 Châtellerault - France

1) INTRODUCTION

The photothermal measurement technique uses a periodic heating of the sample by a pump beam and the detection of the thermal distribution induced. In the mirage effect and optical reflectance, this detection is achieved by a second beam called the probe beam (see fig. 1).

Since the first "Mirage" experiment run in the laboratory of ESPCI in 1979 [1], this method has been used by many other laboratories for the determination of optical and thermal properties and for non destructive evaluation [2] [3] [4].

Despite of the fact that numerous applications deal with industrial problems, up to now there is no set up introduced close to a production line in a factory, to the best of our knowledge. Nevertheless, the technique is sufficiently reliable and mature, and its introduction in an industrial environment is now possible. In this paper we will describe two examples dealing with very weak thermal and optical losses determinations.

Finally two applications of detections at micron scale show that with a rather similar technology, these methods are able to detect fatigue cracks or grain boundaries.

2) THE "INDUSTRIAL MIRAGE SENSOR"

The "Mirage" cell constitues the "heart" of the experimental set up. The sample surface is heated periodically by a modulated light beam and this sensor allows the measurement of the sample periodic temperature, through the deviation measurement of a probe beam crossing the heated area.

The cell derives from the realization we have described in ref [5], which was built for laboratory applications. Despite of its high sensitivity, this sensor required careful alignments of the sample and (or) of the probe beam. For industrial applications, we have chosen either to maintain tightly the sample surface at a fixed position with respect to the heating and probe beams (section 3) or to introduce an automatic positioning of the sample-probe beam distance (section 4). Moreover to avoid acoustic vibrations effects and specific noises, we have replaced the He-Ne laser by a diode laser coupled to a microscope objective. This last laser probe, despite of its larger noise level up to 200 Hz, is almost insensitive to acoustic and mechanical perturbations and exhibits less drifts associated to laser modes problems.

3) OPTICAL LOSSES DETERMINATION IN MIRROR COATINGS

Among the optical components which take place in the laser manufacturing, the mirrors play a major part. The dielectric coatings characteristics are especially important for laser-gyro performances , a typical absorption losses level being of the order of a few tens per million.

It was particularly important for SFENA company to use, close to a production line, a set-up allowing a fast, precise and sensitive measurement of absorption losses.

The compact mirage cell was designed in order to allow colinear mirage detection [6]. Colinear detection is possible as shown in fig. (1) because the multidielectric selective mirrors under test, highly reflective at He-Ne pump beam wavelength of 633 nm, are transparent at the wavelength of the laser diode probe beam (780 nm). The heat deposited in the coating diffuses into the substrate (i.e. silica), and the radial temperature gradient is probed within the bulk of this transparent substrate.

If the modulation frequency is chosen low enough in order to neglect the thermal diffusion in the coating, a simple calibration allows the direct measurement of the absorption level. The calibration is obtained by using a coating deposited on the same substrate but with an important absorption level so that it can be measured by classical reflection transmission photoelectric experiments.

On the one hand, in this mirage experiment, the signal increases linearly with the incident power of the pump beam ; on the other hand, the noise level, with a good laser diode can reach the photon noise of the position sensor at modulation frequency higher than 200 Hz. So that In our set-up, using a 20 mW He-Ne as pump beam, the signal equivalent to noise corresponds to an absorption level of one per a million.

The set-up is reliable and the experiments can be reproduced within a few percents.

In practice, the sample holder can be easily removed from the mirage cell and its design defines a reference plane for the coating under test with a very good precision without mechanical adjustments.

It is important to note that it takes few seconds for the non specialist operator to put the sample in its holder and to get a numerical result from the lock-in and the microcomputer acquisition devices. Actually, after the coating operation each sample is systematically tested by this method which constitues the first in-line industrial application of the mirage detection.

Among all the methods that exist to measure optical losses with a high sensitivity, let us recall three principal classes:

The first ones are flying-time decrease measurements methods. These methods allow global losses measurement (Absoption + Transmission + Diffusion) and are limited in practice for high reflectivity mirors (>99%). But it remains delicate to make use of theese methods without precise and difficult optical adjustments. This situation is often incompatible with in-line sample tests after manufacturing.

The second ones are calorimetric methods. They allow absorption losses measurement but they take too long time to be compatible with industrial production requirements.

The third group of methods uses photothermal phenomenons. In this group, we have chosen the mirage detection for its simplicity and high sensitivity: the same absorption losses measurements are possible with a photoacoustic cell, for example, but in that case, sensitivity is smaller by one or two orders of magnitude than the one we have obtained.

4) THERMAL RESISTANCE LOSSES DETERMINATION IN TITANIUM HEAT EXCHANGER PIPES.

The second application deals with the detection of very thin air slices in metallic systems. Some heat exchangers are built with titanium pipes made of two coaxial cylinders. When theese two cylinders are not enough tighted together, a residual air slice appears between them (see Fig. 2).

This defect decreases considerably the heat exchange efficiency. In our case, a thickness of 1 µm leads to a loss larger than 10 %. Practically the problem is to measure the thickness of the slice located approximately at 1 mm below the surface. The phase of the normal mirage deflection Φ_n have been calculated using the geometrical and thermal properties of the three dimensional problem (see Fig. 3). The different curves are related to the thickness of air slices varying from 0 to 1 µm by steps of 0.1 µm. Let us underline that these curves depend upon the distance z between the probe beam and the sample surface. For modulation frequency lower than 9 Hz this phase is strongly sensitive to the thickness of the air slice : a variation of 0.1 µm of this thickness induces a phase shift of about 1 degree. The compact bench previously described can easily detect such phase shifts (its experimental phase precision is of the order of 0.2 degree). On the contrary, at frequencies higher than 9 Hz, the phase shift of Φ_n is only sensitive to the distance z between the probe beam and the sample surface.

Experimentally the measurements are performed for each point on the pipe at two modulation frequencies.

- Firstly at $f \approx 25$ Hz, the measurement phase of Φ_n is used to determine the distance z, between the probe beam and the sample surface.

- In the second step, the modulation frequency is decreased down to 3 Hz. With the knowledge of the distance z, the measured phase shift can be compared with theoretical calculation and finally gives the thickness of the defect.

For reasons of simplicity and time saving, only the theoretical results, calculated at a given distance $z = z_0$, are stored in the computer memory, so that the phase measurement must be performed at this specific distance z_0. To achieve this experimental condition, the mirage bench can be displaced with respect to the sample . At $f \approx 25$ Hz, the measured phase (which depends only upon the distance z) is compared to the one related to the distance z_0 and the bench is moved until these two values become equal.

The machine (Fig. 4) is totally automatized. The pipe is heated by the light beam (≈ 1 W) of an iodine lamp modulated successively at f≈ 25 Hz and f≈ 3 Hz. The results appear on the screen as images in false colors or experimental points on the theoretical curves. The complete quantitative test takes less than 2 minutes per point.

To detect small slices of air inside a metal, at least two methods are possible: Eddy currents and ultrasonics. Comparative measurements with these two probes and the mirage cell were done at the manufacturing plant, where Eddy currents and ultrasonics are both currently used in production quality control. It is important to note that the mirage detection has appeared fastly to be the quantitative standard of the comparative test as well as for the precision and the repeatibility of the measurements. Moreover, the thermal theoretical modelisation of the defect measurement remain simpler than electromagnetic or stress and strain modelisation. For all these reasons the mirage detection was introduced not to replace other methods, but because, it was particularly well adapted to the defect geometry and to the physical phenomenon concerned: heat exchange tested with heat probe.

5) PHOTOTHERMAL MICROSCOPY

To achieve detection at micron scales, the two beams are highly focused in an optical microscope. The sample is heated by an argon laser and the defects are detected by an He-Ne laser. The two beam spots are about 1 micron wide and can be separated by distances varying between 0 to 5 microns.

The modulation frequency can be varied up to 10 MHz. At this frequency the thermal diffusion length is of the order of 1 micron in the materials analysed.

The experimental arrangement is very similar to the one used by Opsal et al [7] except that we do not expand the two beams before the microscope.

Two kinds of microstructures have been observed (fig. 7) : fatigue cracks and grain boundaries. In the first case a large amplitude change and phase shift of π are observed when the probe beam is on the crack. This effect is probably due to the thermal expansion of the crack produced by the pump beam.

In the second case, the amplitude and the phase shift exhibit variations identical to the ones calculated by Grice [8] et al and Mc Donald et al [9] as due to thermal resistances.

These different behaviours allow a very easy differentiation between simple thermal resistances (which can be present prior to a crack) and actual open cracks.

6) CONCLUSION

The two macroscopic systems we have described were chosen among other industrial solutions, first because they present the qualities required for industrial testing :

- non destructive (with such low powers pump)
- reliable and highly sensitive
- automatic and robust
- insensitive to environment perturbations

secondly, they present two specific qualities : elementary to use and quantitative.

In Germany, the Mirage Sensor is commercially available at Phototherm - Dr. Petry, GmbH [10]. This Society can also develop appropriate solutions for specific problems.

The microscopic system is of course a laboratory set up with precise optical components. Its good resolution allows micro nondestructive testing.

ACKNOWLEDGMENT

We would like to thank S.T.C.A.N. from French National Navy for constant encouragement and financial support allowing to realize the machine described in section 4.

REFERENCES

1. A.C. Boccara, D. Fournier, and J. Badoz, Appl. Phys. Lett. 36 130 (1980).
2. F. Lepoutre, D. Fournier, and A.C. Boccara, J. Appl. Phys. 57 1009 (1985).
3. K.R. Grice, L.J. Inglehart, L.D. Favro, P.K. Kuo, and R.L. Thomas, J. Appl. Phys. 54 6245 (1983).
4. L.C. Aamodt, and J.C. Murphy, J. Appl. Phys. 54 581 (1983)
5. F. Charbonnier, D. Fournier, Rev. Sci. Instrum., 57 1127 (1986).
6. A.C. Boccara, D. Fournier, W.B. Jackson, N. Amer, Optics Letters 5 377 (1980).
7. J. Opsal, M.W. Taylor, W. Lee Smith and A. Rosencwaig. J. Appl. Phys. 61 240 (1987).
8. K.R. Grice, L.J. Inglehart, L.D. Favro, P.K. Kuo and R.L. Thomas, J. Appl. Phys. 54 6245 (1984).
9. F.A. Mc Donald and G.C. Wetsel Jr., in Physical Acoustics edited by Warren P. Mason, Vol. XVIII, p. 167 (1988) Academic Press (New York).
10. Phototherm, Dr. Petry GmbH, Altenkesseler Str. 17, D-6600 Saarbrücken 5, West Germany.

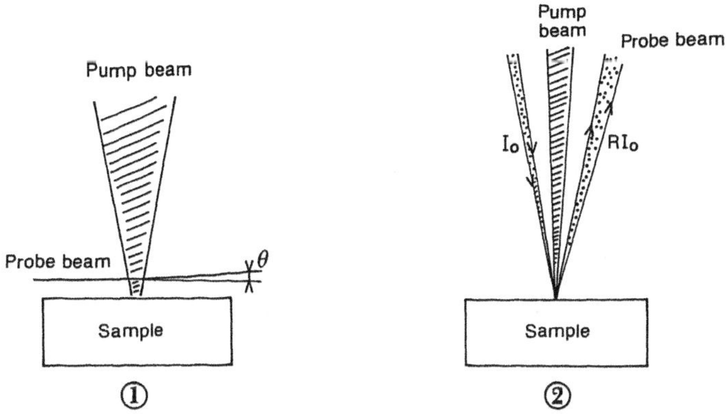

Fig. 1. Schematic arrangement of the beams in mirage (1) and photothermal reflectance (2) experiments.
The sample is periodically heated by a modulated pump beam. The periodic temperature induced is sensitive to the presence of defects in the material.
(1) : The pump and probe beams are perpendicular. The probe beam is deflected (angle Θ) by the thermal gradient close to the heated area of the sample. The periodic temperature is deduced from the measurement of Θ.
(2) : The pump and probe beams are colinear. The periodic temperature induces a periodic variation of the optical reflectance of the sample. The periodic temperature is deduced from the measurement of the periodic part of the reflected probe beam (RI_o).

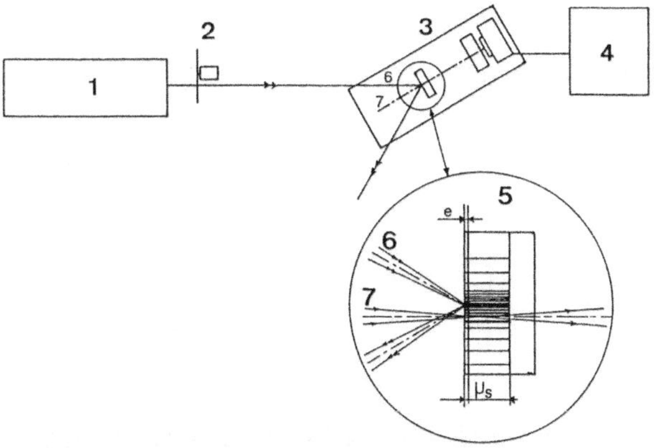

Fig. 2. Experimental set up for optical losses determination in coatings using the colinear mirage detection. -1- 20 mW He-Ne pump laser.-2- Mechanical chopper.-3- Compact mirage cell. -4- Lock-in and microcomputer acquisition devices.
-5- Coating on substrate under testing.-6- Focused He-Ne pump beam absorbed and reflected by the coating.-7- Laser diode probe beam transparent for the coating and deviated into the substrat bulk.- The coating thickness (e) is much smaller than the thermal diffusion length (μs)-

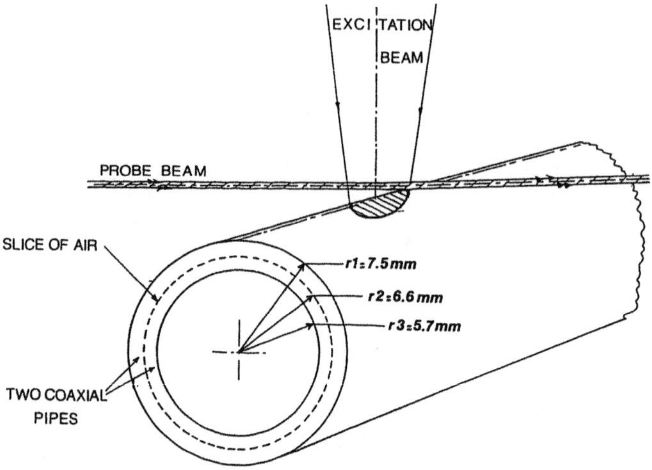

Fig. 3. Titanium pipes sample study ; disposition of the two beams of the Mirage detection.

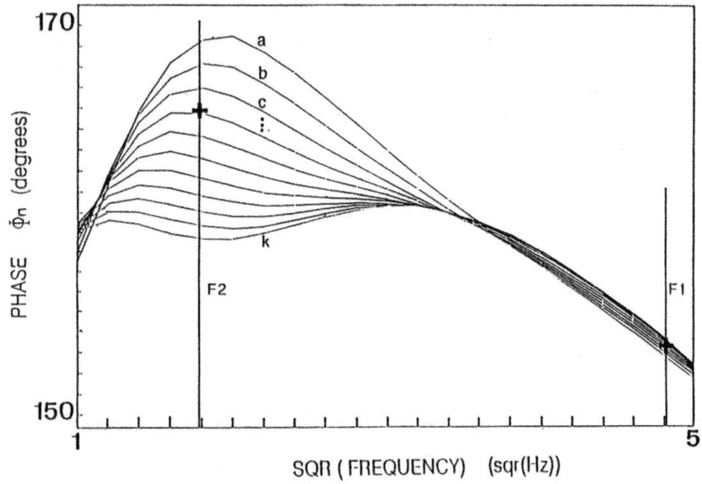

Fig. 4. Titanium pipes characterization : Phase of the normal Mirage deflection versus the square root of the frequency. The different curves are deduced from calculation of the phase for different thicknesses of the air slice :

- curve a corresponds to 0 micron, curve k corresponds to 1 micron, step between each curve: 0.1 micron
- F1 and F2 represent the square roots of the modulation frequencies used for the test. For instance the two experimental points (+) correspond to a 0.3 micron thick air slice.

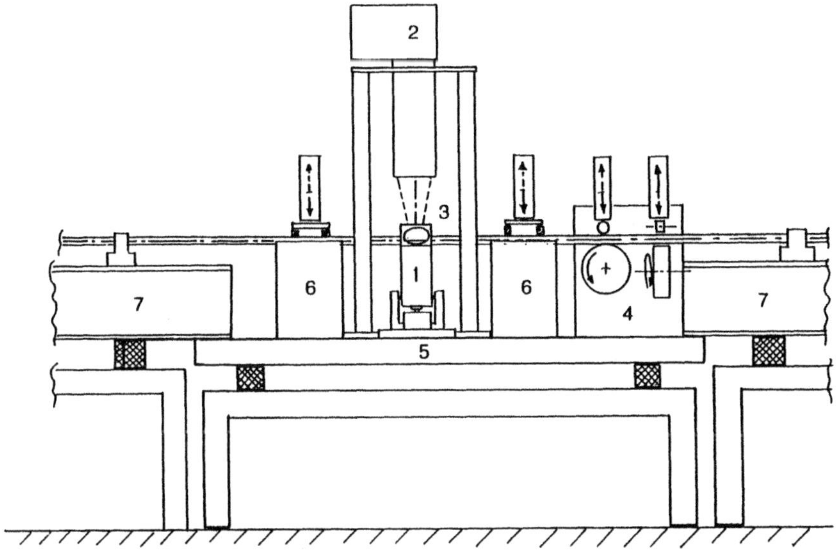

Fig. 5. Central part of the machine for the test of titanium pipes.
-1- Mirage block cell.
-2- Modulated iodine lamp.
-3- Titanium pipe under testing.
-4- Mechanical system for automatic displacement of the pipe.
-5- Marble.
-6- Flanges.
-7- Rail to sustain the pipe

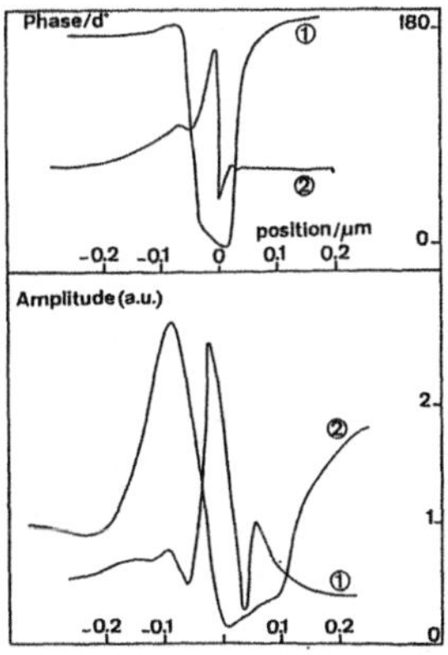

Fig. 6. Phase shifts and amplitude variations of the periodic signals obtained on two kinds of microdefects :
(1) fatigue crack in aluminium. The phase shift of π on the crack is probably characteristic of thermal expansions of the edges of the crack.
(2) grain boundary in a ceramic. This experimental behaviour is characteristic of a thermal resistance.

Closing Comments

R. S. SHARPE

National Nondestructive Testing Centre,
Building 149,
Harwell Laboratory,
Oxfordshire. OX11 ORA

They tell me this 'spot' in the programme was reserved for the member of the Conference Advisory Board who proved to be the least successful in drumming up Conference attendees. From the size of the UK contingent at the Conference, it might appear that we are a 'third world' nation as regards materials characterisation. I don't think this is in fact the case. I obviously just didn't do my job well enough and so I now have to pay the penalty!

Actually I am under no real illusion about the implication of the invitation. It has become abundantly clear to me during the week that a research man's career follows an inevitable progression.

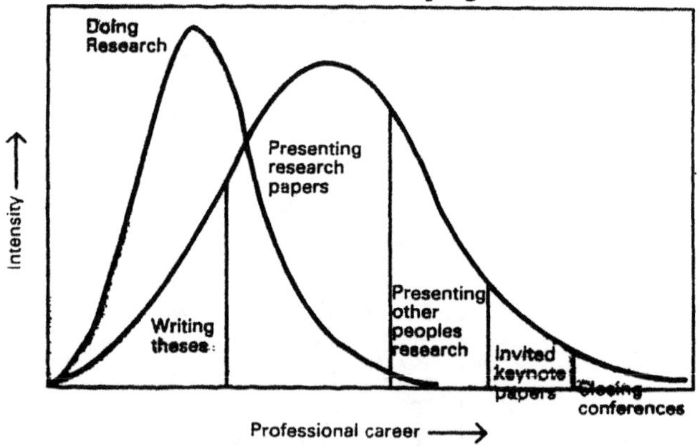

National NDT Centre

In the early days of one's career papers abound that are littered with equations. They are topical and many are the required passport to attend conferences. By nature they are highly specialised and hence understandable to the minority and in my view form the ideal material for poster sessions.

Moving along one's career, a conference attracts papers by heads of research groups or supervisors summarising the work of their research students or their departmental programmes. These are generally informative to a wider audience and have the added benefit of a certain amount of 'filtering out of detail' and a broader 'interpretation' content.

Then we come to keynote papers intended to introduce experience and judgement and to put progress in a particular topic into a wider context. Unfortunately the timing of the parallel sessions at this Conference has meant that many of these have clashed.

Finally, and fortunately few and far between, there are those like myself who are asked to close conferences. Such people are basically a danger to the organisers because, by the nature of their remit, they could not prepare or submit anything beforehand - and some have been known to be dangerously provocative!

I never cease to be amazed how the decidedly down-market subject that I stumbled on some 40 years ago (at about the same time as Friedrich Förster who I am particularly pleased to see at the Conference) has managed to blossom into a subject supporting so many international and national gatherings:-

- * WORLD CONFERENCE SERIES ON NDT
- * QUANTITATIVE NDE CONFERENCES
- * INTERNATIONAL NUCLEAR NDT CONFERENCES
- * INTERNATIONAL INST. WELDING ANNUAL ASSEMBLIES
- ** NDE MATERIALS EVALUATION CONF. SERIES
- * EUROPEAN NDT CONFERENCES
- * PAN-PACIFIC NDT CONFERENCES
- * INTERNATIONAL AC EMISSION CONFERENCES
- * ULTRASONICS INTERNATIONAL CONF. SERIES
- * NATIONAL SOCIETY CONFERENCES

It is indeed not surprising that we keep meeting each other, complaining about jet lag and the pressures of keeping abreast of day-to-day business back 'at the ranch'in the rare gaps in the conference calendar!

Is it all a confidence trick? Have we stopped to think where all this erudition and all these outpourings is leading? It has become much clearer to me this week that there is a logic and there is an objective. Whether or not it is a generally accepted objective or that you see things the same way as me is of course open to debate. But in accord with current good practice I have prepared a model of where I think we are now:

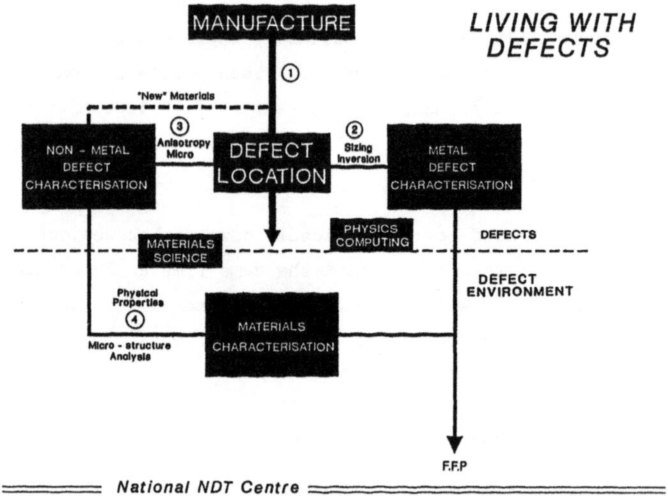

From this you can see that 'lemming-like' the NDT scientists all first moved into sizing defects and solving the 'inversion' problem. Then new materials came into fashion and the challenge of anisotropy and micro defects moved them all into non-metal defect characterisation. Then suddenly 'defects' lost their glamour and everyone has now moved into a study of the 'defect environment' and are using a much greater variety of nondestructive techniques to monitor physical properties, microstructural changes, anisotropy and stress fields. This is a good 'circle of activity' because the combination of quantitative information on defects and detailed information on the defect environment makes it much easier to establish whether a product is truly fit for the purpose for which it was designed and manufactured.

Somebody asked me this week to predict where we 'lemmings' should go next! So I have extended my model.

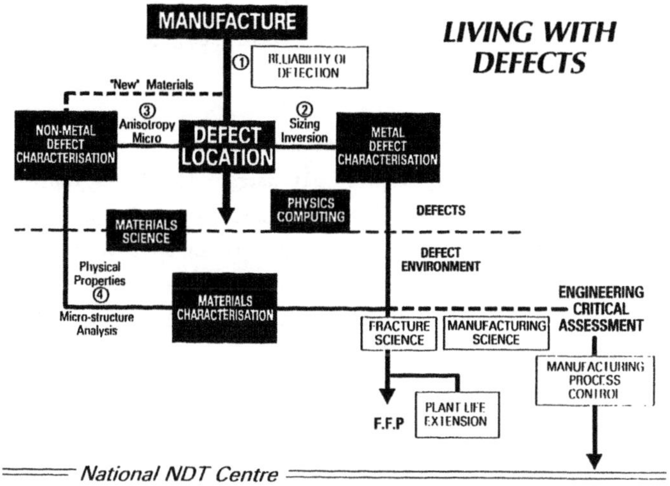

There are opportunities to go more deeply into 'fracture science' incorporating all of the new knowledge we have now accumulated on defects and materials characteristics, and pursue this still further into the economically important subject of plant life extension. The next logical move is that of turning manufacture into a 'science' and encouraging the incorporation of 'manufacturing process control' into the engineering critical assessment studies from which future manufacturing practices will develop. By **controlling** the process during manufacture we are well on the way to a 'defect avoidance' philosophy that will make much of our present researchers obsolescent!

As a cautionary note I would say that our **present** edifice of erudition has a particularly weak link still left in it - namely the detection of defects in the first place: This is not nearly as reliable as it should be. However I suspect that reliability of traditional practices is not going to be a subject for research with any great academic 'pull'!

Bearing this model in mind, what have I found as particularly interesting at this Conference? It can of course only be a very personalised selection form those papers I heard:

1. Confucius, he say 'one picture is worth 1000 words, - and as far as I am concerned '10,000 equations'. Having researched in the field myself I was particularly gratified therefore to see the quality and detail in the micro-computerised, tomo-denistometer pictures.

2. Staying with radiography (and its nice to see it's revival after so many years of subjugation by ultrasonics!) I was impressed by the elegance and possibilities of X-ray scanning microscopy for in-depth resolution of structure. It is a brave man - and often a subsequently proven-wrong man who says he has a 'new' technique, but I think the authors are justified in making that claim in this instance.

3. I was interested in the venture into neural network pattern recognition. We also have started on this road at Harwell and it is nice to see NDT - regarded so long as a traditionally backward subject-keeping abreast of such up-to-date developments in computing philosophy. However I must confess that my personal neural network, that I carry around on top of my shoulders, found it difficult to interpret all of the detail in the slides

4. I was glad to see papers on weld interface monitoring which is a first step into the deeper study of defect formation - and hence defect avoidance.

5. It was nice to see positron annihilation getting an airing having ourselves, at Harwell, been something of lone researchers in the subject over so many years.

6. The interest in laser generation and detection of ultrasound is important as it provides one of the necessary 'remote' sensing techniques that will be required as work on manufacturing process control develops.

7. Finally the extended session devoted to process control suggests that my 'model' of future 'lemming territory' is reasonably justified!

May I be permitted a few words about the Conference arrangements. One of the speakers said 'If you understand the subject there is no need to use equations'! My corollary to this is that if a paper is peppered with equations put it into the poster session. The posters at this Conference were **excellent** and in my view a major feature of the Conference. Indeed there is not shadow of doubt in my mind that 'posters, a few keynote summary papers and a large coffee room' is the preferred way of presenting any scientific conference in this age of extreme specialisation.

It is no particular reflection on this Conference, but at all Conferences these days I continually find myself asking: Are there any real conclusions to some of the papers? Perhaps, more importantly, do the conclusions only say 'This is what I have done' rather than 'This is the advantage to society at large of what I have done'!

This leads to another problem that has come clearly into focus for me this week. Are we still too introverted in our communications? The ministers and mayors and the distributors of 'megabucks and megamarks' all seemed to drift away into the woodwork after the opening session of the Conference. Wouldn't it be an achievement if the next Conference overtly tried to bridge the gap more effectively and concentrated on the cost effectiveness and practical objectives of the research? In this way we could invite, communicate with, maintain the interest of, and perhaps even excite, the non-scientist, the man wrestling with his problems in industry and the man with the money. The organisers of Conferences of this type should not lightly gloss over the fact that the objective of the specialist may not be in harmony with, or indeed be appreciated by, the funding source or the person with a problem. To communicate vertically is far more of a challenge and far more rewarding than merely communicating horizontally amongst one's peers and should be the pattern of conferences in the future.

No conference of course is complete without discussion periods and these have been excellent during this week. To the younger scientist the incisive questioning of some among you must indeed be awesome and forbidding. What we have missed, of course, this week is the perhaps more descriptively enquiring, but equally penetrating, questioning from the floor by Paul Höller. His absence - and in particular the circumstances of his absence - has certainly provided a cloud over the whole proceedings. He and I have been sparring partners over the past 20 years, running similar-sized organisations with similar objectives and competing often for the same funds. I know how much he would have welcomed the opportunity, that he had planned, to host this meeting personally. We all wish him as speedy and as complete a recovery as medical science can achieve.

May I conclude by thanking the IZfP staff (in particular Gerd Dobmann and Herbert Griess) who have been responsible for the local organisation, and the staff of the Germany NDT Society, (secretary Wolfgang Bock) and their many advisors and helpers for the overall planning of the Conference. It has been an excellent and intimate meeting and made more so by the professionalism and experience of the translators.

Mr Chairman, ladies and gentlemen, my words may not be 'conclusive'(as in the programme) but I fear that, as far as my own personal involvement as 'conference closer' in this particular series of conferences is concerned, they are indeed likely to be the last I shall be called on to make. I wish the series continuing success in the future.

GPSR Compliance

The European Union's (EU) General Product Safety Regulation (GPSR) is a set of rules that requires consumer products to be safe and our obligations to ensure this.

If you have any concerns about our products, you can contact us on ProductSafety@springernature.com

In case Publisher is established outside the EU, the EU authorized representative is:

Springer Nature Customer Service Center GmbH
Europaplatz 3
69115 Heidelberg, Germany

Batch number: 10844661

Printed by Printforce, the Netherlands